ADVANCES IN CHEMICAL PHYSICS

VOLUME LXXXI

Advances in
CHEMICAL PHYSICS

Edited by

I. PRIGOGINE

University of Brussels
Brussels, Belgium
and
University of Texas
Austin, Texas

and

STUART A. RICE

Department of Chemistry
and
The James Franck Institute
The University of Chicago
Chicago, Illinois

VOLUME LXXXI

AN INTERSCIENCE® PUBLICATION
JOHN WILEY & SONS, INC.
NEW YORK · CHICHESTER · BRISBANE · TORONTO · SINGAPORE

Library of Congress Catalog Number: 58-9935

ISBN 0-471-54570-8

Printed and bound in the United States of America
by Braun-Brumfield, Inc.

10 9 8 7 6 5 4 3 2 1

CONTRIBUTORS TO VOLUME LXXXI

VLADIMIR R. BELOSLUDOV, Institute of Inorganic Chemistry, Academy of Sciences of USSR, Siberian Branch, Novosibirsk, USSR

I. B. BERSUKER, Laboratory of Quantum Chemistry, Institute of Chemistry, Academy of Sciences of SSRM, Kishinev, USSR

S. A. BORSHCH, Laboratory of Quantum Chemistry, Institute of Chemistry, Academy of Sciences SSRM, Kishinev, USSR

STEPHEN E. BRADFORTH, Department of Chemistry, University of California, Berkeley, California

ROSS BROWN, Centre de Physique Moléculaire Optique et Hertzienne, Université de Bordeaux I, Talence, France

M. W. EVANS, Center for Theory and Simulation in Science and Engineering, Cornell University, Ithaca, New York

FRIEDRICH HUISKEN, Max Planck Institut, für Strömmungsforschung, Göttingen, Federal Republic of Germany

PHILEMON KOTTIS, Centre de Physique Moléculaire Optique et Hertzienne, Université de Bordeaux I, Talence, France

RICHARDO B. METZ, Department of Chemistry, University of California, Berkeley, California

DANIEL M. NEUMARK, Department of Chemistry, University of California, Berkeley, California

EVALDAS E. TORNAU, Semiconductor Physics Institute, Lithuanian Academy of Sciences, Vilnius, Lithuania, USSR

VLADIMIR E. ZUBKUS, Semiconductor Physics Institute, Lithuanian Academy of Sciences, Vilnius, Luthuania, USSR

INTRODUCTION

Few of us can any longer keep up with the flood of scientific literature, even in specialized subfields. Any attempt to do more and be broadly educated with respect to a large domain of science has the appearance of tilting at windmills. Yet the synthesis of ideas drawn from different subjects into new, powerful, general concepts is as valuable as ever, and the desire to remain educated persists in all scientists. This series, *Advances in Chemical Physics*, is devoted to helping the reader obtain general information about a wide variety of topics in chemical physics, which field we interpret very broadly. Our intent is to have experts present comprehensive analyses of subjects of interest and to encourage the expression of individual points of view. We hope that this approach to the presentation of an overview of a subject will both stimulate new research and serve as a personalized learning text for beginners in a field.

ILYA PRIGOGINE
STUART A. RICE

CONTENTS

TRANSITION STATE SPECTROSCOPY OF BIMOLECULAR REACTIONS USING NEGATIVE ION PHOTODETACHMENT

RICARDO B. METZ,* STEPHEN E. BRADFORTH,
and DANIEL M. NEUMARK†

Department of Chemistry, University of California, Berkeley, California

CONTENTS

*University Fellow, University of California,
†Alfred P. Sloan Fellow and NSF Presidential Young Investigator.

Advances in Chemical Physics, Volume LXXXI, Edited by I. Prigogine and Stuart A. Rice.
ISBN 0-471-54570-8   © 1992 John Wiley & Sons, Inc.

I. INTRODUCTION

A primary objective of experimental and theoretical investigations in chemical physics has been to gain a detailed understanding of the transition state region in a chemical reaction. The transition state region is the region of the reactive potential energy surface where chemical change occurs, and studies which probe this region offer the promise of qualitative improvements in our understanding of the microscopic forces that govern the course of a chemical reaction. Although much of the effort in this area has concentrated on the extraction of properties of the transition state region through scattering experiments, a parallel and considerably smaller effort has focused on the development of spectroscopic methods which can be applied directly to the transition state. These "transition state spectroscopy" experiments are complicated by the short-lived nature of the transition state (10^{-15}–10^{-12} sec). In addition, the transition state region must be accessed in a sufficiently well-defined manner so that one can observe meaningful structure associated with the transition state.

A number of frequency and time-resolved techniques have been applied to the transition state in recent years. These have been admirably reviewed by Brooks[1] and Zewail.[2] Zewail and co-workers[3,4] have carried out a particularly noteworthy set of experiments in which chemical reactions and photodissociation are monitored in real time using femtosecond lasers. This chapter focusses on a complementary type of transition state spectroscopy experiment performed in our laboratory in which photodetachment of a stable negative ion is used to probe the transition state region in bimolecular chemical reactions.

The principle behind photodetachment-based transition state spectroscopy is that if the geometry of a stable anion is similar to that of the neutral transition state, then one can photodetach the anion and prepare the unstable transition state complex in a well-defined manner. The idea of using photodetachment to create an unstable neutral species has been in the literature for some time. In 1968 Golub and Steiner[5] reported the total photodetachment cross section of the anion $OH^-(H_2O)$; the resulting $OH \cdot H_2O$ complex dissociates rapidly. More recently, Brauman and co-workers[6] measured total cross sections for the photodetachment of a series of anions $ROHF^-$ in which the neutral $ROHF$ complex dissociates to $HF + RO$.

To obtain more detailed information on the dissociating neutral complex, one can use techniques such as negative ion photoelectron spectroscopy[7,8] and threshold photodetachment spectroscopy.[9] These methods can yield vibrational structure associated with the transition state even if dissociation occurs in as little as 10^{-14} sec. These techniques can be applied to unimolecular decomposition, as in the photoelectron spectrum of HCO^- obtained by Lineberger and co-workers,[10] which shows resolved transitions to vibrationally excited levels of HCO which lie up to 4000 cm^{-1} above the H + CO dissociation limit. One can also study unimolecular isomerization, as in Lineberger's photoelectron spectrum of the vinylidene anion $C_2H_2^-$.[11] Although the neutral vinylidene radical rapidly isomerizes to acetylene, vibrational features attributed to the vinylidene structure are observed in the spectrum.

During the last few years, we have applied photodetachment spectroscopy to the study of transition states in *bimolecular* chemical reactions. We have studied several hydrogen transfer reactions A + HB → HA + B, where A and B are atomic or polyatomic species, via photodetachment of the stable anion AHB^-.[12] In many cases, the AHB^- anion is strongly hydrogen-bonded with a dissociation energy of ~1 eV or higher, and its overall size and geometry are such that one obtains good overlap with the neutral transition state region upon photodetachment. One then obtains resolved vibrational progressions which reveal the spectroscopy and dissociation dynamics of the neutral AHB complex, and from this one can learn about the features of the A + HB potential energy surface near the transition state.

One of the primary motivations for conducting these studies was that they appeared to hold high promise for experimentally observing *reactive resonances*. In the early 1970s, quantum mechanical scattering calculations on model collinear potential energy surfaces for the H + H$_2$ reaction by Truhlar and Kuppermann[13] and Wu and Levine[14] revealed pronounced oscillatory structure in the reaction probability as a function of translational energy. This structure was attributed[15] to reactive resonances which occur because of quasi-bound states with high probability density in the transition state region. In the following years, resonances were found in collinear scattering calculations for the F + H$_2$ reaction.[16–19] Especially sharp resonance structure due to long-lived states of the collision complex was predicted in calculations by Pollak[20] and Bondi et al.[21] for heavy + light-heavy reactions such as Cl + HCl or I + HI. This is the mass combination relevant to most of the experiments performed in our laboratory.

Clearly, the experimental observation of resonances would be an important step in understanding the vibrational structure of the transition state.[22] However, while resonances have been shown to exist in scattering calculations on three-dimensional (rather than collinear) potential energy surfaces,[23] their experimental detection in a scattering experiment is complicated by the contribution of many partial waves to the total reaction cross-section. This has been predicted[24,25] to cause substantial blurring of resonances in total cross-section measurements, such as those recently performed on the H + H$_2$ reaction.[26] Differential cross-section measurements on the F + H$_2$ reaction[27] show evidence for reactive resonances, but the ultimate interpretation of these results requires a more accurate potential energy surface than is currently available.

The photodetachment experiment offers an excellent opportunity for the observation of reactive resonances. Long-lived quasi-bound states on the A + HB potential energy surface should appear as sharp structure in the AHB$^-$ photoelectron spectrum, provided these quasi-bound states have good Franck–Condon overlap with the anion. In our experiments, the anions are produced in a free-jet expansion which should result in substantial rotational cooling. This means that, subsequent to photodetachment, the total angular momentum available to the reaction is much more restricted than in a reactive scattering study of the same system. Resonances should therefore be more pronounced in the photodetachment experiment. The results and analysis presented below will show that resonances are responsible for some, but not all, of the structure seen in the AHB$^-$ photodetachment spectra.

The organization of this chapter is as follows. In Section II the experimental methods used in this work are described. Section III presents a general discussion of the theoretical methods useful in analyzing these spectra. Section IV covers the symmetric Br + HBr and I + HI reactions, which are studied by photodetaching the bihalide anions BrHBr$^-$ and IHI$^-$. Finally, several asymmetric hydrogen exchange reactions are discussed in Section V.

II.   EXPERIMENTAL METHODS

The spectra shown below were obtained using two negative ion photodetachment methods: "fixed-frequency" photoelectron spectroscopy and threshold photodetachment spectroscopy. Both experiments involve the generation of an internally cold, mass-selected negative ion beam, but different photodetachment and electron detection schemes are employed. The two instruments are described briefly in this section; more detailed descriptions may be found elsewhere.[9,28]

Figure 1 illustrates the principles behind the two techniques. In photoelectron spectroscopy (Fig. 1a), the negative ions are photodetached with a fixed-frequency laser and the kinetic energy distribution of the ejected photoelectrons is measured. All energetically accessible levels of the neu-

a) 'fixed-frequency' photoelectron spectroscopy
(8 meV resolution)

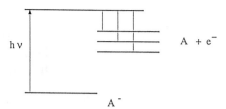

b) threshold photodetachment spectroscopy (0.4 meV)

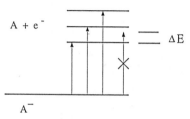

Figure 1.   Energy levels and laser excitation scheme for (*a*) photoelectron spectroscopy and (*b*) threshold photodetachment spectroscopy.

tral are populated according to their Franck–Condon overlap with the initial state of the anion. For a single photodetachment event, the electron kinetic energy (eKE) is given by

$$eKE = h\nu - E_b^{(-)} - E_i^{(0)} + E_i^{(-)} \tag{1}$$

Here $h\nu$ is the photon energy and $E_b^{(-)}$ is the energy difference between the anion and neutral ground states. In the case of IHI$^-$ photodetachment, where the IHI complex is unstable with respect to dissociation, $E_b^{(-)}$ is the energy required to remove an electron from the ground state of IHI$^-$ to form I + HI($v = 0$). $E_i^{(-)}$ and $E_i^{(0)}$ are the internal energies of the anion and neutral, respectively. For IHI$^-$ photodetachment, $E_i^{(0)}$ is measured relative to I + HI($v = 0$).

Thus, the electron kinetic energy distribution exhibits peaks resulting from transitions between ion and neutral energy levels. In nearly all cases, the energy resolution of photoelectron spectroscopy is insufficient to discern rotational structure and one only learns about vibrational energy levels of the neutral and (occasionally) the anion. In our spectrometer,

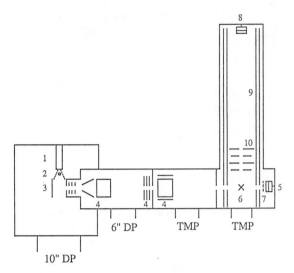

Figure 2. Schematic diagram of time-of-flight photoelectron spectrometer. Salient features are described in text (Ref. 28).

for example, the energy resolution of our time-of-flight analyzer is 8 meV at eKE = 0.65 eV and degrades at higher electron kinetic energies as $(eKE)^{3/2}$. This is the instrumental resolution function used in the simulations of Sections IV and V.

Figure 2 shows a schematic of the negative ion photoelectron spectrometer used in these studies.[28] A similar instrument is described by Posey et al.[29] Because many of the AHB⁻ anions of interest have high electron binding energies (3.80 eV for IHI⁻, 4.27 eV for BrHBr⁻), a pulsed UV laser is the most appropriate photodetachment light source, and the entire instrument is based on pulsed technology. An excellent review of pulsed methods in ion spectroscopy is provided by Johnson and Lineberger,[30] and several components of our instruments are described in more detail therein.

In the photoelectron spectrometer, negative ions are generated by expanding a mixture of neutral gases through a pulsed molecular beam valve (1) and crossing the molecular beam with a 1-keV electron beam (2) just outside the valve orifice. The fast electrons produce relatively slow secondary electrons by ionization. These slow electrons, which rapidly thermalize due to the high gas density in front of the beam valve, efficiently produce negative ions through low-energy attachment processes. To make BrHBr⁻, for example, a 5% HBr/Ar mixture is used. The likely mechanism is formation of Br^- via dissociative attachment to HBr, and formation of $BrHBr^-$ via three-body clustering reactions:

$$Br^- + HBr + M \rightarrow BrHBr^- + M \qquad (2)$$

The ions are formed in the continuum flow region of the supersonic expansion and cool internally as the expansion progresses.

A time-of-flight mass spectrometer of the Wiley–Maclaren design[31] is used to select the mass of the ion of interest. A small volume of negative ions is extracted at 90° from the molecular beam by applying a negative pulse to the electrode (3). The ions then pass through a 1-kV potential drop. As they traverse the 140-cm drift region (4), the ions separate into bunches according to mass. The photodetachment laser pulse is timed so that it crosses the ion beam at (6) and interacts with the ion bunch of the desired mass. The spectra shown below were obtained with the fourth harmonic (266 nm, 4.660 eV) or fifth harmonic (213 nm, 5.825 eV) of a Nd:YAG laser. A small fraction (10^{-4}) of the ejected photoelectrons is detected (8) at the end of a 100-cm field free-flight tube (9) and energy analyzed by time of flight.

Considerably higher resolution can be obtained with threshold photodetachment spectroscopy. The principle of this method is shown in Fig. 1b. In this experiment, negative ions are photodetached with a tunable pulsed laser. At a given laser wavelength, only electrons produced with nearly zero kinetic energy are detected. The zero-kinetic energy spectrum plotted as a function of laser wavelength consists of a series of peaks, each corresponding to an ion → neutral transition. The width of the peaks is determined by the ability of the instrument to discriminate against photoelectrons produced with high kinetic energy. By adapting the methods developed by Müller–Dethlefs et al. for threshold photoionization of neutrals,[32] we have achieved a resolution of $3 \, cm^{-1}$ (0.37 meV) with this instrument and were able to obtain a spectrum of SH^- in which transitions between individual rotational levels of the anion and neutral were resolved.[9]

Figure 3 shows a schematic of the threshold photodetachment spectrometer.[9] The ion source (1, 2) is similar to that shown in Fig. 2, but the ions pass through a skimmer and into a differentially pumped region prior to being accelerated to 1 keV. A coaxial time-of-flight mass spectrometer (3) of the design proposed by Bakker[33] is used for mass separation. The anions are photodetached using an excimer-pumped dye laser with frequency-doubling capability.

In order to selectively detect zero-kinetic-energy photoelectrons, the region in which detachment occurs (4) is initially field-free. A weak extrac-

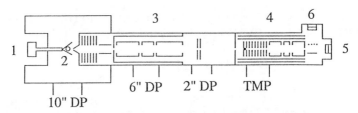

Figure 3. Schematic diagram of threshold photodetachment spectrometer: (1) pulsed beam valve, (2) electron beam, (3) coaxial time-of-flight mass spectrometer, (4) photodetachment region, (5) ion detector, (6) electron detector. Laser–ion beam interaction region is marked by ● (Ref. 9).

tion pulse is applied along the ion beam axis 200–300 ns after the photodetachment laser pulse. The purpose of this delay is to allow the higher energy electrons to spatially separate from the zero-kinetic-energy electrons. The higher-energy electrons that scatter perpendicularly to the ion beam axis are discriminated against since they will not pass through one of the many apertures between the detachment region and the electron detector (6). The higher-energy electrons that scatter along the beam axis will be in different regions of the extraction field when it is applied. They therefore emerge from the extraction region with different kinetic energies from the electrons produced with zero kinetic energy, arrive at the electron detector at different times, and are discriminated against by gated detection of the electron detector signal. The combination of spatial and temporal filtering is essential to achieving the ultimate resolution of this detection scheme.

III. THEORETICAL METHODS

In most negative-ion photodetachment experiments, a stable neutral species is formed by photodetaching a negative ion. Photoelectron spectra are typically analyzed by invoking the Born–Oppenheimer and Condon approximations. The electronic and vibrational degrees of freedom in the anion and neutral are assumed to be separable, so that the cross-section for photodetachment from anion vibrational level v'' to neutral vibrational level v' is given by

$$\sigma(v' \leftarrow v'') \propto |\tau_e|^2 |\langle \psi_{v'}^{(0)} | \psi_{v''}^{(-)} \rangle|^2 \tag{3}$$

Here τ_e is the dipole matrix element between the electronic state of the anion and the (neutral + e^-) continuum, which is assumed constant for a given electronic band. The approximations in Eq. (3) have been used in all theoretical treatments of photodetachment-based transition state spectroscopy experiments to date. In general, neither the anion or neutral potential energy surface is sufficiently well known to warrant a more sophisticated treatment.

The application of Eq. (3) to the spectra presented below is problematic since they all involve formation of an unbound, short-lived neutral complex. Thus, the vibrational wavefunction $\psi_{v'}^{(0)}$ in Eq. (3) must be replaced by a scattering wavefunction. One can then simulate the photoelectron spectrum within a time-independent formalism, in which the Franck–Condon overlap between the anion and neutral scattering wavefunctions is calculated as a function of energy. Alternatively, the spectrum can be simulated using time-dependent wavepacket analysis. Both types of simulations are discussed below.

A. Time-Independent Analysis

The time-independent analysis of these photodetachment experiments has been summarized nicely by Schatz.[34] Equation (3) is replaced by

$$\sigma(E) \propto |\langle \psi^{(0)}(E)|\psi_{v''}^{(-)}\rangle|^2 \tag{4}$$

where $\psi^{(0)}(E)$ is the scattering wavefunction supported by the neutral potential energy surface at energy E. In most scattering calculations, E is measured relative to the neutral potential energy minimum in the asymptotic region for either reactants or products and differs from $E_i^{(0)}$ in Eq. (1) by the zero-point energy of the neutral fragments.

The accurate simulation of a photoelectron spectrum using Eq. (4) therefore requires the determination of A + HB scattering wavefunctions over the energy range for which $\sigma(E)$ is nonnegligible. This is a formidable task which, so far, has only been attempted for triatomic systems, in which case a three-dimensional scattering wavefunction must be determined. Schatz has simulated the ClHCl⁻ and IHI⁻ photoelectron spectra in this manner[35,36] using model London–Eyring–Polanyi–Sato (LEPS) potential energy surfaces for the Cl + HCl and I + HI reactions. The neutral wavefunctions are calculated with a coupled-channel method using hyperspherical coordinates.[37] The calculations assume the anions to be ground-state harmonic oscillators and are restricted to total angular momentum $J = 0$ for the anion and neutral. Both of these assumptions are justified by the low anion temperatures attained in the free-jet ion source, although the small rotational constants for ClHCl⁻ ($B_e = 0.09685 \text{ cm}^{-1}$)[38] and IHI⁻ mean that several rotational levels will be populated even at 10 K.

Using a different method for calculating the scattering wavefunctions, Zhang and Miller[39] have performed a three-dimensional simulation of the

FH_2^- photoelectron spectrum, again assuming $J = 0$. The calculation uses the *ab initio* $F + H_2$ surface of Steckler et al.[40] and the *ab initio* anion geometry of Kendall et al.[41] This simulation has recently been compared to the experimental FH_2^- spectrum.[42]

A more approximate method for simulating AHB^- photoelectron spectra in three dimensions has been developed by Gazdy and Bowman[43] and used to simulate the $ClHCl^-$ and IHI^- photoelectron spectra. They use an L^2 basis to obtain the neutral wavefunctions. This involves the imposition of boundary conditions which result in replacing three-dimensional continuum wavefunctions with bound-state wavefunctions. The L^2 method is well suited for determining the position of resonances, since the resonance wavefunctions are highly localized and are similar to bound-state wavefunctions. In regions of the $ClHCl^-$ and IHI^- photoelectron spectra dominated by direct scattering, the L^2 method yields a many-line stick spectrum while the "exact" three-dimensional simulation[35,36] typically yields a smaller number of peaks with varying widths. Nonetheless, the envelope of the L^2 stick spectrum matches the exact simulation reasonably well. The L^2 method is therefore a useful first test of a model three-dimensional potential energy surface for a reaction of interest.

The simulation of AHB^- photoelectron spectra is further simplified using Bowman's adiabatic bend approximation.[44] This allows one to reduce the three-dimensional potential energy surface for a three-atom system to a two-dimensional surface by an approximate treatment of the bending motion. For example, the three-dimensional potential energy function for $Br + HBr$ is given by $V(Q_1, Q_3, \gamma)$, where

$$Q_1 = \frac{1}{\sqrt{2}}(r_1 + r_2)$$

$$Q_3 = \frac{1}{\sqrt{2}}(r_1 - r_2) \tag{5}$$

are the symmetric and antisymmetric stretch symmetry coordinates (r_1 and r_2 are the two H–Br bond lengths) and γ is the Br–H–Br bending angle. By assuming the bending motion is separable from the two stretches, $V(Q_1, Q_3, \gamma)$ is reduced to a two-dimensional "effective" collinear potential energy surface $V(Q_1, Q_3)$ via

$$V(Q_1, Q_3) = V(Q_1, Q_3, \gamma = \pi) + \epsilon_0(Q_1, Q_3) \tag{6}$$

Here $\epsilon_0(Q_1, Q_3)$ is the bending zero-point energy for the linear configuration of the nuclei specified by Q_1 and Q_3. $V(Q_1, Q_3)$ is therefore an approximate potential surface for the BrHBr complex in its ground-bending vibration. This approximation should be reasonably accurate if the reaction is collinearly dominated, so that photodetachment of a linear ion such as BrHBr^{-} [45,46] should not produce substantial bending excitation in the neutral complex.

Once the effective collinear surface is obtained, one can simulate the AHB^{-} photoelectron spectrum via Eq. (4), where $\psi^{(0)}(E)$ is now a two-dimensional scattering wavefunction. One can obtain $\psi^{(0)}(E)$ from an exact two-dimensional calculation,[47] by using an L^2 basis,[48] or by making the DIVAH approximation[49] in which $\psi^{(0)}(E)$ is assumed to be a product of two one-dimensional wavefunctions. The DIVAH approximation is quite accurate for heavy + light-heavy reactions[21] and will be discussed in more detail in Section IV, where it is used to obtain an effective collinear surface for the Br + HBr reaction. One can also perform an exact two-dimensional simulation using the time-dependent methods described in the following section.

B. Time-Dependent Analysis

An alternative conceptual framework for analyzing and interpreting the photodetachment experiments is provided by time-dependent wavepacket analysis. This approach, initially developed by Heller,[50] has been extended to a broad range of problems in spectroscopy and dynamics.[51] Particularly relevant applications of this method to the material in this chapter include the work of Bisseling et al.,[52,53] who have used time-dependent analysis to study resonances in the F + DBr reaction, and the work of Lorquet,[54] who first used the time-dependent formalism to interpret the photoelectron spectrum of neutrals in which a dissociating ionic species is formed.

The basic idea behind the application of time-dependent analysis to photodetachment is as follows. In the time-independent analysis described above, the stationary state A + HB scattering wavefunctions are calculated at many values of energy and the photoelectron spectrum is simulated via Eq. (4). The time-dependent perspective is based on the equivalent equation in which the photoelectron spectrum $\sigma(E)$ is expressed as the Fourier transform of a time-autocorrelation function $C(t)$,

$$\sigma(E) \propto \int_{-\infty}^{\infty} \exp(iEt/\hbar)C(t)\, dt \qquad (7)$$

where

$$C(t) = \langle \phi(t = 0) \mid \phi(t) \rangle \tag{8}$$

Here $\phi(t = 0)$ is the initial wavepacket formed on the A + HB potential energy surface which, within the Condon approximation, is given by the stationary state wavefunction for the anion. The wavepacket $\phi(t)$ at later times is found by propagating $\phi(0)$ on the A + HB potential energy surface,

$$|\phi(t)\rangle = \exp[-i\hat{H} t/\hbar] |\phi(0)\rangle \tag{9}$$

where $\hat{H}$ is the Hamiltonian on the neutral surface. In our implementation of time-dependent analysis,[55] we restrict ourselves to two-dimensional wavepacket propagation on "effective" collinear potential energy surfaces. We use the propagation scheme of Kosloff and Kosloff[56] in which the propagator is approximated by a second-order differencing scheme and the kinetic energy at each time t is evaluated by the Fourier method. Although simulated photoelectron spectra using Eq. (6) are identical to those obtained using Eq. (4), the time-evolving wavepacket $\phi(t)$ and the autocorrelation function $C(t)$ provide additional insight concerning the AHB$^-$ photoelectron spectra. The connection between observed features in the spectra and dissociation dynamics of the AHB complex is particularly clear using the time-dependent formalism.

IV.   SYMMETRIC HYDROGEN EXCHANGE REACTIONS

A.   Photoelectron Spectrum of BrHBr$^-$

1.   *Summary of Results*

The 213-nm photoelectron spectra[28] of BrHBr$^-$ and BrDBr$^-$ (Fig. 4) show a series of well-resolved peaks with widely varying widths. The widths of the peaks labelled 0, 2, and 4 in the BrHBr$^-$ spectrum are 170, 80, and 20 meV, respectively, and the widths of peaks 0, 2, 4, and 6 in the BrDBr$^-$ spectrum are 175, 100, 64, and 20 meV, respectively. The discussion in Sections IV.A.1–3 focusses on the peak positions and intensities. The peak widths will be discussed in Sections IV.A.4–6.

We first consider whether the peaks in these spectra correspond to transitions to levels of the neutral BrHBr complex which are bound or unbound with respect to Br + HBr ($v = 0$). This can be determined with

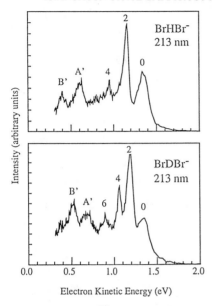

Figure 4. BrHBr⁻ and BrDBr⁻ photoelectron spectra taken at 213 nm photodetachment wavelength ($h\nu = 5.825$ eV). Peaks are labelled with v_3' quantum number (Ref. 28).

the energy level diagram, Fig. 5. This figure shows that peaks with electron kinetic energy *less* than $E_0^{\text{ke}} \equiv h\nu - D_0(\text{BrHBr}^-) - \text{EA(Br)}$ result from levels of the neutral complex that lie *above* Br + HBr ($v = 0$). Here, $h\nu$ is the photon energy (5.825 eV), $D_0(\text{BrHBr}^-)$ is the energy needed for the dissociation reaction $\text{BrHBr}^- \rightarrow \text{Br}^- + \text{HBr}$, and EA(Br) is the electron affinity of Br. $D_0(\text{BrHBr}^-)$ is given approximately by the enthalpy of dissociation of BrHBr⁻, 0.91 ± 0.05 eV,[57] and with EA(Br) = 3.365 eV,[58] we find $E_0^{\text{ke}} = 1.55 \pm 0.05$ eV. The peak at highest electron kinetic energy (Peak 0) appears at 1.353 eV for BrHBr⁻, so *all* the observed peaks in the BrHBr⁻ spectrum result from transitions to unbound states of the neutral complex.

We next consider the isotope effects observed in the spectra. Peak 0 occurs at the same electron kinetic energy in the BrHBr⁻ and BrDBr⁻ spectra. The remaining peaks shift to higher electron energy, resulting in a smaller peak spacing. These two observations indicate that peak 0 is the origin of a vibrational progression in the *neutral complex*, and that the active mode involves H atom motion. None of the peaks appear to be "hot bands" originating from levels of the anion with excitation in the active mode, as these would shift to lower electron kinetic energy upon deuteration.

The active mode could be either the bend (v_2) or antisymmetric stretch (v_3) of the complex. BrHBr⁻ is predicted to be linear,[45,46] and, based on *ab initio* surfaces calculated for the Cl + HCl reaction,[59,60] the minimum

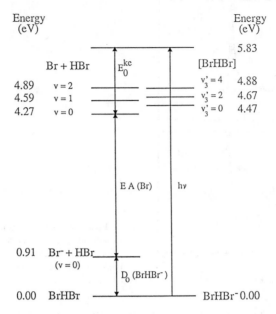

Figure 5.   Energy level diagram for BrHBr⁻/BrHBr system (Ref. 28).

energy path for the Br + HBr reaction is likely to be nearly collinear. An extended progression in the bend is therefore unlikely, and we assign the peaks to a progression in the v_3 mode. Symmetry considerations show that only transitions to even v'_3 levels of the complex are allowed from the $v''_3 = 0$ level of the ion. The peaks in each spectrum are labelled by their v'_3 quantum number. As discussed in more detail below, the two broad peaks at lowest energy in the BrHBr⁻ and BrDBr⁻ spectra (A' and B') do not appear to belong to the same progression as the higher-energy peaks and are assigned to transitions to an electronically excited Br + HBr surface.

The energies of the v'_3 states for the BrHBr complex and the Br + HBr asymptotic vibrational energy levels are shown in Fig. 5. The vibrational spacing in the BrHBr complex is significantly smaller than that in free HBr. For example, peaks 0 and 2 in the BrHBr⁻ spectrum are separated by 0.194 eV (1565 cm⁻¹), whereas the $v = 0$ and $v = 1$ levels of HBr are separated by 0.317 eV (2557 cm⁻¹). The intuitive explanation for this is that the strong HBr bond in free HBr is replaced by two much weaker bonds in the BrHBr complex. This "red shift" indicates that we are probing the transition state region of the potential energy surface, where the hydrogen is interacting strongly with both bromine atoms.

2. Analysis: Preliminary Considerations

The goal of the analysis of the BrHBr$^-$ and BrDBr$^-$ photoelectron spectra is to construct a potential energy surface for the Br + HBr reaction which reproduces the experimental spectra in a simulation. This is accomplished using a potential energy surface with many adjustable parameters. By varying these parameters until the experimental spectra are reproduced, one can, in principle, map out the true potential energy surface in the region probed by photodetachment. Because of the large number of simulations required in this iterative procedure, the analysis was limited to constructing the best "effective" collinear surface (see Eq. (6)) that reproduced the experiments. This reduces the hardest part of the simulation, the calculation of the three-dimensional scattering wavefunctions in Eq. (4), to a much simpler two-dimensional problem.

In the past, LEPS potential energy surfaces have often been used to model the results from kinetics and reactive scattering experiments. However, while a LEPS surface can provide a qualitative framework for understanding a photoelectron spectrum of the type under consideration here, we have found that a more flexible function for the effective collinear surface is needed to reproduce the spectrum. The actual functional form used in our simulations is discussed in detail in Ref. 28.

The simulations also require wavefunctions for the anion. In contrast to FHF$^-$ [61] and ClHCl$^-$,[38] no gas-phase spectroscopic data exist for BrHBr$^-$. *Ab initio* calculations by Peyerimhoff[45] and Nomura[46] predict a $D_{\infty h}$ equilibrium geometry for BrHBr$^-$ with an interhalogen distance, R_e(BrBr), of 3.43 and 3.36 Å, respectively. Nomura predicts the ν_1 frequency to be 200 cm^{-1} and the ν_3 frequency to be 837 cm^{-1}; the matrix isolation spectroscopy values for these frequencies are 164 and 728 cm^{-1}, respectively.[62] In our analysis, we assumed R_e(BrBr) to be 3.50 Å. (Nomura's value became available subsequent to publication of our results.) We use harmonic oscillator wavefunctions with the matrix isolation values for the frequencies. Since all the peaks in the photoelectron spectra appear to originate from the ground state of the anion, this is assumed to be the initial anion vibrational state in all the simulations.

The end result of our analysis is the "best fit" effective collinear surface in Fig. 6. The method by which this surface is obtained is described in Section IV.A.4; for now, we assume it is the correct surface. The barrier height on this surface is 46.8 kJ/mol, and the saddle point occurs at $R_{Br-Br} = 3.29$ Å. This surface is plotted using modified hyperspherical coordinates ρ and z, which, for the Br–H–Br mass combination are approximately proportional to the symmetry coordinates Q_1 and Q_3 (Eq. (5)):

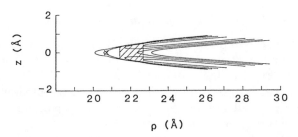

Figure 6. Contour plot of "best-fit" effective collinear potential energy surface for Br + HBr reaction. Coordinates are defined in text (Eq. 10). The Franck–Condon region (see text) is shaded. Contours are at −300, −325, −350 , and −370 kJ/mol with respect to three-atom dissociation (Ref. 28).

$$\rho = \sqrt{\frac{\mu_{Br,HBr}}{\mu_{HBr}}} R_{Br-Br} = 6.30 R_{Br-Br} = 8.91 Q_1$$

$$z \approx \frac{Q_3}{\sqrt{2}} \tag{10}$$

The approximations in Eq. (10) are valid for the heavy + light-heavy mass combination. The acute skew angle in Fig. 6 is characteristic of a heavy + light-heavy reaction such as Br + HBr and is given by

$$\phi_m = \tan^{-1}\{[m_H(m_H + 2m_{Br})/m_{Br}^2]^{1/2}\} = 9.1° \tag{11}$$

The modified hyperspherical coordinates can be expressed in terms of the more familiar mass-scaled coordinates x and y. The latter are defined for the reaction $Br' + HBr \rightarrow HBr' + Br$ by

$$x = \sqrt{\frac{\mu_{Br,HBr}}{\mu_{HBr}}} R_{Br',HBr} \tag{12}$$

$$y = R_{H-Br}$$

where $R_{Br',HBr}$ is the distance between the Br' atom and the HBr center of mass. The two sets of coordinates are related by

$$\rho = (x^2 + y^2)^{1/2}$$

$$z = \rho \left(\arctan \frac{y}{x} + \frac{\phi_m}{2} \right) \tag{13}$$

The shaded region in Fig. 6 indicates the area on the neutral surface that is directly probed by our photodetachment experiment. The center of the shaded region, at $z = 0$, $\rho = 22.05$ Å, corresponds to $R_e(BrBr) = 3.50$ Å, the assumed interhalogen distance in $BrHBr^-$. The extent of the shaded region along the ρ coordinate represents the zero-point amplitude for the symmetric stretch in the ground vibrational state of the ion. Our spectra are very sensitive to the details of this Franck–Condon region of the potential energy surface, which lies quite close to the saddle point on the surface.

3. One-Dimensional Simulation of BrHBr⁻ Spectrum

In Section IV.A.1, the first few peaks in the $BrHBr^-$ and $BrDBr^-$ spectra were assigned to a progression in the v_3 antisymmetric stretch mode of the neutral BrHBr complex on the basis of the observed isotope shift. This assignment is supported by a simple one-dimensional analysis based on the potential energy surface in Fig. 6. The z coordinate in Fig. 6 is proportional to the antisymmetric stretch coordinate in the complex. By taking a cut along the z axis through the center of the Franck–Condon region, one obtains an approximate antisymmetric stretch potential for the complex at the geometry probed by photodetachment. This is a double minimum potential with a barrier at $z = 0$. We should therefore be able to simulate the peak positions and intensities in the photoelectron spectrum by calculating the Franck–Condon overlap between the anion $v_3'' = 0$ level and the eigenfunctions supported by a double minimum potential.

The harmonic antisymmetric stretch potential for $BrHBr^-$ is shown at the bottom of Fig. 7. The lower double minimum potential in Fig. 7 is the potential assumed for the v_3 mode in the ground electronic state of the neutral complex.[63] The $v_3'' = 0$ levels for $BrHBr^-$ and $BrDBr^-$ are shown, as well as the first few eigenvalues for the BrHBr and BrDBr complexes. As discussed above, all the peaks in the spectra originate from the $v_3'' = 0$ anion level, and from this level, only transitions to even v_3' levels of the complex have nonzero intensity. The resulting one-dimensional simulations of the $BrHBr^-$ and $BrDBr^-$ spectra are superimposed on the experimental spectra. Reasonable agreement is obtained with the positions and intensities of the $v_3' = 0$, 2, and 4 peaks in the $BrHBr^-$ spectrum and the $v_3' = 0$, 2, 4, and 6 peaks in the $BrDBr^-$ spectrum. Note that the one-dimensional simulation yields a stick spectrum; to facilitate comparison with the experimental intensities, the sticks have been convoluted with Gaussians to match the experimental peak widths.

The two broad low-energy peaks (A′ and B′) do not seem to belong to the same vibrational progression as the higher-energy peaks. It is likely

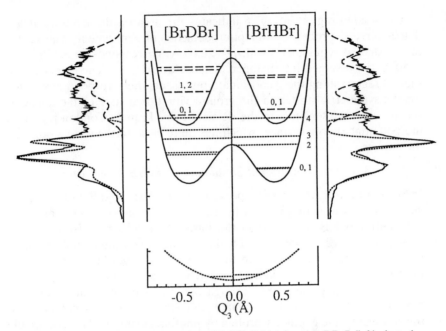

Figure 7. One-dimensional simulation of BrHBr⁻ (right) and BrDBr⁻ (left) photoelectron spectra. Anion v_3 harmonic oscillator potential and $v_3'' = 0$ levels of BrHBr⁻ and BrDBr⁻ are shown at bottom. Double minimum v_3 potentials at middle and top are for ground and electronically excited neutral BrHBr (BrDBr) complex, respectively. The first few v_3 energy levels are shown and v_3 quantum numbers are indicated for several levels including several nearly degenerate pairs. Results of one-dimensional simulations are compared with experimental spectra (solid) at the sides of the figure. For comparison with the experiment, the simulated peaks have been convoluted with Gaussians to match the experimental peak widths. Each tick mark on the vertical axis is 0.1 eV.

that peaks A′ and B′ are part of a second vibrational progression arising from a low-lying excited electronic state of the complex. The large isotope shift of peak A′ indicates that it is not the origin of this progression. However, the entire spectrum for each isotope can be simulated by assuming peaks A′ and B′ result from transitions to the $v_3' = 2$ and $v_3' = 4$ levels of the upper double-minimum potential shown in Fig. 7. The simulation from this potential is also shown in Fig. 7. The high barrier between the wells makes the $0 \leftarrow 0$ transition relatively weak.

The $v_3 = 0$ levels of the two double minimum potentials in Fig. 7 are separated by 0.49 eV. This is slightly larger than the $^2P_{1/2}-^2P_{3/2}$ spin-orbit splitting in Br (0.45 eV) and suggests that the upper potential corresponds to an electronically excited state of the complex which asymptotically correlates to Br*($^2P_{1/2}$) + HBr. Based on diatomics-in-molecules (DIM) calculations[64] on F + HF and Cl + HCl, this excited state interaction is expected to be considerably more repulsive than the ground state. The high barrier between the wells of this potential is consistent with such a repulsive interaction.

The contribution of transitions to multiple potential energy surfaces in these spectra presents a complex and interesting problem. For example, the photoelectron spectrum of ClHCl⁻, which would at first seem like an ideal model system for study by this technique, shows barely resolved, very broad vibrational features.[28,65] Recent *ab initio* calculations by Yamashita and Morukuma[66] suggest that the broad peaks result from overlapping transitions to both the Cl + HCl ground state and a very low-lying excited state. We have also observed transitions to excited electronic states in the photoelectron spectrum of IHI⁻ and several asymmetric bihalide anions. These are discussed in more detail below.

In summary, one can use the simple one-dimensional analysis scheme described above to obtain a good first-order understanding of the BrHBr⁻ photoelectron spectrum. The analysis supports the assignment of the peaks in the spectrum to a progression in the v_3 antisymmetric stretch mode of the BrHBr complex. However, this is clearly only part of the story. The effective collinear surface in Fig. 6 has no wells in the transition state region, so one expects the BrHBr complex formed by photodetachment to dissociate rapidly. The dissociation dynamics of the complex affect the observed peak widths in the photoelectron spectrum, and this is not treated at all in our one-dimensional analysis.

The v_3 vibration treated in the one-dimensional analysis is largely decoupled from the dissociation coordinate of the BrHBr complex. This can be seen from Fig. 6; motion along the z coordinate is nearly perpendicular to the minimum energy path that leads to reaction along the reactant and product valleys. To understand the dissociation dynamics of the BrHBr complex, we must at least include motion along the $\rho(v_1)$ coordinate on the Br + HBr potential energy surface in our analysis. With reference to Fig. 6 again, the ρ coordinate is clearly strongly coupled to dissociation. Thus, a two-dimensional analysis is necessary to incorporate dissociation of the complex.

4.   Two-Dimensional Simulation (Time-Independent)

An exact two-dimensional simulation of the BrHBr⁻ photoelectron spectrum can be generated by determining the Franck–Condon overlap between the ground state of the anion and the scattering wavefunctions supported by the effective collinear potential $V(Q_1, Q_3)$ or $V(\rho, z)$. In the (ρ, z) coordinate system, the Schrodinger equation is

$$-\frac{1}{2\mu_{HBr}}\left\{\frac{\partial^2}{\partial\rho^2}+\frac{1}{\rho}\frac{\partial}{\partial\rho}+\frac{\partial^2}{\partial z^2}+V(\rho,z)\right\}\psi(\rho,z)=E\psi(\rho,z) \qquad (14)$$

Rather than solve the two-dimensional Schrödinger equation (9), we invoke an adiabatic approximation (the DIVAH approximation) that has been used by Römelt, Manz, and others[21,49] in studies of heavy + light-heavy reactions. This approximation is based on the separation of time scales between the fast motion along the the z coordinate, which primarily involves H atom motion, and the much slower motion along the ρ coordinate which corresponds to displacement of the heavy halogen atoms.

The mathematics of this approximation are virtually identical to the Born–Oppenheimer approximation for diatomic molecules. The two-dimensional scattering wavefunction $\psi(\rho,z)$ is assumed to be a product of two wavefunctions

$$\psi(\rho,z)=\theta(z;\rho)R(\rho) \qquad (15)$$

Here $\theta(z;\rho)$ is the wavefunction for the antisymmetric stretch coordinate z. This function varies slowly with ρ. $R(\rho)$ is the wavefunction for motion along the ρ coordinate. θ and R satisfy the one-dimensional differential equations

$$\left[-\frac{1}{2\mu_{HBr}}\frac{\partial^2}{\partial z^2}+V(\rho,z)\right]\theta(z;\rho)=\epsilon_{v_3}(\rho)\theta(z;\rho) \qquad (16)$$

and

$$\left[-\frac{1}{2\mu_{HBr}}\frac{\partial^2}{\partial\rho^2}+U_{v_3}(\rho)\right]R(\rho)=ER(\rho) \qquad (17)$$

where

$$U_{v_3}(\rho)=\epsilon_{v_3}(\rho)-\frac{1}{8\mu_{HBr}\rho^2}-\frac{1}{2\mu_{HBr}}Q_{v_3v_3} \qquad (18)$$

Equation (16) is the one-dimensional Schrodinger equation for a particle of mass μ_{HBr} with potential energy $V(\rho,z)$ at a fixed value of ρ. (The appropriate reduced mass for the Br + DBr reaction is μ_{DBr}). It is the same equation used in the one-dimensional analysis to determine the antisymmetric stretch energy levels and wavefunctions at the value of ρ corresponding to the equilibrium geometry of BrHBr⁻. For each value of ρ at which Eq. (16) is solved, one obtains a set of antisymmetric stretch

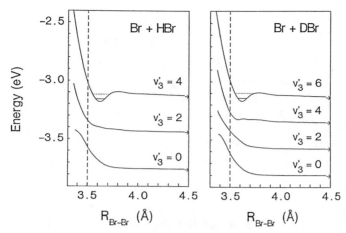

Figure 8. Effective potentials $U_{v_3}(\rho)$ (Eq. 18) for Br + HBr and Br + DBr derived from effective collinear surface in Fig. 6. Only potentials with even v_3 are shown (see text). Quasibound states are indicated by solid and dashed lines (solid lines denote states that can decay only by vibrational predissociation). The Br + HBr ($v = 0, 1, 2$) and Br + DBr ($v = 0$–3) asymptotic levels are indicated by arrows. The dashed vertical line at $R_{\text{Br-Br}} = 3.50$ Å is the assumed value of $R_e(\text{BrBr})$ in BrHBr$^-$ (Ref. 28).

eigenvalues $\epsilon_{v_3}(\rho)$ and eigenfunctions $\theta_{v_3}(z; \rho)$. One therefore generates a set of "vibrationally adiabatic" curves $\epsilon_{v_3}(\rho)$ showing how the energy of the v_3 antisymmetric level varies with ρ, which is proportional to the interhalogen separation (Eq. (10)). In the limit of infinite ρ, the $v_3 = 2m$ and $v_3 = 2m + 1$ levels become degenerate and correlate to the Br + HBr ($v = m$) vibrational energy level. The associated wavefunctions θ_{2m} and θ_{2m+1} have even and odd parity, respectively.

From the vibrationally adiabatic curves $\epsilon_{v_3}(\rho)$, one uses Eq. (18) to obtain the "effective potentials" $U_{v_3}(\rho)$. The DIVAH correction to $\epsilon_{v_3}(\rho)$, $Q_{v_3 v_3}$ in Eq. (18), is defined in Ref. 49. It is very small for a heavy + light-heavy reaction except near avoided crossings, and these do not occur for symmetric reactions. Figure 8 shows several of the lowest-lying effective potentials $U_{v_3}(\rho)$ with even v_3 obtained from the potential energy surface in Fig. 6.

The significance of the effective potentials can be seen from Eq. (17), the one-dimensional Schrodinger equation for a particle with total energy E and potential energy $U_{v_3}(\rho)$. For each value of v_3, $U_{v_3}(\rho)$ governs the motion of the BrHBr complex along the ρ coordinate. The effective potentials $U_0(\rho)$ and $U_2(\rho)$ in Fig. 8 are purely repulsive; each set of the solutions $R_0(\rho)$ and $R_2(\rho)$ of Eq. (12) using these potentials is a continuum of one-dimensional scattering wavefunctions which vary smoothly as a function of E.

However, the potential $U_4(\rho)$ has a small well and can support states of the BrHBr complex which are *quasi-bound* along the ρ coordinate in addition to a continuum of scattering wavefunctions; the energies of these quasi-bound states are indicated in Fig. 8. In a scattering calculation on this surface, these quasi-bound states would appear as resonances. Both states supported by the U_4 potential can decay by vibrational predissociation due to coupling with the repulsive effective potentials. In addition, the higher of the two quasi-bound states lies above the Br + HBr ($v = 2$) asymptote and can decay by tunneling through the barrier on the U_4 potential.

The effective potentials provide an intiuitive explanation of the wide variation in peak widths observed in the BrHBr$^-$ and BrDBr$^-$ photoelectron spectra (Section IV.A.1). Since the U_0 and U_2 effective potentials are repulsive, transitions to the $v_3' = 0$ and $v_3' = 2$ levels of the complex should be quite broad. Specifically, the $v_3 = 0$ peak should be broader than the $v_3 = 2$ peak since the slope of the U_0 potential is steeper than that of the U_2 potential near $R_{Br-Br} = 3.50$ Å, the assumed interbromine distance in the anion. On the other hand, transitions to the quasi-bound states supported by the U_4 effective potential should yield narrow peaks in the photoelectron spectrum. The width of each peak is determined, in principle, by the lifetime of the quasi-bound state. For the BrDBr$^-$ photoelectron spectrum, one expects the $v_3 = 0$, 2, and 4 peaks to involve transitions to repulsive effective potentials, and the $v_3 = 6$ peak to result from a transition to a resonance supported by the $v_3 = 6$ effective potential. An examination of the effective potentials for Br + HBr and Br + DBr shows that as v_3 increases, the effective potentials become less repulsive and eventually develop wells. This general trend has been noted for several reactions.

We can now simulate the BrHBr$^-$ and BrDBr$^-$ photoelectron spectra by calculating the Franck–Condon overlap between the ground vibrational level of the anion, $\psi_{v_1''=0}^{(-)}(\rho)\psi_{v_3''=0}^{(-)}(z)$, and the neutral wavefunction in Eq. (10). The full spectrum is generated by summing over even values of v_3':

$$\sigma(E) \propto \sum \left| \int d\rho \left\{ \int dz \ \theta_{v_3'}(z; \rho) \psi_{v_3''=0}^{(-)}(z) \right\} R_{v_3'}^{(E)}(\rho) \psi_{v_1''=0}^{(-)}(\rho) \right|^2 \qquad (19)$$

Here $R_{v_3'}^{(E)}(\rho)$ is the solution to Eq. (17) at total energy E. Note that the integral is zero for odd v_3' due to the opposite parities of the anion ($v_3'' =$

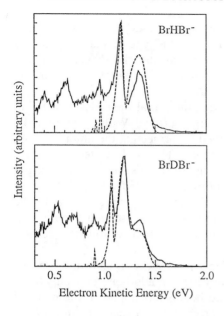

Figure 9. Results of two-dimensional simulation of BrHBr⁻ and BrDBr⁻ photoelectron spectra using Br + HBr surface in Fig. 6. The simulated spectra (– – – –) are convoluted with the instrumental resolution and are superimposed on the experimental spectra (Ref. 28).

0) and neutral (odd v_3') antisymmetric stretch wavefunctions. This is why only the effective potentials with even v_3' are shown in Fig. 8.

Figure 9 shows the simulated BrHBr⁻ and BrDBr⁻ spectra superimposed on the experimental spectra. The simulations are obtained by convoluting the results of Eq. (19) with the instrumental energy resolution. This has only a small effect on the broad peaks resulting from transitions to repulsive effective potentials, but has a large effect on transitions to the quasi-bound states. Within the DIVAH approximation, in which coupling between the effective potentials is neglected, quasi-bound states which can only decay by vibrational predissociation have infinite lifetimes; the peak widths of transitions to these states are determined *entirely* by the instrumental resolution. The comparison of the simulated and experimental spectra shows that the peak positions, intensities, and widths are within reasonable agreement for the ground state progression in both spectra.

5. *Time-Dependent Analysis of the BrHBr⁻ Photoelectron Spectrum*

The discussion in the previous section focused on the relationship between the effective potentials in Fig. 8 and the features in the BrHBr⁻ photoelectron spectrum. This was phrased in terms of Franck–Condon overlap between the anion ground state and the scattering wavefunctions supported by the Br + HBr potential energy surface in Fig. 6. In this section, a complementary analysis and simulation of the BrHBr⁻ spectrum is presented in which an exact two-dimensional time-dependent wavepacket analysis is carried out on the same surface. This type of analysis provides an explicit connection between the spectrum and the dynamics of the BrHBr complex formed by photodetachment. The methods used in the simulation were outlined in Section III.2.

The analysis is carried out by allowing the initial wavepacket, $\phi(0)$, to evolve on the Br + HBr potential energy surface. Within the Franck–Condon approximation, $\phi(0)$ is obtained by projecting the anion ground-state wavefunction onto the neutral surface. This is illustrated in Fig. 10, $t = 0$. The potential energy surface in Fig. 10 is the same as in Fig. 6, but is plotted using the mass-weighted coordinates x and y (Eq. 12). These coordinates are more convenient for wavepacket propagation than the hyperspherical coordinates discussed in the previous section.[53] Note that x and y are nearly parallel to the v_1 and v_3 normal coordinates, respectively, in the anion.

The subsequent time evolution of this wavepacket is shown in Fig. 10. We observe rapid oscillation along the y axis. In addition, after only 60 fs, most of the wavepacket bifurcates and begins moving down the reactant and product valleys. However, even after 300 fs, a small fraction of the original wavepacket remains in the transition state region. This remnant has four nodes along the y-coordinate and is due to quasi-bound states of the complex with $v_3 = 4$. These are the resonances which, in the time-independent approach, occur because of the well in the $v_3 = 4$ effective potential (Fig. 8).

The time-dependent function $\phi(t)$ represents the evolution of $\phi(0)$, which is in turn a coherent superposition of scattering eigenfunctions ψ_E created at $t = 0$. While $\phi(t)$ is uniquely related to the photoelectron spectrum via Eq. (8), it is important to emphasize that, in contrast to a short-pulse laser absorption experiment, photodetachment by a 10-ns laser pulse does not actually create the localized wavepacket $\phi(0)$. Instead, each photodetachment event results in the formation of a well-defined neutral scattering state ψ_E with probability $|\langle \phi(0) | \psi_E \rangle|^2$. Nonetheless, the plots of $\phi(t)$ are useful as they show what would occur if the initial wavepacket were created on the neutral potential energy surface; as such they give an overall picture of the dynamics of the BrHBr complex formed by photodetachment.

The time evolution of the initial wavepacket relevant to the photoelectron spectrum is described by the autocorrelation function $C(t)$ defined in Eq. (7). The modulus of this function, $|C(t)|$, which is plotted in Fig. 11, shows a rapidly decaying oscillation with a period of about 20 fs. This recurrence is due to the v_3 vibration of the complex, and the substantial decay of $C(t)$ after about 40 fs represents movement of most of the

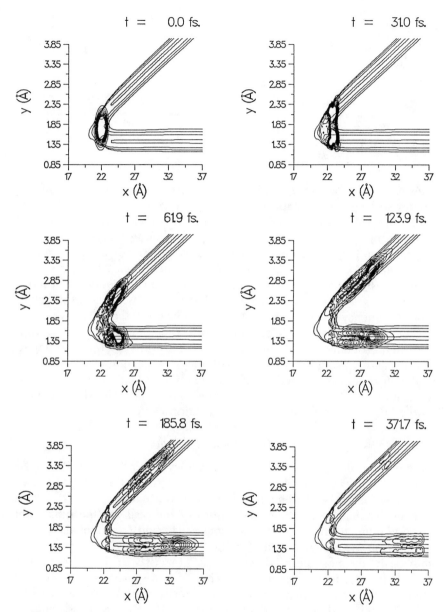

Figure 10. Time evolution of initially prepared wavepacket $\phi(0)$ on best-fit Br + HBr surface. The surface and contours are the same as in Fig. 6, but mass-scaled coordinates (Eq. 12) with an expanded y axis are used.

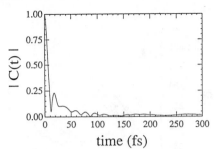

Figure 11. Modulus of time–autocorrelation function $|C(t)|$ (Eq. 8) obtained from wavepacket propagation on Br + HBr surface in Fig. 6.

wavepacket out of the Franck–Condon region of the neutral potential energy surface. Roughly speaking, the time scale of this fast decay determines the width of the broad features in the photoelectron spectrum.

Thus, on this surface, the wavepacket dynamics primarily involve vibrational motion along the v_3 coordinate as the BrHBr complex rapidly dissociates. *It is this vibrational motion occurring on the same (or slightly faster) time scale as dissociation that leads to resolved vibrational features in the photoelectron spectrum.* For $t > 120$ fs, we observe a set of small, lower-frequency recurrences due to the resonances described above which are quasi-bound along the v_1 coordinate. The simulated photoelectron spectrum is obtained by taking the Fourier transform of $C(t)$ according to Eq. (8) and convoluting the result with the experimental resolution. This is therefore an exact two-dimensional simulation. One obtains the same simulation as with the time-independent method, confirming the suitability of the DIVAH approximation used in Section IV.A.4.

6. *Discussion of Results*

We now consider factors affecting the accuracy of our "best-fit" Br + HBr potential energy surface. The first is the geometry of the $BrHBr^-$ anion. The effect of this can be understood with reference to the effective potentials in Fig. 8. The positions and widths of the peaks in the simulation corresponding to transitions to the repulsive U_0 and U_2 potentials depend strongly on the assumed value for $R_e(BrBr)$ in the anion. As mentioned previously, the most recent *ab initio* value[46] for $R_e(BrBr)$ is 0.14 Å smaller than that used in the simulations. Thus, our assumed value may well contribute to error in the best fit surface. In any case, an accurate experimental determination of the ion geometry would be of great assistance in these simulations. This could be obtained from the gas-phase vibration–rotation spectrum of $BrHBr^-$ measured by, for example, velocity-modulated infrared spectroscopy.[67]

The other vital issue in assessing the accuracy of our potential energy surface is the validity of the adiabatic bend approximation. Using this approximation, the surface must be quite repulsive along the minimum energy path in the Franck–Condon region in order to obtain satisfactory agreement with experiment, particularly with the widths of the peaks due to direct scattering. Such a surface yields steeply repulsive effective potentials for the lowest v_3 levels of the complex which are, in turn, necessary for the $v_3 = 0$ and 2 peaks in the simulated spectra to be sufficiently broad.

Figure 12a shows an effective collinear LEPS surface which has the same barrier height and saddle-point geometry as our best-fit surface. The simulation of the photoelectron spectrum on this surface, shown in Fig. 12b, yields much narrower peaks because the slope of the minimum energy path is significantly less in the Franck–Condon region. In comparison with the best-fit surface, the effective potentials on the LEPS surface are less repulsive for low v_3 and the higher v_3 potentials exhibit deeper wells with more resonances.

Similarly, in the time-dependent picture, making the Br + HBr surface repulsive in the Franck–Condon region insures that, on the best-fit surface, the initial wavepacket leaves this region rapidly and moves into the reactant and product valleys. This appears as a fast decay of the autocorrelation function $C(t)$, leading to broad peaks in the simulated photoelectron spectrum. Figure 12c shows $|C(t)|$ resulting from wavepacket propagation on the LEPS surface in Fig. 12a. Clearly, the decay of $|C(t)|$ is much less rapid than in Fig. 11. In addition, at long times ($t > 100$ fs), Fig. 12c shows the amplitude of the wavepacket in the Franck–Condon region remains high, leading to pronounced resonance features in the simulated photoelectron spectrum.

The key question is whether the steeply repulsive nature of the best-fit surface in the Franck–Condon region is real or merely an artifact that results from simulating the spectrum within the adiabatic bend approximation. Schatz[35] has simulated the ClHCl$^-$ spectrum in an exact three-dimensional calculation using the LEPS surface of Bondi et al.[21] for the Cl + HCl reaction. In order to check the validity of the adiabatic bend approximation, we have compared his simulation to that obtained on the same model surface using this approximation. The results (after convolution with the experimental resolution) are similar.[28] In particular, the widths of the $v_3 = 0$ peak, which is the only peak due to direct scattering, are quite close: 50 meV in the three-dimensional simulation and 35 meV in our simulation.

However, the extent of agreement between the two methods depends on the model potential energy surface used in the simulation. We will see in the next section that three-dimensional simulations of the IHI$^-$ photo-

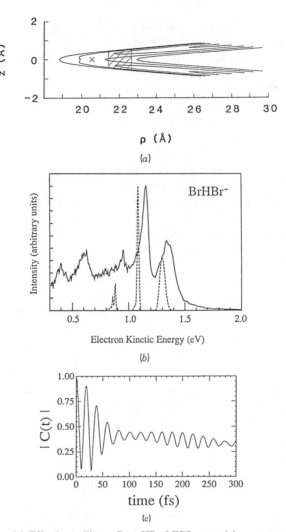

Figure 12. (*a*) Effective collinear Br + HBr LEPS potential energy surface with same barrier height and saddle-point geometry as surface in Fig. 6 (Ref. 28). (*b*) Simulation of BrHBr⁻ spectrum (———) using LEPS surface in (*a*) superimposed on experimental spectrum (solid) (Ref. 28). (*c*) $|C(t)|$ obtained from two-dimensional wavepacket propagation on LEPS surface in (*a*).

electron spectrum[36] using the LEPS-A surface of Manz and Römelt[68] for the I + HI reaction show less agreement between the two methods. After describing the experimental and theoretical results for IHI⁻, we will return to the implications these have on our analysis of the BrHBr⁻ spectrum.

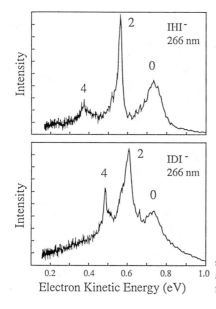

Figure 13. IHI^- and IDI^- photoelectron spectra taken at 266 nm ($h\nu = 4.660$ eV). Peaks are labelled with v_3' quantum numbers (adapted from Ref. 69).

B. Photoelectron and Threshold Photodetachment Spectra of IHI^-

1. *Photoelectron Spectroscopy of IHI^- and IDI^-*

The photoelectron spectra of IHI^- and IDI^- at 266 nm are shown in Fig. 13.[69] Each spectrum shows a progression of three clearly resolved peaks. Several similarities to the $BrHBr^-$ and $BrDBr^-$ spectra are evident. In the first place, the spectra show a large isotope shift; the peak spacing in the IDI^- spectrum is clearly smaller than in the IHI^- spectrum. The progressions are therefore assigned to the v_3 antisymmetric stretch of the neutral IHI complex. As in the $BrHBr^-$ spectra, only transitions to even v_3 levels of the complex occur. For each peak, the v_3 quantum number in the neutral complex is indicated in Fig. 13. Secondly, the frequency of the v_3 mode in the complex is substantially less than in free HI. The spacings between peaks 0 and 2 in the IHI^- and IDI^- spectra are 0.169 ± 0.012 eV $(1360 \pm 100$ cm$^{-1})$ and 0.126 ± 0.012 eV $(1020 \pm 100$ cm$^{-1})$, respectively. The vibrational frequencies in HI and DI are 2309 and 1633 cm^{-1}, respectively. The lower frequency in the complex is attributed to the weakening of the HI bond near the transition state of the I + HI reaction. Finally, we note that, just as in the $BrHBr^-$ spectra, the peak widths in the IHI^- and IDI^- spectra show considerable variation. The $v_3' = 4$ peak in the IDI^- spectrum is only 0.013 eV wide. The peak widths in these spectra will be discussed in more detail below.

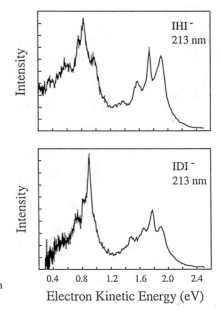

Figure 14. IHI⁻ and IDI⁻ photoelectron spectra taken at 213 nm.

Electron Kinetic Energy (eV)

All the peaks in both spectra occur at electron kinetic energies less than $E_0^{ke} = 0.86 \pm 0.13$ eV, the energy corresponding to formation of I + HI ($v = 0$, $j = 0$). (The error bars are entirely due to the uncertainty in the enthalpy of dissociation for IHI⁻: 0.74 ± 0.13 eV.[57]) This means all the peaks are due to transitions to states of the IHI complex with sufficient energy to dissociate. Manz, Pollak, and co-workers[70] and Clary and Connor[71] have predicted the existence of "vibrationally bound" levels of IHI which actually lie below I + HI($v = 0$, $j = 0$), based on the LEPS-A surface[68] proposed for the I + HI reaction. Our spectra show no evidence for truly bound IHI states; the discrepancy between experiment and theory is most likely due to inaccuracies in the LEPS surface used in the calculations.

Figure 14 shows the IHI⁻ and IDI⁻ photoelectron spectra taken at 213 nm. The progressions at the 266 nm spectrum appear in Fig. 14 at higher electron kinetic energy (and lower resolution). In the 213-nm spectrum, these progressions include a fourth, small peak at lower electron energy (1.38 eV in the IHI⁻ spectrum, 1.47 eV in the IDI⁻ spectrum); at 266 nm, the electrons corresponding to this peak are too low in energy to pass through our time-of-flight analyzer. The most prominent feature in the 213-nm spectra is the appearance of a second, intense band. In the IHI⁻ spectrum, the separation between the origins of the two bands is

0.94 eV, which is essentially identical to the spin-orbit splitting in I. It therefore appears that we are observing transitions to the ground state I + HI surface and an excited-state surface that asymptotically correlates to HI + I*($^2P_{1/2}$). The two bands are better separated than the corresponding bands in the BrHBr$^-$ spectrum, which is expected since the spin-orbit splitting in I is about twice that in Br.

The ground state IHI$^-$ photoelectron spectrum has been simulated by our group using the methods described in Sections IV.A.4–5. In addition, three-dimensional simulations (with $J = 0$) have been performed by Gazdy and Bowman,[43] using an L^2 basis for the I + HI scattering wavefunctions, and by Schatz,[36] using exact I + HI scattering wavefunctions. The simulated spectra from all three methods are shown in Fig. 15(a)–(c). All the simulations assume IHI$^-$ to be linear and centrosymmetric with an equilibrium interiodine distance $R_e = 3.88$ Å and use the LEPS-A potential energy surface for the I + HI reaction. In comparison to our best-fit surface for the Br + HBr reaction, the LEPS-A surface for I + HI has a considerably smaller barrier (4.8 kJ/mol) and is much less repulsive along the minimum energy path in the Franck–Condon region. If we calculate the effective potentials on the LEPS-A surface using the adiabatic bend approximation, we find that they have deeper wells and support more resonances than the analogous curves in Fig. 8. Thus, one might expect a considerable amount of resonance structure in simulations of the IHI$^-$ spectrum.

Indeed, all the simulations show that the $v_3' = 2$ feature in the IHI$^-$ spectrum appears not as a single broad peak, but rather as a progression of peaks spaced by about 12 meV (100 cm^{-1}). The closely spaced peaks represent transitions to quasi-bound symmetric stretch levels of the complex. These levels are analogous to those discussed in the Br + HBr analysis and are responsible for resonances in the I + HI reaction. The simulation by Schatz in Fig. 15c, which is the most realistic, shows three closely spaced peaks associated with the $v_3' = 2$ level of [IHI] resulting from transitions to the $v_1' = 0$, 1, and 2 symmetric stretch levels of the complex. The linewidths of these peaks are 2–5 meV, indicating resonance lifetimes of several hundred femtoseconds. Our simulation as well as that of Gazdy and Bowman (Fig. 15b) also predict resonance structure underlying the $v_3' = 4$ peak in the IHI$^-$ spectrum. Resonances are also observed in simulations of the IDI$^-$ spectrum.[34,69]

On the other hand, the simulations show that the $v_3' = 0$ feature in the IHI$^-$ spectrum is dominated by overlap with direct scattering wavefunctions rather than transitions to quasi-bound IHI states. Nonetheless, this feature is structured even in Schatz's "exact" simulation, where the

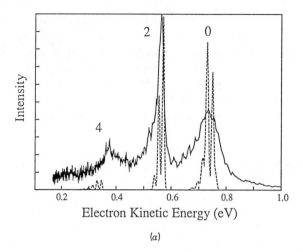

(a)

Figure 15. (a).

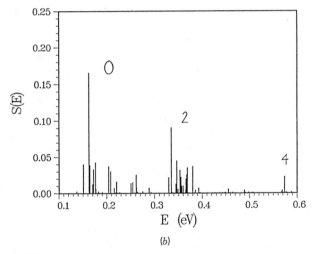

(b)

Figure 15. (b).

$v_3' = 0$ feature consists of a series of peaks spanning 0.1 eV, rather than the single broad peak in the experimental spectrum. The origin of this structure is discussed in Section IV.B.2.

To summarize, the simulated IHI^- spectrum shows considerably more structure than the experimental spectrum. However, the LEPS-A surface used in these simulations is approximate at best. Thus, one must ask if the resonance and structured direct-scattering features in the simulations

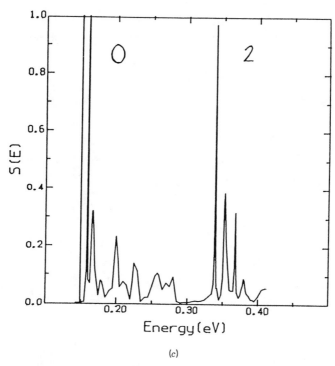

(c)

Figure 15. (a) Simulation of IHI⁻ photoelectron spectrum using effective collinear LEPS-A I + HI surface from Ref. 68. The simulation (———) is superimposed on the 266-nm experimental spectrum. The simulated spectrum is shifted down by 0.100 eV to match the experimental spectrum. (b) Three-dimensional L^2 simulation of IHI⁻ spectrum (Ref. 43). (c) Exact three-dimensional simulation of IHI⁻ spectrum (Ref. 36). Note that different energy scales are used in the simulations. In (b) and (c), E is measured relative to the bottom of the HI well in the asymptotic region of the LEPS-A surface. E and eKE (a) are related by $E = 1.00 - \text{eKE}$, assuming that $E_b^{(-)} = 3.80$ eV for IHI⁻ (Eq. 1).

are real and are not seen in the IHI⁻ photoelectron spectrum due to the limited resolution of the experiment, or whether the simulated features are artifacts resulting from inaccuracies in the LEPS-A surface. In order to answer this question, higher-resolution studies of IHI⁻ were performed using our threshold photodetachment spectrometer. These experiments are described in the next section.

2. Threshold Photodetachment Spectroscopy of IHI⁻

The threshold photodetachment spectra of the $v_3' = 2$, 4, and 0 peaks are shown in Figs. 16–18.[72] The horizontal axis at the bottom of each plot shows the photodetachment wavelength λ. For convenient comparison with Fig. 13, the top axis shows the corresponding electron kinetic energy

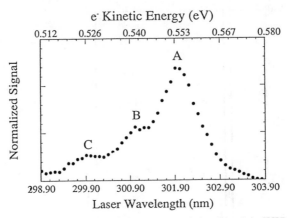

Figure 16. Threshold photodetachment spectrum of $v_3' = 2$ peak in IHI⁻ photoelectron spectrum (see Fig. 13). The points are spaced by 0.1 nm ($\sim 10\,\text{cm}^{-1}$) (Ref. 72).

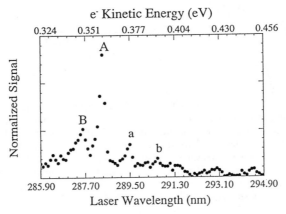

Figure 17. Threshold photodetachment spectrum of $v_3' = 4$ peak in IHI⁻ photoelectron spectrum. Point spacing is 0.1 nm (Ref. 72).

(eKE) that would result from a fixed-frequency photoelectron spectrum at 266 nm; the two axes are related by eKE = 1240(1/266 − 1/λ).

The threshold photodetachment spectrum of the $v_3' = 2$ peak (Fig. 16) reveals three partially resolved peaks spaced by approximately 12 meV. The peak widths are about 12 meV, and the intensity of the peaks decreases towards shorter wavelength. The experimental peak spacings and intensities are quite similar to the simulations of the $v_3' = 2$ feature in Fig. 15. The major difference is that in the simulation by Schatz,[36] the peaks in the symmetric stretch progression are narrower (2 meV) and are therefore fully resolved. Nonetheless, the correspondence between the experimental and simulated spectra strongly suggests we are seeing a progression in quasi-bound IHI symmetric stretch states.

Figure 17 shows the threshold photodetachment spectrum of the $v_3' = 4$ feature in the IHI$^-$ photoelectron spectrum. Four well-resolved peaks are evident in this spectrum. Peaks A, a, and b are evenly spaced by an interval of 16.0 meV (129 cm^{-1}), while peaks A and B are separated by 12.5 meV (101 cm^{-1}). The peaks are substantially narrower than those in Fig. 16; peaks A and a are 3.7 meV wide, and peak B is 5.6 meV wide.

The presence of two intervals among the peaks in Fig. 17 suggests the presence of two progressions. Matrix isolation studies[12] on IHI$^-$ have yielded a value of 129 cm^{-1} for the symmetric stretch frequency in the ion, in excellent agreement with the spacing between peaks A, a, and b. We therefore assign peaks a and b to hot-band transitions originating from the $v_1'' = 1$ and $v_1'' = 2$ symmetric stretch levels of the ion and terminating in the same level of the neutral as peak A. The a/A intensity ratio yields a vibrational temperature of 100 K for the IHI$^-$ ions.

Peaks A and B are assigned to transitions between the $v_1'' = 0$ level of the ion and two different symmetric stretch levels of the IHI complex. This assignment is supported by the peak spacing of 101 cm^{-1}, and with the noticeably different peak widths which imply that the transitions are to two distinct levels of the complex with different lifetimes. The widths yield lower bounds of 180 and 120 fs for the upper-state lifetimes of A and B, respectively. Since the symmetric stretch vibration of [IHI] is strongly coupled to the dissociation coordinate of the complex, it seems

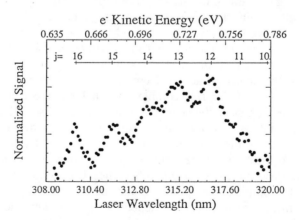

Figure 18. Threshold photodetachment spectrum of $v_3' = 0$ peak in IHI$^-$ photoelectron spectrum. Point spacing is 0.1 nm. The energetic thresholds for the asymptotic levels I + HI ($v = 0$, $j = 11$–16) are shown (Ref. 72).

reasonable that the higher-lying symmetric stretch level of the complex (peak B) should have a shorter lifetime. Similar trends have been observed in collinear scattering calculations on the $Cl + HCl$[73] and $F + DBr$[74] reactions.

The observation that the peaks associated with $v_3' = 4$ in Fig. 17 are narrower than the $v_3' = 2$ peaks in Fig. 16 is of interest. The $v_3' = 4$ peaks lie 0.49 eV above the $I + HI$ ($v = 0$) asymptote compared to 0.30 eV for the levels with $v_3' = 2$, and one might therefore expect the levels with higher total internal energy (the $v_3' = 4$ levels) to dissociate more rapidly. Our result shows that the v_3 mode really is poorly coupled to dissociation of the complex, as discussed in Section IV.A.3. The narrower peaks for $v_3' = 4$ may be due to the deeper wells in the effective potentials for the higher v_3 value (see Fig. 8), or because dissociation of the complex through bending motion is less facile for $v_3' = 4$ than for $v_3' = 2$. The possible coupling between the antisymmetric stretch and bending vibration of the complex is discussed further below.

Figure 18 shows the threshold photodetachment spectrum for the $v_3' = 0$ levels of the complex. This feature, which appears as a single broad peak in the photoelectron spectrum, actually consists of a series of peaks spaced by an interval that increases from 20 to 25 meV towards lower laser wavelength. This is quite distinct from the 12-meV interval seen in the other spectra. A possible assignment of these peaks is suggested by the approximate correspondence between the peak positions and the energetic thresholds for formation of the asymptotic levels $I + HI$ ($v = 0$, $j = 11$–16) from photodetachment of IHI^-. These thresholds, shown in Fig. 18, are drawn assuming the threshold for $I + HI$ ($v = 0, j = 0$) is at 3.79 eV, well within the error bars of the value of $E_b^{(-)}$ for IHI^-, 3.80 ± 0.13 eV, used above.

Our results therefore indicate that photodetachment is enhanced near the thresholds for formation of HI ($v = 0$) in high rotational states. This suggests that the peaks in the spectrum correspond to transitions to hindered rotor levels of the IHI complex which look like nearly free HI rotational states. Because the peak widths are comparable to their spacing, the spectrum implies that internal rotation of the HI is occurring on the same time scale as dissociation of the IHI ($v_3 = 0$) complex. In other words, subsequent to photodetachment, the H atom, which initially lies between the two I atoms, rotates and becomes associated with one of the I atoms as the complex falls apart. Tannor and co-workers[75] are currently working on three-dimensional time-dependent wavepacket studies of IHI^- photodetachment which should provide further insight into the dissociation dynamics of IHI ($v_3 = 0$).

In his three-dimensional simulation of the IHI^- photoelectron spectrum, Schatz[36] observes peaks in the transitions to $v'_3 = 0$ levels of the complex which correspond approximately to asymptotic I + HI rotational energy levels. However, his results show significant intensity down to the I + HI ($j = 0$) level, suggesting that much lower-lying hindered rotor states of the complex are accessed in the simulation than in the experimental spectrum. Schatz's calculations also show that these hindered rotor levels of IHI dissociate in a rotationally adiabatic manner.[34] That is, the IHI level formed near the energy asymptote for I + HI ($v = 0, j$) predominantly dissociates to I + HI ($v = 0, j$). This is physically reasonable if the HI is already undergoing nearly free rotation in the complex. Hence, our experiment suggests that a more highly rotationally excited HI ($v = 0$) distribution results from dissociation of the complex formed by photodetachment on the real I + HI potential energy surface than on the LEPS-A surface used in the simulations. This is likely to be intimately connected with differences in the bending potential and, possibly, the barrier height on the real and model surfaces.

In summary, the IHI$^-$ threshold photodetachment results show that, at higher resolution, the $v'_3 = 2$ and 4 peaks seen in the photoelectron spectrum are qualitatively different from the $v'_3 = 0$ peak. The threshold photodetachment spectrum of the $v'_3 = 2$ and 4 features reveals vibrational progressions in quasi-bound symmetric stretch levels of the IHI complex, while the spectrum of the $v'_3 = 0$ peak shows what appears to be a progression in internal rotor levels of the IHI complex. The lack of resonance structure in the $v'_3 = 0$ peak is reasonable in light of the expectation discussed above that resonances become more pronounced for higher v_3 levels of the IHI complex.

The absence of the rotational structure underlying the $v'_3 = 0$ peak in the $v'_3 = 2$ and 4 peaks is more intriguing. It suggests that the bending motion of the complex resembles internal rotation only for the $v'_3 = 0$ level. If the effective bend potential is stiffer for the $v'_3 = 2$ and 4 levels, and more closely resembles the bending normal vibration in a symmetric triatomic molecule, one might expect photodetachment from IHI$^-$ ($v''_2 = 0$) to have good Franck–Condon overlap only with the lowest bending level associated with these antisymmetric stretch levels of the IHI complex. This implies the effective bending potential depends strongly on the v_3 quantum number. One certainly might expect coupling between the bending and antisymmetric stretch vibrations in the complex, since both are essentially H atom vibrations. This has been explored in a recent paper by Kubach,[76] in which the eigenvalues and eigenfunctions of the two-dimensional antisymmetric stretch/bend Hamiltonian are determined as a function of interiodine distance.

We now consider the implications of the higher resolution IHI$^-$ results for the analysis of the BrHBr$^-$ photoelectron spectrum. A comparison of the IHI$^-$ simulations in Fig. 15 shows that the adiabatic bend approximation used in the BrHBr$^-$ analysis does a reasonably accurate job reproducing the positions of resonances. However, this approximation is clearly inadequate if the complex dissociates in a noncollinear manner, and it therefore fails to reproduce the overall width and underlying rotational structure of the $v_3' = 0$ peak in the exact three-dimensional simulation.

As discussed in Section IV.B, the best-fit Br + HBr surface was designed to reproduce the broad $v_3' = 0$ and 2 experimental peaks assuming these were single, homogeneous peaks. If these peaks consist of unresolved progressions of the type predicted (and experimentally observed) for the $v_3' = 0$ peak in the IHI$^-$ spectrum, then some revision of the proposed Br + HBr surface will be required. In particular, it should be possible to fit the data with a surface which is less repulsive in the Franck–Condon region. It is clearly worthwhile to obtain the threshold photodetachment spectrum of BrHBr$^-$ to see if any of the broad peaks exhibit underlying structure at higher resolution, and these experiments will begin soon in our laboratory.

V. ASYMMETRIC HYDROGEN EXCHANGE REACTIONS

A.   General Considerations

This section describes the application of photodetachment spectroscopy to studies of the transition state region for reactions A + HB → HA + B, where A ≠ B. A and B can be atomic or polyatomic species. These asymmetric reactions are of considerable chemical interest. In particular, studies of reactions in which A and B are (unlike) halogen atoms have led to the formulation of many fundamental concepts in reaction dynamics. Several of the first product state-resolved experimental studies in chemical dynamics were performed on these reactions.[77,78] On the theoretical side, classical[79–83] and quantum mechanical[74,84,85] scattering calculations have been carried out on model potential energy surfaces in order to understand the experimental results. Studies by Manz and co-workers[53,74,86] on the F + DBr reaction predict sharp resonance structure, just as in the symmetric reactions.

Asymmetric hydrogen exchange reactions are therefore a tempting target for photodetachment studies, since asymmetric AHB$^-$ anions are often as strongly hydrogen-bonded as the symmetric BrHBr$^-$ and IHI$^-$ anions. However, for an asymmetric reaction, one must consider the geometric overlap between the anion and neutral transition state more carefully than for a symmetric reaction. This point has been emphasized by Brauman and co-workers[6] in their study of ROHF$^-$ anions.

For the Br + HBr/BrHBr$^-$ system, the BrHBr$^-$ anion is linear and centrosymmetric, and the major concern from the point of view of geometric overlap is how close the equilibrium interhalogen distance in the anion is to the saddle point geometry on the neutral surface. For an asymmetric reaction such as Br + HI $\rightarrow$ HBr + I, one must also consider the location of the H atom in BrHI$^-$, as this will determine whether photodetachment primarily accesses the reactant or product valley on the neutral potential energy surface. To first order, the H atom location is determined by the relative proton affinities of Br$^-$ and I$^-$. Since the proton affinity of Br$^-$ is 0.40 eV higher than that of I$^-$,[87] one expects the ion to look like I$^-$··HBr, and a vertical photodetachment transition from the equilibrium geometry of the anion should land in the I + HBr product valley of the reaction. The photoelectron spectra of BrHI$^-$, FHI$^-$, and FHBr$^-$ will be discussed in Sections V.B and V.C as examples of asymmetric bihalide systems.

We have also studied the reaction F + CH$_3$OH $\rightarrow$ HF + CH$_3$O by photoelectron spectroscopy of CH$_3$OHF$^-$. In this case, the proton affinity of CH$_3$O$^-$ is 0.42 eV higher than that of F$^-$,[87] the anion should look like CH$_3$OH··F$^-$, and photodetachment should access the reactant valley and saddle-point region of the neutral potential energy surface. The photoelectron spectrum of CH$_3$OHF$^-$ is presented in Section V.D.

B. Photoelectron Spectroscopy of BrHI$^-$

1. Results and Qualitative Analysis

Figure 19 shows the photoelectron spectra of BrHI$^-$ and BrDI$^-$ taken at 213 nm.[55] Each spectrum shows two progressions of approximately evenly spaced peaks. The peaks labelled A and A* occur at the same electron kinetic energy in both spectra and are taken to be band origins of the progressions. The peak spacing within each progression in Fig. 19 is noticeably less in the BrDI$^-$ spectrum than in the BrHI$^-$ spectrum. The direction of this isotope shift shows we are observing progressions in the neutral BrHI complex in a vibrational mode primarily involving H atom motion. This is assigned to the v_3 stretching mode of the BrHI complex. The A–A* separation in each spectrum is 0.90 ± 0.02 eV (7300 ± 200 cm^{-1}). This is slightly less than the spin-orbit splitting in atomic I (7600 cm^{-1}) and suggests that the two progressions with band origins A and A* correspond to two electronic states of the BrHI complex which asymptotically correlate to HBr + I($^2P_{3/2}$) and HBr + I*($^2P_{1/2}$), respectively. The arrows at 2.07 and 1.36 eV show the electron kinetic energies corresponding to formation of I + HBr ($v = 0$) ground-state prod-

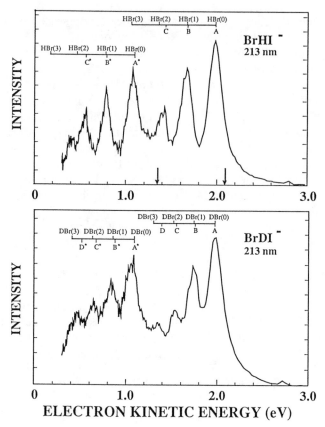

Figure 19. Photoelectron spectra of BrHI$^-$ and BrDI$^-$ at 213 nm. Arrows at 2.07 and 1.36 eV show thresholds for formation of ground-state products I($^2P_{3/2}$) + HBr ($v = 0$) and ground-state reactants Br($^2P_{3/2}$) + HI ($v = 0$), respectively. These thresholds are referred to in the text as product and reactant asymptotes. The spacing between the first few HBr and DBr vibrational levels is shown at the top of each spectrum.

ucts and Br + HI ($v = 0$) ground-state reactants, respectively. All the peaks are due to states of the neutral complex with sufficient energy to dissociate to ground-state products.

A comparison of the BrHI$^-$ and BrDI$^-$ spectrum with the symmetric bihalide spectra discussed previously shows two important differences. In the symmetric bihalide spectra, large variations in the peak widths were observed. In contrast, the peaks in Fig. 19 are all quite broad. In the BrHI$^-$ spectrum, for example, the peak widths in the ground state progression are ~170 meV, somewhat wider than the peaks in the excited state progression (~140 meV). Secondly, the symmetric XHX$^-$ spectra showed

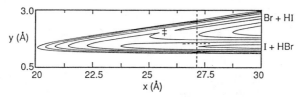

Figure 20.   Effective collinear LEPS potential energy surface for Br + HI reaction obtained from three-dimensional surface of Ref. 83. The surface is plotted using mass-weighted coordinates defined in Eq. (20). The skew angle (see Eq. 11) is 8.2°. The saddle point is marked by ‡. The estimated BrHI⁻ equilibrium geometry (see text) is at the intersection of the dashed lines (Ref. 55).

a substantial "red shift" in the v_3 frequency relative to diatomic HX. Figure 19 shows that the peak spacings in the BrHI⁻ and BrDI⁻ spectra closely parallel the first few vibrational energy levels HBr and DBr. The peak spacings are, on the average, only 200 cm⁻¹ less than the vibrational frequency of HBr or DBr.

The differences between the BrHI⁻ and BrDI⁻ spectra and the symmetric bihalide spectra will now be explored within the framework of the time-independent and time-dependent methods discussed in the previous sections, with the ultimate goal of simulating the ground-state progressions in the spectra. The BrHI⁻ and BrDI⁻ anions are not well characterized; only the v_3 fundamentals have been measured in a matrix[88] (920 and 728 cm⁻¹, respectively). No high-level *ab initio* calculations have been performed on BrHI⁻ to determine the ion geometry, although the ions are expected to be linear, based on *ab initio* studies of FHCl⁻ and FHBr⁻.[89] A LEPS potential energy surface for the Br + HI reaction has been proposed by Broida and Persky[83] (the BP surface). This surface was constructed on the basis of quasi-classical trajectory calculations which reproduce the reaction rate constant[90] and HBr product vibrational distribution.[77] We use the BP surface in our simulations without modification and vary the anion geometry (see below) to achieve optimal agreement with the experimental photoelectron spectra. This yields a qualitative understanding of the features in the photoelectron spectra. In order to actually extract new information on the Br + HI surface from our spectra, further characterization of the ion geometry is required.

The BP surface has a collinear minimum energy path and a 0.88 kJ/mol barrier which lies in the Br + HI reactant channel. Figure 20 shows the effective collinear potential energy surface obtained from the three-dimensional BP surface. The surface is plotted using the mass-weighted coordinates x and y, where

$$x = \sqrt{\frac{\mu_{I,HBr}}{\mu_{HBr}}} R_{I,HBr} \approx 7.0 R_{I-Br} \tag{20}$$

$$y = R_{H-Br}$$

Here $R_{I,HBr}$ is the separation between the I atom and the HBr center of mass. Note that for the heavy + light-heavy mass combination, $x^2 \gg y^2$, and therefore $x \approx \rho$ (see Eq. 13).

The mass-weighted coordinates x and y are assumed to be proportional to the normal coordinates for the v_1 and v_3 modes, respectively, for the anion. The v_3 stretch potential for the anion is approximated by a Morse potential:

$$U(y) = D_e(1 - \exp\{-\beta[y - R_e(HBr)]\})^2 \tag{21}$$

Here $R_e(HBr)$ is the equilibrium H–Br separation in the anion. D_e and β are fixed by the matrix isolation values for the v_3 frequency in $BrHI^-$ and $BrDI^-$.[88] The v_1 potential is assumed to be a harmonic oscillator potential with frequency $100\ cm^{-1}$ centered at $R_e(IBr)$, the equilibrium interhalogen distance in the anion. The values of $R_e(IBr)$ and $R_e(HBr)$ affect the peak spacings and intensities, respectively, in the simulated photoelectron spectrum, and were adjusted in order to optimize agreement between a one-dimensional simulation and experiment.[55] The best agreement is obtained for $R_e(IBr) = 3.88\ Å$ ($x_e = 27.1\ Å$, Eq. 20) and $R_e(HBr) = 1.55\ Å$. This assumed equilibrium geometry for the anion is indicated in Fig. 20 at the intersection of the dashed lines.

The anion equilibrium geometry shown in Fig. 20 indicates that photodetachment primarily accesses the I + HBr product valley, as expected based on the relative proton affinities of I^- and Br^-. In this region of the potential energy surface, the nascent HBr bond is nearly fully formed. This is consistent with the experimental observation that the peak spacing in the v_3 progression is only slightly less than the vibrational frequency in diatomic HBr. Good geometric overlap with the product valley is also consistent with the overall intensity distribution in the spectra; the most intense peak occurs near the I + HBr product asymptote (arrow at 2.07 eV in Fig. 19), and the peak intensity decreases as the Br + HI reactant asymptote (arrow at 1.36 eV) is approached.

These points can be illustrated further with the aid of Fig. 21, which shows the v_3 potential for the anion (Eq. 22) and the approximate v_3 potential for the neutral resulting from taking a cut through the BP potential energy surface at $x = x_e$. The latter is an asymmetric double minimum

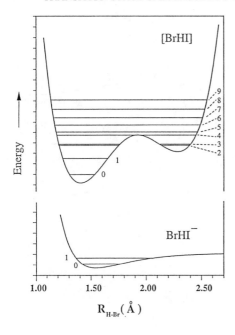

Figure 21. Anion and neutral v_3 potentials used in one-dimensional analysis of BrHI$^-$ spectrum. Eigenstates are labelled by v_3 quantum number. Tick marks on vertical axis are spaced by 0.2 eV (Ref. 55).

potential with the minimum in the I + HBr valley lying 0.58 eV below the minimum in the Br + HI valley. The first few eigenvalues of both potentials are shown. A one-dimensional Franck–Condon simulation (see Ref. 55) shows that the $v_3'' = 0$ level of the anion has the best Franck–Condon overlap with the neutral $v_3' = 0$, 1, and 3 levels; these are localized in the I + HBr valley and look like perturbed HBr vibrational levels. Transitions to the $v_3' = 0$, 1, and 3 levels result in the three highest energy peaks in the BrHI$^-$ photoelectron spectrum. The one-dimensional analysis predicts the $v_3' = 0$ peak to be the most intense, in agreement with experiment. Although the $v_3' = 2$ and 3 levels are nearly degenerate, the $v_3' = 2$ eigenfunction is localized in the Br + HI well and has very little overlap with the BrHI$^-$ ($v_3'' = 0$) level.

We next consider the origin of the uniformly broad peaks in the BrHI$^-$ and BrDI$^-$ spectra. In contrast to the photodetachment of symmetric bihalides, photodetachment of BrHI$^-$ accesses a more repulsive region of the potential energy surface, namely, the exit valley of an exothermic reaction. One therefore expects rapid dissociation of the BrHI complex, consistent with the broad peaks observed in the photoelectron spectra.

This point can be explored in more detail with reference to Fig. 22, which shows the vibrationally adiabatic potentials $\epsilon_{v_3}(\rho)$ obtained by solving Eq. (16) for the potential energy surface in Fig. 20.[91] (Recall $\rho \approx x$ for this reaction.) These curves correlate asymptotically to various I +

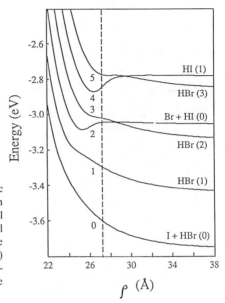

Figure 22. Vibrationally adiabatic curves $\epsilon_{v_3}(\rho)$ (Eq. 16) for Br + HI reaction derived from surface in Fig. 20. The vertical dashed line indicates ρ_e for BrHI$^-$. The level numbering is the same as in Fig. 21. The Br + HI($v = 0, 1$) and I + HBr($v = 0$–3) asymptotic levels are indicated. The apparent intersections of some of the curves are actually weakly avoided crossings.

HBr(v) and Br + HI (v) levels, as indicated in Fig. 22. The dashed vertical line in Fig. 22 indicates ρ_e for BrHI$^-$. The intersections of the adiabatic curves with this line are the v_3 energy levels of the BrHI complex shown in Fig. 21, and the curves are numbered accordingly. As discussed above, the $v_3'' = 0$ level of BrHI$^-$ has good Franck–Condon overlap with the v_3' = 0, 1, and 3 levels. The corresponding adiabatic curves are repulsive, and transitions to these curves should yield broad peaks in the photoelectron spectrum. Note that these curves correlate diabatically with the I + HBr ($v = 0$, 1, and 2) product states. While several of the adiabatic curves (2, 4) have wells which could support resonance states, the v_3 wavefunctions near ρ_e for these curves are localized in the Br + HI valley and have little overlap with the ground state of BrHI$^-$. The curves with wells in the transition state region all correlate diabatically to Br + HI reactant states.

2.   *Time-Dependent Simulations of the Spectra*

In this section, the two-dimensional time-dependent wavepacket analysis discussed in Sections III.B and IV.A.5 will be used to simulate the ground electronic state progressions in the BrHI$^-$ and BrDI$^-$ photoelectron spectra and to probe the underlying dynamics.[55] Figure 23 shows the initial wavepacket, $\phi(t = 0)$, which results from projecting the ground-state vibrational wavefunction of BrHI$^-$ onto the BP effective collinear potential

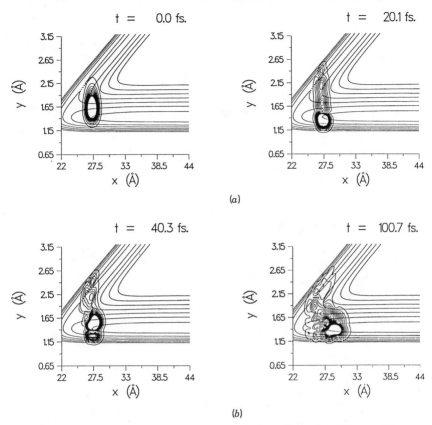

Figure 23. Evolution of initially prepared wavepacket $\phi(0)$ on Br + HI potential energy surface shown in Fig. 20 (Ref. 55).

energy surface. This shows that photodetachment primarily accesses the product side of the Br + HI surface, as discussed in the previous sections. However, $\phi(0)$ does have some amplitude at the saddle-point region of the potential energy surface and therefore will have small but nonzero overlap with states localized in the Br + HI valley.

Figure 23 also shows the time evolution of the initial wavepacket on the Br + HI surface. At early times ($t < 60$ fs), we observe rapid oscillation of the wavepacket along the y (v_3) coordinate, and slower movement of the wavepacket along the x coordinate. While the wavepacket in this time interval has some amplitude in the Br + HI reactant valley, the bulk of the wavepacket remains in the I + HBr product valley. This is in contrast to the wavepacket simulations on our best-fit Br + HBr surface (Fig. 10). In those simulations, due to the higher symmetry of the anion and neutral,

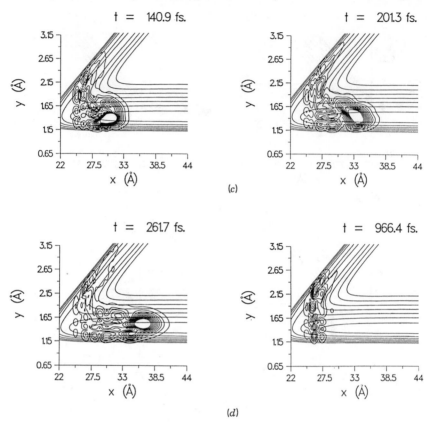

Figure 23.   (*Continued*).

the wavepacket splits evenly between the two valleys of the $Br + HBr$ potential energy surface.

At longer times, $t > 60$ fs, several trends are noticeable. By this time, most of the wavepacket has moved out of the initial region accessed by photodetachment, so we expect $\langle \phi(t) | \phi(0) \rangle$ to be small. Most of the wavepacket remains confined to the $I + HBr$ product valley. The spreading of this portion of the wavepacket along the x coordinate is quite evident for $t > 200$ fs; from the leading edge (at higher x) to the tail there is an increasing number of nodes along the y coordinate. This suggests that those states localized in the product valley with higher v_3 excitation dissociate more slowly. This effect is consistent with the observation that the repulsive adiabatic curves in Fig. 22 are less steep at higher levels of excitation.

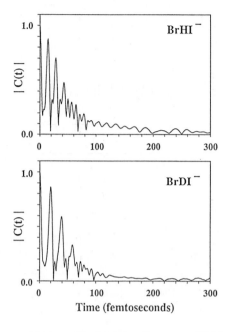

Figure 24. Modulus of time-autocorrelation function $|C(t)|$ obtained from wavepacket propagation on Br + HI surface shown in Fig. 20 (Ref. 55).

Even at the longest time, $t = 966$ fs (Fig. 23d), the wavepacket still has some amplitude near the transition state region. This occurs because the BP potential energy surface does support resonances, and some of the quasi-bound states responsible for these resonances have nonzero overlap with the ground state of the anion. The wavefunction in Fig. 23d represents a superposition of these quasi-bound states, but the most noticeable feature of this wavefunction is the four nodes along the y coordinate. This indicates that $v_3 = 4$ for the dominant resonance state.

Figure 24 shows the modulus of the autocorrelation function $|C(t)|$ (Eq. 8) obtained from propagating $\phi(0)$ on the BP surface. The correspondence between the wavefunctions in Fig. 23 and $|C(t)|$ is quite clear. At times $t < 60$ fs, $|C(t)|$ exhibits a fast oscillation whose period corresponds to the v_3 motion of the BrHI complex. This decays after about 60 fs, indicating that dissociation of the complex is occurring on this time scale. Thus, as in the Br + HBr analysis, the simulations show fast vibrational motion, primarily involving the H atom, occurring as the complex dissociates. At longer times ($t > 200$ fs), we see that $|C(t)|$ does not become zero, but instead shows a persistent oscillation. This results from the small resonance contribution to the simulated spectrum; Fig. 24 shows that even at $t = 300$ fs, there is some residual overlap with the initial wavepacket due to quasi-bound states localized near the transition state region.

Finally, we simulate the BrHI⁻ and BrDI⁻ photoelectron spectra by

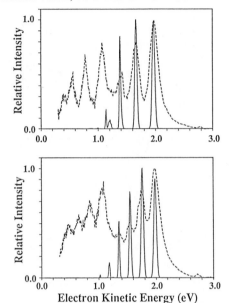

Figure 25. Simulated photoelectron spectra (solid) for (top) BrHI⁻ and (bottom) BrDI⁻ from two-dimensional calculation. The simulations are shifted so that the 0–0 transitions line up with the experimental spectra (———) and have been convoluted with the instrumental resolution function (Ref. 55).

taking the Fourier transform of $C(t)$ (Eq. 7). The results convoluted with our experimental resolution are shown in Fig. 25. The five peaks in the BrHI⁻ simulation result from transitions to the $v'_3 = 0, 1, 3, 4$, and 5 levels of the complex. The intensity of the transition to the $v'_3 = 2$ level is too small to be seen. Figure 22 shows that the vibrationally adiabatic curve for the $v'_3 = 4$ level, which diabatically correlates to Br + HI ($v = 1$), has a well which should support resonances. Indeed, if we Fourier transform $C(t)$ assuming a somewhat higher instrumental resolution of 4 meV,[55] we find the weak $v'_3 = 4$ peak at 1.2 eV does consist of a progression of closely spaced v_1 levels of the complex. However, the overall contribution of resonances to the simulation is quite small, as expected based on our wavepacket dynamics calculations.

The first three peaks in the BrHI⁻ simulation (those at highest electron kinetic energy) line up well with the experimental peaks, as do the first four peaks in the BrDI⁻ simulation. The excited state progression in the experimental spectra makes it difficult to compare the remaining simulated peaks with experiment, although shoulders appear on the high-energy side of the first excited-state peak in each spectrum near the energies of the lowest energy peaks in the simulations. As discussed above, the agreement between the peak positions and intensities in the experimental and simulated spectra results from assuming the BP surface is correct and optimizing the anion geometry. The analysis therefore shows that we can reproduce experimental peak positions and intensities using a reasonable ion geometry and neutral potential energy surface.

However, the peak widths in the simulations are considerably narrower than the experimental peak widths. Here, as in the $BrHBr^-$ analysis, one must consider if this discrepancy results from deficiencies in the BP surface, or if the effective collinear approximation artificially constrains the peak widths as it only allows for collinear decay of the complex. This can clearly be answered by performing three-dimensional simulations of the $BrHI^-$ spectrum using the same anion geometry and neutral potential energy surface.

3. Excited Electronic States in the $BrHI^-$ Photoelectron Spectrum

As discused in Section V.B.1, the photoelectron spectra of $BrHI^-$ and $BrDI^-$ show an additional progression which has been assigned to an excited state of the complex that asymptotically correlates to spin-orbit excited $I^*(^2P_{1/2})$ + HBr (DBr). This section discusses the role of excited electronic states in these photoelectron spectra in more detail. The effects of spin-orbit excitation in the Br + HI and F + HBr reactions have been studied previously by Bergmann[92] and Hepburn,[93] respectively.

Figure 26 shows an electronic correlation diagram for the Br + HI reaction, assuming $C_{\infty v}$ symmetry. Let us consider the product side of this diagram, since this is the region of the Br + HI surface most relevant to our experiment. In the asymptotic region of the product valley, the ground- and first-excited electronic state are separated by the spin-orbit splitting between the $I(^2P_{3/2})$ and $I^*(^2P_{1/2})$ states, $7600\ cm^{-1}$. Closer to the transition state region, the HBr perturbs the spherical symmetry of the I atom. In this region, Hund's case (c) coupling is appropriate: J, the electronic and spin-angular momentum of the I atom, is no longer a good quantum number, but Ω, the projection of J on the internuclear axis, is a good quantum number. The interaction of HBr with a ground state I atom splits the degenerate $I(^2P_{3/2})$ state into the $X(\Omega = 1/2)$ and $I(\Omega = 3/2)$ states, and the $II(\Omega = 1/2)$ state results from the interaction of HBr with $I^*(^2P_{1/2})$. At even smaller interhalogen distances, where all three atoms strongly interact, Hund's case (a) coupling is more appropriate. The $X(1/2)$ and $II(1/2)$ states are mixed, yielding the $^2\Sigma_2$ and $^2\Pi_{1/2}$ states in the diagram, while the $I(3/2)$ state is more appropriately labelled as $^2\Pi_{3/2}$. The two $^2\Pi$ states correlate diabatically to excited-state reactants and should lie much higher than the $^2\Sigma_{1/2}$ state in this region.

Since the energy interval between the two progressions in the $BrHI^-$ and $BrDI^-$ photoelectron spectra, $7300\ cm^{-1}$, is very close to the asymptotic spin-orbit splitting in atomic I, it appears that Hund's case (c) coupling is appropriate for describing the electronic states probed in our experi-

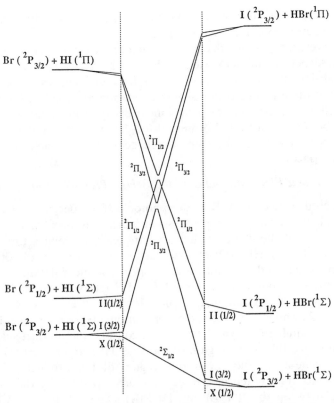

Figure 26. Electronic correlation diagram for Br + HI reaction assuming $C_{\infty v}$ symmetry. The region between the dashed lines is where Hund's case (a) is appropriate (Ref. 55).

ment. In other words, the electronic state structure of the BrHI complex in the geometry resulting from photodetachment of BrHI⁻ is clearly recognizable as that of an I atom slightly perturbed by its interaction with HBr. On the other hand, the vibrational structure in the ground- and excited-state progressions is quite similar. This indicates that, in the geometry probed by photodetachment, the distortion of the HBr bond is approximately independent of the I atom electronic state, and the ground- and spin-orbit excited-state potential energy surfaces are not very different in the I + HBr exit valley.

The observation of two, rather than three, electronic states in the photoelectron spectra indicates that the splitting of the $I(^2P_{3/2})$ state by the HBr is too small to resolve in our experiment. However, the peaks in the ground-state progression are, on the average, 0.020 eV broader than in the excited state progression (0.170 eV vs. 0.150 eV). This could

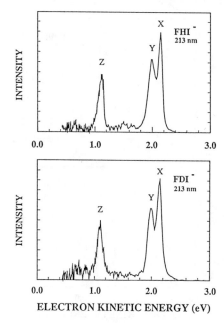

Figure 27. 213-nm photoelectron spectra of FHI⁻ and FDI⁻ (Ref. 55).

mean that each ground-state peak is actually composed, for example, of two 0.150-eV wide peaks split by a 0.020-eV X(1/2)–I(3/2) interval; the sum of two such peaks would yield a single peak in the photoelectron spectrum. A stronger interaction between the halogen atom and HX molecule should yield three distinct electronic states in the photoelectron spectrum. Examples in which this is the case are presented in the next section.

C. Photoelectron Spectroscopy of FHI⁻ and FHBr⁻

The photoelectron spectra of FHI⁻, FHBr⁻, and their deuterated analogs are shown in Figs. 27 and 28. The proton affinity of F⁻ is substantially higher than that of Br⁻ and I⁻ (2.07 and 2.47 eV, respectively[87]), and the F + HI and F + HBr reactions are highly exothermic (2.815 and 2.111 eV, respectively). Therefore, photodetachment should access the repulsive HF + I (Br) product valley in both cases, just as in BrHI⁻ photodetachment. However, although the spectra in Figs. 27 and 28 show resolved features, they are qualitatively different from the BrHI⁻/BrDI⁻ spectra in that no isotope shifts are observed. The FHI⁻ and FDI⁻ spectra each show three peaks labelled X, Y, and Z. These peaks occur at identical energies in both spectra. The X–Y spacing is 0.151 eV (1220 cm⁻¹), and the X–Z spacing is 1.045 eV (8430 cm⁻¹). Three peaks clearly analogous to these appear in the FHBr⁻ and FDBr⁻ spectra. In both spectra, the X–

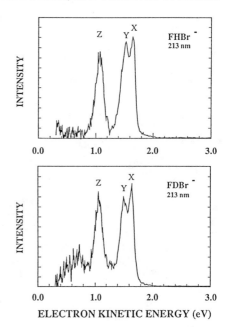

Figure 28. 213-nm photoelectron spectra of FHBr⁻ and FDBr⁻.

ELECTRON KINETIC ENERGY (eV)

Y and X–Z spacings are 0.121 eV (980 cm⁻¹) and 0.600 eV (4820 cm⁻¹), respectively.

The absence of an isotope shift in these spectra suggests that peaks X, Y, and Z are due to transitions to three distinct electronic states of the neutral complex. In the FHI⁻ spectrum, the X–Z spacing is 0.102 eV larger than the spin-orbit splitting in atomic I, while the X–Z spacing in the FHBr⁻ spectrum is 0.150 eV larger than the $^2P_{3/2}$–$^2P_{1/2}$ splitting in atomic Br (0.450 eV). With reference to the correlation diagram (Fig. 26), the peak pattern in the FHI⁻ and FHBr⁻ spectra is what would be expected if the neutral complex formed by photodetachment can be described by a Br or I atom interacting with HF. The larger deviation from the atomic spin-orbit splitting in these spectra compared to the BrHI⁻ spectrum indicates that this electronic interaction is stronger in the FHBr and FHI complexes. This is consistent with the observation of three distinct electronic states in the FHI⁻ and FHBr⁻ spectra compared to only two in the BrHI⁻ spectrum. The closely spaced X and Y peaks are assigned to the two states that result when degeneracy of the Br or I($^2P_{3/2}$) ground state is noticeably split by its interaction with HF.

The FHI⁻ and FHBr⁻ spectra also differ from the BrHI⁻ spectrum in that they do not exhibit a vibrational progression in the v_3 mode; only $\Delta v_3 = 0$ transitions are observed. For this to occur, the v_3 potentials for the anion and neutral (see Fig. 21) must be similar near the anion equilibrium

geometry. A recent *ab initio* calculation of the FHBr$^-$ equilbrium geometry by Sannigrahi and Peyerimhoff[89] yields R_e(HF) = 0.945 Å and
R_e(HBr) = 1.824 Å. The H–F separation is very close to R_e = 0.917 Å in
diatomic HF, while the H–Br separation is considerably greater than R_e =
1.413 Å in diatomic HBr.[94] This means that FHBr$^-$ can be thought of as
an essentially intact HF molecule complexed to Br$^-$. Near the equilibrium
geometry, R_{H-F} is the appropriate v_3 coordinate for FHBr$^-$, and the v_3
potential for FHBr$^-$ is very similar to the potential function for HF. The
highly asymmetric *ab initio* structure is reasonable as the proton affinity
of F$^-$ is much higher than that of Br$^-$. A similar highly asymmetric
geometry and v_3 potential are expected for FHI$^-$.

The lack of a v_3 progression in the photoelectron spectra therefore
means that the neutral v_3 potentials for the FHBr and FHI complexes must
also resemble the potential for diatomic HF, at least near the mimimum in
the potential function. This is a reasonable expectation. The F + HI and
F + HBr reactions are highly exothermic, so the double minimum v_3
potential should be highly asymmetric with the X + HF product valley
side much lower in energy than the F + HX reactant valley side. One
therefore expects the lowest few v_3 levels to look like isolated HF vibrational energy levels. The neutral v_3 wavefunctions will begin to deviate
from diatomic HF wavefunctions only near the energy of the second
(reactant) minimum in the v_3 potential; these highly excited levels have
no Franck–Condon overlap with the $v_3'' = 0$ levels of FHBr$^-$ and FHI$^-$.

In summary, the photoelectron spectra show that in the neutral complex
resulting from photodetachment of FHI$^-$ and FHBr$^-$, the HF interacts
with the I or Br atom strongly enough to noticeably perturb its electronic
structure. One is still in the Hund's case (c) regime, since the spin-orbit
splitting in the halogen atom is larger than the X(1/2)–I(3/2) splitting, but
the deviation from the energy levels of the isolated halogen atom is
noticeably larger than in the BrHI complex. Although the electronic interactions in the FHI and FHBr complexes are stronger than in the BrHI
complex, the HF bond is less perturbed in the former systems than the
HBr bond is in the latter, since the BrHI$^-$ spectra shows a vibrational
progression while the FHI$^-$ and FHBr$^-$ spectra do not. Finally, we note
that although the calculations by Manz and co-workers[53,74,96] on the
F + DBr reaction predict strong resonances, the highly excited v_3 states
responsible for the resonances have nearly zero overlap with the anion
ground state.

D. Photoelectron Spectroscopy of CH$_3$OHF$^-$ and CH$_3$ODF$^-$

The photoelectron spectra of the asymmetric bihalide anions discussed in
Sections V.B and V.C are sensitive to features of the product valley of
the potential energy surface for the corresponding neutral reaction, and

are relatively insensitive to the entrance valley and saddle-point region. This is the case for all of the asymmetric bihalides, not just the ones discussed above. The nature of the product valley in an exothermic reaction is of considerable interest since it plays a major role in the partitioning of excess energy among product degrees of freedom. On the other hand, it was pointed out in Section V.A that photodetachment of CH_3OHF^- should access the reactant valley saddle-point region of the reaction $F + CH_3OH \rightarrow HF + CH_3O$. This is arguably the most important region of the potential energy surface as it includes the barrier along the minimum energy path and the classical transition state. The $CH_3OHF^-/F + CH_3OH$ system is also of interest as it presents an opportunity to extend our method to reactions involving polyatomic species. In this section, the photoelectron spectrum of CH_3OHF^- is presented and compared to the asymmetric bihalide spectra.

The $F + CH_3OH$ reaction has been the subject of both kinetics[95] and product state-resolved experiments.[96-98] Hydrogen abstraction by the F atom can yield either CH_3O or CH_2OH. Although CH_2OH production is more exothermic (176.6 kJ/mol vs. 134.7 kJ/mol at 298 K), the CH_3O channel dominates at room temperature.[95] In the CH_3OHF^- anion, F^- will preferentially bind to the hydroxyl hydrogen since this is considerably more acidic than the methyl hydrogens in CH_3OH.[99] Thus, only the transition state region for the CH_3O channel is probed in our experiment.

The photoelectron spectra of CH_3OHF^- and CH_3ODF^- at 213 nm are shown in Fig. 29. The spectra show resolved features, although they are more complex than the spectra of the triatomic anions discussed previously. The product and reactant asymptotes are indicated by the arrows at electron kinetic energies of 2.59 and 1.17 eV, respectively. These are based on the $CH_3OH \cdot\cdot F^-$ binding energy of 1.28 eV determined by Larson and McMahon.[99]

The most prominent feature in the CH_3OHF^- spectrum is the set of four steps with onsets at 2.69, 2.24, 1.85, and 1.54 eV. The interval between the two highest energy steps is 0.45 eV (3630 cm^{-1}) and decreases to 0.31 eV (2500 cm^{-1}) between the two lowest energy steps. In the CH_3ODF^- spectrum, a clear isotope effect is apparent; six more closely spaced steps occur in the same energy range. Thus, the steps appear to correspond to a vibrational progression in the neutral CH_3OHF complex. The isotope shift and the magnitude of the interval between the steps suggests this progression involves a vibrational mode analogous to the v_3 mode seen in the bihalide photoelectron spectra in which the light H atom is vibrating between the much heavier F atom and CH_3O group in the neutral complex. The steps occur at energies where only the $CH_3O + HF$ products are accessible, but the intervals between the steps are consider-

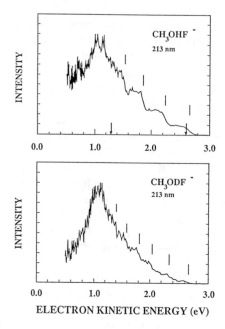

Figure 29. 213-nm photoelectron spectra of CH_3OHF^- and CH_3ODF^-. Arrows indicate product and reactant asymptotes (see text). Step onsets are indicated with short vertical lines.

ably less than the HF fundamental ($3958\ cm^{-1}$). This shows we are close to the transition state region on the $F + CH_3OH$ surface.

In contrast to the $BrHI^-$ spectrum, the intensity in the CH_3OHF^- spectrum is small at the product asymptote and builds as the reactant asymptote is approached. This is a consequence of the improved Franck–Condon overlap between the anion and reactant valley of the neutral potential energy surface. If one approximates $CH_3OHF^-/F + CH_3OH$ as a three-atom $AHB^-/A + HB$ system, the neutral potential energy surface for the collinear geometry will resemble the surface in Fig. 20, and the effective v_3 potential will look like the double minimum potential in Fig. 21. However, the minimum in the anion v_3 potential will lie somewhere under the reactant minimum of the neutral potential. The anion ground state will therefore have poor overlap with the lowest neutral v_3 levels, as these are localized in the product minimum, but better overlap with the higher v_3 levels is expected. Thus, the overall intensity distribution in the CH_3OHF^- spectrum is reasonable in light of this simple one-dimensional picture.

The step structure in the CH_3OHF^- photoelectron spectrum is quite different from the well-resolved peaks in the bihalide spectra. This raises the question of homogeneous vs. inhomogeneous features in these spectra. Since we are photodetaching a polyatomic anion it is possible that other modes in addition to the "v_3 mode" may be active in the photoelectron spectrum, and that the steps represent clumps of unresolved transitions.

Another possibility is that the broad step structure results from the dynamics on the $F + CH_3OH$ potential energy surface. The steps occur between the product and reactant asymptotes, and therefore correspond to states of the complex that dissociate to $HF + CH_3O$ but not $F + CH_3OH$. The dissociation energy of CH_3OHF^- (1.28 eV) is considerably larger than that of $BrHI^-$ (0.70 eV),[57] FHI^- (0.65 eV),[100] and $FHBr^-$ (0.74 eV).[100] The more tightly bound anion leads one to expect the neutral complex to be more compact. Treating CH_3OHF^- as a triatomic linear anion, this means the dashed vertical line in Fig. 20, which in this case corresponds to R_{O-F} at the equilibrium geometry of the anion, should be moved in to a smaller value of x than is appropriate for $BrHI^-$ photodetachment. At such a value, the product valley is more repulsive, leading one to expect that levels of the complex localized in the product valley will dissociate more rapidly.

On the other hand, since the $F + CH_3OH$ reaction is thought to have a small barrier, the reactant valley and saddle-point region of the potential energy surface should be quite flat, and states of the complex localized in this region should have relatively long lifetimes. Indeed, several distinct peaks appear in the CH_3OHF^- photoelectron spectrum at electron energies where $F + CH_3OH$ reactants are accessible as dissociation products of the complex. The observation of relatively narrow peaks in one region of the spectrum and broad steps in another may therefore be an indication of state-specific dissociation dynamics in the CH_3OHF complex.

The explanation of the CH_3OHF^- spectrum in terms of dissociation dynamics on a collinear three-atom potential energy surface is somewhat of an oversimplification, and it seems likely that the polyatomic nature of the system contributes to the complexity of the spectrum. However, we have recently obtained the photoelectron spectrum of FHO^- and found it to be remarkably similar to the CH_3OHF^- spectrum.[101] This suggests that three-atom dynamics may explain a good fraction of the CH_3OHF^- spectrum. This is clearly a subject that warrants further investigation in the near future.

VI. SUMMARY

This chapter has summarized several studies of the transition state region in bimolecular reactions using negative ion photodetachment. The results presented here show how photoelectron spectroscopy and threshold photodetachment spectroscopy can provide a detailed picture of the spectroscopy and dissociation dynamics of a complex formed in the transition state region of a reactive potential energy surface. The higher resolution threshold photodetachment results are of particular interest as they represent the definitive observation of reactive resonances in the I + HI reaction. Although much of the work so far has focused on hydrogen exchange between two halogen atoms, the results in Section V.D show that these methods can be applied to polyatomic systems as well. Future applications of these methods will focus on studies of other polyatomic reactions as well as experiments in which reactions are initiated by photodetaching bihalide anions embedded in clusters of known size.[102]

ACKNOWLEDGMENTS

The photoelectron spectroscopy experiments are supported by the Air Force Office of Scientific Research, Grant. No. AFOSR-91-0084. The threshold photodetachment studies are supported by the Chemistry Division of the Office of Naval Research, Grant. No. N0014-87-K-0495. The authors thank George Schatz, Joel Bowman, and David Tannor for many stimulating discussions. DMN thanks the other members of his research group who contributed to the work presented here: Don Arnold, Caroline Chick, Theofanis Kitsopoulos, Dr. Irene Waller, and Alexandra Weaver.

REFERENCES

1. P. R. Brooks, *Chem. Rev.* **88**, 407 (1988).

2. A. H. Zewail, *Science* **242**, 1645 (1988); M. Gruebele and A. H. Zewail, *Physics Today* **43**, 24 (1990).

3. N. F. Scherer, L. R. Khundkar, R. B. Bernstein, and A. H. Zewail, *J. Chem. Phys.* **87**, 1451 (1987); N. F. Scherer, C. Sipes, R. B. Bernstein, and A. H. Zewail, *J. Chem. Phys.* **92**, 5239 (1990).

4. M. Dantus, M. J. Rosker, and A. H. Zewail, *J. Chem. Phys.* **87**, 2395 (1987); R. M. Bowman, M. Dantus, and A. H. Zewail, *Chem. Phys. Lett.* **56**, 131 (1989); M. Dantus, R. M. Bowman, M. Gruebele, and A. H. Zewail, *J. Chem. Phys.* **91**, 7437 (1989).

5. S. Golub and B. Steiner, *J. Chem. Phys.* **49**, 5191 (1968).

6. C. R. Moylan, J. D. Dodd, C. Han, and J. I. Brauman, *J. Chem. Phys.* **86**, 5350 (1987).

7. M. W. Siegel, R. J. Celotta, J. L. Hall, J. Levine, and R. A. Bennett, *Phys. Rev, A* **6**, 607, 631 (1972).

8. H. Hotop, R. A. Bennett, and W. C. Lineberger, *J. Chem. Phys.* **58**, 2373 (1973).

9. T. N. Kitsopoulos, I. M. Waller, J. G. Loeser, and D. M. Neumark, *Chem. Phys. Lett.* **159**, 300 (1989).

10. K. K. Murray, T. M. Miller, D. G. Leopold, and W. C. Lineberger, *J. Chem. Phys.* **84**, 2520 (1986).

11. S. M. Burnett, A. E. Stevens, C. S. Feigerle, and W. C. Lineberger, *Chem. Phys.*

58   RICARDO B. METZ, STEPHEN E. BRADFORTH, AND DANIEL M. NEUMARK

Lett. **100**, 124 (1983); K. M. Ervin, J. Ho, and W. C. Lineberger, *J. Chem. Phys.* **91**, 5974 (1989).

12. B. S. Ault, *Acc. Chem. Res.* **15**, 103 (1982) and references therein.

13. D. G. Truhlar and A. Kuppermann, *J. Chem. Phys.* **52**, 384 (1970).

14. S.-F. Wu and R. D. Levine, *Mol. Phys.* **22**, 991 (1971).

15. G. C. Schatz and A. Kuppermann, *J. Chem. Phys.* **59**, 964 (1973); S.-F. Wu, B. R. Johnson, and R. D. Levine, *Molec. Phys.* **25**, 609 (1973).

16. G. C. Schatz, J. M. Bowman, and A. Kuppermann, *J. Chem. Phys.* **63**, 674 (1975).

17. S. L. Latham, J. F. McNutt, R. E. Wyatt, and M. J. Redmon, *J. Chem. Phys.* **69**, 3746 (1978).

18. J. N. L. Connor, W. Jakubetz, and J. Manz, *Molec. Phys.* **35**, 1301 (1978).

19. J. M. Launay and M. Le Dourneuf, *J. Phys. B* **15**, L455 (1982).

20. E. Pollak, *J. Chem. Phys.* **78**, 1228 (1983).

21. D. K. Bondi, J. N. L. Connor, J. Manz, and J. Römelt, *Mol. Phys.* **50**, 467 (1983).

22. A. Kuppermann, in *Potential Energy Surface and Dynamics Calculations*, D. G. Truhlar, Ed. Plenum, New York, 1981, pp. 375–420.

23. G. C. Schatz and A. Kuppermann, *Phys. Rev. Lett.* **35**, 1266 (1975).

24. M. J. Redmon and R. E. Wyatt, *Chem. Phys. Lett.* **63**, 209 (1979).

25. J. Z. H. Zhang and W. H. Miller, *J. Chem. Phys.* **91**, 1528 (1989); D. E. Manolopoulos and R. E. Wyatt, *J. Chem. Phys.* **92**, 810 (1990); M. Zhao, D. G. Truhlar, D. W. Schwenke, and D. J. Kouri, *J. Phys. Chem.* **94**, 7074 (1990).

26. J.-C. Nieh and J. J. Valentini, *Phys. Rev. Lett.* **60**, 519 (1988).

27. D. M. Neumark, A. M. Wodtke, G. N. Robinson, C. C. Hayden, and Y. T. Lee, *J. Chem. Phys.* **82**, 3045 (1985).

28. R. B. Metz, A. Weaver, S. E. Bradforth, T. N. Kitsopoulos, and D. M. Neumark, *J. Phys. Chem.* **94**, 1377 (1990).

29. L. A. Posey, M. J. DeLuca, and M. A. Johnson, *Chem. Phys. Lett.* **131**, 170 (1986).

30. M. A. Johnson and W. C. Lineberger, in *Techniques in Chemistry*, Vol. 20, J. M. Farrar and W. H. Saunders, Jr., Eds., Wiley, New York, 1988, pp. 591–635.

31. W. C. Wiley and I. H. Maclaren, *Rev. Sci. Instrum.* **26**, 1150 (1955).

32. K. Müller-Dethlefs, M. Sander, and E. W. Schlag, *Z. Naturforsch.* **39a**, 1089 (1984); *Chem. Phys. Lett.* **12**, 291 (1984).

33. J. M. B. Bakker, *J. Phys. E* **6**, 785 (1973); **7**, 364 (1974).

34. G. C. Schatz, *J. Phys. Chem.* **94**, 6157 (1990).

35. G. C. Schatz, *J. Chem. Phys.* **90**, 1237 (1989).

36. G. C. Schatz, *J. Chem. Phys.* **90**, 4847 (1989).

37. G. C. Schatz, *Chem. Phys. Lett.* **150**, 92 (1988).

38. K. Kawaguchi, *J. Chem. Phys.* **88**, 4186 (1988).

39. J. Z. H. Zhang and W. H. Miller, *J. Chem. Phys.* **92**, 1811 (1990).

40. R. Steckler, D. G. Truhlar, and B. C. Garrett, *J. Chem. Phys.* **82**, 5499 (1985).

41. R. A. Kendall, J. A. Nichols, and J. Simons, *J. Phys. Chem.* **94**, 1074 (1991).

42. A. Weaver, R. B. Metz, S. E. Bradforth, and D. M. Neumark, *J. Chem. Phys.* **93**, 5352 (1990); A. Weaver and D. M. Neumark, *Faraday Discuss. Chem. Soc.* **91**, (1991).

43. B. Gazdy and J. M. Bowman, *J. Chem. Phys.* **91**, 4615 (1989).

44. J. M. Bowman, *Adv. Chem. Phys.* **61**, 115 (1985).

45. A. B. Sannigrahi and S. D. Peyerimhoff, *J. Mol. Struct.* **165**, 55 (1988).

46. S. Ikuta, T. Saitoh, and O. Nomura, *J. Chem. Phys.* **93**, 2530 (1990).

47. J. M. Bowman, B. Gazdy, and Q. Sun, *J. Chem. Soc. Faraday Trans.* **86**, 1737 (1990).

48. J. M. Bowman and B. Gazdy, *J. Phys. Chem.* **93**, 5129 (1989).

49. J. Römelt, *Chem. Phys.* **79**, 197 (1983).

50. E. J. Heller, *J. Chem. Phys.* **68**, 3891 (1978).

51. R. Kosloff, *J. Phys. Chem.* **92**, 2087 (1988).

52. R. H. Bisseling, R. Kosloff, and J. Manz, *J. Chem. Phys.* **83**, 993 (1985).

53. R. H. Bisseling, P. L. Gertitschke, R. Kosloff, and J. Manz, *J. Chem. Phys.* **88**, 6191 (1988).

54. A. J. Lorquet, J. C. Lorquet, J. Delwiche, and M. J. Hubin-Franskin, *J. Chem. Phys.* **76**, 4692 (1982).

55. S. E. Bradforth, A. Weaver, D. W. Arnold, R. B. Metz, and D. M. Neumark, *J. Chem. Phys.* **92**, 7205 (1990).

56. D. Kosloff and R. Kosloff, *J. Comput. Phys.* **52**, 35 (1983).

57. G. Caldwell and P. Kebarle, *Can. J. Chem.* **63**, 1399 (1985).

58. H. Hotop and W. C. Lineberger, *J. Phys. Chem. Ref. Data* **14**, 731 (1985).

59. P. Botschwina and W. Meyer, *Chem. Phys. Lett.* **44**, 449 (1976).

60. B. C. Garrett, D. G. Truhlar, A. F. Wagner, and T. H. Dunning, Jr., *J. Chem. Phys.* **78**, 4400 (1983).

61. K. Kawaguchi and E. Hirota, *J. Chem. Phys.* **87**, 6838 (1987).

62. V. Bondybey, G. C. Pimentel, and P. N. Noble, *J. Chem. Phys.* **55**, 540 (1971); D. E. Milligan and M. E. Jacox, *J. Chem. Phys.* **55**, 2550 (1971).

63. See Ref. 28 for the explicit form of this one-dimensional potential.

64. N. C. Firth and R. Grice, *J. Chem. Soc. Faraday Trans. 2.* **87**, 1023 (1987).

65. R. B. Metz, T. Kitsopoulos, A. Weaver, and D. M. Neumark, *J. Chem. Phys.* **88**, 1463 (1988).

66. K. Yamashita and K. Morokuma, *J. Chem. Phys.* **93**, 3716 (1990).

67. R. J. Saykally, *Science* **239**, 157 (1988).

68. J. Manz and J. Römelt, *Chem. Phys. Lett.* **81**, 179 (1981).

69. A. Weaver, R. B. Metz, S. E. Bradforth, and D. M. Neumark, *J. Phys. Chem.* **92**, 5558 (1988).

70. J. Manz, R. Meyer, E. Pollak, and J. Römelt, *Chem. Phys. Lett.* **93**, 184 (1982); E. Pollak, *Chem. Phys. Lett.* **94**, 85 (1983); J. Manz, R. Meyer, E. Pollak, J. Römelt, and H. H. R. Schor, *Chem. Phys.* **83**, 333 (1984).

71. D. C. Clary and J. N. L. Connor, *Chem. Phys. Lett.* **94**, 81, 1983; *J. Phys. Chem.* **88**, 2758 (1984).

72. I. M. Waller, T. N. Kitsopoulos, and D. M. Neumark, *J. Phys. Chem.* **94**, 2240 (1990).

73. G. C. Schatz, private communication.

74. P. L. Gertitschke, J. Manz, J. Römelt, H. H. R. Schor, *J. Chem. Phys.* **83**, 208 (1985).

75. D. Tannor, private communication.

76. C. Kubach, *Chem. Phys. Lett.* **164**, 475 (1989); C. Kubach, G. N. Vien, and M. Richard-Viard, *J. Chem. Phys.* **94**, 1929 (1991).

77. K. G. Anlauf, P. J. Kuntz, D. H. Maylotte, P. D. Pacey, and J. C. Polanyi, *Disc. Faraday Soc.* **44**, 183 (1967); D. H. Maylotte, J. C. Polanyi, and K. B. Woodall, *J. Chem. Phys.* **57**, 1547 (1972).

78. N. Jonathan, C. M. Melliar-Smith, S. Okuda, D. H. Slater, and D. Timlin, *Molec. Phys.* **22**, 4 (1971).

79. P. J. Kuntz, E. M. Nemeth, J. C. Polanyi, S. D. Rosner, and C. E. Young, *J. Chem. Phys.* **44**, 1168 (1966); C. A. Parr, J. C. Polanyi, and W. H. Wong, *J. Chem. Phys.* **58**, 5 (1973).

80. I. W. M. Smith, *Chem. Phys.* **20**, 437 (1977).

81. J. C. Brown, H. E. Bass, and D. L. Thompson, *J. Phys. Chem.* **81**, 479 (1977).

82. P. Beadle, M. R. Dunn, N. B. H. Jonathan, J. P. Liddy, and J. C. Naylor, *J. Chem. Soc. Faraday Trans. 2* **74**, 2170 (1978).

83. M. Broida and A. Persky, *Chem. Phys.* **133**, 405 (1989).

84. M. Baer, *J. Chem. Phys.* **62**, 305 (1975).

85. J. A. Kaye and A. Kuppermann, *Chem. Phys. Lett.* **92**, 574 (1982).

86. R. A. Fischer, P. L. Gertitschke, J. Manz, and H. H. R. Schor, *Chem. Phys. Lett.* **156**, 100 (1989); J. Manz and J. Römelt, *J. Chem. Soc. Faraday Trans.* **86**, 1689 (1990).

87. S. G. Lias, J. E. Bartmess, J. F. Leibman, J. L. Holmes, R. D. Levin, and W. G. Mallard, *J. Phys. Chem. Ref. Data* **17**, Suppl. No. 1 (1988).

88. C. M. Ellison and B. S. Ault, *J. Phys. Chem.* **83**, 832 (1979).

89. A. B. Sannigrahi and S. D. Peyerimhoff, *Chem. Phys. Lett.* **112**, 267 (1984); **164**, 348 (1989).

90. C. C. Mei and C. B. Moore, *J. Chem. Phys.* **70**, 1759 (1979); D. A. Dolson and S. R. Leone, *J. Chem. Phys.* **77**, 4009 (1982).

91. These curves are the eigenvalues of Eq. (11); the DIVAH correction (Eq. 18) is **not** included. Manz and co-workers (Ref. 86) have shown the $Q_{v_3 v_3}$ term in Eq. (18) becomes large near avoided crossings of the adiabatic curves $\epsilon_{v_3}(\rho)$.

92. K. Bergmann, S. R. Leone, and C. B. Moore, *J. Chem. Phys.* **63**, 4161 (1975).

93. J. W. Hepburn, K. Liu, R. G. Macdonald, F. J. Northrup, and J. C. Polanyi, *J. Chem. Phys.* **75**, 3353 (1981).

94. K. P. Huber and G. Herzberg, *Molecular Spectra and Molecular Structure IV. Constants of Diatomic Molecules*, Van Nostrand, New York, 1979.

95. J. A. McCaulley, N. Kelly, M. F. Golde, and F. Kaufman, *J. Phys. Chem.* **93**, 1014 (1989).

96. R. G. MacDonald, J. J. Sloan, and P. T. Wassell, *Chem. Phys.* **41**, 201 (1979).

97. B. Dill and H. Heydtmann, *Chem. Phys.* **54**, 9 (1980); M. A. Wickramaaratchi, D. W. Setser, H. Hildebrandt, B. Köbitzer, and H. Heydtmann, *Chem. Phys.* **94**, 109 (1985).

98. B. S. Agrawalla and D. W. Setser, *J. Phys. Chem.* **88**, 657 (1984); *J. Chem. Phys.* **90**, 2450 (1986).

99. J. W. Larson and T. B. McMahon, *J. Am. Chem. Soc.* **105**, 2944 (1983).

100. J. L. Larson and T. B. McMahon, *Inorg. Chem.* **23**, 2029 (1984).

101. S. E. Bradforth, R. B. Metz, A. Weaver, and D. M. Neumark, *J. Phys. Chem.* **95** (in press).

102. D. M. Neumark, in *The Physics of Electronic and Atomic Collisions: XVI International Conference (AIP Conference Proceedings 205)*, A. Dalgarno, R. S. Freund, P. M. Koch, M. S. Lubell, and T. B. Lucatorto, Eds., American Institute of Physics, New York, 1990, pp. 33–48.

INFRARED VIBRATIONAL PREDISSOCIATION SPECTROSCOPY OF SMALL SIZE-SELECTED CLUSTERS*

FRIEDRICH HUISKEN

Max Planck Institut für Strömungsforschung, Göttingen, Federal Republic of Germany

CONTENTS

ABSTRACT

Infrared predissociation spectroscopy has been shown to provide detailed information on the structure and dissociation dynamics of small van der Waals and hydrogen-bonded clusters. This information, however, is only obtained if the measured spectral features can be properly assigned. In many cases this assignment is very complicated or even impossible, because many different cluster sizes whose abundance is unknown contribute

---

* This chapter has been presented as habilitation thesis for obtaining the "venia legendi" at the Physics Department of the University of Göttingen, FRG.

Advances in Chemical Physics, Volume LXXXI, Edited by I. Prigogine and Stuart A. Rice.
ISBN 0-471-54570-8   © 1992 John Wiley & Sons, Inc.

to the spectrum. Then it is extremely important to provide a selection technique with which the spectra can be measured separately for each cluster size. This article describes infrared predissociation experiments which have been combined with a cluster selection technique, taking advantage of the constraints imposed on the scattering of a cluster beam with an atomic rare gas beam by the conservation laws of momentum and energy. The clusters are generated by adiabatic expansion of the desired gas into vacuum. After being stabilized by collisions and evaporation, they are subjected to the line-tunable radiation of a pulsed CO_2 laser. Subsequent scattering by the secondary rare gas beam disperses the cluster beam and allows the unequivocal detection of selected cluster species off-axis by a rotatable mass spectrometer. Predissociation spectra are recorded by monitoring the decrease in cluster signal as a function of laser frequency and laser energy and as a function of cluster size. Results are presented for the van der Waals clusters $(C_2H_4)_n$, $(C_6H_6)_n$, and $(SF_6)_n$. The investigated cluster sizes are between $n = 2$ (dimer) and $n = 6$ (hexamer). For $(C_2H_4)_n$ and $(C_6H_6)_n$ the absorption profiles are rather featureless and their dependence on the cluster size is weak. Nevertheless, valuable structural information has been obtained for $(C_6H_6)_n$ by fluence dependence measurements. In the case of $(SF_6)_n$, a dynamical dipole–dipole interaction results in a splitting of the absorption lines which is strongly dependent on the cluster size. The spectra of the hydrogen bonded systems $(NH_3)_n$ and $(CH_3OH)_n$ exhibit a completely different behavior. The dimer spectra are characterized by two well separated peaks revealing the existence of two nonequivalent monomers in the dimer. In contrast, the trimer and tetramer spectra are characterized by single peaks, which is in accordance with the cyclic structures of these species.

I. INTRODUCTION

In recent years there has been increased interest in the investigation of atomic and molecular clusters. These species can be considered as intermediates between isolated noninteracting atoms or molecules and liquid or solid phase particles. Generally the motivation for their investigation is the desire to understand how physical and chemical properties vary if one goes from molecular to condensed phase systems. Several journals have already devoted special issues to this exciting topic.[1-5]

Neutral clusters may be distinguished according to the type and strength of their binding. It is useful to introduce the following classification:[6] (1) Van der Waals (vdW) clusters are bound by rather weak dispersive forces. Typically their dissociation energy D is below 0.2 eV. Examples of this type of binding are the rare gas clusters R_n and closed-shell molecules with no permanent dipole moment such as $(H_2)_n$, $(CO_2)_n$, $(SF_6)_n$, and $(C_2H_4)_n$. (2) The next group with moderate binding energies ($D \sim 0.3$–0.5 eV) constitutes the hydrogen-bonded clusters. These species are characterized by charge transfer and molecules having a permanent dipole moment. Examples are $(HF)_n$, $(H_2O)_n$, $(NH_3)_n$, and $(CH_3OH)_n$. (3) Ionic clusters such as $(NaCl)_n$ and $(CaF_2)_n$ are characterized by rather strong ionic bonds with binding energies $D \sim 2$–4 eV. (4) Conventional chemical bonds are encountered in valence clusters. These species, examples of which are C_n, S_8, and As_4, are also characterized by large binding energies $D \sim 1$–4 eV. Finally, (5) there are metallic clusters with moderate to strong bonding ($D \sim 0.5$–3 eV). Examples of this type of binding are Na_n, Al_n, Cu_n, and W_n. In this chapter we will not consider strong interaction clusters of types (3–5). Instead, the discussion will be limited to vdW and hydrogen-bonded complexes.

A convenient and, therefore, most widely used technique to produce vdW and hydrogen-bonded clusters is by adiabatic expansion of the desired gas into vacuum. Because of the weak intermolecular forces holding these species together, very low temperatures are needed for their production and stabilization. In addition, a collisionless environment must be provided to ensure sufficiently long lifetimes. These requirements are met, in an ideal manner, in a supersonic molecular beam.[7,8] By expanding a gas at high pressure through a nozzle into vacuum, very low temperatures are achieved and, at the resulting high collision frequencies, large numbers of clusters can be readily produced.[9,10] Typically, the central part of the jet is sampled by a conical skimmer, and, thus, a supersonic cluster beam is formed. In this beam the density has decreased so greatly that no further collisions take place, and the clusters can exist for hundreds of microseconds until they hit a wall. For a recent review about molecular beams of clusters the reader is referred to Ref. 11. Although with this technique it is easily possible to produce large clusters comprised of thousands of molecules, we shall focus our interest on small species containing no more than 10 molecules. Such small clusters are often referred to as microclusters.[6]

Much work in cluster research has been devoted to the interaction of clusters with photons of various wavelengths, that is, to their spectroscopy. Such investigations may provide extremely valuable information on the structure and energetics of these species. If the clusters are excited, how and at what timescale the excitation energy is distributed among the various degrees of freedom can be further studied. Thus, insight into dynamical processes is obtained.

Spectroscopic studies of vdW and hydrogen-bonded clusters have been carried out in many wavelength regions using various techniques. Some applicable examples are mentioned. The rare gas dimers Xe_2 and Kr_2 have been investigated in supersonic jets by fluorescence excitation spectroscopy using tunable VUV radiation.[12,13] Spectroscopic constants for the ground state and several excited states have been obtained. In a pioneering work, Levy and co-workers[14-16] studied the excitation and dissociation of vdW complexes of iodine with rare gases by laser-induced fluoresence (LIF). Aside from structural information, valuable insight into the dissociation dynamics has been gained. The energetics and dynamics of large vdW molecules of several aromatic molecules bound to rare gas atoms have been studied by Jortner and co-workers.[17,18] Very recently, laser spectroscopy on benzene dimers and larger clusters has been performed using LIF[19] and resonance enhanced multiphoton ionization (REMPI).[20,21]

Considerable work has been done in the infrared (IR) region of the spectrum. Howard and co-workers have studied complexes of C_2H_2, OCS, N_2O, and CO_2 (as well as their combinations with each other and with rare gases) using infrared absorption spectroscopy in pulsed supersonic jets.[22-24] Although performed in a static cell, the important study on $(HF)_2$ by Pine, Lafferty, and Howard should also be mentioned in this context.[25] Using a molecular beam electric resonance spectrometer, Klemperer's group has obtained rotational spectra of complexes of NH_3 bound to various other molecules.[26] An important result of these studies was the conclusion that the structure of the NH_3 dimer must be somewhat different from the classical hydrogen bonded structure.[27] Very recently it has become possible to probe vdW modes directly. Saykally and co-workers[28,29] have developed a very sensitive laser system operating in the $35\,cm^{-1}$ wavenumber region. With their technique they were able to directly excite the vdW vibrations of ArHCl and to investigate the spectroscopy of this complex with very high resolution.

Complementary information to the IR spectroscopy of vdW molecules and clusters can be obtained by Raman techniques. Rotational and vibrational Raman spectra of Ar_2 have been measured by Godfried and Silvera[30,31] using an intracavity supersonic jet design. Recently, nonlinear Raman techniques like coherent anti-Stokes Raman scattering (CARS) and stimulated Raman loss spectroscopy (SRLS) have also been applied to determine vibrational frequency shifts in complexes of HCN,[32] CO_2,[33] NH_3,[34] H_2O,[35] C_2H_4,[36] N_2,[37] and HCl.[38]

Perhaps the greatest part of work in IR cluster spectroscopy has been done using a technique exploiting the collimation and directional properties of a molecular beam, and the dissociative behavior of loosely bound complexes upon absorption of a single IR photon. This technique is termed molecular beam depletion spectroscopy (MBDS). It was first proposed by

Klemperer[39] and later experimentally realized by Gough, Miller, and Scoles.[40] In their experiment, Gough et al.[40] produced a molecular beam containing N_2O monomers and dimers. This beam is crossed by the radiation of a tunable IR laser and directed to a liquid-helium-cooled silicon bolometer, or alternatively, to a mass spectrometer detector. In the region around 2223.5 cm^{-1} the ν_3 vibration of both N_2O monomers and dimers can be excited. Since the excitation energy is larger than the dimer binding energy and since there is sufficient coupling between the excited intramolecular mode and the intermolecular vdW mode, every dimer excitation results in its prompt dissociation. The fragments of the dissociated complex are expelled from the cluster beam and, thus, the occurrence of dissociation, or equivalently absorption, is readily detected by monitoring the decrease in cluster signal. This new technique constitutes an elegant alternative to the more common direct absorption spectroscopy.

Since its first realization in 1978, this molecular beam depletion technique via IR predissociation has been applied to a large variety of weakly bound complexes. The experiments have been carried out by various research groups using all types of available laser sources. The studies were done on vdW or hydrogen-bonded clusters of N_2O,[40] CO_2,[41] SF_6,[42] by Scoles, on HF,[43-45] H_2O,[44-47] C_6H_6,[45,48] by Lee, on C_2H_4,[49,50] by Janda, on C_2H_4,[51] OCS,[52] NH_3,[53] CH_3OH,[54] C_6H_6,[55] by Gentry, on NH_3,[26] by Klemperer, on SF_6,[56-60] SiH_4,[59] SiF_4,[59-61] C_2H_4,[62,63] NH_3,[64,65] CF_3Br,[61] and other halogenated methanes,[66] by Reuss, on C_2H_4,[67-69] C_2H_2,[69,70] N_2O,[71,72] CO_2,[72-74] C_3H_4,[69] H_2O,[75] HF[76] by Miller, and on C_2H_4,[77] HF,[78] C_6H_6[79] by Lisy and their co-workers (this list does not include heterogeneous clusters). Further studies were conducted on C_2H_4,[80-82] SF_6,[83,84] C_6H_6,[85-87] HF,[88] CO_2,[89] and C_2H_2.[90] For a more comprehensive account also considering heterogeneous clusters the reader is referred to the review articles by Janda[91] and Miller.[92]

A serious problem encountered in cluster-depletion experiments is the fact that typical molecular beams do not contain only one specific cluster size; instead one obtains a cluster mass distribution. Since cluster-specific detectors are not available, this distribution is normally not known. Bolometers are not mass specific, but neither do mass spectrometers provide cluster-specific detection because larger clusters fragment extensively during the ionization process.[93] Thus, the contribution of larger clusters to the dissociation spectra constitutes a serious problem in interpreting the data. Only in some few cases in which the measured spectra can be compared with calculations, is it possible to assign spectral features unambiguously. This has been done in special cases when rotationally resolved high-resolution spectra were obtained, for example, for CO_2,[72-74,89] HF,[76] and C_2H_2.[90] Thus far, these high resolution measurements have been restricted to the frequency range between 3000 and 4000 cm^{-1} accessible with a color center laser.

In all cases where cluster identification by spectroscopic means is not possible, any other technique to distinguish between different cluster sizes should be extremely helpful. Such a technique has recently been introduced by Buck and Meyer.[93,94] Their method is based on the different kinematic behavior of different clusters in the scattering process with rare gas atoms. Due to scattering by a secondary beam, the cluster beam is dispersed, that is, lighter clusters are scattered over larger angular ranges, whereas the heavier species are confined to smaller angles. With a rotatable mass spectrometer, it is thus possible to select one specific cluster size. Detection is not disturbed by fragmentation artifacts from larger clusters.

If this selection technique is combined with an ordinary molecular beam depletion experiment, cluster specific spectroscopy may be performed. The contribution of each cluster size to the measured absorption spectrum can be determined and all spectral features can be unambiguously assigned. It is possible to carry out the corresponding experiment in two alternative configurations. The laser–molecular-beam interaction zone can be placed either in front of or behind the scattering center. In the first case the laser photons interact with internally cold clusters cooled by the adiabatic expansion, whereas in the second configuration the laser probes clusters excited by the collisions with the rare gas atoms. In our institute, cluster selective predissociation is being performed using both approaches. Buck and co-workers have chosen the second configuration and investigate the spectroscopy of internally "warm" clusters. Systems so far studied are $(C_2H_4)_n$,[95–98] $(CH_3OH)_n$,[99,100] $(N_2H_4)_n$,[101] $(CH_3CN)_n$,[102] and $(CH_3NH_2)_n$.[103] A survey of this work is given in Ref. 104.

In this chapter we shall give a complete account of the experiments which have been performed with the laser interaction zone located anterior to the scattering center, that is, on vibrational predissociation experiments with *size-selected cold* clusters. Results are presented for the vdW clusters $(C_2H_4)_n$, $(C_6H_6)_n$, and $(SF_6)_n$ and for the hydrogen-bonded systems $(NH_3)_n$ and $(CH_3OH)_n$. The investigated cluster sizes are between $n = 2$ (dimer) and $n = 6$ (hexamer). Some results on $(C_2H_4)_n$,[105] $(C_6H_6)_n$,[106] $(SF_6)_n$,[107] and $(NH_3)_n$,[108] and preliminary results on $(CH_3OH)_2$[109] have already been published.

This chapter is organized as follows. Section II will be devoted to some kinematic aspects relevant to the cluster selection technique. Advantages and limitations of the technique will be discussed. In Section III the experimental setup and in Section IV the data collection and analysis are described. The results for the investigated systems $(C_2H_4)_n$, $(C_6H_6)_n$,

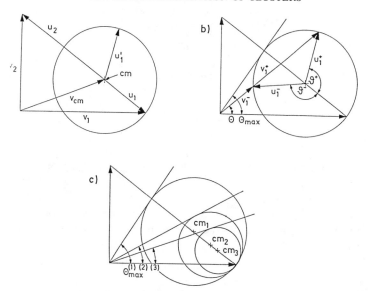

Figure 1. Newton diagrams illustrating the cluster-selection technique.

$(SF_6)_n$, $(NH_3)_n$, and $(CH_3OH)_n$ are presented and discussed in Section V. Finally, in Section VI, some concluding remarks are given.

II. KINEMATIC CONSIDERATIONS

The purpose of this section is to discuss the principles on which the cluster selection by molecular beam scattering[93] is based. We consider two colliding particles, indexed 1 and 2, with masses m_1 and m_2 and velocities $\mathbf{v}_1$ and $\mathbf{v}_2$. The collision process is governed by the conservation laws of Newton's mechanics and can be best visualized in a velocity vector diagram which is also termed a Newton diagram.[110,111] In Fig. 1a such a Newton diagram is drawn for the special case of perpendicularly crossing molecular beams, a situation encountered in most experiments. Here $\mathbf{v}_1$ and $\mathbf{v}_2$ are the initial velocity vectors in the laboratory (lab) reference frame, whereas $\mathbf{u}_1$ and $\mathbf{u}_2$ designate the corresponding (initial) velocities in the center of mass (cm) system. The relative velocity g is obtained as the hypotenuse of the right triangle generated by $\mathbf{v}_1$ and $\mathbf{v}_2$. Thus, we have

$$g = \sqrt{|\mathbf{v}_1|^2 + |\mathbf{v}_2|^2} = |\mathbf{u}_1| + |\mathbf{u}_2| \tag{1}$$

The velocities $u_1 = |\mathbf{u}_1|$ and $u_2 = |\mathbf{u}_2|$ in the cm system are calculated from the conservation laws to be

$$u_1 = \frac{m_2}{m_1 + m_2} g \qquad (2)$$

and

$$u_2 = \frac{m_1}{m_1 + m_2} g \qquad (3)$$

In the example of Fig. 1, m_1 is larger than m_2 and hence u_1 smaller than u_2. The small cross labelled with cm and dividing g into u_1 and u_2 is the center of mass and the vector pointing to it, $\mathbf{v}_{cm}$, is the center of mass velocity. For any velocity $\mathbf{u}$ in the cm system the corresponding velocity $\mathbf{v}$ in the lab system is obtained as the sum

$$\mathbf{v} = \mathbf{u} + \mathbf{v}_{cm} \qquad (4)$$

Now we consider the scattering of particle 1 by particle 2. If the scattering process is elastic, that is, if there is no energy transfer between internal and translational modes, only the direction of $\mathbf{u}_1$ can be changed and not the amount u_1. Thus, the possible velocity vectors $\mathbf{u}_1'$ of the scattered particles 1 lie on a sphere about cm with radius u_1. The section of this sphere with the scattering plane gives the circle shown in Fig. 1a. Assume that particle 1 is scattered at the angle ϑ^+. Its cm velocity then is $\mathbf{u}_1^+$ (see Fig. 1b), and its velocity in the lab system is according to (4) $\mathbf{v}_1^+ = \mathbf{u}_1^+ + \mathbf{v}_{cm}$. Thus, in the lab system this particle is deflected by the angle Θ and, in order to observe it in the experiment, the detector must be set to Θ.

A slight complication now arises, because not only particles with velocity $\mathbf{v}_1^+$, but also particles with velocity $\mathbf{v}_1^-$ will be detected at the same lab deflection angle Θ. As can be seen in Fig. 1b, a deflection in the cm system of either ϑ^- or ϑ^+ results in the same laboratory scattering angle Θ. Note that the situation of two contributing velocities is only encountered if $m_1 > m_2$. In some favorable cases, the two velocities v_1^+ and v_1^- can be resolved in the experiment, if a velocity analysis of the scattered beam is performed (see for example Ref. 112). In many cases, however, a separation is not possible, namely when the difference between v_1^+ and v_1^- is too small, or when the initial velocities v_1 and v_2 are not sufficiently well defined, or when the system is too inelastic. The inelastic collisions are associated with a change in velocity, and in that case the velocity distributions about v_1^+ and v_1^- may be broadened and shifted so much that they merge into one single peak. It is important to note that, in the laboratory reference frame, there is a maximum scattering angle Θ_{max} which is defined by the tangent to the Newton circle (see Fig. 1b). Beyond this angle no scattered particles will be detected.

Now let us assume that the primary beam (1) contains a mixture of monomers, dimers, and trimers all having the same velocity v_1. (Equal velocities are easily obtained if the clusters are generated in expansions of sufficiently dilute mixtures in a rare gas.) The corresponding kinematics are illustrated in Fig. 1c. Since the masses are different (m_1, $2m_1$, and $3m_1$), the scattering of each particle is associated with a different center of mass ($cm^{(1)}$, $cm^{(2)}$, and $cm^{(3)}$). Their positions are determined by calculating the cm velocities for the different cluster sizes n according to (2), that is,

$$u_1^{(n)} = \frac{m_2}{nm_1 + m_2} g \tag{5}$$

It is important to realize that, with increasing cluster size, (1) the centers of mass move towards the primary beam and (2) the cm velocities $u_1^{(n)}$ and, thus, the radii of the Newton circles become smaller. Consequently, the maximum scattering angles $\Theta_{max}^{(n)}$ and therefore the allowed scattering ranges decrease with increasing cluster size n. Dimers can only be detected at angles smaller than $\Theta_{max}^{(2)}$, while trimers are confined to an angular range below $\Theta_{max}^{(3)}$.

It is now conceivable that the cluster size distribution can be analyzed, if one uses a mass spectrometer detector which can be rotated about the scattering center. In the angular range between $\Theta_{max}^{(3)}$ and $\Theta_{max}^{(2)}$, for example, the only observable cluster size is the dimer. All larger clusters are confined to smaller angles. Thus, this angular range is ideally suited to study dimer properties, for example its spectroscopy, without interference from larger clusters fragmenting in the ionizer.

Let us consider three different fragmentation behaviors. First, there is the possibility that there is no or only moderate fragmentation. Thus, every cluster has a large probability of being detected at its parent ion mass even if there is some contribution from fragmenting larger clusters. This is the most favorable case in which the scattering technique is ideally suited to discriminate against larger cluster fragments. For example, if one wants to observe the pure trimer, the mass spectrometer must be tuned to the trimer ion mass and positioned to a scattering angle $\Theta_{max}^{(3)} > \Theta > \Theta_{max}^{(4)}$. Then all detected signals can be unambiguously assigned to the trimer.

Second, we consider the case in which there is strong fragmentation but the main fragment ion mass is still larger than the parent ion mass of the next lighter cluster. Such fragmentation behavior is encountered in many hydrogen-bonded systems. For example, in $(NH_3)_n$ and $(CH_3OH)_n$ clusters the fragmentation channels

$$(NH_3)_n^+ \rightarrow (NH_3)_{n-1}H^+ + \text{rest}$$

and

$$(CH_3OH)_n^+ \rightarrow (CH_3OH)_{n-1}H^+ + \text{rest}$$

are observed. Despite the strong fragmentation, the situation is similarly favorable as in the first case, for the mass spectrometer can be properly tuned and positioned to discriminate against both lighter and heavier clusters.

The most unfavorable situation is encountered in the third case, where very strong or even complete fragmentation occurs. Here it may happen that all clusters must be detected on one single fragment ion mass. An example is $(SF_6)_n$ where, for all clusters, a significant signal is only observed at the SF_5^+ ion mass. In this case the mass spectrometer does not help to discriminate against the lighter clusters, but the scattering technique is still useful to discriminate against the heavier species. Especially in this situation, the merits of the new technique become apparent. First, the mass spectrometer is moved to an angle $\Theta > \Theta_{max}^{(3)}$. Here the dimer can be studied without interference from larger clusters. Then the detector is rotated to a smaller angle $\Theta_{max}^{(3)} > \Theta > \Theta_{max}^{(4)}$ where trimers are also scattered. Here, both dimers and trimers are present, but any new feature compared to the previous measurement must be assigned to the trimer. If the detector is moved one step further ($\Theta < \Theta_{max}^{(4)}$), the tetramer contribution can be studied. The procedure can be continued and even larger clusters may be investigated at still smaller angles.

Now we want to discuss the limitations of the cluster-scattering technique and direct our attention to the problem of resolution. As can be seen in Fig. 1c, the Newton circles become continually smaller with increasing cluster size, as do the scattering ranges or intervals $[\Theta_{max}^{(n+1)}, \Theta_{max}^{(n)}]$ in which selective detection of cluster size n is possible. Thus, it is obvious that there is a maximum cluster size n_{max} beyond which no cluster separation is possible. The size n_{max} depends very much on the system. In particular, it depends (1) on the mass m_1 of the molecule from which the clusters are formed, (2) on its velocity v_1, (3) on the mass m_2 of the scattering partner and (4) its velocity v_2, and (5) finally on the system's inelasticity. Increasing m_1 or v_1 deteriorates the resolution, whereas increasing m_2 or v_2 improves it. Inelastic scattering leads to a

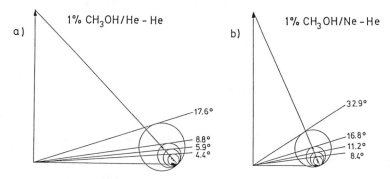

Figure 2. Newton diagrams for the scattering of methanol clusters by He. (*a*) He is used as carrier gas; (*b*) Ne is used.

diminution of the system's translational energy and therefore results in a reduction of both Newton circles and scattering ranges. Another effect of inelastic scattering is that the transitions between the different scattering regions become washed out.

In a typical scattering experiment, if the system is not very inelastic, clusters containing up to $n = 6$ molecules can be well separated. As already mentioned, the resolution can be further increased, if smaller v_1 or greater m_2 or v_2 is chosen. However, any improvement in resolution is combined with a loss in intensity and, since the main interest was so far directed to the study of small clusters, no attempt has been made to optimize the conditions for the investigation of larger species. But it is believed that the technique can be applied to clusters with n as large as $n = 10$.

In Fig. 2 the effect of choosing a heavier carrier gas on the resolution is demonstrated. Fig. 2*a* shows the Newton diagram for methanol clusters seeded in He and scattered by He. Since the methanol clusters are relative fast, they are only slightly deflected in the lab frame and, consequently, the resolution is rather poor. As shown in the Newton diagram of Fig. 2*b*, the situation is already improved if, instead of He, Ne is used as carrier gas. For example, the range where exclusively methanol dimers can be investigated, $\Theta_{max}^{(2)} - \Theta_{max}^{(3)}$, increases from 2.9° in Fig. 2*a* to 5.6° in Fig. 2*b*. The only reason for this improvement is the lower velocity of the methanol cluster beam. It is, however, important to realize that the methanol cluster size distribution may be different in both cases. A further improvement in resolution can be achieved if Ne instead of He is chosen as the scattering partner.

When the cluster-selection technique is combined with a laser-induced molecular beam depletion experiment, there are, as already mentioned, principally two possible configurations. The interaction of the clusters with

a)

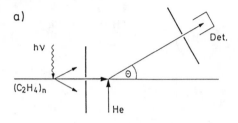

b)

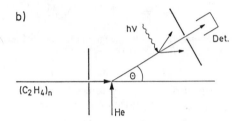

Figure 3. Possible configurations for dissociation experiments with size-selected clusters. (*a*) The laser interacts with all clusters. (*b*) Only size-selected species are subjected to the laser radiation.

the laser radiation can be located either anterior or posterior to the scattering center. These two configurations are depicted in Fig. 3.

In the arrangement of Fig. 3*a*, the cluster beam is first subjected to the laser field and then scattered by a He beam. Assume that the laser is tuned so that its radiation can be absorbed by the clusters. This will result in their dissociation and, since the dissociation products are expelled from the beam, they will be prevented from reaching the scattering center. Further assume that the detector (Det.) is set to a scattering angle Θ with $\Theta_{\max}^{(n+1)} < \Theta < \Theta_{\max}^{(n)}$ and tuned to a mass larger than $(n-1)\,m_1$ so that only clusters of size n are observed. Although the laser interacts with all the clusters in the beam (and may dissociate them), only the effect of the laser radiation on the cluster size n is observed by the detector. Thus, true cluster-specific spectroscopy is performed. If the mass spectrometer is kept at the same angle but tuned to a smaller mass, as for example $k m_1$ with $k < n$, then, of course, the laser-induced depletion signal contains information on various clusters whose sizes range from k to n.

In the configuration of Fig. 3*b*, first the cluster selection is performed by scattering with the He beam and then the selected cluster species are subjected to the laser radiation. In this case the laser interacts only with scattered clusters, that is, with cluster sizes smaller than n. As in case *a*, the effect of the laser field is only observed for cluster size n, provided that the mass spectrometer is tuned to a mass larger than $(n-1)\,m_1$.

Now we want to discuss a problem of configuration (*a*) which is related to the fact that all clusters are subjected to the laser field and which might arise if large clusters are investigated. Assume that the dissociation of a cluster X_n of size n is studied at the appropriate mass spectrometer position by monitoring the decrease in signal at mass nm_1. Now it is possible that clusters of size $n + 1$ are also excited at the same laser frequency and that the excess energy is released by monomer abstraction $X_{n+1}^* \rightarrow X_n + X$. In this unimolecular reaction the monomer will be expelled with a much larger cm velocity than X_n and, as a result, X_n will hardly be deflected from the original direction. These X_n clusters may reach the scattering center and may then be scattered into the detector. Thus, at the same laser frequency, we have two effects which cannot be distinguished, a positive contribution due to the creation of X_n clusters and a negative contribution due to the destruction of X_n clusters. This effect has actually been observed in the dissociation of large methanol clusters.[113] Therefore, it appears that configuration (b) is more suitable for the investigation of large clusters, since in this configuration the laser interacts only with the scattered (selected) clusters.

The main difference between configurations a and b, however, has its origin in the fact that the scattering selection technique is not a completely nonintrusive method. Due to the collisions with rare gas atoms, the clusters may be internally excited and, thus, after scattering, the clusters may be in a different state than before. In the expansion the clusters are formed in internal states characterized by very low temperatures. They are referred to as cold clusters. In the collisions the clusters are internally excited (especially their vdW modes) and, therefore, the scattered clusters are internally "warm". Correspondingly, in configuration a the spectroscopy of cold clusters is studied, while in configuration b internally excited or warm clusters are investigated.

Each method has its own merits. Configuration a has the advantage that the clusters are investigated in exactly that internal state in which they are formed. Due to the cooling in the expansion, the internal state distribution of the clusters is very narrow and not very far from the ground state. Hence, the spectra are expected to be much simpler and easier to interpret than if many internal states are contributing. Another advantage of configuration a is that here the spectrum obtained for any cluster size does not depend on the observation angle. Consequently, spectra in which different cluster sizes contribute can be properly disentangled by successively subtracting spectra obtained at larger scattering angles. The procedure is shown in the following example. Assume that the trimer spectrum has been measured at the proper trimer angle $\Theta_3(\Theta_{max}^{(4)} < \Theta_3 < \Theta_{max}^{(3)})$ but, for reasons of intensity, only on the dimer mass. Hence, this spectrum contains also contributions from the dimer. If the dimer spectrum is known from a measurement at $\Theta_2(\Theta_{max}^{(3)} < \Theta_2 < \Theta_{max}^{(2)})$, this spectrum can be subtracted from the composed trimer–dimer spectrum to obtain a pure trimer spectrum. In general, this method cannot be applied, if the spectra have been measured in configuration b, because here the dimer spectrum at Θ_3 is in principle different from that at Θ_2 because dimers of different internal energies are sampled.

Now we want to discuss the advantages of configuration b. It has already been pointed out that configuration b may be superior if large clusters are to be investigated. The fact that the laser interacts with warm clusters can also be seen as an advantage, for the measured spectra contain also information on the vdW modes which have been excited in the collisions with the rare gas atoms. By taking data at different scattering angles, the influence of different internal temperatures can also be studied. As discussed in context with the Newton diagrams of Fig. 1, larger angles correspond to lower internal temperature or smaller energy transfer, whereas smaller angles correspond to higher internal temperature or larger energy transfer. The most valuable information, however, is obtained if the warm spectra measured in configuration b can be compared with cold spectra obtained in configuration a. Then all spectral features which are observed in the warm spectrum and which are absent in the cold spectrum can obviously be assigned to hot bands from excited vdW modes.

III. EXPERIMENTAL SETUP

The molecular beam machine in which the experiments have been carried out is a further development of the famous construction by Lee.[114] A cross-sectional view of the apparatus is shown in Fig. 4.

The primary cluster beam is produced in the source chamber (P) by expanding a mixture of the desired gas in He or Ne through a nozzle of 0.035–0.100 mm diameter. The nozzle can be heated and cooled, but has been kept at room temperature in the experiments presented here. The cluster beam is collimated by a 0.4-mm diameter skimmer, travels 6 cm through a pressure-reducing differential chamber (R) and enters the main chamber (M) through a cone-shaped collimator of 2 mm diameter. Here the cluster beam is crossed by the secondary rare gas beam which is produced in a similar manner in source chamber (S). To increase the intensity, a differential chamber is not used for the secondary beam. Both molecular beam sources are mounted on xyz translational stages and can be positioned during the experiment relative to the fixed skimmers. An alignment of the skimmers is neither necessary nor possible, since all relevant parts fit snugly into their counterparts, thus ensuring that the skimmer position is the same after every dis- and reassembly. Each source chamber is mounted on a frame equipped with casters and can be easily

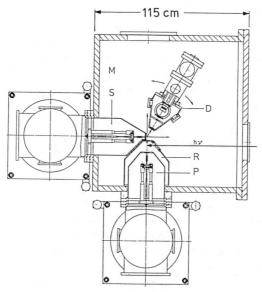

Figure 4. Cross-sectional view through the molecular beam machine. (M) = main chamber, (P) = primary source chamber in which the cluster beam is generated, (R) = differential chamber, (S) = source chamber for secondary beam, (D) = rotatable detector chamber with quadrupole mass spectrometer, and ($h\nu$) = laser radiation.[105]

detached from the main chamber. Both have an unbaffled 6000 l/s oil diffusion pump (Leybold Heraeus DI 6000), a 350 m³/h roots blower (Alcatel RSV 350), and a 65 m³/h rotary pump (Alcatel 2063). All distances and some other relevant parameters are presented in Table I.

The main chamber (M) is a rectangular welded stainless steel box with inside dimensions of $108 \times 108 \times 59$ cm³ and 4-cm thick walls. It is pumped by a baffled 5300 l/s oil diffusion pump (Varian VHS 10) and a 35 m³/h rotary pump (Alcatel 2033). With the aid of a liquid nitrogen trap an operating pressure of $5 \cdot 10^{-7}$ mbar is obtained.

The detector unit (D) is mounted inside the main vacuum chamber and

TABLE I
Molecular Beam Machine Parameters

	Primary Beam	Secondary Beam
Nozzle diameter (mm)	0.035, 0.100	0.030
Skimmer diameter (mm)	0.4	0.4
Nozzle-skimmer distance (mm)	10	8
Nozzle-collision center distance (mm)	145	50

has three nested chambers which are connected rigidly to a 76-cm diameter flange on top of the apparatus. The first region serves as a buffer chamber only and is pumped by a 300 l/s turbopump (Balzers, TPU 330). The second region encloses the third stage, the quadrupole mass spectrometer (Extranuclear, model 4–324–9), and the electron multiplier (EMI, model 9642/3B) which is used as the ion detector. This region is pumped by a 220 l/s ion pump (Varian Triode VacIon). The third and innermost chamber is surrounded by a liquid nitrogen trap and contains the electron bombardment ionizer (Extranuclear, model 041-1). With liquid nitrogen and a 110-l/s ion pump (Varian), a total background pressure of $1 \cdot 10^{-10}$ mbar is obtained in the ionizer chamber. All the pumps for the detector are mounted on the flange on top of the main chamber. Together with this flange, the entire detector assembly can be rotated about the intersection zone of the two molecular beams. This allows us to measure the angular dependence of selected scattered cluster species.

In order to sustain ultra-high vacuum (UHV) in the detector chamber at all times, the entrance can be closed by a slide valve. Thus, it is possible to vent the main chamber (M) for maintenance purposes without affecting the detector unit. Three different apertures are fixed to the slide valve, each of which can be positioned from outside onto the detector axis. For regular scattering experiments, a 3-mm diameter aperture is used. However, if very high intensities are encountered as, for example, in the direct beam or at very small deflection angles, a 0.2-mm diameter pinhole is placed in front of the detector entrance. The particles passing through the entrance hole are ionized in the electron bombardment ionizer, mass-selected in the quadrupole mass filter, and finally detected by the electron multiplier. The distance between scattering center and ionizer is 24 cm.

The signal from the electron multiplier is fed to a preamplifier (Ortec, model 9301), followed by an amplifier (Ortec, model 574) and a discriminator (Ortec, model 436). Here the pulses are normalized to NIM and TTL level. The TTL pulses are fed into a ratemeter (Ortec, model 449-2), thus providing a convenient analog reading. The NIM pulses are fed to a CAMAC crate where digital processing is accomplished under control of a minicomputer (DEC Micro/PDP-11).

The crossed molecular beam apparatus was constructed as a universal machine for elastic, inelastic, and reactive scattering experiments as well as for cluster research. In the experiments presented here, four different types of measurements were carried out. These include (1) the measurement of mass spectra, (2) the determination of the angular distribution of the scattered cluster beam, (3) the measurement of the velocity distribution of both the direct and deflected cluster beam, and finally (4) the determination of the dissociation yield upon irradiation with a CO_2 laser (MBDS experiment).

In order to obtain background-corrected mass spectra, the cluster beam is modulated by a lock-in chopper mounted inside the differential pumping chamber. Thus, it is possible to alternately measure signal plus background and background alone. The events are counted by a 100-MHz CAMAC quad scaler (Borer, model 1004 A) and the appropriate delays and gates are provided by two programmable dual preset counters (Borer, model 1008) triggered by the chopper. Another preset counter of the same type is used to control the integration time by counting the chopper cycles. The quadrupole mass filter is externally tuned by a CAMAC D/A converter (Kinetic systems, model 3112) which supplies voltages between 0 and 10 V to the quadrupole control unit.

Cluster size distributions, unperturbed by ionizer fragmentation artifacts, are determined by scattering the cluster beam with a rare gas beam and measuring the angular distribution of the deflected molecules. Again the cluster beam is modulated by the lock-in chopper and the same previously described technique for background-corrected data collection is applied. Typically, signal plus background and signal alone are counted for 200–20,000 chopper cycles. With gates of 2 ms duration, this corresponds to integration times of 0.8–80 s for each data point. From one angular position to the next, the detector unit is rotated by a dc motor which is controlled by a homemade microprocessor-supported CAMAC module.

In order to carry out IR photodissociation experiments, the light of an IR laser ($h\nu$) is introduced into the differential chamber (R), as depicted in Fig. 4. The radiation source is a pulsed, line-tunable, high-power CO_2 laser (Lambda Physik, EMG 201E). Its linewidth is 2.1 GHz and its maximum repetition rate is 25 cps. The laser beam enters the main chamber through a ZnSe window and is then slightly focused onto the molecular beam axis by a cylindrical ZnSe lens with a 30-cm focal length. In the initial stage of the experiment the dimensions of the laser beam in the interaction region were 29.4 mm long and 4.6 mm high. Later they were changed to 25×4 mm^2. The laser energy is measured outside the molecular beam machine with a Gentec ED-500 energy meter and with the laser running at 1 cps. Values thus obtained are corrected to account for a 21% decrease in energy if the laser is operated at 25 cps instead of 1 cps. This correction is determined with a Scientech power meter (model 38–0202) which can also be employed at 25 cps. To obtain a well-defined energy on every laser line, the power is adjusted with a 12-cm long attenuation cell filled with propylene at variable pressure.

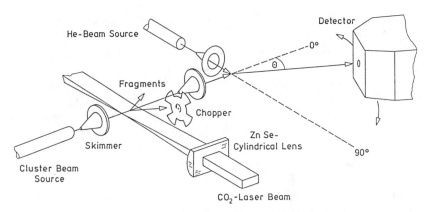

Figure 5.   Perspective view of the experimental setup.

A perspective schematic view of the experimental setup used in the IR photodissociation experiments is given in Fig. 5. In this setup the cluster beam is also modulated by the lock-in chopper. From its light barrier a reference signal is derived which, after being delayed and divided by 8, externally triggers the laser during the open phase of the chopper. The chopper frequency is adjusted so that the laser repetition rate is near its maximum of 25 cps. If the laser is properly tuned, certain clusters may be dissociated and their fragments expelled from the beam. Thus, a "hole" is burned in the cluster beam by the pulsed radiation. This hole propagates to the scattering center and can therefore be observed at all angular positions where the dissociated clusters would have been scattered without laser irradiation. Simultaneously to the laser firing, a multichannel time-of-flight (TOF) analyzer is started. Thus, the effect of the pulsed laser on the cluster beam can be detected time-resolved as a function of cluster size.

The TOF analyzer is incorporated into a CAMAC module and communicates with a LeCroy histogramming memory module (model 3588) where the clusters' arrival times are stored. Details of this TOF analyzer have been published recently.[115] Synchronization between the chopper and the laser is such that the cluster beam is blocked shortly after the hole burned by the laser has passed the chopper wheel. Using this method, it is possible to determine the background in the detector after the hole has arrived. A typical TOF dissociation spectrum will be presented in the next section (see Fig. 7).

In order to determine TOF distributions of direct and scattered cluster beams, the lock-in chopper is replaced by a fast-spinning TOF chopper. Using a wheel of 145 mm diameter with two 1-mm wide slits rotating at

TABLE II
Cluster Beam Conditions

Species	C_2H_4	C_6H_6	SF_6	NH_3	CH_3OH
Carrier gas	He	He (Ne)	Ne	He	Ne
Percentage of species	10%	4% (5%)	5%	5%	1.9%
Bath temperature (°C)	–	0.2 (11.3)	–	–	–15
Total pressure (bar)	4	0.7 (1.0)	1.5	1.75	1.5
Nozzle diameter (μm)	35	100	100	100	100
Most probable velocity (m · s^{-1})	1430	1332 (769)	683	1650	781
Scattering partner	He	Ne	Ne	He	He

a frequency of 400 Hz, the conventional single-shot TOF technique is applied. The data collection scheme is exactly the same as in the previously described laser dissociation experiments. The calibration of the TOF analyzer is discussed in Ref. 115. Typically, TOF measurements are carried out for all cluster beams, in order to determine their velocities, which must be known for the construction of the Newton diagram. In addition, TOF distributions are also measured at every scattering angle where dissociation spectra are taken. Thus, the beam parameters required for the deconvolution of the dissociation TOF spectra are obtained.

The systems reported in this chapter include ethene (C_2H_4), benzene (C_6H_6), sulfur hexafluoride (SF_6), ammonia (NH_3), and methanol (CH_3OH). The conditions under which the respective cluster beams have been generated are tabulated in Table II.

IV. DATA COLLECTION AND ANALYSIS

In this section we will present and discuss typical measurements which are carried out to characterize the distribution of clusters in the beam and to determine their absorption and predissociation cross-sections. As already mentioned, conventional mass spectra obtained in the direct beam suffer from the fragmentation of larger clusters in the ionizer. Although we have carried out such experiments, we shall not discuss them because their information content is very limited. The next type of measurement is the determination of the angular dependence of scattered cluster species. Typical angular distributions of $(C_2H_4)_n$ clusters scattered by He, as measured on various mass settings of the mass spectrometer, are shown in Fig. 6. We shall use these measurements to discuss what information can be obtained from cluster-scattering experiments and how they can be utilized to overcome the problem of fragmentation of larger clusters. It should be noted that Fig. 6 is different from an earlier published result.[105] For the measurements shown in Fig. 6 we used Ne as carrier gas.[116] This

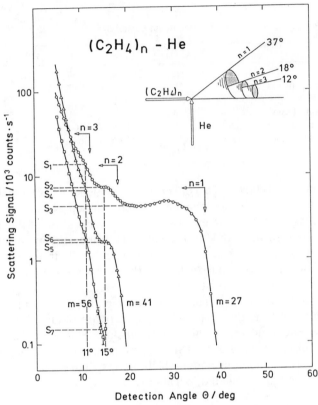

Figure 6. Angular distributions of $(C_2H_4)_n$ clusters scattered from He measured at various mass settings of the quadrupole mass spectrometer. The ethene clusters were generated in a Ne expansion. The arrows indicate maximum deflection angles and scattering ranges for various cluster species n.[116]

results in a slower cluster beam and, consequently, a better resolution (see Section II).

At the top of Fig. 6 the kinematics of the scattering process are sketched. According to their masses, différent $(C_2H_4)_n$ cluster species are scattered into different angular ranges. Taking the most probable velocities of the beams as measured by TOF analysis, we calculate that the monomers are scattered into a laboratory angular range between $-2°$ and $37°$. For dimers ($n = 2$), the maximum deflection angle is only $18°$. Larger $(C_2H_4)_n$ clusters are confined to still narrower ranges with maximum scattering angles of $12°$, $9°$, and $7.5°$ for $n = 3$, 4, and 5. The respective scattering ranges and maximum angles for $n = 1$, 2, and 3 are indicated by arrows above the upper curve.

The curve with circular data points shows the angular distribution as measured on mass $m = 27$ amu ($C_2H_3^+$). This mass has been chosen for monomer detection instead of $m = 28$ because of the much lower background. As expected from the Newton diagram, the curve rises at about $\Theta = 37°$. It goes through a broad maximum and is rather flat around 22°. At $\Theta = 18°$ we observe a significant increase in the signal because (1) we have entered the dimer scattering range and (2) dimers fragment in the ionizer to give $C_2H_3^+$ ions. From the relative increase we deduce that the contribution of dimers to the total $m = 27$ signal is $(S_2 - S_3)/S_2 = 40\%$ in this angular range. Thus, it appears that the fragmentation of dimers to $m = 27$ amu is a rather strong effect. At $\Theta = 12°$ another increase in signal is observed because now we are starting to detect trimers on $m = 27$ amu. Their contribution is calculated to be $(S_1 - S_2)/S_1 = 47\%$.

If the mass spectrometer is tuned to $m = 41$ amu ($C_2H_4 \cdot CH^+$), C_2H_4 monomers can of course no longer be observed. Correspondingly, the signal sets in only after 18° (triangular data points). It is constant at $\Theta = 15°$ where we detect solely dimers, and increases strongly at 12° since now trimers are starting to contribute.

With the mass spectrometer tuned to the dimer parent ion mass $m = 56$ amu, the curve with square data points is determined. In principle, the same cluster species can be detected, but the signal in the dimer scattering range around 15° is very low. This indicates that the signal at $m = 56$ amu is hardly produced by dimers. Comparison with the $m = 41$ curve shows that it is $S_5/S_6 \approx 10$ times more likely for the dimer to form a $C_2H_4 \cdot CH^+$ ion. This result is also confirmed by Buck et al.,[117] who have measured complete mass spectra of scattered $(C_2H_4)_n$ clusters in the entire angular range. These data provide accurate information on the fragmentation ratios of ethene clusters.

Assume that the selection technique is not used and that the $(C_2H_4)_n$ clusters are detected in the direct beam. From our measured angular distributions it is evident that the signal measured at $m = 56$ amu contains hardly any dimer contribution, but is instead mainly produced by fragmented larger clusters. Earlier predissociation experiments on ethene dimers were carried out with the mass spectrometer tuned to $m = 56$ amu.[51] Our results indicate that the dimer dissociation spectra measured at this setting must be strongly disturbed by larger $(C_2H_4)_n$ clusters. But even if the mass spectrometer is tuned to $m = 41$ amu,[49] the dimer spectrum may contain contributions from larger clusters. Thus, ethene appears to be an important candidate to combine the cluster selection technique with a predissociation experiment and to perform cluster-specific spectroscopy, which will be discussed now.

If the mass spectrometer is set to $\Theta = 15°$ and tuned to $m = 41$ amu, solely $(C_2H_4)_2$ dimers are detected. Monomers are rejected by the mass filter and trimers and larger clusters are deflected to smaller angles. When the cluster beam is subjected to the radiation of an IR laser, only its effect on dimers is observed. Hence, a pure dimer spectrum can be measured. It is also possible to study the dimer dissociation on $m = 27$ amu and $\Theta = 15°$. However, here dimers and monomers are present and a 100% destruction of the dimer beam would result only in a $(S_2 - S_3)/S_2 = 40\%$ attenuation of the scattering signal.

In order to study the pure trimer dissociation, it would be best to set the detector to $\Theta = 11°$ and to tune the mass filter to $m = 56$ amu. Here the signal (S_6) is dominated by trimers. The 10% dimer contribution (S_7) may be neglected or appropriately considered (see below). It is equally possible to measure the trimer dissociation at $m = 41$ amu. Here the absolute trimer signal $(S_4 - S_5)$ is stronger, but the dimer contribution (S_5) to the total signal (S_4), which is S_5/S_4, is also larger. A pure trimer spectrum is obtained when the previously measured and properly weighted (by S_5/S_4) dimer spectrum is subtracted from the total spectrum. The same procedure can be applied to investigate the tetramer dissociation. At $\Theta = 8.5°$ a total dissociation spectrum, preferentially on $m = 56$ amu, is measured and then the dimer and trimer contributions are subtracted.

As described in the experimental section, the effect of the pulsed-laser radiation on the cluster beam is measured time-resolved using a multichannel TOF analyzer.[115] In Fig. 7 such a TOF spectrum is shown. The normalized $(C_2H_4)_3$ trimer signal, as measured at $\Theta = 11°$ and $m = 56$ amu, is plotted versus the cluster flight time. At time $t = 0$, the CO_2 laser is fired and a certain fraction of trimers is dissociated. Thus, a hole is burned in the trimer beam. Corresponding to the flight time of the molecules from the laser interaction region to the ionizer, this hole arrives at the detector approximately 270 μs later, as manifested in the negative peak in the spectrum. Shortly after the laser has been fired, the cluster beam is blocked by the lock-in chopper which is synchronously running with fixed phase with respect to the laser system. Thus, at the end of the spectrum, the detector background is measured; this must be known for a precise determination of the absolute trimer signal.

The shape of the negative peak is basically determined by the TOF distribution of the scattered trimers. Its amplitude is a measure of the fraction of trimers dissociated by the laser field. In order to obtain exact dissociation yields, we have to deconvolute the measured spectrum, taking into account the TOF distribution of the scattered molecules. We assume that the hole burned by the laser in the cluster beam can be described by the square function

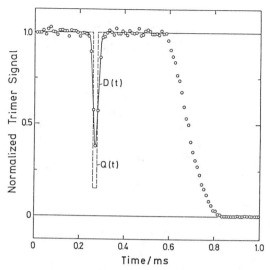

Figure 7. Measured real-time spectrum for $(C_2H_4)_3$ trimer dissociation. The solid curve $D(t)$ is fitted to the data point. The dashed square function $Q(t)$ represents the deconvoluted dissociation signal.[105,115]

$$Q(t) = \begin{cases} 1 - A & \text{for } -\tau/2 \leq t \leq \tau/2 \\ 1 & \text{else} \end{cases} \tag{6}$$

Since the laser pulse is very short (120 ns), the width τ is determined by the time required for the particles to travel through the 25–30 mm long interaction region ($\tau \approx 15$–$40\,\mu s$). The amplitude $A(0 \leq A \leq 1)$ corresponds to the actual fraction of clusters dissociated by the laser field. The velocity of the cluster beam and the TOF distributions of the scattered cluster species are determined by TOF analysis. In most cases they can be very well represented by a Gaussian distribution

$$F(t) = \beta \pi^{-1/2} \exp[-\beta^2(t - t_0)^2] \tag{7}$$

where t_0 is the most probable flight time and β is given by

$$\beta^2 = 4 \ln 2/\gamma^2 \tag{8}$$

γ is the full width at half maximum (FWHM). Carrying out the convolution, we obtain for the signal $D(t)$ observed at the detector

$$D(t) = \int_{-\infty}^{+\infty} Q(t') F(t - t') \, dt'$$

$$= 1 - A \int_{-\tau/2}^{+\tau/2} F(t - t') \, dt' \tag{9}$$

and finally

$$D(t) = 1 - \frac{A}{2} \left[\text{erf}(z_1) - \text{erf}(z_2) \right] \tag{10}$$

erf is the error function and

$$z_{1,2} = \beta(t - t_0 \pm \tau/2) \tag{11}$$

Using Eq. (10), $D(t)$ is fitted to the measured spectrum by a least-squares fit procedure treating A and t_0 as free parameters. The fitted curve is shown in Fig. 7 by the solid line. As result of the analysis, we obtain the square function $Q(t)$ on which the convolution is based and which is represented by the dashed line. Its amplitude A corresponds to the actual fraction of dissociated trimers, which is 0.85 in this specific example.

In order to determine exact dissociation yields, three quantities must be known very precisely. These are (1) the length of the laser molecular beam interaction region, (2) the molecular beam velocity, and (3) the width γ of the TOF distribution of the scattered cluster beam. It turns out that the most critical quantity is the width γ since, in many cases, it cannot be determined experimentally but must be estimated. This is especially true for larger clusters with strong fragmentation behavior when their TOF distribution cannot be measured without contributions from smaller clusters. For example, if the width γ has been overestimated, the dissociation yield is determined to be too large. Therefore, in some cases, the dissociation yield values may be affected with a larger uncertainty than indicated by the statistical error bars. But note that in every measurement all data points are equally affected and that only the absolute values are influenced by this uncertainty.

In the laser dissociation experiments the relative attenuation of the cluster beam is measured as a function of either laser frequency or laser fluence. For every laser adjustment a complete real-time spectrum as depicted in Fig. 7 is measured. Accumulation times are between 5 and 120 min, where the longer times are typically needed for dimers. The measured real-time spectrum is then deconvoluted according to the procedure described above. Thus, the actual fraction of clusters dissociated by the laser field can be plotted as a function of frequency and fluence.

Now we want to discuss a simple model referred to as the two-level-plus-decay model. In this model the system under study is represented by only two levels $|1\rangle$ and $|2\rangle$. After the expansion the vdW molecules are in level $|1\rangle$, which is usually the lowest vibrational state. The laser then excites the molecules from $|1\rangle$ to $|2\rangle$. If stimulated and spontaneous emission back into level $|1\rangle$ can be neglected, integration of the rate equations[49,51] yields for the population $N_1(\tau)$ in level $|1\rangle$ after interaction with the laser field

$$N_1(\tau) = N_1(0) \exp[-\sigma F/(h\nu)] \qquad (12)$$

Here τ is the temporal pulse width of the laser, $N_1(0)$ the initial population of state $|1\rangle$, σ the absorption cross section, F the laser fluence, and $h\nu$ the photon energy. If we assume that, on the way to the detector, all excited molecules $N_1(0) - N_1(\tau)$ dissociate and leave the beam, the relative attenuation or dissociated fraction P_{diss} measured at the detector can be written

$$P_{diss} = \frac{N_1(0) - N_1(\tau)}{N_1(0)} = 1 - \exp[-\sigma F/(h\nu)] \qquad (13)$$

Taking the logarithm, we have

$$\ln(1 - P_{diss}) = -\sigma F/(h\nu) \qquad (14)$$

Thus, if the logarithm of the undissociated fraction $\ln(1 - P_{diss})$ is plotted versus the laser fluence, a straight line with slope $-\sigma/(h\nu)$ is expected in this model.

For the frequency dependence of the dissociation cross section Casassa et al.[49] have derived

$$\sigma(\tilde{\nu}) = L(\tilde{\nu}) = A_0 \frac{\Gamma^2}{4(\tilde{\nu} - \tilde{\nu}_0)^2 + \Gamma^2} \qquad (15)$$

where the amplitude A_0 contains the squared transition moment $\langle \mu \rangle^2$ and other constant numerical factors. Equation (15) represents a Lorentzian line-shape function $L(\tilde{\nu})$ with peak position $\tilde{\nu}_0$ and full width at half maximum (FWHM) Γ both measured in cm^{-1}. In the two-level-plus-decay picture the linewidth Γ is related to the predissociative lifetime according to

$$\tau_{\text{life}} = (2\pi c\Gamma)^{-1} \qquad\qquad (16)$$

From Eqs. (14) and (15) it follows that, when $-\ln(1 - P_{\text{diss}})$ is plotted as a function of wavenumber, a Lorentzian line-shape function is obtained. If the dissociated fraction P_{diss} is plotted directly, a very good approximation of a Lorentzian is also obtained. This is because $-\ln(1 - x) \approx x$ for small values of x.

In some cases the measured dissociation spectrum is characterized by a single featureless line, in others more complicated structures may be encountered. Correspondingly, we try to represent the experimental spectra by fitting a single Lorentzian line shape function $L(\tilde{\nu})$ or a set of several Lorentzians to the data. It should be noted that there is no implication connected with the choice of the Lorentzian profiles. In particular, we do not assume the validity of the two-level-plus-decay model nor do we imply that the linewidth is caused by predissociation lifetime. It just turned out that the Lorentzians provide a good representation of the experimental data needed for a quantitative characterization of the spectra and for further analysis.

V.   RESULTS

A.   Ethene

One of the first and probably most widely studied molecules in vibrational predissociation spectroscopy is the ethene dimer $(C_2H_4)_2$. Extensive investigations have been carried out in the spectral region of the CO_2 laser by the groups of Janda,[49,50] Gentry,[51] Lee,[118] and Reuss.[119] The observed spectra revealed a single Lorentzian-like line suggesting that homogeneous broadening was the dominant broadening mechanism. The exceptionally large linewidth between 10 and 17.5 cm^{-1} was associated with a lifetime in the range from 0.3 to 0.5 ps. Using F-center lasers and optical parametric oscillators, Fischer et al.[67,69] and Liu et al.[77] investigated the spectral range above 2900 cm^{-1}. The spectra of Fischer et al. were interpreted to be in accord with the CO_2 laser results, whereas Liu et al. reported nonhomogeneous absorption lines. On the basis of current theories,[120–123] it is difficult to understand the exceptionally short lifetimes.

So far, all experiments were carried out in the direct unselected beam where, especially at stronger stagnation conditions, larger clusters than dimers could also be present. If these larger clusters fragment in the ionizer to the same ion mass as that which has been chosen to monitor the dimer dissociation, then of course the measured beam attenuations would not yield pure dimer spectra. Thus, it could very well be that the observed broad linewidths were caused by fragmenting larger clusters which contributed to the dimer signal. In order to clarify this question, it was desirable to carry out the dissociation experiments with a pure dimer beam.

For this reason we chose ethene as our first candidate to apply the cluster-selection technique to an IR predissociation experiment. In order to find the optimal conditions for ethene dimer production and detection, we measured angular distributions of $(C_2H_4)_n$ clusters scattered from He at various conditions. A typical example has already been presented in Fig. 6 of the previous section. From these investigations it became apparent that probably all previously reported ethene dimer dissociation spectra were disturbed by fragmented larger clusters. For example, we were able to show that the ethene dimer could hardly be detected at its nominal mass $m = 56$ amu (compare Fig. 6).

The first dissociation experiment with size-selected $(C_2H_4)_n$ clusters was carried out at slightly different conditions than the scattering experiment presented in Fig. 6. For, in order to obtain stronger signals in the dimer scattering range, we used as carrier gas He instead of Ne. As discussed in Section II, this results in a higher cluster-beam velocity and a slightly lower resolution, but an enhanced dimer signal. The experiment was carried out in configuration b sketched in Fig. 3b, that is, the clusters were first scattered by a He beam and then subjected to the CO_2 laser radiation. In this manner the first dissociation spectrum of size-selected ethene dimers was obtained.[95,97]

The result is displayed in Fig. 8a. The absorption profile is considerably broader than observed before.[49,51] In addition, a pronounced structure could be resolved. Although the dissociation spectrum is clearly not homogeneous, its overall behavior can be well characterized by a Lorentzian line-shape function of full width at half maximum (FWHM) $\Gamma = 30.1$ cm^{-1}, if finer structural details are neglected.

Due to the scattering with He atoms, the ethene dimers contain an average internal energy of 29 meV. Thus, the spectrum of Fig. 8a reflects the dissociation of internally excited or "warm" dimers. Since it was not clear whether the broadened linewidth and the structure in the spectrum were the result of the internal excitation or whether these features had been obscured in earlier experiments by the presence of larger ethene clusters, it was desirable to carry out the cluster-specific dissociation experiment with internally cold C_2H_4 dimers. Therefore, we have performed another experiment in which the laser–molecular beam interaction occurs before the scattering center where the clusters are still cold, that is, in configuration a of Fig. 3a.

The result of this experiment[105] is shown in Fig. 8b. Now the dissoci-

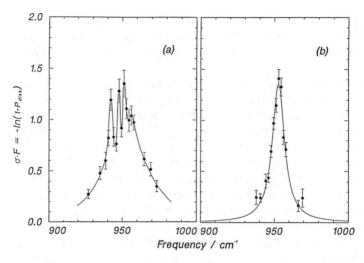

Figure 8. $(C_2H_4)_2$ dimer dissociation spectra. The left spectrum (a) has been obtained with internally excited (warm) dimers, whereas the right spectrum (b) reflects the dissociation of cold ethene dimers.[96]

ation profile is considerably narrower. It is structureless and can be well represented by a Lorentzian of FWHM $\Gamma = 10.4 \, \mathrm{cm}^{-1}$ and peak position $\tilde{\nu}_0 = 952.9 \, \mathrm{cm}^{-1}$. This result is in sharp contrast to the spectrum observed with warm dimers, but in almost perfect agreement with the earlier on-beam experiments.[49,51] Thus, it must be concluded that the features observed in Fig. 8a are the result of the internal excitation during the scattering process.

The structures in the warm dimer spectrum are most probably due to hot bands from collisionally excited vdW modes. Using the azimuthal and vibrational close-coupled infinite-order sudden approximation (AVCC IOS), Peet[124] has calculated the transition frequencies for the combination bands involving the ν_7 fundamental and bending modes of the ethene monomers about their C–C axes. This calculation yields the combination bands at 905 and 995 cm^{-1}, which were previously observed by Snels et al.,[62] but also several lines around 950 cm^{-1} separated by approximately 5 cm^{-1}.[125] Since the separations between the maxima in Fig. 8a are 4 and 6 cm^{-1}, it is very appealing to assign the structure in this spectrum to transitions from excited vdW modes.

Rudolph[125,126] has reinvestigated the dissociation of warm C_2H_4 dimers using the same configuration (b) but employing a cw CO_2 laser with a bandwith of 3 MHz which is 3 orders of magnitude narrower than that of our pulsed laser. Using N_2O gas as lasing medium, he obtained also twice as many laser lines in the spectral region of interest. In this experiment, Rudolph was able to confirm the existence of structure on the broad, 30.1 cm^{-1} wide absorption profile, although the features were less pronounced. However, this could be due to the narrow laser bandwidth and the possibility that the laser frequency sometimes matched a strong absorption by accident and sometimes not. It should be noted that similar coarse structure has also been observed by Snels et al.[62] The cold dimer spectrum in Fig. 8b does not show any structure. This is also not expected, since the vdW modes are probably not excited; however, some lines could be hidden in the flanks of the band.

The observed coarse structure reduces the upper limit of the experimental linewidth to less than 4 cm^{-1}. According to Eq. (16), this corresponds to a lifetime of 1.3 ps. However, this is still 6 orders of magnitude shorter than predicted by the recent coupled-channel calculations of Peet et al.[127,128] which yield lifetimes on the order of 5 μs.

First experimental evidence for much narrower lines than reported here have been found by Snels et al.[62] These investigators supplied their CO_2 laser grating with a piezo drive and, thus, were able to tune the laser with very high resolution within one CO_2 laser transition. In this experiment, Snels et al. found very sharp, 3.5-MHz wide absorption lines superimposed on a broad background. From the width of these lines a lower limit of 45 ns for the lifetime was obtained, reducing the discrepancy between theory and experiment to only 2 orders of magnitude.

Independently, the narrow lines were also observed by Baldwin and Watts[81] and Buck et al.[98] The latter experiment was carried out with size-selected $(C_2H_4)_n$ clusters, proving that the sharp features were actually due to the dimer. An important result of the high-resolution studies of Reuss and co-workers[62,63] and Buck et al.[98] was not only the detection of narrow lines, but also the observation that they were superimposed on a broad homogeneous background. The coexistence of broad and narrow features has been subject of many controversial discussions. It is presumed that an energetically closed channel which involves the ν_{10} vibration could be important; however, the problem has not yet been solved. For the latest discussion see Ref. 129.

Besides the dimer spectra shown in Fig. 8, we have also measured dissociation spectra for larger ethene clusters. The results obtained for cold C_2H_4 trimers and tetramers are shown in Fig. 9. The spectrum in Fig. 9a has been measured with the detector positioned to $\Theta = 6.6°$ and the mass spectrometer tuned to $m = 56$ amu. As discussed in Section IV, with this adjustment predominantly trimers are detected. A 4% dimer contribution to the total signal can be neglected. The solid line represents a Lorentzian line-shape function fitted to the data points. It peaks at $\tilde{\nu}_0 =$

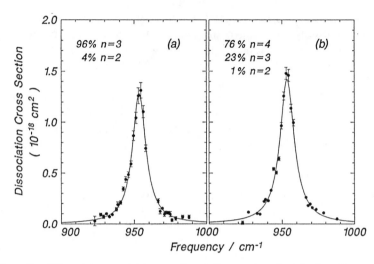

Figure 9. Ethene trimer and tetramer dissociation spectra. Contributions from smaller clusters are indicated in the figure. The solid curves represent Lorentzians fitted to the data points.[105]

953.4 cm^{-1} and has a FWHM of $\Gamma = 11.7$ cm^{-1}. The spectrum in Fig. 9b, measured at $\Theta = 5°$ and $m = 56$ amu, is dominated by the dissociation of tetramers. The trimer and dimer contributions are given in Table III. The combined absorption profile is again well described by a single Lorentzian with $\tilde{\nu}_0 = 953.7$ cm^{-1} and $\Gamma = 12.2$ cm^{-1}. All results of the least-squares fits for ethene are summarized in Table IV. We have also investigated the pentamer dissociation at $\Theta = 4.2°$ and $m = 56$ amu. The maximum dissociation yield is slightly larger, but the shape can be described by the same Lorentzian.

Comparing the various spectra for cold ethene clusters (Figs. 8b and 9; Table IV), it can be seen that—within the range of the experimental error—all have the same width $\Gamma = 12$ cm^{-1} and the same peak position $\tilde{\nu}_0 = 953$ cm^{-1}. Thus, we have the important result that the frequency

TABLE III
Ethene Cluster Contributions at Various Detector Positions

Detector Position	I	II	III
Scattering angle Θ (deg)	9.5	6.6	5.0
Mass (amu)	41	56	56
Dimer contribution (%)	100	4	1
Trimer contribution (%)	–	96	23
Tetramer contribution (%)	–	–	76

TABLE IV
Results of Least-Squares Fits for Ethene Cluster Dissociation

Detector Position	Line Position $(\tilde{\nu}_0/cm^{-1})$	Linewidth (FWHM) (Γ/cm^{-1})
I	952.9	10.4
II	953.4	11.7
III	953.7	12.2

dependence of the dissociation spectra is essentially the same for all investigated $(C_2H_4)_n$ clusters ($n = 2, 3, 4$, and 5).

Previous on-beam investigations on the predissociation of ethene clusters upon excitation of their ν_7 monomer vibration were performed under very different conditions.[49,51,119] Since the cluster distribution in the beam is expected to depend strongly on the source conditions, the cluster-beam composition should have been quite different in each experiment. Nevertheless, all authors report essentially the same frequency dependence. In particular, Casassa et al.[49] report that they varied source conditions over a wide range and that no change in line shape was observed. Geraedts[119] operated the beam source at low temperature ($T_0 = 169$ K), which favors heavy clustering, but the position and shape of the absorption profile remained essentially the same. All these results suggest that there is almost no dependence on cluster size and so far this is in good agreement with our findings. On the other hand, we have strong indications that the previous on-beam experiments were carried out in cluster beams containing not only dimers but also larger $(C_2H_4)_n$ clusters. Evidence for this is given by the reported large-absorption cross sections. Because the larger clusters have the same frequency dependence, their contribution to the dimer dissociation spectrum has obviously not been noticed.

At the top of the absorption profile at $952.9 \ cm^{-1}$ we have measured the fluence dependence of the dissociation yield for all three $(C_2H_4)_n$ species with $n = 2, 3$, and 4. If the logarithm of the nondissociated fraction $[\ln(1 - P_{diss})]$ is plotted versus laser fluence, straight lines are obtained. The dimer curve had to be corrected to account for a counteracting effect arising from trimers and tetramers dissociating at the same frequency (see Section II). Assuming the validity of the two-level-plus-decay picture (compare Section IV), transition probabilities can be derived from the slope of the fluence dependence curves. Within the experimental error, the sum rule is confirmed, that is, the total cluster transition probability is equal to the sum of the transition probabilities of each constituent. In other words, the squared transition moment $\langle \mu \rangle_n^2$ of a $(C_2H_4)_n$ cluster is equal to $n \cdot \langle \mu \rangle_1^2$, where $\langle \mu \rangle_1^2$ is the transition probability for one C_2H_4

unit in the cluster. The transition probability for each monomer unit is $\langle\mu\rangle_1^2 = 21 \cdot 10^{-3} \, D^2$ and somewhat smaller than for a free ethene molecule which is $\langle\mu\rangle^2 = 35.3 \cdot 10^{-3} \, D^2$.[130]

In the experiment carried out in configuration b and involving warm ethene clusters, we have also studied the dissociation of larger $(C_2H_4)_n$ clusters with $n = 3, 4, 5,$ and 6.[97] While the maximum position is nearly the same for all cluster sizes ($\tilde{\nu}_0 = 951.6 \, cm^{-1}$), the overall width (FWHM) decreases from $\Gamma = 30.1 \, cm^{-1}$ for the dimer to $\Gamma = 12.2 \, cm^{-1}$ for the hexamer. This is in contrast to the result with cold ethene clusters where the width does not depend on the cluster size. The difference is obviously due to the internal excitation by the collision with He. This excitation affects the dimer most strongly. For the larger clusters the energy transferred by collisions can be distributed among the increasing number of degrees of freedom of internal motion. Thus, for the hexamer, there is only little internal excitation per vdW bond, resulting in a similar dissociation spectrum as for cold clusters. These results were confirmed in the recent predissociation experiments carried out by Buck et al.[131] with a cw CO_2 laser.

B. Benzene

Benzene C_6H_6 is the simplest 6π aromatic molecule and of great importance in solvent chemistry. Correspondingly, the study of its mutual interaction in small gas-phase clusters has attracted much interest. The determination of the structure of small benzene clusters has been the subject of many investigations, both theoretical and experimental.

The dimer structure has been the subject of some controversy in the literature. Molecular-beam electric resonance experiments by Janda et al.[132] showed that the benzene dimer has a permanent dipole moment. These studies suggest a T-shaped structure which is also known to be the nearest-neighbor configuration in crystalline benzene. Two-color photoionization studies by Hopkins et al.[20] also seem to favor this structure, whereas similar experiments by Law et al.[133] indicate a parallel stacked and displaced configuration. Recent photoionization studies by Börnsen et al.,[134] however, are in agreement with a dihedral (V-shaped) structure. As suggested by Law et al.,[133] the different experimental findings could be reconciled if the benzene dimer were actually to exist in different geometries with similar energies. The relative population in each configuration would be determined by the dimer temperature and this in turn by the molecular-beam expansion conditions. Hence, it is conceivable that each experiment was sensitive to a different conformation and, thus, revealed a different dimer structure.

As far as theoretical studies are concerned, the situation is similarly controversial. Parallel displaced,[135] dihedral,[136,137] and T-shaped[138,139] structures have been reported. The dimer binding energies vary between 600 and 1000 cm^{-1}. However, the energy differences predicted for the different structures are generally quite small within a specific investigation.

Infrared vibrational predissociation experiments on small benzene clusters have been carried out in the frequency ranges around 1000 and 3000 cm^{-1}. In the region around 3000 cm^{-1} a Nd:YAG laser-pumped optical parametric oscillator (OPO)[48,79] or a color-center laser[87] was used to study the predissociation upon excitation of the C–H stretch. Here three well-resolved bands could be observed. In the 1000-cm^{-1} frequency range Nishiyama and Hanazaki,[85] Gentry and co-workers,[55] and Stace et al.[86] employed pulsed CO_2 laser radiation to excite the ν_{18} in-plane C–H bend. The spectra reveal a single featureless band at 1038 cm^{-1} which seemed to be well represented by a Lorentzian line-shape function with full width at half maximum of 2 cm^{-1},[55] or 2.5 cm^{-1}.[86] Stace et al.[86] also investigated the dissociation of larger $(C_6H_6)_n$ clusters with n up to 27. These spectra did not show any significant change in either band position or linewidth. Because of this observation and the experience that benzene clusters fragment strongly in the ionizer, the question arose whether and to what extent fragmentation of larger clusters may have influenced the previously measured dimer dissociation spectra. Therefore, we felt it worthwhile to reinvestigate the dissociation of small benzene clusters using the kinematic selection technique. First results have recently been published.[106]

In order to determine the optimum detector positions for our dissociation experiments, we first measured the angular distributions of benzene clusters scattered from Ne at various mass settings of the quadrupole. The clusters were generated by expanding a 4% mixture of C_6H_6 in He. The corresponding stagnation conditions were total stagnation pressure $p_{tot} = 0.7$ bar, bath temperature $T_{bath} = 0.2°C$, nozzle temperature $T_0 = 21°C$, and nozzle diameter $D = 0.1$ mm. Angular distributions were determined with the mass spectrometer tuned to $m = 78$ amu [$C_6H_6^+$], $m = 156$ amu [$(C_6H_6)_2^+$], and $m = 234$ amu [$(C_6H_6)_3^+$]. We obtained the following results. The monomer angular distribution was rather flat and did not show pronounced shoulders corresponding to the onsets of the various cluster-scattering ranges. This indicates that fragmentation of larger clusters to the monomer ion mass is not a very important process. The dimer curve measured on $m = 156$ amu, instead, was characterized by pronounced shoulders corresponding to the dimer, trimer, and tetramer onsets. On the other hand, the signal on $m = 234$ amu [$(C_6H_6)_3^+$] was very low throughout the entire angular range. These results indicate that there is strong fragmentation of the $(C_6H_6)_3$ trimer to the dimer ion mass. Even

TABLE V

Benzene Cluster Contributions at Various Detector Positions and Beam Conditions[a]

	$p_0 = 0.7$ bar			$p_0 = 1.4$ bar		
Scattering angle Θ (deg)	10.25	7.50	6.25	10.25	7.50	6.25
Dimer contribution (%)	100	54	25	100	31	12
Trimer contribution (%)	—	46	49	—	69	45
Tetramer contribution (%)	—	—	26	—	—	43

[a]The mass spectrometer was tuned to $m = 156$ amu.

in the tetramer scattering range, the signal on $m = 234$ amu was not sufficient for later dissociation experiments.

In summary, the angular distribution measurements indicate that, as a result of the strong fragmentation, all dissociation experiments have to be carried out with the mass spectrometer tuned to the dimer mass $m = 156$ amu. This in turn implies that it is not possible to measure pure trimer or tetramer spectra and that contributions from smaller clusters, that is, dimers and trimers, must be accepted. It should be noted, however, that the technique can clearly discriminate against larger clusters. Based on the calculated Newton diagram and the shape of the $m = 156$ angular distribution, we decided to measure the dimer, trimer, and tetramer dissociation at their respective scattering angles $\Theta = 10.25°$, $\Theta = 7.50°$, and $\Theta = 6.25°$. The contributions of smaller clusters to the signal at 7.50° and 6.25° have been determined from their angular distribution by extrapolation and are tabulated in Table V.

The results of the dissociation measurements, which have been carried out with a laser fluence of $F = 142$ mJ $\cdot$ cm^{-2}, are shown in Fig. 10. Here the relative attenuation of the scattering signal as measured at the respective deflection angle with the mass spectrometer tuned to $m = 156$ amu is plotted as a function of laser frequency. Figure 10a represents a pure dimer spectrum. The solid line is a Lorentzian line-shape function fitted to the data points. Again this functional form has been chosen without associating it with lifetime broadening. In Fig. 10b a combined dimer–trimer spectrum is shown. Since the dimer dissociation spectrum (Fig. 10a) and the dimer contribution (Table V) are known, the dissociation signal due to the trimer can be properly determined, if again a Lorentzian is assumed. The result is the lower-amplitude dashed curve designated with $n = 3$. For the tetramer dissociation (Fig. 10c) the situation is similar. If the dimer and trimer contributions are properly considered, the frequency dependence of the benzene tetramer can be determined. The result is the dashed Lorentzian designated with $n = 4$.

All parameters of the Lorentzian line-shape functions as determined

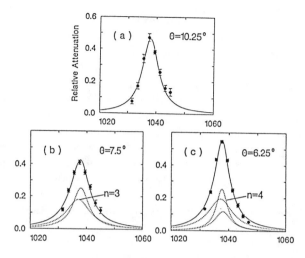

Figure 10. $(C_6H_6)_n$ dissociation spectra for $n = 2$, 3, and 4. The clusters were generated by expanding a 4% mixture of C_6H_6 in He. (*a*) A pure dimer spectrum. (*b*) Dimers and trimers. (*c*) Tetramers also contribute. The various contributions are indicated by the dashed Lorentzians.[106]

by the fit procedure are collected in Table VI. The figures for maximum attenuation refer to a 100% beam of the respective cluster species. Within the experimental error, all absorption profiles have the same line position. The linewidths, however, are distinctively different for each cluster size. It is interesting to note that no trend is observed and that the trimer absorption profile is much broader than the dimer and tetramer lines. As far as the dissociation amplitude or maximum attenuation is concerned, normally a gradual increase with growing cluster size is expected, simply because the absorption cross section becomes larger. Here again the trimer is an exception. Its dissociation cross section is even smaller than the dimer cross section. On the other hand, the tetramer attenuation is roughly twice as large as the dimer attenuation, as expected from the doubled number of absorbers in the tetramer.

TABLE VI
Results of Least-Squares Fits for Benzene Cluster Dissociation

	Line Position $(\tilde{\nu}_0/cm^{-1})$	Linewidth (FWHM/cm^{-1})	Maximum Attenuation (%)
Dimer	1037.9	7.1	46
Trimer	1037.0	10.4	39
Tetramer	1037.5	4.9	97

It should be noted that the exceptional behavior of the trimer cannot be an artifact of our data analysis. Just by comparing the measured dimer and trimer–dimer dissociation spectra of Fig. 10a and b, it is obvious that the lower amplitude and the broader linewidth in Fig. 10b must have been caused by the trimers contributing to the spectrum. If we anticipate the results obtained later from the dependence of the dissociation signal on the laser fluence, that is, that the benzene trimer must absorb two CO_2 laser photons in order to dissociate, its exceptional behavior can be easily explained. The smaller dissociation cross section is simply the result of the two-photon process, whereas the broadening of the dissociation profile may be due to the fact that the two photons do not necessarily have the same frequency.

In addition to the measurements presented in Fig. 10, we have carried out dissociation experiments with 4% C_6H_6 seeded in He but expanded at a higher stagnation pressure ($p_0 = 1$ bar) and with 5% C_6H_6 seeded in Ne ($p_0 = 1$ bar). The peculiar trimer behavior could be confirmed; in all cases the trimer dissociation spectrum was smaller in amplitude and broader than the corresponding dimer spectrum.

From the simple shape of the dissociation profiles, no conclusions about the structure of the clusters can be drawn. In particular our measurements cannot discriminate between a parallel, dihedral, or T-shaped dimer structure. Especially in the T-shaped dimer, the two monomers are in non-equivalent positions. Hence, each monomer should absorb at a slightly different frequency and as a result we expect two peaks in the dissociation spectrum. However, our observing of only one peak cannot be taken as argument against this structure since, in principle, two unresolved lines could be hidden behind the measured, $7 \, \text{cm}^{-1}$ broad absorption profile. If, however, the frequency shift could be calculated and if it turned out to be larger than $7 \, \text{cm}^{-1}$, then our measurement could possibly be evidence against a T-shaped dimer structure.

Now we want to compare our results with previous dissociation experiments carried out in the same frequency range. Johnson et al.[55] and Stace et al.[86] determined dimer linewidths of 2.1 and $2.5 \, \text{cm}^{-1}$, respectively. However, they used Ar as carrier gas and, thus, employed different stagnation conditions. In contrast we obtain values of $7.1 \, \text{cm}^{-1}$ for C_6H_6 seeded in He and $5.6 \, \text{cm}^{-1}$ for C_6H_6 seeded in Ne. The different results can be easily reconciled if one assumes that at least our absorption lines are not homogeneously broadened, but instead merely reflect the internal state distribution. With Ar as carrier gas, probably the lowest dimer temperatures are achieved and, thus, with Ar the narrowest absorption profiles are measured. The fact that the linewidth obtained with Ne falls in between those measured with Ar and He supports this interpretation.

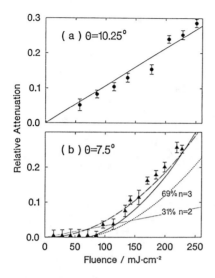

Figure 11. Fractional benzene dimer (*a*) and trimer (*b*) dissociation as a function of laser fluence. The dimer (*a*) shows a linear fluence dependence. In *b* dimers and trimers contribute. If the dimer contribution is subtracted, the parabola-like solid curve is obtained as trimer result.[106]

At the top of the absorption profile at 1037.43 cm^{-1} ($9\mu P30$ laser line), we have measured the dependence of the dissociation yield on the laser fluence for all three $(C_6H_6)_n$ species $n = 2$, 3, and 4. In order to increase the fraction of trimers and tetramers in the beam, the source was operated at a higher stagnation pressure ($p_0 = 1.4 \text{ bar}$). The new relative contributions of the various cluster species to the scattering signal on mass $m = 156$ amu are also tabulated in Table V. In Fig. 11*a* the fluence dependence of the benzene dimer dissociation is shown. The solid curve is a straight line fitted to the data points. In Fig. 11*b* the data measured at the trimer angle are plotted. The functional dependence of this data is clearly nonlinear. Instead it can be well represented by a parabola, as is shown by the chain-dotted curve. Since the fluence dependence of the dimer (Fig. 11*a*) and its contribution to the signal at $\Theta = 7.5°$ (Table V) are known, the dimer dissociation can be properly considered (dotted curve designated with 31% $n = 2$) and subtracted from the measured curve. As a result the dashed curve designated with 69% $n = 3$ is obtained. It reflects the fluence dependence of a pure (100%) trimer beam, if the y values are scaled and divided by 0.69 (solid curve).

In Fig. 12 the results obtained at the tetramer scattering angle ($\Theta = 6.25°$) are shown. As in the case of the dimer, the measured data can be well represented by a straight line. In order to determine a pure tetramer fluence dependence, the dimer and trimer contributions shown by the dotted curves must be subtracted. The result is the dashed curve desig-

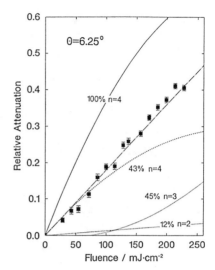

Figure 12. Fluence dependence of the dissociation signal measured at $\Theta = 6.25°$. Consideration of the dimer and trimer contributions yields the 100% tetramer curve (solid line).[106]

nated with 43% $n = 4$. If divided by 0.43, this curve represents the fluence dependence of a 100% tetramer beam (solid curve).

If we assume for a moment that the simple two-level-plus-decay model[49,51,53,54] (see Section IV) is adequate to describe the excitation and subsequent dissociation of C_6H_6 clusters properly, the measured relative attenuation can be converted into a dissociation cross section σ with Eq. (14). When $-\ln(1 - P_{diss})$ is plotted versus the laser fluence F, a straight line with slope $\sigma/(h\nu)$ is obtained, if the relation (Eq. 14) is valid.

If the dimer curve (Fig. 11a) and the final results (solid curves) for the trimer (Fig. 11b) and tetramer (Fig. 12) are formally transformed, corresponding to Eq. (14), and plotted as a function of laser fluence, Fig. 13 is obtained. We note that the dimer dependence is, to a very good approximation, still linear. In contrast, the trimer dissociation shows a nonlinear behavior, whereas the tetramer dissociation also depends linearly on the laser fluence. Although we are aware that the two-level-plus-decay model may not be valid, these results suggest a very appealing interpretation. The benzene dimer dissociation energy is believed to be below $1000\,\mathrm{cm}^{-1}$. Correspondingly, one absorbed CO_2 laser photon is sufficient to dissociate the dimer. As a result, a linear fluence dependence is expected, and this is exactly what we observe in our experiment. For the trimer the fluence dependence is determined to be clearly stronger than linear and approximately quadratic. This suggests that one CO_2 laser photon is not sufficient to dissociate the trimer. A plausible explanation for this behavior would be the assumption of a ring structure for the trimer.

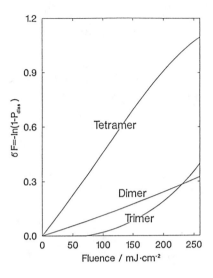

Figure 13. Benzene dimer, trimer, and tetramer dissociation as a function of laser fluence.[106]

Then at least two bonds would have to be broken, and correspondingly two CO_2 laser photons would have to be absorbed, if the binding energy of each bond is larger than 500 cm^{-1}. The linear fluence dependence observed for the tetramer suggests that again one photon is sufficient to dissociate this cluster, and using the same arguments it can be concluded that the tetramer exists predominantly in a noncyclic structure.

In order to provide further information to either support or reject this interpretation, we have carried out theoretical calculations. Utilizing a computer program written by Schmidt[131] and using the nonbonded potential functions of Williams,[136] the potential energy was minimized with respect to molecular translational and rotational coordinates. For the dimer we obtained a dihedral structure with an angle of $\varphi = 26°$ between the two benzene planes. The total energy is -918 cm^{-1}. The trimer was calculated to have a cyclic "pinwheel" structure with a total energy of -2466 cm^{-1} or -822 cm^{-1} per bond. The monomers in the ring are in almost equivalent positions. For the tetramer the energy minimization yields a structure which consists of a trimer in combination with a dimer, that is, a monomer is bound to a trimer ring in such a way that the angle between the planes of the molecule in the ring and the attached monomer is close to the dimer angle. The total energy of the tetramer is -3656 cm^{-1}.

The results of our calculations are in perfect agreement with the interpretation of our experimental fluence dependence data given above. According to the calculation, the trimer has a cyclic structure with a total energy difference of $2466 - 918 = 1548 \text{ cm}^{-1}$ to the dimer. Correspondingly, two CO_2 laser photons must be absorbed to dissociate this ring. In order to destroy the tetramer, it is sufficient to detach the monomer, which is bound to the trimer ring with an energy of $3656 - 2466 = 1190\,cm^{-1}$. Thus, taking into account the zero-point vibrational energy and the possibility that the clusters may not be completely cooled, it is conceivable that the tetramer can be dissociated upon absorption of a single CO_2 laser photon.

In this context, an important conclusion about the dissociation dynamics can be drawn from our fluence dependence measurements. Comparing the dimer and tetramer curves in Fig. 13, it can be seen that the tetramer dissociation cross section is more than twice as large as the dimer cross section. This might be regarded as obvious since the number of absorbers are doubled, but it also bears the following implication. Apparently it does not matter which benzene molecule in the tetramer is initially excited. The energy flows very fast into the bond holding the attached monomer and causes the rupture of this bond. Otherwise, the large tetramer cross section cannot be explained.

Very recently, in their photoionization experiment, Börnsen et al.[140] found evidence for a "zig-zag" rather than cyclic benzene trimer structure. This result is in sharp contrast to our findings. However, both results can be reconciled if one remembers that the structure of a cluster may depend on its internal temperature. Börnsen et al. have employed different stagnation conditions (0.7% C_6H_6 in He, $p_0 = 5$ bar) and it could very well be that their benzene trimers were warmer than ours. Thus, it is conceivable that they have predominantly trimers with zig-zag structure in their molecular beam, whereas in our beam the trimers are colder and have a cyclic structure.

C. Sulfur Hexafluoride

During the past few years, several experiments have been carried out to study the IR photodissociation of $(SF_6)_n$,[42,56–59] and SF_6Ar_m,[141–143] vdW clusters. In these experiments, the radiation of a line-tunable CO_2 laser was employed to excite the ν_3 vibration of SF_6 in the cluster. All studies were carried out in the direct undispersed beam. Either a mass spectrometer[56–58,143] or a liquid-helium-cooled bolometer[42,59,141,142] was used as the detector. The isotopic selectivity of the dissociation has also been exploited for isotope separation purposes.[83]

Reuss and co-workers[57,58] were the first to show that the $(SF_6)_2$ dimer dissociation spectrum was characterized by two isolated lines separated by $20\,cm^{-1}$. They were able to explain them as being due to the excitation of the ν_3 vibration parallel and perpendicular to the dimer axis. Although a mass spectrometer was used to monitor the dimer dissociation, contributions from larger $(SF_6)_n$ clusters could not be completely ruled out, since these species are known to fragment extensively in the electron bombardment ionizer. Therefore, we felt it worthwile to reinvestigate the dissociation of $(SF_6)_2$ dimers employing our cluster selection technique. In addition, it was desirable to obtain information on the dissociation of larger $(SF_6)_n$ clusters and to compare the results with model calculations carried out by Reuss and co-workers.[57,58,144]

In this section, we report on our predissociation experiments on size-selected $(SF_6)_n$ clusters with n between 2 and 4. Due to the extensive fragmentation of the $(SF_6)_n$ clusters in the ionizer, it was not possible to measure a pure trimer spectrum. Unfortunately, contributions from dimers had to be tolerated. However, by comparing the spectra taken at two different scattering angles, the spectral features of the trimer could be determined (compare Section II). In any case, our technique is capable of discriminating against fragmenting larger $(SF_6)_n$ clusters. Some important results have already been published.[107]

For stagnation pressures around 1.5 bar, where all our dissociation experiments have been carried out, we have measured conventional mass spectra in the direct undispersed $(SF_6)_n$ cluster beam (using 5% SF_6 in Ne). The most intense mass peaks could be assigned to the ion fragments SF_i^+ and $SF_6SF_i^+$ with $0 \leq i \leq 5$. SF_6^+ and $(SF_6)_2^+$ ions were not observed. The largest signal was recorded for SF_5^+ on $m = 127$ amu. The most intense cluster ion peak, which could be assigned to $SF_6SF_5^+$ ($m = 273$ amu), was only 4.8% of the SF_5^+ peak.

When we take, as is frequently done,[145] the $(SF_6)_{n-1} SF_5^+$ ion peak intensities to characterize the neutral $(SF_6)_n$ cluster distribution, we obtain a monomer–dimer ratio of 100/4.8 if different ionization probabilities are neglected. This assignment, however, is not correct, since possible fragmentation of larger clusters is not considered. Indeed, as we shall see below, the fragmentation of ionized $(SF_6)_n$ clusters is very strong and therefore $(SF_6)_{n-1}SF_5^+$ peak heights must not be related to neutral $(SF_6)_n$ cluster intensities.

We measured angular distributions of SF_6 monomers and clusters scattered by Ne at several mass settings of the quadrupole mass spectrometer. The results obtained for the two most prominent mass settings, $m = 127$ and 273 amu, are shown in Fig. 14. In this experiment, a 5% mixture of SF_6 in Ne was expanded at a stagnation pressure of 1.5 bar through a 0.1-mm diameter nozzle. In the upper part of the figure the Newton diagram for the elastic scattering of $(SF_6)_n$ clusters from Ne is sketched. From this velocity diagram it is deduced that monomers are scattered into an angular range which extends up to $\Theta = 21.0°$, whereas dimers, trimers, and tetramers are confined to smaller ranges with respective maximum angles

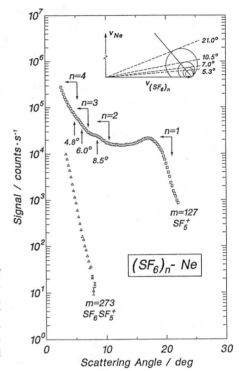

Figure 14. Angular distributions of $(SF_6)_n$ clusters scattered by Ne. Scattering ranges and maximum deflection angles are indicated above the $m = 127$ curve for $n = 1, 2, 3, 4$. The angular positions where the photodissociation experiments have been carried out are indicated below the curve.[107]

of $\Theta = 10.5°$, $7.0°$, and $5.3°$. These maximum angles and the respective scattering ranges are indicated in the figure by vertical and horizontal arrows above the curves.

The upper curve with square data points has been measured at mass $m = 127$ amu (SF_5^+), where the strongest monomer signal was observed. As expected from the Newton diagram, the signal sets in at around 21°. After running through a maximum due to a kinematic effect, the angular distribution is rather flat around $\Theta = 12°$. At $\Theta = 10.5°$ we observe a pronounced increase in signal, because we have entered the dimer scattering range and because dimers fragment in the ionizer and appear as SF_5^+ ions. At $\Theta = 7°$ we reach the trimer scattering range, which is again correlated with a signal enhancement showing that also trimers contribute to the signal at $m = 127$ amu.

At $\Theta = 8.5°$ (this position is marked by an arrow below the curve), the only cluster which can contribute to the SF_5^+ signal is the dimer, since all larger clusters are confined to smaller angles. In order to determine the contribution of dimers to the total scattering signal at this angle, we have measured angular distributions for various stagnation pressures. With decreasing pressure, the dimer shoulder at $\Theta = 8.5°$ becomes less pronounced and finally disappears. By normalizing the various curves in the monomer scattering range at $\Theta = 12°$ and extrapolating the signal at 8.5° to zero pressure, the dimer concentration at this angle can be determined as a function of stagnation pressure.[107] As a result we obtain that, at $p_0 = 1.5$ bar and $\Theta = 8.5°$, the dimer contribution to the SF_5^+ signal is $(28 \pm 2)\%$. This means that 6.8 kHz from the total count rate of 24.4 kHz must be attributed to dimers.

The curve with triangular data points is obtained when the mass spectrometer is tuned to $m = 273$ amu ($SF_6SF_5^+$). From the very steep rise it can already be deduced that the signal at this mass is mainly produced by fragmenting larger $(SF_6)_n$ clusters and not by dimers. In fact, comparing the signals in the dimer scattering range at $\Theta = 8.5°$, we obtain 7 counts/s at $m = 273$ amu, whereas at $m = 127$ amu the dimer contribution was 6800 counts/s. Assuming equal quadrupole mass spectrometer transmission on both mass settings, we obtain the result that it is 1000 times more likely for the ionized $(SF_6)_2$ dimer to appear as SF_5^+ than as $SF_6SF_5^+$ ion. This shows clearly that it is not correct to relate $(SF_6)_{n-1}SF_5^+$ ion peak intensities to the neutral $(SF_6)_n$ cluster distribution in the beam.[145] Even trimers do not contribute very much to the $SF_6SF_5^+$ signal. They are detected as SF_5^+ ions more than 60 times as efficiently.

From the measured angular distributions at $m = 127$ and 273 amu we know that for the smaller $(SF_6)_n$ clusters the most efficient fragmentation channel upon electron bombardment ionization is the dissociation into the SF_5^+ product. Consequently, these species are preferably detected at $m = 127$ amu, where the monomer is also present. Nevertheless the photo-dissociation of the $(SF_6)_2$ dimer can be selectively studied at this mass at $\Theta = 8.5°$ since the SF_6 monomer is not affected by the laser radiation at low fluences. If the dissociation of trimers or tetramers is studied at the same mass, contributions of the respective lighter species must be accepted. It should be noted, however, that contributions from larger clusters are excluded. The angular positions where dissociation experiments have been carried out are indicated in Fig. 14 by arrows below the SF_5^+ curve. The contributions of the various cluster sizes to the $m = 127$ amu signal at these positions are tabulated in Table VII.

The dimer dissociation spectrum as measured at $\Theta = 8.5°$ is shown in Fig. 15a. Each data point has been obtained by integrating over 15 min. The laser fluence was adjusted to $5.5\ mJ \cdot cm^{-2}$. The spectrum is characterized by two absorption peaks at 934.4 and 955.6 cm^{-1}. Compared to the band origin of the SF_6 monomer at 948 cm^{-1},[146] one line is blue shifted by 7.6 cm^{-1}, whereas the other one is red shifted by 13.6 cm^{-1}. The blue

TABLE VII
Sulfur Hexafluoride Cluster Contributions at Various Detector Settings

Detector Setting	I	II	III	IV
Scattering angle Θ (deg)	8.5	6.0	4.8	4.8
Mass (amu)	127	127	127	273
Monomer contribution (%)	72	52	39	—
Dimer contribution (%)	28	20	15^a	—
Trimer contribution (%)	—	28	21^a	b
Tetramer contribution (%)	—	—	25^a	b

[a]Estimated contribution.
[b]Contribution not known.

absorption maximum is approximately twice as high as the red one. The solid curves are Lorentzian line-shape functions fitted to the data points. The entire spectrum is reproduced very well by two Lorentzians of equal width FWHM $\Gamma = 3.0\,\mathrm{cm}^{-1}$. All results of the least-squares fit are collected in Table VIII. The maximum attenuation of 0.244 in the blue absorption peak actually means that 87% of the dimers are dissociated at this laser frequency.

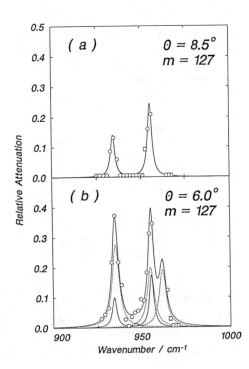

Figure 15. SF$_6$ dimer and trimer dissociation spectra. The upper panel (a) shows a pure dimer spectrum. In spectrum b the trimer absorption lines are represented by the dashed curves.[107]

TABLE VIII
Results of Least-Squares Fits for Sulfur Hexafluoride Cluster Dissociation

Species	Detector Setting	Peak Position $(\bar{\nu}_0/cm^{-1})$	Maximum Attenuation[a]	FWHM (Γ/cm^{-1})
Dimer	I	934.4	0.139 (0.496)	3.0
		955.6	0.244 (0.871)	
Trimer	II	935.2	0.276 (0.986)	4.8
		[946.6]	[0.016 (0.057)]	
		955.1	0.200 (0.714)	
		962.0	0.192 (0.686)	
Tetramer	III	935.3	0.156 (0.624)	9.5
		958.3	0.112 (0.448)	
	IV	936.1	0.999	11.6
		959.1	0.759	

[a]The figures in parentheses refer to pure cluster beams as calculated from the contributions given in Table VII.

In order to ensure that the dimer dissociation spectrum is not affected by monomer multiphoton dissociation, we have carried out the same experiment at a larger scattering angle ($\Theta = 14°$) where only monomers are present. The laser was tuned around the 10μP20 line at $944\,cm^{-1}$ where the maximum multiphoton dissociation yield has been observed by Schulz et al.[147] Even at 40 times higher laser fluence, compared to the dimer dissociation experiment, no attenuation of the SF_5^+ signal could be observed proving that monomer multiphoton dissociation does not occur at low laser fluences.

When the detector is moved into the scattering range of the $(SF_6)_3$ trimer at $\Theta = 6.0°$, the effect of the laser radiation on both dimers and trimers can be studied. The corresponding spectrum obtained at the same laser fluence of $5.5\,mJ \cdot cm^{-2}$ is shown in Fig. 15b. Since the spectroscopy of the dimer is known from Fig. 15a, its contribution to the composed spectrum can be properly considered. This is shown by the lower solid curve obtained by multiplying the curve in Fig. 15a with 20/28 (compare Table VII).

Once the dimer contribution is known, the spectral features of the trimer absorption can be determined. This is accomplished by fitting four independent Lorentzian line-shape functions with the same FWHM to the difference spectrum. The result is shown by the dashed curves. The solid curve through the data points corresponds to the sum of the four dashed curves plus the dimer curves. It nicely reproduces the measured spectrum. This good agreement is only obtained if the trimer linewidth is chosen somewhat larger than the dimer linewidth. All line positions and amplitudes are collected in Table VIII.

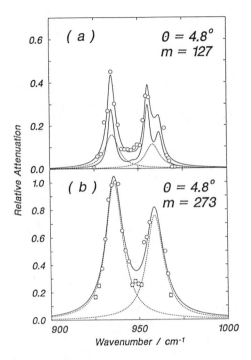

Figure 16. (SF$_6$)$_4$ tetramer absorption lines (dashed curves) derived from dissociation spectra measured at $m = 127$ amu (*a*) and $m = 273$ amu (*b*).[107]

The central line at $\tilde{\nu}_0 = 946.6$ cm^{-1}, which has only very low amplitude, is near to the monomer absorption at 947.97 cm^{-1}.[146] It is probably due to the dissociation of SF$_6$Ne$_m$ vdW clusters. Support for this assumption is supplied by the fact that this line is between the absorption lines observed for SF$_6$He$_m$ at 947.7 cm^{-1} [119] and SF$_6$Ar$_m$ at 946.5 cm^{-1}.[42]

In order to investigate the dissociation of (SF$_6$)$_4$ tetramers, the detector was moved to $\Theta = 4.8°$. With the mass spectrometer tuned to $m = 127$ amu, the effect of the laser radiation on a mixture of tetramers, trimers, and dimers was measured. The result is shown in Fig. 16a. Compared to the detector setting II at $\Theta = 6.0°$, we estimate that the trimer and dimer contributions are reduced by 25%, but assume that their ratio has not significantly changed. Correspondingly, we insert the complete dimer–trimer combination spectrum of Fig. 15b with an amplitude reduced by 25% into Fig. 16a and ascribe the difference to the tetramer. Thus, we obtain two 9.5 cm^{-1} broad tetramer lines at $\tilde{\nu}_0 = 935.3$ cm^{-1} and $\tilde{\nu}_0 = 958.3$ cm^{-1}.

Since the dimer and trimer contributions at $\Theta = 4.8°$ are not exactly known, the procedure for determining the tetramer absorption lines is not very accurate. Therefore, we have repeated the same measurement with the mass spectrometer tuned to $m = 273$ amu ($SF_6SF_5^+$). Although the signal was very low, a complete spectrum could be recorded. The result is shown in Fig. 16b. From what has been said in the discussion of the angular distributions (Fig. 14), it is clear that any contribution from dimers can now be neglected; but also the trimer contribution must be very low. On the other hand, the angular position discriminates against $(SF_6)_n$ clusters with $n > 4$. Thus only tetramers and a few percent of trimers are contributing to the dissociation spectrum of Fig. 16b. It is our most reliable tetramer spectrum and confirms our earlier result that the tetramer absorption is dominated by two absorption lines.

Having investigated the frequency dependence of the dissociation, we measured the fluence dependence at fixed laser frequency. For the dimer we determined the relative attenuation as a function of laser fluence in both absorption peaks, at 934.9 and 956.2 cm^{-1} (10μP30 and 10μP6 laser lines). The curves rise very steeply and go into saturation at quite low fluence values. The maximum attenuations, which are already achieved at $F = 20$ mJ $\cdot$ cm^{-2}, are 0.28 in the blue and 0.155 in the red dimer absorption peak. This means that actually 100% of the dimers are dissociated at 956.2 cm^{-1} whereas at 934.9 cm^{-1} only 55% are affected by the laser radiation. Apparently, at this frequency, not all dimers can be addressed. This result is in contrast to recent observations by Reuss and co-workers,[148,149] who found in an intracavity experiment that, at high laser fluence, both dimer peaks became equal in height. At present, we have no plausible explanation for this discrepancy.

The fluence dependence curves for the larger $(SF_6)_n$ clusters have been measured on the 10μP30 laser line at 934.9 cm^{-1}. With detector settings II and III, saturation values of 0.42 and 0.50 are obtained. From these maximum attenuation values it can be deduced that, above 20 mJ $\cdot$ cm^{-2}, practically all SF_6 trimers and tetramers are dissociated.

Calculations based on Eq. (13) and the simple two-level-plus-decay model[49,51,53,54] do not agree with the measured fluence dependences. If the theoretical curves are fitted to the data points to reproduce (1) the low fluence dependence and (2) the saturation region, large discrepancies are obtained in the intermediate fluence range. This shows that the simple model is not adequate to describe the absorption and dissociation of $(SF_6)_n$ clusters properly.[107]

The IR photodissociation of the $(SF_6)_2$ dimer was first studied by Geraedts et al.[56-58] These authors found two well separated peaks which they assigned to the excitation of the ν_3 vibration parallel and perpendicular to the dimer axis. In our present study on size-selected $(SF_6)_n$ clusters we have shown that the two peaks are actually due to $(SF_6)_2$ dimer absorptions.

If we compare our dimer results with previous measurements,[58,59,42] we find very good agreement on the position of the absorption lines, their widths, and relative amplitudes. All measured line positions agree within $1\ cm^{-1}$. However, it appears that our red absorption peak is slightly more red shifted, whereas the blue absorption peak is slightly more blue shifted. As will be discussed later, this observation can be very well explained, if one assumes that the dimers formed in our molecular beam are somewhat colder. As far as the linewidth is concerned, our absorption lines are slightly broader than observed in the low-power cw CO_2 laser experiments. This can be explained by power broadening effects. The amplitude ratio of the blue peak to the red peak is in all experiments between 1.7 and 1.9.

Very recently, Heijmen et al.[60,150] have employed an elegant pump and probe technique using two CO_2 lasers to determine the $(SF_6)_3$ trimer dissociation spectrum. They found three trimer lines at 934.9, 955.2, and $963.3\ cm^{-1}$, which is in excellent agreement with our observations (see Table VIII). Unfortunately, our $(SF_6)_4$ tetramer results cannot be compared with other experiments since no data is available.

According to the model of the dynamical dipole–dipole interaction by Geraedts et al.,[57,58] the degeneracy of the $3n$ degenerate states of a $(SF_6)_n$ cluster is partially lifted upon absorption of one ν_3 quantum. As a result, m levels are obtained whose separations $\Delta_{n,m}$ from the monomer absorption line at $948\ cm^{-1}$ [146] are expressed in units of

$$\lambda = (2\pi\epsilon_0)^{-1}\mu_{01}^2\langle R^{-3}\rangle \tag{17}$$

where R describes the nearest-neighbor distance and $\mu_{01} = 0.388\ D$ [151] is the monomer transition dipole moment. For the dimer, two lines with $\Delta_{2,1} = -2\lambda$ and $\Delta_{2,2} = \lambda$ are predicted. In order to be consistent with their experimental data, Geraedts et al. had to take $\lambda = 6.8\ cm^{-1}$ as their unit value. This value corresponds to one specific averaged intermolecular separation R. For the trimer three different transition frequencies were determined, whereas for the tetramer only two contributions are obtained.

A comparison between our experimental results and the model calculations of Geraedts et al.[57,58] is shown in Fig. 17. The Lorentzian curves are the final results of our experiment for each cluster size, whereas the solid sticks are the results of the dynamical dipole–dipole interaction calculation. As far as the dimer spectrum is concerned, very good agreement is found. The separation between the peaks is slightly larger in our experiment. This can be explained by assuming that our dimers are somewhat colder than those investigated by Geraedts et al. For colder

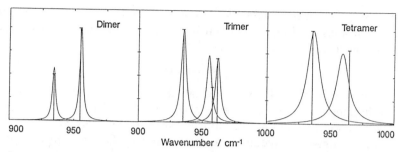

Figure 17. Comparison of the experimental results with the model of the dynamical dipole–dipole interaction of Geraedts et al.[57,58] (solid sticks).[107]

dimers, we expect a smaller intermolecular separation R and therefore, corresponding to Eq. (17), larger shifts from the monomer line.

For the trimer, we find three absorption lines in good agreement with the results of the model calculations. The positions and amplitudes of the low- and high-frequency lines are both very well predicted, whereas the position and amplitude of the central absorption line deviate slightly. It should be noted, however, that our result is in excellent agreement with the experimental findings of Heijmen et al.[60,150] For the tetramer only two lines are predicted by the model calculation. This is again in agreement with the experimental findings. In view of the simplicity of the model, the overall agreement between the calculations and our experimental results must be considered as surprisingly good.

In order to explain the experimental observations in the case of $(SiH_4)_2$,[59] Snels and Reuss[144] have introduced a refined model which, in addition to the dominant dipole–dipole term, also includes induction effects. Comparing both model calculations with our experimental findings,[107] it is clearly seen that the simple dipole–dipole interaction model gives better agreement for the SF_6 trimer and tetramer. It seems that, in addition to the induction, another interaction must be included for this model to give better results for SF_6.

When considering the positions of the dimer absorption lines and their relation to the dimer temperature, the very recent double resonance experiments of Heijmen et al.[60,150] should be mentioned. Employing a pump and probe technique with two CO_2 lasers, they were able to demonstrate the inhomogeneous nature of the dimer absorption lines. They showed that the red wing of the low-frequency dimer line correlates with the blue wing of the high-frequency line. The inner wings are related to warmer dimers, whereas the outer wings are associated with colder dimers having smaller intermolecular separation R (compare Eq. 17). These findings agree nicely with the previously presented interpretation of the dimer line positions. In addition, these observations show that the width of the absorption lines are not determined by lifetime broadening, but instead merely reflect the internal state distribution of the cluster.

D. Ammonia

The ammonia dimer $(NH_3)_2$ was originally thought to be hydrogen bonded with a binding energy of more than $1000\ cm^{-1}$.[152] Correspondingly, one CO_2 laser photon would not be sufficient to dissociate this complex. The first dissociation experiments were carried out by Gentry and co-workers.[53] They measured a single broad non-Lorentzian absorption profile near $980\ cm^{-1}$ and found that conditions producing colder beams resulted in a lower dissociation signal on the dimer mass. These observations were interpreted as an indication that only hot dimers having sufficient internal energy could be dissociated by a single CO_2 laser photon.

Klemperer and co-workers[26,27] studied the IR predissociation of $(NH_3)_2$ in their molecular beam electric resonance apparatus in which they could discriminate against the larger $(NH_3)_n$ clusters. They found that the absorption band at $980\ cm^{-1}$ consisted of two well-separated peaks. Employing the microwave infrared double-resonance technique, they were able to assign them to two torsional sublevels of the complex. In addition, they could show that the dimer dissociated upon absorption of a single CO_2 laser photon and that its binding energy was less than $1000\ cm^{-1}$.

More recently, Snels et al.[64] investigated the dissociation of $(NH_3)_n$ clusters employing a liquid-helium-cooled bolometer detector. In the region near $1004\ cm^{-1}$, which became accessible with their $^{13}CO_2$ laser, they found another absorption band of $(NH_3)_2$ which had not been previously observed. The existence of two absorption bands was explained with the excitation of the two nonequivalent NH_3 molecules in the complex.

In order to prove whether the two absorption bands observed by Snels et al.[64] were really due to dimers, we reinvestigated the dissociation of ammonia clusters using our cluster-selection technique. All dimer features reported so far could be confirmed to be actually due to the dimer. In addition, we have measured dissociation spectra for larger $(NH_3)_n$ clusters with $n \leq 8$, which had not been previously observed. Valuable information about the structure and dissociation energies of these species was obtained.

Using a mixture of 5% NH_3 in He, we first measured conventional mass spectra in the direct undispersed cluster beam. Since an intracluster proton transfer reaction occurs rapidly upon ionization,[153] $(NH_3)_n$ clusters are predominantly detected as $(NH_3)_m H^+$ ions with $m \leq n - 1$. At a stagnation pressure of $p_0 = 1.75\ bar$, at which the dissociation experiments were carried out, only cluster ions with $m \leq 4$ were observed. Disregarding fragmentation effects, one would conclude that the largest neutral $(NH_3)_n$

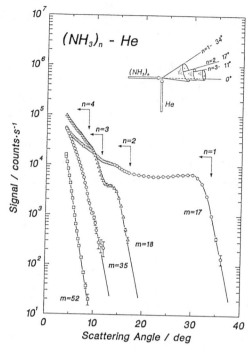

Figure 18. Angular distributions of the scattered $(NH_3)_n$ cluster beam measured at different mass settings of the mass spectrometer. The arrows above the $m = 17$ curve indicate the scattering ranges and maximum deflection angles for the various cluster species.[108]

in the beam was that with $n = 5$. However, as we shall see below, this conclusion is incorrect because fragmentation effects play an important role. Thus, $n = 5$ is only the smallest upper bound of neutral $(NH_3)_n$ clusters in the beam.

Angular distributions of ammonia monomers and clusters scattered by He and detected at various mass settings of the quadrupole mass spectrometer are shown in Fig. 18. At the top of the figure the kinematics of the scattering process are illustrated. This diagram shows the maximum deflection angles and scattering ranges for $n = 1, 2,$ and 3 which are also indicated above the measured curves.

The upper curve with circular data points has been measured at $m = 17\,\text{amu}$ (NH_3^+). It is dominated by NH_3 monomer scattering. However, as can be seen by the pronounced shoulders below $\Theta = 17°$ and $\Theta = 11°$, larger $(NH_3)_n$ clusters also fragment in the ionizer and appear as NH_3^+ ions.

If the mass spectrometer is tuned to the main dimer fragmentation channel $m = 18\,\text{amu}$ (NH_3H^+), the curve with triangular data points is obtained. At the small plateau around $\Theta = 12.5°$ only dimers are detected. This is the ideal position to study dimer features undisturbed by fragmentation artifacts of larger clusters. The strong increase in signal below $\Theta = 11.4°$ is due to trimer fragments. Thus, it appears that at $\Theta = 8.7°$ 83% of the signal at $m = 18$ amu originates from trimers. This proves that, at our stagnation conditions, most of the signal on the nominal dimer mass observed at small scattering angles and particularly in the direct beam must be ascribed to larger clusters.

With the mass spectrometer tuned to $m = 35$ amu $[(NH_3)_2H^+]$ only trimers and fragmented higher clusters are observed. However, at $\Theta = 8.7°$ only trimers can be detected. Comparison of the $m = 18$ and $m = 35$ curves at this angle shows that it is 10 times more likely for the ionized trimer to appear as an $(NH_3)H^+$ ion than an $(NH_3)_2H^+$ ion. This indicates that one must be very careful in interpreting direct-beam mass spectra. The weak signal at $m = 35$ amu cannot be considered as an indication that the fraction of trimers is negligibly small. A similar branching ratio for the trimer fragmentation was determined by Buck and Lauenstein.[154]

If the mass spectrometer is tuned to $m = 52$ amu $[(NH_3)_3H^+]$, the curve with square data points is obtained. Now the smallest cluster size is the tetramer. The very steep rise signals that here also strong fragmentation of larger clusters to $m = 52$ amu occurs.

The Newton diagram and the measured angular distributions show that cluster-specific detection of small ammonia clusters is best performed if the mass spectrometer settings are $\Theta = 12.5°$ and $m = 18$ amu for the dimer, $\Theta = 8.7°$ and $m = 35$ amu for the trimer, and $\Theta = 7.2°$ and $m = 52$ amu for the tetramer. Because of the strong fragmentation, however, it appears that, for all clusters, much higher signals are obtained when the mass spectrometer is always tuned to $m = 18$ amu (NH_3H^+). In this case some selectivity is lost, but we can proceed as discussed in Section II. Starting at the dimer angle and measuring the dissociation spectra at successively smaller deflection angles, the spectral features of each cluster size can be revealed. In any case, the technique is capable of discriminating against fragmenting larger clusters. For a proper analysis of the measured spectra, the fractional contribution of each cluster size to the $m = 18$ signal must be known. These numbers are obtained from the angular distribution curves as discussed in Section IV. They are collected in Table IX.

At detector setting I we observe, as already stressed, solely $(NH_3)_2$ dimers. Consequently, with this setting a pure dimer dissociation spectrum can be measured. The result is shown in Fig. 19a. Each data point has been obtained by integrating over 2 h. The spectrum is characterized by two absorption bands at 977 and 1003.4 cm^{-1}. In order to measure the higher-frequency band, the laser had to be operated with $^{13}CO_2$ isotope.

TABLE IX
Ammonia Cluster Contributions at Various Detector Settings[a]

Detector Setting	I	II	III	IV
Scattering angle Θ (deg)	12.5	8.7	7.2	5.8
Dimer contribution (%)	100	17	7	4
Trimer contribution (%)	—	83	40	29
Tetramer contribution (%)	—	—	53	35
Pentamer contribution (%)	—	—	—	32
Laser fluence (mJ · cm^{-2})	179	179	89	89

[a]The mass spectrometer was tuned to $m = 18$ amu.

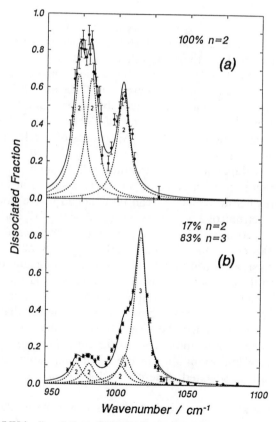

Figure 19. $(NH_3)_n$ dissociation spectra for $n = 2$ and $n = 3$. (a) A pure dimer spectrum. (b) Spectrum contains contributions from dimers and trimers.[108]

TABLE X
Results of Least-Squares Fits for Ammonia Cluster Dissociation

Species	Detector Setting	Peak Position $(\bar{\nu}_0/\mathrm{cm}^{-1})$	Maximum Attenuation[a]	FWHM $(\Gamma/\mathrm{cm}^{-1})$
Dimer	I	972.5	0.66	10.0
		981.3	0.64	
		1003.4	0.58	
Trimer	II	1006.1	0.16 (0.19)	10.0
		1016.3	0.80 (0.96)	
Tetramer	III	1036.4	0.12 (0.21)	8.8
		1044.7	0.21 (0.40)	6.1
Pentamer	IV	999.6	0.06 (0.18)	11.3
		1044.8	0.20 (0.63)	27.8

[a]The figures in parentheses refer to pure cluster beams as calculated from the contributions given in Table IX.

The dashed curves are Lorentzian lineshape functions fitted to the data points and the solid curve represents their sum. Careful inspection of the lower-frequency band reveals a splitting which can be reproduced by assuming two absorption lines centered at 972.5 and 981.3 cm^{-1}. The entire spectrum is reproduced very well by three Lorentzians of equal FWHM $\Gamma = 10\,\mathrm{cm}^{-1}$. At a laser fluence of 179 mJ $\cdot$ cm^{-2}, the maximum dissociated fraction amounts to 0.66. All results of the least-squares fits are collected in Table X.

The frequency dependence of a mixture of 17% ammonia dimers and 83% trimers measured with detector setting II is shown in Fig. 19b. The spectral features below 1000 cm^{-1} can be assigned to the dimer. In fact, if the dimer spectrum of Fig. 19a is weighted with 0.17 and drawn into the combined spectrum of Fig. 19b, a perfect fit is obtained for wavenumbers below 1000 cm^{-1}. On the other hand, the strong absorption line at 1016.3 cm^{-1} must be attributed to the trimer. Unfortunately, the line center falls into the gap between two laser bands so that the line position and amplitude can only be determined by a least-squares fit with the assumption of a certain lineshape function. The shoulder that we observe at 1006.1 cm^{-1} can only partly be assigned to the dimer, the rest must be attributed to trimer dissociation. The 1016.3 cm^{-1} line, however, appears to be much more important. In the top of this line approximately 96% of all trimers are dissociated, whereas at 1006.1 cm^{-1} only 19% can be addressed.

In order to study the ammonia tetramer dissociation, the detector is moved to position III. The result is shown in Fig. 20a. Comparing this spectrum with the spectrum of Fig. 19b, it is immediately seen that the

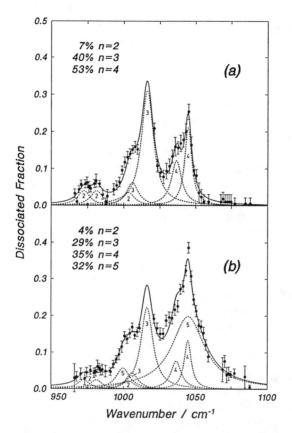

Figure 20. $(NH_3)_n$ dissociation spectra including tetramer (a) and pentamer (b) dissociation. The various contributions of each cluster size n are indicated in the figure.[108]

contributions from dimers and trimers can be separated very easily. It appears that the spectral range below $1030 \, cm^{-1}$ is governed by dimer and trimer dissociation, whereas the tetramers exclusively dissociate at wavenumbers above $1030 \, cm^{-1}$. Inserting the measured dimer and trimer absorption profiles with variable amplitude (but fixed amplitude ratio for one specific cluster size), the spectrum below $1030 \, cm^{-1}$ is easily reproduced. From this fit also the fractional dimer (7%) and trimer (40%) contributions are obtained. The spectral region above $1030 \, cm^{-1}$ is fitted by two Lorentzians with peak positions at $\tilde{\nu}_0 = 1036.4 \, cm^{-1}$ and $\tilde{\nu}_0 = 1044.7 \, cm^{-1}$. These absorption lines must be assigned to the dissociation of tetramers. With a resulting tetramer contribution of 53%, the maximum dissociated fraction of tetramers is determined to be 0.40.

The spectrum in Fig. 20b has been obtained with detector setting IV.

Besides dimers, trimers, and tetramers, now pentamers are also included. The main difference between the two spectra in Fig. 20 is observed at wavenumbers above $1050 \, cm^{-1}$ where spectrum b shows a much higher dissociation signal. As described above, the known frequency dependences of dimers, trimers, and tetramers are inserted into spectrum b and, thus, their contributions can be determined. The remaining dissociation signal must be ascribed to the pentamer. So we obtain the result that the pentamer contributes with two Lorentzians positioned at $\tilde{\nu}_0 = 999.6 \, cm^{-1}$ and $\tilde{\nu}_0 = 1044.8 \, cm^{-1}$. The high-frequency pentamer absorption profile is with $\Gamma = 27.8 \, cm^{-1}$, much broader than those observed for the smaller clusters. The existence of the small pentamer absorption peak at $\tilde{\nu}_0 = 999.6 \, cm^{-1}$ is not very certain, but its inclusion clearly improves the fit.

We have also measured dissociation spectra for even larger $(NH_3)_n$ clusters. A spectrum taken in the direct undispersed beam at $m = 69$ amu ($n \geq 5$) shows a very broad ($\Gamma > 70 \, cm^{-1}$) featureless profile centered at $\tilde{\nu}_0 = 1040 \, cm^{-1}$. Strong dissociation signals are observed above $1075 \, cm^{-1}$ showing that (1) the absorption profiles of larger ammonia clusters are further blue shifted and that (2) the cluster beam contains much larger clusters than expected from the conventional mass spectrum.

In order to gain information about the nature of the photon-induced transition in the cluster prior to dissociation, we have investigated its dependence on the laser fluence.[108] For the dimer the measurements were carried out at the maxima of both absorption bands. The curves very quickly become saturated and at fluence values around $200 \, mJ \cdot cm^{-2}$ maximum dissociation is reached. In the red absorption band more than 90% of the dimers can be dissociated, whereas in the blue absorption peak only 58% can be affected. However, these numbers are difficult to compare, since at least the red absorption band contains two, perhaps even more, overlapping transitions. The fact that a certain fraction of dimers remains undissociated even at high laser fluences is in accordance with the double resonance experiments of Heijmen et al.,[65,150] which reveal the inhomogeneous character of the dimer absorption peaks.

For tetramer absorption the saturation value of the dissociation is measured to be $P_{diss} = 0.68$. This value is only slightly larger than the maximum dissociation obtained in the blue dimer peak ($P_{diss} = 0.58$). A comparable study for the trimer was not possible, since the frequency of the absorption maximum was not accessible to our CO_2 laser.

Unfortunately, we obtained no information from the shape of the measured fluence dependence curves as to whether we are dealing with a one- or two-photon transition, since they are dominated by saturation effects. Some information can be inferred from the height of the tetramer dissociation cross section, which is only slightly larger than that for the dimer, although the absorption cross section is expected to be approximately twice as large. The difference between absorption and dissociation cross section is possibly an indication that the tetramer cannot be dissociated with a single CO_2 laser photon, that is, that this species has a dissociation energy larger than $1050 \, cm^{-1}$ (see discussion below).

Now we want to compare our ammonia dissociation spectra with earlier results. The first studies were carried out by Howard et al.,[53] who observed a single broad absorption profile at $980 \, cm^{-1}$. Their observation that conditions which produced colder beams resulted in a lower dissociation signal on the "dimer mass" $m = 18$ amu was interpreted as indication that only hot dimers having enough internal energy can be dissociated by a single IR photon. In our study, we observe that the fragmentation channel of the ionized ammonia trimer is also $m = 18$ amu. Therefore, it could very well be that Howard et al. had some nonnegligible contribution from larger clusters in their molecular beam, although the signal observed on the "trimer mass" $m = 35$ amu was very low. This leads us to offer an alternative explanation for the observed behavior. At conditions which produce colder beams, larger ammonia clusters are also formed. These clusters contribute to the signal at $m = 18$ amu but cannot be dissociated at $980 \, cm^{-1}$. Therefore, the relative attenuation of the $m = 18$ signal is lowered if colder beams are produced, thus falsely indicating a decrease in dimer dissociation efficiency.

Howard et al. observed only the red dimer absorption band. The linewidth of $\Gamma = 18 \, cm^{-1}$ is the same as we observe. There is even some indication of a splitting in the top of the absorption profile. Since the blue dimer absorption band and trimer dissociation fall outside the tuning range of their $^{12}CO_2$ laser, they could not be observed by Howard et al.

A doublet structure in the red dimer absorption band was first observed by Fraser et al.[26] Using the microwave–IR double resonance technique, they were able to show that the two peaks correspond to different torsional sublevels of the dimer. From the measured rotational and quadrupole coupling constants and from the measured dipole moment, Nelson et al.[27] deduced valuable information about the dimer structure. In particular, they found that the ammonia dimer structure is somewhat different than the theoretically predicted structure which has a nearly linear N–H . . . N arrangement.

Fraser et al.[26] found two peaks in the red dimer absorption band at 977.2 and $980.9 \, cm^{-1}$ with linewidths of about $5 \, cm^{-1}$. Compared with our results (see Table X), we find a slight inconsistency in the position of the low-frequency peak. We explain this as probably being due to different beam conditions, that is, different dimer temperatures. The overall agreement is very good.

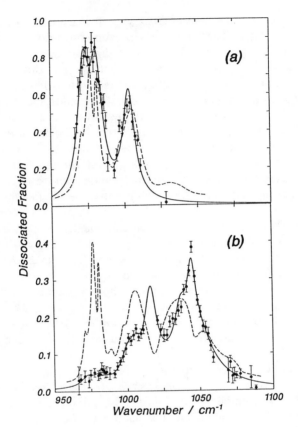

Figure 21. Comparison between our results (data points) and those of Snels et al.[64] (dashed curves). The upper spectra are pure dimer dissociation spectra. The bottom spectra include contributions from larger $(NH_3)_n$ clusters.[108]

The first experimental study reporting the existence of two dimer absorption bands at 979 and 1004 cm^{-1} was published by Snels et al.[64] They employed a liquid-helium-cooled bolometer detector and a cw CO_2 laser operated with different CO_2 isotopes and N_2O gas. The two bands were assigned to the excitation of the two nonequivalent NH_3 units in the dimer. A comparison of our results with those obtained by Snels et al. is shown in Fig. 21. The dashed curves are adapted from Ref. 64. The upper spectrum (*a*) shows the comparison of the dimer results. Because the bands at 979 and 1004 cm^{-1} showed the same temperature dependence upon varying source conditions, Snels et al. assigned them both to dimer absorptions. Our spectrum, obtained with size-selected dimers, proves that this assignment is correct. Although there is a slight deviation in the fine structure of the red absorption band, the overall agreement must be considered as very statisfactory. As we will show below, the blue dimer band corresponds to the excitation of the more strongly perturbed monomer, whereas the red doublet must be assigned to the more weakly interacting submolecule.

The dashed curve of the lower spectrum (b) was obtained by Snels et al. at reduced temperature. Correspondingly, the broad peak at 1035 cm^{-1} was assigned to the absorption of larger $(NH_3)_n$ clusters. Comparison with our results makes it evident that this broad peak cannot be due to trimers or tetramers. Apparently it must be assigned to the dissociation of pentamers or even higher ammonia clusters. The dissociation of trimers and tetramers was not observed by Snels et al. In contrast, they observe pronounced minima at 1016.3 and 1044.7 cm^{-1}, the region in which we observe the dissociation of trimers and tetramers. On the other hand, however, it is hard to believe that Snels et al. had exclusively dimers and larger ammonia clusters in their beam, that is, no trimers and tetramers. In order to reconcile the contradictory results, we postulate that the dissociation energies of the trimer and tetramer are larger than 1000 cm^{-1}.

Since Snels et al.[64] employed a low-power CO_2 laser, two-photon transitions are not expected. Consequently, trimers and tetramers cannot be dissociated in their experiment, if the dissociation energies are larger than 1000 cm^{-1}. However, as we know from our experiment, their trimers and tetramers are excited at 1016.3 and 1044.7 cm^{-1}, respectively. In contrast to dissociation, which is detected by a decrease in signal, this internal excitation results in a positive change of the bolometer signal. Assume that the ammonia pentamer can be dissociated with a single CO_2 laser photon. If the molecular beam is composed of a mixture of dimers, trimers, tetramers, and pentamers, we would then expect, besides the dimer features, a broad pentamer dissociation profile with dips at 1016.3 and 1044.7 cm^{-1} caused by absorbing but nondissociating trimers and tetramers. This is exactly what we see in the comparison of the two spectra in Fig. 21b. Our spectrum is obtained at high laser fluence (89 mJ · cm^{-2}) where two-photon transitions can play an important role.

The explanation above has been confirmed by the recent double resonance experiments of Heijmen et al.[65,150] They observed that the ammonia trimer and tetramer can indeed absorb a CO_2 laser photon without being dissociated. From the experimental observation that the ammonia trimer and tetramer dissociation energies are larger than 1000 cm^{-1}, important conclusions about the structure of these species can be drawn. The experimental findings strongly suggest that trimer and tetramer have cyclic structures. Because in cyclic clusters at least two bonds must be broken, the dissociation energies are expected to be roughly twice as large as the dimer dissociation energy. This conclusion is also in perfect agreement with the molecular beam electric deflection studies by Odutola et al.[155] who showed that the permanent dipole moment of the $(NH_3)_n$ clusters is smaller than 0.3 D, for all clusters from $n = 3$ to $n = 6$. Such a constraint of an almost zero dipole moment is fulfilled by a symmetric cyclic geometry in which all NH_3 units are equivalent.

Additional support for cyclic trimer and tetramer structures is provided by the shape of our measured dissociation profiles. For the dimer, two bands are observed which are separated by $26.5\,cm^{-1}$ and which are assigned to the excitation of the two nonequivalent monomer units. For the trimer and tetramer a comparable splitting has not been observed. In contrast, the absorption bands of these clusters essentially consist of one single line (see Figs. 19b and 20a). This observation strongly suggests cyclic trimer and tetramer geometries, for in symmetric cyclic structures all monomer units are equivalent and correspondingly only one absorption band is expected.

A closer examination, however, reveals some structure in the trimer and tetramer absorption bands. Both show a red-shifted shoulder. The separations are $10.2\,cm^{-1}$ for the trimer and $8.3\,cm^{-1}$ for the tetramer. Since the splitting is nearly the same as that observed in the red dimer absorption band, we were tempted to assign these features to some combination bands within the given ring structure or to another slightly different cyclic geometry. Another possible explanation is based on the recent double resonance experiments by Heijmen et al.[65,150] which reveal that two photons are needed to dissociate the ammonia trimer and tetramer. In addition, they show that the second photon which excites the cluster beyond the dissociation limit is red-shifted by $10\,cm^{-1}$ for the trimer and $17\,cm^{-1}$ for the tetramer. We are able to dissociate the trimers and tetramers with two successively absorbed photons of equal frequencies. Consequently, the respective absorption lines for the two transitions must overlap and the resulting dissociation spectra should be described by the product of two absorption lines separated by 10 and $17\,cm^{-1}$, respectively. In Fig. 22 we have plotted the experimental tetramer data from Fig. 20a together with two Lorentzians $L_1(\tilde{\nu})$ and $L_2(\tilde{\nu})$ (dashed curves) as well as their product $L_1 \cdot L_2$ (solid curve). The red-shifted shoulder is obtained because the lower-frequency line is assumed to be broader. Comparing the data points with the solid curve it becomes clear that the tetramer absorption can be equally well described by the product of two Lorentzians. With the parameters given in the figure caption, a perfect fit is obtained for the tetramer absorption. The separation $\Delta\tilde{\nu} = 12\,cm^{-1}$ between the two lines is somewhat smaller than observed by Heijmen et al.

In contrast to the smaller $(NH_3)_n$ clusters, the $(NH_3)_5$ pentamer dissoci-

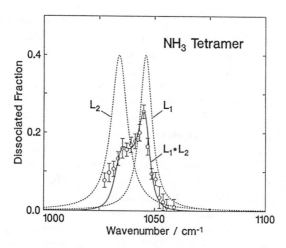

Figure 22. The $(NH_3)_4$ tetramer dissociation profile can be described by the product of two Lorentzians L_1 and L_2 (solid curve). The parameters of the Lorentzians are $\tilde{\nu}_{01} = 1046\,\text{cm}^{-1}$, $\Gamma_1 = 6.6\,\text{cm}^{-1}$, $\tilde{\nu}_{02} = 1034\,\text{cm}^{-1}$, and $\Gamma_2 = 8.8\,\text{cm}^{-1}$.

ation is characterized by an absorption profile which with $\Gamma = 28\,\text{cm}^{-1}$ is remarkably broader (see Fig. 20b). This sudden spectral change must be related to a significant structural change when going from the tetramer to the pentamer geometry. The pentamer structure seems to be much less rigid than the tetramer structure. This allows for additional internal motions which tend to broaden the spectrum. This explanation is also supported by the observation that the dissociation energy of the pentamer is smaller than for the tetramer.

With increasing cluster size a progressive blue shift of the ν_2 absorption is observed. However, all absorptions are found to lie between the absorption of the free, noninverting NH_3 monomer at $950\,\text{cm}^{-1}$ [156] and the band positions observed in liquid and solid ammonia which are between 1059 and $1063\,\text{cm}^{-1}$.[157]

In the cyclic trimer none of the NH_3 molecules has three free H atoms. For this reason we assign the blue dimer band which is very near to the trimer absorption to the excitation of the more strongly perturbed molecules. The red dimer band corresponds to the excitation of the more weakly interacting submolecule or "proton acceptor." This assignment is in agreement with the conception that the ν_2 vibration of the NH_3 molecule with three nonbonded H atoms is certainly less disturbed and should, therefore, show a smaller shift from the ν_2 vibration of the free noninverting monomer at $950\,\text{cm}^{-1}$.

Finally, we want to compare our results with theoretical predictions.

In their recent *ab initio* calculation, Frisch et al.[158] obtained dissociation energies between 700 and 800 cm^{-1} for two different $(NH_3)_2$ dimer geometries. However, neither of the two structures are compatible with the measured dipole moment of 0.74 D.[26,27] Frisch et al. further calculated the shifts of the vibrational frequencies upon dimerization. For the ν_2 vibration in the dimer they obtained two bands shifted by 37 and 46 cm^{-1} from the monomer transition. Thus, the theory predicts two dimer bands at 987 and 996 cm^{-1} which are close to the observed positions. The perhaps best overall agreement with various experimental observations is obtained by Sagarik et al.[159] They predict an asymmetric cyclic equilibrium structure with a nonzero dipole moment of $\mu = 0.48$ D. The binding energy of 1081 cm^{-1} is compatible with the experimental observation that the ammonia dimer can be dissociated by a single CO_2 laser photon at $\tilde{\nu} = 977$ cm^{-1}, if one considers the zero-point vibrational energy of the hydrogen bond and the possibility that the dimers may not be completely cooled in the experiments.

Very recently, the dimer potential of Sagarik et al.[159] has been adapted to cluster calculations by Greer et al.[160] They assumed that the total cluster energy is the sum of the pairwise interactions between the ammonia molecules and determined cluster geometries and binding energies by minimizing the potential energy with respect to molecular positions. The results of these calculations are in almost perfect agreement with our experimental findings. (1) For the ammonia trimer and tetramer Greer et al. find cyclic structures. (2) The dissociation energies are 1871 and 1629 cm^{-1} and, thus, much larger than the energy of a single CO_2 laser photon. (3) For the pentamer, a completely asymmetric structure with many possible orientations of the NH_3 molecules within the cluster is found. This explains why the pentamer absorption profile is much broader than the trimer and tetramer spectra. (4) Finally, Greer et al. calculate a pentamer dissociation energy which is, with 1339 cm^{-1}, considerably smaller than for the trimer and tetramer. The experiments show that the pentamer can even be dissociated by a single photon of frequency $\tilde{\nu} = 1045$ cm^{-1}.

E. Methanol

A typical example of a hydrogen-bonded complex is the methanol cluster $(CH_3OH)_n$. Since methanol plays an important role in solvent chemistry, the investigation of the interactions between the methanol molecules in the cluster is of fundamental interest. The first dissociation experiments on methanol clusters were carried out in Gentry's group by Hoffbauer et al.[54] These investigators observed a broad single-peaked absorption profile which they assigned to the dissociation of methanol dimers. In our study on size-selected methanol clusters, we could show that the dimer spectrum consists of two distinct peaks separated by $25.1\,\mathrm{cm}^{-1}$ and that the broad spectrum observed by Hoffbauer et al. is most probably due to the dissociation of larger methanol clusters. First results of this experiment have already been published.[109]

In our study, the $(CH_3OH)_n$ cluster beam was generated by bubbling Ne or He at 1.5 bar through a reservoir of methanol kept at a constant temperature of $-15°C$. The resulting mixture of 1.9% methanol in Ne(He) is then expanded through a 0.1-mm diameter room-temperature nozzle. After the interaction with the CO_2 laser, which is tuned to excite the ν_8 vibration (C–O stretch) of a monomer unit in the adiabatically cooled complex, the cluster beam is dispersed by scattering it with a secondary He beam and finally detected off-axis by the rotatable mass spectrometer.

Angular distributions of $(CH_3OH)_n$ clusters scattered from He and measured at various mass settings of the quadrupole mass spectrometer are shown in Fig. 23. These measurements provide valuable information about the cluster distribution in the beam as well as fragmentation patterns in the ionizer. As usual, the arrows above the upper curve designate the maximum deflection angles and scattering ranges of the various $(CH_3OH)_n$ clusters as calculated from the Newton diagram. The upper curve with triangular data points has been measured on $m = 31$ amu (CH_2OH^+) where the strongest monomer signal was observed. Several pronounced shoulders near the maximum deflection angles indicate that dimers and trimers are in the beam and that these species also fragment in the ionizer and contribute to the CH_2OH^+ ion signal.

If we tune the mass spectrometer to $m = 33$ amu, where the dimers are most efficiently detected as $(CH_3OH)H^+$ ions, the curve with the circular data points is obtained. Corresponding to the kinematics, the maximum scattering angle for dimers is $\Theta_{\max}^{(2)} = 16.8°$. The steep increase in signal at $\Theta = 11.2°$ is due to the onset of trimer scattering, indicating that we have strong contributions from trimers on the typical dimer mass. At still smaller angles even larger clusters contribute. The curves measured on $m = 65$ amu $[(CH_3OH)_2H^+]$ and $m = 97$ amu $[(CH_3OH)_3H^+]$ reflect the angular dependences of trimers and tetramers, but again include contributions from larger species.

When the detector is set to $\Theta = 13.5°$ and the mass spectrometer tuned to $m = 33$ amu, that is, on the shoulder of the $m = 33$ curve, only dimers are observed. Larger clusters are excluded by the kinematic constraints and lighter species are discriminated against by the mass spectrometer. Thus, with these settings, pure dimer spectroscopy may be performed. Trimers can be studied when the detector is moved to $\Theta = 9.0°$. If the mass filter stays on $m = 33$ amu, a mixture of dimers and trimers will be

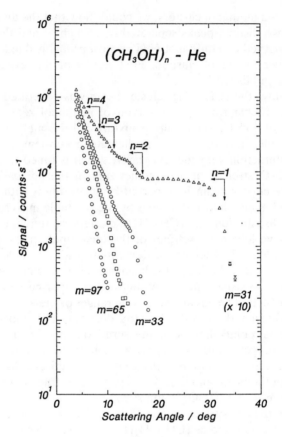

Figure 23. Angular distributions of scattered $(CH_3OH)_n$ clusters measured at various mass settings of the quadrupole. Scattering ranges and maximum deflection angles are indicated in the figure.[109]

detected; if it is tuned to $m = 65$ amu exclusively, trimers are observed. Similarly, tetramers can be investigated at $\Theta = 7.5°$

In the region of the C–O stretch vibration, the dependence of IR photon absorption on laser frequency was investigated as a function of cluster size. The result is shown in Fig. 24. Here we have plotted the depletion signal as measured on $m = 33$ amu for three different detector positions. The laser fluence was adjusted to $F = 82 \text{ mJ} \cdot \text{cm}^{-2}$. The upper spectrum a shows the frequency dependence of the methanol dimer. The middle spectrum b, measured at $\Theta = 9.0°$, is dominated by the dissocation of trimers. Since the mass spectrometer was tuned to $m = 33$ amu, some

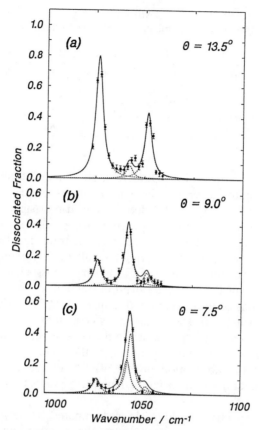

Figure 24. Dissociation spectra for $(CH_3OH)_n$ clusters ($n = 2, 3, 4$) obtained at different scattering angles Θ with the mass spectrometer tuned to $m = 33$ amu. The main cluster sizes in the spectra are dimers (a), trimers (b), and tetramers (c).[109]

dimer contribution must be taken into account. The bottom spectrum c has been obtained at $\Theta = 7.5°$. Again it is a composite spectrum where the main contribution is made by tetramers, but dimers and trimers are also present. The fractional contributions of the different cluster sizes at the various positions are given in Table XI.

The upper spectrum a features the frequency dependence of the methanol dimer absorption. It shows two distinct peaks separated by 25.1 cm^{-1}. The solid curves are Lorentzians fitted to the data points. Their peak positions and full widths at half maximum (FWHM) are collected in Table XII. The small absorption peak between the two dimer lines is assigned

FRIEDRICH HUISKEN

TABLE XI

Methanol Cluster Contributions at Various Detector Settings

Detector Setting	I	II	III	IV	V
Scattering angle Θ (deg)	13.5	9.0	7.5	9.0	5.3
Mass (amu)	33	33	33	65	65
Dimer contribution (%)	84	26	12.5	–	–
Trimer contribution (%)	16	74	41.7	69	[a]
Tetramer contribution (%)	–	–	45.8	31	[a]

[a]Contribution not known.

to a small trimer contribution. Unfortunately, our resolution is not good enough to discriminate completely against trimers. As can be seen in the $m = 65$ curve in Fig. 23, we still detect some trimer signal at $\Theta = 13.5°$. Careful analysis of the angular distributions measured at $m = 33$ and 65 amu yields a 16% trimer contribution to the total signal at $m = 33$ amu and $\Theta = 13.5°$. Anticipating the trimer result (56% dissociation at $\tilde{\nu}_0 = 1042.2\ cm^{-1}$), we thus expect an additional dissociated fraction of $0.16 \cdot 0.56 = 0.09$ at the trimer absorption frequency.

Spectrum b has been obtained with a mixture of 26% dimers and 74% trimers. Comparing with spectrum a it is clear that the outer peaks are due to the dissociation of dimers, while the central peak must be assigned to the trimer. The Lorentzian fit yields $\tilde{\nu}_0 = 1042.2\ cm^{-1}$ and $\Gamma = 4.1\ cm^{-1}$. On the top of the absorption profile, 41.5% of the observed clusters (dimers and trimers) are dissociated. This corresponds to a trimer dissociation of 56%.

The bottom spectrum in Fig. 24c is dominated by the dissociation of $(CH_3OH)_4$ tetramers. Taking into account the measured dimer and trimer frequency dependences, the tetramer absorption profile is determined by a least-squares fit, assuming a Lorentzian line shape function. The result is shown in Fig. 24c by the dashed curve with the largest amplitude. This

TABLE XII

Results of Least-Squares Fits for Methanol Cluster Dissociation

Species	Detector Setting	Peak Position ($\tilde{\nu}_0/cm^{-1}$)	Maximum Attenuation[a]	FWHM (Γ/cm^{-1})
Dimer	I	1026.5	0.80 (0.95)	4.5
		1051.6	0.42 (0.50)	4.2
Trimer	II	1042.2	0.42 (0.56)	4.1
	IV	1043.1	0.62 (0.62)	4.3
Tetramer	III	1044.0	0.40 (0.88)	4.0

[a]The figures in parentheses refer to pure cluster beams as calculated from the contributions given in Table XI.

curve peaks at $\tilde{\nu}_0 = 1044.0 \, cm^{-1}$ and has a width of $\Gamma = 4.0 \, cm^{-1}$. Thus, the tetramer absorption is slightly blue shifted with respect to the trimer line. It should be noted that, besides the tetramer dissociation profile, also the relative contributions of dimers and trimers are determined by the least-squares fit.

A closer look to the spectra in Fig. 24 reveals that the blue dimer line at $\tilde{\nu}_0 = 1051.6 \, cm^{-1}$ is not reproduced very well by the fits in spectra b and c. From the amplitude of the red dimer line, we would expect larger dissociation yields around $1050 \, cm^{-1}$. The observed deviation is explained by an effect which has been discussed in Section II. Around $1050 \, cm^{-1}$ larger methanol clusters are also dissociated. Their dissociation may result in the production of trimers and tetramers with very small perpendicular velocity. These newly formed trimers and tetramers can reach the scattering center and are deflected into the respective scattering ranges. Because of their strong fragmentation, these species will produce a positive signal on $m = 33$ amu and, thus, reduce the dimer depletion signal.[113]

We have also tried to measure a pure trimer spectrum. For this purpose the mass spectrometer was tuned to $m = 65$ amu so that contributions from dimers could be excluded. However, in the trimer scattering range at $\Theta = 9.0°$, the signal was very weak and, therefore, long integration times had to be accepted. The result is shown in Fig. 25a. Comparison with spectrum b of Fig. 24 proves that the trimer dissociation is indeed characterized by a single line. Because of the limited angular resolution, it is possible that some tetramers are contributing and causing the slight blue shift with respect to the trimer line of Fig. 24b.

In order to determine the dissociation spectra of larger methanol clusters, the detector was moved to $\Theta = 5.3°$. With the mass spectrometer tuned to the nominal trimer mass $m = 65$ amu $[(CH_3OH)_2H^+)]$, the spectrum in Fig. 25b has been obtained. From the Newton diagram we expect that $(CH_3OH)_n$ clusters with n between 3 and 6 contribute to the spectrum. The two low-amplitude lines are assigned to the trimer and tetramer, respectively, whereas the two lines with large amplitude are probably due to the dissociation of pentamers and hexamers. Their line positions are $\tilde{\nu}_0 = 1044.9$ and $1049.7 \, cm^{-1}$. Thus, it appears that the absorption lines of the larger methanol clusters become progressively more blue shifted with increasing cluster size. In their very recent investigation on size-selected internally excited methanol clusters carried out in configuration b, Buck and co-workers[99,100,125] found that the hexamer absorption profile consists of two lines at $\tilde{\nu}_0 = 1041.2$ and $1052.5 \, cm^{-1}$. With this knowledge one is tempted to presume that our pentamer line in Fig. 25b is partly due to hexamers as well. However, our data allows no decision as to whether this assignment is correct.

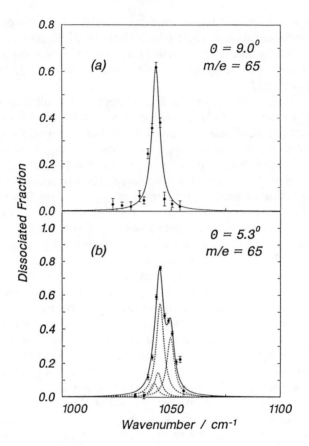

Figure 25. (a) $(CH_3OH)_3$ trimer dissociation spectrum without dimer contributions. (b) Spectrum contains contributions from methanol trimers to hexamers.[113]

In another experiment,[113] we have used He instead of Ne as carrier gas for the methanol cluster beam. Because of the different kinematics, the detector has to be positioned to $\Theta = 8.0°$ if a pure dimer spectrum is to be measured. The result is shown in the lower panel *b* of Fig. 26. For comparison, the corresponding dimer spectrum obtained with Ne as the carrier gas (same as spectrum in Fig. 24*a*) is displayed in the upper panel of the same figure. Comparing the two dimer spectra, it is noticed that the positions of the absorption lines and their amplitude ratio are the same in both spectra but that the linewidths are approximately a factor of 2 larger ($\Gamma = 7.8$ and $8.2 \, \text{cm}^{-1}$) when the dimers are produced in a He expansion.

This behavior can be explained if we assume that the Ne expansion

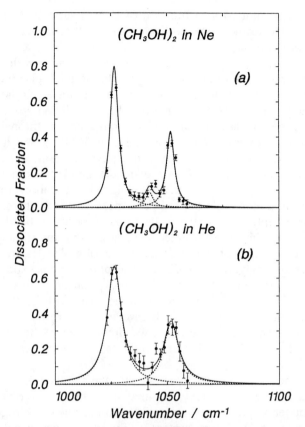

Figure 26. Dissociation spectra of methanol dimers for two different internal temperatures. (*a*) Spectrum reflecting the lower temperatures with Ne used as carrier gas. (*b*) Spectrum with He used as carrier gas.[113]

provides better cooling for the methanol dimers than the He expansion. As a result, dimers produced in He have a higher internal temperature. This then leads to a broadening of the absorption lines since more internal states are populated. Strong support for this interpretation is provided by the dissociation experiments on hot methanol dimers by Buck et al.[99,100] In these experiments the two dimer peaks are so broad that they are hardly separated.

Finally, we have measured the dependence of the dissociation yield on the laser fluence. For the dimer these measurements have been carried out in the top of the red absorption peak at $1027.4 \, \text{cm}^{-1}$ (9μP40 laser line). The trimer fluence dependence has been measured on the 9μP24 laser line at $1043.2 \, \text{cm}^{-1}$. When the logarithm of the nondissociated fraction $1 - P_{diss}$ is plotted as a function of laser fluence, straight lines are obtained.[109] However, from the linear fluence dependence observed for both dimers and trimers, we cannot conclude that the two-level-plus-decay model (compare Eq. 14) is valid and that we are dealing with one-photon processes. Arguments will be given below that in particular the methanol trimer cannot be dissociated upon absorption of a single CO_2 laser photon.

Our study was the first to show that the dissociation spectrum of the methanol dimer consists of two well-separated peaks. In analogy to the results for the ammonia dimer (see Section V.D), we assign the two peaks to the ν_8 excitation of the two nonequivalent monomer units in the hydrogen-bonded complex. With the same arguments as in the case of ammonia, the blue absorption peak is assigned to the more strongly per-turbed proton donor while the red peak must be due to the excitation of the proton acceptor. In contrast, the trimer absorption band consists of one single line. This we interpret as indication that the trimer has a ring structure. In a cyclic configuration all methanol molecules are equivalent and, correspondingly, only one single absorption band is expected in low-resolution experiments like ours. The absorption lines observed for the tetramer and pentamer also indicate cyclic structures for these species. For the hexamer, Buck et al.[99,100] observed two absorption lines separated by $11.3\,cm^{-1}$. This surprising result was interpreted as being due to the excitation of molecules at different positions in a nonplanar hexamer ring structure.

Our findings are in good agreement with the results of molecular beam electric resonance experiments on methanol clusters,[161] which indicate that the dimer is polar and compatible with a linear hydrogen-bonded structure, whereas the larger clusters are found to be consistent with cyclic geometries. Very recently, Buck et al.[100] have calculated the structures of small methanol clusters employing an energy minimization program. These calculations perfectly confirm the structures found experimentally.

In addition, Buck and Schmidt[162] have calculated the splitting of the ν_8 vibration upon dimerization. They determined shifts from the monomer absorption at $1034\,cm^{-1}$ of $+22\,cm^{-1}$ for the proton donor and $-4.1\,cm^{-1}$ for the acceptor. This is in reasonable agreement with our experimental values of $+17.6$ and $-7.5\,cm^{-1}$. Note that the calculation also confirms our assignment of the two dimer lines made earlier. This assignment was based on the assumption that the absorption of the proton donor should be near the trimer absorption line since in the cyclic trimer structure all molecules have their H atom in the hydrogen bond. Additional support for this assignment is supplied by the very recent predissociation experiments on deuterated methanol dimers by LaCosse and Lisy.[163]

Finally, we want to address the question of dissociation energies. Various theoretical estimates of the dimer binding energy yield values between 1950 and 2400 cm^{-1}.[164-166] Thus, a single 1020 cm^{-1} photon would not be sufficient to dissociate the complex even if a zero point energy of approximately 400 cm^{-1} is assumed. On the other hand, the measured linear fluence dependence suggests a one-photon process. However, this argument is not very convincing since it can be shown that also more-photon processes may have a linear power dependence if certain assumptions about the coupling of the excited states to the internal modes are made.[167] In fact, very recent double resonance photodissociation experiments by Reuss and co-workers[167] have shown that cold methanol dimers as produced by seeding in Ne must absorb two photons in order to dissociate. Only warm dimers as, for example, obtained with He as a carrier gas can be dissociated with a single CO_2 laser photon. For the dissociation of the methanol trimer even three photons are needed. The higher dissociation energy of the trimer can be easily understood if a cyclic trimer structure is assumed because at least two bonds must be broken.

VI. CONCLUDING REMARKS

It has been shown that the technique of dispersing the clusters by molecular beam scattering provides an ideal means to perform cluster-specific spectroscopy. In some cases, contributions from smaller cluster species must be accepted; however, the technique is always capable of discriminating against larger clusters which may seriously affect on-beam experiments by their extensive fragmentation in the ionizer.

We have studied the photodissociation of several vdW and hydrogen-bonded complexes. For each system, the new technique could provide important contributions to our understanding of their basic properties, such as their structures, energetics, and dissociation dynamics. In many cases, essential features could be revealed which are as yet unknown. It could be shown, for example, that the $(C_2H_4)_n$ dissociation profiles were practically the same for all investigated cluster sizes $n = 2, 3, 4, 5$. For the benzene trimer a quadratic dependence of the dissociation yield on the laser fluence was observed. This observation suggests that the benzene trimer has a cyclic structure which can only be dissociated if at least two CO_2 laser photons are absorbed. Strong support for this interpretation was supplied by structure calculations. For the benzene tetramer an acyclic structure was found and important conclusions about the dissociation dynamics could be made. In the case of SF_6, trimer and tetramer absorption spectra could be measured for the first time. These measurements provide an ideal test for the dynamical dipole–dipole interaction model.[57,58] Similarly, for ammonia, the dissociation spectra of trimers,

TABLE XIII
Absorption Frequencies for all Investigated Clusters

Molecule	Vibrational Mode	Monomer ($\bar{\nu}_0/cm^{-1}$)	Dimer ($\bar{\nu}_0/cm^{-1}$)	Trimer ($\bar{\nu}_0/cm^{-1}$)	Tetramer ($\bar{\nu}_0/cm^{-1}$)
C_2H_4	ν_7 CH_2 out	949	952.9	953.4	953.7
C_6H_6	ν_{18} CH bend	1038	1037.9	1037.0	1037.5
SF_6	ν_3 bending	948	934.4	935.2	936.1
			955.6	955.1	959.1
				962.0	
NH_3	ν_2 umbrella	950	976.9	1006.1	1036.4
			1003.4	1016.3	1044.7
CH_3OH	ν_8 CO stretch	1034	1026.5	1042.2	1044.0
			1051.6		

tetramers, and pentamers could be determined for the first time. By comparison with earlier low-fluence experiments carried out without size selection, important information about the structure and dissociation energies of these species could be obtained. Finally, in contrast to earlier results, with the cluster-selection technique it was found that the methanol dimer spectrum consists of two well-separated lines which must be assigned to the two nonequivalent subunits in the hydrogen-bonded complex. For the methanol trimer and tetramer single lines are observed, which is in accordance with a cyclic structure of these species. An important result of our studies, the line positions of dimers, trimers, and tetramers, is collected in Table XIII for all investigated systems.

So far, all dissociation experiments with size-selected clusters have been carried out using a CO_2 laser. This laser has the advantage of high power, but the drawback of limited tunability. In addition, only a few molecules can be excited with this laser. Thus it is desirable to carry out similar experiments in another frequency range. For this purpose the molecular beam machine has recently been equipped with a Nd:YAG laser-pumped optical parametric oscillator (OPO).[168,169] With this setup it is possible to carry out the dissociation experiments in the 3-μ region in which many molecules absorb. However, not only will many new molecular clusters be amenable to their experimental investigation, but also some of the systems of the present CO_2 laser study may be reinvestigated in the different frequency range (C_2H_4, C_6H_6, NH_3, CH_3OH). This comparison will be particularly interesting since it will reveal important information about the dependence of the dissociation dynamics on the vibrational mode which is initially excited. First results have already been obtained for the dissociation of $(CH_3OH)_2$ dimers upon excitation of the O–H stretch around 3681 cm^{-1}.[170]

ACKNOWLEDGMENTS

It is a great pleasure for me to thank Dr. T. Pertsch, Mr. M. Stemmler, and Mr. A. de Meijere for their enthusiasm in carrying out the experiments. Many thanks go to Professor H. Pauly, who continuously supported this work. I'm grateful to Professor U. Buck for reading the manuscript and for many fruitful discussions. Finally, I would like to thank Mrs. V. Rosenthal for typing the manuscript. The computer calculations have been carried out at the "Gesellschaft für wissenschaftliche Datenverarbeitung Göttingen" (GDWG).

REFERENCES

1. *Ber. Bunsenges. Phys. Chem.* **88**, (1984).

2. *Surface Science* **156**, (1985).

3. *Z. Phys. D* **3**, (1986).

4. *Chem. Rev.* **86**(3), (1986).

5. *J. Phys. Chem.* **91**, (1987).

6. J. Jortner, *Ber. Bunsenges. Phys. Chem.* **88**, 188 (1984).

7. J. B. Anderson, in *Molecular Beams and Low Density Gasdynamics*, P. P. Wegener, Ed., Dekker, New York, 1974, pp. 1–91.

8. D. R. Miller, in *Atomic and Molecular Beam Methods*, G. Scoles, Ed., Oxford, New York, 1988, pp. 14–53.

9. O. F. Hagena, in *Molecular Beams and Low Density Gasdynamics*, P. P. Wegener, Ed., Dekker, New York, 1974, pp. 93–181.

10. S. B. Ryali and J. B. Fenn, *Ber. Bunsenges. Phys. Chem.* **88**, 245 (1984).

11. M. Kappes and S. Leutwyler, in *Atomic and Molecular Beam Methods*, G. Scoles, Ed., Oxford, New York, 1988, pp. 380–415.

12. R. H. Lipson, P. E. LaRocque, and B. P. Stoicheff, *J. Chem. Phys.* **82**, 4470 (1985).

13. P. E. LaRocque, R. H. Lipson, P. R. Herman, and B. P. Stoicheff, *J. Chem. Phys.* **84**, 6627 (1986).

14. W. Sharfin, K. E. Johnson, L. Wharton, and D. H. Levy, *J. Chem. Phys.* **71**, 1292 (1979).

15. J. E. Kenny, K. E. Johnson, W. Sharfin, and D. H. Levy, *J. Chem. Phys.* **72**, 1109 (1980).

16. K. E. Johnson, W. Sharfin, and D. H. Levy, *J. Chem. Phys.* **74**, 163 (1981).

17. U. Even, A. Amirav, S. Leutwyler, M. J. Ondrechen, Z. Berkovitch-Yellin, and J. Jortner, *Faraday Discuss. Chem. Soc.* **73**, 153 (1982).

18. S. Leutwyler and J. Jortner, *J. Phys. Chem.* **91**, 5558 (1987).

19. P. R. R. Langridge-Smith, D. V. Brumbaugh, C. A. Haynam, and D. H. Levy, *J. Phys. Chem.* **85**, 3742 (1981).

20. J. B. Hopkins, D. E. Powers, and R. E. Smalley, *J. Phys. Chem.* **85**, 3739 (1981).

21. A. Kiermeier, B. Ernstberger, H. J. Neusser, and E. W. Schlag, *Z. Phys.* **D10**, 311 (1988).

22. B. J. Howard, in *Structure and Dynamics of Weakly Bound Molecular Complexes*, A. Weber, Ed., Reidel, Dordrecht, 1987, pp. 69–84.

23. M. A. Walsh, T. H. England, T. R. Dyke, and B. J. Howard, *Chem. Phys. Lett.* **142**, 265 (1987).

24. D. G. Prichard, R. N. Nandi, J. S. Muenter, and B. J. Howard, *J. Chem. Phys.* **89**, 1245 (1988).

25. A. S. Pine, W. J. Lafferty, and B. J. Howard, *J. Chem. Phys.* **81**, 2939 (1984).

26. G. T. Fraser, D. D. Nelson, A. Charo, and W. Klemperer, *J. Chem. Phys.* **82**, 2535 (1985).

27. D. D. Nelson, G. T. Fraser, and W. Klemperer, *J. Chem. Phys.* **83**, 6201 (1985).

28. D. Ray, R. L. Robinson, D.-H. Gwo, and R. J. Saykally, *J. Chem. Phys.* **84**, 1171 (1986).

29. K. L. Busarow, G. A. Blake, K. B. Laughlin, R. C. Cohen, Y. T. Lee, and R. J. Saykally, *J. Chem. Phys.* **89**, 1268 (1988).

30. H. P. Godfried and I. F. Silvera, *Phys. Rev.* **A27**, 3008 (1983).

31. H. P. Godfried and I. F. Silvera, *Phys. Rev.* **A27**, 3019 (1983).

32. M. Maroncelli, G. A. Hopkins, J. W. Nibler, and T. R. Dyke, *J. Chem. Phys.* **83**, 2129 (1985).

33. G. A. Pubanz, M. Maroncelli, and J. W. Nibler, *Chem. Phys. Lett.* **120**, 313 (1985).

34. H. D. Barth and F. Huisken, *J. Chem. Phys.* **87**, 2549 (1987).

35. S. Wuelfert, D. Herren, and S. Leutwyler, *J. Chem. Phys.* **86**, 3751 (1987).

36. H. D. Barth, C. Jackschath, T. Pertsch, and F. Huisken, *Appl. Phys.* **B45**, 205 (1988).

37. R. Beck and J. W. Nibler, *Chem. Phys. Lett.* **148**, 271 (1988).

38. A. Furlan, S. Wülfert, and S. Leutwyler, *Chem. Phys. Lett.* **153**, 291 (1988).

39. W. Klemperer, *Ber. Bunsenges. Phys. Chem.* **78**, 128 (1974).

40. T. E. Gough, R. E. Miller, and G. Scoles, *J. Chem. Phys.* **69**, 1588 (1978).

41. T. E. Gough, R. E. Miller, and G. Scoles, *J. Phys. Chem.* **85**, 4041 (1981).

42. T. E. Gough, D. G. Knight, P. A. Rowntree, and G. Scoles, *J. Phys. Chem.* **90**, 4026 (1986).

43. J. M. Lisy, A. Tramer, M. F. Vernon, and Y. T. Lee, *J. Chem. Phys.* **75**, 4733 (1981).

44. M. F. Vernon, J. M. Lisy, D. J. Krajnovich, A. Tramer, H.-S. Kwok, Y. R. Shen, and Y. T. Lee, *Faraday Discuss. Chem. Soc.* **73**, 387 (1982).

45. J. M. Lisy, M. F. Vernon, A. Tramer, H.-S. Kwok, D. J. Krajnovich, Y. R. Shen, and Y. T. Lee, in *Laser Spectroscopy V*, A. R. W. McKellar, T. Oka, and B. P. Stoicheff, Eds., Springer Ser. Opt. Sci. 30, Springer, Berlin, 1981, pp. 324–332.

46. M. F. Vernon, D. J. Krajnovich, H. S. Kwok, J. M. Lisy, Y. R. Shen, and Y. T. Lee, *J. Chem. Phys.* **77**, 47 (1982).

47. R. H. Page, J. G. Frey, Y. R. Shen, and Y. T. Lee, *Chem. Phys. Lett.* **106**, 373 (1984).

48. M. F. Vernon, J. M. Lisy, H. S. Kwok, D. J. Krajnovich, A. Tramer, Y. R. Shen, and Y. T. Lee, *J. Phys. Chem.* **85**, 3327 (1981).

49. M. P. Casassa, D. S. Bomse, and K. C. Janda, *J. Chem. Phys.* **74**, 5044 (1981).

50. M. P. Casassa, C. M. Western, and K. C. Janda, *J. Chem. Phys.* **81**, 4950 (1984).

51. M. A. Hofflauer, K. Liu, C. F. Giese, and W. R. Gentry, *J. Chem. Phys.* **78**, 5567 (1983).

52. M. A. Hoffbauer, K. Liu, C. F. Giese, and W. R. Gentry, *J. Phys. Chem.* **87**, 2096 (1983).

53. M. J. Howard, S. Burdenski, C. F. Giese, and W. R. Gentry, *J. Chem. Phys.* **80**, 4137 (1984).

54. M. A. Hoffbauer, C. F. Giese, and W. R. Gentry, *J. Phys. Chem.* **88**, 181 (1984).

55. R. D. Johnson, S. Burdenski, M. A. Hoffbauer, C. F. Giese, and W. R. Gentry, *J. Chem. Phys.* **84**, 2624 (1986).

56. J. Geraedts, S. Setiadi, S. Stolte, and J. Reuss, *Chem. Phys. Lett.* **78**, 277 (1981).

57. J. Geraedts, M. Waayer, S. Stolte, and J. Reuss, *Faraday Discuss. Chem. Soc.* **73**, 375 (1982).

58. J. Geraedts, S. Stolte, and J. Reuss, *Z. Phys.* **A304**, 167 (1982).

59. M. Snels and R. Fantoni, *Chem. Phys.* **109**, 67 (1986).

60. B. Heijmen, A. Bizarri, S. Stolte, and J. Reuss, *Chem. Phys.* **132**, 331 (1989).

61. J. Geraedts, M. Snels, S. Stolte, and J. Reuss, *Chem. Phys. Lett.* **106**, 377 (1984).

62. M. Snels, R. Fantoni, M. Zen, S. Stolte, and J. Reuss, *Chem. Phys. Lett.* **124**, 1 (1986).

63. B. Heijmen, C. Liedenbaum, S. Stolte, and J. Reuss, *Z. Phys.* **D6**, 199 (1987).

64. M. Snels, R. Fantoni, R. Sanders, and W. L. Meerts, *Chem. Phys.* **115**, 79 (1987).

65. B. Heijmen, A. Bizzarri, S. Stolte, and J. Reuss, *Chem. Phys.* **126**, 201 (1988).

66. C. Liedenbaum, B. Heijmen, S. Stolte, and J. Reuss, *Z. Phys.* **D11**, 175 (1989).

67. G. Fischer, R. E. Miller, and R. O. Watts, *Chem. Phys.* **80**, 147 (1983).

68. G. Fischer, R. E. Miller, and R. O. Watts, *J. Phys. Chem.* **88**, 1120 (1984).

69. G. Fischer, R. E. Miller, P. F. Vohralik, and R. O. Watts, *J. Chem. Phys.* **83**, 1471 (1985).

70. R. E. Miller, P. F. Vohralik, and R. O. Watts, *J. Chem. Phys.* **80**, 5453 (1984).

71. R. E. Miller, R. O. Watts, and A. Ding, *Chem. Phys.* **83**, 155 (1984).

72. R. E. Miller and R. O. Watts, *Chem. Phys. Lett.* **105**, 409 (1984).

73. K. W. Jucks, Z. S. Huang, D. Dayton, R. E. Miller, and W. J. Lafferty, *J. Chem. Phys.* **86**, 4341 (1987).

74. K. W. Jucks, Z. S. Huang, R. E. Miller, G. T. Fraser, A. S. Pine, and W. J. Lafferty, *J. Chem. Phys.* **88**, 2185 (1988).

75. D. F. Coker, R. E. Miller, and R. O. Watts, *J. Chem. Phys.* **82**, 3554 (1985).

76. Z. S. Huang, K. W. Jucks, and R. E. Miller, *J. Chem. Phys.* **85**, 3338 (1986).

77. W. L. Liu, K. Kolenbrander, and J. M. Lisy, *Chem. Phys. Lett.* **112**, 585 (1984).

78. D. W. Michael and J. M. Lisy, *J. Chem. Phys.* **85**, 2528 (1986).

79. K. D. Kolenbrander and J. M. Lisy, *J. Chem. Phys.* **85**, 6227 (1986).

80. D. S. Bomse, J. B. Cross, and J. J. Valentini, *J. Chem. Phys.* **78**, 7175 (1983).

81. K. G. H. Baldwin and R. O. Watts, *Chem. Phys. Lett.* **129**, 237 (1986).

82. K. G. H. Baldwin and R. O. Watts, *J. Chem. Phys.* **87**, 873 (1987).

83. J.-M. Philippoz, J.-M. Zellweger, H. van den Bergh, and R. Monot, *J. Phys. Chem.* **88**, 3936 (1984).

84. B. B. Brady, G. B. Spector, and G. W. Flynn, *J. Phys. Chem.* **90**, 83 (1986).

85. I. Nishiyama and I. Hanazaki, *Chem. Phys. Lett.* **117**, 99 (1985).

86. A. J. Stace, D. M. Bernard, J. J. Crooks, and K. L. Reid, *Mol. Phys.* **60**, 671 (1987).

87. G. Fischer, *Chem. Phys. Lett.* **139**, 316 (1987).

88. R. L. DeLeon and J. S. Muenter, *J. Chem. Phys.* **80**, 6092 (1984).

89. A. S. Pine and G. T. Fraser, *J. Chem. Phys.* **89**, 100 (1988).

90. G. T. Fraser, R. D. Suenram, F. J. Lovas, A. S. Pine, J. T. Hougen, W. J. Lafferty, and J. S. Muenter, *J. Chem. Phys.* **89**, 6028 (1988).

91. K. C. Janda, *Adv. Chem. Phys.* **60**, 201 (1985).

92. R. E. Miller, *J. Phys. Chem.* **90**, 3301 (1986).

93. U. Buck, *J. Phys. Chem.* **92**, 1023 (1988).

94. U. Buck and H. Meyer, *Phys. Rev. Lett.* **52**, 109 (1984).

95. F. Huisken, H. Meyer, C. Lauenstein, R. Sroka, and U. Buck, *J. Chem. Phys.* **84**, 1042 (1986).

96. U. Buck, F. Huisken, C. Lauenstein, T. Pertsch, and R. Sroka, in *Structure and Dynamics of Weakly Bound Molecular Complexes*, A. Weber, Ed., Reidel, Dordrecht, 1987, p. 477.

97. U. Buck, F. Huisken, C. Lauenstein, H. Meyer, and R. Sroka, *J. Chem. Phys.* **87**, 6276 (1987).

98. U. Buck, C. Lauenstein, A. Rudolph, B. Heijmen, S. Stolte, and J. Reuss, *Chem. Phys. Lett.* **144**, 396 (1988).

99. U. Buck, X. J. Gu, C. Lauenstein, and A. Rudolph, *J. Phys. Chem.* **92**, 5561 (1988).

100. U. Buck, X. J. Gu, C. Lauenstein, and A. Rudolph, *J. Chem. Phys.* **92**, 6017 (1990).

101. U. Buck, X. J. Gu, M. Hobein, and C. Lauenstein, *Chem. Phys. Lett.* **163**, 455 (1989).

102. U. Buck, X. J. Gu, R. Krohne, and C. Lauenstein, *Chem. Phys. Lett.* **174**, 247 (1990).

103. U. Buck, X. J. Gu, R. Krohne, C. Lauenstein, and H. Linnartz, *J. Chem. Phys.* **94**, 23 (1991).

104. U. Buck, in *Dynamics of Polyatomic van der Waals Complexes*, N. Halberstadt and K. C. Janda, Eds., NATO ASI Series, Plenum, London, 1991.

105. F. Huisken and T. Pertsch, *J. Chem. Phys.* **86**, 106 (1987).

106. A. de Meijere and F. Huisken, *J. Chem. Phys.* **92**, 5826 (1990).

107. F. Huisken and M. Stemmler, *Chem. Phys.* **132**, 351 (1989).

108. F. Huisken and T. Pertsch, *Chem. Phys.* **126**, 213 (1988).

109. F. Huisken and M. Stemmler, *Chem. Phys. Lett.* **144**, 391 (1988).

110. H. Pauly and J. P. Toennies, in *Advances in Atomic and Molecular Physics*, Vol. 1, D. R. Bates and I. Estermann, Eds., Academic Press, New York, 1965, pp. 195–334.

111. U. Buck, in *Atomic and Molecular Beam Methods*, G. Scoles, Ed., Oxford, New York, 1988, pp. 449–471.

112. U. Buck and H. Meyer, *J. Chem. Phys.* **84**, 4854 (1986).

113. F. Huisken and M. Stemmler, to be published.

114. Y. T. Lee, J. D. McDonald, P. R. LeBreton, and D. R. Herschbach, *Rev. Sci. Instrum.* **40**, 1402 (1969).

115. F. Huisken and T. Pertsch, *Rev. Sci. Instrum.* **58**, 1038 (1987).

116. F. Huisken and T. Pertsch, unpublished results.

117. U. Buck, C. Lauenstein, H. Meyer, and R. Sroka, *J. Chem. Phys.* **92**, 1916 (1988).

118. D.J. Krajnovich, Ph.D. thesis, University of California, Berkeley, 1983.

119. J. Geraedts, Ph.D. thesis, University of Nijmegen, Nijmegen, 1983.

120. J. A. Beswick and J. Jortner, *Adv. Chem. Phys.* **47**(I), 363 (1981).

121. G. E. Ewing, *J. Chem. Phys.* **72**, 2096 (1980).

122. G. E. Ewing, *Chem. Phys.* **63**, 411 (1981).

123. G. E. Ewing, *Faraday Discuss. Chem. Soc.* **73**, 325 (1982).

124. A. C. Peet, *Chem. Phys. Lett.* **132**, 32 (1986).

125. A. Rudolph, Ph.D. thesis, University of Göttingen, Göttingen, 1989.

126. U. Buck, C. Lauenstein, and A. Rudolph, *Z. Phys. D.* **18**, 181 (1991).

127. A. C. Peet, D. C. Clary, and J. M. Hutson, *Chem. Phys. Lett.* **125**, 477 (1986).

128. A. C. Peet, D. C. Clary, and J. M. Hutson, *Faraday Discuss. Chem. Soc.* **82**, 327 (1986).

129. S. R. Hair, J. A. Beswick, and K. C. Janda, *J. Chem. Phys.* **89**, 3970 (1988).

130. T. Nakanaga, S. Kondo, and S. Saëki, *J. Chem. Phys.* **70**, 2471 (1979).

131. R. Ahlrichs, S. Brode, U. Buck, M. de Kieviet, C. Lauenstein, A. Rudolph, and B. Schmidt, *Z. Phys. D.* **15**, 341 (1990).

132. K. C. Janda, J. C. Hemminger, J. S. Winn, S. E. Novick, S. J. Harris, and W. Klemperer, *J. Chem. Phys.* **63**, 1419 (1975).

133. K. S. Law, M. Schauer, and E. R. Bernstein, *J. Chem. Phys.* **81**, 4871 (1984).

134. K. O. Börnsen, H. L. Selzle, and E. W. Schlag, *J. Chem. Phys.* **85**, 1726 (1986).

135. M. Schauer and E. R. Bernstein, *J. Chem. Phys.* **82**, 3722 (1985).

136. D. E. Williams, *Acta Cryst.* **A36**, 715 (1980).

137. B. W. van de Waal, *Chem. Phys. Lett.* **123**, 69 (1986).

138. J. Pawliszyn, M. M. Szczęśniak, and S. Scheiner, *J. Phys. Chem.* **88**, 1726 (1984).

139. P. Čársky, H. L. Selzle, and E. W. Schlag, *Chem. Phys.* **125**, 165 (1988).

140. K. O. Börnsen, S. H. Lin, H. L. Selzle, and E. W. Schlag, *J. Chem. Phys.* **90**, 1299 (1989).

141. T. E. Gough, D. G. Knight, and G. Scoles, *Chem. Phys. Lett.* **97**, 155 (1983).

142. T. E. Gough, M. Mengel, P. A. Rowntree, and G. Scoles, *J. Chem. Phys.* **83**, 4958 (1985).

143. G. Delacrétaz, J.-D. Ganière, P. Melinon, R. Monot, R. Rechsteiner, L. Wöste, H. van den Bergh, and J.-M. Zellweger, in *Proceedings of Thirteenth International Symposium on Rarefied Gas Dynamics*, O. M. Belotserkovskii, M. N. Kogan, S. S. Kutateladze, and A. K. Rebrov, Eds., Plenum Press, New York, 1985, p. 1173.

144. M. Snels and J. Reuss, *Chem. Phys. Lett.* **140**, 543 (1987).

145. O. Echt, A. Reyes Flotte, M. Knapp, K. Sattler, and E. Recknagel, *Ber. Bunsenges. Phys. Chem.* **86**, 860 (1982).

146. C. Brodbeck, I. Rossi, E. Strapelias, and J.-P. Bouanich, *Chem. Phys.* **54**, 1 (1980).

147. P. A. Schulz, Aa. S. Sudbø, E.R. Grant, Y. R. Shen, and Y. T. Lee, *J. Chem. Phys.* **72**, 4985 (1980).

148. J. Reuss, private communication.

149. C. Liedenbaum, Ph.D. thesis, University of Nijmegen, Nijmegen, 1989.

150. B. Heijmen, Ph.D. thesis, University of Nijmegen, Nijmegen, 1988.

151. K. Fox and W. B. Person, *J. Chem. Phys.* **64**, 5218 (1976).

152. K. D. Cook and J. W. Taylor, *Intern. J. Mass. Spectrom. Ion Phys.* **30**, 345 (1979), and references therein.

153. S. T. Ceyer, P. W. Tiedemann, B. H. Mahan, and Y. T. Lee, *J. Chem. Phys.* **70**, 14 (1979).

154. U. Buck and C. Lauenstein, *J. Chem. Phys.* **92**, 4250 (1990).

155. J. A. Odutola, T. R. Dyke, B. J. Howard, and J. S. Muenter *J. Chem. Phys.* **70**, 4884 (1979).

156. G. Herzberg, *Molecular Spectra and Molecular Structure* Vol. II, Van Nostrand, New York, 1945.

157. A. Bromberg, S. Kimel, and A. Ron, *Chem. Phys. Lett.* **46**, 262 (1977), and references therein.

158. M. J. Frisch, J. A. Pople, and J. E. DelBene, *J. Phys. Chem.* **89**, 3664 (1985).

159. K. P. Sagarik, R. Ahlrichs, and S. Brode, *Mol. Phys.* **57**, 1247 (1986).

160. J. C. Greer, R. Ahlrichs, and I. V. Hertel, *Chem. Phys.* **133**, 191 (1989).

161. J. A. Odutola, R. Viswanathan, and T. R. Dyke, *J. Am. Chem. Soc.* **101**, 4787 (1979).

162. U. Buck and B. Schmidt, *J. Mol. Liquids* **46**, 181 (1990).

163. J. P. LaCosse and J. M. Lisy, *J. Phys. Chem.* **94**, 4398 (1990).

164. J. E. del Bene, *J. Chem. Phys.* **55**, 4633 (1971).

165. W. L. Jorgensen, *J. Phys. Chem.* **90**, 1276 (1986).

166. G. Pálinkás, E. Hawlicka, and K. Heinzinger, *J. Phys. Chem.* **91**, 4334 (1987).

167. A. Bizzarri, S. Stolte, J. Reuss, J. G. C. M. van Duijneveldt-van de Rijdt, and F. B. van Duijneveldt, *Chem. Phys.* **143**, 423 (1990).

168. S. J. Brosnan and R. L. Byer, *IEEE J. Quantum Electron.* **QE-15**, 415 (1979).

169. D. W. Michael, K. Kolenbrander, and J. M. Lisy, *Rev. Sci. Instrum.* **57**, 1210 (1986).

170. F. Huisken, A. Kulcke, C. Laush, and J. M. Lisy, *J. Chem. Phys.* in press (1991).

THE DYNAMICS OF TRIPLET EXCITONS IN MIXED MOLECULAR CRYSTALS

ROSS BROWN and PHILEMON KOTTIS

*Centre de Physique Moléculaire Optique et Hertzienne,
Université de Bordeaux I, Talence, France*

CONTENTS

Advances in Chemical Physics, Volume LXXXI, Edited by I. Prigogine and Stuart A. Rice.
ISBN 0-471-54570-8   © 1992 John Wiley & Sons, Inc.

INTRODUCTION

The general theory of the dynamics of electronic excitations in disordered molecular solids has been widely developed during the last 25 years. Silbey[1] has reviewed much of this work. A common feature of most approaches is the tight-binding electronic Hamiltonian,

$$H_e = \sum_i E_i B_i^+ B_i + \sum_{i,j} V_{ij} B_i^+ B_j$$

where E_i is the transition energy of an excited state created on site i by the operator B_i^+ and V_{ij} is the exchange integral between sites i and j. Disorder appears either in the energies E_i, diagonal disorder, or in the couplings V_{ij}, off-diagonal disorder.

Neither Bloch waves nor pure-site excitations are stationary states of H_e: The exchange coupling mixes site states, while disorder in either part of H_e causes scattering between Bloch waves. When disorder is sufficiently strong, interference of the scattered waves causes Anderson localization[2] of the electronic excitation on groups of sites which happen to be resonant: $|E_i - E_j| \lesssim |V_{ij}|$. Discussion of transport properties commonly tackles disorder by perturbation of one or other limit wavefunction.

Besides disorder, coupling to phonons is a further cause of dephasing of Bloch waves. It was introduced originally in a semiclassical way, through stochastic perturbations $\delta E_i(t)$ and $\delta V_{ij}(t)$ of the tight-binding Hamiltonian, representing fluctuations due to thermal agitation.[3] Subsequent work, by introducing quantitized phonon modes, allowed for asymmetrical relaxation toward the eigenstates of lowest energy.

This chapter differs from the work described above. It is not a general theory of exciton dynamics, and the tools used below are not new. What we believe is new, is an attempt at applying them to a nonphenomenological discussion of triplet excitons in widely studied, yet we suggest misunderstood, isotopically mixed crystals of naphthalene. Mixed aromatic crystals at low temperatures are interesting for a first-principles approach because much more is known about their structure, electronic states, intermolecular forces, exciton–phonon coupling, and so on than in say glasses. They are thus useful systems for testing a variety of concepts like coherence, incoherence, the Anderson transition, hopping transport, and fractals. This approach may help to bridge the gap between the more formal, general work mentioned above, and phenomenological application of theory, which may lead to wrong interpretations.

A distinction with much earlier work is that the discussion will be based on the eigenbasis of the tight-binding Hamiltonian, rather than perturbation of either Bloch waves or single site states. The density matrix at equilibrium of the electronic excited states is diagonal in the eigenbasis of H_e, making it the appropriate choice for discussing relaxation. However, connections with other approaches will be pointed out. Of course, there is no analytical expression for the eigenstates of the disordered Hamiltonian, so we shall have to resort to numerical diagonalization. This is a nuisance, but avoids the largely unknown approximations inherent in mean-field methods for disordered media.[4]

Nonphenomenological modelling contributes to the ten-year-old debate on the wavelike or the hopping character of exciton transport in disordered molecular solids. This originally was set as a dichotomy, percolation, or the Anderson transition, but as we shall see, the problem, which remains open, needs restating and careful numerical analysis. It seems useful to return to this problem, because it turns out that triplet excitons on deep traps in mixed crystals of naphthalene are a case of extreme localization. Evidence of coherent processes, examined here, is thus surprising but important for other strongly disordered systems, such as glasses, where hopping, percolation, and fractals are the present paradigms: Significant "coherent" processes may occur in real systems through the overlap of the exponential tails of the localized, but not point-like, states.

We shall develop the subject as follows: Section I discusses the experimental context of this work and earlier interpretations of triplet energy transfer in mixed aromatic crystals. As it is clear that temperature has a strong influence on the results, we have developed a description based on the total Hamiltonian of the sample, including exciton–phonon coupling, and on the master equation for the electronic density matrix.

In Section II we therefore discuss the different terms in the total Hamiltonian of a mixed crystal needed for a nonphenomenological description of transport: The electronic states, the phonons of the crystal, and coupling between excitons and phonons.

In Section III we provisionally accept the hopping model of exciton transport and a simple Golden Rule approach to the calculation of the temperature dependence of the rate of hopping assisted by various mechanisms of exciton–phonon coupling. Simulations of hopping transport with these dependences are compared with the experimental thermal threshold behavior at 5 K. The threshold is due to coupling to two optical phonons. However the temperature-independent results below the threshold are typical of the absence of diffusion. Moreover, samples at different guest concentrations cross over to different limits at different temperatures.

In order to investigate this and to provide a unified description of the results we need the electronic density matrix and its pilot equation, introduced in Section IV. This section begins with a simple discussion of the meaning of coherence and its dependence on the choice of representation. A simple perturbative calculation shows how laser excitation builds up coherence. Derivation of the pilot equation is sketched for two cases: The eigenbasis of the electronic Hamiltonian and the site basis. The master equation approximation, with vanishing memory, corresponding to weak coupling to a broad heat bath, is discussed. This also shows the origin of memory terms in generalized master equations: elimination of variables, either bath variables or coherences.

Section V draws these results together in a unified description of the hopping and "coherent" components of transport of triplet excitons below 20 K, using the overlap of the eigenstates of H_e. Section VI gives an estimate of the coherent component, based on the exact guest eigenstates of mixed crystals of naphthalene. In our opinion the current assumption of two-dimensional transport must be dropped, because the detailed numerical analysis of the two-dimensional models leads to unbelievably small couplings in the hopping model or amplitudes in the diffusionless model.

Qualitatively, however, the change from temperature independence to thermally activated transport is due to the freezing out of diffusion, revealing "coherent" transport through the overlap of the exponential tails of the wavefunctions. Suggestions are made for further investigations.

Section VII returns to the starting point, hopping transport, and to the currently popular phenomenological models of hopping on fractals, used to explain nonexponential relaxation. The direct numerical simulation used here enables us to include the real intermolecular forces explicitly and to take account of the correlation between the local geometry in a disordered solid and the distribution of hopping times. It shows up the limitations of phenomenological application of hopping on fractals with step-function interactions: Introduction of a step-function interaction enables one to cast the hopping transport problem in terms of percolation and the special laws of hopping on percolation clusters, which are known to be fractals. This was a major step forward in understanding the importance of restricted geometry on hopping transport. But this representation fails to describe the full dynamics with distance-dependent hopping rates. This failure is explained and corrected by introducing hopping on a hierarchical structure, an ultrametric space. Such simulations also underline how the timing of excitation and observation in real experiments smooths out the effects of disorder.

Section VIII is a restatement of the main conclusions of the preceding sections.

I.   THE EXPERIMENTAL PROBLEM

The dynamics of triplet excitons in mixed aromatic crystals continues to attract interest, because they have a special kind of disorder: Contrary to glasses, where the intermolecular distances and orientations fluctuate widely, with consequent, uncontrolled variations of the couplings, the lattice structure of mixed crystals generally produces discrete disorder. Moreover, the components often may be chosen with similar sizes and electronic structures, with the important result that substitution of host molecules by guests is random, practically the only situation for which the theory can be developed. Isotopically mixed crystals fulfill this condition very well. In these systems, part of the molecules have some or all of their hydrogen atoms replaced by deuterium isotopes. Replacement of one atom raises the electronic transition energies by about $10\,cm^{-1}$ $(10^{-3}\,eV)$. This introduces "diagonal" disorder in the site Hamiltonian, or fluctuations in the site transition energies, which take the values E_{Host} on hosts and E_{Guest} on guest sites. Intermolecular couplings are only very slighty affected, as is shown by the nearly equal bandwidths of the pure host and guest crystals.[5] We thus have a nearly perfect case of diagonal disorder. As a rule, deuterated host crystals are used, because at low temperatures, $k_B T \ll E_H - E_G$, excitations are localized in the guest system. Any transport is then due to guest–guest interactions. Moreover, the spectral separation of the hosts and the guests enables selective, laser excitation of the guests, producing a well-defined initial state. Finally, triplet lifetimes of aromatics can be very long, up to 10 s, a convenience for temporal measurements.

Transport in the guest states can be revealed in two main ways: Spectral diffusion and delayed fluorescence. In the first case, emission from traces of an initially nonexcited acceptor species is a sign of transport on length scales of the order of the mean distance between the acceptor molecules. The acceptor may be a chemically different molecule or an aggregate of the main, "donor" guest species. It is convenient to choose deep acceptors to reduce back transfer. This method can be applied to the singlet and the triplet states. Figure 1.1 shows the principle of the method and Fig. 1.2 shows some typical results for triplet and singlet excitons in benzene and naphthalene.

Transport in the triplet state also may be detected by the delayed fluorescence due to the fusion of excitons which though initially far apart, may meet after transport. Delayed fluorescence has the same spectral composition as prompt fluorescence, but a much slower decay. It was observed first in rigid solutions[6] and then in crystals. Its bimolecular origin is shown by a quadratic dependence of the intensity on the incident flux.

Sternlicht, Nieman, and Robinson[7] made the first low-temperature study of delayed fluorescence, in mixed crystals of benzene C_6H_6 in C_6D_6. Colson and Robinson[8] followed by Colson and his colleagues[9,10] were among the first workers to study transport in mixed crystals by sensitized luminescence.

Kopelman and his team made a study of sensitized luminescence of 2-methylnaphthalene fed from naphthalene-H_8 in a naphthalene-D_8 host crystal.[11–13] Figure 1.3 shows the energy levels of these crystals. The main features of the triplet state are a host band with Davydov splitting $\sim 10 \, \text{cm}^{-1}$, due to exchange coupling with the four nearest inequivalent neighbors of each site[5,14,15] and a narrow band of localized states for each guest species. The acceptor band is essentially a single line, because most acceptors are isolated from one another at the trace concentrations discussed here. The donor band is structured, Fig. 1.3b, by exchange coupling in random clusters of nearest-neighbour donors.[16–18]

At low donor concentrations the acceptor emission has some low value, owing to direct excitation of the acceptors in these experiments and to direct donor to acceptor transfer from donors which happen to be near

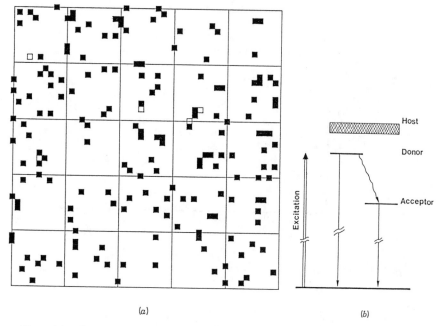

(a)   (b)

Figure 1.1.   Detection of transport of excitons by sensitized luminescence. (*a*) Simulation of the random distribution of donors (■, concentration 7%) and acceptors (□, 0.2%) on a host square lattice (grid lines every 10 units). Strong emission from the acceptors after excitation of the donors shows transport in the donor states. Note that random clustering produces a spread of the relaxation rates. (b) Sensitized luminescence at low temperatures, following selective excitation, is a clear sign of transport. Indirect excitation, or higher temperatures, which allow detrapping to the host band, lead to emission from the acceptors, complicating the interpretation in this case.

enough to an acceptor molecule. As the donor concentration is increased, the acceptor emission increases. This must be due to transport in the donor guest band. The usual measure of the efficiency of donor–donor transport is the yield of acceptor emission,

$$r = I_A/(I_A + I_D) \qquad (1.1)$$

where $I_A(I_D)$ is the steady-state intensity of the acceptor (donor) emission. Kopelman interpreted these results as reflecting the crossing of a percolation threshold.

Percolation may be defined as the long-range propagation of a signal in a set of randomly distributed sites with short-range interactions, when the concentration of the active species is sufficiently high.[19-22] This hap-

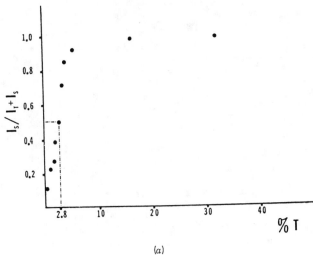

(a)

Figure 1.2. (a).

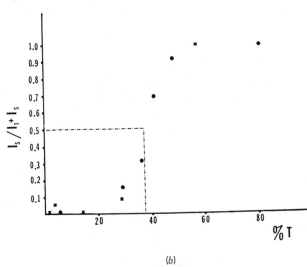

(b)

Figure 1.2 (b).

pens because the size of the clusters of connected sites diverges as the concentration approaches a critical value, the percolation threshold. Below the threshold, the size of the clusters grows with increasing concentration, but remains finite. The size of the typical clusters diverges at the threshold and an infinite cluster appears. Above the threshold the percolating cluster mops up the finite clusters, which shrink in number

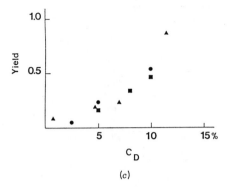

(c)

Figure 1.2. Dependence of the yield of sensitized luminescence on the donor concentration c_D in mixed crystals below 2 K, $r = I_A/(I_A + I_D)$ (see text). Triplet (a) and singlet (b) sensitized luminescence of pyrazine acceptors in mixed benzene-H_6 donor and benzene-D_6 host crystals.[10] The acceptor concentrations are, respectively, 1.3×10^{-3} and 2.5×10^{-2} m/m. (c) The same results for the triplet state of 2-methylnaphthalene acceptors and naphthalene-H_8 donors in a naphthalene-D_8 host;[11,13] acceptor concentration 0.1%. Data drawn from references: ▲ (Ref. 12a); ■ (Ref. 115b); ● (Ref. 47).

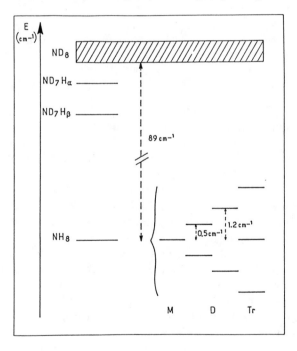

(a)

Figure 1.3. (a).

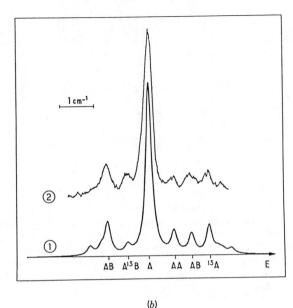

(b)

Figure 1.3. (a) Level diagram of the T_1 states of isotopically mixed crystals of naphthalene. (b) Total luminescence excitation spectrum of the donor band. Besides the central line due to monomer or isolated donors, structures appear due to resonance coupling by electron exchange between sites in random clusters of nearest-neighbor donors.

and size. Figure 1.4 illustrates long range propagation by percolation of nearest neighbour interactions on a square lattice.

The short-range interactions here are the excitation exchange interactions, which are much stronger for nearest-neighbor molecules in the (ab) plane of naphthalene crystals than for next-nearest-neighbors or for molecules in different planes. The results on transport in the singlet state can thus be modelled phenomenologically by assuming nearest-neighbor hopping between donors, until either natural decay or trapping by an acceptor intervenes. The yield of acceptor emission rises in this model because the clusters of connected donors grow with increasing concentration and include acceptors with higher probability. The infinite, percolating cluster, which must contain acceptors, soon dominates the samples at concentrations above the threshold, leading to ~100% yield of acceptor emission. The percolation threshold for nearest-neighbor interactions on a square lattice is indeed 59%, in good agreement with the experimental concentration at which acceptor emission predominates the fluorescence of the samples at 1.6 K. Triplet interactions are also strongest for nearest neighbors in the (a, b) plane, but the acceptor emission predominates at

35%   50%   60%

Figure 1.4.   Nearest-neighbor interactions between randomly occupied sites (■) on a square lattice percolate at a threshold concentration $\bar{c} \sim 59\%$ of active sites. At 35 and 50%, all clusters of connected sites are finite. The cluster size diverges at the threshold, so that long-range propagation of signals by nearest-neighbor interactions is possible, for example, at 60%, just above the threshold.

a much lower concentration, 10%. This was supposed to be due to the much longer triplet lifetime, which allows weaker interactions to compete with natural decay. Nonneighboring guests may then exchange an excitation by tunneling or superexchange interactions. These interactions appear in perturbative development of the true mixed crystal eigenstates about the site localized states. One obtains a coupling of the form

$$v = G \cdot V (V/\Delta)^n \tag{1.2}$$

for two guests of depth Δ below the host band, separated by n host sites. V is the nearest-neighbor exchange interaction. G is a geometrical factor, accounting for the number of paths of n hosts between the guests. The hopping time was found from the Rabi formula for the recurrence time in coherent motion,[12a] $t = h/v$. Comparison with the lifetime fixes the hopping range of triplet excitons. Using the values $V \sim 1\,\text{cm}^{-1}$ and $\Delta \sim 100\,\text{cm}^{-1}$ one obtains $n = 4$ to 5 host sites. Nonnearest neighbor hopping on randomly occupied lattices also displays percolation. The percolation threshold for hops of radius 5 on a square lattice is about 7%[23,24] in good agreement with the experimental increase of the acceptor emission.

Klafter and Jortner[25] proposed that the experimental increase of acceptor emission is due to an Anderson transition from localized to extended donor states as the donor concentration is raised.

This theory examines the nature of the eigenstates of a pure molecular solid with the tight-binding Hamiltonian

$$H_e = \sum_i E_i |i\rangle\langle i| + \sum_{i,j} V_{ij} |i\rangle\langle j| \qquad (1.3)$$

The site transition energy E_i contains an intramolecular term, the same for all sites of a given species and stabilization terms due to interactions with neighboring molecules. These are all the same in an ideal crystal, but in a disordered material there will be some spread of the values or diagonal disorder, δ. The reader will recall that the eigenstates of a pure crystal form a band of delocalized Bloch waves, with equal amplitude on all sites. The bandwidth is $B = 2zV_0$, where z is the coordination number and V_0 is the order of magnitude of the exchange couplings. As diagonal disorder is increased, excentric sites are uncoupled from the bulk of the material, giving rise to localized states in the band tails. Eventually, at a critical disorder, all states are localized and the solid becomes an insulator:[2] There is some constant η such that if $B < B_c = \delta/\eta$, all states are localized. Klafter and Jortner first applied this idea using a fictive superlattice with the effective, tight-binding, nearest-neighbor Hamiltonian,

$$H_{\mathrm{eff}} = \sum_p E_p |p\rangle\langle p| + \sum_{p,p'} J |p\rangle\langle p'| \qquad (1.4)$$

The spread of site energies E_p was estimated from high-resolution excitation spectra of molecular crystals. The exchange integral J was set equal to the superexchange coupling at the average donor–donor separation, in the disordered mixed crystals

$$J = v(\langle r \rangle)$$
$$\langle r \rangle = \Gamma(3/2)(\pi c_D)^{-1/2} \text{ lattice units,}$$

for donor concentration c_D in two dimensions and

$$v(r) = V_0 \exp(-\beta r)$$

where r is in lattice units and $\beta = \ln(\Delta/V_0)$.

Klafter and Jortner deduced an Anderson transition at the concentration

$$\bar{c}_D = \frac{\Gamma(3/2)^2/\pi}{[\ln(2z\eta V_0/\delta)/\ln(\Delta/V_0) + 1]^2} \tag{1.5}$$

or about 10% using $\delta = 0.1$ to $4\,\text{cm}^{-1}$, $\eta = 2.7$, and $z = 4$. This is about right for naphthalene.

Monberg and Kopelman[26] subsequently pointed out that the average coupling is not representative of tunneling interactions in a disordered system. Spatial disorder is amplified by the exponential law to give a wide spread of couplings, or off-diagonal disorder. They expected this off-diagonal disorder to prevent formation of extended states in the mixed crystals in the experimental concentration range. Klafter and Jortner improved their model in several ways. First they replaced the coupling at the average distance by the average coupling calculated from the distribution of guest–guest distances.[27a] This markedly increased the coupling in the effective Hamiltonian. Secondly, they showed that off-diagonal disorder in the principle species, guest monomers, should only slightly lower the concentration at which the Anderson transition occurs in their model.[27b] Finally,[27c] they introduced the localization length defined by Lukes.[28] The amplitude of localized states in disordered solids generally has an exponential radial decay about some cluster of sites at a point $\vec{R}_0$:

$$|\psi(\vec{R})| \underset{|\vec{R}-\vec{R}_0|\to\infty}{\sim} \exp(-|\vec{R} - \vec{R}_0|/\lambda_\psi)$$

When λ_ψ, the localization length, exceeds a few lattice units, it is natural to talk of extended states. The above relation is of course an average behavior, each state showing fluctuations due to chance resonances and to clustering. The localization length in the superlattice model of mixed crystals was found to be

$$\lambda(E, c_D) = \langle r\rangle \left\{ (2/3)\ln\left[\frac{\delta\exp(E/\delta)}{2\eta\langle v\rangle}\right]\right\}^{-3/4} \tag{1.6}$$

for states at energy E with respect to the band center. Assuming thermal equilibrium in the narrow donor band ($\delta < 1\,\text{cm}^{-1}$ at $T > 1.6\,\text{K}$), transfer to acceptors will be strong when the localization length is of the order of the average donor–acceptor distance, $\langle r_{AD}\rangle = (\pi c_A)^{-1/2}$. This enables comparison with experiments on the effect of changing the acceptor concentration[13] to distinguish between the hopping model and the Anderson transition: As the acceptor concentration rises, the required donor localization length drops and a given yield of acceptor emission can be achieved at lower donor concentrations. This is indeed observed, but the predicted effect considerably exceeds the experimental one.

Ahlgren and Kopelman[29] later published a scaling plot of the yield of acceptor emission as a function of the donor concentration. This produced the critical exponents for two-dimensional percolation. While this is interesting concerning the dimensionality of the transport (providing step-function interactions are appropriate, see Section VII below) it does not necessarily eliminate the Anderson transition, which is also a percolation phenomenon, because the existence of extended states depends on the presence of large clusters of sites which are resonant to within the exchange coupling.[30-32]

The nature of exciton transport in mixed crystals remains an open question for a number of reasons: The off-diagonal disorder, an important parameter in the localization model, was equated with the width of the excitation spectra, but these give only an upper limit to the diagonal disorder on the scale of 200 Å, involved in the transport experiments. The spectra may be the envelope of a set of narrower spectra of more homogeneous domains. The localization lengths would then be longer. One may also ask to what extent exchange coupling contributes to the apparent linewidth.

Klafter and Jortner initially assumed thermal equilibrium in the narrow donor band would empty nearly all donor states to acceptors, once states in the middle of the band were sufficiently extended. A narrow band compared to thermal energy is not however a sufficient condition for strong coupling. As a simple counter example, remember that the triplet spin sublevels of aromatic molecules are commonly out of equilibrium at temperatures of 2–3 K, much larger than the splitting, about $0.1\,\mathrm{cm}^{-1}$.

Can one replace the guest Hamiltonian in the mixed crystal by the effective superlattice Hamiltonian? Is even the average coupling representative of the wide spread of superexchange interactions? It may be that the localization idea is right, but that the numerical treatment was too approximate.

The determination of the hopping radius by equating the hopping time with the coherent recurrence time[12a] is incorrect. The hopping time must be found by explicit inclusion of coupling to phonons.

It is unfortunate that no quantitative estimations of the effect of the acceptor concentration were made in the hopping model.

Brenner and his colleagues recently reported on the effect of pressure on the experimental yield of acceptor emission in mixed crystals of naphthalene.[33,34] As they pointed out, increased pressure will have several effects. It must increase the exchange couplings by reducing the intermolecular distances. It should also modify the off-diagonal disorder, by changing the intermolecular stabilization. Both these effects must change the hopping rates in the percolation model and the localization length in the transition model, but it is difficult to say by how much, without an exact calculation, as we shall show below. Both models should be worked out in more detail before trying to draw conclusions.

We concentrate here on triplet excitons in naphthalene, but similar questions could be asked about other aromatic isotopically mixed crystals such as phenazine,[35] dibromonaphthalene;[36] chemically mixed crystals like 1-bromo,4-chloronaphthalene/1,4-dibromonaphthalene,[37] p-dichloroben-zene/p-dibromobenzene[38,39] and phenanthrene/biphenyl;[40] orientationally disordered crystals like 2,3-dimethylnaphthalene,[41] 2,3-dimethyl-anthracene,[42] dibenzofurane,[43] and 1-bromo,4-chloronaphthalene;[44] or glassy aromatic solids.[45,46]

As a rule, phenomenological models of hopping transport have been applied to these systems and most of the work centered on the concentration dependence of the transport. The temperature dependence, which received less attention, can provide much more interesting and perhaps more reliable information on the nature of the dynamics.

Thus, returning to isotopically mixed naphthalene, one of the arguments against the model of transport by extended states was that activated hopping between localized states would tend to mask the Anderson transition,[26,27a] but data were available at two temperatures only, 1.6 and 4.2 K. Acceptor emission was indeed higher at the higher temperature,[29] but measurements at only two temperatures are insufficient to unambiguously identify the nature of the transport.

The temperature dependence of transfer to acceptors and of delayed fluorescence is in fact the following,[47,48] (Fig. 1.5):

- At temperatures below 5 K, all transport sensitive emissions are independent of the temperature or only slightly dependent on temperature. They are, however, concentration-dependent.

- At ~5 K there are sharp thermal thresholds, dependent on the concentration.

- Delayed fluorescence drops above 5 K, and rises again at temperatures above 15 K.

This behavior suggests trapping to acceptors from more or less extended donor states below 5 K. Thermally activated hopping between these states could be responsible for the rise in yields at about 5 K. These complicated effects of temperature require a detailed, nonphenomenological analysis, identifying the nature of the electronic states and the mechanisms by which they relax by coupling to phonons. This is attempted below.

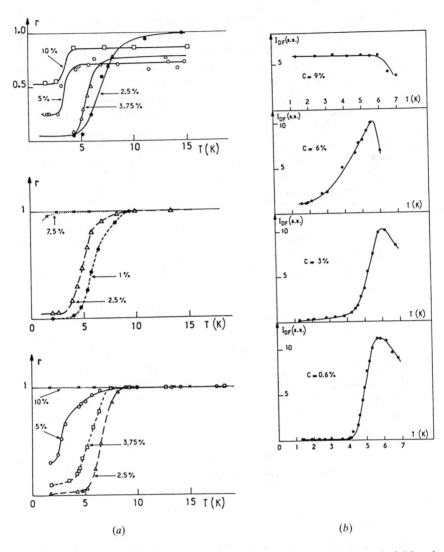

(a)   *(b)*

Figure 1.5. The complex temperature dependence of acceptor emission and of delayed fluorescence in isotopically mixed crystals of naphthalene suggests several kinds of transport, rather than either an Anderson transition or hopping transport. (*a*) Yield of emission of 2-methylnaphthalene acceptors (0.1%) in a naphthalene-D_8 host at several concentrations of naphthalene-H_8 donors.[47] (*b*) Intensity of the delayed fluorescence of naphthalene-H_8/naphthalene-D_8 mixed crystals.[48]

II. HAMILTONIAN

The first step in a nonphenomenological model of triplet exciton transport in mixed aromatic crystals is to set up the model Hamiltonian. We shall discuss successively the electronic Hamiltonian, the phonon Hamiltonian, and exciton-phonon coupling, responsible for relaxation between electronic states.

A. Overview

1. *Molecular Hamiltonian*

Readers will recall that the total wavefunctions $\Psi(\vec{r}, \vec{R})$ of the electronic and nuclear coordinates ($\vec{r}$ and $\vec{R}$) of an isolated molecule are solutions of the wave equation

$$H\Psi(\vec{r}, \vec{R}) = E\Psi(\vec{r}, \vec{R})$$

with Hamiltonian

$$H = T_e + V_{en} + V_{ee} + V_{nn} + T_n \tag{2.1}$$

$T_e = \Sigma_e \vec{p}_e^2/2m_e$ and $T_n = \Sigma_n \vec{p}_n^2/2M_n$ are respectively the kinetic energies of the electrons, e, and of the nuclei, n, and

$$V_{ne} = \sum_e \sum_n \frac{-Z_n e^2}{4\pi\epsilon_0|\vec{r}_e - \vec{R}_n|}$$

and

$$V_{ee} = \sum_e \sum_{e' \neq e} \frac{e^2}{4\pi\epsilon_0|\vec{r}_e - \vec{r}_{e'}|}$$

are, respectively, the electrostatic attraction between the nuclei and the electrons and the repulsion between the electrons. Finally,

$$V_{nn} = \sum_n \sum_{n' \neq n} \frac{Z_n Z_{n'} e^2}{4\pi\epsilon_0|\vec{R}_n - \vec{R}_{n'}|}$$

is the repulsion of the nuclei.

The electrostatic forces on the electrons and the nuclei are of the same order of magnitude, but the much smaller mass of the electrons causes their motion to be faster and their states to be wider split apart. This allows Born–Oppenheimer separation of the electronic and the nuclear degrees of freedom:[49] (1) The electronic motion can be solved for any given nuclear coordinates; (2) the electronic motion is supposed to instantaneously adapt itself to the much slower changes in the nuclear configuration; (3) electronic states corresponding to slightly different nuclear configurations are similar to one another, so that a single set of electronic states, ψ^f, can be defined parametrically for all nuclear configurations near equilibrium. The time-independent Schrödinger equation can then be written as follows for the electronic wavefunctions:

$$H_e(\vec{R})\Psi^f(\vec{r}; \vec{R}) = E^f(\vec{R})\Psi^f(\vec{r}; \vec{R}) \tag{2.2}$$

where $\vec{r}$ and $\vec{R}$ stand respectively, for the electronic and the nuclear coordinates and $H_e(\vec{R})$ is the electronic Hamiltonian for a given configuration of the nuclei,

$$H_e(\vec{R}) = T_e + V_{en}(\vec{R}) + V_{ee} + V_{nn}(\vec{R}) \tag{2.3}$$

Approximate solutions of Eq. (2.2) can be obtained by the various methods of quantum chemistry, leading in general to a set of molecular orbitals, the lowest ones being doubly occupied in the ground state by electrons of opposite spin, in the usual case of a singlet ground state. The highest occupied orbital in the ground state will be written below φ_{-1}, the next lower one φ_{-2}, and so on. Successively higher unoccupied orbitals will be called φ_{+1}, φ_{+2}, and so on. The ground state is then the (antisymmetric) Slater determinant for $2N$ electrons:

$$|\varphi^0\rangle = |\varphi_{-1\uparrow}, \varphi_{-1\downarrow}, \varphi_{-2\uparrow}, \varphi_{-2\downarrow}, \ldots, \varphi_{-N\uparrow}, \varphi_{-N\downarrow}\rangle$$

Excited states are created in the two electron approximation by promoting an electron to an unoccupied orbital, such as the singlet state built on transition $\varphi_{-1} \rightarrow \varphi_{+1}$:

$$|\psi_s^*\rangle = \frac{1}{\sqrt{2}}(|\varphi_{-1\downarrow}, \varphi_{+1\uparrow}\rangle - |\varphi_{-1\uparrow}, \varphi_{+1\downarrow}\rangle)$$

The total wavefunction can then be approximated by products of the form

$$\Psi(\vec{r}, \vec{R}) = \psi^f(\vec{r}; \vec{R})\chi^f(\vec{R}) \tag{2.4}$$

where χ^f is a vibrational state corresponding to the electronic potential in state ψ^f. Carrying Eq. (2.4) into Eq. (2.2) one obtains

$$(E^f(\vec{R}) + T_n)\chi^f(\vec{R}) = E\chi^f(\vec{R}) \tag{2.5}$$

where E is the total energy of the molecule. In going from Eq. (2.1) to Eq. (2.5), terms due to T_n, of the form

$$\langle \psi^f | \partial/\partial\vec{R} | \psi^f \rangle$$

an operator on $\chi^f(\vec{R})$ and

$$\langle \psi^f | \partial^2/\partial\vec{R}^2 | \psi^f \rangle$$

a scalar, are neglected in the adiabatic approximation. They are responsible for transitions like internal conversion between the trial wavefunctions (Eq. 2.4).

The potential energy $E^f(\vec{R})$ can be developed as a power series about the equilibrium coordinates $\vec{R}_{f_0}$:

$$E^f(\vec{R}) = E^f(\vec{R}_{f0}) + (\vec{R} - \vec{R}_{f0})^t \cdot \frac{\partial^2 E^f}{\partial^2 \vec{R}}\bigg|_{\vec{R}_{f_0}} \cdot (\vec{R} - \vec{R}_{f0}) + \cdots \tag{2.6}$$

The second-order term is the potential of a set of coupled harmonic oscillators, which can be uncoupled by introducing the normal modes of vibration, Q_p, which diagonalize the tensor $\partial^2 E^f/\partial\vec{R}^2$. The nuclear eigenstates then appear as products of states of the independent oscillators $\chi^f_{v_p}(Q_p)$, where v_p is the number of vibrational quanta in mode p:

$$\chi^f(\vec{R}) = \prod_p \chi^f_{v_p}(Q_p) \tag{2.7}$$

2. Born–Oppenheimer Separability in Molecular Solids

The states of a solid of N molecules, each of M atoms, also may be described by the Born–Oppenheimer separation of nuclear and electronic coordinates. The total Hamiltonian can be written as before (Eq. 2.1) and we can carry through the previous steps:

$$\Psi(\vec{r}, \vec{R}) = \psi^f(\vec{r}; \vec{R})\omega^f(\vec{R})$$
$$H_e(\vec{R}) = T_e + V_{en}(\vec{R}) + V_{nn}(\vec{R}) + V_{ee} \tag{2.8}$$

$$(E^f(\vec{R}) + T_n)\omega^f(\vec{R}) = E\omega^f(\vec{R})$$

Here $\omega^f(\vec{R})$ describes the motion of the 3MN nuclear coordinates in the electronic potential $E^f(\vec{R})$ created by all the electrons in the sample. The decomposition (Eq. 2.8) once more corresponds to the assumption that the electronic state instantaneously adjusts to the nuclear configuration. However, distinction of the intramolecular and intermolecular nuclear coordinates, $\vec{Q}$ and $\vec{R}$, respectively, generally is possible in molecular solids because their spectra strongly resemble those of the isolated molecules: The same vibronic lines appear with phonon sidebands, proof that the intramolecular modes retain their identity. The electronic potential can thus be split into monomolecular and bimolecular parts:

$$E^f(\vec{Q}, \vec{R}) = E_f^{(1)}(\vec{Q}) + E_{fv}^{(2)}(\vec{R})$$

where the notation $E_{fv}^{(2)}(R)$ expresses the fact that the intermolecular potential may differ for different intramolecular vibronic states $|fv\rangle$. Approximate total wavefunctions of the form

$$\Psi(\vec{r}, \vec{Q}, \vec{R}) = \psi^f(\vec{r}; \vec{Q})\chi_v^f(\vec{Q})\lambda^{fv}(\vec{R}) \tag{2.9}$$

then distinguish 3M-6 internal coordinates, $\vec{Q}$ per molecule, and 6N intermolecular coordinates $\vec{R}$. The corresponding wave equation for the intermolecular motion is

$$(T_{\vec{R}} + E_{fv}^{(2)}(\vec{R}))\lambda^{fv}(\vec{R}) = E\lambda^{fv}(\vec{R}) \tag{2.10}$$

$E_{fv}^{(2)}$ can be developed about the equilibrium configuration to introduce the intermolecular vibrational modes, which we shall discuss in more detail below because they are needed to calculate the electronic transport properties of the mixed crystal. We note in passing that the intermolecular adiabatic approximation does not always hold, for example, when the molecules are flexible or when they have small intramolecular quanta. The lowest internal modes of anthracene, for example, are significantly coupled to the intermolecular phonons.[50]

B.   Hamiltonian of a Molecular Crystal

1.   Electronic Hamiltonian of a Rigid Crystal

The conservation of the identity of the intramolecular states in molecular solids makes the monoexcited oriented gas states a meaningful basis for the study of their excited states at low densities of excitation. Assuming for simplicity that each molecule m has only one excited state, ψ_m^*, the oriented gas states are: (1) an (approximate) ground state, where all molecules are in their ground state, ψ_m^0,

$$|\Phi_0\rangle = \prod_m |\psi_m^0\rangle \tag{2.11}$$

and (2), mono-excited states where only molecule m is excited (say transition $\varphi_{-1} \to \varphi_{+1}$):

$$|\Phi_m^*\rangle = |\psi_m^*\rangle \prod_{n \neq m} |\psi_n^0\rangle \tag{2.12}$$

The total electronic Hamiltonian for a given nuclear configuration, H, containing all interactions between electrons and between electrons and nuclei, stabilizes and mixes these states, which are not crystal eigenstates:

$$H|\Phi_0\rangle = \left(\sum_m E_m^0 + D_m^0\right)|\Phi_0\rangle + \sum_m \langle\Phi_m^*|H|\Phi_0\rangle|\Phi_m^*\rangle$$

$$H|\Phi_m^*\rangle = (E_m^* + D_m^*)|\Phi_m^*\rangle + \sum_{n \neq m} V_{mn}|\Phi_n^*\rangle + \langle\Phi_0|H|\Phi_m^*\rangle|\Phi_0\rangle \tag{2.13}$$

$E_m^0(E_m^*)$ is the ground (excited) state energy of molecule m, containing all interactions in molecule m. $D_m^0(D_m^*)$ is the stabilization of the ground (excited) state of molecule m in the "crystal field" due to all other molecules:

$$D_m^0 = \sum_{n \neq m} D_{m,n}^0 = \sum_{n \neq m} \langle\psi_m^0\psi_n^0|H_{m,n}|\psi_m^0\psi_n^0\rangle$$

$$D_m^* = \sum_{n \neq m} D_{m,n}^* = \sum_{n \neq m} \langle\psi_m^*\psi_n^0|H_{m,n}|\psi_m^*\psi_n^0\rangle \tag{2.14}$$

where $H_{m,n}$ is the part of H referring to electrons in orbitals on molecules m and n:

$$H_{m,n} = v_{mn} + V_m(n) + V_n(m) + U_{mn} \tag{2.15}$$

with v_{mn} the repulsion between electrons, $V_n(m)$ the attraction of the nuclei of molecule n on electrons in orbitals on molecule m, and U_{mn} nuclear repulsion between the molecules. The stabilizations are, in general, multipolar interactions. Further "van der Waals" stabilizations or dispersion forces appear in the second-order perturbation of the energies by the intermolecular part of the Hamiltonian. These terms are neglected here, as is the mixing of the approximate ground and excited states. The stabilization representing the cohesion of the solid is of the order of the sum of the latent heats, about $5000 \, \text{cm}^{-1}$ for aromatic crystals. The terms coupling the ground and the excited states are in general much smaller than the transition energy $E_m^* - E_m^0$, so that mixing is negligible. Mixing of the resonant excited states by the exchange interactions $V_{m,n}$ is on the other hand important, giving rise to collective excitations, spread over all sites in perfect crystals, or over random groups of resonant sites in disordered solids.

Exchange interactions are of the form

$$V_{m,n} = \langle \Phi_m^* \Phi_n^0 | H | \Phi_m^0 \Phi_n^* \rangle = 2J_{m,n} - K_{m,n} \tag{2.16}$$

where $J_{m,n}$ is the Coulomb or multipolar interaction between the electronic distributions of molecules m and n, for example,

$$J_{m,n} = \int \int d\vec{r}_1 \, d\vec{r}_2 \bar{m}_{+1}(\vec{r}_1) \bar{n}_{-1}(\vec{r}_2) \frac{1}{r_{12}} \bar{m}_{-1}(\vec{r}_1) \bar{n}_{+1}(\vec{r}_2)$$

where $\bar{m}_{+1}$ is the $+1$ orbital on molecule m and so on. The specifically quantum term $K_{m,n}$ is due to exchange of electrons between the molecules when their orbitals overlap:

$$K_{m,n} = \int \int d\vec{r}_1 \, d\vec{r}_2 \bar{m}_{+1}(\vec{r}_1) \bar{n}_{-1}(\vec{r}_2) \frac{1}{r_{12}} \bar{n}_{-1}(\vec{r}_1) \bar{m}_{+1}(\vec{r}_2)$$

J is of the order of $10–50 \, \text{cm}^{-1}$ for exchange of singlet excitations between nearest neighbors in aromatic crystals. K, the only term in the triplet state, is of the order $10^{-2} - 1 \, \text{cm}^{-1}$ and decays exponentially with the separation. Next-nearest interactions are estimated to be $\sim 10^{-3}$ times smaller than nearest-neighbour couplings.[51]

Introducing the ground-state energy,

$$\mathscr{E}_0 = \sum_m E_m^0 + D_m^0 \tag{2.17}$$

and the stabilizations of the molecular transitions,

$$D_m = D_m^* - D_m^0 \qquad (2.18)$$

Eq. (2.13) can be written as

$$H|\Phi_0\rangle = \mathscr{E}_0|\Phi_0\rangle \qquad (2.19)$$

and

$$H|\Phi_m^*\rangle = (\mathscr{E}_0 + E_m)|\Phi_m^*\rangle + \sum_{n \neq m} V_{m,n}|\Phi_m^*\rangle$$

or

$$H = \mathscr{E}_0 B_0^+ B_0 + \sum_m (\mathscr{E}_0 + E_m) B_m^+ B_m + \sum_m \sum_{n \neq m} V_{m,n} B_n^+ B_m \qquad (2.20)$$

In Eq. (2.20) state $|m\rangle = |\Phi_m^*\rangle$ is created (annihilated) by the second quantification operator[52] $B_m^+(B_m)$. In these relations E_m is the "site energy" of the localized excitation on site m, including the crystal field, $E_m = E_m^* - E_m^0 + D_m$. The creation and annihilation operators are assumed to satisfy

$$[B_m^+, B_n]_+ = \delta_{m,n} \qquad \text{and} \qquad [B_m, B_n] = 0$$

meaning that each molecule has only one excited state.

The stabilization D_m is of the order of 100–500 cm^{-1} between the vapor and the crystal of aromatic compounds.[53] It is sensitive to defects like vacancies, dislocations, distortions due to impurities, and so on. These contribute to "diagonal disorder" of the Hamiltonian in the site basis, a source of inhomogeneous or statistical broadening of electronic transitions. The effect may be violent enough to break off states from the main exciton band forming X-traps.[54,55] "Off-diagonal" disorder of the exchange interactions must exist too, but is more difficult to reveal, because the smaller values are in general masked by the diagonal disorder.

2. Phonons in Molecular Crystals

Thermal effects are essential to the discussion of the nature of exciton transport at low temperatures. It may therefore be useful to refresh our memory with a brief reminder about phonons, before introducing exciton–phonon coupling. Readers are referred to standard works for a complete treatment.[56,57]

Consider then the ground electronic state of a crystal of N lattice cells with σ molecules per unit cell ($\sigma = 2$ for naphthalene). Cells are numbered by Latin letters and sites in the cell by Greek letters. As noted above, intermolecular degrees of freedom generally are well separated from intra-molecular coordinates, so that the intermolecular motion is described by the solutions of

$$(T_{\vec{R}} + E_0^{(2)}(\vec{R}))\lambda(\vec{R}) = E\,\lambda(\vec{R}) \tag{2.21}$$

where $E_0^{(2)}(\vec{R})$ is the intermolecular potential in the electronic ground state and $T_{\vec{R}}$ is the kinetic energy operator of the coordinates of the intermolecular motion and E is the total energy. The potential can be put in terms of the stabilization energies,

$$E_0^{(2)}(\vec{R}) = \sum_m \sum_{n \neq m} D_{m,n}^0(\vec{R}) \tag{2.22}$$

where $D_{m,n}^0$ stands for the stabilization of molecule m by molecule n. Developing the potential to second order about the ground-state equilibrium positions we have

$$E_0^{(2)}(\vec{R}) = E_0^{(2)}(\vec{R}_0) + (\vec{R} - \vec{R}_0)^t \cdot \left.\frac{\partial^2 E_0^{(2)}}{\partial \vec{R}^2}\right|_{\vec{R}_0} \cdot (\vec{R} - \vec{R}_0) \tag{2.23}$$

where the second term is a tensor product

$$(\vec{R} - \vec{R}_0)^t \cdot \left.\frac{\partial^2 E_0^{(2)}}{\partial \vec{R}^2}\right|_{\vec{R}_0} \cdot (\vec{R} - \vec{R}_0)$$

$$= \sum_{\substack{mn\ xy \\ \alpha\beta}} (R_{m,x}^\alpha - {}^{(0)}R_{m,x}^\alpha) \cdot \left.\frac{\partial^2 E_0^{(2)}}{\partial R_{m,x}^\alpha \partial R_{n,y}^\beta}\right|_{{}^{(0)}R_{m,x}^\alpha,\, {}^{(0)}R_{n,y}^\beta} \cdot (R_{n,y}^\beta - {}^{(0)}R_{n,y}^\beta)$$

$$\tag{2.24}$$

In this expression $R_{m,x}^\alpha$ is the xth coordinate (out of three rotations and three translations) of the αth molecule in the mth cell, and the prefix (0) stands for the equilibrium value, set to zero below as a matter of convenience. In pure crystals the second derivative depends only on the relative displacement of the molecules, so that Fourier transformation can be used to obtain

$$H_R = E_0 + \sum_{\vec{q},\alpha,x} \frac{1}{2M_{\alpha x}} P^\alpha_{\vec{q},x} P^\alpha_{\vec{q},x} + \frac{1}{2} \sum_{\vec{q}} \sum_{\alpha,\beta} \sum_{x,y} R^\alpha_{\vec{q},x} V^{xy}_{\alpha\beta}(\vec{q}) R^\beta_{-\vec{q},y} \quad (2.25)$$

where

$$V^{xy}_{\alpha\beta}(\vec{q}) = \sum_m \exp(-i\vec{q} \cdot \vec{r}_m) \frac{\partial^2 E_0^{(2)}}{\partial R^\alpha_{0,x} \partial R^\beta_{m,x}}\bigg|_{\text{equilibrium}} \quad (2.26)$$

and $P^\alpha_{\vec{q},x}$ is the momentum operator of the transformed coordinate

$$R^\alpha_{\vec{q},x} = \frac{1}{\sqrt{N}} \sum_m \exp(i\vec{q} \cdot \vec{r}_m) R^\alpha_{m,x} \quad (2.27)$$

The wave vector sums cover the first Brillouin zone, for example,

$$q_a = k\pi/N_a a; \qquad k = 0, 1, 2, \ldots, N_a - 1$$

for a cubic crystal with side N_a lattice units in the a direction. $M_{\alpha x}$ is the mass (moment of inertia) of the xth coordinate of the αth molecule in the unit cell. Equation (2.25) is the Hamiltonian of N sets of 6σ coupled harmonic oscillators. The 6σ modes corresponding to a given wavevector are solutions of the eigenvalue problem

$$(D(\vec{q}) - \omega^2 I)\vec{e}_{\vec{q}} = 0 \quad (2.28)$$

where I is the unit matrix and D is defined by

$$D_{\alpha x, \beta y}(\vec{q}) = \frac{1}{\sqrt{M_{\alpha x} M_{\beta y}}} V^{xy}_{\alpha\beta}(\vec{q}) \quad (2.29)$$

Vector $e_{\vec{q}}$ contains the 6σ amplitudes of the mode of frequency ω projected on the local coordinates x. These projections largely are determined by the symmetry of the lattice. In centro-symmetric crystals rotations and translations are completely uncoupled at the center of the zone, $|\vec{q}| = 0$, but for general $\vec{q}$, modes are mixed rotational (librational) and translational states. There are 6σ phonon branches with mode properties varying continuously with $\vec{q}$, of which three are "acoustic" branches for which the frequency vanishes at small wavevectors because all the molecules in the crystal are in phase. The remaining $6\sigma - 3$ "optical" branches have

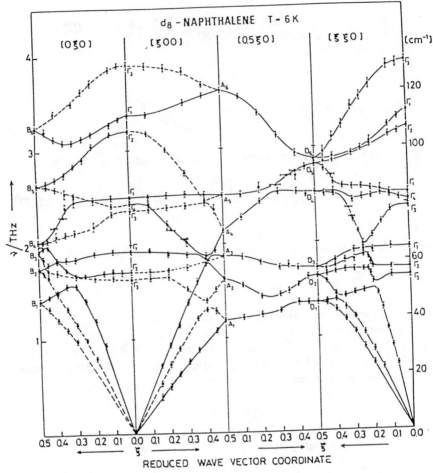

Figure 2.1. The phonon dispersion curves of deuterated naphthalene deduced from neutron scattering at room temperature.[58] Note the high density of modes, $\rho(E) \propto (1/dE/dq)$ at 50 cm^{-1} where the optical branches flatten out.

finite frequencies at long wavelengths, typically 50–150 cm^{-1} in aromatic crystals. Complete dispersion curves for lattice phonons can be obtained by a mixture of Raman scattering,[58] neutron scattering at special points in the Brillouin zone, and fitting dynamical models of the lattice to these data. Figure 2.1 shows the dispersion curves of naphthalene.[59]

The phonon Hamiltonian can be quantitized by introduction of the creation and annihilation operators $b_{\vec{q},p}^{+}$ and $b_{\vec{q},p}$ of phonons (quanta) in mode p of wavevector $\vec{q}$. These are defined by:[60]

$$b_{\vec{q},p}^{+} = \alpha U_{\vec{q},p} - i\beta \Pi_{\vec{q},p}^{+}$$

and   (2.30)

$$b_{\vec{q},p} = \alpha U_{\vec{q},p}^{+} + i\beta \Pi_{\vec{q},p}$$

with

$$\alpha = \sqrt{\omega_p(\vec{q})/2\hbar} \quad \text{and} \quad \beta = \sqrt{\frac{1}{2\hbar\omega_p(\vec{q})}} \tag{2.31}$$

$$U_{\vec{q},p} = \sum_{\alpha,x} \sqrt{M_{\alpha x}} \, e_{\vec{q},x,p} R_{\vec{q},x}^{\alpha} \tag{2.32}$$

$$\Pi_{\vec{q},p} = \sum_{\alpha,x} \frac{1}{\sqrt{M_{\alpha x}}} \overline{e_{\vec{q},x,p}^{\alpha}} P_{\vec{q},x}^{\alpha}$$

Then

$$H_R = E_0 + \sum_{\vec{q},p} \hbar\omega_p(\vec{q}) \left(b_{\vec{q},p}^{+} b_{\vec{q},p} + \frac{1}{2} \right) \tag{2.33}$$

where the creation and annihilation operators obey Bose statistics. This model corresponds to second order development of the intermolecular potential. Higher-order terms contribute to transitions between these modes, such as fission of phonons. The masses and intermolecular forces in isotopically mixed crystals like deuterated and protonated naphthalene change only slightly, so that the phonon modes are very similar to those of a pure crystal. Prasad and Kopelman[61] have thus reported small changes in the frequencies of the optical modes of mixed crystals of naphthalene, varying linearly with the concentration of protonated substituents. In the general case, such as heavy impurities in a light matrix, local vibrational modes may be associated with the impurities and the host modes may be strongly distorted.

3.   Exciton-Phonon Coupling

Intermolecular vibrational modes in the adiabatic approximation depend on the electronic potential $E_{fv}^{(2)}(R)$, and generally differ a little for different vibronic states $|fv\rangle$. The equilibrium position may for example be shifted or the frequency may be different and we are faced with working out the phonon modes for each electronic state. Changes in the electronic state of the crystal, such as absorption of a photon in the ground state, producing a site excitation, or exchange of site excitation between neighbors, which do not involve the intermolecular phonon coordinates, are possible provided the total energy is conserved and the initial and final phonon states are not orthogonal. The probability of such an event contains the electronic matrix element, weighted by an intermolecular Frank–Condon overlap factor. An alternative and sometimes more convenient way to write this is to develop the total Hamiltonian entirely in terms of the phonon modes of the electronic ground state. As a first step, write the total Hamiltonian as

$$H = T_{\vec{R}} + \mathscr{E}_0(\vec{R})B_0^+ B_0 + \sum_m (\mathscr{E}_0(\vec{R}) + D_m(\vec{R}) + E_m^* - E_m^0)B_m^+ B_m$$

$$+ \sum_{m,n} V_{m,n}(\vec{R})B_m^+ B_n \qquad (2.34)$$

$$= \mathscr{E}_0(0)B_0^+ B_0 + \sum_{p,\vec{q}} \hbar\omega_p(\vec{q})(b_{p,\vec{q}}^+ b_{p,\vec{q}} + 1/2)$$

$$+ \sum_m (\mathscr{E}_0(0) + D_m(\vec{R}) + E_m^* - E_m^0)B_m^+ B_m + \sum_{m,n} V_{m,n}B_m^+ B_n$$

$$\qquad (2.35)$$

Note that the intermolecular exchange integrals, V_{mn} also depend parametrically on the intermolecular coordinates. They thus introduce off-diagonal exciton–phonon coupling in the site representation. However, exchange terms in the triplet state are much smaller than the stabilization energies, so that off-diagonal coupling will be neglected here.

Now develop the intermolecular stabilization as a power series about the equilibrium position of the ground state. Its form is

$$D^*(\vec{R}) - D^0(\vec{R}) = D^*(0) + \frac{\partial D^*}{\partial \vec{R}}\bigg|_0 \cdot \vec{R} + \frac{1}{2}\vec{R}^t \frac{\partial^2 D^*}{\partial \vec{R}^2}\bigg|_0 \cdot \vec{R} + \cdots$$

$$- D^0(\vec{R}) - \frac{\partial D^0}{\partial \vec{R}}\bigg|_0 \cdot \vec{R} - \frac{1}{2}\vec{R}^t \cdot \frac{\partial^2 D^*}{\partial \vec{R}^2}\bigg|_0 \cdot \vec{R} - \cdots$$

$$= D^*(0) - D^0(0) + \frac{\partial D^*}{\partial \vec{R}}\bigg|_0 \cdot \vec{R} + \frac{1}{2}\vec{R}^t \cdot \left(\frac{\partial^2 D^*}{\partial \vec{R}^2}\bigg|_0 - \frac{\partial^2 D^0}{\partial \vec{R}^2}\bigg|_0\right) \cdot \vec{R} + \cdots$$

$$\qquad (2.36)$$

The second and third terms are respectively linear and quadratic, local or diagonal (in the site basis) couplings between excitons and ground-state

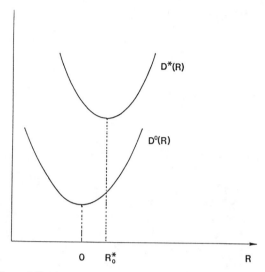

Figure 2.2. The stabilizations of ground and excited states by the crystal field depend on the intermolecular coordinates. In general the excited state potential has a different minimum position leading to linear exciton–phonon coupling and a different curvature, causing different phonon frequencies, or quadratic exciton–phonon coupling. These effects correspond physically to neighbors around the excited molecule moving over to make room for its distended electronic cloud and to generally increased strain energy when the excited molecule vibrates.

phonons. They correspond respectively to a shift of the equilibrium position and a change in the frequency of the excited state phonons relative to the ground state ones. Let $\vec{R}_0^*$ be the equilibrium position in the excited state. Thus, cf. Figure 2.2,

$$D^*(\vec{R}) - D^0(\vec{R}) = D^*(\vec{R}_0^*) + \frac{1}{2}(\vec{R} - \vec{R}_0^*)^t \cdot \frac{\partial^2 D^*}{\partial \vec{R}^2}\bigg|_{\vec{R}_0^*} \cdot (\vec{R} - \vec{R}_0^*)$$

$$- D^0(0) - \frac{1}{2}\vec{R}^t \cdot \frac{\partial^2 D^0}{\partial \vec{R}^2}\bigg|_0 \cdot \vec{R} + \cdots$$

and

$$D^*(0) = D^*(\vec{R}_0^*) + \vec{R}_0^{*t} \cdot \frac{\partial^2 D^*}{\partial R^2}\bigg|_0 \cdot \vec{R}_0^* + \cdots$$

whence

$$D^*(\vec{R}) - D^0(\vec{R}) = D^*(0) - D^0(0)$$

$$- \frac{1}{2} \left(\vec{R}_0^{*t} \cdot \left. \frac{\partial^2 D^*}{\partial \vec{R}^2} \right|_{\vec{R}_0^*} \cdot \vec{R} + \vec{R}^t \cdot \left. \frac{\partial^2 D^*}{\partial \vec{R}^2} \right|_{\vec{R}_0^*} \vec{R}_0^* \right)$$

$$+ \frac{1}{2} \vec{R}^t \cdot \left(\left. \frac{\partial^2 D^*}{\partial \vec{R}^2} \right|_{\vec{R}_0^*} - \left. \frac{\partial^2 D^0}{\partial \vec{R}^2} \right|_0 \right) \cdot \vec{R} + \cdots \quad (2.37)$$

These relations show up the linear coupling as due to a change in the equilibrium position and the quadratic term as due to a change in the curvature of the electronic potential between the ground and the excited states: Neighbours must move over to make room for the slight expansion of the electronic cloud of the excited molecule and the strain energy of intermolecular vibrations may increase. Most of the external vibrational frequencies of anthracene for example increase slightly in the triplet state.[62]

Finally, $\vec{R}$ can be expressed in terms of the creation and annihilation operators of the ground-state phonons:

$$R_{m,x}^\alpha = \sum_{p,\vec{q}} \sqrt{\frac{\hbar}{2NM_{\alpha x}\omega_p(\vec{q})}} \exp(i\vec{q} \cdot \vec{r}_m) e_{\vec{q},x,p}^\alpha (b_{\vec{q},p}^+ + b_{-\vec{q},p}) \quad (2.38)$$

The total Hamiltonian then has the form

$$H = \mathscr{E}_0(0) + \sum_{m,\alpha} (E_{m\alpha}^* - E_{m\alpha}^0 + D_{m\alpha}(0))B_{m\alpha}^+ B_{m\alpha} + \sum_{\substack{m,\alpha \\ n,\beta}} V_{m\alpha,n\beta}(0)B_{m\alpha}^+ B_{n\beta}$$

$$+ \sum_{p,\vec{q}} \hbar\omega_p(\vec{q}) \left(b_{p,\vec{q}}^+ b_{p,\vec{q}} + \frac{1}{2} \right)$$

$$+ \sum_{m,\alpha} \sum_{p,\vec{q}} F_1(m\alpha, p, \vec{q}) B_{m\alpha}^+ B_{m\alpha} (b_{p,\vec{q}}^+ + b_{p,-\vec{q}})$$

$$+ \sum_{m,\alpha} \sum_{p,\vec{q}} \sum_{p',\vec{q}'} F_2(m\alpha, p, \vec{q}, p', \vec{q}') B_{m\alpha}^+ B_{m\alpha}$$

$$\times (b_{p,\vec{q}}^+ + b_{p,-\vec{q}})(b_{p',\vec{q}'}^+ + b_{p',-\vec{q}'}) \quad (2.39)$$

where the coupling constants are

$$F_1(m\alpha, p, \vec{q}) = \sum_{n,\beta,x} A(p, \vec{q}, n\beta, x) \frac{\partial D^*_{m\alpha}}{\partial R^\beta_{nx}}\bigg|_{R^\beta_{nx}=0}$$

and   (2.40)

$$F_2(m\alpha, p, \vec{q}, p', \vec{q}') = \sum_{\substack{n,\beta,x \\ n',\beta',x'}} A(p, \vec{q}, n\beta, x)A(p', \vec{q}', n'\beta', x')$$

$$\times \left(\frac{\partial^2 D^*_{m\alpha}}{\partial R^\beta_{nx} \partial R^{\beta'}_{n'x'}}\bigg|_{\substack{\text{excited state} \\ \text{equilibrium}}} - \frac{\partial^2 D^0_{ma}}{\partial R^\beta_{nx} \partial R^{\beta'}_{n'x'}}\bigg|_{\substack{\text{ground state} \\ \text{equilibrium}}} \right)$$

and the amplitudes are defined by

$$A(p, \vec{q}, n\beta, x) = \sqrt{\frac{\hbar}{2NM_{\beta x}\omega_p(\vec{q})}} \exp(-i\vec{q} \cdot \vec{r}_n)e^\beta_{\vec{q},x,p} \qquad (2.41)$$

The first two lines of Eq. (2.39) correspond to independent electronic excitations and phonons. The third and fourth lines are respectively linear and quadratic coupling. Higher-order terms in Eq. (2.39), not expressed here, would correspond to higher-order differences in the ground-state and excited-state intermolecular potentials. This Hamiltonian thus contains just the same information as the definition of the harmonic phonon modes for each electronic state. If the reader were presented with Eq. (2.39), with no prior knowledge of the problem, a suitable canonical transformation would remove the coupling, "revealing" shifted modes with altered frequencies in the excited state.[63]

Note that the diagonal coupling in the site basis will couple the electronic eigenstates, allowing for relaxation. This is discussed in Section IV with the help of the electronic density matrix.

III. TEMPERATURE DEPENDENT HOPPING IN NARROW IMPURITY BANDS

A. Quantitative Consequences of Hopping Transport of Site Excitations

As pointed out in Section I, the quantitative predictions of the hopping model were not worked out in the debate about the nature of the transport. We carried out calculations, in order to understand the temperature dependence of the yield of exciton transfer.[64] Hopping transport generally is modelled by introducing the Pauli master equations for the diagonal elements of the electronic density matrix, or populations of the sites involved. Postponing a detailed analysis of the validity of this representation, let us assume that it is possible to write such equations for the site-occupation probabilities $p_i(t)$:

$$\frac{dp_i}{dt} = -p_i/\tau_i - \left(\sum_{j \neq i} k_{ji}\right) p_i + \sum_{j \neq i} k_{ij} p_j + s_i(t) \qquad i = 1, \ldots, M \quad (3.1)$$

for M molecules with intrinsic lifetimes τ_i, connected by transfer rates k_{ij} between sites j and i. The last term, $s_i(t)$, gives the pumping rate of site i. The hopping rates in this model are deduced from the Golden Rule and are of the form

$$k_{ij} = |v_{ij}|^2 \xi(E_i - E_j, T) \qquad (3.2)$$

In this expression v_{ij} is the electronic matrix element between sites i and j and ξ, the thermal factor, depends on the density of final phonon states which balance the energy mismatch between sites i and j. It depends on the temperature and on the particular mechanism involved. We assume here, and prove later, that the rates are symmetric for donor–donor transfer, because of the narrowness of the donor band compared to experimental temperatures. Further, donor–acceptor transfer is assumed to be irreversible because of the great depth of the acceptor: $E_D - E_A \sim 240 \, \mathrm{cm}^{-1} \gg k_B T$, where k_B is the Boltzmann constant.

We study here the steady-state yield of acceptor emission, defined by

$$r = \varphi_A P_A / (\varphi_A P_A + \varphi_D P_D) \qquad (3.3)$$

where P_D (respectively P_A) is the total probability of a donor (acceptor) exciton surviving anywhere in the sample, under steady state, homogeneous illumination of all the donors,

$$P_D = \sum_{\substack{i \in D \\ \lim t \to \infty}} p_i(t) \qquad \text{with } s_i(t) = s \qquad (3.4)$$

The phosphorescence quantum yields of the donors and acceptors are assumed to be equal, as are their lifetimes, after Birks.[65] Under steady-state conditions the matrix form of the master equations can be written as

$$d\vec{P}/dt = k\vec{P} + \vec{S} = 0 \qquad (3.5)$$

where $\vec{P}$ is the vector of components $p_i(t)$ and $\vec{S}$ is the vector of source terms $s_i(t) = s$. The kinetic matrix K is given by

$$K_{ij} = -\delta_{ij}\left(\frac{1}{\tau_i} + \sum_{l \neq i} k_{li}\right) + (1 - \delta_{ij})k_{ij} \qquad (3.6)$$

The steady-state populations can be found by solution of the linear system

$$K\vec{P} = -\vec{S} \qquad (3.7)$$

Equation (3.7) corresponds to one particular configuration of the guests on the scale of the transport measured by trapping, typically $\langle r_{AA} \rangle \sim c_A^{-1/2}$. Experimental samples are much larger, so that observable quantities are configurational averages. Numerical methods are preferred here to analytical methods of configurational averaging,[66–68] because very little is known about the accuracy of mean field methods. Full numerical details are discussed in Ref. 64, but the main points are as follows.

Donors and acceptors, randomly distributed in independent (a, b) crystallographic planes, are coupled by exchange interactions between translationally equivalent $(0.5\,\text{cm}^{-1})$ and inequivalent $(1.2\,\text{cm}^{-1})$ neighbours. The depths of the donor and acceptor excited states below the host band are respectively $\Delta_D \sim 100\,\text{cm}^{-1}$ and $\Delta_A \sim 340\,\text{cm}^{-1}$. Nonnearest-neighbor guests can interact by tunneling or superexchange interactions, due to renormalization of the guest subspace by interactions with the host sites (see Section B below). The effective or superexchange interaction may be expressed by the approximate formula

$$v = GV(V/\Delta_D)^n \quad \text{between donors}$$

and

$$v = GV(V/\Delta_D)^{n-1}(V/\Delta_A) \quad \text{between donors and acceptors}$$

where G is a geometrical factor for the number of shortest paths, through n host sites, between the two guests. These interactions are well described in naphthalene by the approximate form:[16]

$$v \sim 136 \exp(\sim(\alpha/2)R)\,\text{cm}^{-1}$$

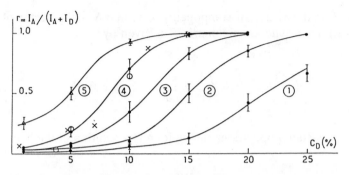

Figure 3.1. Concentration dependence of the yield of donor–acceptor (acceptor concentration 0.1%) transfer in mixed crystals of naphthalene, calculated from Eq. (3.1) with different values of the interaction radius: (1) 14 Å; (2) 16.6 Å; (3) 19.1 Å; (4) 21.7 Å; (5) 21.7 Å with a higher acceptor concentration, 0.5%. The error bars are 95% confidence limits.

where R is in units of the distance between inequivalent nearest neighbors (5.095 Å) and $\alpha = 2\ln(\Delta/V) \sim 9.2$.

The transfer rates can then be written in the form

$$k_{ij} = \frac{1}{\tau_D}\exp(-\alpha(r_{ij} - \bar{R}))\xi(|E_i - E_j|, T) \tag{3.8}$$

The transfer radius $\bar{R}$, at which transfer and decay are equiprobable, is determined in the following way. We assume the thermal factors ξ to be the same for all donor pairs (narrow donor band) and equal to the temperature-independent donor–acceptor factor at some temperature T_0 in the low-temperature region where the yield of transfer is independent of temperature, $T \lesssim 5$ K. Donor–acceptor transfer is supposed to be temperature independent here because it involves emission of large phonons to conserve energy. The transfer radius can then be adjusted by setting $\xi(T_0) = 1$ and fitting the calculated yield of acceptor emission to the experimental results. Figure 3.1 shows this for $T_0 = 1.6$ K. The best fit is obtained with $\bar{R} = 21.7$ Å $= 4.3$ units. The fifth curve of Fig. 3.1 shows the effect of increasing the acceptor concentration.

Figure 3.2 shows the next step, which is to determine by how much the thermal factor must change in order to account for the experimental change in the acceptor yield between 1.6 and ~5 K. The thermal factor clearly must change by 6–8 orders of magnitude to account for the experimental results. This is rather surprising for such a small temperature

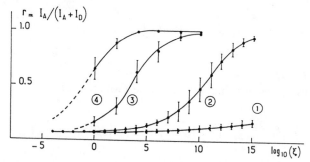

Figure 3.2. Influence of the thermal factor in Eq. (3.2) on the yield of donor–acceptor transfer ($c_A = 0.1\%$). Curves 1–4 are the calculated yields at donor concentrations 1, 2.5, 5 and 10%. Comparison with Fig. 1.5 shows that the thermal factor must change by a factor of 10^6–10^8 between 2 and 5 K to account for the threshold behavior of Fig. 1.5(a).

interval and shows the need for a closer look at the mechanisms of exciton–phonon coupling, pursued in Section B below.

B. Hopping Rates Between Nearly Resonant Guests

1. *General Relations*

In this section we calculate the hopping rates k_{ij} appearing in Eqs. (3.1) and (3.2) by application of the Golden Rule.[69] The starting point is the total Hamiltonian, expressed in the site excitation basis. We consider here just two guests, labelled 1 and 2, and linear or quadratic coupling to phonons. Thus:

$$
\begin{aligned}
H = {} & \sum_m E_m B_m^+ B_m + \sum_{m,n} V_{mn} B_m^+ B_n + \sum_{\vec{q}} \hbar\omega(\vec{q}) \left(b_{\vec{q}}^+ b_{\vec{q}} + \frac{1}{2} \right) \\
& + \sum_{m,\vec{q}} F_1(m,\vec{q}) B_m^+ B_m (b_{\vec{q}} + b_{-\vec{q}}) \\
& + \sum_{m,\vec{q},\vec{q}'} F_2(m,\vec{q},\vec{q}') B_m^+ B_m (b_{\vec{q}}^+ + b_{-\vec{q}})(b_{\vec{q}}^+ + b_{-\vec{q}'}) \qquad (3.9)
\end{aligned}
$$

This Hamiltonian is divided into a principle part, H_0, under which the site excitations (Eq. 2.12) are stationary states and a perturbation, V, responsible for transitions between sites. The partition for site excitations is

$$H_0 = E_1 B_1^+ B_1 + E_2 B_2^+ B_2 + \sum_{\vec{q}} \hbar\omega(\vec{q}) \left(b_{\vec{q}}^+ b_{\vec{q}} + \frac{1}{2} \right) \qquad (3.10)$$

and $V = H - H_0$.

Transition rates are then given by the Golden Rule:

$$k_{21} = \frac{2\pi}{\hbar} \sum_{\alpha,\beta} p(\alpha)\langle 1\alpha| \mathscr{T}(z = E_{1\alpha})|2\beta\rangle^2 \delta(E_{1\alpha} - E_{2\beta}) \qquad (3.11)$$

where the "T-matrix," $\mathscr{T}$, is given by

$$\mathscr{T} = V + VG_0V + VG_0VG_0V + \cdots \qquad (3.12)$$

and $G_0(z) = 1/(z - H_0)$ is the unperturbed propagator.

Here α and β stand respectively for the initial and final bath states and $p(\alpha)$ is the probability of initial state α, supposing the bath to be in thermal equilibrium at temperature T. Nonnearest neighbors are coupled by developing $\mathscr{T}$ to a sufficiently high order in the electronic coupling to cause the superexchange interaction to appear. This interaction will be written as v. Further, the main contribution to the probability amplitude comes from the first term to contain this interaction, with phonon interactions only on the initial and final guest sites, because phonon interactions on intermediate hosts are attenuated by unfavorable energy denominators. Each phonon operator introduces a factor $(n + 1)^{1/2}$ for creation and $n^{1/2}$ for annihilation, where n is the number of phonons in the mode in the initial bath state α. Let $\delta = E_1 - E_2$ be the energy gap between the guests. Excitation spectra show it is less than or of the order of $1\ \mathrm{cm}^{-1}$ for pairs of donors. It may be very much smaller on the scale of the transport path (cf. Section I). It is about $240\ \mathrm{cm}^{-1}$ for donor-to-acceptor transfer.

2. Transition Amplitudes for One- and Two-Phonon Processes

a. One Phonon Processes

The one-phonon process stems from the term VG_0V in which the electronic coupling appears once with the phonon coupling, but in either order, giving

$$\left\{ \begin{array}{c} \sqrt{n+1} \\ \sqrt{n} \end{array} \right\} F_1(1, p, \vec{q}) \frac{1}{\pm E} v + v \frac{1}{\delta} F_1(2, p, \vec{q}) \left\{ \begin{array}{c} \sqrt{n+1} \\ \sqrt{n} \end{array} \right\} \qquad (3.13)$$

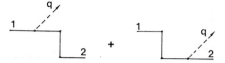

Figure 3.3.   Diagrams contributing to hopping assisted by one-phonon processes. Horizontal lines are unperturbed propagators, vertical lines are electronic couplings, and dots are exciton–phonon coupling.

where $E = \hbar\omega_p(\vec{q})$ is the energy of the phonon of wavevector $\vec{q}$ in branch p. On account of conservation of energy, $0 = \delta \pm E$, only one process is possible for a given δ. If $\delta > 0$, the process is emission, for example, giving the amplitude

$$\sqrt{n+1}\,\frac{v}{\delta}(F_1(1,p,\vec{q}) - F_1(2,p,\vec{q})) \qquad (3.14)$$

Figure 3.3 shows the corresponding diagrams.

b.   Two-Phonon Processes

Quasi-resonant transfer in the narrow donor band can occur by a number of two-phonon processes: Emission or absorption of two phonons, or an absorption–emission process. Either linear coupling, taken twice, or quadratic coupling may be responsible. Only absorption–emission processes are studied here, since it is easily seen that it is the only possible one for coupling to optical phonons. It is also the major contribution of acoustic phonons because absorption–emission is possible for any pair of phonons $\vec{q}, \vec{q}'$ such that $\delta + E' - E = 0$, whereas the first two processes require respectively $\delta + E + E' = 0$ and $\delta - E - E' = 0$, implying $|E|, |E'| < \delta$. Hence double absorption or emission contributes only in small regions around the center of the Brillouin zone, where the density of acoustic modes and the corresponding matrix elements vanish. The absorption–emission process, on the other hand, contains contributions from all the acoustic modes. The diagrams corresponding to two-phonon processes are shown in Fig. 3.4. The corresponding scattering amplitudes are given respectively by the expressions

$$\sqrt{n(n'+1)}\,\frac{v}{EE'}(F_1(1,p,\vec{q}) - F_1(2,p,\vec{q}))(F_1(1,p',\vec{q}') - F_1(2,p',\vec{q}'))$$

$$(3.15)$$

for double linear coupling and

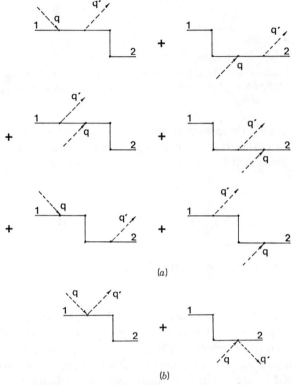

Figure 3.4. Diagrams for two-phonon processes, with the same conventions as Fig. 3.1. (a) Second-order linear coupling. (b) Quadratic coupling.

$$\sqrt{n(n'+1)}\,\frac{v}{\delta}[F_2(1,p,\vec{q},p',\vec{q}') - F_2(2,p,\vec{q},p',\vec{q}')] \qquad (3.16)$$

for quadratic coupling. The coupling strengths are given by Eq. (2.40).

c. Discussion

Expressions (3.14)–(3.16) show that transfer between sites 1 and 2 is possible only if the local coupling strengths differ. This can come about in several ways. Molecules 1 and 2 may have different weights or electronic structures and thus different participation in the perturbed phonon modes of the host. This effect of mass is expected to be negligible for protonated aromatic molecules in a deuterated host. Different local couplings can then occur if the guests occupy different sites in their respective lattice cells, or if they are on the same site but in cells which are out of phase.

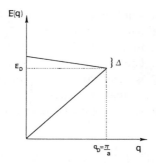

Figure 3.5. The calculation of the hopping rates assumes isotropic, linear dispersion of the active phonon modes (cf. Fig. 2.1). E_D is the Debye energy of the acoustic modes. Δ is the dispersion of the lowest optical modes.

Using Eqs. (2.14, 2.16, 2.18, 2.40 and 2.41), the difference in the linear coupling between the sites can be written as

$$
F_1(1, p, \vec{q}) - F_2(2, p, \vec{q}) = \sqrt{\frac{\hbar}{2N\omega_p(\vec{q})}} \sum_x \frac{1}{\sqrt{M_x}}
$$

$$
\times \left\{ \left[\sum_{m\beta} e^x_{\vec{q},\beta_1,p} \frac{\partial D^*(1, m\beta)}{\partial R^x_{1\beta_1}} + \vec{e}^x_{\vec{q},\beta,p} \frac{\partial D^*(1, m\beta)}{\partial R^x_{m\beta}} \exp(-i\vec{q} \cdot (\vec{r}_m - \vec{r}_1)) \right] \right.
$$

$$
\times \exp(-i\vec{q} \cdot \vec{r}_1)
$$

$$
- \left[\sum_{m\beta} \vec{e}^x_{\vec{q},\beta_2,p} \frac{\partial D^*(2, m\beta)}{\partial R^x_{2\beta_2}} + \vec{e}^x_{\vec{q},\beta,p} \frac{\partial D^*(2, m\beta)}{\partial R^x_{m\beta}} \exp(-i\vec{q} \cdot (\vec{r}_m - \vec{r}_2)) \right]
$$

$$
\left. \times \exp(-i\vec{q} \cdot \vec{r}_2) \right\} \tag{3.17}
$$

where M_x is the moment of inertia of the local coordinate x.

All the quantities relating to phonons in Eq. (3.17) are known for systems like naphthalene, for which the complete phonon dispersion curves and modes have been determined.[58] However, such sophistication will be dropped here, because of other approximations like two-dimensional transport. We shall suppose that the three acoustic modes have linear dispersion up to a common Debye energy, E_D, and that the lowest optical mode has linear dispersion between E_D and $E_D + \Delta$ (see Fig. 3.5).

Finally, $\|\vec{e}_q\| = 1$, so we shall drop reference to the site in the cell in the order of magnitude estimations below.

Both sites in the naphthalene lattice have the same surroundings, connected by a screw axis. The contents of the square brackets are therefore equal. This is the most we can say for coupling to a general mode. The sum in the square bracket may be expected to be small, due to interference. The coupling strength to optical phonons is then

$$F_1(1, p, \vec{q}) - F_1(2, p, \vec{q}) \sim \left(\frac{\hbar}{2N\omega_p(\vec{q})}\right)^{1/2} \sum_x \frac{1}{\sqrt{M_x}} \frac{\partial D_1^*}{\partial R_1^x} (e^{-i\vec{q} \cdot \vec{r}_1} - e^{-i\vec{q} \cdot \vec{r}_2})$$

$$(3.18)$$

Better can be done for acoustic phonons of long wavelength, at the zone center, $|\vec{q}| = 0$. These are the only populated modes at low temperatures. Further, they are pure translational motions. But $D^*(1, m)$ depends only on the difference $\vec{r}_1 + \vec{R}_1 - \vec{r}_2 - \vec{R}_2$, so that

$$\frac{\partial D^*(1, m)}{\partial R_1^x} = -\frac{\partial D^*(1, m)}{\partial R_m^x} \tag{3.19}$$

Carrying this into Eq. (3.17) we have

$$F_1(1, p, \vec{q}) - F_1(2, p, \vec{q}) \sim \left(\frac{\hbar}{2N\omega_p(\vec{q})}\right)^{1/2} \left(\sum_x \frac{1}{\sqrt{M_x}} \frac{\partial D_1^*}{\partial R_1^x}\right)(qa)$$

$$\times (e^{-i\vec{q} \cdot \vec{r}_1} - e^{-i\vec{q} \cdot \vec{r}_2}) \tag{3.20}$$

for acoustic phonons of long wavelength, where a is the magnitude of the lattice constant. Similar arguments hold for quadratic coupling, yielding

$$F_2(1, 2, p, \vec{q}, p', \vec{q}') \sim \frac{\hbar}{2N}\left(\frac{1}{\omega_q(\vec{q})\omega_{p'}(\vec{q}')}\right)\left(\sum_{x,x'} \frac{1}{\sqrt{M_x M_{x'}}} \frac{\partial^2 (D_1^* - D_1^0)}{\partial R_1^x \partial R_1^{x'}}\right)$$

$$\times (qa)(q'a)(e^{-i(\vec{q}+\vec{q}') \cdot \vec{r}_1} - e^{-i(\vec{q}+\vec{q}') \cdot \vec{r}_2}) \tag{3.21}$$

for long wavelength acoustic phonons and

$$\frac{\hbar}{2N} \left(\frac{1}{\omega_p(\vec{q})\omega_{p'}(\vec{q}')} \right)^{1/2} \left(\sum_{x,x'} \frac{1}{\sqrt{M_x M_{x'}}} \frac{\partial^2 (D_1^* - D_1^0)}{\partial R_1^x \partial R_1^{x'}} \right)$$

$$\times (e^{-i(\vec{q}+\vec{q}') \cdot \vec{r}_1} - e^{-i(\vec{q}+\vec{q}') \cdot \vec{r}_2}) \tag{3.22}$$

for optical phonons.

3. Transfer Rates

The transfer rates can now be found by summing the squares of the amplitudes of all the processes discussed above. Discrete summation is replaced by integration over a sphere of radius q_D in spherical coordinates

$$\sum_{\vec{q}} \delta(E(\vec{q}) - E_0) f(\vec{q}) = \frac{N}{\frac{4\pi}{3} q_D^3} \int_0^{q_D} dq \, q^2 \int_0^\pi d\theta \sin \theta$$

$$\times \int_0^{2\pi} d\psi f(q, \theta, \psi) \frac{1}{\dfrac{dE}{dq}\Big|_{q_0}} \delta(q - q_0) \tag{3.23}$$

where $\vec{q}_0$ is a root of $E_0 = \hbar\omega(\vec{q})$. Approximation of the polyhedral Brillouin zone is acceptable here because (1) the thermal weighting of the acoustic modes annihilates contributions from the zone boundaries; (2) the approximation has very little effect on the temperature dependence of the coupling to optical phonons. Finally, we use the relation $\hbar/M_x E_D \sim 1/\hbar\delta R_x^2$, where δR_x is the amplitude of the zero-point motion of coordinate x, to obtain the following estimations of the transfer rates in the narrow donor band, $v \ll \delta \ll k_B T$:

1. One-acoustic-phonon processes

$$k_{21} \sim \frac{1}{\hbar} |v(r_{12})|^2 \left(\frac{r_{12}}{a} \right)^2 \left(\frac{\partial D^*}{\partial R} \cdot \delta R \right)^2 \delta^2 k_B T / E_D^6 \tag{3.24}$$

2. Absorption-emission of acoustic phonons

$$k_{21} \sim \frac{1}{\hbar} |v(r_{12})|^2 \left(\frac{r_{12}}{a} \right)^4 \left(\frac{\partial D^*}{\partial R} \cdot \delta R \right)^4 (k_B T)^7 / E_D^{12} \tag{3.25}$$

for double linear coupling

$$k_{21} \sim \frac{1}{\hbar} |v(r_{12})|^2 \left(\frac{r_{12}}{a}\right)^2 \left(\frac{\partial^2 (D^* - D^0)}{\partial R^2} \cdot \delta R^2\right)^2 (k_B T)^9 / E_D^{12}$$

$$(3.26)$$

for quadratic coupling

3. Absorption–emission of optical phonons

$$k_{21} \sim \frac{1}{\hbar} |v(r_{12})|^2 \left(\frac{\partial D^*}{\partial \theta} \delta \theta\right)^4 \frac{e^{-E_D/k_B T}}{E_D^4 \Delta} \qquad (3.27)$$

for double linear coupling

$$k_{21} \sim \frac{1}{\hbar} |v(r_{12})|^2 \left(\frac{\partial^2 (D^* - D^0)}{\partial \theta^2} \delta \theta^2\right)^2 \frac{\exp(-E_D/k_B T)}{E_D^2 \Delta} \qquad (3.28)$$

for quadratic coupling

Coupling to acoustic phonons thus leads to power-law dependence of the hopping rates on temperature. Hamilton, Selzer, and Yen[70] noted that the exact power depends rather strongly on the dispersion assumed for the acoustic modes. We add here that cancelling of the derivatives for crystals in Eq. (3.19), which appears to have gone unnoticed, adds a factor T^4 to double linear coupling and a factor T^2 to quadratic coupling. It could help to distinguish between these mechanisms. In any case, coupling to acoustic phonons is favored by their thermal populations, compared to those of optical modes, but they are disserved by a lower density of modes, varying in $|\vec{q}|^2$ and vanishing matrix elements. The activation behavior of coupling to optical modes reflects the thermal population of optical phonons at temperature T. Note that the rates in Eqs. (3.24) and (3.28) are independent of the sign of the energy mismatch, confirming the initial assumption in the simulations.

C. Application to Naphthalene

1. *Application to the Temperature Dependence of Donor-Acceptor Transfer*

Figure 3.6 shows the result of using the laws given above of exciton–phonon coupling to scale the thermal factor $\xi(T)$ in the simulations of Fig. 3.2. Only the absorption–emission process of coupling to optical phonons explains the very sharp rise in the yield of acceptor emission at about 5 K. Further confirmation of the likelihood of this law comes from Arhenius plots of the experimental yield of acceptor emission (Fig. 3.7). Experiments show a low temperature constant value followed by activated

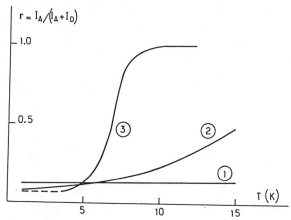

Figure 3.6. Temperature dependence of the yield of acceptor emission calculated with different kinds of exciton–phonon coupling in the donor band: (1) One-acoustic phonon process, $k_{hop} \propto T$. (2) Two-acoustic phonon process, $k_{hop} \propto T^7$. (3) Coupling to two optical phonons, $k_{hop} \propto \exp(-E_D/k_B T)$. The donor and acceptor concentrations are, respectively, 2.5 and 0.1%.

emission. If the low-temperature constant values are subtracted before plotting, the plots are straight lines from which the activation energy 35 ± 5 cm^{-1} can be deduced. This is quite close to the energy of the optical phonons in naphthalene. Delayed fluorescence of crystals containing only the donor guests also shows activation, at the same temperatures and with the same activation energy.[48] Note that this activation is not due to transport through the host band. Firstly, the activation energy is wrong. Secondly, the delayed fluorescence of the same crystals shows clear activation with energy 90 cm^{-1}, the depth of the guest below the host band, at temperatures between 15 and 20 K, (Fig. 1.5).[71]

Increasing the temperature has little effect on the calculated yield at donor concentrations below 1%. This is because of the large average distance and consequently weak couplings at these concentrations. The yield probably would be higher if account was taken of out of plane couplings (see discussion below).

2. *Difficulties of the Hopping Model*

In this section we have put the hopping model of exciton transport on a more quantitative footing, by assuming that a set of master equations can be written for the site-occupation probabilities (Eq. 3.1), in which the transition probabilities are deduced from the Golden Rule. Two-dimensional transport was assumed. Extension to three dimensions is conceptually simple, but poses formidable numerical problems. Such an extension

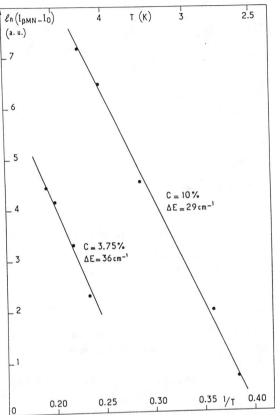

Figure 3.7. Arrhenius plot of the experimental yield of acceptor emission after subtraction of the part independent of temperature below the threshold. The activation energies are of the right order of magnitude for coupling to optical phonons, Fig. 2.1.

is desirable because even though the direct exchange interaction between planes is estimated[51] to be about 10^{-4}–10^{-3} cm^{-1}, due to poor overlap of the π orbitals, it is larger than the superexchange interaction in the plane at distances over R_{3d} = 2–3 lattice units. Fitting the data to a three-dimensional model should therefore lower the hopping radius to more reasonable values. The two dimensional fit, $\bar{R}$ = 21.7 Å, implies that couplings of the order of 10^{-7} cm^{-1} are relevant to the transport problem, a value we view with scepticism.

An equally serious limitation is that we considered only hopping between monomer donors. As the concentration rises, there are in fact more and more clusters of neighboring sites, with energy levels determined by the resonance exchange couplings. The lower levels of these clusters act at low temperatures as traps,[5] with depths below the monomer levels in the range of $5 \, cm^{-1}$. Back transfer from such shallow traps occurs by absorption of an acoustic phonon.[72,73] There is thus a mixture of temperature dependences, determined by the fluctuations of the local donor concentration and singularly complicating the analysis.

Next, we must point out a difficulty of the hopping model, independent of its dimensionality, when confronted with the temperature dependence of the yield of acceptor emission. Hopping transport is predicted to be temperature-dependent at all temperatures, due to contributions of several kinds of exciton-phonon scattering (Eqs. 3.24–3.28). The steep radial dependence of the hopping rates (superexchange interactions), requires a strongly temperature-dependent thermal factor due to coupling to optical phonons, to create the required connectivity of the donor sites. The transition between temperature-independent behavior and the thermal activation could be thought to be due to the transition between these two kinds of scattering, maintaining the hopping model over the entire experimental temperature range. This is not correct, because the activation behavior is observed at different temperatures for different concentrations. Thus scattering on optical phonons is the predominant mechanism. The constant yield of acceptor emission then corresponds to the freezing out of hopping transport. The constant value might be thought to be due to direct trapping of donor excitons from donors which happen to be close to an acceptor. The problem is that the yield of direct, one-step donor–acceptor trapping should be independent of the donor concentration. Its value is determined by the branching ratio for donor emission in the presence of traps, averaged over all donors:

$$r = 1 - \left\langle \frac{1/\tau_D}{1/\tau_D + 1/\tau_{\text{trapping}}} \right\rangle = 1 - \frac{1}{N_D} \sum_{i \in D} \frac{1}{1 + \Sigma_{j \in A} k_{ji}}$$

This is the value, about 6% in Fig. 3.2, reached by the yield when the thermal factor is very small. The experimental results show on the contrary that the constant yield increases with the donor concentration. This behavior is what would be expected from direct trapping from donor states (temperature independence) whose extension increases with the donor concentration due to delocalization. Both behaviors being present in a small temperature region and at the same temperature for different donor concentrations, it seems necessary to include both in a unified model, including both hopping and delocalization. This is achieved in the next section by introducing and studying the donor eigenstates and their relaxation, described by the pilot equation for the electronic density matrix.

IV. TRANSPORT OF EXCITONS DESCRIBED BY THE ELECTRONIC DENSITY MATRIX

A. Introduction

Much has been said about the coherence or the incoherence of triplet excitons in mixed crystals, but in view of the continuing confusion, we feel it is useful to discuss the problem here, starting from a first-principles discussion of site coherence, using the electronic density matrix and the pilot equation. Our reason is to distinguish below between the ideas of coherence and absence of diffusion. The problem is simplified here because we can neglect macroscopic coherence, the in-phase motion of all the transition moments of the sample, because the sharpest optical excitation used to date had a width of around $10^{-3}\,\mathrm{cm}^{-1}$. Straightforward precession will disperse macroscopic coherence of this set of states in about $10^{-8}\,\mathrm{s}$, much shorter than experimental time scales.

The Hamiltonian (Eq. 2.39) can in principle be used in the Schrödinger equation to predict the evolution of the sample, but we do not know the exact initial state, nor even the distribution of traps in the mixed crystal. Two kinds of average are needed here. First, an average over initial states of the phonon bath and second, an average over all possible configurations of the traps. This chapter deals with the first average. The second can be tackled by simulation or by mean field theory.

1. *The Density Matrix*

Consider a system for which we have no detailed knowledge of the initial state, but just the probability, p_s, of finding it in state ψ_s at time $t = 0$, with $\Sigma_s p_s = 1$. Imagine now an ensemble of replicas of the system, with initial states drawn from this distribution of ψ_s. A system in state ψ_s evolves according to the Schrödinger equation:

$$i\hbar \frac{\partial |\psi_s(t)\rangle}{\partial t} = H(t)|\psi_s(t)\rangle \tag{4.1}$$

The quantum average value of an observable G measured in this state at time t is

$$G_s(t) = \langle \psi_s(t)|G|\psi_s(t)\rangle \tag{4.2}$$

The same is true of all the other replicas, so the ensemble average is

$$\langle G \rangle(t) = \sum_s p_s G_s(t) \tag{4.3}$$

This average can be written as

$$\langle G \rangle(t) = \mathrm{Tr}(\rho(t)G) \tag{4.4}$$

where the density matrix $\rho(t)$ is the weighted sum of the projection operators on the states reached by all the systems in the ensemble,[74,75]

$$\rho(t) = \sum_s p_s |\psi_s(t)\rangle\langle\psi_s(t)| \tag{4.5}$$

Referred to a basis $\{|u_i\rangle\}_i$, the diagonal elements of $\rho(t)$ represent occupation probabilities, whereas the off-diagonal elements or coherences describe the persistence of quantum interference effects when the initial state is a superposition of states in this basis. Thus, inserting the identity operator in Eq. (4.4), the average value of G is

$$\langle G \rangle(t) = \mathrm{Tr}(\rho(t)G) = \sum_i \langle u_i|\rho(t)|u_i\rangle\langle u_i|G|u_i\rangle$$
$$+ \sum_i \sum_{i\neq j} \langle u_i|\rho(t)|u_j\rangle\langle u_j|G|u_i\rangle \tag{4.6}$$

which should be compared with its value for one system in a superposition of states u_i:

$$|\psi\rangle = \sum_i c_i |u_i\rangle$$
$$\langle G \rangle_\psi = \sum_i |c_i|^2 \langle u_i|G|u_i\rangle + \sum_i \sum_{j\neq i} \bar{c}_i c_j \langle u_i|G|u_j\rangle$$

The kinetic equation for the density operator, the Liouville equation, follows from Eqs. (4.1) and (4.5):

$$i\hbar \frac{\partial \rho(t)}{\partial t} = [H(t), \rho(t)] \tag{4.7}$$

whence we have for closed systems, with Hamiltonians independent of time,

$$\rho(t) = \exp(-iHt/\hbar)\rho(0)\exp(iHt/\hbar) \tag{4.8}$$

The populations of the eigenbasis states of H are then constant and the coherences precess at the Bohr frequencies:

$$\rho_{kk}(t) = \rho_{kk}(0)$$

$$\rho_{kk'}(t) = \exp(-i\omega_{kk'}t)\rho(0) \qquad (4.9)$$

$$\hbar\omega_{kk'} = E_k - E_{k'}$$

It can be shown[74,75] that if the system is put into thermal equilibrium by contact with a bath at temperature T, then the equilibrium density matrix is

$$\rho(t \to \infty) = Z^{-1}\exp(-H/k_BT) \qquad (4.10)$$

where Z is the partition function. This gives the eigenbasis of H a special meaning as the natural basis for describing the relaxation of the system after action of external forces. The equilibrium populations deduced from Eq. (4.10) are

$$p_k = e^{-E_k/k_BT} \Big/ \sum_m e^{-E_m/k_BT}$$

whereas the coherences vanish. In other words, repeated interaction with the bath pushes the populations towards equilibrium and destroys the memory of the initial phase which would otherwise have produced interference effects. We also see from Eq. (4.10) that the eigenbasis of a system in equilibrium with a heat bath verifies the random-phase approximation. In any other basis, such as the site basis for our present problem, the coherences are nonzero. Nonrandom phases in the eigenbasis may result from driving the system with a periodical force, as we shall see below.

2. Optical Coherence, Exciton Coherence, and Site Coherence

Resonant exchange interactions in the triplet state occur by short-ranged electron exchange interactions. Moreover, the coupling to the radiation field is very weak, because the transition is spin-forbidden to first order. We will thus neglect the delayed light field and use a tight-binding Hamiltonian, the electronic part of Eq. (2.39) for N sites (ground-state energy set to zero):

$$H = \sum_m^N E_m |m\rangle\langle m| + \sum_m^N \sum_{n \neq m} V_{mn} |m\rangle\langle n| \qquad (4.11)$$

Introducing the eigenbasis of H, we can also write

$$H = \sum_k E_k |k\rangle\langle k|$$

and   (4.12)

$$|k\rangle = \sum_m U_{km} |m\rangle$$

where U is the unitary transformation matrix between the bases. Disorder of the site energies leads to two kinds of states in mixed crystals. First, perturbed band states, which do not concern us here, and second, trap states. The trap states are localized: The amplitudes U_{km} decay exponentially around some center of gravity of large values, identifiable with an aggregate of strongly coupled, nearest-neighbor traps.

The *site coherences* and *populations*, or density operator elements in the basis of perfectly localized, oriented gas excitations, are not observable quantities, but the evolution of the system is often discussed in the site representation. The first reason is that the site basis shows up interference effects which disturb our classical intuition. The second is that stationary site or local coherence is a measure of the purity of the quantum state in the relaxation basis. As an example, consider the evolution of the site coherence deduced from Eqs. (4.10) and (4.12):

$$\rho_{mn}(t) = \sum_{kk'} U_{km} \bar{U}_{k'n} \rho_{kk'}(0) \exp(-i\omega_{kk'}t) \qquad (4.13)$$

This shows that local coherences will be stationary only if there are no *exciton coherences* in the eigenbasis. This is the case for systems in equilibrium with a heat bath. Thus

$$\rho_{mn}(t) = \sum_k U_{km} \bar{U}_{kn} \rho_{kk}(0) \qquad (4.14)$$

At high temperatures, that is, when $k_B T \gg (E_k - E_{k'})$, all the N eigenstates are equally populated so that

$$\rho_{mn}(t) = \delta_{mn}/N \qquad (4.15)$$

The site populations are thus all equal and there is no local coherence.

At sufficiently low temperatures, on the other hand, relaxation brings the system into its lowest eigenstate, which being in the band tail, generally will be a quantum-localized state. Site coherence is then stationary and strongest for sites in the center of gravity of the state.

Before introducing relaxation and dephasing, let us look simply at how electronic coherence can be created in a disordered solid. The following perturbational approach shows heuristically how longer-ranged local coherence can be induced by selectively preparing a $|k\rangle$ state in the middle of the band, where the amplitudes U_{km} decay slower than in the band tail. Monochromatic optical excitation forces all the transition dipoles of the sample into coherent oscillation. The interaction of these induced dipoles with the light field then builds up population in the excited state. Suppose, for example, that we irradiate the crystal with an electromagnetic pulse described by a Lorentzian wave packet of width Γ_L at mean frequency ω_L:

$$\vec{E}(\vec{r}, t) = \vec{E}_0 \int_{-\infty}^{+\infty} d\omega\, g(\omega) \cos(\omega t - \vec{k} \cdot \vec{r}) \qquad (4.16)$$

$$g(\omega) = \frac{\Gamma_L/2\pi}{(\omega - \omega_L)^2 + \Gamma_L^2} \qquad (4.17)$$

As the states we are interested in are localized in regions smaller than the wavelength of optical transitions, we shall neglect the spatial phase. This field perturbs the crystal, for which we may write a semi-classical, time-dependent Hamiltonian,

$$H(t) = H + V(t)$$
$$V(t) = -\vec{d} \cdot \vec{E}(t) \qquad (4.18)$$

where $\vec{d}$ is the polarization operator

$$\vec{d} = \sum_m \vec{d}_m(|0\rangle\langle m| + |m\rangle\langle 0|) \qquad (4.19)$$

Alternatively, use Eq. (4.12) to define the transition dipole $\vec{d}_k$ of state $|k\rangle$:

$$\vec{d}_k = \sum_m \bar{U}_{km}\vec{d}_m$$

Suppose that before turning on the field the crystal was in the ground state so that the initial density operator was

$$\rho_{-\infty} = \rho_{00}|0\rangle\langle 0|, \qquad \text{with } \rho_{00} = 1 \qquad (4.20)$$

The unperturbed motion in the Liouville equation can be removed by changing to interaction representation of operators A

$$A^I(t) = e^{iHt/\hbar} A e^{-iHT/\hbar}$$

whence

$$\rho^I(t) = \rho_{-\infty} + \int_{-\infty}^{t} dt' [V^I(t'), \rho^I(t')] \qquad (4.21)$$

Iterating leads to the perturbative development of the density matrix. The first-order approximation to the "optical" coherence (i.e., at optical frequencies) between the ground and the excited states is

$$\begin{aligned}
\rho_{k0} &= \frac{1}{2} \frac{(\vec{E}_0 \cdot \vec{d}_k) e^{-i\omega_L t} e^{\Gamma_L t}}{E_k - E_L - i\hbar\Gamma_L}, \qquad t < 0 \\
&= \frac{1}{2} \frac{(\vec{E}_0 \cdot \vec{d}_k) e^{-i\omega_L t} e^{-\Gamma_L t}}{E_k - E_L - i\hbar\Gamma_L} + \frac{i\hbar\Gamma_L (\vec{E}_0 \cdot \vec{d}_k) e^{-i\omega_k t}}{(E_k - E_L)^2 + (\hbar\Gamma_L)^2}, \qquad t > 0 \quad (4.22)
\end{aligned}$$

Recalling that the polarization of the sample at time t is

$$\begin{aligned}
\vec{D}(t) &= \mathrm{Tr}(\vec{d}\rho(t)) \\
&= \sum_k \vec{d}_k (\rho_{k0}(t) + \rho_{0k}(t))
\end{aligned} \qquad (4.23)$$

we can see two effects:

- Forced oscillation of the dipoles $\vec{d}_k$ at the frequency ω_L of the source. These oscillations are maximum at the same time as the field at $t = 0$.
- Natural oscillations of the dipoles at frequencies ω_k. These are strongest for states resonant with the source. They last indefinitely because we neglected dephasing and relaxation by coupling to the bath.

There are thus neither exciton populations nor coherences. These appear in the second-order term of $\rho(t)$. Exciton populations and coherences are built up by the interaction of the optical coherence with the field. As above we receive a driven term which decays with the applied field and a persistent term

$$\rho_{kk'}(t) \underset{t \to \infty}{=} \frac{1}{4} \cdot \frac{(\vec{E}_0 \cdot \vec{d}_k)(\vec{E}_0 \cdot \vec{d}_{k'})}{[(E_k - E_L)^2 + (\hbar\Gamma_L)^2][(E_{k'} - E_L)^2 + (\hbar\Gamma_L)^2]}$$
$$\cdot [(E_K - E_L)(E_{k'} - E_L) + 7(\hbar\Gamma_L)^2 + 3i\hbar\Gamma_L(E_k - E_{k'})] \cdot e^{-i\omega_{kk'}t}$$
$$(4.24)$$

The coherences precess at the Bohr frequencies, with amplitudes weighted by the spectral overlap of the source and of the electronic transitions. Note however that the transitions are not saturated when the pulse is very long. This is a shortcoming of a semiclassical, simple, perturbative approach. There are two extreme cases:

1. Selective excitation of one $|k\rangle$ state, with a source narrower than the separation of the eigenfrequencies: $E_L = E_k$ and $\hbar\Gamma_L \ll |E_k - E_{k'}|$. We then have

$$\rho_{kk}(t) \propto |\vec{E}_0 \cdot \vec{d}_k|^2/(\hbar\Gamma_L)^2 \qquad \text{as } t \to \infty \qquad (4.25)$$
$$\rho_{kk'}(t) = 0$$

Exciton coherence is thus absent, and site coherence such as it may be for state $|k\rangle$, is conserved:

$$\rho_{mn}(t) = U_{mk}\bar{U}_{kn}(\vec{E}_0 \cdot \vec{d}_k)^2/(\hbar\Gamma_L)^2 \qquad (4.26)$$

The total polarisation can be written as

$$\langle\vec{D}\rangle(t) = \sum_m U_{km}\vec{d}_m e^{-i\omega_k t} \qquad (4.27)$$

The site polarizations are different, but they all precess at the same frequency, so the state may reasonably be called "coherent."

2. Broad-band excitation: $\hbar\Gamma_L \gg (E_k - E_{k'})$. In this case site coherences and populations show interference effects due to preparation of a superposition of states.

$$\rho_{mn}(t) \propto \sum_k \frac{(\vec{E}_0 \cdot \vec{d}_k)^2}{(\hbar\Gamma_L)^2} U_{km}\bar{U}_{kn} + \sum_{kk'}' \frac{(\vec{E}_0 \cdot \vec{d}_k)(\vec{E}_0 \cdot \vec{d}_{k'})}{(\hbar\Gamma_L)^2} U_{km}\bar{U}_{k'n}\, e^{-i\omega_{kk'}t}$$

$$\rho_{mm}(t) \propto \sum_k \frac{(\vec{E}_0 \cdot \vec{d}_k)^2 |U_{km}|^2}{(\hbar\Gamma_L)^2} + \sum_{kk'}' \frac{(\vec{E}_0 \cdot \vec{d}_k)(\vec{E}_0 \cdot \vec{d}_{k'})}{(\hbar\Gamma_L)^2} U_{km}\bar{U}_{k'm}\, e^{-i\omega_{kk'}t} \tag{4.28}$$

We consider for the sake of simplicity identical molecules with the same transition moments, $\vec{d}$. The site populations are then

$$\rho_{mm}(t) = \frac{(\vec{E}_0 \cdot \vec{d})^2}{(\hbar\Gamma_L)^2} + \sum_{kk'}' \frac{(\vec{E}_0 \cdot \vec{d}_k)(\vec{E}_0 \cdot \vec{d}_{k'})}{(\hbar\Gamma_L)^2} U_{km}\bar{U}_{k'm}\, e^{-i\omega_{kk'}t} \tag{4.29}$$

If the sites absorbed independently, only the constant term would appear. The oscillatory terms, with vanishing average value over time scales longer than the reciprocal Bohr frequencies, have a simple meaning. They are due to the internal motion on each cluster of resonant molecules, due to excitation of a superposition of stationary states by the broad-band source. Thus, summing $\rho_{mm}(t)$ over all sites in a cluster of N_A nearest-neighbor traps, fulfilling the resonance condition,

$$\sum_{m=1}^{N_A} \rho_{mm}(t) = N_A \frac{(\vec{E}_0 \cdot \vec{d})^2}{(\hbar\Gamma_L)^2} + \sum_{kk'}' \left(\sum_{m=1}^{N_A} U_{km}\bar{U}_{k'm} \right)$$

$$\times \frac{(\vec{E}_0 \cdot \vec{d}_k)(\vec{E}_0 \cdot \vec{d}_{k'})}{(\hbar\Gamma_L)^2} e^{-i\omega_{kk'}t} \tag{4.30}$$

Because of the exponential localization of the wavefunctions, the major terms in the sum in brackets come from states $|k\rangle$ and $|k'\rangle$ centered on the same cluster. Moreover,

$$\sum_{m=1}^{N_A} U'_{km}\bar{U}_{k'm} \approx \delta_{kk'} \tag{4.31}$$

Since $k = k'$ is excluded, the double sum vanishes and the sum of the occupation probabilities of the cluster is constant.

The stationary part of the site coherence is large only when both U_{km} and U_{kn} are large. This generally is possible when m and n are in the same cluster. The oscillatory part may contain large contributions when m and n are in different clusters, A_m and A_n:

$$\rho_{mn}(t)_{\text{oscillatory}} = \sum_{k \in A_m} \sum_{k' \in A_n} \frac{(\vec{E}_0 \cdot \vec{d}_k)}{(\hbar \Gamma_L)^2} (\vec{E}_0 \cdot \vec{d}_{k'}) U_{km} \bar{U}_{k'n} e^{-i\omega_{kk'}t} \quad (4.32)$$

But, either (1) States $|k\rangle$ and $|k'\rangle$ are nonresonant and the corresponding term is averaged to zero in a short time of the order of $(\omega_{kk'})^{-1}$. If m is a monomer site and n a dimer site, $E_k = E_M$ and $E_{k'} = E_M \pm V$, where the exchange coupling is of the order of 1 cm^{-1}. Thus the coherence decay is about 10^{-11} s; or (2) states $|k\rangle$ and $|k'\rangle$ are resonant (e.g., two monomer sites). The oscillatory term may be nearly stationary over long periods. Coherence could then be very long-ranged. In practice this coherence is destroyed by thermal fluctuations of the energies of states k and k', so that these cross terms are rapidly averaged out.

A simple way to see this is to think of the site energies E_m and the couplings V_{mn} as the averages of stochastic variables $E_m(t)$ and $V_{mn}(t)$, whose fluctuations reflect thermal agitation. The Hamiltonian then contains an unperturbed, time-independent part and a time-dependent perturbation:

$$H(t) = H_0 + V(t)$$
$$V(t) = \sum_m (E_m(t) - E_m)|m\rangle\langle m| + \sum_{m,n} (V_{mn}(t) - V_{mn})|m\rangle\langle n| \quad (4.33)$$

In much the same way as the electronic states corresponding to close nuclear configurations can be indexed by a single set of indices, we can here introduce the time-dependent eigenstates of $H(t)$, $\{|k(t)|, E_k(t), U_{km}(t)\}_k$. The local coherence derived from an initally pure state $|k\rangle$ is

$$\rho_{mn}(t) = U_{km}(t)\bar{U}_{kn}(t) \quad (4.34)$$

$V(t)$ can be seen as an additional source of disorder in H_0, tending to localize the eigenstates so that the amplitudes $U_{km}(t)$ will have faster radial decay than in the rigid crystal.[76] Sign changes in the U matrix, which are easier to create for small amplitudes, also tend to reduce the time-averaged coherence $\rho_{mn}(t)$. Alternatively, and more rigorously, one can look at transitions between the eigenstates of H_0 caused by the perturbation $V(t)$. Early approaches of this kind were made by Haken and Strobl[77] and by Haken and Reineker[78] (see also Ref. 75). When bath fluctuations are fast, the electronic density matrix obeys the equation

$$i\hbar\rho = [H_0, \rho] - \frac{i\Gamma}{\hbar}\rho \qquad (4.35)$$

The commutator describes the "coherent," unperturbed electronic motion, which conserves the site coherence present in an initially pure state, while the relaxation operator Γ damps coherence. Aslangul and Kottis[79] show how to derive this relation by superoperator algebra in Liouville space. A disadvantage of this model is that the relaxation constants are phenomenological and independent of the energy differences between the states because the phonon bath is treated classically. Quantum treatment of the phonons recognizes that emission is more probable than absorption at a given temperature, so that the rates become asymmetrical and lead to irreversible relaxation towards the lowest eigenstates. This relaxation can be extracted from the Liouville equation for the total density matrix by projection operators which yield the pilot equation of the electronic density matrix:

$$i\hbar\frac{d\sigma_E(t)}{dt} = [H_0, \sigma_E(t)] - \frac{i}{\hbar}\int_0^{+\infty} R(\tau)\sigma_E(t - \tau)\,d\tau \qquad (4.36)$$

The relaxation operator depends on the details of exciton–phonon coupling, the temperature, the phonon spectrum, and so on. The apparent memory in the pilot equation is a mathematical consequence of removing the phonon variables in the Liouville equation, which is of course a memory-free, ordinary differential equation. Equation (4.36) begins by trying to say that the electrons are a closed system. This they are not and the state now and its evolution must contain corrections, "memory," because of earlier interactions with the bath, which Eq. (4.36) hides. These corrections are in the relaxation operator. In the common case of weak coupling to a broad phonon spectrum, the relaxation operator decays very fast with a characteristic correlation time τ_c. The integral is then close to $-i\Gamma\sigma_E(t)$ where

$$\Gamma = \int_0^{+\infty} R(\tau)\,d\tau \qquad (4.37)$$

The pilot equation is then equivalent to the phenomenological model (Eq. 4.35), with the advantage that we know how to calculate the asymmetric relaxation rates. The differential equations thus derived are Pauli master equations. Examples of this approach are the works of Grover and Silbey,[80] Munn and Silbey,[81] Silbey[1] (review paper), Capek,[82] and Capek and Szocs.[83] Schmid and Reineker[84] applied the method to E.S.R. linesh-apes in naphthalene. We sketch the essential mathematical steps below with a view to estimating later whether the master equation formalism is valid for triplet excitons in mixed naphthalene crystals at low tempera-tures.

B. The Pilot Equation for the Electronic Density Matrix

1. Derivation and Meaning

The first step towards the pilot equation is to remove the commutator in the Liouville equation by changing to the Liouville space, $\mathscr{L}$, of operators on ordinary state space, $\mathscr{E}$: Operators A acting on $\mathscr{E}$, $A: \mathscr{E} \to \mathscr{E}$ are represented by vectors $|A\rangle\rangle$ in $\mathscr{L}$. Liouville space has a Hilbert structure for the scalar product

$$\langle\langle A|B\rangle\rangle = \mathrm{Tr}(A^+ B) \tag{4.38}$$

where A^+ is the Hermitian conjugate of A. The Liouville operator L, defined by

$$L|A\rangle\rangle = |[H, A]\rangle\rangle \tag{4.39}$$

can be used to write the Liouville equation as

$$i\hbar \frac{d|\rho\rangle\rangle}{dt} = L|\rho\rangle\rangle \tag{4.40}$$

with the immediate solution

$$|\rho(t)\rangle\rangle = \exp(-iLt/\hbar)|\rho(0)\rangle\rangle \tag{4.41}$$

The density matrix $\rho(t)$ nonetheless contains a lot of uninteresting information on the phonon states of the system. This can be eliminated by introducing the reduced electronic density matrix, $\sigma_E(t)$. Thus, the Hamiltonian (Eq. 2.39) consists of parts describing independent excitons and phonons and their coupling. Let $\mathscr{E}_E$, $\mathscr{E}_R$, and $\mathscr{E}$ stand for the electronic, the phonon, and the total state spaces.

$$\mathscr{E} = \mathscr{E}_E \otimes \mathscr{E}_R = \{|m\rangle \otimes |\alpha\rangle\}_{m,\alpha} \tag{4.42}$$

where the $|m\rangle$ and the $|\alpha\rangle$ are eigenbases of $\mathscr{E}_E$ and $\mathscr{E}_R$. This decomposition is always possible. It is physically significant here because the electronic and the nuclear motions can be separated.

Electronic properties can be found from a reduced density operator σ_E, defined on $\mathscr{E}_E$.

The average of some electronic operator G_E is

$$\langle G_E \rangle = \mathrm{Tr}(\rho G_E \otimes Id_R)$$
$$= \mathrm{Tr}(\sigma_E G_E) \tag{4.43}$$

where the reduced density operator σ_E is a partial trace on $\mathscr{E}_R$:

$$\langle m|\sigma_E|n\rangle = \sum_\alpha \langle m\alpha|\rho|n\alpha\rangle \tag{4.44}$$

The motion of the reduced density operator is easily deduced from that of ρ by projection operator methods.[85,86] Introduce a projection operator $P:\mathscr{L}\to\mathscr{L}$ defined by

$$P|A\rangle\rangle = |\mathrm{Tr}_R(A)\otimes\sigma_R(0)\rangle\rangle \tag{4.45}$$

Then $P|\rho\rangle\rangle = |\sigma_E(t)\otimes\sigma_R(0)\rangle\rangle$, that is, σ_E multiplied by a constant vector of $\mathscr{L}_R$, the phonon density matrix at $t = 0$. Suppose further that the initial density matrix is factorized, $|\rho(0)\rangle\rangle = |\sigma_E(0)\rangle\rangle \otimes |\sigma_R(0)\rangle\rangle$, and introduce $Q = Id_{\mathscr{E}} - P$. Information on phonons is in $Q|\rho\rangle\rangle$ and does not interest us directly here. These projection operators correspond to decomposition of $H(L)$ into a principle part $H_0(L_0)$ describing independent excitons and phonons and a perturbation $V(L_1)$ coupling them.

$$H = H_0 + V$$
$$L = L_0 + L_1 \tag{4.46}$$

where $L_0 = [H_0, .]$, $L_1 = [V, .]$, $H_0 = H_E + H_R$ and $V = H_{ER}$. Suppose the phonon bath initially is in thermal equilibrium at temperature T, so that

$$\sigma_R(0) = \frac{\exp(-H_R/k_B T)}{Z} \tag{4.47}$$

where Z is the partition function. Then H_R and $\sigma_R(0)$ commute and $PL_0 =$

$L_0 P$ and $PL_0 Q = 0$. We can also assume that $PL_1 P = 0$, because for any product of operators A_E and A_R acting respectively on $\mathscr{E}_E$ and $\mathscr{E}_R$,

$$PL_1 P|A_E \otimes A_R\rangle\rangle = \text{Tr}_R(A_R) | [\text{Tr}_R(H_{ER}, \sigma_R(0)), A_E] \otimes \sigma_R(0)\rangle\rangle \tag{4.48}$$

Now $\text{Tr}_R (H_{ER}\sigma_R(0))$ is an operator on $\mathscr{E}_E$ representing the average potential of the phonon system acting on the excitons. It can simply be included in the renormalized electronic Hamiltonian, just as emissive states are shifted by the radiation field.

The problem is thus to find

$$P \exp(-iLt/\hbar)|\rho(0)\rangle\rangle = P \exp(-iLt/\hbar)(P + Q)|\rho(0)\rangle\rangle$$

Let us express the evolution operator as a path integral in the complex plane:

$$\exp(-iLt/\hbar) = \frac{1}{2i\pi} \int_{C^+} dz e^{-(izt/\hbar)} G(z), \qquad \text{valid for } t > 0 \tag{4.49}$$

where $G(z) = 1/(z - L)$ is the Green's function and C^+ is the contour from $+\infty$ to $-\infty$, just above the real axis and closed anticlockwise in the lower half plane. PGP follows from inserting $Id = P + Q$ in the identity $G(z - L) = Id$.

$$PG(z)P = \frac{P}{z - PLP - PR(z)P} \tag{4.50}$$

where

$$R(z) = L_1 + L_1 Q \frac{1}{z - QLQ} QL_1$$

Carrying these relations into Eq. (4.49) and differentiating, one finds

$$ i\hbar \frac{d}{dt} P|\rho(t)\rangle\rangle = i\hbar P|\rho(0)\rangle\rangle\delta(t) + PL_0 P|\rho(t)\rangle\rangle $$

$$ + PLQ \exp(-iQLQ\, t/\hbar)Q|\rho(0)\rangle\rangle $$

$$ + \frac{1}{2i\pi}\int_{C^+} dz\, e^{-izT/\hbar} PR(z)P \frac{P}{z - PL_0 P - PR(z)P} $$

$$ \times \left(1 + PL_1 Q \frac{1}{z - QLQ} Q\right)|\rho(0)\rangle\rangle \qquad (4.51) $$

Define now an operator

$$ R(t) = -\frac{1}{2i\pi}\int_{C^+} dz\exp(-izt/\hbar)R(z) $$

and take the inverse Fourier transform to obtain the pilot equation for the reduced electronic density operator

$$ i\hbar \frac{dP|\rho\rangle\rangle}{dt} = i\hbar P|\rho(0)\rangle\rangle\delta(t) + PL_0 P|\rho(t)\rangle\rangle $$

$$ + PL_1 Q \exp(-iQLQt/\hbar)Q|\rho(0)\rangle\rangle $$

$$ + \int_{-\infty}^{+\infty} d\tau\, PR(t - \tau)P|\rho(\tau)\rangle\rangle \qquad (4.52) $$

The first term describes sudden preparation of the system in a state described by the initial density matrix $P|\rho(0)\rangle\rangle$. The second term describes unperturbed electronic motion. Since $PL_R = L_R P = 0$,

$$ PL_0 P|\rho(t)\rangle\rangle = |[H_E, \sigma_E(t)] \otimes \sigma_R(0)\rangle\rangle \qquad (4.53) $$

The populations of the eigenstates are constant and their coherences precess at the Bohr frequencies. Site coherence will be observed provided the conditions of selective excitation set out above are met.

The third term gives the influence of the part of the initial condition which was orthogonal to the space $\mathscr{E}_E \otimes |\sigma_R(0)\rangle\rangle$. It vanishes if the initial density matrix is factorized, as is the case here. Exciton phonon coupling is absent in the electronic ground state. Sudden optical excitation creates an excited electronic state, but has little effect on the vibrational state, so that just after excitation the density matrix still is factorized. It develops thereafter under the influence of dissipative exciton–phonon coupling in the excited state.

The last term describes electronic relaxation due to interaction with the phonon bath. It shows an apparent memory, due to elimination of the phonon variables. Operator R is nonhermitian. It introduces dissipation into the electronic motion, through the nonuniform distribution of probabilities of different initial phonon states. The electronic populations relax towards the lowest states and the coherences are damped out. Note however that as pointed out by Aslangul,[87] even the conservative part, L_0, generally gives irreversible electronic motion. Equation (4.13) for the matrix elements of the density matrix shows they are aperiodic whenever the system has incommensurable Bohr frequencies. Even under the influence of dispersive forces only, an excitation initially localized on one site becomes diluted over all sites and is never reconstituted in a finite time on the initial site.

Equation (4.52) is exact, but in general insoluble. Fortunately, it can be solved perturbatively when exciton phonon coupling is weak, $L_{ER}/L_0 \ll 1$. Set $z = \hbar\omega + i\epsilon$ ($\epsilon \to 0^+$) in Eq. (4.51) and develop PRP to second order in L_{ER} in interaction representation. Remembering that $PL_{ER}P = 0$, the pilot equation becomes

$$i\hbar \frac{d}{dt} P|\rho^I(t)\rangle\rangle = \frac{1}{2} \exp(iL_0 t/\hbar) \int_{-\infty}^{+\infty} d\tau \, P$$

$$\times \int_{-\infty}^{+\infty} d\omega \, e^{-i\omega t} \frac{1}{\hbar\omega + i\epsilon - L_0} L_{ER}$$

$$\times P \exp\left(-\frac{iL_0(t-\tau)}{\hbar}\right) P|\rho(t-\tau)\rangle\rangle \qquad (4.54)$$

From the identities $I = \exp(-iL_0 t/\hbar)\exp(iL_0 t/\hbar)$ and $P = I_E \otimes |\sigma_R(0)\rangle\rangle\langle\langle 1_R|$, where $|1_R\rangle\rangle = \Sigma_\beta |\beta\beta^+\rangle\rangle$, the pilot equation can be written to second order in Schrödinger notation as

$$i\hbar \frac{d\sigma_E}{dt} = [H_E, \sigma_E(t)] - \frac{i}{\hbar} \text{Tr}_R \int_{-\infty}^{+\infty} d\tau$$

$$\times [ER, [E^I(-\tau)R^I(-\tau), \sigma_R(0)\sigma_E(t-\tau)]] \qquad (4.55)$$

$$E'(-\tau) = \exp\left(\frac{-iH_E\tau}{\hbar}\right) E \exp\left(\frac{iH_E\tau}{\hbar}\right)$$

$$R'(-\tau) = \exp\left(\frac{-iH_R\tau}{\hbar}\right) R \exp\left(\frac{iH_R\tau}{\hbar}\right)$$

Use is made here of the product form of the coupling in Eq. (2.39): $H_{ER} = E \cdot R$ where E and R are, respectively, electronic and phonon operators. Finally, developing the commutators and introducing the scalar $g(\tau) = \text{Tr}_R(\sigma_R(0)Re^{-iH_R t/\hbar}Re^{iH_R t/\hbar})$, the second-order pilot equation becomes

$$i\hbar\frac{d\sigma_E}{dt} = [H_E, \sigma_E(t)]$$

$$-\frac{i}{\hbar}\int_{-\infty}^{+\infty} d\tau\{EE'(-\tau)\sigma_E(t-\tau)g(\tau) + \sigma_E(t-\tau)E'(-\tau)E\bar{g}(\tau)$$

$$- E'(-\tau)\sigma_E(t-\tau)Eg(\tau) - E\sigma_E(t-\tau)E'(-\tau)\bar{g}(\tau)\}$$

Now $E'(-\tau)R'(-\tau)$ is the force exerted on the exciton by the phonon bath at time $-\tau$. It is replaced above by $E'(-\tau)$ weighted by $g(\tau)$, the thermal average of $RR'(-\tau)$ [or $R'(\tau)R$]. The correlation function $g(\tau)$ is a sum of contributions from all phonon modes coupled to the exciton, each term being weighted by the probability of the phonon state in the initial distribution.

$$g(\tau) = \sum_{\alpha\beta} P_T(\alpha)|\langle\alpha|R|\beta\rangle|^2 e^{i\omega_{\alpha\beta}t} \tag{4.56}$$

where $\hbar\omega_{\alpha\beta} = E_\alpha - E_\beta$ and $P_T(\alpha) = \exp(-E_\alpha/k_BT)$ is the thermal population of state α at temperature T. Alternatively, we can define the spectral density of the coupling,

$$\bar{g}(\omega) = 2\pi \sum_{\alpha\beta} P_T(\alpha)|\langle\alpha|R|\beta\rangle|^2\delta(\omega - \omega_{\alpha\beta}) \tag{4.57}$$

and put

$$g(\tau) = \int_{-\infty}^{+\infty} d\omega e^{i\omega\tau}\bar{g}(\omega) \tag{4.58}$$

The correlation function, defined as the Fourier transform of the power spectrum, must decay in a correlation time τ_c inversely proportional to the width of the spectrum. This time is of the order of the time for changes of sign in the energies and couplings in the stochastic Hamiltonian (Eq. 4.33). When the spectrum is broad, we may neglect the perturbed evolution of $\sigma_E(t')$ in Eq. (4.55) between $t - \tau_c$ and t and replace $\sigma_E(t - \tau)$ by $\sigma_E(t)$. This short memory approximation turns the integro-differential pilot equation into a differential equation:

$$\frac{d\sigma_E}{dt} = \frac{1}{\hbar}[H_E, \sigma_E] - \Gamma \cdot \sigma_E \qquad (4.59)$$

where the relaxation operator is defined by

$$\Gamma \cdot \sigma_E(t) = \frac{1}{\hbar^2}\left\{\left(\int_0^{+\infty} d\tau\, EE'(-\tau)g(\tau)\right)\sigma_E(t)\right.$$

$$+ \sigma_E(t)\left(\int_0^{+\infty} d\tau\, E'(-\tau)E\bar{g}(\tau)\right)$$

$$- \int_0^{+\infty} d\tau\, E\sigma_E(t)E'(-\tau)\bar{g}(\tau)$$

$$\left. - \int_0^{+\infty} d\tau\, E'(-\tau)\sigma_E(t)Eg(\tau)\right\} \qquad (4.60)$$

This equation is the same as that of Haken and Strobl,[77] with the advantage that the relaxation rates can be calculated from first principles, as discussed in the next section.

2. Relaxation Rates in the Electronic Eigenbasis

Equation (4.59) is a linear differential system coupling all the populations and coherences of the electronic density matrix. The order of magnitude of the matrix Γ of relaxation rates follows from inserting the identity operator twice in Eq. (4.60) and collecting terms. Consider the population relaxation rate beween two states i and j with Bohr frequency ω_{ij}:

$$\Gamma_{ji} \sim \frac{2}{\hbar^2} \sum_{\alpha\beta} P_T(\alpha) |\langle i|E|j\rangle|^2 |\langle \alpha|R|\beta\rangle|^2 \int_0^{+\infty} d\tau \cos \frac{(E_{i\alpha} - E_{j\beta})\tau}{\hbar}$$

$$= \frac{4\pi}{\hbar} \sum_{\alpha\beta} P_T(\alpha) |\langle i|E|j\rangle|^2 |\langle \alpha|R|\beta\rangle|^2 \delta(E_{i\alpha} - E_{j\beta})$$

$$\sim \frac{4\pi}{\hbar} |\langle i|E|j\rangle|^2 |\langle \alpha|R|\beta\rangle|^2 \rho(E_i - E_j) \qquad (4.61)$$

where $\rho(E_i - E_j) \sim \tau_c/\hbar$, is the density of pairs of initial and final bath states conserving energy in the transition. It is inversely proportional to the width of the spectrum of coupled states. Thus,

$$\Gamma \sim \langle H_{ER}^2 \rangle_T \tau_c / \hbar^2$$

where the average is to be taken at the temperature of the bath, T. The relaxation times are of the order of $1/\|\Gamma\|$. The short-memory approximation is valid if the relaxation times are longer than the correlation time, beyond which the correlation function vanishes. This requires that

$$\frac{1}{\|\Gamma\|} \gg \tau_c \Leftrightarrow \langle H_{ER}^2 \rangle_T \tau_c^2 / \hbar^2 \ll 1$$

that is, weak coupling to a broad spectrum of phonon states.

The pilot equation with the short memory approximation can be written in the electronic eigenbasis as

$$\frac{d\sigma_{ij}}{dt} = -i\omega_{ij} + \sum_{k,l} \Gamma_{ij,kl}\sigma_{kl} \quad \text{with} \quad \Gamma_{ij,kl} = \frac{1}{\hbar} \int_{-\infty}^{+\infty} d\tau\, R_{ijkl}(\tau) \quad (4.62)$$

The matrix elements of R are

$$R_{ijkl}(\tau) = \frac{1}{2\pi} \int_{-\infty}^{+\infty} d\omega\, e^{-i\omega\tau} \langle\langle ij^+ 1_R | L_{ER}$$

$$\times \frac{1}{\hbar\omega + i\epsilon - L_0} L_{ER} | kl^+ \sigma_R(0) \rangle\rangle \qquad (4.63)$$

where $|ij^+\rangle\rangle$ stands for the vector in $\mathscr{L}_E$ corresponding to the operator $|i\rangle\langle j|$ on $\mathscr{E}_E$. Development of the matrix element is tedious but not difficult. The result is

$$R_{ijkl}(\tau) = \frac{1}{i\hbar} \left\{ \delta_{jl} \sum_{\substack{\alpha \\ s\gamma}} P_T(\alpha) V_{k\alpha}^{s\gamma} V_{s\gamma}^{i\alpha} e^{-i(\omega_{sl} + \omega_{\gamma\alpha})\tau} \right.$$

$$+ \delta_{ik} \sum_\alpha P_T(\alpha) V_{s\gamma}^{l\alpha} V_{j\alpha}^{s\gamma} e^{-i(\omega_{ks} + \omega_{\gamma\alpha})\tau}$$

$$- \sum_{\alpha\beta} P_T(\alpha) V_{k\alpha}^{i\beta} V_{j\beta}^{l\alpha} (e^{-i(\omega_{il} + \omega_{\beta\alpha})\tau}$$

$$\left. + e^{-i(\omega_{kj} + \omega_{\alpha\beta})\tau}, \right. \tag{4.64}$$

where

$$V_{k\alpha}^{s\gamma} = \langle s\gamma | H_{ER} | k\alpha \rangle$$

The relaxation of population from state $|j\rangle$ to state $|i\rangle$ is determined by

$$R_{iijj}(\tau) = \frac{-1}{i\hbar} \sum_{\alpha,\beta} P_T(\alpha) |V_{j\alpha}^{i\beta}|^2 \cos(\omega_{ij} + \omega_{\alpha\beta})t \tag{4.65}$$

and hence the rate is

$$k_{ij} = \frac{2\pi}{\hbar} \sum_{\alpha,\beta} P_T(\alpha) |V_{j\alpha}^{i\beta}|^2 \delta(E_{i\beta} - E_{j\alpha}) \tag{4.66}$$

that is, the Golden Rule.

All terms of the density matrix are in general coupled, but when the transition frequencies lie in well-separated groups of nearly resonant values, the groups of transitions are independent. By well separated is meant that if two groups of transitions $\{|i\rangle\langle j| : \omega_{ij} \sim \omega_0\}$ and $\{|k\rangle\langle i| : \omega_{kl} \sim \omega_0'\}$ have relaxation rates such that

$$\|\Gamma_{\omega_0 \cdot \omega_0}\| \quad \text{and} \quad \|\Gamma_{\omega_0' \cdot \omega_0'}\| \ll |\omega_0 - \omega_0'| \tag{4.67}$$

then $R_{ij,kl}(\tau)$ oscillates faster than the internal relaxation rates of each group so that the cross relaxation is averaged out in the expression for $\Gamma_{ij,kl}$.

The populations (vanishing Bohr frequency) are a special case of this, so we finally receive a Pauli master equation

$$\frac{dp_i}{dt} = -\left(\sum_{j\neq i} k_{ji}\right)p_i + \sum_{j\neq i} k_{ij}p_j \tag{4.68}$$

where $p_i = \sigma_{ii}$ is the population of state i.

The coherences at Bohr frequency ω_0 obey

$$\frac{d\sigma_{ij}}{dt} = -i(\omega_{ij} + \Delta_{ij})\sigma_{ij} - \Gamma'_{ij}\sigma_{ij} + \sum_{\substack{kl\neq ij \\ \omega_{kl}\sim\omega_0}} \Gamma_{ijkl}\sigma_{kl} \tag{4.69}$$

where Δ_{ij} is a shift in the transition frequency due to coupling to the bath and

$$\Gamma'_{ij} = \Gamma^{(1)}_{ij} + \Gamma^{(2)}_{ij} \tag{4.70}$$

with

$$\Gamma^{(1)}_{ij} = \frac{1}{2}\left(\sum_{m\neq i} k_{mi} + \sum_{m\neq j} k_{mj}\right) \tag{4.71}$$

and

$$\Gamma^{(2)}_{ij} = \frac{2\pi}{\hbar}\sum_{\alpha\beta} P_T(\alpha)\delta(E_\alpha - E_\beta)$$
$$\times \{\tfrac{1}{2}|\langle i\beta|H_{ER}|i\alpha\rangle|^2 + \tfrac{1}{2}|\langle j\beta|H_{ER}|j\alpha\rangle|^2 - \langle i\beta|H_{ER}|i\alpha\rangle\langle j\alpha|H_{ER}|j\beta\rangle\} \tag{4.72}$$

$\Gamma^{(1)}_{ij}$ represents "T_1" loss of coherence through relaxation of the population of each level of the transition, in which inelastic scattering on bath states occurs. $\Gamma^{(2)}_{ij}$ is the "T_2" or pure dephasing term in which scattering is elastic. The last term in Eq. (4.69), $\Gamma_{ij.kl}$ appears only for degenerate transitions. It is given by

$$\Gamma_{ijkl} = \frac{2\pi}{\hbar}\sum_{\alpha,\beta} P_T(\alpha)\langle i\beta|H_{ER}|k\alpha\rangle\langle l\alpha|H_{ER}|j\beta\rangle\delta(E_{k\alpha} - E_{i\beta})$$

and represents inelastic scattering. In Section V we apply this formalism to triplet excitons in naphthalene. We should like now to underline the convenience of the eigenbasis by showing how difficult it can be to judge the validity of the master equation in the site basis.

3. Derivation and Limits of Validity of the Master Equation for the Populations in the Site Basis

Transport of excitations in disordered molecular solids at low temperatures generally is assumed to be incoherent, meaning that site coherences or local coherence (see Section IV.A above) can be neglected and that one can write Pauli master equations for the site populations. We adopted this point of view in Section III by writing

$$\frac{dp_i}{dt} = - \sum_{j \neq i} k_{ji} p_i + \sum_{j \neq i} k_{ij} p_j, \qquad i = 1, \ldots, N \qquad (4.73)$$

for the site occupation probability $p_i(t)$, with hopping rates

$$k_{ij} = \frac{2\pi}{\hbar} |v_{ij}|^2 \rho(E_i - E_j, T) \qquad (4.74)$$

where V_{ij} is the electronic coupling and ρ is the density of bath states at temperature T which conserve energy in the transition. It is worthwhile here to look at the assumptions underlying this representation. Complete treatment of this problem is very difficult. Here we shall simply point out some of the difficulties involved in justifying the master equation in the site representation. The evolution of the site populations can be found quite simply by introducing a suitable projection operator P', similar to P of Section IV.B.1.

$$P': \mathcal{L} \to \mathcal{L} \qquad P'|A\rangle\rangle = |\text{diag}(\text{Tr}_R(A)) \otimes \sigma_R(0)\rangle\rangle$$
$$\text{or} \quad P' = \sum_i |ii^+\rangle\rangle\langle\langle ii^+| \otimes |\sigma_R(0)\rangle\rangle\langle\langle 1_R| \qquad (4.75)$$

whence

$$P'|\rho(t)\rangle\rangle = |\sigma_E^d(t) \otimes \sigma_R(0)\rangle\rangle$$

where σ_E^d is the diagonal part of the electronic density matrix.

The total Hamiltonian $H = H_E + H_R + H_{ER}$ must now be cut up in such a way that the "principal" part is stable under the projection P': $H = H_0' + H_1'$ where $P'L_0' = L_0'P'$ and $P'L_0'Q' = 0$ where $L_0' = [H_0', .]$. Set therefore

$$H_0' = H_E^d + H_R$$

and

$$H_1' = H_E^{nd} + H_{ER}$$

where the superscripts d and nd stand, respectively, for the diagonal and nondiagonal parts of the electronic Hamiltonian in the site basis. Defining $Q' = I - P'$ as above, all the steps in the derivation of the pilot equation can be carried through to receive

$$i\hbar \frac{dP'|\rho(t)\rangle\rangle}{dt} = i\hbar P'|\rho(0)\rangle\rangle + P'L_1'Q' \exp(-iQ'LQ't/\hbar)Q'|\rho(0)\rangle\rangle$$

$$+ \int_{-\infty}^{+\infty} d\tau\, P'R'(t - \tau)P'|\rho(\tau)\rangle\rangle \qquad (4.76)$$

Note that the second term disappears if the initial density matrix was diagonal in the site basis, that is, if site coherence initially was absent. This does not mean that coherences can be neglected. They are active occultly in the memory of the relaxation operator R', as we shall see below. Using the definition of $R'(z)$, it is easy to show that

$$\sum_j \langle\langle ii^+1_R|R'(z)|jj^+\sigma_R(0)\rangle\rangle = 0 \Leftrightarrow \sum_j \langle\langle ii^+1_R|R'(\tau)|jj^+\sigma_R(0)\rangle\rangle = 0$$

The pilot equation then takes the form of a generalized master equation (master equation with time-dependent transition rates)

$$\frac{dp_i}{dt} = p_i(0)\delta(t) + \int_{-\infty}^{+\infty} d\tau \sum_j (w_{ij}(t - \tau)p_{ji}(\tau) - w_{ji}(t - \tau)p_i(\tau))$$

$$(4.77)$$

where the memory functions

$$w_{ij}(\tau) = \frac{1}{i\hbar} \langle\langle ii^+1_R|R'(\tau)|jj^+\sigma_R(0)\rangle\rangle \qquad (4.78)$$

obey detailed balance

$$w_{ii} = -\sum_{j \neq i} w_{ij} \qquad (4.79)$$

The memory functions are easily found by inverse Fourier transformation of $R'(z)$ expressed as a power series in L_1'/L_0'. Letting $K, L, M, N, \ldots$ stand for intermediate, virtual states (bath and electronic coordinates), the result is of the form

$$R(t) = \sum_K \langle ii^+ 1_R | L_1' | K \rangle\rangle\langle\langle K | L_1' | jj^+ \sigma_R(0) \rangle\rangle \, e^{-i\omega_K t}$$

$$+ \sum_{K,L} \frac{\langle\langle ii^+ 1_R | L_1' | K \rangle\rangle\langle\langle K | L_1' | L \rangle\rangle\langle\langle L | L_1' | jj^+ \sigma_R(0) \rangle\rangle}{E_L - E_K} (e^{-i\omega_K t} - e^{-i\omega_L t})$$

$$+ \sum_{K,L,M} \langle\langle ii^+ 1_R | L_1' | K \rangle\rangle\langle\langle K | L_1' | L \rangle\rangle\langle\langle L | L_1' | M \rangle\rangle\langle\langle M | L_1' | jj^+ \sigma_R(0) \rangle\rangle$$

$$\times \left(\frac{e^{-i\omega_M t}}{(E_M - E_K)(E_M - E_L)} + \frac{e^{-i\omega_L t}}{(E_L - E_K)(E_L - E_M)} \right.$$

$$\left. + \frac{e^{-i\omega_K t}}{(E_K - E_L)(E_K - E_M)} \right) + \cdots \tag{4.80}$$

It is thus a sum of oscillatory terms of several kinds, the first of which are terms containing only L_E^{nd}. These have frequencies $\omega_{lm} \sim |E_l - E_m|/h$, where E_l and E_m are site energies. Their integrated contribution to the memory vanishes at times beyond ω_{lm}^{-1}, that is, they introduce a long memory if the sites are resonant. The short-memory approximation to coupling to bath states is thus not a sufficient condition for obtaining a master equation. The lowest-order term of this kind is

$$w_{ij}^{EE}(t') = \frac{2}{\hbar^2} |V_{ij}|^2 \cos \omega_{ij} t' \tag{4.81}$$

and nth order terms are of the form

$$\sum_{\omega_{electronic}} e^{-i\omega_e t'} x \frac{n \text{ couplings } V}{n - 2 \text{ energy differences}}$$

Convergence of this perturbation series requires that the couplings should be smaller than the energy differences. These terms produce the nonlocal memories noticed by Kenkre.[88] Their importance increases with the delocalization of the electronic eigenstates, which was neglected in the local electronic basis.

Terms containing L_{ER} once only, vanish because of the presence of the trace of an off-diagonal operator. The lowest order nonzero term in L_{ER} is second order

$$w_{ij}^{RR}(t') = \frac{2}{\hbar^2} \sum_{\alpha\beta} P_T(\alpha)|\langle i\beta|H_{ER}|j\alpha\rangle|^2 \cos(\omega_{ij} - \omega_{\alpha\beta})t' \qquad (4.82)$$

which for coupling to a broad spectrum of bath states yields the first secular term:

$$w_{ij}^{RR}(t') = \delta(t')x\frac{2\pi}{\hbar} \sum_{\alpha\beta} P_T(\alpha)|\langle i\beta|H_{ER}|j\alpha\rangle|^2\delta(E_{i\beta} - E_{j\alpha}) \qquad (4.83)$$

Note, however, that this term involves the nonlocal electron–phonon coupling $\partial V_{ij}/\partial R$, not V_{ij}.

The first secular term of the form $w_{ij} = |V_{ij}|^2\rho(E_i - E_j, T)$ is in fact fourth order in L_1'. It is

$$k_{ij}(T) = \frac{2\pi}{\hbar}\frac{|V_{ij}|^2}{|E_i - E_j|^2} \sum_{\alpha\beta} P_T(\alpha)\left|\langle i\alpha|\frac{\partial D}{\partial R}|i\beta\rangle - \langle j\alpha|\frac{\partial D}{\partial R}|j\beta\rangle\right|^2 \delta(E_{i\alpha} - E_{j\beta})$$

$$(4.84)$$

This shows that transfer is possible only if the local coupling is different on sites i and j (cf. Section III.B.2). Moreover, it is the dominant secular term only if

$$\frac{\partial V_{ij}}{\partial R} \ll \frac{V_{ij}}{E_i - E_j}\left|\frac{\partial D_i}{\partial R} - \frac{\partial D_j}{\partial R}\right| \qquad (4.85)$$

On account of the condition $V_{ij}/(E_i - E_j) \ll 1$ required for convergence of the perturbation series above, Eq. (4.76) is a much stronger condition than the usually assumed one,

$$\frac{\partial V_{ij}}{\partial R} \ll \frac{\partial D_i}{\partial R}, \frac{\partial D_j}{\partial R} \qquad (4.86)$$

Thus, elimination of the bath variables and of the site coherences in the Liouville equation leads to a generalized master equation for the site populations, with time dependent hopping rates. This equation may be replaced by a master equation for the site-occupation probabilities, provided several conditions are met:

- The time scale of interest must be much longer than the Bohr frequencies.
- Diagonal disorder should be strong compared to the exchange coupling.
- Local exciton–phonon coupling should be much stronger than off-diagonal coupling.
- Oscillatory, nonsecular terms like Eq. (4.72) should have higher frequencies than the relaxation rates.

The presence of the perturbation series in the exchange coupling makes rigourous discussion very difficult. It is simpler to use the electronic eigenbasis and to convert the results to the site basis if an intuitive, diffusive picture is required. This is the approach applied below to triplet excitons in naphthalene.

V. COHERENT AND INCOHERENT TRANSPORT IN DISORDERED SOLIDS

A. Introduction

In Section III we saw indications of both hopping transport and extended states in the temperature dependence of the yields of acceptor emission and delayed fluorescence of isotopically mixed crystals of naphthalene. Section IV was a reminder of the density matrix and its pilot equation and the difficulties in assuming hopping transport described by master equations for the site populations. In this section we estimate from first principles whether master equations can be used to describe relaxation between triplet eigenstates, not site states. The estimation justifies this approach, which we use to give a unified description of the experimental results.

1. Electronic Eigenstates

a. Notation
The first step is to rewrite the electronic tight-binding Hamiltonian in the electronic eigenbasis $\{|k\rangle\}_k$ with creation (annihilation) operators $B_k^+(B_k)$:

$$H = \sum_m E_m B_m^+ B_m + \sum_{m,n} V_{mn} B_m^+ B_n = \sum_k E_k B_k^+ B_k \qquad (5.1)$$

This is a difficult eigenvalue problem because of the disorder of the site energies, E_m, which has two components: (1) Isotopic substitution leads to randomly distributed guest sites at a depth Δ below the host levels. (2) Values of E_m for hosts and guests fluctuate about the values 0 and $-\Delta$ because of disorder in the crystal field (stabilization terms D_m in Section II). This disorder depends on how well the host matrix accepts the guest, on the presence of crystallographic defects, other impurities, and so on. Well-grown Bridgman or sublimation crystals of isotopically mixed naphthalene have a total inhomogeneous width $\delta \lesssim 0.1 \text{ cm}^{-1}$ much smaller than the trap depth $\Delta \sim 100 \text{ cm}^{-1}$. This may, however, be an average over a set of locally more homogeneous domains.

The exchange interactions V_{mn}, are much smaller than the trap depth: $V_{mn} \sim V_0 \ll \Delta$. The eigenstates then split into two bands, a set of host states, perturbed by the presence of the guests, and a set of guest states, which may be more or less extended, depending on concentration. Both kinds of state have projections on the other species, determined by the ratio V_0/Δ.

Diagonalization of H amounts to finding the unitary transformation between the site basis and the eigenbasis:

$$B_k^+ = \sum_m c_{km} B_k^+ \tag{5.2}$$

In this expression do not confuse the kth eigenstate of the mixed crystal, for which no momentum can be defined, with the Bloch states of a pure crystal, with wave vector $\vec{k}$. Since the transformation is unitary, the inverse relation is

$$B_m^+ = \sum_k \bar{c}_{km} B_k^+ \tag{5.3}$$

Putting these relations in the exciton–phonon coupling (Eq. 2.39), we have

$$
\begin{aligned}
H_{\text{ex-phon}} = {} & \sum_{k,k'} \sum_{\vec{q}} G_1(k, k', \vec{q}) B_k^+ B_{k'} (b_{\vec{q}}^+ + b_{-\vec{q}}) \\
& + \sum_{k,k'} \sum_{\vec{q},\vec{q}'} G_2(k, k', \vec{q}, \vec{q}') B_k^+ B_{k'} (b_{\vec{q}}^+ + b_{-\vec{q}})(b_{\vec{q}'}^+ + b_{-\vec{q}'})
\end{aligned}
\tag{5.4}
$$

where

$$G_1(k, k', \vec{q}) = \sum_m \bar{c}_{km} c_{k'm} F_1(m, \vec{q}) \tag{5.5}$$

and

$$G_2(k, k', \vec{q}, \vec{q}') = \sum_m \tilde{c}_{km} c_{k'm} F_2(m, \vec{q}, \vec{q}') \qquad (5.6)$$

b. Example

As an example, consider a pair of sites whose site energies differ by δ, coupled by an exchange interaction $v \ll \delta$. The electronic Hamiltonian is

$$H_{el} = \delta/2(B_1^+ B_1 - B_2^+ B_2)_2 + v(B_1^+ B_2 + B_2^+ B_1) \qquad (5.7)$$

with approximate eigenstates

$$|\tilde{1}\rangle \sim |1\rangle + v/\delta|2\rangle \quad \text{and} \quad |\tilde{2}\rangle \sim |2\rangle - v/\delta|1\rangle \qquad (5.8)$$

In Section III.B the coupling v is responsible for the jump and the exciton phonon coupling makes up the energy difference. Here, v is absorbed by the unitary transformation

$$U \sim \begin{bmatrix} 1 & v/\delta \\ -v/\delta & 1 \end{bmatrix} \qquad (5.9)$$

while the linear exciton–phonon coupling, for example, becomes

$$H_{\text{ex-phon}}^{\text{lin}} = \sum_{\vec{q}} F_1(1, \vec{q}) B_1^+ B_1 + F_1(2, \vec{q}) B_2^+ B_2$$
$$- v/\delta(F_1(2, \vec{q}) - F_1(1, \vec{q}))(B_{\tilde{1}}^+ B_{\tilde{2}} + B_{\tilde{2}}^+ B_{\tilde{1}}) \qquad (5.10)$$

The diagonal part renormalizes the exciton energies while the off-diagonal part causes relaxation between them. These ideas are used below to calculate the relaxation rates between the crystal eigenstates.

2. Electronic Relaxation

Following the formal discussion of the density matrix in Section IV, relaxation of the electronic eigenstates can be described by a set of master equations in which the populations are uncoupled from the coherences, given the conditions fulfilled below:

$$\frac{dp_k}{dt} = -p_k/\tau_k - \sum_{k' \neq k} \Gamma_{k'k} p_k + \sum_{k' \neq k} \Gamma_{kk'} p_{k'} \qquad k = 1, 2, \ldots \quad (5.11)$$

with relaxation rates due to linear and to quadratic coupling

$$\Gamma_{k',k}^{L} = \frac{2\pi}{h} \sum_{\alpha,\beta} P_T(\alpha) \left| \Sigma_I \langle k\alpha | H_{\text{ex-phon}}^{L} \frac{1}{E_I} H_{\text{ex-phon}}^{L} | k'\beta \rangle \right|^2 \delta(E_{k\alpha} - E_{k'\beta})$$

(5.12)

and

$$\Gamma_{k,k'}^{Q} = \frac{2\pi}{h} \sum_{\alpha,\beta} P_T(\alpha) |\langle k\alpha | H_{\text{ex-phon}}^{Q} | k'\beta \rangle|^2 \delta(E_{k\alpha} - E_{k'\beta})$$ (5.13)

Only two phonon processes are considered here. They are absorption–emission of optical phonons in the donor–donor eigenstate transport (cf. Section III.C) and emission of two large phonons in relaxation between donor and acceptor states.

The amplitudes of the second-order linear and the quadratic processes for donor–donor scattering are, respectively,

$$\sqrt{\nu(\nu'+1)} \sum_{m,n} \bar{c}_{km} C_{k'm} (|c_{k'n}|^2 - |c_{kn}|^2)$$

$$\times \left(\frac{F_1(m,\vec{q})F_1(n,\vec{q}')}{-\hbar\omega(\vec{q})} + \frac{F_1(m,\vec{q}')F_1(n,\vec{q})}{\hbar\omega(\vec{q}')} \right)$$

(5.14)

and

$$\sqrt{\nu(\nu'+1)} \sum_{m} \bar{c}_{km} c_{k'm} F_2(m,\vec{q},\vec{q}')$$ (5.15)

These amplitudes depend on the spatial disorder through the overlap of the states $c_{km} c_{k'm}$ and on the temperature through the probability of occupation ν, ν' of the active modes at temperature T, $P_T(\alpha)$.

The amplitude of the quadratic process between donors and acceptors has the same form as Eq. (5.15), but the bilinear amplitude is

$$\sqrt{(\nu+1)(\nu'+1)} \sum_{m,n} (|c_{km}|^2 \bar{c}_{kn} c_{k'n} + |c_{k'n}|^2 \bar{c}_{km} c_{k'm})$$

$$\times \left(\frac{F_1(m,\vec{q})F_1(n,\vec{q}')}{-\hbar\omega(\vec{q})} + \frac{F_1(m,\vec{q}')F_1(n,\vec{q})}{\hbar\omega(\vec{q}')} \right)$$

(5.16)

Note that the factor $\sqrt{(\nu+1)(\nu'+1)}$ will give negligible temperature dependence in the final result, since the populations of the active modes are very small at low temperatures.

The total monomolecular lifetime of the kth eigenstate differs negligibly from that of the isolated molecule, because the radiative lifetime is about 20 times longer than the intramolecular relaxation rate. Although collective emission of strongly coupled molecules in general shortens the radiative lifetime, the aggregates of such sites are much too small for this effect to be competitive at the concentrations below 20% considered here. We assume therefore that $\tau_k \sim \tau_{\text{molecular}}$ for all states.

3.   *Validity of the Master Equation in the Eigenbasis*

The master equation (Eq. 5.1) is valid provided a number of conditions are met. Let us examine them step by step, for two monomer states with energy difference $10^{-2}\,\text{cm}^{-1}$. Other values used are $V_0 = 1\,\text{cm}^{-1}$, $\Delta = 100\,\text{cm}^{-1}$, and for the active optical modes an average energy $30\,\text{cm}^{-1}$, bandwidth $10\,\text{cm}^{-1}$, and thermal population $\bar{\nu} = 10^{-6}$ at 1.6 K.

1. The initial bath density matrix is supposed to commute with the bath Hamiltonian. This is true for a bath initially in thermal equilibrium at temperature T: $\sigma_R(0) \sim \exp(-H_{\text{phon}}/k_B T)$.

2. The density matrix is supposed to be factorised at $t = 0$. This corresponds to the absence of exciton–phonon coupling in the ground state before optical excitation.

3. We supposed that the average force exerted by the bath on the electronic system vanishes: $\text{Tr}_R(\sigma_R(0)H_{ER}) = 0$. This is certainly true for linear coupling which is off-diagonal in the phonon basis. Quadratic coupling contains a diagonal part which will renormalize the exciton energies and an off-diagonal part which poses no problem.

4. The exciton-phonon coupling is factorized. This is indeed the form of Eq. (5.4).

These conditions suffice to write down the pilot equation. The master equations (Eq. 5.1) require the further conditions:

5. The small parameter in the perturbative development of the kernel of the pilot equation should be small:

$$\epsilon \sim |H_{ER}|\tau_c/h \ll 1 \qquad (5.17)$$

The bath correlation time is of the order of the inverse of the dispersion of the optical phonons coupled to the electronic states, $10\,\text{cm}^{-1}$, whence $\tau_c \sim 10^{-12}\,\text{s}$. $H_{\text{ex-phon}}$ is of the order of magnitude

$$H_{\text{ex-phon}} \sim |c_{km}||c_{k'm}| \frac{\partial^2 D}{\partial \theta^2} \Delta\theta^2 \sqrt{\nu} \tag{5.18}$$

where $\Delta\theta \sim 10^{-1}$ rad is the amplitude of the zero-point libration of the optical modes. The stabilizations D^* and D^0 are of the same order of magnitude as their change when a molecule is turned over. Assuming $D \sim D_0 \cos\theta$, $\partial D/\partial \theta \sim D$. The part of Eq. (5.18) dependent on the bath is thus of the order of $5 \cdot 10^{-2}$ cm^{-1} at 1.6 K for phonon modes of energy 30 cm^{-1}. It is weighted by the overlap factor. States centred on sites with an energy difference δ, coupled by the superexchange interaction

$$v \sim V_0 \exp(-\beta(r-1))$$

with

$$\beta = \ln(\Delta/V_0)$$

and r in lattice units, have an overlap $v/\delta \ll 1$. With the values given above,

$$\epsilon < 0.5 e^{-\beta(r-1)} \ll 1$$

6. The secular approximation allowing separation of the populations and the coherences requires that the relaxation rates should be slower than the precession. This means that $|E_k - E_{k'}| \gtrsim \delta \gg h\Gamma_{kk'}$. Using the overlap v/δ

$$\hbar\Gamma_{kk'}/\delta \sim \frac{V_0^2}{\delta^3} e^{-2\beta(r_{kk'}-1)} \left(\frac{\partial^2 D^*}{\partial \theta^2} \Delta\theta^2\right) \tau_c \bar{\nu}/\hbar \ll 1 \tag{5.19}$$

where $r_{kk'}$ is the separation of the centers of gravity of the states. Hence,

$$\hbar\Gamma_{kk'} \sim \frac{V_0^2}{\delta^2} e^{-\alpha(r_{kk'}-1)} \left(\frac{\partial^2 D^*}{\partial \theta^2} \cdot \Delta\theta^2\right)^2 \tau_c \bar{\nu}/\hbar \tag{5.20}$$

7. Finally, the relaxation times should be much longer than the bath correlation time τ_c so that the memory in the pilot equation can indeed be neglected. With the values given above,

$$\Gamma_{kk'} \sim \Gamma_0 \exp(-2\beta r)$$

with $\Gamma_0 = 3 \cdot 10^{15} \, s^{-1}$. This is indeed slower than the bath fluctuations. This first-principles estimate of Γ_0 is only two decades off the single adjusted parameter $\exp(2\beta\bar{R})/\tau_D$ in the simulations of Section III (Eq. 3.8), a fair fit, considering the approximate nature of the model, in which the real, inhomogeneous, nature of the sample, with a spread of splittings δ, was neglected.

Thus, transport of triplet excitons in isotopically mixed naphthalene crystals at the temperature of liquid helium can reasonably be approximated by a set of master equations for the relaxation between the electronic eigenstates, in which the populations and the coherences are uncoupled.

4. Approximate Form of the Relaxation Rates

The relaxation rates between eigenstates depend on their overlap, weighted by the local exciton–phonon coupling. Putting Eq. (5.15) in Eq. (5.13), the relaxation rate by quadratic coupling between two nearly resonant donor eigenstates D and D' is

$$\Gamma^Q_{DD'} = \sum_{\vec{q},\vec{q}'} \bar{\nu}(\vec{q})(\nu^-(\vec{q}')+1) \sum_{m,n} \bar{c}_{Dm}c_{D'm}c_{Dn}\bar{c}_{D'n}F_2(m,\vec{q},\vec{q}')\bar{F}_2(n,\vec{q},\vec{q}')$$

$$\sim \sum_{\vec{q},\vec{q}'} \bar{\nu}(\vec{q})(\bar{\nu}(\vec{q}')+1) \sum_{m,n} \bar{c}_{Dm}c_{Dm}c_{Dn}\bar{c}_{D'n}e^{i(\vec{q}+\vec{q}')(\vec{m}-\vec{n})}|\tilde{F}(\vec{q},\vec{q}')|^2$$

$$(5.21)$$

Note that $\tilde{F}(\vec{q},\vec{q}') = F_2(m,\vec{q},\vec{q}')e^{i\vec{q}\cdot\vec{m}}$ is independent of the lattice cell m. We shall neglect its dependence on the site in the cell. Suppose also that the phonon dispersion is isotropic and that $E_D \neq E_{D'}$, so that $|\vec{q}| \neq |\vec{q}'|$. The sum then vanishes when $m \neq n$. Since the first branches of optical phonons have energy about E_D (the Debye energy), the (symmetric) donor–donor relaxation rate takes the final approximate form

$$\Gamma_{DD'} \sim \Gamma_0 R_{DD'} \exp(-E_D/k_B T) \qquad (5.22)$$

where the overlap is

$$R_{DD'} = \sum_m |c_{Dm}|^2|c_{D'm}|^2 \qquad (5.23)$$

The same approximations applied to linear and quadratic coupling to acoustic modes gives the complete result

$$\Gamma_{DD'} \approx \Gamma_0(T)R_{DD'}$$

where

(5.24)

$$\Gamma_0(T) = \gamma_1 T + \gamma_2 T^7 + \gamma_3 \exp(-E_D/k_B T)$$

The constants γ can be estimated from first principles as above or regarded as adjustable parameters.

Quadratic coupling with emission of two large phonons in trapping between donor and acceptor states is of the same form, except that the rate is independent of temperature:

$$\Gamma_{AD} \sim \Gamma'_0 R_{DA}$$

with

(5.25)

$$R_{DA} \sim \sum_m |c_{Dm}|^2 |c_{Am}|^2$$

Second-order linear coupling gives a slightly different overlap factor:

$$R'_{kk'} = \sum_{m,n} |c_{km}|^2 |c_{k'm}|^2 (|c_{kn}|^2 - |c_{k'n}|^2)^2 \qquad (5.26)$$

Noting that c_{kn} and $c_{k'n}$, cannot both be large for states with different centers of gravity, we see that

$$R'_{DD'} \sim R_{DD'}L$$

where

(5.27)

$$L = L(k') \sim L(k) = \sum_m |c_{km}|^4$$

is the inverse participation ratio of the states.

L varies between 1 for a perfectly localized site excitation and $1/N$ for an excitation shared with equal amplitude by N sites. The inverse participation ratio is widely used in studies of Anderson localization and of quantum percolation. As shown below, the k states are well localized with exponential tails and L is close to unity for all states in the narrow guest band, so we shall include it in constant Γ_0.

B. Interpretation of the Temperature and Concentration Dependence of Exciton Transport

We can now discuss the experimental results using the kinetic scheme

$$\frac{dp_D(t)}{dt} = -p_D(t)/\tau_D - \sum_{D' \neq D} \Gamma_{D'D}(T)p_D(t) + \sum_{D' \neq D} \Gamma_{DD'}(T)p_{D'}(T)$$

$$- \sum_A \Gamma_{AD}p_D(T)$$

$$\frac{dp_A(t)}{dt} = -p_A(t)/\tau_A + \sum_D \Gamma_{AD}(T)p_D(t) \qquad (5.28)$$

where D and A stand for any donor or acceptor eigenstates and $\tau_D(\tau_A)$ is the monomolecular-donor (acceptor) decay rate, unaffected by collective emission at the concentrations concerned here. We can see two ways to understand the experimental temperature and concentration dependence of the yield of acceptor emission.

One way to interpret the results is to suppose that scattering on optical phonons is the predominant process at $T \gtrsim 1.6$ K, the lowest temperature in the data. This means that couplings in the donor band vary very fast because of the exponential factor in Eq. (5.24). Figure 5.1 shows the shape of the distribution of donor–donor relaxation rates at two donor concentrations c_1 and $c_2 > c_1$ at some temperature T_0. The plot is of the distribution of the logarithm of $\Gamma_{DD'}$ rather than of $\Gamma_{DD'}$ because the exponential overlap factor leads to a wide spread of the rates, typically 6–8 decades or more.

This approximate distribution is deduced from the exponential radial decay of the overlap of the eigenstates with the separation of their centers of gravity, and from the distribution of nearest donor–donor distances. Using

$$\Gamma = \Gamma_0 \exp(-\alpha r) \qquad (\alpha = 2\beta = 2\ln(\Delta/V_0)$$

and

$$\varphi(r) = 2\pi c_D r \exp(-\pi c_D r^2)$$

we have

$$\psi(\ln \Gamma) = \frac{2\pi c_D}{\alpha^2} (\ln \Gamma_0 - \ln \Gamma) \exp\left(\frac{-\pi c_D}{\alpha^2}(\ln \Gamma_0 - \ln \Gamma)^2\right) \quad (5.29)$$

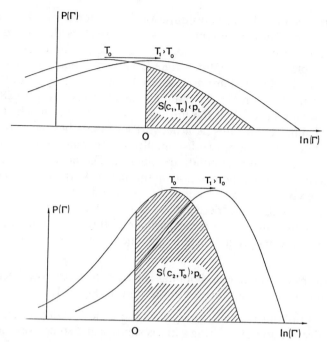

Figure 5.1. Distribution of the relaxation rates between donor eigenstates at two concentrations, c_1 and $c_2 > c_1$ and two temperatures T_0 and $T_1 > T_0$. The hatched region shows rates faster than the intramolecular decay. A logarithmic rate scale is chosen because the exponential radial decay of the overlap between states produces a wide spread of the relaxation rates.

We assume here that all the donors are monomers and hence the center of gravity of some state, which is not quite true because of clustering in groups of donors coupled by nearest-neighbor, direct, exchange interactions. At $c_D = 10\%$, about 45% of the donors are in such clusters.

The relaxation rates at concentration c_2 are on average faster than at c_1. The distribution is also narrower, as is that of the nearest-neighbor distances. When the temperature increases from T_0 to T_1, all the donor–donor rates are multiplied by $\exp(E_D/k_B(1/T_0 - 1/T_1))$, corresponding to a horizontal shift of all the distribution in Fig. 5.1.

The yield of acceptor emission stems from two sources (neglecting direct population of the acceptors or assuming selective excitation of the donors). The first is direct donor–acceptor trapping, between sufficiently close donor and acceptor states. This is responsible for the temperature-independent part of the yield. It is due to quantum overlap of the tails of the localized states. There is no diffusion in this temperature range. We refrain from calling it coherent transport for reasons developed below. The second source is by migration between donor states until a state suitably coupled to an acceptor is found. This amounts to hopping transport because the donor states are localized. The efficiency of this process depends on the fraction $S(c, T)$ of donor–donor rates in Eq. (5.28) which are substantially faster than the intrinsic decay rate, the hatched area of Fig. 5.1. Think of the rates as a network of bonds between points, the states in state space. Bonds (substantially) faster than the intrinsic decay rate $1/\tau_D$ are open. Slower bonds are blocked. The high yield of acceptor emission corresponds to percolation on this network when $S(c, T)$ is larger than the percolation concentration. Let T_c stand for the temperature at which the bond structure percolates for a given concentration c. Identifying this temperature with the experimental threshold temperature, we obtain the following predictions, in agreement with experiment:

- T_c drops as c increases; hopping is harder to freeze out because the histogram starts out further to the right in Fig. 5.1.

- The threshold becomes sharper as the concentration rises, because the distribution of hopping rates is sharper and so crosses the cut-off rate $1/\tau_D$ over a smaller range of temperatures.

- At given temperature, the yield of acceptor emission increases with c_D, due to greater connectivity.

- Finally, below the threshold temperature at a given concentration hopping is inefficient, but direct trapping to acceptors occurs through the exponential overlap of the eigenstates. Qualitatively, this overlap must rise with the donor concentration. Section VI shows that in the experimental range $c_D \lesssim 15\%$, this is not due to an Anderson transition, but to an increase in the localization length of the localized donor states.

Let us pause here to distinguish between coherence and the absence of diffusion. In the present interpretation, there is no diffusion in the donor band at temperatures below T_c. This does not mean, however, that the site coherence will be large or long-ranged. Site coherence would be strongest for selective excitation of a single eigenstate. The largest coherences will be between donors in the cluster of nearest neighbors at the center of the state, coupled by direct exchange interactions. Site coherence between clusters will be smaller, because of the strong localization of the states. In practice, site coherence may even vanish because the indirect excitation of the singlet states, or insufficiently selective excitation of the triplet states, creates a superposition of states. Although the state populations may be constant because of the absence of diffusion, the site coherence will vanish by interference as discussed in Section IV.A.

A second way to interpret the results would be to suppose that the transition from weak to strong temperature dependence is due to crossing over from scattering on acoustic phonons to scattering on optical phonons. Let T_{tr} be the temperature below which scattering on acoustic phonons predominates. Then the argument in the first interpretation can be repeated for sufficiently low concentrations, such that $T_c > T_{tr}$. Since T_c decreases with increasing c, this will be true up to some limiting $\bar{c}$ such that $T_{\bar{c}} = T_{tr}$. The constant yield below T_{tr} is then the result of residual hopping assisted by acoustic phonons. We should expect that the threshold temperatures would cease decreasing with increasing concentration above the value such that $T_{\bar{c}} = T_{tr}$. The current data give no firm support to this idea, Fig. 1.5. There are further difficulties in identifying this limit. First, the width of the distribution of relaxation rates decreases more and more slowly as the concentration is increased, so the differences in the threshold temperatures must become smaller. Secondly, we assume here transport between nearly resonant states on different clusters of donors, activated by coupling to optical phonons. This is a simplification, however, because it neglects coupling between non-resonant levels on the same aggregate. These are due to exchange coupling and produce levels as deep as $5\,\text{cm}^{-1}$ below the monomer levels. These states are always much more localized than the monomer states in the middle of the donor band and are uncoupled from them: They have large weights on different sets of sites, as shown in Part VI below. At helium temperatures, trapping of excitons in these aggregate levels will hinder transport. The larger energy gaps favor coupling to acoustic phonons and the spread of depths will lead to a spread of activation energies and to a complication of the relaxation scheme.

Our understanding of the nature of the transport in the temperature-independent region could be improved in two ways. First, experiments could be carried to much lower temperatures in a dilution cryostat. If hopping assisted by acoustic phonons is responsible for the weak temperature dependence of transport below 5 K, then we would expect a significant change in the yield of acceptor emission between T_{tr} and say 0.1 K. If trapping from the tails of the donor states is the cause of the temperature independence, then the yield should be the same right down to the absolute zero. A second way would be to look at triplet dephasing by optically detected spin locking in the temperature range 1.6–5 K. This method is insensitive to spin-spectral diffusion and can reach dephasing rates as slow as the triplet decay rate.[89] Aggregates have definite spin quantitization axes[90] so that hopping transport should be accompanied by dephasing rates faster than the intramolecular decay. These experiments could help to distinguish between the contributions of hopping and of direct trapping below 5 K because the predictions are different. We stress, however, that neither point of view is adequate by itself. Transport in a given sample may be predominantly of one kind or the other at different temperatures. Different regions of the sample will have different behavior too, owing to natural fluctuations of concentration.

Brenner et al.[33,34] recently reported on the effect of pressure on the yield of acceptor emission. However, pressure must affect both the extension of the donor eigenstates and the hopping rates in complicated ways, since it will give rise to several effects: A probable increase in the resonant exchange interactions, an increase in the diagonal disorder, and modified phonon modes and exciton–phonon coupling. The reader may by now be convinced of the complexity of the effects in even the simple case of transport in isotopically mixed crystals and see that unravelling these effects along the lines suggested by the present analysis is a formidable task. In the meantime we may also look at the numerical predictions of the first interpretation, based on direct trapping from the tails of the donor states. Section VI below calculates the donor states and estimates the yield of acceptor emission due to this process. This completes the numerical comparison of the hopping model and the delocalization model on the same footing.

VI. DIFFUSIONLESS TRANSFER

In this section we complete the numerical comparison of the two models of exciton transport in the triplet state of isotopically mixed crystals of naphthalene. The first point is exact calculation of the donor eigenstates. It is believed now that the Anderson transition never occurs in two dimensions.[91] All states are localized. What does change with the degree of disorder is the localization length. The present numerical results confirm that all the donor states are localized in the two-dimensional model of isotopically mixed crystals of naphthalene, at concentrations below 20% and discusses for the first time the nature of the tails of the states. Donor–acceptor transfer is estimated from the overlap of the exponential tails of the localized wavefunctions.

A. Donor Eigenstates

1. Hamiltonian

Consider the tight-binding Hamiltonian of the isotopically mixed crystal:

$$H = \sum_{m=1}^{N} E_m |m\rangle\langle m| + \sum_{m=1}^{N} \sum_{n \neq m} V_{mn} |m\rangle\langle n| \qquad (6.1)$$

or, in the eigenbasis,

$$H = \sum_{k=1}^{N} E_k |k\rangle\langle k|$$

with

$$|k\rangle = \sum_{m}^{N} c_{km} |m\rangle$$

Site energies in Eq. (6.1) will be reckoned relative to the transition of the host molecules, for which we set $E_m = 0$. Let Δ be the depth of the traps below the host band. Both the host and the trap energies have some spread about their average values, because of fluctuations of stabilization forces, due to missing sites, dislocations, strains induced during crystal growth, and so on. Narrow-band excitation spectra of the phosphorescence or of delayed fluorescence[14] have a width $\delta < 0.25 \, \text{cm}^{-1}$, but this is only an upper limit to the microscopic disorder. Trapping by acceptors monitors transport on a scale $L \sim (c_A)^{-1/d}$ where d is the spatial dimension. Assuming two-dimensional transport and, typically, $c_A \gtrsim 10^{-5} \, \text{m/m}$ the scale here is $L \lesssim 300$ lattice constants. It is possible that the experimental lineshape is the envelope of many small such domains with smaller local inhomogeneous broadening. Here we regard the microscopic, inhomogeneous broadening as unknown and vary δ as a parameter. The distribution of site energies used below is rectangular, with width δ.

The resonance exchange interactions also have some spread, but the effect on localization is smaller than the spread of the diagonal terms, so we shall neglect it here: Superexchange coupling between nonnearest-neighbor traps is of the order of the diagonal fluctuations, so that a change of a few percent in the nearest-neighbor exchange couplings is unimportant.

Numerical values used below are $\Delta = 100 \, \text{cm}^{-1}$, $V_{ij} = V = 1.2 \, \text{cm}^{-1}$ for inequivalent neighbors in the (a, b) plane, $V_{ij} = v = 0.5 \, \text{cm}^{-1}$ for equiva-

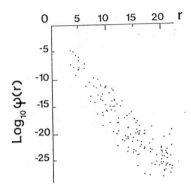

Figure 6.1. Exponential decay of the amplitude of a dimer state in a model crystal with $c = 10\%$ and $\delta = 5 \cdot 10^{-3} \, \text{cm}^{-1}$. The localization length is 0.3 lattice constants, but note the large fluctuations at a given lattice distance, due to azimuthal fluctuations of concentration and to chance resonances. Only the projections on the donor states are shown, for clarity.

lent neighbors along the b axis and $V_{ij} = 0$ otherwise. δ was varied from 10^{-6} to $10^{-2} \, \text{cm}^{-1}$.

Eigenvalues and eigenvectors of the Hamiltonian H were calculated for models of several thousand sites, containing a random distribution of guests with average concentration c. Using Gerschgorin's theorem on the spectral radius,[92] the spectrum can be seen to consist of two bands of width at most $4V + 2v$ centered respectively on $E = 0$ and $E = -\Delta$, corresponding respectively to perturbed band states of the host matrix and to trap states. The trap states which are of interest to us here were calculated by Ritz iteration.[93,94]

2. Guest Wavefunctions

We saw above that there are in the mixed crystal clusters of nearest-neighbor guests, with couplings larger than the inhomogeneous broadening, δ, and the superexchange coupling between clusters. These clusters give rise to the satellite lines around the main monomer transition in the optical spectra,[16] (cf. Fig. 1.3). Each of the corresponding wavefunctions has a "nucleus" of strong projections on such a cluster, with amplitude $c_{km} \sim 1/\sqrt{N_A}$ where N_A is the number of sites in the cluster. The nucleus is surrounded by a tail of projections on host and guest sites with amplitudes decaying exponentially with distance (Fig. 6.1). Monomer states generally have a single-site nucleus, but it can happen that several monomer sites form an extended nucleus (see Fig. 6.5 below). A fit to the radial decay of the wavefunction generally yields

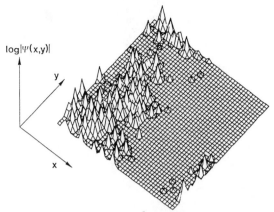

$\log|\psi(x,y)|$

y

x

Figure 6.2. Plot of the amplitudes over 10^{-5} on guest sites, of a monomer state in the guest band center at $c = 15\%$ and $\delta = 10^{-4}\,\text{cm}^{-1}$, illustrating the azimuthal fluctuations underlying the average, exponential radial decay of the wavefunctions. Periodic boundary conditions are assumed. The mesh represents the square lattice.

$$\psi(r) \sim \exp(-r/l_\psi) \tag{6.2}$$

where l_ψ is the localization length of the state.

Localization lengths of the order of 0.5–2 lattice constants showed three known effects. They were longest for states in the middle of the guest band. They also increased with the guest concentration and decreased with increasing diagonal disorder, δ. Note, however, that the average, exponential, radial decay of the wavefunction smears out very large azimuthal fluctuations at a given distance, caused by the random distribution of the guests. Figure 6.2 gives an idea of the strong disorder of the wavefunctions. Clearly, we should be very cautious before calculating observables with the mean properties of the wavefunctions.

Figure 6.3 illustrates the increase of the localization length between the edge of the band and the center. Figure 6.4 shows the influences of the concentration and of the inhomogeneous broadening on localization.

At all concetrations studied here, $c < 15\%$, all the states were localized and there was no Anderson transition, as originally proposed by Klatfer and Jortner,[25] to explain the rise in supertrap emission below this concentration. This agrees with the conclusion drawn by Lemaistre and Blumen[95,96] from their study of the inverse participation ratio,

$$L(k) = \sum_{m=1}^{N} |c_{km}|^4$$

This ratio lies between $1/N$ for a state uniformly delocalised on N sites

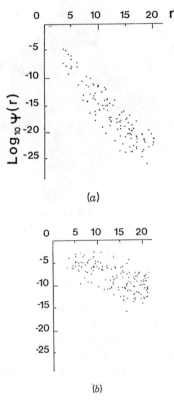

Figure 6.3. Comparison of: (*a*) a state in the edge of the band—a trimer—with (*b*) a monomer state in the center; $c = 10\%$ and $\delta = 10^{-4} \text{cm}^{-1}$. The localization length increases from 0.3 to 0.7 lattice units.

and 1 for a perfectly localized state. It is thus a useful criterion for detecting the Anderson transition, but in the present case it is sensitive only to the nucleus of the wavefunctions and misses the increase in the exponential tails, because of the fourth-power weighting of the projection on the site basis.

Figure 6.5 shows an example of another interesting fact revealed by the numerical results—the uncoupling of states with different energies although they may be close together in space. This is an important effect, because transitions between states occur with rates determined by their overlap (see Section V.A.4).

Finally, Fig. 6.6 sums up the change in the localization length with concentration. The values obtained in the numerical work are much

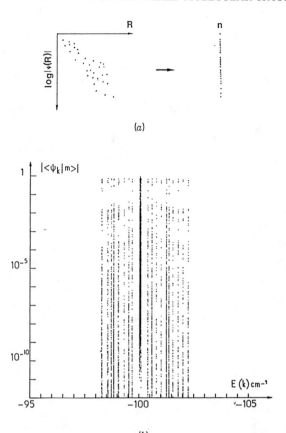

Figure 6.4. Influence of the concentration c and the diagonal disorder δ on localisation. (a) How to read part (c): Each state $|n\rangle$ is projected on to the amplitude axis. States are plotted in (c) by increasing energy, but are equally spaced for clarity, because of the bunching caused by the disparity of the nearest-neighbor exchange coupling and the next-nearest-neighbor or longer, superexchange coupling (b), where bands corresponding to monomers, dimers, trimers, and other small nearest-neighbor clusters appear. (c) Combined influence on the localization of the eigenstates of the diagonal disorder and of the concentration.

smaller than those deduced by Klafter and Jortner[27c] from Lukes's formula (Eq. 1.6). This formula gives for example $L(E = 0, c = 5\%) = 60$ lattice units, whereas even when δ is as small as 0.005 cm^{-1} the simulation gives 0.5 units. This disagreement illustrates the inapplicability of mean-field reasoning to strongly disordered systems.[4]

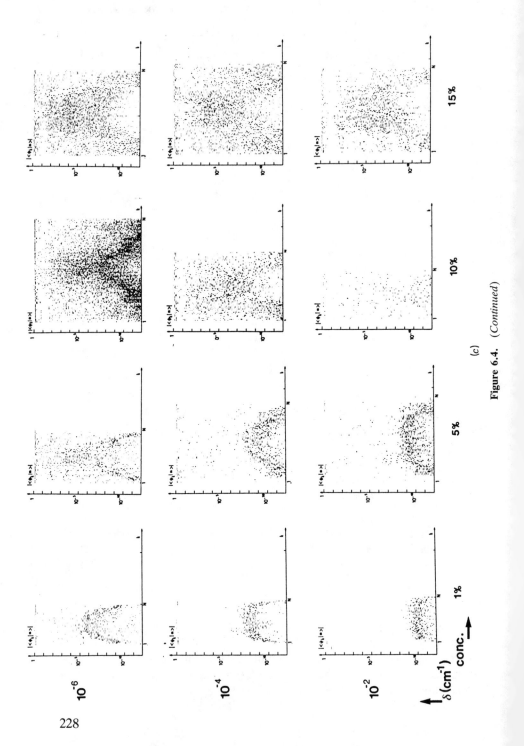

Figure 6.4. (*Continued*)

228

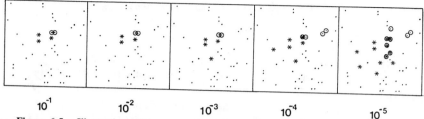

Figure 6.5. Illustration of the uncoupling of nonresonant states, which generally have largest amplitudes on distinct sets of sites. In this example, guest sites participating in a monomer state (*) and a dimer state (○), are distinct down to amplitude 10^{-5}, despite the proximity of the states; $c = 10\%$ and $\delta = 10^{-2} \, cm^{-1}$, 30×30 model.

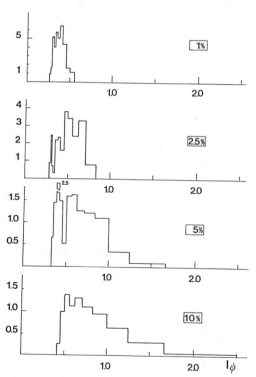

Figure 6.6. Histograms showing the increase of the localization lengths of the donor states with donor concentration between 1 and 10%, with diagonal disorder $\delta = 5 \times 10^{-3} \, cm^{-1}$.

B.   Direct Donor–Acceptor Trapping

1.   *Introduction*

The rate of trapping from donors to acceptors can now be found by inserting the exact donor eigenstates in the expressions for trapping by overlap in Section V.

Donor excitons in the experiments were prepared in three ways: (1) relaxation after excitation of the singlet manifold of the mixed crystal, (2) excitation of the host triplet band, and (3) direct excitation of the donor band. We assume that all these methods lead to equiprobable excitation of all the donor states because preparation of guest singlet or triplet excitations by trapping from the host exciton band depends on the overlap of the host and donor wavefunctions. The overlap is insensitive to the donor state because the traps are deep. Laser excitation of the donors[47] was done with a linewidth larger than the total bandwidth of the donors. Granted that the donor states all have about the same radiative lifetime (see Section V), they will be equiprobable. The yield of acceptor emission deduced from Eq. (5.28) without donor–donor diffusion is then

$$r = 1 - \frac{1}{N_D} \sum_{D=1}^{N_D} \frac{1/\tau_D}{1/\tau_D + \Sigma_A \Gamma_{AD}} \tag{6.3}$$

for N_D donor states coupled to acceptor states A by trapping rates Γ_{AD} as in Eq. (5.25). Exact evaluation of Eq. (6.3) and configurational averaging would require much too much numerical work. We suggest an approximate method below.

2.   *Approximate Yield of Acceptor Emission*

An approximation to Eq. (6.3) can be found by looking for the proportion P of donor states D which are not coupled to at least one acceptor state A by a trapping rate faster than the monomolecular decay rate τ_D. Assuming that trapping is then inevitable,

$$r = 1 - P$$

The next step is identification of the principal terms in the overlap of the states D and A. Consider the order of magnitude of the tails of a donor state D and an acceptor state A, along a cut between their centers of gravity on sites 0 and n. The magnitude of the overlap terms along this cut is

$$\frac{V^{2n}}{\Delta_A^{2n-2}|\Delta_A - \Delta_D|^2}$$

on site $m = 0$,

$$\frac{V^{2n}}{\Delta_D^{2m}\Delta_A^{2n-m}}$$

on sites m, $0 < m < n$, and

$$\frac{V^{2n}}{\Delta_D^{2n-2}|\Delta_A - \Delta_D|^2}$$

for $m = n$.

With the numerical values $\Delta_D \sim 100\ \mathrm{cm}^{-1}$, $\Delta_A \sim 340\ \mathrm{cm}^{-1}$, and $V \sim 1\ \mathrm{cm}^{-1}$, the dominant term is thus $m = n$. We shall therefore approximate the overlap by the weight of the donor state on the acceptor site m_A:

$$\Gamma_{AD} \sim \Gamma_0'|c_{Dm_A}|^2 \tag{6.4}$$

Further, the weight of the donor state on the acceptor site must be of the same order as the weight if site m_A was a host site, because $\Delta_D \sim |\Delta_D - \Delta_A|$. But the latter weight was calculated in Section A. Transfer will thus occur if an acceptor occupies a host site m, which in the crystal with donors only (Section A) had an amplitude c_{Dm} greater than ϵ given by

$$\epsilon = (\Gamma_0'\tau_D)^{-1/2} \tag{6.5}$$

Let $N(\epsilon, D)$ be the number of host sites fulfilling this condition for donor state D. From the probability that none of them is an acceptor $(1 - c_A)^{N(\epsilon,D)}$, we find the yield of transfer is

$$r \sim 1 - \frac{1}{N_D}\sum_D (1 - c_A)^{N(\epsilon,D)} \tag{6.6}$$

Such sites are counted easily by inspection of the eigenstates in Section A. Constant Γ_0' is then to be adjusted by comparing the calculated

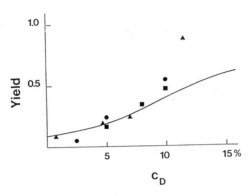

Figure 6.7. Yield of diffusionless transfer as a function of the donor concentration in mixed crystals of naphthalene, deduced from Eq. (6.6) with $\Gamma_0' = 6 \times 10^{20}\,\text{s}^{-1}$.

yields with the experimental results at 1.6 K. Figure 6.7 shows this comparison for $\Gamma_0' = 6.10^{20}\,\text{s}^{-1}$ and $c_A = 0.1\%$. We note that the calculated curve underestimates the yield at concentrations above 10%. This is not surprising, because the temperature dependence of the yield shows that hopping contributes significantly at concentrations over $c_D = 10\%$ (see Fig. 1.5 and Section V.5) but the main problem is elsewhere. The limiting amplitude in Eq. (6.5) is about 10^{-10}. We disbelieve that such small amplitudes can be responsible for observable effects. On the other hand, the hopping radius required to explain the same data was $\bar{R} = 4.2$ (see Section III.C), corresponding to a superexchange coupling of only $10^{-7}\,\text{cm}^{-1}$. The real problem with both models is neglect of the third dimension. Both radii would be reduced to more believable values by including three-dimensional transport.

VII. DISPERSIVE HOPPING TRANSPORT

Hopping transport in disordered media has received much attention. Many recent contributions have addressed the nonexponential dynamics of reactions in disordered systems by introducing the concept of hopping on fractals. Although the success of phenomenological models of hopping on percolation clusters in the triplet state of mixed crystals of naphthalene has contributed to the popularity of fractals, we should like to question here whether these models are consistent with what generally is known about hopping transport in bulk materials. Our conclusion is that fractals capture one important point about nonexponential relaxation, the need for a low spatial dimension. On the other hand, fractals are introduced into bulk relaxation by percolation of the interactions, which implies modelling distance-dependent hopping rates by a step function. This is impossible for slowly varying interactions, such as multipolar forces. When applied to interactions with a strong distance dependence, such as tunneling, it completely misses the hierarchical nature of the dynamics and produces numerical results which are misleading. We shall illustrate this below with reference to hopping in naphthalene, but let us begin with a reminder on fractals.

A. Hopping and Diffusion-Limited Reactions on Fractals

One of the main ideas on hopping transport which has emerged in the last decade is that nongaussian transport and nonexponential relaxation are to be expected whenever the spatial dimension of the set of accessible sites is low. Causes of this might be dispersion of the site energies, with insufficient thermal energy to reach higher neighbors around a given site, or a combination of a low concentration of the active sites and a hopping rate decaying sharply with distance. In such situations, the accessible sites are not distributed homogeneously around a given point. Fractals are one way to model this. Suppose the number of active sites within radius R grows slower than the volume in which they are distributed

$$M(R) \propto R^{\bar{d}} \tag{7.1}$$

where $\bar{d}$ is smaller than d, the spatial dimension. The fractal dimension, $\bar{d}$ may not be an integer, but a fraction, whence "fractal." Equation (7.1) is characteristic of self-similar objects with dilation symmetry: They look the same at all scales of length. Figure 7.1 illustrates this for the Sierpinski gasket in two dimensions. It is obtained by repeated division of a triangle into four smaller ones and removal of the middle. Each triangle in the nth stage of subdivision is similar to the original, but has all its elements 2^n times smaller. It is clear that because the middle is removed at each stage of subdivision, the masses of portions of the gasket of sides R and $2R$ verify

$$M(2R) = 3M(R)$$

This is consistent with Eq. (7.1), which implies

$$M(2R) = 2^{\bar{d}}M(R)$$

on condition that $2^{\bar{d}} = 3$ whence $\bar{d} = \log 3/\log 2 < 2$. The fractal dimension of the gasket is thus lower than that of the plane in which it is built. This

Figure 7.1. A two-dimensional, deterministic fractal, the Sierpinski gasket, produced by repeated subdivision of an equilateral triangle, according to the rule in the inset.

means physically that the fractal has holes of all sizes, as suggested by inspection of Fig. 7.1. The Sierpinski gasket is regular, but many natural objects obey Eq. (7.1) in a statistical sense.[97] Examples are trees, clouds, river basins, ria coastlines, and diffusion-limited aggregates.

Fractals in fact require more than one dimension to describe them. Of importance here is the spectral dimension $\tilde{d}$, which appears in the elastic properties of fractals. Consider a large set of particles of mass m, connected by a mesh of springs. Let k_{ij} be the elastic constant of the spring between particles i and j. On projecting the equations of motion of the particles onto the x axis, we have for example, assuming independent x,

y and z movements (scalar model)

$$m\ddot{x}_i = -\sum_{j\neq i} k_{ij}(x_i - x_j) \tag{7.2}$$

for the ith displacement x_i. On looking for eigenmodes of the form

$$x_i = X_i \exp(i\omega t)$$

we receive the eigenvalue problem

$$(K - m\omega^2 Id)\vec{X} = 0$$

where $\vec{X}$ is a vector with components X_i and K is the dynamic matrix

$$K_{ij} = \left(\sum_{j\neq i} k_{ij}\right)\delta_{ij} - k_{ij}(1 - \delta_{ij}) \tag{7.3}$$

Nontrivial solutions correspond to the condition

$$\det(K - m\omega^2 Id) = 0 \tag{7.4}$$

The spectral density $\rho(\omega)$ can thus be found by determining the eigenvalues of K.

If the mesh is periodic, with identical springs $k_{ij} = k$, then we have the ball and spring model of the vibrations of a crystal. The frequency spectrum of the vibrational modes has a cutoff at about $\omega_{\max}$ such that

$$\omega_{\max}^2 \sim k/m$$

while the low frequency spectral density obeys[57]

$$\rho(\omega) \sim \omega^{d-1} \tag{7.5}$$

where d is the spatial dimension. A similar relation holds for fractal assemblies of balls and identical springs,

$$\rho(\omega) \sim \omega^{\beta} \tag{7.6}$$

and can be used to define the "spectral" dimension by

$$\beta = \tilde{d} + 1$$

The Sierpinski gasket with identical masses in the corners of the triangles and identical springs along their sides has for example a spectral dimension $\tilde{d} = 2 \log 3/\log 5 \sim 1.36$. In general, the following inequality holds for the spatial, the fractal, and the spectral dimensions, with equality for the usual, solid, objects of ordinary geometry.

$$\tilde{d} < \bar{d} < d \qquad (7.7)$$

The spectral dimension is of interest here because it governs random walks on the fractal, at long times. Suppose we replace the balls by sites and the elastic constants by hopping rates. Then hopping transport is described by the equation for the vector of site occupation probabilities, $p_i(t)$ or $\vec{P}$,

$$\frac{d\vec{P}}{dt} = W\vec{P} \qquad (7.8)$$

where

$$W = -K$$

whence

$$\vec{P}(t) = \exp(Kt)\vec{P}(0) \qquad (7.9)$$

The probability of return to the origin (including never leaving) at time t is, for example,

$$P_0(t) = \frac{1}{N} \sum_{i=1}^{N} \langle i|\exp(Wt)|i\rangle$$

$$= \frac{1}{N} \sum_{i} \sum_{v} |\langle i|v\rangle|^2 e^{\lambda_v t}$$

$$= \frac{1}{N} \sum_{v} e^{\lambda_v t}$$

$$= \int_0^{+\infty} d\lambda \, e^{\lambda t} \varphi(\lambda) \qquad (7.10)$$

where λ_v is the eigenvalue of W corresponding to the eigenvector $|v\rangle$ and

$\varphi(\lambda)$ is the spectral density of W. From conservation of probability and the relation $\lambda \propto \omega^2$ we have

$$\varphi(\lambda) \underset{\lambda \to 0}{\propto} \lambda^{\bar{d}/2-1} \qquad (7.11)$$

and

$$P_0(t) \underset{t \to \infty}{\propto} t^{-\bar{d}/2} \qquad (7.12)$$

Now if sites are revisited frequently,[98,99] $P_0(t)$ is inversely proportional to the number of distinct sites visited at time t, $S(t)$:

$$P_0(t) \underset{t \to \infty}{\sim} 1/S(t) \qquad (7.13)$$

If the random walk occurs on a fractal substrate, these sites are distributed in a volume of size $l(t)$ such that (cf. Eq. 7.1),

$$S(t) \sim l(t)^{\bar{d}}$$
$$\underset{t \to \infty}{\sim} t^{\bar{d}/2} \qquad (7.14)$$

Hence, the mean squared displacement $r^2(t)$ will be

$$r^2(t) \underset{t \to \infty}{\sim} t^{\bar{d}/d} \qquad (7.15)$$

According to Eq. (7.7), the power in the last relation is less than unity for diffusion on fractals. Diffusion is thus "anomalous." Alternatively, we could introduce a time-dependent diffusion constant

$$r^2(t) \sim D(t)t$$
$$D(t) \underset{t \to \infty}{\sim} t^{\bar{d}/d-1} \qquad (7.16)$$

Note that Eq. (7.14) states that the volume explored by the random walker grows slower than the time. This is called compact exploration.[100] The walker returns repeatedly to points it has already visited, like an "ant in a labyrinth."[101] Consequently, reactions dependent on diffusion, such as trapping by acceptor molecules or fusion with other walkers, are slowed down compared to usual diffusion, resulting in non-exponential decay kinetics.

Consider for example a particle A hopping on a fractal substrate until it finds a particle of type B, with which it reacts:

$$A + B \rightarrow B$$

This reaction is a simple way to model trapping by acceptors with negligible lifetime. Let R_n be the number of distinct sites visited in a particular walk of n steps. The walker survives these steps if none of the sites is an acceptor. Assuming random distribution of the acceptors, with average concentration p, the probability of survival of the A particle after n steps is

$$\Phi_n = \langle (1 - p)^{R_n - 1} \rangle$$
$$= e^{\lambda} \langle e^{-\lambda R_n} \rangle \qquad (7.17)$$

with $\lambda = -\ln(1 - p) \sim p$ for $p \ll 1$.

The average is over all possible walks of n steps. The lowest-order approximation to this average, the first term in a cumulant expansion, is

$$\Phi_n \sim e^{\lambda} \exp(-\lambda S_n) \qquad (7.18)$$

Clearly, compact exploration will result in nonexponential decay of the population of A particles. Under the present assumption of a single hopping time, τ (Eq. 7.18) is in fact a stretched exponential or Williams–Watts or Kohlrausch decay: Setting $t = n\tau$ and $S_n = \kappa n^{\tilde{d}/d}$ in Eqs. (7.14) and (7.18),

$$\Phi(t) \sim \exp(-\kappa p t^{\beta})$$
$$\beta = \tilde{d}/d < 1 \qquad \text{if } \tilde{d} < 2 \qquad (7.19)$$

Nonexponential decay is to be expected whenever diffusion occurs in a space of dimension lower than three, fractals being a special case. Thus, the number of distinct sites visited by nearest-neighbor hopping on periodic structures of dimension d have the following forms at long times:[102]

$$S_n = \begin{cases} a_1 n + a_2 \sqrt{n} + \cdots & d = 3 \\ a_1 \log(n)/n + \cdots & d = 2 \\ a_1 \sqrt{n} + a_2/\sqrt{n} & d = 1 \end{cases} \tag{7.20}$$

where the constants depend on the lattice structure.

Blumen, Klafter, and Zumofen have determined the validity of the approximate form (Eq. 7.18). It turns out that corrections are strongest in spaces of low spatial dimension, but the conclusion that hopping in restricted geometries leads to compact exploration and hence to nonexponential kinetics remains true. See Ref. 103 for a systematic review of diffusion-limited reactions on substrates of low dimension.

Several authors have used a generalization of Eq. (7.19) when both A and B particles diffuse. Differentiating Eq. (7.19) with respect to time, we have

$$\frac{d\Phi(t)}{dt} = -\kappa p \frac{dS(t)}{dt} \Phi(t)$$

which suggests a generalization:

$$\frac{dn_A(t)}{dt} = \frac{dn_B(t)}{dt} = -k(t)n_A(t)n_B(t) \tag{7.21}$$

with a time-dependent reaction rate

$$k(t) \sim \kappa \frac{dS(t)}{dt}$$

This equation states that A (B) particles sweep a volume $dS(t) = k(t)\,dt$ in time dt and that the probability of presence of the other species in this volume is, for very low concentrations, $dS(t)n_X(t)$. Argyrakis and Kopelman studied trapping according to Eq. (7.21). de Gennes[100] derived a similar relation, with $k(t) \sim S(t)/t$, equivalent to Eq. (7.21) when $S(t)$ is a power of t. In these expressions, $n_X(t)$ is the total concentration of species X. As pointed out by de Gennes, Eq. (7.21) will be appropriate if the reactants remain well mixed throughout the reaction. This is true of trapping and of homogeneous fusion, $A + A \to 0$, but is now known to be false for heterogeneous fusion, $A + B \to 0$, owing to segregation of the surviving reactants.[103-108] The reaction then proceeds only at the boundaries of regions containing either A or B particles.

The decay of mobile A particles due to trapping at fixed B particles deduced from Eq. (7.21) is

$$\frac{n_A(t)}{n_A(0)} \sim \exp(-\kappa n_B(0)S(t)), \qquad n_A(0) \ll n_B(0) \qquad (7.22)$$

while decay due to homogeneous fusion follows

$$\frac{n_A(t)}{n_A(0)} \underset{t \to \infty}{\sim} \frac{1}{1 + \kappa n_A(0)S(t)} \sim 1/S(t) \sim t^{-\gamma}, \qquad \gamma = \tilde{d}/2 \qquad (7.23)$$

Several authors have applied these ideas to nonexponential luminescence decays in disordered solids. Consider, for example, reversible hopping transport on a set of randomly distributed active sites, described by a set of master equations like Eq. (7.8) for the site occupation probabilities. The argument has two steps. First, hopping rates are decreasing functions of distance, $w(r)$, so that if the lifetime of an excitation, or the experimental apparatus available, sets an upper limit to the observable time, say τ, it seems reasonable to define a cutoff in the interactions at r_L such that $w(r_L) \sim 1/\tau$. Next one can look at the clusters of sites at distances less than r_L from one another. At low concentrations these clusters are small. As the concentration is raised, they grow and coalesce, until a percolating cluster eventually appears (cf. Section I). Now it is known that percolation clusters are fractals, so we may expect hopping transport and diffusion-limited reactions at the percolation threshold $\bar{c}$ to be described by relations such as Eqs. (7.13)–(7.23). At concentrations close to the percolation threshold $c \lesssim \bar{c}$, the correlation length ξ, a measure of the linear size of the largest cluster, diverges according to

$$\xi \propto |c - \tilde{c}|^{-\nu}$$

where ν depends only on the spatial dimension. Similarly, the number of sites in the largest cluster at concentrations over the threshold grows as

$$P_\infty \propto |c - \bar{c}|^{\beta}$$

where β depends again on the dimension and not on the details of the problem, such as the connectivity or the range of the interactions. According to the scaling theory of percolation,[109] the fractal dimension of the percolating cluster is

$$\bar{d} = d - \beta/\nu \tag{7.24}$$

The value in two dimensions is about 1.89, with $\beta \sim 0.14$ and $\nu \sim 1.35$. Gouker and Family[110] checked that ν really is independent of the interaction radius. Their data also provide values of the percolation concentration as a function of the connectivity. The independence of percolation exponents with respect to the details of the problem should give $\bar{d}$ a meaning independent of the experimental context. If measurement of some macroscopic observable element, such as the conductivity or the emission of traces of acceptors, shows percolation, then the active sites are part of a fractal of dimension $\bar{d}$, no matter what the ranges of experimental concentration or time. Now according to a conjecture of Alexander and Orbach,[99] the spectral dimension of percolating clusters should be

$$\tilde{d} = 4/3$$

in all dimensions (whereas $\bar{d} \sim 1.89$ for $d = 2$ and $\bar{d} \sim 2.7$ for $d = 3$). Alexander,[111] and Aharony and Stauffer[112] later proposed that

$$\tilde{d} = 4/3 \quad \text{if } \bar{d} > 2$$

and $\tag{7.25}$

$$\tilde{d} = \frac{2\bar{d}}{\bar{d} + 1} \quad \text{if } \bar{d} \leq 2$$

The difference between these estimates is very small for $d = 2$, with simulations supporting Eq. (7.25).[113,114]

Evesque,[115] Evesque and Duran,[116] and Klymko and Kopelman[117] applied these results to triplet excitons in mixed crystals of naphthalene. Hopping occurs here with rates determined by the exponential tunneling or superexchange interaction, up to times of the order of the triplet lifetime. According to Evesque and Duran, trapping and fusion of triplet excitons should give a delayed fluorescence signal,

$$I_{DF}(t) \sim n_{D*}(n_{D*} + n_{A*})S(t)/t$$
$$\sim S(t)/t \sim t^{-1/3} \tag{7.26}$$

if the donor and acceptor populations n_{D*}, and n_{A*} are approximately constant in their experiments with periodic excitation. Klymko and Kopelman[107] and Kopelman[117] compared the delayed fluorescence to the phosphorescence:

$$I_{DF}(t) \sim \gamma \frac{dS(t)}{dt} (n_{D*})^2$$

$$I_{PH}(t) \sim n_{D*}(t) \tag{7.27}$$

$$\frac{I_{DF}(t)}{I_{PH}(t)^2} \sim dS/dt \sim t^{-1/3}$$

Figure 7.2 shows the power-law behavior of the data, in fair agreement with Eqs. (7.26) and (7.27).

Two points are however unclear. The first is that the laws given above correspond to a single, sharp excitation, whereas pulsed exitation and shuttered continuous excitation were used in the experiments. We can expect this to modify these laws. In the second case time is measured from the closing of the shutter, but the sample then contains excitations with all ages up to the exciton lifetime, which already have explored part of the cluster of accessible sites defined by the cutoff in the interaction. Secondly and more fundamentally, the model described above neglects the spread of hopping rates which the interaction law produces in a disordered sample. Is this reasonable? Does it influence the dependence on the experimental conditions? Only solution of the complete master equations can answer these questions, as illustrated below.

B. Simulation of Variable Range Hopping

Section III.A presented simulations of the steady-state populations of donors and acceptors described by the master equations (Eq. 3.1). Brown et al.[118] calculated the kinetics of the same model, which we restate briefly for convenience. Donor and acceptor molecules were distributed at random, with respective average concentrations c_D and c_A on a host square lattice. Reversible hopping between donors occcurs by tunneling through the host, with hopping rates of the form

$$w_{ij} = w_0 \exp(-\alpha r_{ij}) \tag{7.28}$$

where r_{ij} is the distance between donors i and j and α is determined by the depth Δ of the donor states below the host and the order of magnitude V of the nearest-neighbour exchange interactions. The appropriate value for protonated naphthalene in deuterated host crystals is[16] $\alpha \sim 9.2$, but

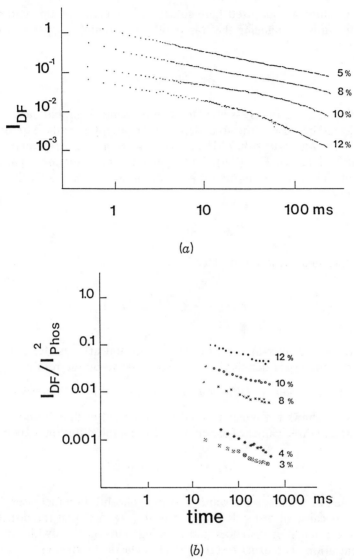

(a)

(b)

Figure 7.2. (a) Decay of the delayed fluorescence of isotopically mixed crystals of naphthalene at 1.6 K and several guest concentrations, after Ref. 114. (b) Delayed fluorescence normalized by the squared intensity of the phosphorescence,[117] I_{DF}/I^2_{PHOS}. Power law behavior is shown by the straight log–log plots.

several values will be used here to see the influence of this parameter. This amounts to changing the trap depth, for example by partial isotopic substitution, since

$$\alpha = 2\ln(\Delta/V)$$

Trapping to acceptors is assumed to be of the same form, but back transfer is neglected because of the great depth of the acceptors in the experiments: $E_D - E_A \sim 240\ cm^{-1} \gg k_B T$. The master equations (Eq. 3.1) can be solved in matrix notation, Eq. (7.9). The probability of observing a particle at site j at time t, following its preparation at site i at $t = 0$, is the matrix element

$$P(j, t \mid i, 0) = [\exp(Wt)]_{ji} \tag{7.29}$$

while the mean-squared displacement is

$$r^2(t) = \frac{1}{N}\sum_{i,j} r_{ij}^2 P(j, t \mid i, 0) \tag{7.30}$$

The average is over all possible departures and arrivals. Reference 118 gives the numerical details. Here we concentrate on the results.

1. Anomalous Diffusion by Tunneling in Two Dimensions

Figure 7.3 shows the mean-squared displacement at three donor concentrations and three values of exponent α. In general, diffusion is anomalous,

$$r^2(t) \sim t^\eta \qquad \text{with } \eta < 1 \tag{7.31}$$

The value of η increases towards 1 as the concentration increases. This is normal because disorder decreases. Recall (Fig. 5.1) that the distribution of nearest-neighbor distances and hopping rates narrows with increasing concentration. The exact result at $c = 1$ is classical diffusion,

$$r^2(t) = Dt$$

with a diffusion coefficient

$$D(t) = \sum_j r_{0j}^2 w(r_{0j})$$

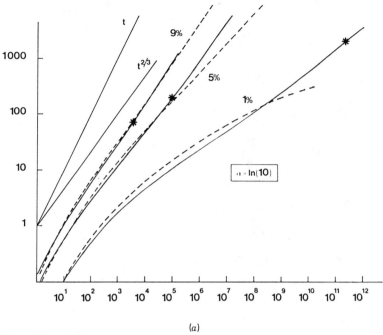

9%

t

$t^{2/3}$

5%

1%

$\alpha = \ln(10)$

1000

100

10

1

10^1 10^2 10^3 10^4 10^5 10^6 10^7 10^8 10^9 10^{10} 10^{11} 10^{12}

(a)

Figure 7.3. (a).

Classical diffusion is expected[66–68,119a] and observed,[119b] at very long times, beyond those reported here.

Increasing α on the other hand enhances disorder, by increasing the spread of the hopping rates. This lowers the value of η. Large values of α produce steps in the law (Eq. 7.31), because the hopping rates for successive possible separations on the square lattice are then very different. The hopping law translates a discrete set of distances R_i into a set of widely separated hopping times, $\tau_i \sim 1/w(R_i)$ (see Fig. 7.4). Excitations explore the sample by stages, first sampling neighbors at distances less than R_1, then any additional sites which can be reached by jumps of length R_2 and so on. Figure 7.5 illustrates this hierarchical exploration by showing some clusters of mutually accessible sites at different times. Similar steps are seen in the probability of return to the origin, corresponding to local equilibrium of the probabilities of occupation on successively larger clusters. Sawteeth appear for the same reasons in the spectral density of the transport matrix.

2. Inapplicability of Hopping on Percolation Clusters

As pointed out above, for a given interaction radius r_L there is a critical

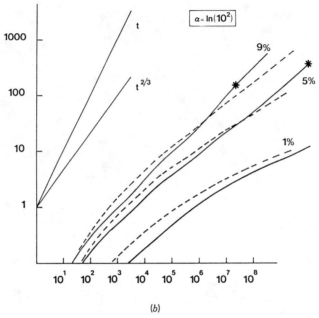

Figure 7.3. (*b*).

concentration $\bar{c}$ at which the clusters of connected sites percolate. Conversely, for a given concentration, there is a critical interaction radius, beyond which the clusters of connected sites percolate. Shante and Kirkpatrick[19] discussed an empirical relation between the coordination number of the interaction z and the percolation threshold for bond and site percolation on lattices:

$$z\bar{c} = K \tag{7.32}$$

where constant K depends on the spatial dimension. On a square lattice, the coordination number is easily found by enumeration of the neighbors, and is approximately

$$z(r_L) = \pi r_L^2 - 1$$

The value of K is 4.15 ± 0.05 for long range ($r_L \gtrsim 3$) site percolation on a square lattice.[120] The steep radial dependence of the tunneling rates suggests neglect of hopping at distances over r such that $w(r)t = 1$ at time t: A slight increase in the distance corresponds to a large increase in the characteristic hopping time. Combining this relation with Eq. (7.32) we can predict the time at which the bond structure percolates:

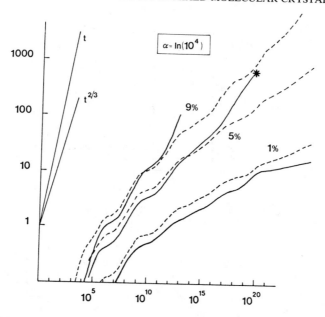

Figure 7.3. Anomalous diffusion by tunneling between random guests on a square lattice, with rates $w(r) = w_0 \exp(-\alpha r)$. (a) $\alpha = \ln(10)$; (b) $\alpha = \ln(10^2)$; (c) $\alpha = \ln(10^4)$ (as for naphthalene). Displacement is in lattice units. Asterisks mark the times at which the time-dependent bond structure percolates, where the $t^{2/3}$ law could be expected to hold.

$$t_c \sim w_0^{-1} \exp(\alpha (K\pi c)^{1/2}) \qquad (7.33)$$

Before t_c, the bond structure is split up into small clusters. An infinite cluster appears at time t_c and subsequently dominates the sample. Figure 7.5 shows examples of the clusters of accessible sites before and after t_c. The largest cluster at t_c should be a fractal with a fractal dimension given by Eq. (7.34). Figure 7.6 confirms the statistical dilation symmetry of the largest cluster at t_c, with a fractal dimension 1.9 in agreement with the prediction of scaling theory.[109,121]

Figure 7.7 also confirms that hopping by a uniform long-range interaction also obeys the law of anomalous diffusion predicted by scaling. In this simulation the tunneling interaction was replaced by a step function with width equal to the critical radius for a given concentration.

It might be expected that this law would be close to the exact law in Fig. 7.3 at time t_c. This is not the case. Changing the concentration or the tunneling exponent α results in powers η larger or smaller than the value $\tilde{d}/\bar{d} = 0.69 \pm 0.01$ in two dimensions. Anomalous diffusion was generally slower than predicted by percolation. When it was faster, this was true

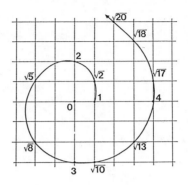

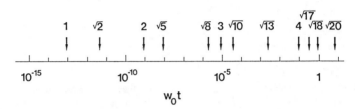

Figure 7.4. The tunneling law on a square lattice translates small differences in the length of jumps (upper) to large differences in the jump times (lower). Here $\alpha = \ln(10^4)$ as for naphthalene.

over a wide range of times and not a distinguishing feature of t_c. Percolation alone does not fix the exponent of anomalous diffusion when there is a spread of the hopping rates.

The cause of this failure is the hierarchical nature of the clusters of connected sites. Consider, for example, hopping with a step function interaction $w(r) = w_0$ (hopping time $t_0 = 1/w_0$) up to the critical radius for percolation. At time $t = t_0/2$, random walks are breaking new ground all over the clusters of sites connected by rates $w \geq w_0$. Random walks with a spread of hopping rates (Eq. 7.28) have on the other hand completely and repeatedly visited all sites accessible by jumps faster than $t_0/2$. This slows down the diffusion, because new sites are on the periphery of the already visited volume.

3. *Tunneling Represented as a Walk on an Ultra-Metric Space*

That tunneling leads to local equilibrium on sets of sites which grow with time seems first to have been pointed out by Zvyagin and Esser.[122,123] This idea can be expressed formally in the language of ultra-

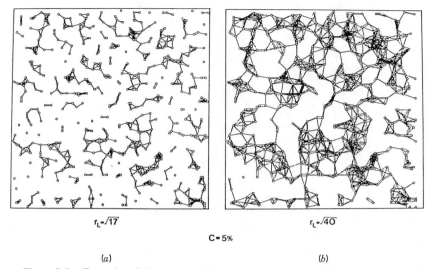

$r_L = \sqrt{17}$ $r_L = \sqrt{40}$

$c = 5\%$

(a)   (b)

Figure 7.5. Examples of the clusters of connected sites at times before (a) and after (b) percolation of the interactions at $r_L = \sqrt{26}$ and $c = 5\%$. The corresponding times in Fig. 7.3(a) are, respectively, $w_0 t = 10^{4.12}$ and $10^{6.35}$.

metric spaces,[103,124,126] illustrated here by an approximate calculation of the decay of donors due to trapping limited by diffusion.[120]

Consider donors D and acceptors A distributed at random on a d-dimensional lattice, with respective concentrations c_D and c_A. Let $p = c_A/(c_D + c_A)$ be the proportion of acceptors. Now, because the hopping lengths can have only discrete values, $R_1, R_2, \ldots, R_i, \ldots$, with well-separated jump times, $\tau_1 \ll \tau_2 \ll \cdots \ll \tau_i \ll \cdots$, the time taken to get from one point to another depends less on the distance between them than on the longest hop required. We can in fact define an ultra-metric distance on the set of hopping centers,

$$d(A, B) = \text{Minimum over all} \quad (\text{Maximum hopping time on the path})$$
$$\text{paths between A and B}$$

Thus,

$$d(A, C) = \max[d(A, B), d(B, C)] \le d(A, B) + d(B, C)$$

the usual triangle inequality for ordinary distances. Hopping over potential barriers is an example of this kind of distance. The time to get from A to

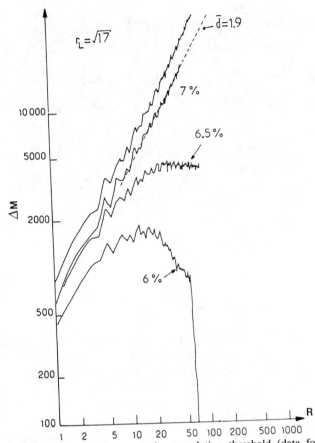

Figure 7.6. The largest cluster at the percolation threshold (data for $c = 5\%$ and $r_L = \sqrt{26}$) obeys dilation symmetry (Eq. 7.1), checked here in a plot of the mass of rings of width 1 and internal radius R, $\Delta M(R) \propto R^{d-1}$. The value of $\bar{d}$ is 1.9, in agreement with scaling theory. Clusters below the percolation threshold are finite, so the density decreases at large radii (e.g., the bottom curve).

C is determined by the higher of the barriers between A and B and B and C.

Figure 7.8 illustrates the ultra-metric structure. Guests connected to the same knot at level i in the hierarchy can all be reached one from another in about the same time, τ_i, and will have been visited many times before the next possible hopping time connects them to other sites. At time τ_i, the sample is split into clusters $C_1^{(i)}, C_2^{(i)}, \ldots, C_n^{(i)}, \ldots$, of $s_1^{(i)}, s_2^{(i)}, \ldots, s_n^{(i)}, \ldots$ sites, which grow and fuse together, as illustrated by Fig. 7.5. Excitations created anywhere in the sample will survive so long

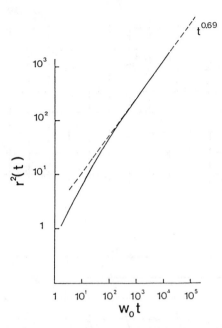

Figure 7.7. Random walks with a step-function hopping rate, $w(r) = w_0$ for $r < r_L$, obey the law predicted by scaling theory, for critical combinations of c and r_L. Data for $c = 5\%$ and $r_L = \sqrt{26}$ on a square lattice. After a brief transitory regime, diffusion follows $r^2(t) \propto t^{0.69}$ in agreement with estimates of $\bar{d}/d = 0.69 \pm 0.01$ for two-dimensional percolation, (see Eq. 7.15).

as their initial site is not included in a cluster containing an acceptor. Thus, cf. Eq. (7.17)

$$\Phi_D(\tau_i) = \frac{1}{N_D} \sum_s \sum_n s\delta(s - s_n^{(i)})$$

$$= \langle (1 - p)^{s-i} \rangle \tag{7.34}$$

where the average is over all clusters. Zumofen and Blumen[102] have shown that this average is well approximated on ultra-metric spaces by the second-order cumulant expansion. Here, introducing the mean cluster size, S_i and its variance, σ_i^2 at time τ_i:

$$\Phi_D(\tau_i) = e^{+\lambda} \exp(-\lambda S_i - \lambda^2 \sigma_i^2) \tag{7.35}$$

This expression can be evaluated by direct counting of clusters in simulations, but the shape of the decay can be deduced from the scaling theory

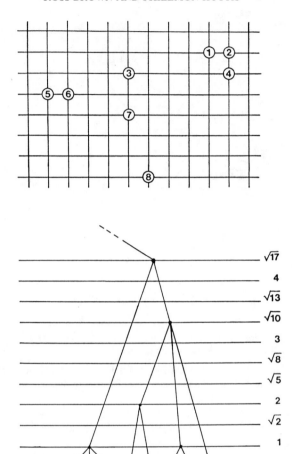

Figure 7.8. Representation of the hierarchy of hopping times due to tunneling between random centers on a square lattice. Small increases in the distance translate into large increases in the hopping times, so that the time to get from site 1 to site 8, for example, is determined by step 1–3 and path 1–3–7–8 is faster than the direct jump 1–8, although the latter is shorter.

of percolation. Thus, S_i is the second moment of the distribution of cluster sizes,

$$S_i = \frac{1}{c_D} \sum_s n_s s^2 \qquad (7.36)$$

which according to scaling theory diverges as[109]

$$\sum_s n_s s^2 \underset{|R-\bar{R}|\to 0}{\sim} \left| \frac{R-\bar{R}}{\bar{R}} \right|^{(\tau-3)/\sigma} \tag{7.37}$$

where τ and σ are the critical exponents for the concentration of clusters of size s and the typical cluster size[109]

$$n_s \sim s^{-\tau}, \qquad \tau > 0$$

and

$$s_{\text{typ}} = A \left| \frac{R-\bar{R}}{\bar{R}} \right|^{-1/\sigma}, \qquad \sigma > 0$$

Combining this relation with Eq. (7.35) we have the following approximate form for the donor decay:

$$\Phi_D(t) \sim e^{-\lambda} \exp(\lambda A (R_c \alpha)^\gamma [\ln(t_c/t)]^{-\gamma}) \tag{7.38}$$

with $\gamma = (\tau - 2)/\sigma$. Figure 7.9 shows the qualitatively correct form of this approximation, using a value $\gamma = 2.2$ deduced from a scaling plot of the cluster sizes in Monte Carlo simulations.[120] This decay is much slower than the stretched exponential which follows from the fractal approximation (Eq. 7.14) with $S(t) \sim t^{2/3}$ in two dimensions, because of the slow exploration discussed above. Note, however, two limitations to Eq. (7.38): (a) it is valid only up to the percolation time t_c; (b) the possible hopping lengths tend to run together at long times, so that the characteristic times are not so well separated. We turn now to the last point brought out by numerical simulation, that experimental measurement of dispersive transport is very sensitive to the excitation and observation conditions.

C. Slow Kinetics and Experimental Conditions

This section shows how the apparent kinetics of a system with spread-out rate constants may depend strongly on the method of experimental excitation and observation. Triplet-state transport in naphthalene will be used as an illustration. The starting point is again the kinetic scheme (Eq. 7.8). The acceptor concentration in these simulations was $c_A \gtrsim 0.1\%$, the highest experimentally possible value, due to segregation in Bridgman tubes, so that quantitative agreement with the data is not to be expected. The simulations did however predict qualitatively some interesting effects of real experimental sources. Figure 7.10 shows the decay of the donor population after a single excitation pulse. Steps in the decay are again due to the discrete set of hopping rates. The decay clearly is nonexponential.

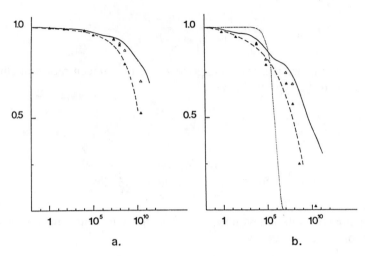

Figure 7.9. Decay of the donor population due to migration and trapping by tunneling on a square lattice, at two acceptor concentrations, (*a*) 0.1% and (*b*) 0.5%. The reduced time is in units of the nearest-neighbor jump time. Solid curves: numerical solution of the full master equations. Full and hollow triangles: first- and second-order approximations in Eq. (7.35), with S_i and σ_i^2 the actual average and variance of the number of connected sites in the ultra-metric approximation. Dashed line: first-order approximation using the scaling relation (Eq. 7.37). The dotted line, an exponential decay with lifetime $10^6 \tau_1$, underlines the absence of a single characteristic time.

Figure 7.11 shows the later part of the decay, in a double logarithmic plot. Stretched exponential behavior would give a straight line with slope β. This is indeed observed over about four decades of time, with deviations, a residue of the discrete structure, which probably could not be measured experimentally, since they amount to at most 10%. At long-times only isolated particles remain. Their monomolecular decay leads to a long-time asymptote with slope $\beta = 1$, that is, exponential decay.

It is unfortunate that nonexponential, slow kinetics can be revealed only if intramolecular decay is slower. This implies weak luminescence. Stronger experimental signals are obtained with periodic or continuous excitation, to build up population. Figure 7.12 shows some excitation–observation sequences which do this. The disadvantage of these methods is that at "time" t, measured from the end of the last excitation, the system contains particles with widely varying ages and histories, produced by the preceding pulses. The experimental signal is in fact the convolution of the single-shot response with the time-dependent source term in Eq. (3.1) or Eq. (3.5):

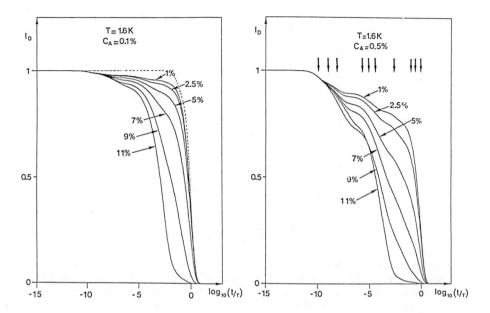

Figure 7.10. Simulated decay of the triplet donor population in mixed crystals of naphthalene, represented by tunneling on a square lattice, at several donor concentrations and two acceptor concentrations. The dashed curve is an exponential, for comparison. The plots include a natural lifetime $\tau = 2.7$ s and the value of α in Eq. (7.28) is $\ln(10^4)$. Arrows show the correlation between steps in the decay and the possible hopping times on the lattice (cf. Fig. 7.4). Time is in units of the natural donor lifetime.

$$\vec{P}(t) = \int_{-\infty}^{t} d\tau \, \exp[W(t - \tau)]\vec{s}(\tau) \qquad (7.39)$$

Distortion results for any form of donor decay except exponential decay. The population vector has the form

(1)   and (2)   Single shot: $\vec{P}(t) = \exp(Wt)\vec{P}(0)$

(3)   Pulsed excitation with period T:

$$\vec{P}(t) = \frac{\exp(Wt)}{\exp(WT) - I} \vec{P}(0)$$

(4)   Continuous excitation:

$$\vec{P}(t) = \exp(Wt)(-W)^{-1}\vec{P}(0)$$

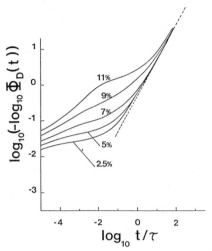

Figure 7.11. An approximately linear double logarithmic plot of the decays of Fig. 7.10 reveals stretched exponential behavior, $P_D = \exp(-At^\beta)$, with A a constant and $\beta < 1$. The dashed line with slope $\beta = 1$ is the result of the natural, monomolecular decay of the remaining isolated excitations, at long times.

Note that the effective population on which the exponential acts is just the equilibrium value after a long continuous excitation (cf. Eq. 3.7).

Figure 7.13, for example, compares the decays after using a continuous source, such as a shuttered lamp, switched off at time "$t = 0$" and a single-shot excitation. Integration of the signal over a finite interval would introduce further distortion. Both decays could be adequately represented by a stretched exponential, but the power β changes from 0.31 in the single-shot experiment to 0.78 in the experiment with a shuttered continuous source. The continuous source gives a decay closer to exponential form because the equilibrium population reached just before the power is turned off is deficient in the short-lived excitations created on donors with strong conducting paths to an acceptor. Heber et al.[126] have seen such effects on chromium ions in lanthanum aluminate. Parus and Kopelman[127] recently reported similar effects on the delayed fluorescence of naphthalene.

As another illustration, consider the delayed fluorescence of mixed crystals of naphthalene, owing to fusion of triplet excitons. A general description of delayed fluorescence, at high densities of excitation, requires the hierarchy of 1,2,3,4-particle distribution functions,[128,129] not just the site-occupation probabilities (one-particle functions) and master equations (Eq. 3.1), valid for low densities of excitation and negligible reactions between the particles. However, we may note that fusion and

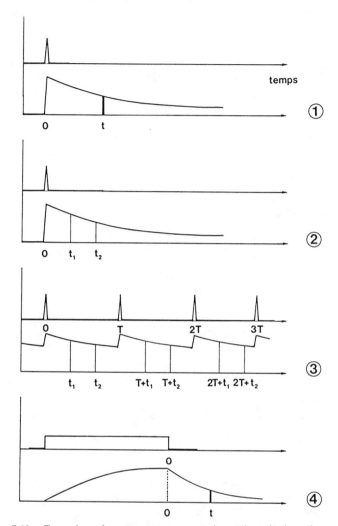

Figure 7.12. Examples of some common experimental excitation–observation sequences, which may distort the true kinetics of a system with a spread of relaxation rates. (1) Ideal single shot and observation with perfect temporal resolution. (2) A single shot, followed by integration of the signal over a finite interval $[t_1, t_2]$. (3) Pulsed excitation with period T and averaging over a window, as with a boxcar. (4) Continuous excitation until equilibrium is reached, as with a lamp and a shutter.

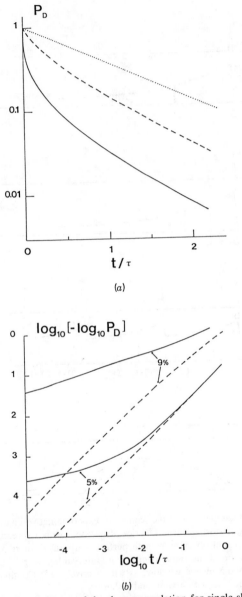

Figure 7.13. Simulated decays of the donor population for single-shot excitation (solid curves) and continuous excitation (dashed curves). (*a*) Semilogarithmic plot of the decay at $c_D = 9\%$, with the natural decay (dotted line) for comparison. (*b*) Double log plot for 5 and 9%, showing the change in the apparent exponent of the stretched exponential behavior.

trapping deplete the donor population, so that at long times the dominant contribution to delayed fluorescence must eventually be fusion between an already trapped excitation and a mobile excitation. The rate of fusion should thus be proportional to the trapping rate, which is limited by the anomalous diffusion. The traps might be residual or introduced impurities, such as 2-methylnaphthalene in naphthalene, or X traps, or even, at low temperatures, the lower levels of clusters of strongly coupled protonated donors. Since fusion reactions have the same distance dependence as the trapping rates,[7] the rate of fusion at time t is thus

$$F(t) \sim \sum_{i \in D} \sum_{A} w_{Ai} p_i(t) \tag{7.40}$$

The equations for the donor population (Eq. 7.28) are unchanged because both donor–acceptor fusion and trapping remove one donor excitation from the system. Equation (7.40) also gives the intensity of delayed fluorescence, because the singlet lifetime, 120 ns, is shorter than the experimental timescale. Thus, after an initial rise due to its finite lifetime, the population, of the singlet state follows Eq. (7.40) at long times. Figure 7.14 compares Eq. (7.40) with the data of Evesque and Duran[116] and of Kopelman.[117] Among the limitations of this calculation are the assumption of two-dimensional transport, the concentration of acceptors, 0.1%, probably higher than in the experiments (leading to steeper calculated decays), and neglect of diffusionless transport (cf. Section VI). The decays are nonetheless close to power laws over the relevant timescale, with appropriate values of the power, increasing with the concentration of guests:

$$I_{DF}(t) \sim t^{-\gamma} \quad \gamma \sim 0.5$$

Figure 7.15 shows the effect of changing the laser repetition rate in pulsed excitation. In agreement with Evesque,[115b] the power γ decreases as the repetition rate increases. A very distorted result is obtained with a shuttered lamp, in the top curve. Clearly, one should be very careful in comparing experimental results with uncorrected theoretical results such as Eq. (7.22) and Eq. (7.23), which assume single-pulse excitation.

VIII. CONCLUSION

Our purpose has been determination of the nature of the transport of triplet excitons in isotopically mixed crystals of naphthalene. Is it hopping transport or diffusionless transport? This problem was tackled in two ways.
The first problem was to go beyond the phenomenological hopping

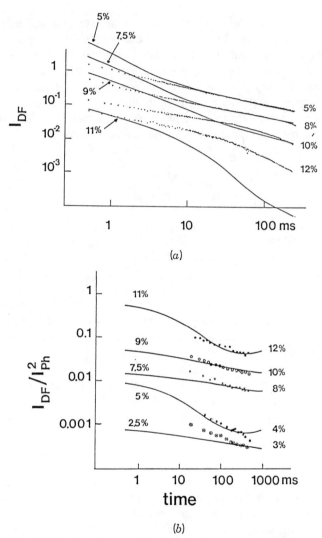

Figure 7.14. (*a*) Simulated decay of the delayed fluorescence (donor concentrations on the left) and measurements (concentrations on the right, redrawn from Ref. 116). Fusion at trap sites is assumed to be the dominant process. (*b*) Simulated, normalized signal (cf. Eq. 7.27) (concentrations on the left) and data redrawn from Ref. 117 (concentrations on the right).

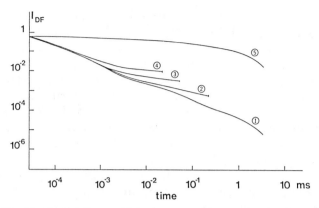

Figure 7.15. Simulated influence of the laser repetition rate on the decay of delayed fluorescence in Fig. 7.13a. The donor and acceptor concentrations are respectively 5 and 0.1%. (1) Single shot; (2) 5 Hz ; (3) 20 Hz; (4) 50 Hz; (5) after turning off a continuous source.

model, putting it on a quantitative footing, by explicit calculation of the hopping rates in the mixed crystals, since numerical simulation of hopping transport had shown that the rates must increase by a factor of the order of 10^6 between 2 and 5 K. Several mechanisms of exciton-phonon coupling were examined. Hopping in the narrow trap band is assisted mainly by coupling to optical phonons: Each hop is accompanied by the emission of a phonon and the absorption of another of nearly the same energy to conserve the total energy in the transition. This leads to activation of the hopping rates with an energy of the order of the Debye energy. Coupling to acoustic phonons, which has a much slower temperature dependence, is weaker because of a lower density of modes and smaller coupling constants.

The two-dimensional hopping model, including explicitly all the distance-dependent hopping rates, was found capable of explaining the dependence of the yield of acceptor emission on donor concentration at fixed low temperature. The "hopping radius", the distance at which hopping is as probable as monomolecular decay, required to fit the data was then 22 Å and the smallest electronic coupling by superexchange or tunneling was 10^{-7} cm^{-1}. However, hopping of perfectly localized site excitations does not account for the temperature dependence of the yield. Specifically, the temperature independence of the yield below the thermal threshold corresponds in this model to residual, direct, one-step trapping to acceptors, after all hopping has been frozen out. The yield should then depend only on the acceptor concentration and not on the donor concentration, contrary to experiment.

This is evidence in favor of trapping between more or less extended donor states. Increasing the concentration should produce more extended states, with higher probability of trapping by overlap with acceptor states. On the other hand, the original proposition of an Anderson transition of the donor states of these crystals, based on an effective, superlattice Hamiltonian, has been shown here to be wrong. At all the relevant experimental concentrations, all the states are strongly localised. There is however a significant change in the localization lengths. The yield of diffusionless transfer has been estimated by numerical simulation, on the same basis as the hopping model. We conclude that two-dimensional, diffusionless transfer, calculated explicitly from the overlap of the states, can account for the yield of acceptor emission up to donor concentrations of 10%. Beyond this value, the temperature dependence shows that hopping has not yet been frozen out. The trapping radius is estimated to be about 24 Å.

Both the hopping radius above and the trapping radius in diffusionless transfer correspond to very weak couplings, or equivalently, to small amplitudes of the wavefunctions. The problem here is in our opinion neglect of the third dimension. Out of plane couplings, known only from calculation of the overlap of the wavefunctions, are estimated to be of the order of 10^{-3} cm^{-1}. Since this is larger than the in-plane tunneling interaction through two host sites, the unreasonably large interaction radii in both the above models would be reduced in a three-dimensional model. The essential point is that out-of-plane coupling would introduce more neighbors and a larger connectivity or coordination number, at a given level of the coupling or the amplitude.

The exact simulation used here to compare these models shows in any case that current arguments, based on average properties, prove nothing for this system. Disorder is far too strong to be represented reliably by average values.

In view of the mixed nature of transport suggested by the experimental temperature dependence, we have developed a description of relaxation of triplet excitons in the eigenbasis of the electronic Hamiltonian. This incorporates the diffusionless component of transport in the extent of the eigenstates, which increases with concentration. Hopping between states contributes to the temperature dependence. High yields of transport depend qualitatively on percolation of the relaxation rates in the space of eigenstates. This can be achieved either by increasing the overlap through an increase of the concentration, or by increasing thermal relaxation, by increasing the temperature. Interpreting the experimental results with this model, we conclude that at 1.6 K, significant diffusionless transport occurs — 50% transfer to acceptors at a concentration of 0.1% (average separation about 70 Å in a three-dimensional distribution) at donor concentration 10%. In view of the strong localization of the wavefunctions, this suggests looking for diffusionless transport in glassy materials, where the ratio of the inhomogeneous broadening to the couplings is hardly less favorable.

Finally, we have examined the validity of the fractal model of hopping transport. This model was introduced into the description of transport in bulk materials, in which the molecular substrate is in no way a fractal, by allowing a cutoff in the reaction rates. The cutoff leads to formulation of a percolation problem and to the expectation that hopping on percolation clusters, a known example of non-Gaussian or dispersive transport, may account for the nonexponential luminescence decays so common in disordered materials. While this approach is intuitively appealing, because it introduces a restricted geometry, it is numerically unreliable. The problem is that it ignores the hierarchical nature of the time-dependent bond structure, induced by the wide spread of relaxation rates due to tunneling interactions. We furthermore draw attention to the need to include experimental timing sequences in theoretical predictions of the luminescence decays, if erroneous interpretation of the distorted experimental decays is to be avoided.

ACKNOWLEDGMENTS

We are indebted to Drs. F. Dupuy and Ph. Pée, whose experimental work and skeptical attitude towards theory have done much to shape this work. Our thanks go also to Dr. M. Orrit, who has taken a constant interest and made many useful remarks.

REFERENCES

1. R. Silbey in *Spectroscopy of Excitation Dynamics of Condensed Molecular Systems*, V. M. Agranovich and R. M. Hochstrasser, Eds., North Holland, New York, 1983.

2. P. W. Anderson, *Phys . Rev.* **109**, 1492 (1958).

3. H. Haken and G. Strobl, in *The Triplet Sate*, A. B. Zahlan, Ed., Cambridge University Press, 1968.

4. M. Orrit and R. Brown, *J. Phys. C: Solid State Phys.* **18**, 5585 (1985).

5. U. Doberer and H. Port, *Z. Naturforsch.* **39a**, 413 (1984).

6. B. Muel, *C. R. Acad. Sci.* **255**, 3149 (1962).

7. M. Sternlicht, C . G. Nieman, and G. W. Robinson, *J. Chem. Phys.* **38**, 1326 (1963).

8. S. D. Colson and G. W. Robinson, *J. Chem. Phys.* **48**, 2550 (1968).

9. F. B. Tudron and S. D. Colson, *J. Chem. Phys.* **65**, 4184 (1976).

10. S. D. Colson, S. M. George, T. Keyes, and V. Vaida, *J. Chem. Phys.* **67**, 4941 (1977).

11. R. Kopelman, E. M. Monberg, F. W. Ochs, and P. N. Prasad: (a) *Phys. Rev. Lett.* **34**, 1506 (1975); (b) *J. Chem. Phys.* **62**, 292 (1975).

12. R. Kopelman, E. M. Monberg, and F. W. Ochs: (a) *Chem. Phys.* **19**, 413 (1977); (b) *ibid*, **21**, 373 (1978).

13. D. C. Ahlgren and R. Kopelman, *J. Chem. Phys.* **70**, 3133 (1979)

14. F. Dupuy, Ph. Pée, R. Lalanne, J-P. Lemaistre, C. Vaucamps, H. Port, and Ph. Kottis, *Mol. Phys.* **35**, 595 (1978).

15. J-P. Lemaistre, A. Blumen, F. Dupuy, Ph. Pée, R. Brown, and Ph. Kottis, *J. Chem. Phys.* **88**, 4655 (1984).

16. Ph. Pée, R. Brown, F. Dupuy, Ph. Kottis, and J-P. Lemaistre, *Chem. Phys.* **35**, 429 (1978).

17. U. Doberer, H. Port, and H. Benk, *Chem. Phys. Lett.* **85**, 253 (1982).

18. U. Doberer, H . Port, D. Rund, and W. Tuffenstammer, *Mol. Phys.* **49**, 1167 (1983).

19. V. K. S. Shante and S. Kirkpatrick, *Adv. Phys.* **20**, 325 (1971).

20. D. Stauffer, *Introduction to Percolation Theory*, Taylor and Francis, 1985.

21. E. Guyon, *C. R. Séances Acad. Sci.*, *Vie Acad.* **294**, XXVII (1982).

22. J. M. Ziman, *Models of Disorder*, Cambridge University Press, 1979.

23. M. Gouker and M. Family, *Phys. Rev. B* **28**, 1449 (1983).

24. P. Argyrakis and R. Kopelman, *Phys. Rev. B* **31**, 6008 (1985).

25. J. Klafter and J. Jortner, *Chem. Phys. Lett.* **49**, 410 (1977).

26. E. M. Monberg and R. Kopelman, *Chem. Phys. Lett.* **58**, 497 (1978).

27. J. Klafter and J. Jortner: (a) *Chem. Phys. Lett.* **60**, 5 (1978); (b) *J. Chem. Phys.* **71**, 1961 (1979); (c) *ibid.* **71**, 2210 (1979).

28. T. Lukes, *J. Non-Cryst. Solids* **8–10**, 461 (1972).

29. D. C. Ahlgren and R. Kopelman, *Chem. Phys. Lett.* **77**, 135 (1981).

30. D. J. Thouless, *Phys. Rep.* **13c**, 93 (1974).

31. A. L. Efros: (a) *Sov. Phys. Usp.* **21**, 746 (1978); (b) *Usp. Phys. Nauk.* **126**, 41 (1978).

32. P. V. Elyutin, *J. Phys. C: Solid State Phys.* **16**, 4151 (1983).

33. B. F. Variano, H. C. Brenner, and W. B. Daniels, *J. Phys. Chem.* **90**, 7 (1986).

34. H. C. Brenner and B. F. Variano, *Mol. Cryst. Liq. Cryst.* **175**, 1 (1989).

35. D. D. Smith, R. D. Mead, and A. H. Zewail, *Chem. Phys. Lett.* **50**, 358 (1977).

36. D. D. Smith, R. C. Powell, and A. H. Zewail, *Chem. Phys. Lett.* **68**, 309 (1979).

37. G. T. Talapatra, D. N. Rao, and P. N. Prasad, *Chem. Phys.* **101**, 147 (1986).

38. T. Kirski, J. Grimm, and C. von Borczyskowski, *J. Chem. Phys.* **87**, 2062 (1987).

39. C. von Borczyskowski and T. Kirski, *J. Phys. Chem.* **93**, 1373 (1989).

40. S. Taen, *J. Phys. Chem.* **92**, 107 (1988).

41. W. Schrof, E. Betz, H. Port, and H. C. Wolf, *Chem. Phys. Lett.* **123**, 300 (1986).

42. W. Wachtel, *Thesis*, University of Stuttgart, 1986.

43. U. Fischer, C. von Borczyskowski, and N. Schwenter, *Phys. Rev B* **41**, 9126 (1990).

44. P. N. Prasad, J. R. Morgan, and M. A. El-Sayed, *J. Phys. Chem.* **85**, 3569 (1981).

45. R. Richert and H. Bässler, *J. Chem. Phys.* **84**, 3567 (1986).

46. J. Lange, B. Ries, and H. Bässler, *Chem. Phys.* **128**, 47 (1988).

47. R. Brown, J-P. Lemaistre, J. Mégel, Ph. Pée, F. Dupuy, and Ph. Kottis, *J. Chem. Phys.* **76**, 5719 (1982).

48. Ph. Pée, Y. Rebière, F. Dupuy, R. Brown, Ph. Kottis, and J-P. Lemaistre, *J. Phys. Chem.* **88**, 959 (1984).

49. H. C. Longuet-Higgins, *Adv. Spectry.* **2**, 429 (1961).

50. S. L. Chaplot, G. S. Pawley, B. Dorner, V. K. Jindal, J. Kalus, and I. Natkaniec, *Phys. Stat. Sol.* (b) **110**, 4454 (1982).

51. J. Jortner, S. A. Rice, J. L. Katz, and S. Choi, *J. Chem. Phys.* **42**, 309 (1965).

52. P. A. M. Dirac, *The Principles of Quantum Mechanics*, Oxford University Press, 1930.

53. T. L. Muchnick, R. E. Turner, and S. D. Colson, *Chem. Phys. Lett.* **42**, 570 (1976).

54. J. Grimm, T. Kirski, and C. von Borczyskowski, *Chem. Phys. Lett.* **128**, 569 (1986).

55. J. Kolenda and C. von Borczyskoski, *J. Lumin.* **42**, 217 (1988).

56. J. M. Ziman, *Electrons and Phonons*, Oxford University Press, 1960.

57. C. Kittel, *Introduction to Solid State Physics*, Wiley, New York, 1976.

58. R. Kopelman, F. W. Ochs, and P. N. Prasad, *J. Chem. Phys.* **57**, 5409 (1972).

59. G. S. Pawley, G. A. MacKenzie, E. L. Bokhenkov, E. F. Sheka, B. Dorner, J. Kalus, U. Schmeltzer, and I. Natkaniec, *Mol. Phys.* **39**, 251 (1980).

60. J. M. Ziman, *Elements of Advanced Quantum Theory*, Cambridge University Press, 1969.

61. P. N. Prasad and R. Kopelman, *J. Chem. Phys.* **57**, 863 (1972).

62. H. Port, D. Rund, G. J. Small, and V. Yakhot, *Chem. Phys.* **39**, 175 (1979).

63. M. Orrit and Ph. Kottis, in *Advances in Chemical Physics*, Vol. LXXIV, I. Prigogine and S. Rice, Eds., Wiley, New York, 1988.

64. R. Brown, Ph. Pée, F. Dupuy, and Ph. Kottis, *J. Phys. C: Solid State Phys.* **17**, 5549 (1984).

65. J. Birks, *Photophysics of Aromatic Molecules*, Wiley, New York, 1970.

66. C. R. Gochanour, H. C. Andersen, and M. D. Fayer, *J. Chem. Phys.* **70**, 4254 (1979).

67. K. Godzik and J. Jortner, *J. Chem. Phys.* **72**, 4471 (1980).

68. J. Klafter and R. Silbey, *J. Chem. Phys.* **72**, 843 (1980).

69. T. Holstein, S. K. Lyo, and R. Orbach, in *Topics in Applied Physics*, Vol. 49, *Laser Spectroscopy of Solids*, Springer, 1981.

70. D. M. Hamilton, P. M. Selzer, and W. M. Yen, *Phys. Rev. B* **16**, 1858 (1977).

71. Ph. Pée, Thèse d'Etat, Université de Bordeaux I, 1982.

72. B. J. Botter, Thesis, University of Leiden, 1977.

73. F. Dupuy, Y. Rebière, J-L. Garitey, Ph. Pée, R. Brown, and Ph. Kottis, *Chem. Phys.* **110**, 195 (1986).

74. R. C. Tolmann, *The Principles of Statistical Mechanics*, Oxford University Press, 1938.

75. A. Abragam, *The Principles of Nuclear Magnetism*, Oxford University Press, 1961.

76. D. A. Evensky, R. T. Scatetarr, and P. G. Wolynes, *J. Phys. Chem.* **94**, 1149 (1990).

77. H. Haken and G. Strobl, in *The Triplet Strate*, A. B. Zahlan, Ed., Cambridge University Press, 1968.

78. H. Haken and R. Reineker, *Z. Phys.* **249**, 253 (1972).

79. Cl. Aslangul and Ph. Kottis: (a) *Phys. Rev.* **B10**, 4364 (1974); (b) *ibid.* **13**, 5544 (1976).

80. M. Grover and R. Silbey, *J. Chem. Phys.* **54**, 4843 (1971).

81. R. W. Munn and R. Silbey, *J. Chem. Phys.* **72**, 2763 (1980).

82. V. Capek, *Z. Phys. B* **60**, 101 (1985).

83. V. Szocs and V. Capek, *Phys. Stat. Sol. (b)* **131**, 667 (1985).

84. U. Schmid and R. Reineker: (a) *Chem. Phys. Lett.* **94**, 510 (1985); (b) *Mol. Phys.* **55**, 77 (1985).

85. R. Zwanzig, *Physica* **30**, 1109 (1964).

86. Cl. Cohen-Tannoudji, *Cours au Collége de France*, 1975–1976, unpublished.

87. Cl. Aslangul, Thesis, Université P. et M. Curie, Paris, 1977.

88. V. M. Kenkre, *Phys. Rev. B* **18**, 4064 (1978).

89. J. Grimm, T. Kirski, and C. von Borczyskowski, *Chem. Phys. Lett.* **131**, 522 (1986).

90. Ph. Pée, J. P. Lemaistre, F. Dupuy, R. Brown, J. Mégel, and Ph. Kottis, *Chem. Phys.* **64**, 389 (1982).

91. J. L. Pichard and G. Sarma, *J. Phys. C: Solid State Phys.* **14**, L617 (1986).

92. J. H. Wilkinson, *The Algebraic Eigenvalue Problem*, Oxford University Press, 1965.

93. H. Rutishauser, in *Handbook of Automatic Computing*, Vol. II, *Linear Algebra*, F. L. Bauer, Ed., Springer, 1971.

94. R. Brown, Ph. Pée, F. Dupuy, and Ph. Kottis, *J. Phys. Coll.* **46-C7**, 79 (1985).

95. J. P. Lemaistre and A. Blumen: (a) *Chem. Phys. Lett.* **99**, 291 (1983); (b) in *Structure and Dynamics of Molecular Systems*, R. Daudel, J. P. Korb, J. P. Lemaistre, and J. Marouani, Eds., Reidel, Dordrecht, 1985.

96. A. Blumen, J. P. Lemaistre, and I. Mathlouthi, *J. Chem. Phys.* **81**, 4610 (1984).

97. B. Mandelbrot, *Fractals: Form, Chance and Dimension*, Freeman, San Francisco, 1977.

98. R. Rammal and G. Toulouse, *J. Phys. Lett. (Paris)* **44**, L-13 (1983).

99. S. Alexander and R. Orbach, *J. Phys. Lett. (Paris)* **43**, L-625 (1982).

100. P. G. de Gennes, *J. Phys. Chem.* **76**, 3316 (1982).

101. P. G de Gennes, *C. R. Acad. Sci.* **296–II**, 881 (1983).

102. E. W. Montroll and G. Weiss, *J. Math. Phys.* **6**, 167 (1965) and references therein.

103. A. Blumen, J. Klafter, and G. Zumofen, in *Optical Spectroscopy of Glasses*, I. Zschokke, Ed., Reidel, Dordrecht, 1986, review paper.

104. P. Argyrakis and R. Kopelman, *J. Chem. Phys.* **72**, 3035 (1980).

105. G. Zumofen and A. Blumen, *Chem. Phys. Lett.* **88**, 63 (1982).

106. G. Zumofen, A. Blumen, and J. Klafter, *J. Chem. Phys.* **82**, 3198 (1985).

107. P. W. Klymko and R. Kopelman, *J. Phys. Chem.* **87**, 4565 (1983).

108. J. S. Newhouse and R. Kopelman, *J. Phys. Chem.* **92**, 1538 (1988).

109. D. Stauffer, *Phys. Rep.* **54**, 1 (1979).

110. M. Gouker and F. Family, *Phys. Rev.* **B28**, 1449 (1983).

111. S. Alexander, *Ann. Israel Phys. Soc.* **5**, 149 (1983).

112. A. Aharony and D. Stauffer, *Phys. Rev. Lett.* **52**, 2368 (1984).

113. D. Ben-Avraham and S. Havlin, *J. Phys. A: Math. Gen.* **15**, L691 (1982).

114. A. Keramiotis and R. Kopelman, *Phys. Rev. B* **31**, 4617 (1985).

115. P. Evesque: (a) *J. Phys. (Paris)* **44**, 1217 (1983); (b) Thèse d'Etat, Université P. et M. Curie, Paris, 1984.

116. P. Evesque and J. Duran, *J. Chem. Phys.* **80**, 3016 (1984).

117. R. Kopelman, *J. Stat. Phys.* **42**, 185 (1986).

118. R. Brown, J. L. Garitey, F. Dupuy, and Ph. Pée, *J. Phys. C: Solid State Phys.* **21**, 1191 (1988).

119. (a) T. Schneider, M. P. Sörensen, E. Tosatti, and M. Zanetti, *Europhys. Lett.*, **2**, 167 (1986). (b) R. Brown and C. von Borczyskowski, unpublished results.

120. R. Brown, F. Dupuy, and Ph. Pée, *J. Phys. C: Solid State Phys.* **20**, L649 (1987).

121. M. P. M. den Nijs, *J. Phys. A: Math. Gen.* **12**, 1857 (1979).

122. I. P. Zvyagin, *Phys. Status Solidi (b)* **95**, 227 (1979).

123. B. Esser and I. P. Zvyagin, *Phys. Status Solidi (b)*, **104**, 523 (1981).

124. R. Rammal, G. Toulouse, and M. A. Virasoro, *Rev. Mod. Phys.* **58**, 765 (1987).

125. A. Blumen, J. Klafter, and G. Zumofen, *J. Phys. A: Math. Gen.* **15**, L691 (1986).

126. J. Heber, H. Dornauf, and H. Siebold, *J. Lumin.* **24/25**, 735 (1981).

127. S. J. Parus and R. Kopelman, *Mol. Cryst. Liq. Cryst.* **175**, 119 (1989).

128. N. G. Van Kampen, *Int. J. Quant. Chem., Quant. Chem. Symp.* **16**, 101 (1982).

129. B. Sipp, *Chem. Phys.* **77**, 257 (1983).

130. B. Sipp and R. Voltz, *J. Chem. Phys.* **79**, 434 (1983).

THEORETIC PHYSICOCHEMICAL PROBLEMS OF CLATHRATE COMPOUNDS

VLADIMIR E. ZUBKUS* and EVALDAS E. TORNAU

Semiconductor Physics Institute, Lithuanian Academy of Sciences, Vilnius, Lithuania, USSR

VLADIMIR R. BELOSLUDOV

Institute of Inorganic Chemistry, Academy of Sciences of USSR, Siberian Branch, Novosibirsk, USSR

CONTENTS

* *Present address:* Division of Electronic Organic Materials, Institute of Chemical Physics, Academy of Sciences of USSR, Moscow, Kosygina 4.

Advances in Chemical Physics, Volume LXXXI, Edited by I. Prigogine and Stuart A. Rice.
ISBN 0-471-54570-8 © 1992 John Wiley & Sons, Inc.

INTRODUCTION

Widespread investigation of clathrate compounds (clathrates) over the last 50 years revealed basic structural, physical, and chemical properties of these systems. Clathrates have gained recognition in organic and inorganic chemistry, physical chemistry, crystal chemistry, geochemistry, biology, and other branches of science.[1-4]

Clathrate formation occurs due to the inclusion of molecules called "guests" to the closed cavities or channels constituted of "host" lattice molecules. In accordance with the classical approach,[5-10] guest and host molecules of clathrate combine in a certain way without forming any specific chemical bonds. Only van der Waals interactions are observed between the molecules. Thermodynamic stability of the compounds is maintained thanks to the favorable geometric form of cavities and the convenient arrangement of guest molecules in the cavities. As a result, weak intermolecular interactions lead to gain of sufficient energy for clathrate formation. Usually van der Waals interaction between guest and host components is observed in the compounds argon–, krypton–, and xenon–hydroquinone, methane–, ethane–, and propane–water, and normal structure hydrocarbon-urea.

Sometimes the term clathrate is also used to denote the coordinational systems with various types of chemical bonds and more complicated interactions in comparison with the van der Waals between guest and host molecules. Among these there are monomolecular inclusion compounds of cyclodextrin with iodine, naphthalene, and so on ("cavito-clathrates"[8]); so-called "inclusion phases" of hafnium, tantalium carbides, or nitrides; macromolecular inclusion compounds of clathrasiles and zeolites; and numerous class of biocompounds which have obtained the special name "clathrins".[9]

To determine physical and chemical properties of clathrates, the formation of these nonstoichiometric systems ought to be examined. The information collected thus far is predominantly of the so-called lattice clathrate systems, for example, of gas hydrates, zeolites, compounds of urea and thiourea with paraffins, and hydroquinone compounds with volatile components.

The compounds of crown ethers and cryptates with metal ions also attract great interest. The Nobel prize for chemistry was presented in 1987 to C. J. Pedersen for the synthesis of the first crown ethers[11] and M. J. Lehn for the discovery and investigation of cryptates.[12] These compounds, sometimes called "clathrate complexes,"[8] are assumed to be intermediate between lattice clathrates and coordinational compounds. They are characterized by large molecules (dibenzo-18-crown-6 or cryptands) containing the cations of alkaline metals. The theory of formation of these systems is now in the beginning stage.

The investigation of clathrates is stimulated by practical interests. Here we present the most distinctive examples of practical application of inclusion compounds:

1. The formation of gas hydrates at positive temperatures allows desalination of sea water due to salt-molecule inclusion into the solid carcass of hydrates.

2. The separation of certain guest molecules from the mixture is carried out by inclusion of this molecule into the suitable cavity of clathrate or cryptane. This property allows the use of clathrates for the storage, transport, and treatment of radioactive guest molecules (e.g., radioisotopes and noble gases[13]). The cryptanes are used to exclude the radioactive and poisoning substances from human organisms.

3. The clathrates serve as the storage rooms for volatile components, which can be firmly caged in the host lattice at temperatures considerably higher than the boiling temperature of volatile components.

The clathrates and clathration processes play an important role in the regulation of specific processes of live organisms.

4. According to Pauling[14] general anesthesia is caused by the ability of albumen molecules to form with water the clathrate-like hydrate structures which become stable at the human body temperature when the gases N_2O, Xe, and CH_3Cl_3 are added. These structures influence the movement of nerve impulses.

5. The process of DNA transcription during the growth of cancer cells can be halted, introducing plain aromatic cation heterocycles between DNA helices.

6. The considerable scope for disease treatment offers application of croup compounds which can transport the medicine through the lipid membranes inside the cell.[12]

From the physical point of view the clathrates represent one of the inclusion-type systems, the so-called many-component solid solution, in which one component corresponds to the host and others to the guest molecules. The mobility of the guest component of the solution appears to be rather weak in lattice clathrates, contrary to such solid solutions as hydrogen in metals, alloys, and intercalates. Therefore, clathrates kinetics is less interesting, and various phase diagrams with liquid, liquid–solid, and solid phases, characteristic of the compounds with high mobility of included components, are not observed in clathrates. It was thought that the limited structural and chemical properties precluded the possibility of serious investigations of physical properties of clathrates. However, that opinion has changed as new experimental information has become available. It was experimentally found that β-hydroquinone is unstable and transforms to the stable α-form in the absence of guest molecules. Later the analogous property was found to be characteristic of ice and urea lattices. Thus, it was shown that the guests can transform the host lattice from unstable metastable to thermodynamically stable state.

These results required the investigation of lattice stabilization techniques and the estimation of minimal and/or maximal guest molecules concentration c_k, at which the β form turns out to be stable. The dependence of concentration c_k on external conditions (temperature T and pressure P) has been calculated. The van der Waals and Platteeuw theory[5] published in the late 1950s answered the major problems and summarized the experimental results.

The next step in the understanding of physical and chemical problems of clathrate formation was the reexamination of guest molecule interaction, which was not taken into account earlier. It had been assumed[5] that this interaction did not have a considerable influence on the clathrate formation process. However, forthcoming experimental data on phase transitions in clathrates with polyatomic guest molecules CH_3CN, CH_3OH could not be interpreted in the framework of this theory. As a result, proposals have been made (see, e.g., Refs 15–17) to revise the role of guest–guest interactions.

Phase transitions in hydroquinone with HCN and CH_3OH gases were found by Matsuo et al.[18,19] from the temperature dependencies of specific heat and dielectric constant. The phase transition temperature T_c appeared to be comparable with the temperatures at which the clathrates were usually investigated ($T_c \approx 180\,\text{K}$ in β-hydroquinone with HCN and

$T_c \approx 70$ K in β-hydroquinone with CH_3OH). The transitions were interpreted as orientational, caused by guest molecule ordering in the closed cavities of β-hydroquinone. The estimations[18,19] of interaction potentials V_{gg} of guest molecules have shown that $V_{gg} \approx 300$ K in β-hydroquinone with HCN. It became obvious that such a value cannot be neglected, though the real role of V_{gg} interaction remained rather vague until the beginning of the 1980s.[20-26]

It should be noted that clathrate formation along with the cooperative behavior of guest molecules can be determined by the cooperative behavior of the hosts. At certain concentrations of guest molecules it is thermodynamically unbeneficial for the host lattice to accept guests. As a result, the reconstruction of the crystallic structure of the host occurs together with the change of cavity or channel form.

Now we present a summary of the main distinctive physicochemical characteristics by which clathrates differ from other inclusion compounds:

1. Empty host lattices are metastable. They become stable due to guest molecule inclusion.
2. The mobility of guest molecules in the host lattice is restricted. Guests cannot be transferred from one cavity (channel) to another.
3. The size of cavity or channel is comparable with the size of guest molecule. Therefore, there is room for just one guest molecule in every cavity or across the channel.
4. Clathrate compounds can contain various molecules, including the large and massive ones. The concentration of guest molecules in the host lattice considerably exceeds the concentration of these molecules soluted in the lattices of the nonclathrate phase.
5. External factors (pressure, temperature) influence considerably the total amount of included guest molecules.

Along with these properties some features of intermolecular interactions of lattice clathrates can be distinguished:

1. The interactions of host molecules are the main intermolecular interactions in clathrates. These interactions determine the form of host lattice.
2. Intermolecular interactions cause the difference of lattice energies $\Delta E = E_\alpha - E_\beta$ to be small in comparison with the energy E_α or E_β, where E_α, E_β are the energies of stable nonclathrate α-phase and metastable β-phase host lattices.

3. The energy of guest–host interactions turns out to be small in comparison with the energy of host–host interactions, but can be larger or comparable with ΔE. Guest–host interaction of lattice clathrates is of van der Waals nature.

4. The energy of guest–guest interaction is comparable with ΔE. This interaction can be van der Waals or electrostatic in nature.

These properties of intermolecular interactions allow us to consider various clathrate formation models, which differ by completeness of description. Here we propose the model of clathrate formation that begins with the general microscopic Hamiltonian, which is simplified to account for certain pecularities of clathrates. The calculation of thermodynamic characteristics of clathrates is performed in various approximations, taking into account the experimental data of concrete clathrates.

In this chapter we present and analyze the contemporary state of clathrate formation theory. We consider lattice clathrates in general since the main advances in theory are in descriptions of the properties of these clathrates. Thermodynamic properties of clathrates are investigated in terms of solid solution models wherever possible. The experimental data of phase transitions and cooperative phenomena in clathrates are presented and theoretic consideration of these problems is carried out. The similarities and differences of clathrates and other inclusion-type systems are also discussed, as well as the unsolved problems of clathrate formation. Finally, experiments which could help to clarify some of the physicochemical properties of clathrates are proposed.

I. STRUCTURAL AND CHEMICAL PROPERTIES OF LATTICE CLATHRATES

Lattice clathrates can be classified by the following main features:

1. Type of host–guest interaction.
2. Geometric form of cavity in which the guest molecule is included.
3. Type of chemical bond of host molecules.
4. Stability degree of empty host lattice.
5. Chemical composition of clathrate.

The term "lattice clathrate" is applied to denote the compounds with solely van der Waals guest–host interactions. Layered compounds with prevailing van der Waals interactions are called "coordinate-clathrates" (e.g., intercalates, hexahydrates, urotropin[9,10]). Prevailing chemical inter-

TABLE I
Characteristics of Hydroquinone Modifications

Modification	Symmetry Space Group	Cell Parameters (Å)	Density ρ_{exp} (g/cm³)	Melting Temperature, T_m, (°C)	Stoichiometry Coefficient
α	Rhombohedral R$\bar{3}$	$a = b = 38.46$ $c = 5.65$ $z = 54$ $n = 3$	1.36	172.1	1/18
β	Rhombohedral R$\bar{3}$	$a = b = 16.613$ $c = 5.4746$ $z = 9$ $n = 3$	1.25	165.0	1/3
γ	Monoclinic P2$_1$/c	$a = 8.07$ $b = 5.20$ $c = 13.20$ $\beta = 107$ $z = 4$	1.32	—	—

$^a z$ — Number of neighbors; n — number of cavities.

actions are characteristic to "clathrate-complex" compounds (e.g., crown ethers, cryptands).

In this section attention is paid to the lattice clathrates of hydroquinone, hydrates, urea, and thiourea. We present the structural and chemical properties of the formation of these compounds.

A. Hydroquinone

The description of hydroquinone properties is mainly based on the data of Refs. 27–36.

There are three known (α, β, and γ) crystalline modifications of hydroquinone clathrate $3C_6H_4(OH)_2(Q)$. The properties of Q_γ are not considered owing to the absence of empty cavities. The remaining two modifications Q_α and Q_β are rhombohedrical space group R3 systems (see Table I). Oxygen molecules in hydroquinone are linked up by the hydrogen (H–) bonds and arranged in a regular hexagon, shown in Fig. 1. Three of the six molecules are directed upward at an angle approximately equal to 45° and the remaining three downward. The cavity is formed of the fragments of these molecules arranged accurately one over another 5.5–5.6 Å apart (Fig. 2). Thus, the "floor" and the "ceiling" of the closed Q_α and/or Q_β cavity are formed of hexagons and the "sides" of six benzol rings of hydroquinone. The diameter of the cavity is approximately equal

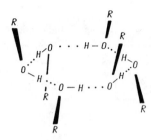

Figure 1. Hexamer usually found in clathrate carcasses of hydroquinone, phenol, and phenol derivatives.

to 4.5 Å. The cavities of α- and β-hydroquinone is almost equal by size, but the modifications just differ in the spatial arrangement of cavities.

The structure of Q_β is more loose in comparison with Q_α in which the molecules are bridged up by H bonds into the form of the carcass. In Q_β two interpenetrating H-bond carcasses are not interconnected by H bonds. In Q_α and Q_β the hexagons joined cavities create one-dimensional chains extended along a three-fold symmetry axis. The distance between the centers of neighboring cavities in the chain is equal to 5.45–5.65 Å. In Q_β the distance between the centers of the nearest cavities of the neighboring chains is almost twice that and is about 9.8 Å. Since the cavity is created of six hydroquinone molecules and every molecule borders two cavities, the ratio of number of cavities per number of hydroquinone molecules (i.e., the stoichiometry coefficient) is equal to 1/3 for Q_β.

In the Q_α modification the cavities are formed by one-third of the molecules (6 molecules per cavity as in Q_β), the remaining 12 molecules forming the double helix. Therefore the stoichiometry coefficient is equal to 1/18. Let us arbitrarily divide the Q_α structure into β- and helix subsystems. Over the β-subsystems the same one-dimensional chains of cavities as in Q_β are implied. However, dense helical formations of hydroquinone

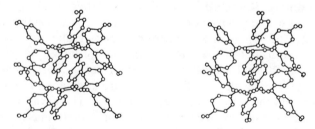

Figure 2. Stereoview of fragment of 12 hydroquinone molecules corresponding to the two hexamers of Fig. 1 located one over another, which form the cavities of α and β modifications.[10]

molecules are located between these chains. Therefore the distance between one-dimensional chains in Q_α is considerably longer than that in Q_β and is equal to 22.2 Å. The density ρ_h of the helical subsystem can be estimated provided the values ρ_α and ρ_β (see Table I) are used. Then one obtains the value $\rho_h = 1.417\,\text{g/cm}^3$, which is rather considerable for compounds composed of light atoms.

Since Q_α contains the same sort of cavities as Q_β, the inclusion of guest molecules into an α-modification carcass is also possible. Even such gases as helium and neon, which cannot stabilize Q_β, can be soluted in Q_α owing to the thermodynamic stability of this modification. Therefore, two competing mechanisms of hydroquinone molecules casting into the crystalline structure are available. Van der Waals interactions tend to pack molecules in the densest way, but the hydrogen bonds require the determined orientation of molecules which does not agree with the dense packing rules. As a result, the formed structure reflects the competition of these mechanisms. In Q_β the H bond (length 2.68 Å) is benefitted at the expense of packing quality (calculated density $1.26\,\text{g/cm}^3$, Ref. 31). In Q_γ the H bond is weakened (2.84 Å), but the packing is better $(1.38\,\text{g/cm}^3)$.[29] In Q_α we can observe the most satisfactory compromise which makes the structure more stable almost till the melting temperatures. The mean length of the H bond is equal to 2.72 Å (from 2.66 to 2.78 Å) and the density $\rho_\alpha = 1.36\,\text{g/cm}^3$.

It should be noted that the mean occupation number (concentration) c of α-hydroquinone varies from zero to one, since Q_α is stable in the absence of guest molecules. It makes Q_α similar to other inclusion-type systems. In Q_β the c values can vary from the fixed values $c_k \neq 0$, which determines the concentration of guests necessary to stabilize the clathrates.

B. Hydrates

We discuss here systems whose carcass and cavities are made of water molecules, taking into account that water molecules cannot be included as the guests. The structural and chemical properties of these compounds was taken from Refs. 7 and 37–47.

Since the reports of Stackelberg and Claussen[48-51] it is common practice to consider the gas hydrates to be described by two cubic structures, CS-I and CS-II. Because of the available theoretic models we shall also pay close attention to these modifications. Table II presents the current information about clathrate carcass structures built of water molecules.

Tetrahedral coordination and flexibility of H-bond length and angle allows us to build a number of loose structures with almost equal energies of water molecules. The most stable of these structures is that of ice, I_h. All the bonds lengths of I_h, as well as the angles between the bonds, are

TABLE II
Structural Characteristics of Hydrate Carcasses[7,37–47,65]

Type, Space Group and Unit Cell Formulas, R_t^a	Cell Parameters (Å) and Empty Carcass Density ρ (g/cm³)	Examples of Hydrates	Structural Stoichiometry (Hydrate Number)
Cubic I (CS-I)^b Pm3n 6 T · 2 D · 46 H$_2$O $R_t = 0.33$	$a = 12$ $\rho = 0.796$	H$_2$S · 6 H$_2$O Cyclo-C$_3$H$_6$ · 7.8 H$_2$O	$5\frac{3}{4}$ $7\frac{2}{3}$
Cubic II (CS-II) Pd3m 8 H · 16 D · 136 H$_2$O $R_t = 2.0$	$a = 17.1$ $\rho = 0.812$	SF$_6$ · 17 H$_2$O Ar · 6 H$_2$O Kr · 6 H$_2$O TGF · 0.5 Pr$_4$NF · 16 H$_2$O	17 $5\frac{2}{3}$ $5\frac{2}{3}$ 16
Hexagonal (HS-I) P6/mmm 2P · 2 T · 3 D · 40 H$_2$O $R_t = 0.75$	$a = 12.4$ $c = 12.4$ $\rho = 0.719$	Br$_2$ · 10 H$_2$O (?) BrCl · 10 H$_2$O (?) iso-Am$_4$NF · 38 H$_2$O	10 10 38
Hexagonal II (HS-II) P6$_3$/mmc 4 H · 8 D · 68 H$_2$O $R_t = 2.0$	$a = 12.1$ $c = 19.7$ $\rho = 0.812$	(CH$_3$)$_2$CHNH$_2$ · 8 H$_2$O^c	17
Tetragonal (TS-I) P4$_2$/mnm 4 P · 16 T · 10 D · 172 H$_2$O $R_t = 0.5$	$a = 23.5$ $c = 12.3$ $\rho = 0.757$	Br$_2$ · 8.6 H$_2$O^d (CH$_3$)$_2$O · 8.6 H$_2$O	8.6 8.6
Rhombic (RS) Pbam 4 H · 4 P · 4 T · 14 D · 148 H$_2$O $R_t = 1.17$	$\alpha = 23.5$ $b = 19.9$ $c = 12.1$ $\rho = 0.782$	Br$_2$ · 12 H$_2$O (?) Cl$_2$ · 12 H$_2$O (?) Bu$_3$PO · 34.5 H$_2$O	$12\frac{1}{3}$
Hexagonal IIIe (HS-III) P6/mmm E · 3 D · 2 D' · 34 H$_2$O $R_t = 5.0$	$a = 12.3$ $c = 10.2$ $\rho = 0.761$	CH$_3$C$_6$H$_{11}$ · 5 Ar · 34 H$_2$O CH$_3$C$_6$H$_{11}$ · 5 H$_2$S · 34 H$_2$O	34
	Structures without D cavities		
Tetragonal II (TS-II) I4/mcm 8 P · 4 d^f · 68 H$_2$O	$a = 15.4$ $c = 12.0$ $\rho = 0.709$	iso-Am$_4$PBr · 32 H$_2$O	
Cubic I43d 16(17-hedron)^f 12(8-hedron) · 156 H$_2$O	$a = 18.81$	(CH$_3$)$_3$CHNH$_2$ · 9.75 H$_2$O	

similar and respectively equal to 2.76 Å and 109.5°. This arrangement induces the packing coefficient to be equal to 0.43. In spite of the very low value of the coefficient, the cavities of the I_h structure are small and can contain only H_2 and He molecules.

Several other structures can be rather easily formed at the slight deviations of H-bonds and angles. These compounds can be more dense than ice, I_h (high-presure ice), so are also more loose. The cavities of molecular size are formed in the loose structures and they are stabilized at the inclusion of chemically inert molecules of appropriate geometry. These carcass structures for the most part have the polyhedric structure. The pentagondodecahedron is the most satisfactory polyhedron formed of water molecules, since the angle between H bonds (108°) slightly deviates from tetrahedral angle and is close to the valence angle (104.5°) characteristic of free-state water molecules. But pentagondodechedron does not occupy the space completely, so it must be combined with the structures containing more energocapacitive cavities, which along with pentagonal have also hexagonal faces. The 14-, 15-, and 16-faced cavities with 12-pentagonal and 2-, 3-, 4-hexagonal faces are the ones most frequently observed (see Table III and Fig. 3). These cavities–polyhedra are called large cavities, contrary to that of dodecahedron, which is called small cavity. Large cavities are not rigorously regular due to distortion of their angles and faces. The 15-faced polyhedron made of the three above-mentioned polyhedra has the largest energy. These cavities with free diameter 5–10 Å can be occupied by molecules of the sizes ranging from Ar to CCH$_4$, dioxane, and even methylcyclohexane.

The information of clathrate carcasses is obtained from the known structure data of gas hydrates, peralkilonium salt hydrates, and so on (see Table II). With the exception of gas hydrate systems, the idealization to

---

$^a R_t$ = number of little cavities/number of large cavities.

^{b}Structure analogous to CS-I has silicone dioxide (the new mineral melanophlogite which can accept the same guest molecules as hydrate CS-I, e.g., CH$_4$,[57] silicon and germanium (in these compounds guests are Na molecules). Structure analogous to CS-II made of silicon dioxide is called dodecasil-3C with guests, for example, (CH$_3$)$_3$N and (CH$_3$)$_2$NH.[58] CS-II type structures with heavy alkaline ions as guests are also found in silicon and germanium.

^{c}In a real structure part of the ideal carcass dodecahedra "widens" until it reaches 14-hedra ($4^2 5^8 6^4$), in which amine molecules are also located.

^{d}Classification by stoichiometry shows that hydrates CHClF$_2 \cdot 8.3$ H$_2$O, CHClF$_2 \cdot 8.5$ D$_2$O, C$_3$H$_8 \cdot 8$ H$_2$O, C$_2$H$_6 \cdot 8.25$H$_2$O, CH$_3$Cl $\cdot 8.2$ H$_2$O could be attributed not necessarily to CS-I, but also to TS-I structure as hydrate Br $\cdot 8.6$ H$_2$O.

^{e}For the first time HS-III was obtained in clathrasil 1 H.[60]

$^f d = 4^2 5^8$; 17-hedron = $4^3 5^9 6^2 7^3$; 8-hedron = $4^4 5^4$.

TABLE III
Types of Cavities–Polyhedra Found in Clathrate Hydrates[41,45]

Cavities	Number of			Free Diameter of Cavity (Å)	Polyhedron Volume (Å)^c
	Apices^a	Edges^b	Faces^c		
D (12-hedron)	20	30	$12(5^{12})$	5.2	168
T (14-hedron)	24	36	$12(5^{12}6^2)$	5.3^d	230
				6.4	
P (15-hedron)	26	39	$15(5^{12}6^3)$	6.1^d	260
				7.0	
H (16-hedron)	28	42	$16(5^{12}6^4)$	6.6	290
E (20-hedron)^e	36	54	$20(5^{12}6^8)$	9.6^d	
				7.3	

^{a}The oxygen atoms of water are located in polyhedra apices.
^{b}The edge corresponds to hydrogen bond O–H··O.
^{c}The designation (e.g., $5^{12}6^3$) means that 15-hedron has 12 pentagonal and 3 hexagonal faces.
^{d}14-, 15-, and 20-hedra can be approximately taken as compressed and extended rotation spheroids with two diameters.
^{e}In this structure an E cavity is found in combination with D and D' ($4^35^66^3$) cavities.

gas hydrate carcass is performed, that is, it is implied that the guest molecule is bonded with water carcass by van der Waals forces only. In the first six carcasses of Table II only the cavities with pentagonal and hexagonal faces are observed. In the remaining carcasses the cavities with more complicated faces (square, sevenangular, etc.) occur. The first group of carcasses is less strained when compared with ice and here we know the examples of carcass realization in gas hydrate version (widely used hydrates CS-I and CS-II and two hydrates TS-I). To stabilize the carcasses of the second group, more strong guest–host interaction is needed, for example, the one observed in the water systems with low molecular amins or peralkilonium salts as guests. The description of the second group of carcasses in terms of idealized polyhedra with mutual faces appears to be

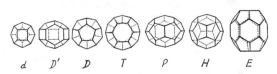

d D' D T P H E

Figure 3. Cavities–polifaces in clathrate hydrate structures. The letters indicate various types of cavities (see also Table 3).[17]

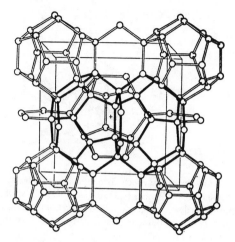

Figure 4.   Unit cell of cubic structure I of clathrate hydrates. The centers of dodecahedral cavities are located in the apices and the center of the cube (the cavity in the center is turned by 90° to those in the apices). The centers of *T*-cavities are located in the faces of the cube.

rather complicated due to increasing discrepancy between idealized models and real structure.

Cubic Structure I. The lattice of CS-I is shown in Fig. 4. Columns made of T cavities coupled by hexagons are arranged parallel to the edges of the cube and bonded as densely as possible by pentagonal faces. The remaining space between the columns represents a pentagonal dodecahedron. Thus, only two types of cavities are characteristic of CS-I, that is, the large T and the little D cavities in a ratio of 3:1 (denotations shown in Table II and Fig. 4). The arrangement of dodecahedra in cubic lattice is pseudo-body-centered. The central dodecahedron is turned by 90° with respect to the dodecahedra in the apices. As is seen from the figure, the distance between T-cavity hexagons is equal to $a/2$. Frequently, when the cation of tetrabutylammonium appears to be the guest, the hydrates crystallize in a cubic lattice with the parameter ~24 Å. This structure corresponds to that of CS-I with doubled parameter. When one sort of guest molecule (M) occupies all the cavities, the stoichiometry amounts to $M \cdot 5\frac{3}{4} H_2O$, but it turns out to be $M \cdot 7\frac{2}{3} H_2O$, when only large cavities are occupied.

Cubic Structure II. At the joining of pentagondodecahedra by the faces slight deviations occur which lead to the formation of the plain floors of

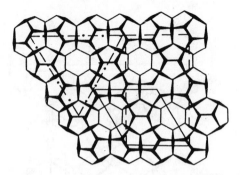

Figure 5. Pentagondodecahedra which form the layer of hexagonal symmetry. These formations are the basis of CS-II, HS-I, HS-II, and HS-III structures. The ground of hexagonal cell I is shown by a solid line, the face of the rhombohedral cell on a basis of HS-I by a dashed line, and dash-dotted line corresponds to the face of the smallest tetrahedron which makes the H cavity in the CS-II carcass.

hexagonal symmetry (see Fig. 5, Table II). These floors are parallel to the faces of the tetrahedron. The remaining space in the middle of the least of them corresponds to a T cavity with hexagonal faces parallel to the faces of tetrahedron. In this structure also two types of cavities, H and D, are observed in a ratio of $1:2$. Cubic lattice CS-II represents a face-centered diamond structure where the carbon sites are substituted for by the centers of H cavities. Stoichiometry is equal to $M \cdot 5\frac{2}{3} H_2O$ when the cavities are completely occupied by one sort of guest molecule. It turns out to be $M \cdot 17 \cdot H_2O$ when only H cavities are occupied. Double hydrates are characteristic of this structure, containing many small cavities. Small molecules occupy small cavities and stabilize the carcass additionally. The limiting stoichiometry in this case is given by Table II— $M_1 \cdot 2M_2 \cdot 17 H_2O$. Only in the case of gas guests were the hydrates of the such structures so far realized. However, recently the authors of Ref. 52, using the hydrates ability to additionally stabilize, synthesized double-hydrate $TGF \cdot 0.5Pr_4NF \cdot 16 H_2O$, in which large cavities are occupied by TGF molecules and all D cavities by cation Pr_4N^+. The cation of this structure is located in a four-sectioned D_4 cavity, which is formed by the exclusion of water molecules from the apex common to four polyhedra and inclusion of the central atom of the cation in the released place without H bonds formation with the neighbors. Hydrocarbon radicals are located in every D section and fluorine ion hydrophilously expels the water molecule from the carcass, forming H bonds with neighbors.

Now we shall discuss the problems of small and large cavities being occupied by guest molecules. The discrepancy (sometimes considerable) between the structural formulas and directly determined composition of gas hydrates was explained by Stackelberg,[50] who proposed the concept of incomplete water carcass occupation. Van der Waals[5] developed this idea, assuming the empty carcass structure of the host to be metastable. At present this viewpoint is held by most investigators (e.g., see Ref. 7). However, as the accuracy of determination of hydrate composition increases, the "homogeneous region" caused by alternating occupation of cavities becomes narrower.

It is convincingly shown for CS-II hydrates that the degree of occupancy of large cavities is close to unity. As is shown in the interesting experiment of Davidson, Handa, and Ripmeester,[53] the ratio of large and small cavities occupation degrees c_T/c_D in the CS-I system is equal to 1.37 at the composition of hydrate Xe · 6.286 H_2O. The ratio of occupation degrees gives the value $c_T = 0.981$. In the framework of ideal clathrate solution theory, the value of chemical potentials difference $\Delta\mu$ was calculated to be equal to 1297 ± 110 J/mol. It follows from this value that CS-I hydrates with the hydrate formator molecules of sizes incapable of occupying D cavities have the limiting hydrate number equal to 7.76.[53] It should be noted that the value of the limiting hydrate number considerably decreased (from 8.47 to 7.76) due to the increase of experimental accuracy and became close to the ideal value 7.67 obtained in the case of complete occupation of large cavities. Complete occupation of large cavities is characteristic also of the systems in which some types of hydrates are formed. The interesting properties were observed in the systems with peralkilonium salts as the guests and in the bromine–water system. It was shown that at the monotonous variation of equilibrium concentration of components in the liquid phase of the guest–host binary system, the changes of this phase composition occur jumpingly due to formation of hydrates with new carcass structures, but not gradual occupation of the same carcass cavities.[54] But it frequently happens, due to insufficient resolving power of the determination method, that the hydrates of similar composition are assumed to be one hydrate of alternating composition. For example, four hydrates with hydrate numbers, 24, 26, 32, and 36 are obtained by varying the composition from 96.0 to 97.3 mol% of water in the system water–tetrabuthylammonium bromide. The hydrates $Br_4NF · 28.6\ H_2O$ and $Br_4NF · 32.4\ H_2O$ occurring in the system with ammonium fluoride have the carcasses CS-I (real structure of the hydrate corresponds to the superstructure CS-I with doubled parameter) and TS-I.[47] An analogous picture is observed in numerous systems with peralkilonium salt as a guest.[54] The cases of the formation of several gas hydrates in one system are less common, but such examples are known, for example, in the systems with trimethylene oxide,[37] cyclopropane,[38] and dimethyl ether.[56] The bromine–water system is assumed to be the most

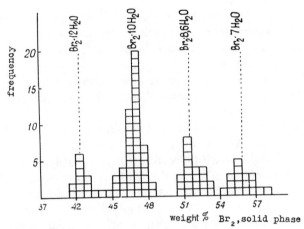

Figure 6. Hystogram of the data of solid-phase composition in a bromide–water system.[54]

illustrative example, in which four hydrate numbers 7, 8.3, 10, and 12 are formed. Only the tetragonal structure I of hydrate $Br_2 \cdot 8.6\ H_2O$, which is frequently found among tetrabuthylammonium polyhydrates, is known. Reasoning from the stoichiometry data, the remaining hydrates are grouped to CS-I, HS-I, and RS hydrates.[54] This attribution is possible because of the existence in these structures of the same cavities as for bromine molecule inclusion, as in TS-I. In any case, the data of crystal phase compositions in a bromine–water system (see hystogram in Fig. 6) do not lead to the conclusion of alternating composition hydrate formation. On the contrary, as it follows from more definite data obtained in bromine and other systems, the hydrates under formation have the composition which corresponds at least to the complete occupation of large cavities.

Thus, two viewpoints of large-cavity occupation are available, those of complete $(c_T = 1)$ and partial $(c_T < 1)$ occupation. As it will be seen below, the theoretical model of gas hydrate formation can be considerably simplified when complete occupation of T cavities occurs.

C. Clathrate Carcasses of Urea and Thiourea

The data of structural and chemical properties of urea and thiourea can be found in Refs. 25 and 61–69.

The packing of urea and thiourea molecules in a free state is the compromise between the requirements of angles for optimal H-bond formation and dense packing. The H-bonds are more rigid in urea, but dispersal interactions are more strong in thiourea. Certain characteristics

TABLE IV

Crystallographic Characteristics of Urea and Thiourea and Their Clathrate Carcasses[65]

Compound	Crystallographic System	Unit Cell Parameters			ρ (g/cm^3)	z	T_m (°C)
		a	b	c			
σ-Urea	P$\bar{4}$21	5.67		4.73	1.335	2	132.7
α-Thiourea		5.50	7.68	8.57	1.405	4	180 (with decomposition)
β-Urea	P6$_1$22 P6$_5$22	8.23		11.005	(0.926)	6	
β-Thiourea	R3	10.0	$\alpha = 104.5°$		(0.840)	6	
	Hexagonal apparatus	15.8		12.5		18	

of these substances and their clathrate carcasses are presented in Table IV.

Urea and thiourea have similar clathrate carcasses, those of hexagonal infinite prisms with channel cavities inside, but the distribution of the host molecules in both prisms turns out to be different.

The carcass of β modification is very loose in comparison with the carcass of the initial α modification. For example, the density of urea clathrate carcass is equal to 0.926 g/cm^3 and it amounts to 0.69 of the initial urea density. This difference appears to be larger in thiourea. Therefore, loose β modifications are particularly unstable in spite of more favorable conditions for H bonds formation in comparison with α modifications.

In Fig. 7 the sections of channels are schematically shown. The diameter of the urea channel is almost invariable and equal to ~5.2 Å. The channel of thiourea has an alternating diameter, which is more wide in the vicinity of sulphur atoms (approximately 7 Å) and more narrow (approximately 6 Å) in the vicinity of amine groups. The size of the channel determines the form and the size of guest molecule. The molecules of hydrocarbon and their normal structure derivatives are available to form urea clathrates. Highly branched hydrocarbons, cyclohexanes and their derivatives, and CCl$_4$ are suitable for thiourea formation. To understand the nature of urea and thiourea more properly, let us consider urea compounds with n paraffins.

It should be noted, that the parameters of urea with n-paraffin carcasses hardly depend on the sort of paraffin, since only the length of paraffin molecules undergoes certain changes. The number (m) of urea molecules per one paraffin molecule must presumably be the linear function of the number of carbon atoms (n) in the paraffin molecule. The statement is

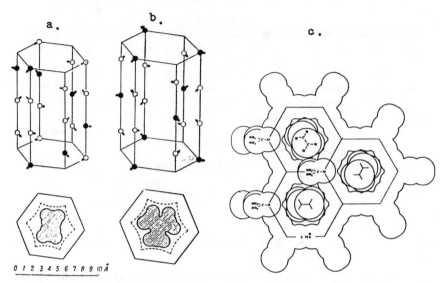

Figure 7. Urea and thiourea structure. Arrangement of urea (*a*) and thiourea (*b*) molecules and channel formation. The circle represents oxygen (sulfur) atom; the arrow is the remaining part of molecule. (*c*) Cross section of channels in urea structure and the projection of *n*-paraffin molecules in channels.

confirmed by experiment, which gives m to be defined with a high degree of accuracy by the expression $m = 0.684n + 1.254$. Thus, it follows that the length of channel per one included molecule can be written as $l_k = 1.254(n - 1) + 3.554$.[68]

If one takes into account the geometry of *n*-paraffin molecules, then the length of it trans-conformer in free state appears to be equal to $l_n = 1.271(n - 1) + 4.06$, i.e. the length of channel is shorter than the length of molecule in ideal extended state (here the values of the angles are $C–C–C = 112°$ and $H–C–H = 108.5°$; lengths are $C–C = 1.533$ Å and $C–H = 1.08$ Å; and intermolecular radii are $r_H = 1.16$ Å, $r_c = 1.71$ Å[69]). This can occur due to superposition of final methyl groups at the dense packing favored by geometry of the channel and certain twisting of the *n*-paraffin molecule.

In Fig. 8 the values are presented of incongruental melting temperatures and decomposition heat (per mol of guest) of urea clathrates. With the increase of guest molecule length the decomposition heat linearly increases. At the transition from tetragonal α modification to that of the clathrate hexagonal form of urea, H bonds become stronger (or, at least, not weaker) and losses of dispersal interactions are completely compensated for by the inclusion of the guest. Clearly, the longer is the guest

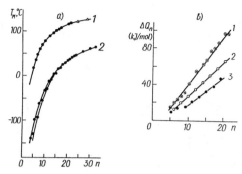

Figure 8.   Decomposition temperature (a) and heat (b) of urea clathrates with n-paraffins (1), Melting temperture (a) and heat (b) of even (2) and odd (3) individual n-paraffins.

molecule, the greater is the number of atom–atom guest–host contacts. It explains the linear dependence of decomposition heat $\Delta Q_n = 5.89n - 16.6$ (in kJ/mol of guest) at the decomposition to the solid α-urea and liquid n-paraffin. With the increase of guest molecule length the decomposition temperature also increases and it can become higher (e.g. 148°C with polyethylene being the guest) than the melting temperature of pure urea (132.7°C).

The dependencies of decomposition heat ΔQ_M per 1 mol of urea and density ρ on the number of carbon atoms in paraffins are given by the formula $\Delta Q_M = Q_0 - \alpha Q_1$ and $\rho = \rho_0 - \alpha \rho_1$ with $Q_0 = 8.612$ kJ/mol, $Q_1 = 40.02$ kJ/mol, $\rho_0 = 1.213$ g/cm^3, $\rho_1 = 0.534$ g/cm^3, and $\alpha = 1/(n + 1.834)$. The analogous dependence of the packing coefficient has the form $k = \Sigma_i v_i/V = k_0 - \alpha k_1$, where $k_0 = 0.629$, $k_1 = 0.257$, and $\Sigma_i v_i$ is the sum of volumes of molecules which constitute the phase of volume V. The volume of the urea molecule is equal to 43.9 Å^3. The data used for calculation (lengths of bonds, angles, molecular radii) were taken from Refs. 68 and 69.

It follows from the given expressions that the parameters represent certain constants characteristic to polyethylene ($n \to \infty$) minus the value proportional to the concentration of guest molecule interfaces. The number of interfaces per mole of urea which is equal to the number of paraffin molecules N is defined by the expression $N = 1/m = 1.462/\alpha$. As a result, decomposition heat, density, and packing coefficient can be written as follows: $\Delta Q_M = Q_0 - 27.37N$, $\rho = \rho_0 - 0.365N$, and $k = k_0 - 0.257N$.

A similar situation is observed in urea clathrates with other guest molecules of normal structure. The violation of guest molecule geometry (branching, large size substituents, presence of double and triple bonds)

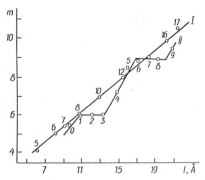

Figure 9. Two variants of thiourea composition dependence on the length of the guest molecule. I — incompressible highly branched hydrocarbons. II — compressible molecules of ω, ω'-dicyclohexils. The digits near the point correspond to the number of hydrocarbon atoms in the longest isoparaffin chain (circles) or in methylene chain of ω, ω'-dicyclohexil (squares).

leads to the decrease of compound stability and makes clathrate formation impossible.

Thiourea has channels of alternating sections in which two types of guests arrangement are observed: fixation in the widened zones and dense, one-dimensional chains of guests characteristic also to urea. The location of ω-dicyclohexils with methylene bridges can serve as an example of the first arrangement (see Fig. 9). If the whole length of extended guest molecule does not considerably exceed the integer value of half lengths of the unit cell along the c axis, the included molecule "tends" to decrease its own length in order to be in accordance with the integer value of half lengths of the unit cell of the thiourea carcass. In these cases the molecules shrink (twist) in the polymethylene bridge, since the thiourea channel is sufficiently wide. If highly branched paraffins with molecules of sizes inappropriate for compression are included, such guests reveal the usual picture characteristic to channel urea compounds. The dense packing is observed also at the distribution of small molecules (see Table V, where the experimental data of stoichiometry and data calculated at the assumption of dense one-dimensional packing are compared). The improvement of clathrate stability is observed at the increase of guest molecule length. In Table V the minimal equilibrium concentration of guests in liquid phase is shown which is necessary to form the clathrate compounds in the three-component system thiourea–guest–acetic acid (acetic acid as a solvent). The most stable clathrate is formed with hexachlorethane owing to optimum transverse size of the guest molecules and the symmetry available to channel. The clathrate is crystallized at the guest concentration of just

TABLE V

Thiourea Clathrate Composition[a] (m = thiourea/guest) and Guest Concentration in Eutectic Point c_0 of Guest–Thiourea–Acetic Acid System at 20°C[65]

Parameter	Guest					
	CH_3Cl_3	CCl_4	$C_2H_2Cl_4$	$CHCl_5$	C_2Cl_6	C_6H_6
m_{exp}	2.25	2.86	2.96	2.94	2.95	2.40
m_{calc}	2.20	2.88	2.95	2.95	2.95	2.40
c_0 (weight %)	46.7	2.6	5.9	0.75	0.73	74.2

[a]It should be noted that the calculated compositions are obtained taking into account the dense packing of guest molecules in the channel.

0.73%, as compared with 47.7% required for chloroform and 74.2% for benzol clathrate formation.

It should be noted that at these minimal liquid-phase guest concentrations initiating clathrate formation, the composition of the solid phase (clathrate) corresponds to the complete occupation of the channel. This is also observed in urea compounds with n-paraffins. Thus, in the whole region of stability (at least at room temperature), urea and thiourea clathrates have the fixed composition corresponding to the completest occupation of channel space. This suggests that the carcass with vacant cavities is absolutely unstable under these conditions.

The character of guest–guest, guest–host, and host–host interactions differs in hydrate, urea, and thiourea clathrates and that is reflected in the intersolubility of host and guest, since the solvent is considerably larger in solid state than in liquid. In the most cases guest–guest interaction is isotropic dispersional. However, host–host interaction shows strictly defined anisotropy occurring due to H bonds. Thus, guest–host interaction turns out to be "unadvantageous" and only dispersional. Therefore, neither water, urea, nor thiourea can be dissolved in the vast majority of liquid hydrophobic guest components. Intersolubility of liquids above the melting temperature of the host is uncommon. The situation is reversed if one considers the solid state composition of clathrate. Reconstructing the initial host modification into that of clathrate with preservation or, presumably, even little gain of H-bond energy is still unadvantageous because of the loss of dispersional interaction. However, this loss is compensated for by inclusion of guest molecules which allows for a large but strictly determined amount of the molecules to be located in the crystal host environment.

II. MICROSCOPIC HAMILTONIAN

Intermolecular interactions determine the specific character of formation and the particular properties of clathrates. Therefore the problem of correct consideration of these interactions becomes of prime importance. To derive the microscopic Hamiltonian of clathrates we use the adiabatic approach,[70] assuming the system to be influenced by electronic degrees of freedom via the ionic core.

Consider the model with the potential energy U written in terms of the pair interaction sum

$$U = \sum V_{ij}(\mathbf{R}_{mi} - \mathbf{R}'_{nj}) \tag{1}$$

where $\mathbf{R}_{mi}$ is the coordinate of the ith ion in the mth lattice.

First, let us divide the crystal into two subsystems, those of guest and host molecules. Then introduce the operator n_r^i, which is equal to 1 when the rth cavity is occupied by the ith guest molecule and to 0 when the cavity is empty. We denote $\boldsymbol{\xi}_R^i$ the shift vector of ith host moelcule with respect to equilibrium state coordinate $\mathbf{R}_i$ and $\mathbf{u}_r^i$ the shift vector of the ith guest molecule in the cavity with coordinate $\mathbf{r}_i$. Then we assume the values $\mathbf{R}_{1i}$ and $\mathbf{r}_{1i}$ to be equal to $\mathbf{R}_{1i} = \mathbf{R}_i + \boldsymbol{\xi}_R^i$, $\mathbf{r}_{1i} = \mathbf{r}_i + \mathbf{u}_r^i$.

The potentials V_{ij} from (1) are presented in a way[71]

$$U = \frac{1}{2} \sum V_{ij}^{(1)}(\mathbf{R}_{1i} - \mathbf{R}'_{1j}) + \sum V_{ij}^{(2)}(\mathbf{R}_{1i} - \mathbf{r}_{1j})n_r^i$$

$$+ \frac{1}{2} \sum V_{ij}^{(3)}(\mathbf{r}_{1i} - \mathbf{r}'_{1j})n_r^i n_{r'}^j \tag{2}$$

The first term of the potential (Eq. 2) expansion over the small shifts $\boldsymbol{\xi}_R^i$ and $\mathbf{u}_r^i$ can be written as

$$U_1 = V_1 + \sum V_{ij}^{(2)}(\mathbf{R}_i - \mathbf{r}_j)n_r^i + \frac{1}{2}\sum V_{ij}^{(3)}(\mathbf{r}_i - \mathbf{r}'_j)n_r^i n_{r'}^j \tag{3}$$

Here $V_1 = \sum V_{ij}^{(1)}$ is the energetic constant leading to renormalization of the potential U_1. Potentials $V_{ij}^{(2)}$ and $V_{ij}^{(3)}$ describe guest–host and guest–guest interactions respectively.

Since the majority of guest molecules in the cavities have dipole moments, the operator which describes the dipole moment direction is to be included in the potential energy (Eq. 3). Consider for simplicity the case of the guest molecule found in two equivalent equilibrium states (corresponding to two directions of dipole moment) with coordinates $\mathbf{r}_i = \mathbf{r}_i^{(0)} \pm \mathbf{b}_i$, where the vector $\mathbf{b}_i$ describes the location of the molecule with respect to the center of molecule gravity. It is convenient to describe two possible values of $\mathbf{r}_i$ by Pauli operator $\sigma_{r_i}^Z = \pm 1$.[72] As a result

$$\mathbf{r}_i = \mathbf{r}_i^{(0)} + \mathbf{b}_i \sigma_{r_i}^Z \tag{4}$$

Substituting Eq. (4) into Eq. (3) we obtain the potential

$$U_1 = \sum V_{ij}^{(2)}(\mathbf{R}_i - \mathbf{r}_i^{(0)} - \mathbf{b}_j \sigma_{r_j}^Z) n_{r^{(0)}}^j$$

$$+ \frac{1}{2} \sum V_{ij}^{(3)}(\mathbf{r}_i^{(0)} - \mathbf{r}_j^{(0)} + \mathbf{b}_i \sigma_{r_i}^Z - \mathbf{b}_j \sigma_{r_j}^Z) n_{r^{(0)}}^i n_{r'^{(0)}}^j \tag{5}$$

where we changed the coordinate $\mathbf{r}$ to $\mathbf{r}^{(0)}$ in the operator n_r^i because of the negligibly small quantity of $|\mathbf{b}|/|\mathbf{r}^{(0)}|$.

If we use the common relation[73]

$$f(x\sigma_1^Z + y\sigma_2^Z) = \sigma_1^+ \sigma_2^+ f(x + y) + \sigma_1^+ \sigma_2^- f(x - y)$$

$$+ \sigma_1^- \sigma_2^+ f(-x + y) + \sigma_1^- \sigma_2^- f(-x - y) \tag{6}$$

with $\sigma^\pm = (1 \pm \sigma^Z)/2$, Eq. (5) takes the form

$$U_1 = \sum W_{ij}^{(1)}(\mathbf{R}_i - \mathbf{r}_j) n_r^j + \sum W_{ij}^{(2)}(\mathbf{R}_i - \mathbf{r}_j) n_r^j \sigma_{r_j}^Z$$

$$+ \frac{1}{2} \sum [V_{ij}(\mathbf{r}_i - \mathbf{r}_j') + J_{ij}(\mathbf{r}_i - \mathbf{r}_j') \sigma_{r_i}^Z \sigma_{r_j}^Z] n_r^i n_{r'}^j \tag{7}$$

The constants W_{ij}, V_{ij}, and J_{ij} in Eq. (7) are expressed via potentials:

$$V_{\alpha\alpha'}^{ij} = V_{ij}^{(n)}(\mathbf{R}_i - \mathbf{r}_j' + \alpha\mathbf{b} - \alpha'\mathbf{b}') \tag{8}$$

taking into account the formulas (Eq. 6), $\alpha, \alpha' = +, -$ and $n = 2, 3$.

The second term of the sum in Eq. (7) is linear with respect to the operator σ_r^Z. Physically, the term describes the action of a certain field. Thus, in the absence of the field the term can be neglected and the expression for U_1 takes the form

$$U_1 = \sum W_{ij}^{(1)}(\mathbf{R}_i - \mathbf{r}_j)n_r^j$$

$$+ \frac{1}{2}\sum [V_{ij}(\mathbf{r}_i - \mathbf{r}_j') + J_{ij}(\mathbf{r}_i - \mathbf{r}_j')\sigma_{r_i}^Z\sigma_{r_j'}^Z]n_r^i n_{r'}^j \qquad (9)$$

and the potential $V_{gg}^{ij}(\mathbf{r}_i - \mathbf{r}_j') = V_{ij}(\mathbf{r}_i - \mathbf{r}_j') + J_{ij}(\mathbf{r}_i - \mathbf{r}_j')$ describes the interaction of guest molecules.

Consider now other terms of the potential (Eq. 2) expansion. They can be written in conventional form as

$$U_2 = \sum V(\boldsymbol{\xi}_R^i) + \sum V(\mathbf{u}_r^i)n_r^i$$

$$+ \sum V(\boldsymbol{\xi}_R^i, \mathbf{u}_r^i)n_r^i + \sum V(\boldsymbol{\xi}_R^i)n_r^i \qquad (10)$$

where $V(\boldsymbol{\xi})$ describes the terms of $\boldsymbol{\xi}\boldsymbol{\xi} + \boldsymbol{\xi}\boldsymbol{\xi}\boldsymbol{\xi}$ type, $V(\boldsymbol{\xi}, \mathbf{u})$ of $\boldsymbol{\xi}\mathbf{u} + \boldsymbol{\xi}\mathbf{u}\mathbf{u}$ type, and so on.

The linear members with respect to the shifts $\mathbf{u}$ and $\boldsymbol{\xi}$ can be neglected in Eq. (10) owing to the absence of an external electric field. The remaining summands in Eq. (10) are of quadratic or higher degree. But if one is not concerned by the phase-transition problems related to the appearance of mean shifts $\langle\boldsymbol{\xi}_R\rangle \neq 0$, all the terms in Eq. (10) can be neglected. Adding to Eq. (9) the summand of kinetic energy of guest molecules we obtain the Hamiltonian of clathrates in the form

$$H = \sum m_i \frac{(\dot{\mathbf{u}}_r^i)^2}{2}n_r^i + U_1 - \sum \mu_i n_r^i \qquad (11)$$

where μ_i and m_i are correspondingly chemical potential and mass of ith guest molecules.

We admitted at the derivation of Eq. (11) the possibility that the guest molecule is oriented along two equivalent directions. When the number of dipole moment directions is $q > 2$, the operator σ_r^Z must be changed by the q-component operator e_r. If the interaction between molecules depends on the number of directions, then $J_{ij} \to J_{ij}^{qq'}(\mathbf{r}_i - \mathbf{r}_j')$.

The specific character of clathrate is taken into account when the terms are neglected in Eq. (10) proportional to the shift $\mathbf{u}_r$ of guest molecule in the cavity. This approximation is valid for clathrates with $r_g/r_h \approx 1$, where r_g and r_h are the effective radii of guest and host molecules. The majority of clathrates meet this condition. When $r_g/r_h < 1$, guest can easily occupy as well as leave the cavity. But then the characteristic features of clathrate are lost and it resembles other inclusion-type systems (e.g., alloys).

III.   CALCULATION OF THERMODYNAMIC FUNCTIONS

In this section the examination of clathrate properties is carried out in accordance with the general principles of statistical physics and solid solution theory.[76,77]

To investigate the conditions of thermodynamic equilibrium, thermodynamic Ω and chemical μ potentials of clathrate must be obtained. Thermodynamic potential Ω per one cavity is expressed as[72]

$$\Omega = -T \ln \mathrm{Sp} \exp(-\beta H) \tag{12}$$

where $\beta \equiv T^{-1}$ and H is determined in Eq. (11). Since the summation over impulses in Eq. (12) just renormalizes the constant $W_{ij}^{(1)}$ in Eq. (9), we can write the Hamiltonian in the general form

$$H = \sum \epsilon_r^i n_r^i + \frac{1}{2} \sum [V_{ij}(\mathbf{r} - \mathbf{r}')$$

$$+ J_{ij}(\mathbf{r} - \mathbf{r}')\sigma_{r_i}^Z \sigma_{r_j}^Z] n_r^i n_{r'}^j - \sum \mu_i n_r^i \tag{13}$$

where ϵ_r^i is the energy required to transport the guest molecule from the cavity to the lattice.

At first, the thermodynamic properties of clathrate consisting of one sort of guest and one sort of host molecule ($i = j = 1$) are to be considered. The calculation of Ω and μ is carried out by the mean-field approximation.

A.   Spherical Guest Molecules

The potential $J_{ij} = 0$ in the case of symmetric guest molecules. Then the Hamiltonian (Eq. 13) takes the more simple form

$$H = \frac{1}{2} \sum V(\mathbf{r} - \mathbf{r}')n_r n_{r'} + (\epsilon - \mu) \sum n_r \tag{14}$$

which can also be rewritten identically as

$$H = \frac{Vc^2}{2} + (\epsilon - \mu + Vc) \sum n_r$$
$$+ \frac{1}{2} \sum V(\mathbf{r} - \mathbf{r}')[(n_r - c)(n_{r'} - c)] \tag{15}$$

Here $V = \sum V(r - r')$, $\epsilon = \sum \epsilon_r$ and $c = \langle n_r \rangle$ is the concentration of guest molecules in clathrate. The Hamiltonian of the mean-field approximation (MFA) is obtained when the last (correlation) term in Eq. (15) is dropped:

$$H_{\text{MFA}}^{(0)} = \frac{Vc^2}{2} + (\epsilon - \mu + Vc) \sum n_r \tag{16}$$

Substituting Eq. (16) into Eq. (12) we obtain

$$\Omega = -\frac{Vc^2}{2} - T\ln(1 + \zeta) \tag{17}$$

where $\zeta = \exp[-\beta(\epsilon - \mu + Vc)]$.

Using the thermodynamic identity

$$c = -\frac{\partial \Omega}{\partial \mu} \tag{18}$$

from Eq. (17) we obtain $\zeta = c/(1 - c)$ or

$$\Omega = -\frac{Vc^2}{2} + T\ln(1 - c) \tag{19a}$$

$$\mu = \epsilon + Vc + T\ln\left(\frac{c}{1 - c}\right) \tag{19b}$$

Free energy $F = \Omega + \mu c$ takes the form

$$F = \epsilon c + \frac{Vc^2}{2} + T\{c \ln c + (1 - c) \ln(1 - c)\} \tag{20}$$

The obtained formula (Eqs. 19 and 20) do not describe the specific features of clathrates and can be applied in principle to any inclusion-type systems. Phase boundary lines between, for example, liquid (L)–solid (S) phases in common inclusion compounds is found from the equalities of chemical and thermodynamic potentials of each phase

$$\mu_L = \mu_S; \qquad \Omega_L = \Omega_S \tag{21}$$

Solving Eq. (21) one can obtain the equilibrium curve in the (T, c) plane.

The problem turns out to be different in clathrates, which do not undergo such transitions and in which equilibrium is reached between other phases. Guest molecules in the cavities form a "liquid" phase (according to the terminology of solid solutions) in the sense that the cavities are occupied with equal probability. Note that here one sort of cavity is initially considered. When one has a lattice with two or more unequivalent cavities with different occupation (e.g., hydrates), the state of guest molecules in the lattice can be treated as "quasi-solid."

If an empty host lattice is stable in the absence of guests, as in zeolite and some other lattices, the problem remains to determine V, ϵ, and T dependence of the optimal concentration c_k, which provides the minimal free energy F of the system. The computer chooses the best set of parameters minimizing F.

Consider the case when an empty lattice (β form) transits to the stable α form in the absence of guests. If the clathrate (Q) is found in equilibrium with the α and gas (G) phases of the guests,

$$G + Q_\alpha \Leftrightarrow \text{clathrate} = Q_\beta G \tag{22}$$

To meet the equilibrium conditions, thermodynamic Ω and chemical μ potentials of both sides of Eq. (22) must be equal, in accordance with Eq. (21). It should be noted that in this case different amounts of cavities are found in α and β forms owing to the specific character of chosen clathrate. Therefore, the stoichiometric coefficient ν_α and ν_β are to be introduced. They determine the number of host molecules forming one closed cavity. As a result, the equilibrium conditions of clathrate take the form

$$\nu_\alpha \Omega_\alpha = \nu_\beta \Omega_\beta \tag{23a}$$

$$\mu = \mu_G \equiv T\ln(P/P_0) \qquad (23b)$$

where μ is determined in Eq. (19), μ_G is the chemical potential of ideal gas, and P is the pressure.

The thermodynamic potential of the α form in the absence of guests is denoted as Ω_α. This potential is still not concretized in theoretic calculations. The potential Ω_Q consists of two terms:

$$\Omega_Q = \Omega + \Omega_\beta \qquad (24)$$

where Ω is determined in Eq. (19) and Ω_β is the thermodynamic potential of the empty host lattice. The potential Ω_β is not calculated, but remains as a phenomenological parameter of the theory.

Substituting Eq. (24) into Eq. (23a) and taking into account Eq. (19a), we obtain the equation which determines the curve of clathrate equilibrium in the (T, c) coordinates

$$-\Delta\Omega = \nu_\beta \left[T\ln(1 - c) - \frac{Vc^2}{2} \right] \qquad (25)$$

with

$$\Delta\Omega = -\nu_\alpha\Omega_\alpha + \nu_\beta\Omega_\beta > 0 \qquad (26a)$$

It should be noted that in clathrate formation theory[5] the values $\nu_\alpha\Omega_\alpha$ and $\nu_\beta\Omega_\beta$ are called chemical potentials of the α and β forms of empty host lattice: $\mu_0^\alpha = \nu_\alpha\Omega_\alpha$, $\mu_0^\beta = \nu_\beta\Omega_\beta$ and correspondingly, $\mu_Q = \nu_\beta\Omega$. If, however, we admit the clathrate to be an inclusion-type system, the terminology of inclusion compound theory must be used. In this theory the potentials Ω and μ are supposed to be the main thermodynamic quantities of inclusion-type systems.[72,76,77] Immediate calculation of these values is needed for a deep insight into similarities and differences of clathrates and other inclusion-type systems. This concept is used throughout the discussion. But in the final expressions we use the chemical potentials μ_0^α, μ_0^β, and μ_Q in order that clathrate terminology not be violated. In this case

$$\Delta\Omega \equiv \Delta\mu = \mu_0^\beta - \mu_0^\alpha \qquad (26b)$$

Now, perform the analysis of Eq. (25). When guest molecules are absent, $V = 0$ and this equation is expressed as

$$-\Delta\mu/T = \nu_\beta \ln(1 - c) \qquad (27)$$

Solving Eq. (27) with respect to c and taking into account the inequality $\Delta\mu > 0$ we obtain

$$c = 1 - \exp(-\Delta\mu/T\nu_\beta) \qquad (28)$$

Equations (27) and (28) were first obtained in Ref. 5. It follows from Eq. (28) that at the given temperature T, the clathrate can have a strictly determined concentration c_k at which the guest–host system appears to be thermodynamically stable. All the remaining concentrations $c \neq c_k$ must lead to nonequilibrium states of clathrate.

Consider the influence of interaction $V \neq 0$ on the clathrate formation process. For convenience we introduce in Eq. (25) the normalized values

$$f = \ln(1 - c) - \frac{Vc^2}{2T} \qquad (29a)$$

$$D = -\Delta\mu/\nu_\beta T \qquad (29b)$$

The inclusion of the term $V \neq 0$ at $|D| \gg |Vc^2/2T|$ cannot lead to new results. But the consideration of this term is necessary in such clathrates as hydroquinone or hydrates with some poliatomic gases, in which $|D| \sim |Vc^2/2T|$.

As is seen from Eq. (29a) the function $f(c)$ has a minimum and maximum in the points

$$c_{\min}^{(1)} = \tfrac{1}{2}[1 - (1 + 4T/V)^{1/2}]$$
$$c_{\max}^{(1)} = \tfrac{1}{2}[1 + (1 + 4T/V)^{1/2}] \qquad (30)$$

which are obtained from the equation $\partial f/\partial c = 0$. It follows from Eq. (30) that $V < 0$ due to $c_{\max}^{(1)} \lesssim 1$, and a guest–guest interaction turns out to be attractive. In the case of the repulsive $V > 0$ interaction, the function $f(c)$ has no extremes. Thus, only one solution c_k of the equation $D = f$ occurs at $T < |V|/4$. If $T > |V|/4$, three solutions c_1, c_2, and c_3 can occur in the

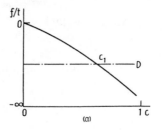

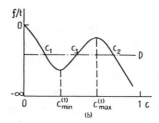

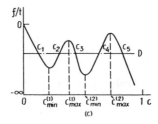

Figure 10. Possible forms of the function f dependence on guest molecules concentration c. (a) In the regions I, I_1 (for labeling of the regions see the phase diagram of Fig. 11). (b) In the regions II, II_1, IV, IV_1. (c) In region III.[78]

system. Let us consider the case in detail. In Fig. 10 the possible types of $f(c)$ dependences at various values of temperature and interaction parameters are shown. Various phase diagram regions are labeled in Fig. 11.

(1) $|D| < |f(c_{max}^{(1)})| = f_{max}$. The right line D intersects the $f(c)$ curve at $c_k < c_{min}^{(1)}$. The clathrate turns out to be a thermodynamically stable compound with the unique concentration value $c_k < 0.5$.

(2) $|D| > |f(c_{min}^{(1)})| = f_{min}$. The right line D intersects the $f(c)$ curve at c_k, but the value of equilibrium concentration is $c_k > 0.5$.

(3) $f_{max} < |D| < f_{min}$. The right line D intersects the $f(c)$ curve in three points—c_1, c_2, c_3. The point c_3 laying between c_1 and c_2 can be dropped,

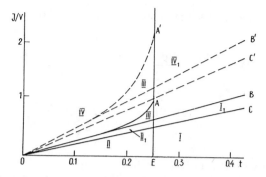

Figure 11. The regions in coordinates $(J/V, t = T/V)$, where one (I, I_1), two (II, II_1, IV, IV_1), and three (III) concentrations of gas in clathrate are realized. Solid curves correspond to Ising model (the number of gas molecule dipole orientations $q = 2$). The dashed line is to $q = 8$.[78]

since the system is unstable at the considered interval of concentrations in accordance with the thermodynamic inequalities.[72] The clathrate is a thermodynamically stable compound in one of the remaining concentration points, which provides minimal chemical potential μ (Eq. 19b), but it is metastable in other points. It is also important that the transition from metastable to unstable state, which is characterized by considerable changes of the β-form structure can proceed just as quickly as slowly. The time of this transition is 1 year in hydroquinone, urea clathrates (see Section VIII).

B.   Non-spherical Guest Molecules

Electrostatic multipole (e.g., dipole) guest–guest interactions as well as van der Waals interactions V_{ij} appear to be important when one considers nonspherical guest molecules. As a result, $J_{ij} \neq 0$ in Eq. (13) and the mean-field Hamiltonian of the system takes the form

$$H_{\text{MFA}} = H_{\text{MFA}}^{(0)} - \frac{Jc^2\sigma^2}{2} + Jc\sigma \sum_r n_r\sigma_r \qquad (31)$$

where $H_{\text{MFA}}^{(0)}$ is determined in Eq. (16), $\sigma = \langle \sigma_r \rangle$ is the average value of the spin (dipole) of the guest molecule, and $J = \Sigma J_{ij}$.

Performing calculations analogous to Eqs. (17)–(20) we obtain

$$\Omega = -\frac{Vc^2}{2} - \frac{Jc^2\sigma^2}{2} + T\ln(1-c) \tag{32a}$$

$$\mu = \epsilon + Vc + T\ln\frac{c}{1-c} - T\ln \text{ch}\left(\frac{Jc\sigma}{T}\right) \tag{32b}$$

which coincide with Eq. (19) at $J = 0$.

The equation determining the equilibrium c_k values takes the form analogous to Eq. (29):

$$D = -\frac{\Delta\mu}{\nu_\beta T} \tag{33a}$$

$$f = \ln(1-c) - \frac{Vc^2}{2T} - \frac{Jc^2\sigma^2}{2T} \tag{33b}$$

Complete analysis of the Eq. (33) can be found in Refs. 22–26 and 78. In Fig. 11 the phase diagram of the model at $J \neq 0$ is plotted. Here we present the main conclusions:

1. The region, where clathrate can be found in two states (stable and metastable), widens at $J \neq 0$. The case of $T < |V|/4$ appears to be possible.

2. Three values of concentrations $c_1 < c_3 < c_5$ (Fig. 11), at which the existence of clathrate is possible, are simultaneously found in the narrow interval of parameters J, V, T in the vicinity of the line $T - |V|/4 \lesssim 0$. Clathrate appears to be metastable at two concentration points and stable at one. Guest molecules can be found in completely disordered ($\sigma = 0$), partially ordered ($\sigma < 1$), and completely ordered ($\sigma = 1$) states at various c_i values.

3. Orientational phase transition to ferroelectric or antiferroelectric states related to the ordering of guest molecules in the cavities can occur at the decrease of temperature.

4. Concentrational phase transition can be observed at the fixed value of temperature, when a jump-like decrease of total concentration of guest molecules occurs. This case is particularly important for controlling the composition of metastable clathrates.

C.   Spherical Guest Molecules in Different Cavities

In this section we examine thermodynamic properties of clathrates with various sorts of cavities. For example, let us consider the hydrate structures CS-I and CS-II in which two sorts of cavities, large and small, are observed with the indices $i, j = 1, 2$ in the Hamiltonian (Eq. 13). Due to the spherical form of guest molecules, $J_{ij} = 0$ in Eq. (13). We denote n_r^a the occupation number of the small (a) cavity with coordinate r and $n_{r'}^b$, that of the large (b) cavity with coordinate r'. Then, the Hamiltonian (Eq. 13) for the case of two nonequivalent cavities takes the form

$$
H = \sum \epsilon_r^a n_r^a + \sum \epsilon_{r'}^b n_{r'}^b + \frac{1}{2} \sum [V_{aa}(r - r')n_r^a n_{r'}^a
$$
$$
+ V_{bb}(r - r')n_r^b n_{r'}^b] + \sum V_{ab}(r - r')n_r^a n_{r'}^b
$$
$$
- \sum \mu_a n_r^a - \sum \mu_b n_{r'}^b
\tag{34}
$$

Generalization of Eq. (34) to lattices of greater number of unequivalent cavities can be performed in an analogous way.

The Hamiltonian (Eq. 34) in mean-field approximation takes the form

$$
H_{\mathrm{MFA}} = -\frac{V_{aa}c_a^2}{2} - \frac{V_{bb}c_b^2}{2} - V_{ab}c_a c_b + \sum n_r^a (V_{aa}c_a + V_{ab}'c_b + \epsilon_a - \mu_a)
$$
$$
+ \sum n_r^b (V_{bb}c_b + V_{ab}''c_a + \epsilon_b - \mu_b)
\tag{35}
$$

Here $c_a = \langle n_r^a \rangle$ and $c_b = \langle n_r^b \rangle$ are the mean occupation numbers of small and large cavities. The potentials $V_{aa} = \Sigma_{r-r'} V_{aa}(r - r')$ and $V_{bb} = \Sigma_{r-r'} V_{bb}(r - r')$. Different values of V_{ab}' and V_{ab}'' are specially introduced in Eq. (35) to describe $V_{ab}(r - r')$ potentials. The introduction is related to the fact that the number z_a of the cavities a interacting with the neighboring cavity b in CS-I and CS-II clathrates is unequal to the number z_b of the cavities b interacting with the neighboring cavity a. For example, $z_a = 4$, $z_b = 12$ in hydrates CS-I and $z_a = 12$, $z_b = 6$ in hydrates CS-II. Therefore the potentials $V_{ab}' = z_b V_{ab}^{(0)}$, $V_{ab}'' = z_a V_{ab}^{(0)}$, $V_{ab} = (z_a + z_b)V_{ab}^{(0)}/2$ and $V_{ab}^{(0)}$ is the potential of the interacting pair of guest molecules located in the neighboring cavities a and b.

Substituting Eq. (35) into Eq. (12), performing the summation over $n_r^{a,b}$, and expressing μ_a, μ_b in terms of c_a, c_b we obtain the thermodynamic potential Ω of clathrate in the form

$$\Omega = \Omega_a + \Omega_b + \Omega_{ab} \tag{36}$$

$$\Omega_a = -\frac{V_{aa}c_a^2}{2} + T\ln(1 - c_a)$$

$$\Omega_b = -\frac{V_{bb}c^2}{2} + T\ln(1 - c_b) \tag{37}$$

$$\Omega_{ab} = -V_{ab}c_a c_b$$

Chemical potentials of guest molecules μ_a and μ_b are expressed as

$$\mu_a = \epsilon_a + V_{aa}c_a + V'_{ab}c_b + T\ln\left(\frac{c_a}{1 - c_a}\right) \tag{38}$$

$$\mu_b = \epsilon_b + V_{bb}c_b + V''_{ab}c_b + T\ln\left(\frac{c_b}{1 - c_b}\right)$$

Introducing chemical potentials of the lattice in the form $\mu_Q^a = v_a\Omega_a$, $\mu_Q^b = v_b\Omega_b$, $\mu_Q^{ab} = v_{ab}\Omega_{ab}$, we obtain the equilibrium conditions of the clathrate with two sorts of cavities

$$-\Delta\mu = \mu_Q^a + \mu_Q^b + \mu_Q^{ab} \tag{39}$$

where $\Delta\mu = \mu_0^\alpha - \mu_0^\beta$ is the difference of chemical potentials of α- and β-form empty lattices. The values $v_a = 1/23$, $v_b = 3/23$, and $v_{ab} = 2/23$ for hydrate I. Correspondingly, $v_a = 2/17$, $v_b = 1/17$, and $v_{ab} = 3/34$ for hydrate II.

It should be noted that the Eqs. (36)–(39) correspond to expressions in Ref. 5 in the absence of guest–guest interactions ($V_{ij} = 0$).

It is rather difficult to perform the analysis Eq. (39) in full measure due to the presence of at least three variational parameters (V_{aa}, V_{bb}, V_{ab}). However, some specific conclusions resulting from the solution of this equation for hydrates with simple gases are discussed in Section VIII.

The consideration of nonspherical guest molecules in unequivalent cavities can be carried out in accordance with the model proposed in Section III.B.

D. Thermodynamics of Clathrates in Cluster Approximation

As the number of nearest neighbors z, interacting via constant V, decreases, so does the accuracy of mean-field approximation.[79] The error of this approximation in the Ising model with $z = 12$ (face-centered cubic lattice) is about 18%, but it amounts to more than 40% for $z = 4$ (square lattice).

The average number of neighbors in hydrates is $z = (z_a + z_b)/2 > 8$. Therefore, the application of mean-field approximation is assumed to be justified for these compounds. But in hydroquinone structures with the cavities extended in quasi-one-dimensional chains ($z = 2$), mean-field approximation ought to distort the true value of quantitative characteristics. In order to obtain the required value of concentration ($c_k \approx 0.99$ at $T \approx 300$ K) for β-hydroquinone with methanol, one must take the value $D \approx -2 \div -2.5$ which differs by 5–6 times from the experimental value.[23] In this situation a notable increase is required of Ω and μ calculation accuracy.

In this section we describe the procedure of Ω and μ calculation by cluster approximation,[79–81] the accuracy of which is at least two times better than that of mean-field approximation.

Let us consider the case when guest molecules in equivalent cavities can be oriented along two equivalent directions (Ising situation). The more general case of $q > 2$ directions was examined in Ref. 23. The calculations of Ω and μ we perform by the simplest two points (cavities) cluster. The Hamiltonian H_2 of this cluster is written as

$$H_2 = (V + J\sigma_1\sigma_2)n_1n_2 + [\varphi(z - 1) + J'c\langle\sigma\rangle](n_1\sigma_1 + n_2\sigma_2)$$
$$+ (n_1 + n_2)[\epsilon + V'c + \psi(z - 1) - \mu] \tag{40}$$

Here V and J are the potentials of the nearest guests molecule interaction in the chosen cluster and subscripts 1, 2 denote both cavities of the cluster. The constants J' and V' describe the next nearest guest interactions, which are taken into account in Eq. (40) by mean-field approximation. Effective fields φ and ψ describe the interaction of cluster guest molecules with the nearest noncluster molecules.

In order to close the problem and perform Ω and μ calculation one must add to Eq. (40) one point (cavity) cluster with the Hamiltonian

$$H_1 = n_1\sigma_1(\psi z + J'c\langle\sigma\rangle) + n_1(\epsilon + V'c + z\psi - \mu) \tag{41}$$

If we use the thermodynamic relation

$$\frac{\partial \Omega}{\partial (1/T)} = \langle H \rangle - \mu c \tag{42}$$

where $\langle H \rangle$ is the average energy of the lattice, the expression of Ω per one cavity takes the form

$$\Omega = -\frac{z}{2} T \ln \text{Sp} \exp(-\beta H_2) - \frac{(z-1)}{2} T \ln \text{Sp} \exp(-\beta H_1) -$$

$$- \frac{V'c^2}{2} - \epsilon c - \frac{J'c^2 \langle \sigma \rangle^2}{2}. \tag{43}$$

where $\beta = 1/T$. The fields φ and ψ are determined by the system of equations

$$\frac{\partial \Omega}{\partial \varphi} = \frac{\partial \Omega}{\partial \psi} = 0 \tag{44}$$

As a result, we obtain the expression for the potential Ω in a two-point cluster approximation in the form

$$\Omega = -\frac{z}{2} T \ln(1 + R_1) + \left(\frac{z-1}{2}\right) T \ln 2 \text{ ch } \beta R_2 + \ln c + \ln(1-c) -$$

$$- \frac{1}{2} V'c^2 - \epsilon c - \frac{1}{2} J'c^2 \langle \sigma \rangle^2 \tag{45}$$

where

$$R_1 = [(2c-1)^2 + 4c(1-c)L]^{1/2}$$
$$R_2 = \psi z + J'c \langle \sigma \rangle$$
$$L = e^{\beta(V+J)}(1 + e^{-2\beta J} \text{ ch } 2\beta R_2)/2$$

Using the thermodynamic relation

$$\frac{\partial \mu}{\partial c} = -\frac{1}{c} \frac{\partial \Omega}{\partial c} \tag{46}$$

we obtain the expression for the chemical potential

$$\mu = \frac{z}{2} T \ln \frac{R_1 + 2c - 1}{(1 - c)L} - \frac{z - 1}{2} T \ln \frac{c}{(1 - c)\,\mathrm{ch}\,\beta R_2} + \\ + V'c + \epsilon + J'c\langle\sigma\rangle^2 \tag{47}$$

The field φ in Eqs. (45) and (47) is eliminated by the first equation (44). Equations (45)–(47) describe the thermodynamics of the clathrate. It should be noted that in the case of a one-dimensional chain ($z = 2$), Eqs. (45) and (47) coincide with the precise expressions obtained by the matrix method.[79] The application of cluster approximation to the concrete calculations of thermodynamic properties of hydroquinone is performed in Section VIII.A.1

E. Guest Molecules in Channel-Type Clathrates

The problems which arise when describing channel-type clathrates differ from those of the clathrate theory of closed cavities. Guest molecules in channel clathrates are as a rule densely packed. But some of empty carcasses, for example, that of urea, cannot exist in β form and transfer to the α modification. Therefore, on one hand we observe a specific clathrate situation when guest molecules stabilize the β form, but on the other, this stabilization is carried out at the dense guest molecules packing in channels.

It is unlikely that clathrate-formation models described in previous sections could be applied to the description of channel-type clathrates. In order to use the models of Sections III.B and III.C one must divide the channels into cells, which would correspond to guest molecule size and be occupied or vacant. The ratio of occupied cells to the full number of cells (c) approaches 1 (dense packing) only when the difference of chemical potentials of α- and β-host phases $\Delta\mu \to \infty$ in Eq. (28). Therefore, it is hardly probable that this model would be available to describe the experimental data.

In Refs. 82 and 83 the model is proposed of channel-type clathrates with an unstable empty host lattice. It is taken into account in Ref. 82 that guest molecules can be oriented along various directions in the channel.

The guests are packed in a one-dimensional channel in such a way that the distance a_i is left between the "head" of ith guest molecule and the "tail" of ith + 1 guest molecule. The interaction energy of the chain of these molecules in the channel is written in the most general form as

$$H = \sum_i \varphi_i(a_i) - \sum_i J(a_i)\sigma_i\sigma_{i+1} - V\sum_i \sigma_i \cos \alpha_i$$
$$- E\sum_i \sigma_i + \sum_i W(a_i - r_i) \tag{48}$$

Here $\sigma_i = \pm 1$ according to guest orientation, $J(a_i)$ describes the interaction of neighboring guest molecules in the channel, $\varphi(a_i)$ is the potential energy of these molecules interaction and $E = J'\langle\sigma\rangle$ describes the interaction of guest molecules of different channels in mean-field approximation. The third term in Eq. (48) takes into consideration the guest–host interaction energy which depends on the angles α_i of guest molecule orientation with respect to host walls. Guest–host interaction energy is denoted as W.

It seems that the form of Eq. (48) notably differs from the form of Hamiltonians used in inclusion-type systems. However, this is not really the case. Let us consider the Hamiltonian (Eq. 7) written in the most general form. Since we have assumed that guest molecules are densely packed in the channels and concentration $c \to 1$, all occupation numbers in Eq. (7) are not variable and $n_r^i = 1$. Then the comparison of Eqs. (7) and (48) yields the identity of these expressions if the following relations are valid:

$$\varphi(a_i) \to V_{ij}(r_i - r_j')$$
$$E + V \cos \alpha_i \to -W_{ij}^{(2)}(R_i - r_j)$$
$$J(a_i) \to J_{ij}(r_i - r_j')$$
$$W(a_i - r_i) \to W_{ij}^{(1)}(R_i - r_j)$$

The concrete calculations of thermodynamic properties with the Hamiltonian (Eq. 48) is carried out at the assumptions of little guest molecule shifts in the channel and $J(a_i) \approx J(a)$, where $a \approx a_1 \approx a_2 \approx \cdots \approx a_N$ is the mean distance between guest molecules. Then one part of the free energy $F = \sum_{\{a_i\}} \exp[-\sum_i \varphi(a_i)]$ can be calculated in quasi-harmonic approximation with $\varphi(a_i)$ presented as a series

$$\varphi(a_i) = \varphi(a) + \frac{1}{2}g_2u^2 + \frac{1}{3!}g_3u^3 + \frac{1}{4!}g_4u^4 + \cdots \tag{49}$$

where u is the difference of mean shifts of two neighboring guest molecules. Assuming that the constants of an harmonic interaction $g_3, g_4 \ll T$, the expression of free energy per one guest molecule can be obtained in the form[84]

$$\frac{F_a}{N_g} = \varphi(a) - \frac{T}{2}\ln\frac{(2\pi T)^2}{g_2} - T^2\left(\frac{1}{2}\frac{g_3^2}{g_2^3} - \frac{g_4}{8g_2^2}\right) \tag{50}$$

where N_g is the number of guest molecules in channel.

The other part of free energy

$$F_\sigma = -T\ln\mathrm{Sp}\exp\left[J(a)\sum_i \sigma_i\sigma_{i+1} + E\sum_i \sigma_i\right] \tag{51}$$

can be calculated as the free energy of a one-dimensional chain of Ising spins in external field E. As a result, one obtains[79]

$$\frac{F_\sigma}{N_g} = -J(a) - T\ln[\mathrm{ch}\,\xi + \sqrt{\mathrm{sh}^2\xi + L}] \tag{52}$$

where $\xi = E/T = J'\langle\sigma\rangle/T$, $L = \exp(-2J(a)/T)$. The average value $\langle\sigma\rangle$ is obtained from the equation

$$\langle\sigma\rangle = \frac{\mathrm{sh}\,\xi}{\sqrt{\mathrm{sh}^2\xi + L}} \tag{53}$$

It follows from Eq. (53) that in the T_c point determined by the equation

$$L = J'/T \tag{54}$$

a second-order orientational phase transition occurs from the state $\langle\sigma\rangle = 0$ to $\langle\sigma\rangle \neq 0$. Consequently, guest molecules in channels turn out to be ordered in one-dimensional chains.

The thermodynamic properties of the system hardly depend on the term $V\sum\sigma_i\cos\alpha_i$ in Eq. (48), because the direction of host fields periodically varies along the channels and averaging over α_i angles leads to the simple J' constant renormalization.

The remaining summand which describes guest-host interaction in Eq. (48) gives the standard contribution to the free energy

$$F_{hg}/N_g = -\epsilon \tag{55}$$

where ϵ is the energy analogous to that of Eq. (13) which is needed to transfer the guest molecule from channel to lattice.

Thus, the free energy F of the channel-type clathrate with the dense packing of guest molecules in channels takes the form

$$F = F_a + F_\sigma + F_{hg} + F_\beta(N_Q, T) \tag{56}$$

where F_β is the free energy of the empty host lattice, N_Q is the number of host molecules, and F_a, F_σ, F_{hg} are determined in Eq. (50), (52), and (55).

To investigate the thermodynamic stability of channel-type clathrates, using Eq. (58), one must know the relation between N_Q and N_g. In accordance with Ref. 83 these values are connected by the relation

$$N_Q = \frac{N_g}{d}[a + l(n - 1)] \tag{57}$$

where n is the number of carbon atoms and l is the distance between carbon atoms in a paraffin molecule, d is the distance between urea carbon atoms along the channel axis. Substituting Eq. (57) in Eq. (56) and using F minimum condition with respect to $\langle \sigma \rangle$ and a, one obtains the system of equations which determines the equilibrium values $a(T)$ and $\langle \sigma \rangle(T)$. These equations are so far investigated just at two limiting cases of high $T \gg T_c$ ($\langle \sigma \rangle = 0$) and low $T \ll T_c$ ($a = $ constant) temperatures with T_c determined from Eq. (54). In general, equilibrium conditions are expressed as $F_\alpha = F$, where F_α is the free energy of urea and F is determined in Eq. (56). In "clathrate" designations it takes the form

$$\Delta\mu = \mu_\alpha - \mu_\beta = \mu \tag{58}$$

where $\mu_\beta = \partial F/\partial N_Q$. The analysis Eq. (58) is performed in Section VIII.C.

In closing this section we should like to make several remarks about the model of linear and nonlinear optical properties of clathrate proposed in Ref. 85. Guest–host interactions in this paper were written for the guest of mass m in the form

$$V_{gh} = \sum_{i=1}^{3} \frac{m\omega_i^2}{2}\left[\frac{2\gamma_i}{\pi}\right]^2 tg^2\left(\frac{\pi x_i}{2\gamma_i}\right) \tag{59}$$

where ω_i is the frequency in the harmonic approximation, when potential width γ_i tends to infinity and x_i are the Cartesian coordinates which take into account the relative location of guest and host.

Guest–guest interaction was neglected in this article. The proposed theory was applied to β-hydroquinone at 100% occupation of cavities by gas ($c = 1$). Since $V_{gg} \ll T$ when the guests are Ar, Kr, and so on, the theory[85] corresponds to the argon inclusion. However, the occupation of hydroquinone cavities by argon amounts to 30–40%.[5,20] Thus, the description of a real situation in hydroquinone with different gases requires the consideration of $V_{gg} \neq 0$ or $c \neq 1$. Contrary to this, in channel-type clathrates $c \to 1$. Therefore, the possibility is not precluded of more successful application of theory[85] to a urea-paraffin system than to that of hydroquinone–argon.

IV. PHASE DIAGRAMS

In order to obtain the phase diagram of clathrate, thermodynamic functions of all phases involved in equilibrium must be known. The equations, which describe a certain equilibrium of phases are obtained by equating thermodynamical and chemical potentials of components of various phases. In 1959 van der Waals and Platteeuw[5] presented the pioneering analysis of phase diagrams in clathrates. The qualitative and quantitative calculation was carried out for hydroquinone with various guest molecules. The equilibriums of gas guest phase, solid stable phase of α-hydroquinone, and clathrate phase were analyzed. The case of hydroquinone α phase not dissolving guests was investigated. The gas phase was assumed to be the ideal gas of the guest component with the chemical potential in the form (Eq. 23b)

$$\mu_G = T \ln(P_G / P_G^0) \tag{60}$$

where P_G indicates pressure of gas. In this case the equilibrium conditions of chemical potentials of host and guest in corresponding phases lead to Eqs. (27) and (23b). Expressing P_G as a function of c we obtain

$$P_G = P_G^0 \frac{c}{1 - c} \tag{61}$$

The value $(P_G^0)^{-1} = C_L$ is also called Langmuir constant. It was assumed at the derivation of Eqs. (23b) and (61) that the total pressure $P \approx P_G$ and partial pressure P_Q can be neglected.

The clathrate phase at given temperature T is formed at the occupation

TABLE VI
Equilibrium Pressure of Hydroquinone at $T = 25°C$

	Pressure (atm)	
Solution	Theory	Experiment
Argon	3.4	3.4
Krypton	0.4	0.4
Xenon	0.06	
Methane	0.88	
Nitrogen	5.2	5.8
Oxygen	2.6	
HCl	0.023	0.0

degree c obtained from Eqs. (27) and (61). These equations determine the curve of monovariant equilibrium $P(T)$. As is seen from Eq. (27), the calculation requires knowledge of the dependences of $\Delta\mu$ vs. T and P. Besides, the parameter ϵ, which determines intermolecular interaction of guest molecules with the host lattice ($P_G^0 \sim \exp(-\epsilon/T)$) must be taken into account. Note that dissolution of the guest in the α phase was neglected at the derivation of Eq. (61). The curve of monovariant equilibrium in Ref. 5 corresponds to the two experimental points obtained at 60 and 120°C. Remaining monovariant equilibria were investigated qualitatively.

Equilibrium values of pressures calculated by Eq. (61) in Ref. 5 are presented in Table VI. As shown they are in satisfactory agreement with the experimental results.

The structural investigations (see Section I.A) have shown that similar-size cavities occur in α and β modifications. The amount of cavities in α-hydroquinone is six times less than in β modification. Consequently, the guest molecules which are capable of being included in β-form cavities can be included in α-form cavities too. Hence the properties of a stable α phase with included molecules can be described in a manner analogous to the β phase.

Consider the case when the α phase can include guest molecules and host molecules can form metastable β- and γ*-clathrate modifications of similar energies. The fragment of the P, T diagram of guest (G)–host(Q) binary system is plotted in Fig. 12. The presence of other phases in the considered system would lead to the occurrence of new nonvariant points

---

*The notation γ phase implies any phase similar by energy (e.g., liquid L phase). This phase is unrelated to the nonclathrate γ phase of hydroquinone mentioned in Section I.A.

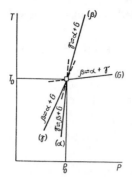

Figure 12. Schematic representation of monovariant equilibrium curves in guest–host system, when the amount (x) of guest in the phases can be expressed as $x_\alpha < x_\beta < x_\gamma < x_G$. The labeling of the phase in brackets shows the monovariant equilibrium in the absence of this phase, that is, (α) corresponds to $\gamma \Leftrightarrow \beta + G$.[26]

and corresponding lines of monovariant equilibrium at the determined values of P and T. We restrict the consideration of equilibrium to the part of the diagram defined in Fig. 12. This approach can be easily generalized for P, T diagrams with other sets of phases in the vicinity of nonvariant points.

The equations which determine pressure P_0, temperature T_0, and the compositions of phases in nonvariant points (see Fig. 12) are obtained from the equalities of chemical and thermodynamic potentials of system components in all phases. These equalities in "clathrate" denotations are as follows:

$$\mu_Q^\alpha = \mu_Q^\beta = \mu_Q^\gamma = \mu_Q^G; \qquad \mu_G^\alpha = \mu_G^\beta = \mu_G^\gamma = \mu_G^G \qquad (62)$$

When one assumes the gas phase to be formed of ideal gas, the expression of guest component chemical potential μ_G^G is given by the Eq. (60) and the one of host by Eq. (60) with index G replaced by Q. The partial pressures of host P_Q and guest P_G components are related with the mole part of the guest in gas phase x_G by the following formula:

$$P_Q = (1 - x_G)P; \qquad P_G = x_G P \qquad (63)$$

The substitution of the corresponding expressions of chemical potentials in Eq. (62) leads to the system of equation with respect to c_α, c_β, c_γ, P, and T. When guest–guest interactions V_{gg} are not taken into account, the system is notably simplified and takes the form

$$\Delta\mu^{\alpha\beta} = kT \ln[(1 - c_\beta)^{\nu_\beta}(1 - c_\alpha)^{-\nu_\alpha}] \tag{64a}$$

$$\Delta\mu^{\beta\gamma} = kT \ln[(1 - c_\gamma)^{\nu_\gamma}(1 - c_\beta)^{-\nu_\beta}] \tag{64b}$$

$$P_Q = A_\gamma(1 - c_\gamma)^{\nu_\gamma} \tag{64c}$$

$$(C_L^\alpha)^{-1}\frac{c_\alpha}{1 - c_\alpha} = (C_L^\beta)^{-1}\frac{c_\beta}{1 - c_\beta} \tag{64d}$$

$$(C_L^\beta)^{-1}\frac{c_\beta}{1 - c_\beta} = (C_L^\gamma)^{-1}\frac{c_\gamma}{1 - c_\gamma} \tag{64e}$$

$$P_G = (C_L^\gamma)^{-1}\frac{c_\gamma}{1 - c_\gamma} \tag{64f}$$

where

$$\Delta\mu^{ij} \equiv \mu_Q^i - \mu_Q^j; \quad A_i \sim T\exp(\mu_Q^i/T); \quad i,j = \alpha, \beta, \gamma \tag{64g}$$

C_L^i are the Langmuir constants of guest molecules in the ith phase. The values P_0, T_0, c_α, c_β, c_γ, and x_G, which characterize invariant equilibrium of the considered binary guest–host system, are obtained as a solution of the six-equation system (Eq. 64). When $V_{gg} \neq 0$, the system of equations becomes more complicated, but its form remains qualitatively unchanged.

The equations, which describe mono- and divariant equilibria, are obtained from the equalities of corresponding potentials of phases involved in equilibrium.

1. The monovariant equilibrium

$$\gamma \Leftrightarrow \beta + G \tag{65}$$

is described by Eq. (64b, c, e and f) with respect to c_β, c_γ, x_G, and T [curve (α)].

2. The curve (β), that is, the equilibrium

$$\beta \Leftrightarrow \alpha + G \tag{66}$$

is described by the following equations:

$$\Delta\mu^{\alpha\gamma} = kT \ln[(1 - c_\gamma)^{\nu_\gamma}(1 - c_\alpha)^{-\nu_\alpha}] \tag{67}$$

$$(C_L^\alpha)^{-1} \frac{c_\alpha}{1 - c_\alpha} = (C_L^\gamma)^{-1} \frac{c_\gamma}{1 - c_\gamma} \tag{68}$$

and Eqs. (64c) and (64f).

3. The curve (γ), that is, the equilibrium

$$\beta \Leftrightarrow \alpha + G \tag{69}$$

is determined by the Eqs. (64a), (64d), and

$$P_Q = A_\beta (1 - c_\beta)^{\nu_\beta} \tag{70}$$

$$P_G = (C_L^\beta)^{-1} \frac{c_\beta}{1 - c_\beta} \tag{71}$$

4. The curve (G), that is, the equilibrium

$$\beta \Leftrightarrow \alpha + \gamma \tag{72}$$

is described by Eqs. (64a), (64b), (64d), (64e).

5. Divariant equilibrium. The conditions of coexistence of α and β phases is described by Eqs. (64a) and (64d); of β and γ phases by Eqs. (64b) and (64e); of γ and G phases by Eqs. (64c) and (64f); of α and γ phases by Eqs. (67) and (68); of β and G phases by Eqs (70) and (71); of α and G phases by equations

$$P_Q = A_\alpha (1 - c_\alpha)^{\nu_\alpha} \tag{73}$$

$$P_G = (C_L^\alpha)^{-1} \frac{c_\alpha}{1 - c_\alpha} \tag{74}$$

The solutions of these equations allow us to plot isobar and isotherm sections of a phase diagram of a binary system. The form of the sections

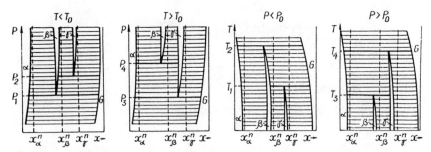

Figure 13. Scheme of isobaric and isothermic sections of guest–host phase diagram.[26]

is schematically shown in Fig. 13. Note that it is more convenient to use as a variable the mole part of the guest in the ith phase x_i instead of c_i

$$x_i = \nu_i c_i / (1 + \nu_i c_i) \tag{75}$$

The limiting value $x_i^n = \nu_i / (1 + \nu_i)$ is obtained from Eq. (75) at $c_i = 1$.

Equations (64), (67), (68), (70), (71), (73), and (74) describe completely a P, T, x diagram (non-, mono-, and divariant equilibria) of a binary guest–host system. To obtain the solutions of the corresponding system of equations, the functions $\mu_G^i(P, T)$, $\Delta\mu^{ij}(P, T)$, $\mu_Q^G(P, T)$, $\mu_G^G(P, T)$, and also the intermolecular potential ϵ (see Eq. 13), which characterizes the interaction of guest molecule with the lattice, must be known. For some systems ϵ can be found with reasonable accuracy by the well-known methods (e.g., Lennard–Jones–Devonshire lattice model,[5] and atomic potentials method[86] (see the discussion in Section VI.A). The values μ_Q^i are included in the equations determining the partial pressure of the host vapor, which is for the most part almost negligible in comparison with the pressure of the guest vapor. This simplifies the solution. A more complicated problem arises with $\Delta\mu^{ij} = \mu_Q^i - \mu_Q^j$, since $\Delta\mu^{ij}$ is, as a rule, a small difference of two large quantities, which cannot be found at present with the required accuracy. Direct experiment determination is also related to certain specific difficulties. Some basic problems also arise when one wants to obtain the metastable phases. This is possible just in the case of small $\Delta\mu^{ij}$ when the activation barrier of the transition is sufficient for the realization of the metastable phase. But the barrier cannot be very high, in order to keep the rate of clathrate formation reaction available. We know reliable data just for $\Delta\mu^{\alpha\beta} = 341 \div 349$ J/mol[5] for hydroquinone at 25°C. At the values of concentration c not vicinous to unity, the reverse situation is more promising when $\Delta\mu^{ij}$ is calculated from the known equations of phases equilibrium in binary systems. In this way the estimations of $\Delta\mu^{ij}$ for clathrate hydrates were made using the ideal model[5] and the experimentally determined values of concentration c. However, the case of hydrates is not beneficial owing to $c \rightarrow 1$. Therefore, even with reliable experimental data and a wealth of statistics one must take into account that this kind of calculation requires the accuracy of one order higher than that obtained in the experiments which define hydrate gas composition.[87]

It seems promising at $c \rightarrow 1$ to make the calculation using the experimentally determined value of pressure which depends both on c and $1 - c$ (see Eq. 64f). Thus, measuring pressure one can almost accurately determine c and $1 - c$. It allows us to obtain $\Delta \mu^{ij}(P, T)$ and to calculate the phase diagram with other guests using Eqs. (64), (67), (68), (70), (71), and (73). Certainly, the calculation is valid for the correct model of clathrate formation.

Qualitative calculation of the P, T, x diagram of hydroquinone with noble gases is presented in Section VIII.A.2.

V. PHASE TRANSITIONS

Three types of phase transitions are observed in clathrates, namely orientational, concentrational, and induced. At the phase transition of the first type the ordering of guest-molecule orientations in cavity or channel occurs. Concentrational transition is characterized by the sudden exit of guest molecules from cavities. Induced phase transition occurs when guest molecules at certain values of concentration induce the reconstruction of the host lattice.

A. Orientational Phase Transitions

These transitions are the best examined. They are characterized by the ordering of guest molecule dipole moments p_d at the transition point. The ordering of CH_3OH, HCl, SO_2 molecules in β-hydroquinone[18,19,24,88] or of trimethylene oxide in hydrates[89] can serve as examples of orientational phase transitions in clathrates.

Consider the case of guest molecules capable of being oriented along two equivalent directions and located in equivalent cavities (the general case of $q > 2$ directions can be found in Refs. 23 and 24). Then thermodynamic potential Ω is expressed in the form of Eq. (32a). The changes of thermodynamic properties related to the occurrence of the state with $\sigma \neq 0$ in Eq. (32a) can be investigated in much the same fashion of phase transition theory as an inclusion-type system (e.g., solid solutions[76,77]). However, contrary to solid solutions, where the concentration varies from zero to one, the value of c in clathrates is determined by the equation of system equilibrium (Eq. 33). This condition imposes certain restrictions on possible changes of thermodynamic properties of clathrates at the orientational phase transitions.

The value $\sigma \neq 0$ is determined from the equation

$$\sigma = \text{th}(Jc\sigma/T) \tag{76}$$

The equation can be obtained either by summation of trace $\sigma = \text{Sp } \sigma_r \exp(-\beta H_{\text{MFA}})/\text{Sp} \exp(-\beta H_{\text{MFA}})$ with H_{MFA} from Eq. (31) or from the free-energy minimum condition $\partial F/\partial \sigma = 0$, where $F = -T \ln \text{Sp} \exp(-\beta H_{\text{MFA}})$. It follows from Eq. (78) that at the point

$$T_c = cJ \tag{77}$$

the phase transition, related with the ordering of guest molecules and occurrence of state $\sigma \neq 0$, takes place.

The required thermodynamic functions (specific heat, entropy, etc.) are calculated from Eqs. (32), (76), and (77). For example, the expression of dielectric susceptibility has the form

$$\epsilon(T) = \epsilon_\infty + 4\pi \left(\frac{dP_s}{dE}\right)_{E \to 0} \tag{78}$$

where E is external electric field connected with the potential $W_{ij}^{(2)}$ (Eq. 7) by the relation $p_d E = W_{ij}^{(2)}$, ϵ_∞ is the value of susceptibility at superhigh frequencies (usually, $\epsilon_\infty \gtrsim 1$). Spontaneous polarization P_s is related to σ in the following way:

$$P_s = p_d \sigma/v_c \tag{79}$$

where $v_c = v/n$, v is the volume of clathrate unit cell and n is the number of cavities in the cell.

Let us turn to the investigation of Eq. (33) since the value of clathrate concentration is determined from this equation. In Fig. 11 the regions are shown where Eq. (33) has one, two, or three physically reasonable solutions. If line D and curve $f(c)$ intersect only at point c_k, the orientational phase transition in point T_c (Eq. 77) occurs with the decrease of temperature. The properties of the transition can be investigated by Eqs. (32) and (76). The situation turns out to be different in the case of two or three solutions of Eq. (33). The extreme values of $f(c)$ at $\sigma \neq 0$ (the regions located above the OC and OC' lines in Fig. 11) are determined from the condition $\partial f/\partial c = 0$ and Eq. (76);

$$\frac{t}{1-c} + c + J_1 c\sigma^2 \frac{ch(J_1 c\sigma/t)}{ch^2(J_1 c\sigma/t) - J_1 c/t} = 0 \tag{80}$$

where the dimensionless temperature $t = T/V$ and $J_1 = J/V$.

The solution of Eq. (80) is rather complicated at arbitrary t values. Thus, consider the value of temperature vicinous to phase transition point $T = T_c - 0$. Excluding σ and c from Eq. (80) with the aid of the expressions (76) and (77) we obtain the formulas of curve which divide "paraphase" and "ferrophase" regions in the form

$$J_1 = 3t - 2[(3t - 2)^2 + 4t]^{1/2} \tag{81}$$

(the curve OB in Fig. 11). It is also easy to verify that function $f(c)$ has two minima and maxima in the regions restricted by the curve

$$J_1 = \frac{4t}{1 + (1 - 4t)^{1/2}} \tag{82}$$

and the straight line $t = 1/4$.

Examine the behavior of clathrates in region III of Fig. 11, namely, in the vicinity of the boundary line EA, described by the expression $t = (1 - \tau)/4$ with $\tau \ll 1$. Order parameter $\sigma = 0$ in the vicinity of $c_{min}^{(1)}$ and $c_{max}^{(2)}$, but it turns out to be $\sigma \neq 0$ at $c > c_{min}^{(2)}$. Let us assume that line D is confined between f_{min} and f_{max} in such a way that

$$D = (f_{min} + f_{max})/2 \tag{83}$$

Then at the given value of τ the line (Eq. 83) and the curve $f(c)$ intersect in the points

$$c_1 = c_{min}^{(1)} - \tilde{\alpha}, \qquad c_3 = c_{max}^{(1)} + \tilde{\alpha}$$
$$c_5 = c_{max}^{(2)} + \tilde{\alpha} \approx 1 - \tilde{\alpha} \tag{84}$$

where $\tilde{\alpha} \ll 1$. Solving Eq. (33) with respect to $\tilde{\alpha}$ with $\sigma = 1 - \sigma_1$, where $\sigma_1 \ll 1$, we obtain

$$\ln \tilde{\alpha} \approx -\ln 2 - (1 + \tau)(J_1 + \tfrac{3}{2}) \tag{85}$$

Let us assume the linear relation between the parameter $b = \epsilon/V$ from Eq. (32b) and c in the form $b = b_0 - b_1 c$. Then equating the chemical potentials μ from Eq. (32b) in the points c_1, c_3, and c_5, we obtain that the state with gas concentration $c = c_5 \lesssim 1$ and $\sigma \lesssim 1$ occurs at

$$2b' - b_1 < \frac{3b_1}{8\sqrt{6}} \tau^{3/4} + \tau \ln 2$$

$$b' = \frac{1}{2}\ln 2 - \frac{1}{8} - \frac{J_1}{4}$$
(86)

In the case of opposite inequality, either the state with c_1 or that with c_3 is realized. It follows from Eqs. (86) and (82) that $b' < 0$. Therefore, if $b_1 \gtrsim 0$, the state with $c_5 \lesssim 1$ is always more beneficial. On the other hand, at $b_1 < 0$ the first-order phase transition can occur from the state with $c_5 \sim \sigma \lesssim 1$ to that of $c_1 < 0.5$, $\sigma = 0$. The point of the transition at $|b_1| \ll 1$ is determined by the simple expression

$$t_k = \frac{1}{4} + \frac{2|b'| - |b_1|}{4 \ln 2}$$
(87)

The stability of clathrate in the remaining regions of the (J_1, t) diagram (Fig. 11) can be investigated in a similar way. It should be noted that the properties of clathrate can be different in these regions, in spite of the similarity of $f(c)$ function behavior. In region II, in the vicinity of $c_{min}^{(1)}$ and $c_{max}^{(2)}$ the order parameter $\sigma = 0$, but it turns out to be nonzero at c_m determined from Eq. (77). If line D and curve $f(c)$ intersect at the points $c_1 \lesssim c_{min}^{(1)}$ and $c_m > c_{max}^{(1)}$, guest molecules are either in a partly ordered (c_m, $\sigma < 1$) or a disordered (c_1, $\sigma = 0$) state. In region IV the parameter $\sigma = 0$ at $c_1 \lesssim c_{min}^{(1)}$, but $\sigma \neq 0$ at $c > c_{min}^{(1)}$. Thus, two values of concentration cannot be found, at which guest molecules are disordered in this region at $|f_{min}| < D < |f_{max}|$. An interesting situation can occur in region IV, where $\sigma \neq 0$, starting from $c < c_{min}^{(1)}$. If line D and $f(c)$ intersect at points $c_1 \lesssim c_{min}^{(1)}$ and $c_3 \gtrsim c_{max}^{(1)}$, guest molecules reside either in partly ordered (c_1, $\sigma < 1$) or ordered (c_3, $\sigma \to 1$) states.

Thus, thermodynamic properties depend considerably on the region of the phase diagram to which the considered clathrate is prescribed.

B. Concentrational Phase Transitions

This type of phase transition can occur when D and $f(c)$ intersect at two points at least. Examine now the case $\sigma = 0$. Consider region II and the line D in the form of Eq. (83), intersecting with $f(c)$ at points c_1 and c_3 from Eq. (84). Solving Eq. (83) with respect to ν, we obtain that at point

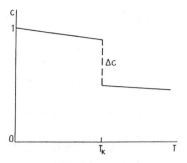

Figure 14. Schematic temperature dependence of guest molecules concentration $c(T)$ in clathrate at the concentrational phase transition.

$$t_k \approx \tfrac{1}{4}[1 - (\tfrac{3}{2})^{2/3}b_1^{4/3}] \tag{88}$$

the concentrational phase transition of the first order occurs from the state $c_3 > 0.5$ to $c_3 < 0.5$ with the value of discontinuity $\Delta c \approx 2\tilde{\alpha}$ (see Fig. 14). The clathrate remains a thermodynamically stable compound, as at the concentration $c_3 - \tilde{\alpha}$, so also at $c_3 + \tilde{\alpha}$, even though part of the guest molecule suddenly exits the cavities of the clathrate.

To determine clathrate occupancy degree we have chosen the value D in the form of Eq. (83). The results do not qualitatively vary when D is written in a slightly changed form (e.g., the value of D is vicinous to $f_{\min}$ or $f_{\max}$). It is just important that the value of D be confined between $f_{\min}$ and $f_{\max}$. If $D < f_{\min}$, only one occupancy degree is realized in clathrate and c increases with the decrease of t. At $D \gtrsim f_{\max}$, the concentration of gas c_1 decreases and c_3 increases with the decrease of t.

C. Induced Phase Transitions

This type of phase transition in clathrates resembles the transition on solid surfaces where the system of chemosorbed or physisorbed atoms induces the reconstruction of substrate at certain concentrations of adsorbed atoms c_k.[71,94,95] But if the adsorption-induced reconstruction of surface occurs quickly, the time of reconstruction in clathrates can be considerably long. The transformation of β-modification into α-modification is assumed to be an example of such a transition. Starting from the concentration of guests $c < c_k$, the β-host lattice appears to be unstable and transforms to the stable α form. The considerable reconstruction time τ_s is related to the specific interactions of clathrates, namely, with the guest molecules inclusion in closed cavities and very similar values of α- and β-phase lattice energies (see Introduction). Note that the difference of reconstructed and unreconstructed phase energies in chemisorbed atoms can attain several eV.

The theory of induced phase transitions for clathrates is not so far proposed. However, certain computer calculations are already carried out for hydrates. We shall return to these results in Section VII.

VI. ESTIMATION OF INTERMOLECULAR INTERACTIONS

Intermolecular interactions of clathrates have a different order of magnitude. Host–host interaction is the largest. This interaction was estimated for ice-type structures (hydrates) for the most part by the simple point-charge method, and has an atomic order of magnitude $V_{hh} \sim 10^4$–10^5 K. Correspondingly, V_{gh} is about 10^3–10^4 K and V_{gg} for polyatomic guest molecules is $\sim 10^2$–10^3 K.

To calculate these interactions, various model potentials are employed. This section is devoted to the estimates of V_{gh} and V_{gg} interactions.

A. Guest–Host Potential

The function $W^{(1)}(R_i - r_j) \equiv W(r)$ in Eq. (9) describes the interaction of guest and host molecules. Here r describes the distance between the centres of guest and host molecules. To relate our denotations and those of Ref. 5 we present the expression which relates the parameter ϵ from Eq. (20) with the parameter h from Ref. 5:

$$\epsilon = \ln h$$

$$h = 4\pi a^3 g\Phi \exp(-W(0)/T)$$

$$\Phi = \sum_k \int \frac{d^3 p_k}{(2\hbar)^3} \exp\left(-\frac{p_k^2}{2mT}\right) \tag{89}$$

$$g = \frac{1}{2\pi a^3} \int d^3r \exp\{-[W(r) - W(0)]/T\}$$

where p_k is the impulse of guest molecule and $\hbar$ is Planck's constant. Note that Langmuir constant $C_L = h/T$. To calculate the functions g and $W(r)$ in Eq. (89) the authors of Ref. 5 used a Lennard–Jones–Devonshire liquid cell model which is based on the assumptions that host molecules are homogenously distributed over the surface of a cavity with radius a and that only the interaction of a guest with the nearest neighbors is taken into account. In this case the guest molecule is located in the potential field of spherical symmetry. The magnitude of the field depends only on a distance between guest and center of cavity. When one takes the energy of guest–host interaction in the form of a Lennard–Jones potential,

$$\varphi(r) = 4\epsilon_L \left\{ \left(\frac{\sigma_L}{r}\right)^{12} - \left(\frac{\sigma_L}{r}\right)^{6} \right\} \qquad (90)$$

the expressions $W(0)$ and $W(r)$ are obtained:

$$W(0) = N\epsilon_L(\alpha^{-4} - 2\alpha^{-2})$$
$$W(r) - W(0) = N\epsilon_L\{\alpha^{-4}l(r^2/a^2) - 2\alpha^{-2}m(r^2/a^2)\} \qquad (91)$$

Here σ_L, ϵ_L are the parameters of the Lennard–Jones potential, N is the number of cavity-forming molecules, $\alpha = (a/\sigma_L)^3/\sqrt{2}$, and the functions $l(x)$ and $m(x)$ are expressed as follows:

$$l(x) = (1 + 12x + 25.2x^2 + 12x^3 + x^4)(1 - x)^{-10} - 1$$
$$m(x) = (1 + x)(1 - x)^{-4} - 1$$

The obtained expression of $W(r)$ (Eq. 91) is commonly used for the investigation of clathrates with the simplest one-atomic guest molecules (e.g., Ar, Kr, Xe, and small spherical methane-type molecules). For the extended and large molecules it is more convenient to use the Kihara potential (e.g., see Ref. 96). This potential, contrary to that of Lennard–Jones, takes into account the influence of length, form, and symmetry of molecules on interaction energy. The nucleus, which is prescribed to every molecule, represents the linear length between the atoms in the case of a two-atomic molecule. The nucleus for the CO_2 is the length between O–O, for the C_2H_4 or C_2H_6 the length between C–C, although it has a spherical form for the CH_4 molecule.[96] The energy of molecule interaction has the usual form of the Lennard–Jones potential (Eq. 90), but the distance between molecules really means the shortest distance between the nuclea of the molecules. The Kihara potential was used to calculate $W(r)$ in the framework of a cell model.[96] The analytic expression of guest and host molecule interaction $W(r)$, similar by form and spherically symmetric as Eq. (91), was obtained averaging over guest-molecule nucleus orientations.

In Ref. 96 the potential $W(r)$ values of various guest molecules were

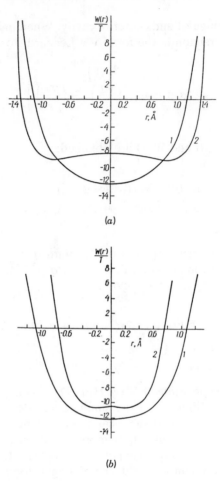

Figure 15. Spherically symmetric potential $W(\mathbf{r})$ of hydrate cavity with (a) guest molecules C_2H_6 and (b) CO_2. Curve 1 corresponds to the Lennard–Jones and curve 2 to the Kihara potential.[96]

calculated for hydrates. In Fig. 15 these values are shown for the guest molecules C_2H_6 and CO_2. As seen from Fig. 15a,b the values of potential $W(0)$ obtained with Lennard–Jones and Kihara potentials differ notably. The difference turns out to be essential at the determination of hydrate dissociation pressure. The choice of Kihara potential for the two-atomic molecules description gives better agreement of theory and experiment (see Section VIII.B).

In Ref. 20 a direct summation of all pair-interaction potentials of vari-

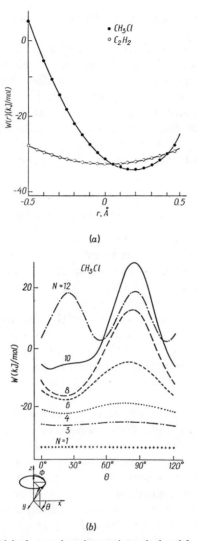

Figure 16. (*a*) Potential of guest–host interaction calculated for CH_3Cl and C_2H_2 molecules in hydroquinone.[20] The axis of the molecule coincides with the *c* axis of the crystal. (*b*) Dependence of interaction potential on the orientation of the CH_3Cl molecule axis in the hydroquinone cavity. The parameter $N = \Phi/7.5°$.[20]

ous guest molecules and atoms of host molecules was carried out in atom-atom approximation for hydroquinone clathrates. The plots were drawn of $W(r)$ potentials dependence on distribution of polyatomic guest molecules and their axes' orientation in a cavity. In Fig. 16*a,b* the potential

$W(r)$ dependencies on guest molecules shifts and its axis orientation angles are shown.[20]

As seen in Fig. 16, the CH_3Cl molecule prefers to be arranged in such a manner that the center of gravity would not coincide with the center of the cavity. The result is of special interest since it was assumed that the centers usually coincide in the hydroquinone molecule.

Note that sometimes considerable errors can occur in the lattice model at low temperatures due to ignorance of the influence of guest molecules on the host lattice.[25] On the other hand, guests with small radii (e.g., Kr, Xe,HCl) weakly influence the vibrational states of the host lattice.[97]

B. Guest–Guest Potential

The problem of calculating guest–guest interaction was recently raised. Therefore, only a few attempts to estimate V_{gg} are so far known. Thus, Sitarski[21] tried to calculate the statistical sum of the clathrate taking into account guest–guest interaction V_{gg} as an addition to guest–host interaction V_{gh}. He included the potential field of guest molecules, located in the centers of neighboring cavities, to the potential field of the lattice. This approach allowed him to obtain the new expression for guest molecules chemical potential and hence the new expression for sorption isotherms. However, the problems related to the influence of guest–guest interaction on the stoichiometry of clathrates were not considered in this paper.

In Ref. 20 the Monte Carlo method was used to calculate the energy and order parameters of the dipole moment orientation along and perpendicular to the c axis of 576 guest molecules located in β-hydroquinone cavities. In the paper the dipole–dipole interaction of guest molecules was taken into account. The calculation was carried out to describe the phase transition to the ferromagnetic state which had been observed in Refs. 19 and 20. The authors of Ref. 20 found the tendency of dipole moment orientations along the c axis to the ordering at the temperature decrease, but the phase transition to ferromagnetic state did not occur in the proposed model. In Refs. 22–26, 74, and 75, V_{gg} interaction was considered as equivalent to V_{gh}, but not as a perturbation of the latter interaction. Here we demonstrate the calculation[65] of pair guest–guest interaction energy for hydroquinone clathrates with the simple guest molecules Ar, Kr, and Xe. It can be suggested in this case that $V_{gg}/V_{gh} \approx 0.1$. The calculation is performed using the formulas obtained in the summation of all pair interactions of guest molecules in the form of the Lennard–Jones potential,

TABLE VII
Potential Energy of Guest–Guest Interaction in α and β Modifications of Hydroquinone

Guest	σ_L (Å)	ϵ_L (K)	$-V_{gg}^{\alpha}$ (kJ/mol)	$-V_{gg}^{\beta}$ (kJ/mol)
Ar	3.40	119.8	0.20	0.24
Kr	3.60	171	0.38	0.46
Xe	4.10	221	0.98	1.19
CH_4	3.82	148	0.45	0.55
SO_2	4.29	252	1.38	1.68

$$V_{gg} = 2\epsilon_L \left[\left(\frac{\sigma_L}{r} \right)^{12} g_{12} - \left(\frac{\sigma_L}{r} \right)^6 g_6 \right]$$

$$g_{12} = \sum P_{ij}^{-12}; \qquad g_6 = \sum P_{ij}^{-6} \tag{92}$$

where rP_{ij} is the distance between atoms i and j, expressed in the nearest distance r dimensions. The parameters g_{12} and g_6 depend on host lattice properties. Thus, $g_{12}^{\alpha} = 2.0005$, $g_6^{\alpha} = 2.0379$, $g_{12}^{\beta} = 2.0041$, and $g_6^{\beta} = 2.2350$ for α- and β-hydroquinone (the following lattice constants were used in the calculation: $a = 38.3$ Å, $c = 5.6$ Å for α-hydroquinone[30] and $a = 16.5$ Å, $c = 5.5$ Å for β-hydroquinone[31]). The values of guest–guest interaction energy and parameters of pair interaction potentials ϵ_L and σ_L are presented in Table VII. The analogous calculation for hydrates provides that $V_{gg} = -2.6$ kJ/mol with Xe and $V_{gg} = -0.98$ kJ/mol with Ar. The values are of one order with the $\Delta\mu$ value from (Eq. 26a) which is about 0.33 kJ/mol for hydroquinone and 1 kJ/mol for hydrates.

The estimates of V_{gg} as pure dipole–dipole interaction are also known.[18,19,98] The calculation yields $V_{gg} \approx -2.6$ kJ/mol for β-hydroquinone–HCN and $V_{gg} \approx -0.22 \div -0.56$ kJ/mol for a β-hydroquinone–CH_3OH system.[19,98] Note that thermodynamic estimation[24] gives the considerably higher value of $V_{gg} \approx -2.29$ kJ/mol for β-hydroquinone with CH_3OH, but also the value $V_{gg} \approx -1.34$ kJ/mol for β-hydroquinone with SO_2, which satisfactorily agrees with the calculations of Eq. (92) (see Table VII).

VII. COMPUTATIONAL METHODS

Numerical calculation of thermodynamic and dynamic properties of clathrates allows us to establish directly the degree of stability of the host lattice and to reveal the role of guest molecules in the stabilization of clathrate. The simulation is performed thus far in hydrates of the CS-I lattice, which is taken to be empty or occupied by simple guest molecules. The two methods of numerical analysis which complement each other are predominantly used, those of molecular[99,100] and lattice[101,102] dynamics.

To make the concrete analysis, the potential of intermolecular interaction must be specified. It is common for hydrates to choose the potential in the Lennard–Jones form (Eq. 90) (interaction oxygen–oxygen, oxygen–guest, guest–guest, etc.) and in the form of simple point-charge potential.[103] The latter includes Coulomb interaction of charged sites on oxygen and hydrogen atoms. The parameters of these potentials for CS-I hydrate are numerically equal to $\sigma_L^{0-0} = 3.16$ Å, $\epsilon_L^{0-0} = 2.6$ kJ/mol, and the charges $q_0 = -0.82|e|$, $q_H = 0.41e$, where e is the electron charge. The selection of potentials is analogous in both methods of numerical analysis, but other means of investigation are different.

A. Molecular Dynamics

Molecular dynamics investigation of concrete systems is performed by numerical solution of molecule motion equations. The system of 20 or more interacting molecules is customarily chosen with the periodic boundary conditions imposed to decrease the influence of the boundary effect on the calculation result. Structural and thermodynamic characteristics of the system can be obtained averaging molecule motion over time.

The investigations of hydrates and ice I_h by the molecular dynamics method were carried out in Refs. 104–109. The water molecule was assumed to be rigid with an O–H bond length of 1.0 Å and an angle H–O–H of 109°28′.

In Ref. 104 this method is used to examine the cluster of 20 water molecules which form the small cavity of hydrate clathrate (pentagondodecahedron). It was found that at temperatures about 230 K the random motion of molecules increases and the momentary structures of clathrate considerably differ from those averaged over time. The fracture of $U(T)$ dependence at $T \approx 200$ K is seen in Fig. 17, where U indicates the internal energy of system. The cluster transits from the open clathrate structure to the more dense and less ordered structure. The authors of Ref. 104 relate the changes with phase transition (probably with melting of system). The small dimensions of the cluster lead to the blurring of the transition. Besides, the notable influence of the boundary effect presumably decreases the transition temperature as compared with crystallic hydrate.

In Refs. 105–107 the crystal structure of empty hydrate CS-I and hy-

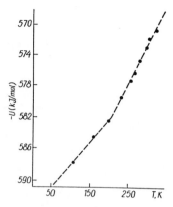

Figure 17. Temperature dependence of energy per molecule for the 20-molecule cluster.[104]

drate with CH_4, Xe, CF_4, C_3H_6, and $(CH_2)_2O$ were examined. In these papers important information about vibrational spectra and hydrate structures were obtained. An instability of hydrate structure was not found over a considered range of temperatures. However, in Ref. 110 the same authors have found the phase transition in ice at pressure 13 kbar and temperature 80 K. The analysis of the considered results strongly suggest that empty cavities are unstable in hydrate clathrates.

In Table VIII and Fig. 18 the vibrational characteristics of various clathrate hydrates are presented. As seen from Figs. 18 and 19, satisfactory

TABLE VIII
Translational Vibrations of Guest Molecules (cm^{-1}) in Various Structure I Clathrate Hydrates[107]

Guest	T (K)	Small Cages	Large Cages
CH_4	103	70	32, 52
$(CH_4)^a$	(145)	(78)	(35, 54)
CF_4	77		21, 37
	20		18, 35
C_3H_6	139		84, 103
$(CH_2)_2O$	66		61, 85, 125

^{a}Results for a spherically symmetric model of CH_4 from Ref. 106.

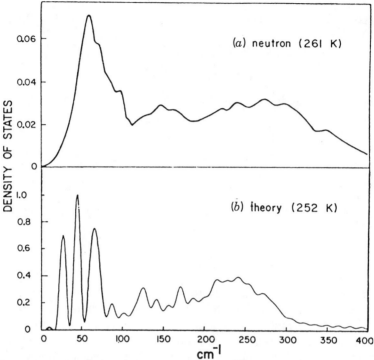

Figure 18.   Translational frequency spectra for ice.[110] (*a*) Experimental result for ice I_h at 261 K. (*b*) Molecular dynamic result for ice I_c obtained by using simple point-charge model at 252 K.

agreement of molecular dynamics and experimental results is obtained for translational frequencies of spectra and radial distribution function of ice.

B.   Lattice Dynamics

The normal vibrational frequencies of crystal and the dependence of frequency on wave vector **q** in lattice dynamics method is obtained from the main equation

$$|D(\mathbf{q}) - \omega^2(\mathbf{q})M| = 0 \qquad (93)$$

Here $D(\mathbf{q})$ is the dynamic matrix of crystal related with the force constants or pair interaction potentials; M indicates matrix consisting of masses and inertia moments of molecules. Vector **q** depends on the first Bruillouin zone and ω_n frequencies correspond to every **q** value (n is the number of unit cell molecules).

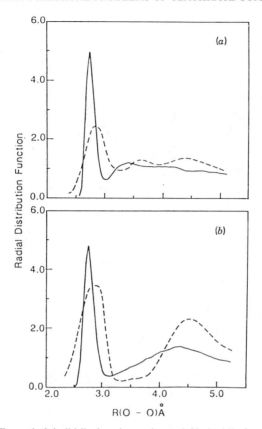

Figure 19. Theoretical (solid line) and experimental (dashed line) oxygen atom radial distribution functions[109] for (a) high-density amorphous ice recovered at zero pressure at approximately 80 K, and (b) low-density amorphous ice obtained by warming of the high-density amorphous sample.

The elements of the dynamic matrix are calculated by formulas presented in Refs. 101 and 102.

The system of Eq. (93) with respect to $\omega(\mathbf{q})$ with the chosen interaction potentials was investigated for ices I_c and I_h, and CS-I structure hydrates.[111,112]

In Fig. 20 the histograms of phonon state density are plotted. It is seen that the range of vibrational frequencies of empty molecules is divided into two zones. At low frequencies $(0\text{--}280\ \text{cm}^{-1})$ the translational vibration of water molecules is predominantly observed. At higher frequencies $(490\text{--}960\ \text{cm}^{-1})$ water molecules execute librational vibrations. Phonon spectra of hydrates with composition $6\,\text{Xe}\cdot 46\,\text{H}_2\text{O}$ and $8\,\text{Xe}\cdot 46\,\text{H}_2\text{O}$ are obtained using the Lennard–Jones potential of interatomic xenon–oxygen

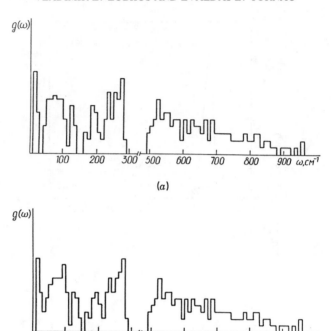

Figure 20. Density of phonon states of CS-I clathrate hydrates (a) $6\,Xe \cdot 46\,H_2O$ (with large cavities filled by xenon) and (b) $8\,Xe \cdot 46\,H_2O$ (large and small cavities filled by xenon).[112]

interaction. The parameters of the potential is calculated by the standard rules (e.g., see Ref. 38). The frequencies of xenon vibration in large cavities amount to $15-18\,cm^{-1}$. The obtained dynamic properties allow us to construct the thermodynamic functions of the crystal. In Fig. 21 the equation of state $V(P)/V_0$, where $V_0 = V(P = 0)$, is plotted for the empty carcass of hydrate CS-I, hydrate CS-II with xenon-occupied large cavities, and with both small and large occupied cavities. The equation of state for ice I_h is also shown in this figure. All the curves end at certain critical pressures, at which the compound becomes dynamically unstable. The empty carcass of CS-I hydrate is dynamically stable at $T = 0$ and $P < 13.5\,kbar$. The stability of hydrate with respect to compression increases as the large cavities become occupied by xenon. The critical pressure amounts to $16\,kbar$ for $6\,Xe \cdot 46\,H_2O$ hydrate and $26\,kbar$ for completely occupied $8\,Xe \cdot 46\,H_2O$ hydrate. The boundaries of thermodynamic stability of the hydrate carcass at the increase of temperature can approximately be found by the Lindeman criterion.[113] The empty carcass is the

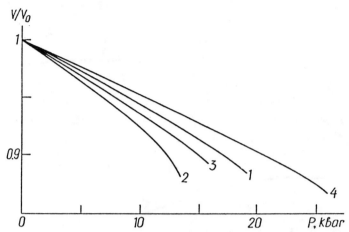

Figure 21. Pressure dependence of volume $V(P)/V_0$ at $T = 0$ for (1) ice, (2) empty CS-I hydrate framework, (3) CS-I hydrate with large cavities filled by xenon ($6\,Xe \cdot 46\,H_2O$), (4) CS-I hydrate with large and small cavities filled by xenon ($8\,Xe \cdot 46\,H_2O$).[112]

least stable as compared with all structures considered. The temperature at which the carcass loses the stability is equal to $-13°C$ at low pressures. The inclusion of guest molecules increases the temperature up to $3°C$. The occupation of all cavities by xenon molecules makes the hydrate stable up to $12°C$. Thus, the calculations show that the carcass of CS-I hydrate is thermodynamically unstable at temperatures about $0°C$ and the reestablishment of stability requires at least the occupation of large cavities. In Fig. 22 the lines of thermodynamical stability limit dependence $T(P)$ are plotted for the empty lattice, $6\,Xe \cdot 46\,H_2O$, and $8\,Xe \cdot 46\,H_2O$. The curve $T(P)$ for hydrates with partly or completely occupied cavities also exhibits the maximum observed in the experiment (curves 6–8). As shown, the two mechanisms lead to clathrate carcass destruction, namely, dynamic instability at low temperatures and high pressures and thermodynamic instability at high temperatures. One might expect that at intermediate temperatures the transition from the one mechanism to another takes place.

Molecular and lattice dynamic methods show the specific instability of the empty CS-I hydrate lattice. If we compare the results of both methods, certain differences can be observed. When we calculate phonon state density and vibrational spectra, the lattice dynamics method agrees with the experiment in the region where anharmonic effects turn out to be notable. This refers in part to those phonon spectra that include longitudinal optic modes. In Fig. 23 lattice dynamics and experimental results

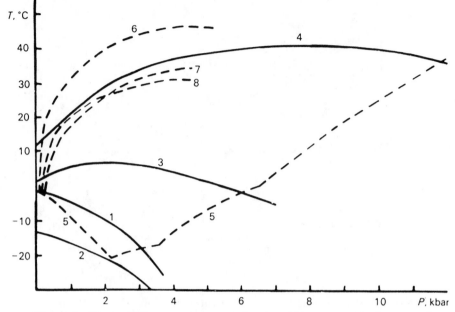

Figure 22. The instability temperature as a function of pressure: theory (curves 1–4); experiment (curves 5–8). (1, 5) Ice; (2) empty CS-I hydrate carcass; (3) CS-I hydrate with large cavities filled by xenon (6 Xe · 46 H$_2$O); (4) CS-I hydrate with large and small cavities filled by xenon (8 Xe · 46 H$_2$O); (6) hydrates CH$_4$; (7) N$_2$; (8) Ar.[112]

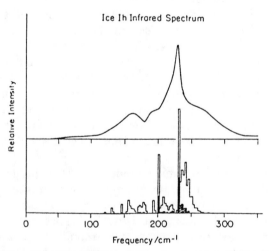

Figure 23. Comparison between the calculated and observed infrared spectrum in the librational mode region. Polarization (induction) effects make a negligible contribution.[108]

obtained in the region mentioned above are shown.[108] As noted, the lattice dynamics method here gives inaccurate results.

VIII. THERMODYNAMIC PROPERTIES OF CONCRETE CLATHRATES

In this section our main focus is on the ability of the theory of clathrate formation to describe the available set of experimental data.

A. Hydroquinone

Prior to comparing theory and experiment, some important remarks must be made concerning the possible amount of guest molecules in β-hydroquinone. For convenience, consider β-hydroquinone–CH_3OH clathrate. Various authors prepared samples of this compound using a wide range concentration c limits. For example, $c = 0.728, 0.897, 0.974,$[19] $c = 0.552,$ $0.664,$ $0.750,$ $0.989,$[114] and recently the samples were prepared with $c \lesssim 0.4.$[115] However, in accordance with the theory presented, the clathrate appears to be a thermodynamically stable compound at a certain temperature and pressure only, and at one concentration point, $c = c_k$. Under real conditions the fluctuation of composition $c = c_k \pm \Delta c$ certainly occurs due to various defects and imperfections of crystal, but $\Delta c \ll c_k$. What is the reason for the discrepancy between the values of hydroquinone concentrations? The analysis of available experimental data of β-hydroquinone–methanol system growth suggests an explanation which is based on the statement that the thermodynamically stable compound occurs at $c_2 \approx 0.99$ and the metastable one at $c_1 \approx 0.47$. All remaining compounds with $c_1 < c < c_2$ are thought to be thermodynamically metastable. In Table IX experimentally obtained data[68] of hydroquinone clathrate compositions are presented. The weighted-average value of the amount of hydroquinone in all the systems corresponds to $c = 0.9944 \pm 0.0006$.

In this subsection it is more convenient to use the common denotation $Q = C_6H_4(OH)_2$ for the consideration of equilibrium conditions of hydroquinone clathrate. In Fig. 24 a solubility isotherm is plotted for the Q_β–CH_3OH–HCl system. It is suggested from the behavior of isotherms that the composition $CH_3OH \cdot 6.33Q^5$ represents the phase which manifests itself rather well in the metastable region of the Q_β–CH_3OH–HCl system. From the data of this isotherm it cannot be concluded that this compound provides a stable section of branch, since at high concentrations of HCl (where the exit from the branch to the stable region could be expected) errors in determining the equilibrium solution and the location of "remain" points increase owing to the enhanced volatility of HCl and, presumably, to the reaction of chloric hydrogen with methanol at this range

TABLE IX

Composition^a of Hydroquinone Clathrate with Methanol in the Systems Hydroquinone (Q)-Methanol-S^b at $20.0 \pm 0.01°C^{68}$

Number of Schreinemakers Rays	General Solution (Weighted Average)	Solution Binary Compound Program (Weighted Average)
	$Q-CH_3OH-HCl$	
26	90.91 ± 0.27 9.09 ± 0.27 −0.006 ± 0.02	91.18 ± 0.12
	$Q-CH_3OH-TBAB$	
30	91.18 ± 0.25 8.83 ± 0.25 −0.01 ± 0.04	91.19 ± 0.09
	$Q-CH_3OH-TEAB$	
17	91.63 ± 0.24 8.45 ± 0.24 −0.08 ± 0.05	91.21 ± 0.10
	$Q-CH_3OH-LiCl$	
30	91.09 ± 0.29 8.91 ± 0.29 0.003 ± 0.02	91.26 ± 0.10
	$Q-CH_3OH-LiBr$	
14	91.15 ± 0.30 8.87 ± 0.30 −0.015 ± 0.05	91.23 ± 0.10
	$Q-CH_3OH-HCl$	
11^c	95.34 ± 0.64 4.66 ± 0.64 0.00 ± 0.04	95.62 ± 0.26
	$Q-CH_3OH-S^d$	
18^c	95.38 ± 0.40 4.61 ± 0.40 0.01 ± 0.03	95.57 ± 0.18

of concentrations.[68] The phase considered was obtained with considerable difficulty in the remaining systems. Three rays were realized in LiBr and four in LiCl systems. These rays emerge from the region of equilibrium solutions with high concentration of the third component. The branch of phase solubility is located in the vicinity of the stable branch, but inside the region of clathrate $CH_3OH \cdot 3Q_\beta$ crystallization. This points out that the compound of composition $CH_3OH \cdot 6.28Q_\beta$ has not, presumably, a stable field of crystallization. The transition of phase to stable state was observed only in single cases. An explanation can be found in the small range of hydroquinone modification energies. The activation barriers of transition from one form to another are rather high. For example, the activation energy of clathrate $CH_3CN \cdot 3Q_\beta$ destruction amounts to 145.6 ± 16 kJ/mol of "guest," whereas the heat of the clathrate decomposition is equal to 31.2 kJ/mol of "guest."

Thus β-hydroquinone–CH_3OH clathrate turns out to be unstable at $0.47 < c < 0.99$. However, the time of the clathrate decomposition is of the order of a year. It should be noted that the time of sample growth must be indicated and close control over composition must be maintained during the experimental investigation of hydroquinone with nonequilibrium composition.

1. Thermodynamics at zero pressure

Since the cavities in Q_β are arranged in quasi-one-dimensional fashion, the mean-field theory (Sections III.A and III.B) can distort qualitative characteristics of the compound when guest–guest interaction potentials appear to be comparable with $\Delta\mu$ (Eq. 39). Therefore, the analysis of $CH_3OH \cdot 3Q_\beta$ properties[23] was carried out by the cluster method (Section III.D), which takes into account the geometry of cavity arrangement.

The intersection of the experimentally found line $D = -0.4 \div -0.5$[5,15] and the curve $f(c)$ in mean-field approximation provides the only point of stable clathrate concentration $c_k < 0.34$. Cluster approximation gives different results, which can be seen from Fig. 25, where $f(c)$ dependence is crossed by the $D = -0.416$ line. The occurrence of three intersection points at c_1, c_2, and c_3 is observed, but the middle value c_3 can be

---

[a]The amount of components in the compound is given in weight % with mean square error.

[b]S = LiCl, LiBr, tetra-n-buthilammonium bromide (TBAB), tetraethilammonium bromide (TEAB) and HCl.

[c]Rays from metastable region.

[d]Here the rays obtained in systems with HCl, LiCL, and LiBr are combined.

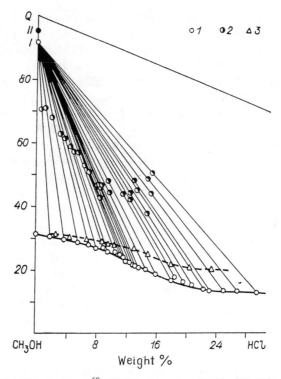

Figure 24. Solubility isotherms[68] (20°C) in the system Q–CH₃OH–HCl. (I) Solid phase of composition $(0.997 \pm 0.013)CH_3OH \cdot 3Q$; (II) Solid phase of composition $CH_3OH \cdot (6.35 \pm 0.43)Q$. The figurative points correspond to (1) saturated solution, (2) "remainder," (3) saturated solution of metastable phase of composition II.

neglected, as discussed in Section III. Thus, the clathrate at $D = -0.416$ is found to exist at two guest concentration values, $c_1 = 0.49$ and $c_2 = 0.99$.[23] A comparison of the theory[23] and the experimental results[68] gives satisfactory agreement, since in experiment hydroquinone–CH₃OH clathrate was found to be thermodynamically stable at $c_2 = 0.99$ and metastable at $c_1 = 0.47$.

Consider now the phase transitions related to the ordering of guest molecule dipole moments p_d. In Table X the experimentally found values of transition temperatures T_c are presented. As noted, T_c is not simply proportional to p_d^2, but has a more complicated dependence on dipole moment. Data matching by mean-field approximation, where $T_c = cJ$, is also rather unsatisfactory. However, one can try to match the T_c data by means of the simple one-dimensional Ising model in the field $J'\langle\sigma\rangle$, where $\langle\sigma\rangle$ indicates the averaged value of the dipole moment. Then the equation

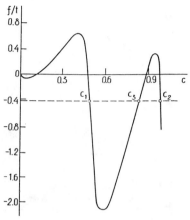

Figure 25. The dependence of f/t vs. c for β-hydroquinone with CH_3OH at room temperature. The point c_3 is thermodynamically unstable.

which determines T_c at $c \rightarrow 1$ takes the form of Eq. (54). At $J'/T \ll 1$, we obtain from Eq. (54) the first approximation with respect to J'/J:

$$T_c \approx J/|\ln(J'/J)| \tag{94}$$

where J and J' are the nearest intra- and interchain guest interaction constants, respectively.

Let us estimate qualitatively the data for β-hydroquinone with SO_2

TABLE X
Values of Orientational Phase-Transition Temperatures T_c in Hydroquinone Clathrates[a]

Guest Molecule	c	p_d (Debye)	T_c (K)
CH_3OH	0.974	1.69	66[19]
CH_3OH	0.99	1.69	68[24,88]
HCN	~1	3.0	178[18]
SO_2	0.9	1.7	44[24,88]
CH_3F	0.868	1.8	110[114]
H_2S	0.95	1.17	7.7[98]
D_2S	0.96	0.9–1.2	7.6[97]

[a]p_d and c denote dipole moments and concentratiosn of guest molecules.

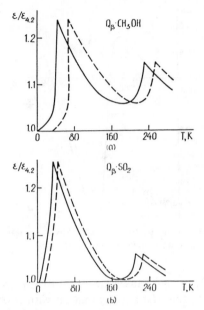

Figure 26. Experimental temperature dependence of dielectric constant $\epsilon(T)/\epsilon(T = 4.2 \text{ K})$ in hydroquinone clathrates. The solid lines were obtained with decreasing temperature, the dashed lines with increasing temperature.

assuming that T_c approaches 50 K at $c \to 1$, since $T_c \approx 44$ K at $c \approx 0.9$.[24] Admit that J in Eq. (94) consists of two terms

$$J = J_{\text{dis}} + J_{\text{dip}} \tag{95}$$

where J_{dis} is the dispersional interaction of SO_2 molecules, which can be calculated in the framework of the Lennard–Jones approximation (Eq. 92). As shown in Table 7, $J_{\text{dis}} \approx 200$ K. Constant J_{dip} in Eq. (95) describing dipole–dipole interactions can be estimated from the formula $J_{\text{dip}} \approx 2p_D/4\pi\epsilon_\infty r^3$.[98] Taking $p_D = 1.7$ Debye, $J_{\text{dip}} \approx 100$ K for SO_2 molecules. Replacing $T_c \approx 50$ K, $J \approx 300$ K in Eq. (94) we obtain $J'/J \approx 0.007$ or $J' \approx 2$ K. Thus, considerable intrachain and rather weak interchain interactions are found in the system. It should be noted that in Q_β with CH_3OH and SO_2, the additional phase transition was observed at $T_c' \approx 230$ K from dielectric constant $\epsilon(T)$ measurements[24,88] (see Fig. 26). The authors relate this transition to reconstruction of the host lattice in the vicinity of $T_c' \approx 230$ K. However, we must also consider the $\epsilon(T)$ anomaly occurring because of short-order changes in the host lattice. Measurements of $\epsilon(T)$ up to $4.2 \, K^{88}$ in $SO_2 \cdot 3Q_\alpha$ have not shown the transition. This can be related to a transition temperature even lower than 4.2 K.

2. Phase P, T, c Diagrams

Concrete calculations of (P, T, c) diagrams have been carried out only for simple guest molecules (e.g., Ar, Kr, N_2, O_2, and CH_4) with $p_d = 0$. The dispersional interactions were accounted for by perturbation theory owing, to small values of J_{dis} in the case of simple guest molecules. Consider the equilibrium in which the coexistence of liquid L phase (guest solution in hydroquinone melt), gas G phase and phases, of α- and β-hydroquinone with soluted guest component are observed. The equations which describe phase equilibrium with the chosen set of phases can be obtained using Section IV formulas and substituting μ_Q^γ for the expression of liquid L-phase chemical potential μ_Q^L and μ_G^γ for the expression of chemical potential μ_G^L of guest in the melt. The expression of solution chemical potentials can be used for μ_Q^L, μ_G^L because of the solubility of noble gas in the hydroquinone melt (see Ref. 72).

Nonvariant Equilibrium. The equations of Section IV can be considerably simplified to take an advantage of the properties of hydroquinone–noble gas system components. Then the following assumptions can be made:

1. The partial pressure of host–component vapor can be neglected $(2.13 \cdot 10^{-3 \, \text{atm}}$ at 446 K).
2. The solubility of the guest in the hydroquinone melt turns out to be negligible.
3. Guest–host interactions of α and β carcasses differ insignificantly. This assumption is based on the structural data. The values of guest–host interactions in the first coordination sphere are almost equal owing to the similarity in size of α- and β-carcass cavities. Guest molecule interaction with the next-nearest hydroquinone molecule is considerably weaker and hardly differs for α and β carcasses because of the similarity of their densities (1.35 and 1.26 g/cm^3, respectively).

Then the system of equations of Section IV takes the form

$$c_0 = 1 - \exp[\Delta\mu_0^{\alpha\beta}/(\nu_\beta - \nu_\alpha)T_0]$$

$$\Delta\mu_0^{L\alpha}/\nu_\alpha = \Delta\mu_0^{\alpha\beta}/(\nu_\beta - \nu_\alpha) \tag{95}$$

$$P_0 = \frac{T_0 c_0}{2\pi a^3 g(1 - c_0)} \exp(W/T_0)$$

where g, a, and W are determined in Eq. (89). To indicate the nonvariant point, the subscript "0" is introduced for P, T, c, $\Delta\mu^{ij}$. To solve Eq. (95) with respect to P_0, T_0, c_0, the values of $\Delta\mu_0^{\alpha\beta}$, $\Delta\mu_0^{L\alpha}$, g, a, and W must be known. The values g, a, W are chosen and calculated as in Ref. 5. The difference $\Delta\mu_0^{\alpha\beta}$ in Eq. (95) is presented as an expansion over $T_0 - T_1$, $P_0 - P_1$,

$$\Delta\mu_0^{\alpha\beta} = \Delta\mu^{\alpha\beta}(T_1, P_1) - \Delta Q^{\alpha\beta}(T_0 - T_1)/T_1 + \Delta v^{\alpha\beta}(P_0 - P_1) \quad (96)$$

where $\Delta Q^{\alpha\beta}$ and $\Delta v^{\alpha\beta}$ are heat and change of volume and T_1 and P_1 are the temperature and pressure at which the concentration is measured. Using Eq. (95) at a given c we determine $\Delta\mu^{\alpha\beta}(T_1, P_1)$ to be equal to -0.35 kJ/mol (the choice of $\Delta Q^{\alpha\beta}$ and $\Delta v^{\alpha\beta}$ values is given below). In the vicinity of the melting point ($T_m = 445.46$ K, $P_m = 0.021$ atm), $\Delta\mu^{L\alpha}$ can be expressed as a linear term of expansion over $T - T_m$, $P - P_m$

$$\Delta\mu^{L\alpha}(T, P) = -\Delta Q_m^{L\alpha}(T - T_m)/T_m + \Delta v_m^{L\alpha}(P - P_m) \quad (97)$$

where $\Delta Q^{L\alpha}$ is the heat of α-hydroquinone melting which is equal to 26.96 kJ/mol; $\Delta v^{L\alpha}$ is the change of volume at the transition from α to L phase, which amounts to 15 cm^3/mol (calculated from densities of α-hydroquinone and it's melt equal to $\rho = 1.1$ g/cm^3 at 447 ± 1 K).[65] Equation (97) together with Eq. (95) give T_0:

$$T_0 = T_m\{1 + [\nu_\alpha\Delta\mu_0^{\alpha\beta}/(\nu_\alpha - \nu_\beta) + \Delta v_m^{L\alpha}(P_0 - P_m)]/\Delta Q_m^{L\alpha}\} \quad (98)$$

It follows from this equation that $T_m < T_0$ in the system considered. Substituting expressions for c_0 (Eq. 95) and T_0 (Eq. 98) into the last of Eq. 95, we obtain nonlinear equation with respect to P_0. Since for the pressure P_0, which meets the inequality

$$P_0 - P_m \ll \Delta Q_m^{L\alpha}/\Delta v_m^{L\alpha} = 1.8 \times 10^{-4} \text{ atm} \quad (99)$$

the third terms in Eq. 98 is smaller than the first, the expression of T_0 (Eq. 98) at $P_0 = P_m$ is taken as a first approximation and P_0 is calculated from Eq. (95).

TABLE XI
Calculated and Experimental Data[5,65,115] at the Nonvariant Point and at $T = 298$ K

Guest (i)	ϵ_L (K)	σ_L (Å)	Calculation (Experiment)				
			T_0 (K)	p (atm)	c_0	p_{298} (atm)	c_{298}
He	10.2	2.56					
Ne	36.2	2.74					
Ar	119.5	3.408	448.0	51.0	0.304	4.57	0.398
						(3.4)	(0.34)
Kr	166.7	3.679	447.1	15.2	0.292	0.59	0.396
			(446.3)	(13.8)	(0.28)	(0.4)	
Xe	255.3	4.069	446.8	5.57	0.289	0.08	0.393
			(446.5)	(5.8)	(0.30)	(0.06)	
N_2	95.05	3.698	448.9	85.2	0.316	6.7	0.401
						(5.8)	
O_2	117.5	3.58	447.9	47.1	0.303	3.3	0.399
CH_4	142.7	3.810	447.3	25.0	0.296	1.02	0.398
HCl	360.0	3.305	446.7	1.6	0.287	0.026	0.397
						(0.01)	

In Table XI the calculated and experimental (in brackets) values of P_0, T_0, and c_0 are presented for various hydroquinone systems. The values of intermolecular interaction parameters ϵ_L and σ_L are also shown.

Monovariant Equilibrium. The equations which describe monovariant equilibrium can be found in Section IV. The assumptions presented above are also taken into account.

1. The equilibrium $\alpha \Leftrightarrow L + G$ is described by the equation

$$c = 1 - \exp[\Delta\mu^{L\alpha}(T, P)/\nu_\alpha T]$$
$$P = \frac{T}{2\pi a^3 g} \frac{c}{1 - c} \exp(W/T)$$

(100)

2. The equilibrium $\beta \Leftrightarrow L + G$ is described by Eq. (100) with index α substituted by β.

3. The equilibrium $\beta \Leftrightarrow \alpha + L$ is described by the Eq. (100) and

$$(\nu_\beta - \nu_\alpha)\Delta\mu^{L\alpha}(T, P) = \nu_\alpha\Delta\mu^{\alpha\beta}(T, P) \qquad (101)$$

4. The equilibrium $\beta \Leftrightarrow \alpha + G$ is described by the Eq. (100) and

$$c = 1 - \exp[\Delta\mu^{\alpha\beta}(T, P)/(\nu_\beta - \nu_\alpha)T] \qquad (102)$$

To plot the monovariant curves in the vicinity of the nonvariant point we take $\Delta\mu^{L\alpha}$ in the form of expansion over $\Delta T = T - T_0$ and $\Delta P = P - P_0$ with just linear terms left:

$$\Delta\mu^{L\alpha}(T, P) = \Delta\mu_0^{L\alpha} - \Delta Q_0^{L\alpha}\Delta T/T_0 + \Delta v_0^{L\alpha}\Delta P \qquad (103)$$

The analogous expressions are valid for $\Delta\mu^{L\beta}$ and $\Delta\mu^{\alpha\beta}$. From the condition of monovariant equilibrium follows the equality

$$\Delta\mu_0^{L\alpha}/\nu_\alpha = \Delta\mu_0^{L\beta}/\nu_\beta = \Delta\mu_0^{\alpha\beta}/(\nu_\beta - \nu_\alpha) \qquad (104)$$

which allows us to determine $\Delta\mu_0^{L\alpha}$ and $\Delta\mu_0^{L\beta}$. Let us ignore the dependencies of $\Delta Q^{L\alpha}$, $\Delta Q^{L\beta}$, $\Delta Q^{\alpha\beta}$, and $\Delta v^{L\alpha}$, $\Delta v^{L\beta}$, $\Delta v^{\alpha\beta}$ on T and P in the region we are investigating. Then $\Delta v_0^{L\alpha} = \Delta v_m^{L\alpha} = 15\, cm^3/mol$; $\Delta v_0^{\alpha\beta} = -5\, cm^3/mol$; $\Delta v_0^{L\beta} = \Delta v_0^{\alpha\beta} + \Delta v_0^{L\alpha} = 10\, cm^3/mol$. It follows from Refs. 116–118, that the value of $\Delta Q^{\alpha\beta}$ is rather small, for example, it is equal to $-0.54\, kJ/mol$,[117] -0.71, -0.75, $-1.08\, kJ/mol$,[118] and -0.67, $-0.21\, kJ/mol$ at $T = 298\, K$.[116] If we assume that $\Delta Q^{\alpha\beta}$ is of the order of value noted and it does not depend on temperature, the calculation of $\beta \Leftrightarrow \alpha + G$ equilibrium leads to notable disagreement with experiment at high temperatures. The reverse calculation can be performed using the data[65] of the phase diagram of this equilibrium. The value $\Delta\mu^{\alpha\beta}(T_1, P_1)$ calculated at 446 K by Eq. (102) is equal to $-0.35\, kJ/mol$ for hydroquinone with Kr or Xe. It agrees with the results of direct measurements at 298 K.[119] Therefore, it is assumed that $\Delta Q^{\alpha\beta} \approx 0$ for the calculations in the whole temperature interval and hence $\Delta Q^{L\alpha} \approx \Delta Q^{L\beta}$.

Substituting the expressions of $\Delta\mu^{\alpha\beta}$ and $\Delta\mu^{L\beta}$ expansions into the equation which describes monovariant equilibria, the P, T diagram of a

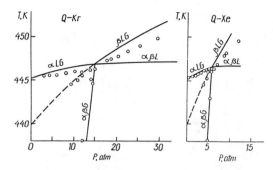

Figure 27. Phase P,T-diagram of hydroquinone–noble gas system. Calculated stable (solid line) and metastable (dashed line) equilibriums and experimental data.[65]

binary system can be numerically plotted. In Fig. 27 the calculated P, T diagrams are presented along with experimental data given for comparison.

Divariant Equilibrium. These equilibria, with the assumptions proposed above taken into account are described by the Eq. (100) (for α and L phases; β and G phases; α and G phases), Eq. (101) (for α and β phases), and $x_G = 1$ (for L and G phases). In Fig. 28 the P, x and T, x sections of the phase diagram are plotted for the hydroquinone–xenon system.

In Table XII the values $\Delta\mu^{\alpha\beta}$ calculated from the data of clathrate composition are presented for monovariant equilibrium. The following assumptions are used for the calculation: (1) α-hydroquinone does not dissolve the guest component and guest–guest interaction in the β phase is negligible. The value of $\Delta\mu_1^{\alpha\beta}$ is determined from Eq. (27). (2) α-hydroquinone dissolves the guest component, but guest–guest interaction is neglected. Then $\Delta\mu_2^{\alpha\beta} = (\nu_\beta - \nu_\alpha)T\ln(1 - c)$, if the condition $c_\alpha = c_\beta = c$ is met. (3) The correction on guest–guest interaction V is introduced for $\Delta\mu_2^{\alpha\beta}$. At the assumption $V_\alpha \approx V_\beta = V$, $\Delta\mu_3^{\alpha\beta} = (\nu_\beta - \nu_\alpha)T\ln(1 - c) - c^2V/2$. For all $\Delta\mu_i^{\alpha\beta}$ the defined dependence of $\Delta\mu^{\alpha\beta}$ vs. T is not observed in the temperature interval of 280–446 K within the limits of experiment accuracy. This fact clearly demonstrates that the value of $\Delta Q^{\alpha\beta}$ does not amount to more than 40–60 J/mol in absolute values. It seems evident that $\Delta\mu_1^{\alpha\beta}$ and $\Delta\mu_2^{\alpha\beta}$ depend on the sort and properties of the guest molecule. The greater the tendency of the guest to van der Waals interaction, the higher is absolute value of $\Delta\mu^{\alpha\beta}$. This suggests the idea that guest–guest interaction makes a notable contribution to the stoichiometry of the clathrate, and therefore it cannot be neglected. However, the dependence of $\Delta\mu_3^{\alpha\beta}$ on guest properties is not found within the limits of experimental error.

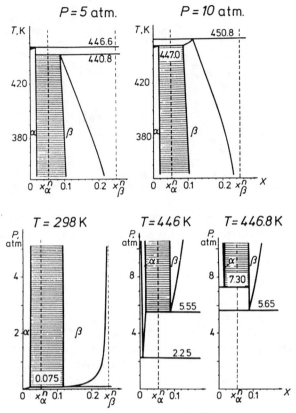

Figure 28. Isobaric and isothermic sections of P, T-diagram of hydroquinone–xenon system.[65]

In the system with Ar (298.2 K) and Kr (298.2–353.2 K), metastable equilibrium is achieved with α-hydroquinone which does not dissolve guests. Calculated at this assumption, values of $\Delta\mu_3^{\alpha\beta}$ (see Table XII, values in brackets) form the block of data which is characterized by the average value -323.6 ± 9.6 J/mol.* The experimental results suggest the possibility of such metastable equilibrium realization. Thus it is natural to assume that the velocity of clathrate formation in the system considered increases with the increase of van der Waals interaction between guest

---

*Thus there is no need to introduce the single value of $\Delta\mu^{\alpha\beta}$ to every concrete system, explaining the variation of $\Delta\mu^{\alpha\beta}$ by β-carcass distortions, as is commonly the case. The assumption of β-carcass distortion in the systems discussed seems artificial, since the sizes of the considered set of guest molecules do not exceed the sizes of β-carcass and hydroquinone cavities.

TABLE XII

Calculation[65] of $\Delta\mu^{\alpha\beta}$ for Hydroquinone from the Data of Cavities Occupancy Degree[a]

Guest	T (K)	P (atm)	c	$\Delta\mu_1^{\alpha\beta}$ (J/mol)	$\Delta\mu_2^{\alpha\beta}$ (J/mol)	$\Delta\mu_3^{\alpha\beta}$ (J/mol)
Ar	298.2	3.4	0.340	-343		(-334)
Kr	446.3	13.8	0.280	-407	-339	-329
	353.2	2.30	0.280	-322		(-311)
	323.0	1.19	0.310	-337		(-324)
	303.2	0.55	0.370	-337		(-321)
	298.2	0.40	0.340	-343		(-327)
Xe	446.5	5.80	0.300	-441	-368	-341
	353.2	0.60	0.370	-452	-377	-336
	328.2	0.26	0.390	-450	-375	-329
CH$_4$	289.3	0.83	0.391	-410	-342	-320
	289.6	0.47	0.399	-400	-334	-311
CH$_3$F	323.0	0.33	0.383	-432	-360	
	298.2	0.12	*0.416	-445	-371	

[a]The correction to $\Delta\mu^{\alpha\beta}$ related to the variation of pressure in the considered interval is not taken into account because of its negligible value. $\Delta\mu_1^{\alpha\beta}$—without regard for solubility of guest component in α modification and without regard for guest–guest interactions; $\Delta\mu_2^{\alpha\beta}$—with this solubility taken into account; $\Delta\mu_3^{\alpha\beta}$—with the solubility and guest–guest interaction taken into account.

and carcass. This assumption is supported by the data of Ref. 116, where it is shown that CH$_3$F reacts with hydroquinone immediately as the required pressure is reached, but the inductive period with methane is approximately 36–48 hours. On the other hand, it is noted in several papers (e.g., see Refs. 5 and 120) that clathrate formation in hydroquinone–guest systems meets with some difficulty at temperatures below 373 K. To avoid the difficulties related with α-hydroquinone destruction inert solvent, vibropomol, and so on (see Refs. 5, 116, and 120) are used. But even at these specific conditions the pressures (concentrations) notably exceeding equilibrium values (e.g., four times for xenon[5]) are needed to initiate the reactions. Such oversaturations cannot be set up for the α phase, because they induce the conditions at which the phase becomes unstable and the β phase is achieved. Therefore, the experiments, which are carried out at high temperatures ($T > 373$ K) with the most "reaction capacitive" heavy guests, seem rather reliable from the viewpoint of real equilibrium establishment.

Since guest–guest interactions are found to be weak in the systems

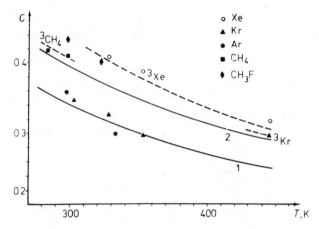

Figure 29. Calculated $c(T)$ curves of the monovariant equilibrium $\beta \Leftrightarrow \alpha + G$. (1) In accordance with Ref. 5. (2) From Eq. (102). (3) From Eq. (102) with the correction on guest–guest interaction. The experimental data are represented by dots.

considered, the correction related to these interactions can be accounted for by the perturbation theory. In the first approximation it follows for the changes at nonvariant point T_0, P_0, c_0 that

$$\Delta c/c_0 = (c_0 - 1)c_0 V/2T$$

$$\Delta P/P_0 = c_0 V/2T$$

(105)

and $\Delta T \sim V_\alpha - V_\beta \approx 0$ (for Xe $V_\alpha \neq V_\beta$ and $\Delta T/T_0 \approx 4.5 \cdot 10^{-5}$). Corrections made by Eq. (105) for Q–Xe system are equal to $\Delta c/c_0 \approx 0.07$, $\Delta P/P_0 \approx 0.1$

The corrections for mono- and divariant equilibria are calculated in an analogous way. In Fig. 29 calculated $c(T)$ curves for monovariant equilibrium $\beta \Leftrightarrow \alpha + G$ are shown together with the experimental data. The experimental values for the clathrates with Xe and Kr at $T = 446.3$ K are satisfactorily described by curve 3. The system of equations does not provide solutions for He and Ne. It looks as if there are no pressures which make the β phase more stable than the α phase. Certainly, the latter conclusion is valid within the limits of the advanced assumptions.

We do not want to make far-reaching conclusions, since the number of assumptions proposed to simplify the solution of the equation system reduces the accuracy of the final results. However, even within the framework of ideal approximation, when guest–guest interaction is neglected, the consideration of guest solubility in the α phase leads to satisfactory qualitative description. The consideration of guest–guest interaction just enhances the quality of description.

B. Hydrates

The thermodynamic properties of hydrates are considerably less investigated than those of hydroquinone. This is because of the more complicated structure of hydrates and certain ambiguity in determining guest occupancy degrees of small c_a and large c_b cavities. Presumably, it is difficult thus far to point out the method of independent determination of the values c_a and c_b. However, estimations of thermodynamic properties of hydrates are given in Ref. 75.

Various phase equilibria of hydrates can be investigated by the relations presented in Section III.C. For example, for three phase equilibrium, gas (G)–hydrate (Q)–water (H_2O) one obtains

$$\mu_Q^a = \mu_Q^b = \mu_G$$
$$\mu_{H_2O} = \mu'_{H_2O}$$
(106)

where μ'_{H_2O} indicates the chemical potential of water in various phases. The system of equations which describes absorption isotherms can be obtained from the first equality (Eq. 106).

Consider the scheme[75] of Langmuir constants and $\Delta\mu_{H_2O} = \mu_{H_2O} - \mu'_{H_2O}$ determination for hydrate structure I. The composition data and degree of cavity occupancy are known for the structure in addition to equilibrium conditions of hydrate formation. In Ref. 121 the composition of xenon hydrate (hydrate number n in formula $Xe \cdot n \cdot H_2O$) is determined as a function of pressure P, that is, at $P > P_0$, where $P_0 \approx 0.148$ MPa is the pressure corresponding to three-phase equilibrium of gas–gas hydrate–water at $T = 273.15$ K. The determination is carried out by the precise weight method at $T \approx 273$ K for the two-phase gas–gas hydrate equilibrium conditions. With the increase of pressure from 0.18 to 0.3 MPa, hydrate number n decreases from 6.39 to 6.13. Ripmeester and Davidson[122] found by the nuclear magnetic resonance method that the small cavities of xenon hydrate are occupied less than the large cavities and $c_b = (1.30 \pm 0.03)c_a$ for the pressure corresponding to the three-phase equilibrium at $T \approx 275$ K. If we assume that the equality $c_b \approx 1.3c_a$ is valid also at $T \approx 273$ K, the joint analysis of the results[121,122] appears possible.

In the framework of an "ideal" hydrate model, that is, when $V_{ab} = 0$ in Eqs. (35)–(39), the values $C_L^a = 17.9 \pm 1$ MPa^{-1}, $C_L^b = 99 \pm 10$ MPa^{-1}, and $\Delta\mu_{H_2O} = 940 \pm 30$ J/mol are obtained in Ref. 75 by the least-squares method, taking into account the error of experimental results.[121,122] In the framework of a "regular" hydrate, the values $C_L^a = 10 \pm 0.2 \, \text{MPa}^{-1}$, $C_L^b = 32 \pm 3 \, \text{MPa}^{-1}$, and $\Delta\mu_{H_2O} = 810 \pm 25 \, \text{J/mol}$ are analogously obtained at $V_{aa} \approx 0$, $V_{bb} \approx -50 \, \text{K}$, $V'_{ab} \approx -14 \, \text{K}$, and $V''_{ab} \approx -25 \, \text{K}$.

As is seen, the "experimental" values of Langmuir constants C_L^a and C_L^b and the value of $\Delta\mu_{H_2O}$ considerably differ in the ideal and regular clathrate models. This means that the thermodynamic models taking into account guest–guest interaction are most beneficial for the description of gas–hydrate phase equilibria.

The differences between hydrate and hydroquinone cavity structure require the careful choice of parameters which determine the potential of interaction of guest with cavity-forming host molecules. In Ref. 96 the dissociation pressure of gas hydrates was evaluated using Lennard–Jones and Kihara potentials in a Lennard–Jones–Devonshire cell model. In Table XIII calculated and experimental values of dissociation pressure and degree of cavity occupancy are presented together with the values of the chosen potential parameters.[5,96] As is shown,[96] the Lennard–Jones potential (Eq. 90) works satisfactorily for the monatomic gases and methane, but poorly for the rod-like molecules C_2H_6, CO_2, N_2, O_2, and C_2H_4. Kihara potential predicts better agreement of dissociation pressure for the hydrates of the rod-like molecules; unlike Lennard–Jones potential, it takes into account the size and shape of interacting molecules. The authors of Ref. 96 suggest that consideration of hydrate lattice distortions and a barrier to internal rotation of the molecule in its cavity would improve the results given both by Lennard–Jones and Kihara potentials. It should be noted that dissociation pressure P_{L-J} calculated with Lennard–Jones potential agrees with the experimental value only for argon.

In Refs. 123–125 the authors consider the theoretic aspects of the stabilizing influence of gas encaged in small cavities and perform the calculation of guest–host interaction energies for argon, xenon, and methane.

The agreement of experimental and calculated equilibrium curves of gas–hydrate–ice for the hydrates with two types of guest molecules is observed for an H_2O–CH_4–C_3H_8 system at $-3°C$.[126] A qualitative consideration of phase diagrams of the systems H_2O–H_2S–CCl_4, H_2O–H_2S–C_3H_6, H_2O–CH_3–CH_3CHF_2, and H_2O–H_2S–$CHCl_3$ is also performed.

C. Urea

Here we return to the practical conclusions of the model proposed in Section III.E.

To obtain the thermodynamic functions of urea clathrate, the temperature dependences of parameter a and order parameter $\sigma = \langle\sigma\rangle$ (Eq. 53)

TABLE XIII

Dissociation Pressure and Occupancy Degree of Gas Hydrates at $T = 273$ K Calculated[5,96] Using Lennard–Jones and Kihara Potentials

Guest	P_{exp} (atm)	Lennard–Jones Potential (Eq. 90)			Kihara Potential[a]			Occupancy Degrees	
		ϵ_L (K)	$\sigma_0 = 2^{1/6} \times \sigma_L$ (Å)	P_{L-J} (atm)	ϵ_K (K)	ρ_K (Å)	P_K (atm)	c_1	c_2
Ar	95.5	119.5	3.83	95.5				0.825[b]	0.841[b]
Kr	14.5	166.7	4.13	15.4				0.832	0.830
Xe	1.15	225.3	4.57	1.0				0.813	0.835
CH$_4$	28	142.7	4.28	19.0	3.81 3.35 3.39	178 205 204	13.0 19.0	0.818	0.836
N$_2$	140	95.1	4.15	90.0	124 132	3.47 3.40	115 115	0.810 0.843[c]	0.845 0.843[c]
O$_2$	100	118.0	3.88	63.0	153	3.14	120	0.821	0.838
CO$_2$	12.47	205 189	4.57 5.04	0.71 1.7	3.70 3.36 2.72	279 309 400	9.0	0.786 0.528	0.861 0.879
N$_2$O	10.0	205 189	4.67 5.15	0.6 1.52	3.4	209	8.2		
C$_2$H$_6$	5.20	243	4.44	1.1	2.59	609	8.4	0.837 0.860	0.827 0.818
C$_2$H$_4$	5.44	199	5.08	0.5	4.0	266	1.3	0.523 0.740	0.879 0.879
CF$_4$	1	152	5.28	1.6	3.22	291	0.6	0.282	0.894 0.852

[a] The argument of Kihara potential is taken to be the shortest distance between molecular cores, $W_K(\rho) = \epsilon_K[(\rho_K/\rho)^{12} - 2(\rho_K/\rho)^6]$, with ϵ_K the potential minimum and ρ_K its position.

[b] The values in this column are for Lennard–Jones potentials.

[c] The values in this column are for Kihara potentials.

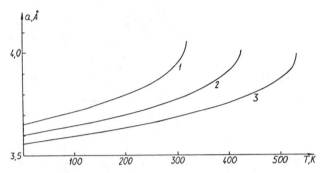

Figure 30. The dependence $a(T)$ for (1) $\Delta\mu = -12.31$ kJ/mol; (2) $\Delta\mu = -16.39$ kJ/mol; and (3) $\Delta\mu = -20.51$ kJ/mol.

must be found. The solution $\sigma \neq 0$ appears at $T < T_c$, where T_c is the phase transition temperature defined by Eq. (54). The solution of the equation (see discussion below), together with the calculations and experiments,[25,82] shows that the transition temperatures of urea with n-paraffins are of rather low value, $T_c \sim 100$ K. It allows us to separate the two temperature regions and find $a(T)$ and $\sigma(T)$ for high and low temperatures.

High Temperatures ($T > T_c$). The dependence $a(T)$ is calculated from Eq. (58) for several values of $\Delta\mu$ by the incompressible molecule approximation. The Lennard–Jones potential (Eq. 90) is chosen as a $\varphi(a)$ function. The value of parameter $\epsilon_L = 148.2$ K is taken to be equal to the value of CH_4–CH_4 interaction. The parameter σ_L is chosen to make $\sigma_L \cdot 2^{1/6}$ equal to 4.06 Å. This value corresponds to the equilibrium distance between hydrocarbon atoms in the crystal lattice of n-paraffins. The distance between urea molecules along a six-fold symmetry axis is $d = 1.834$ Å.

The dependences $a(T)$ at the three values of $\Delta\mu$ are shown in Fig. 30. For every value of $\Delta\mu$ it is found the value of temperature $T = T_0$ above which clathrate cannot exist with solid phase of host, because the equilibrium condition is not fulfilled at any a value. We assume the precise value of $\Delta\mu$ to be the one at which the maximum temperature of stable clathrate T is equal to the destruction temperature of urea clathrate with polyethylene ($T_0 = 148°C$, $\Delta\mu = -16.39$ kJ/mol). The equilibrium distance between neighboring molecules at room temperature appears to be equal to 3.76 Å, which is very close to the experimental value 3.74 Å.[25]

Consider the results of clathrate destruction temperature T_d. If clathrate is destroyed at the single value $a = a_d$, equal for all guests, the destruction temperature would also have to be independent of the guests.

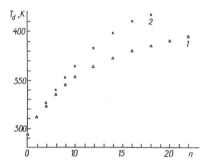

Figure 31. (1) Experimental and (2) theoretical dependences of destruction temperature on carbon atoms number.

It is possible to explain the increase of T_d with the increase of carbon atoms in the molecule if we assume that a_d depends on n (e.g., linearly $a_d = a_0 + \delta n$). The $T_d(n)$ dependence resulting from assumption and the experimental results are shown in Fig. 31. The values a and δ are chosen to give the precise value of destruction temperature at $n = 5.6$.

We assumed so far that guest molecules are incompressible. The decrease of intermolecular distance leads in reality to the compression of molecules. This fact must be reflected in thermodynamic functions of the system as changing the dependence $a(T)$. Several mechanisms can be noted as leading to the compression, for example, the decrease of valence bond length and valence angles and the twisting of molecules. The latter is presumably the main reason for compression, since the least characteristic energy ($\sim$8–12 kJ/mol) is related to the twisting. More precise calculation of $a(T)$ can be performed when one knows the dependence of the potential energy of the molecule vs. the angles of internal rotation and takes into account the interaction of the molecule with channel walls at the twisting. Then the temperature dependence of the expansion would be different for molecules with various n numbers of carbon atoms. This can change the dependence of destruction temperature on n and improve the agreement of calculated $T_d(n)$ results with experiment.

It follows from Eq. (49) that at high temperatures ($\sigma = 0$) $a(T)$ does not correspond to the value a_0, determined by the equation $\varphi(a_0) = 0$, where a_0 is the equilibrium interatomic distance of a one-dimensional chain. Besides, $a < a_0$ since $\Delta\mu < 0$. To stabilize the clathrate phase, guest molecules in clathrate must be in a more compressed state compared with the ideal extended state. Thus, the theory can explain not only dense packing, but also the surprising shortening of guest molecules found in experiments.

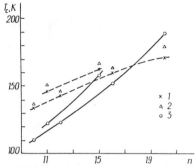

Figure 32. The values of transition temperatures in urea with n-paraffin clathrates as a function of paraffin number. (1) Calculations;[82] (2) by formulas;[54] (3) experiment.[82]

Low Temperature $(T < T_c)$. To solve Eq. (54), the dependence $a(T)$ can be neglected in the considered region of temperatures. The obtained values of T_c almost coincide with the calculations[82] and qualitatively agree with the experiment[82] (see Fig. 32).

It should be noted that the values J and J' calculated by the potential (Eq. 90) may be thought of as only approximate. In Ref. 82 two sets of parameters were chosen with J' differing four times. Thus, these values can be used only to estimate phase transition temperatures. To calculate T_c more precisely, a complete system of equations can be used, taking into account the relation between a and σ.

The guest molecule has properly not two but six orientations in a urea channel. Therefore, the Potts model[127] would be more appropriate for the description of phase transition in urea than the Ising model, especially since the Potts model was used for the description of orientational transitions in β-hydroquinone.[23]

At low temperatures interesting calculations[128] are also performed by the Monte Carlo method. Guest molecules are distributed in a way one calls "head-to-tail." It is shown that T_c notably varies as the transition from the ordering model "head-to-head" to the model "head-to-tail" occurs (see Fig. 33). This result indicates that further development of such

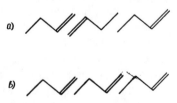

Figure 33. Two versions of alken-1 molecule arrangement in adduct channel. The ordering (a) head to head and (b) and head to tail.

approaches requires the improved methods for estimating the parameters J and J'.

The reliable theoretic models for thiourea are thus far in the formation stage. However, these compounds have been extensively studied using various experimental methods.[25]

CONCLUSIONS

Prior to discussing the unsolved problems and the possible ways of further developing clathrate theory, we want to present some conclusive remarks.

The first models of clathrate formation, proposed almost 30 years ago, may be cited as the typical representatives of ideal solution theories. In the models, guest molecule interaction was not taken into account and guest and host subsystems were assumed to be independent. These restriction allows a description of only a limited number of thermodynamic properties of clathrate compounds. New experimental results in hydroquinone, urea clathrates, and hydrates required that guest–guest interaction be considered. As a result, the clathrates with the unstable empty host lattice were thoroughly investigated. New elaborated concepts allowed us to understand the character of clathrate formation in wide classes of compounds and explain such phenomena as (1) the enhanced value of guest molecule concentration, (2) the existence of two guest molecule concentrations which characterize stable and metastable states of hydroquinone with methanol, (3) the decomposition of solid clathrate solutions, (4) the dense packing of guest molecules in urea, and (5) phase transitions in clathrates.

The existence of clathrates with a labile host carcass leads to the problem of determining the place of clathrate compounds in inclusion-type systems. We think that the host subsystem is always preserved as in labile so in traditional clathrates due to the greater value of host–host rather than guest–host interaction. The difference in bond types and characteristic values of interaction energy leads to a rigid distinction between guest and host subsystems. The separation of the subsystems allows us to consider the clathrates with guest–guest interaction and unstable host lattice as a typical inclusion-type (e.g., solid solution) system. Along with this one must also take into account specific structural and chemical features, without which the concept of clathrate system would make no sense.

We return now to some presently unsolved problems of clathrate formation:

1. *Calculations of intermolecular interactions.* The Lennard–Jones formalism is mostly used for the estimation of interaction potentials in clath-

rates. Unfortunately, the potential does not take into account the structure of complicated molecules. Therefore, the consideration of electrostatic, high-multipolarity interactions (quadrupole, octupole) is required for the calculation of guest–guest interactions, as was the case for CH_4,[129,130] N_3B,[131,132] and others.[133] On the other hand, the pseudopotential method[134] is at present rather reliably developed. The calculations of various systems (e.g., see Refs. 135 and 137) were carried out by this method with the short number of fit parameters determined from independent experiments. We think that the pseudopotential method will find application in the solution of clathrate problems.

The estimations of host–host interactions by the simple point charge (SPC) potential were performed only for hydrates. It would be interesting to see the calculations in these systems by lattice and molecular dynamics methods at variations of SPC potential, since the stability of results depends on the chosen potential. Maybe the calculation could be carried out with various forms of SPC potentials.[109]

The various model potentials[138–140] used in modern theories presumably ought to be more decidedly applied to the calculation of clathrate formation problems. Certainly the criterium of application would be the comparison of theoretic results with experiments.

2. *Composition of clathrates.* This problem appeared because of the lack of experimental information. No investigations have so far been performed to determine the time at which the notable changes in the host lattice of an unstable clathrate occur, for example, we do not even now the time needed for the reconstruction of a hydroquinone–methanol structure. The experimental investigations of clathrate composition dependence on time are very arduous and time-consuming, but they can be a great stimulus for theoretical studies of unstable clathrate composition.

3. *Phase transitions.* The clathrate systems in which ferroelectric and antiferroelectric phases are found at the ordering of guest molecule dipole moments must be intensively studied. The occurrence of these phases is already shown by the first results obtained.[88,115] It would be interesting to verify the conclusions of an experimental result,[24] where the phase transition in β-hydroquinone at $T_c \approx 230$ K was assumed to be independent of the guest. The development of phase transition theory in clathrates may be influenced by improving the methods of calculating guest–guest interaction potentials.

4. *Application.* One might expect the progress of clathrate formation theory, when the models mentioned here will be applied to the investigation of concrete clathrates. The calculations must be carried out by more precise methods than mean-field approximation, for example, by

cluster approximation,[141] Monte Carlo, and molecular dynamics methods. As a result, the comprehensive data would allow us to point out the qualitative relationships between the structural, chemical, and physical properties of clathrates.

This review is primarily devoted to the lattice clathrates in which guest and host molecules are bonded by the van der Waals forces. The next stage of clathrate theory development could be investigation of the intermediate compounds of clathrate-complex and clathrin. Since the compounds have the features of both clathrate and coordinational systems, they can help to determine the influence of anisotropic and chemical bonds on the processes of formation, stability, and destruction of substances, especially since the tendency to form "clathrate regions" is not precluded in most compounds with complicated crystallochemical composition.

ACKNOWLEDGMENTS

The authors would like to thank Yu. A. Dyadin for material on clathrate structure and thermodynamic experiments.

REFERENCES

1. W. S. Fyfe, *Geochemistry of Solids. An Introduction*, McGraw-Hill, New York, 1964, p. 238.

2. J. Needham, *Biochemistry and Morphogenesis*, Cambridge University Press, New York, 1951, p. 661.

3. M. L. Bender and M. Komyjama, *Cyclodextrin Chemistry*, Springer, Berlin, 1978.

4. *Host–Guest Complex Chemistry*, vols. I and II, F. Vogtle, Ed., Springer, Berlin, 1981.

5. J. H. van der Waals and J. C. Platteeuw, *Adv. Chem. Phys.* **2**, 1 (1959).

6. W. C. Child, Jr. *Quart. Rev.* **18**, 321 (1964).

7. D. W. Davidson, *Clathrate Hydrates. Water. A Comprehensive Treatise* **2**, 115 (1973).

8. E. Weberand and H.-P. Josel, *J. Incl. Phenom.* **1**, 79 (1983).

9. J. E. D. Davies, W. Kemula, H. M. Powell, and N. O. Smith, *J. Incl. Phenom.* **1**, 3 (1983).

10. D. D. MacNicol in *Inclusion Compounds*, J. L. Atwood, J. E. D. Davies, and D. D. MacNicol, Eds, Academic Press, London, 1984, vol. 2, p. 1.

11. C. J. Pedersen, *J. Am. Chem. Soc.* **89**, 2495, 7017 (1967).

12. B. Dietrich, J.-M. Lehn, and J. P. Sauvage, *Tetrahedron Lett.* **2885**, 2889 (1969).

13. D. L. Chleck and C. A. Ziegler, *Nucleomics* **17**(9), 131 (1959).

14. L. Pauling and P. Pauling, *Chemistry*, W. H. Freeman and Co., San Francisco, CA, 1975.

15. T. Iwamoto in *Inclusion Compounds*, J. L. Atwood, J. E. D. Davies, and D. D. MacNicol, Eds, Academic Press, London, 1984, vol. 1, p. 29.

16. M. M. Arodž, *Acta Phys. Pol.* **A52**, 555 (1977).

17. K. D. Kleaver, J. D. Davies, and W. J. Wood, *J. Mol. Struct.* **25**, 222 (1975).

18. T. Matsuo, H. Suga, and S. Seki, *J. Phys. Soc. Jpn.* **30**, 785 (1970).

19. T. Matsuo, *J. Phys. Soc. Jpn.* **30**, 794 (1970).

20. P. Sixou, P. Dansas, *Ber. Bensen-Gesellschaft* **80**, 364 (1976).

21. M. Sitarski, *Roczn. Chem.* **49**, 159 (1975).

22. V. R. Belosludov, Yu. A. Dyadin, O. A. Dracheva, and G. N. Cheknova, *Izv. SO AN USSR Ser. Khim. Nauk* **4**, 60 (1979).

23. V. E. Schneider,* E. E. Tornau, and A. A. Vlasova, *Chem. Phys. Lett.* **93**, 188 (1982).

24. V. E. Schneider, E. E. Tornau, A. A. Vlasova, and A. A. Gurskas, *J. Incl. Phenom.* **3**, 235 (1985).

25. N. G. Parsonage and L. A. K. Staveley, *Disorder in Crystals*, Clarendon Press, Oxford, 1978.

26. V. R. Belosludov, Yu. A. Dyadin, G. N. Chekhova, and S. I. Fadeev, *J. Incl. Phenom*, **1**, 251 (1984).

27. D. E. Palin and H. M. Powell, *J. Chem. Soc.* 208 (1947).

28. H. M. Powell, *J. Chem. Soc.* 61 (1948).

29. K. Maartmaan-Moe, *Acta Crystallogr.* **21**, 979 (1966).

30. S. C. Wallwork and H. M. Powell, *J. Chem. Soc. Perkin Trans.* **2**, 641 (1980).

31. S. V. Lindeman, V. E. Shklover, and Yu. T. Struchkov, *Cryst. Struct. Comm.* **10**, 1173 (1981).

32. G. L. Stepakoff and L. V. Coulter, *J. Phys. Chem. Solids* **24**, 1435 (1963).

33. S. J. Allen, Jr., *J. Chem. Phys.* **44**, 394 (1966).

34. J. C. Burgiel, H. Meyer, and P. L. Richards, *J. Chem. Phys.* **43**, 4291 (1965).

35. T. C. W. Mak, *J. Chem. Soc. Perkin Trans.* **2**, 1435 (1982).

36. Tze-Lock Chan and T. C. W. Mak, *J. Chem. Soc. Perkin Trans.* **2**, 777 (1983).

37. D. F. Sargest and L. D. Calvert, *J. Phys. Chem.* **70**, 2689 (1966).

38. D. R. Hafemann and S. L. Miller, *J. Phys. Chem.* **73**, 1392 (1969).

39. D. W. Davidson, Y. P. Handa, C. I. Ratcliffe, J. A. Ripmeester, J. S. Tse, J. R. Dahn, F. Lee, and L. D. Calvert, *Mol. Cryst. Liq. Cryst.* **141**, 141 (1986).

40. D. W. Davidson, S. K. Garg, S. R. Gough, Y. P. Handa, C. I. Ratcliffe, J. S. Tse, and J. A. Ripmeester, *J. Incl. Phenom.* **2**,231 (1984).

41. G. A. Jeffrey in *Inclusion Compounds*, J. L. Atwood, J. E. D. Davies, and D. D. MacNicol, Eds, Academic Press, London, 1984, vol. 2, p. 135.

42. K. W. Allen, *J. Chem. Phys.* **41**, 840 (1964).

43. J. A. Ripmeester and D. W. Davidson, *Mol. Cryst. Liq. Cryst.* **43**, 189 (1977).

44. S. R. Gough, J. A. Ripmeester, and D. W. Davidson, *Can. J. Chem.* **53**, 2215 (1975).

45. J. A. Ripmeester, J. S. Tse, C. I. Ratcliffe, and B. M. Powell, *Nature* **325**, 135 (1987).

46. D. Feil and G. A. Jeffrey, *J. Chem. Phys.* **35**, 1863 (1961).

47. J. A. Ripmeester, J. S. Tse, C. I. Ratcliffe, and B. M. Powell, *Nature* **325**, 135 (1987).

48. W. F. Claussen, *J. Chem. Phys.* **19**, 259 (1951).

*Since 1987 this author uses his new surname, V. E. Zubkus.

49. M. Stackelberg and H. R. Muller, *J. Chem. Phys.* **19**, 1319 (1951).

50. M. Stackelberg and H. R. Muller, *Z. Electrochem.* **58**, 25 (1954).

51. W. F. Claussen, *J. Chem. Phys.* **19**, 1425 (1951).

52. Yu. A. Dyadin, K. A. Udachin, F. B. Jurko, S. B. Bogatyreva, and Yu. I. Mironov, *Izv. SO AN USSR Ser. Khim, Nauk* **1**, 44 (1989).

53. D. W. Davidson, Y. P. Handa, and J. A. Ripmeester, *J. Chem. Phys.* **90**, 6549 (1986).

54. Yu. A. Dyadin and K. A. Udachin, *Zh. Strukt. Khim.* **28**, 75 (1987).

55. G. A. Jeffrey and R. K. McMullan, *Progr. Inorg. Chem.* **8**, 43 (1967).

56. S. L. Miller, S. R. Gough, and D. W. Davidson, *J. Chem. Phys.* **81**, 2154 (1977).

57. H. Gies, H. Gerke, and F. Liebau, *Nenes Jahrb. Mineral Monatsh.* **3**, 119 (1982).

58. H. Gies, F. Liebau, and H. Gerke, *Angew. Chem. Int. Ed. Engl.* **21**, 206 (1982).

59. C. Cros, M. Pouchard, and P. Hagenmuller, *Bull. Soc. Chim. Fr.* **13**, 379 (1971).

60. H. Gerke and H. Gies, *Z. Kristallogr.* **116**, 11 (1984).

61. L. C. Fetterly, in *Non-stoichiometric Compounds*, L. Mandelcorn, Ed., Academic Press, New York and London, 1964, p. 501.

62. A. F. G. Cope, D. J. Gannon, and N. G. Parsonage, *J. Chem. Thermodyn.* **4**, 829 (1972).

63. F. Laves, N. Nicolaides, and K. C. Peng, *Z. Kristallogr. Kristallgeom.* **121**, 258 (1965).

64. H.-U. Lenne, *Acta Crystallogr.* **7**, 1 (1954).

65. V. R. Belosludov, Yu. A. Dyadin, and M. Yu. Lavrentiev, *Theoretic Models of Clathrate Formation*, Nauka, Novosibirsk, 1991 (in Russian).

66. W. Schlenk, *Ann. Chem.* **1145**, 1156, 1179, 1195 (1973).

67. A. E. Smith, *Acta Crystallogr.* **5**, 224 (1952).

68. G. N. Chekhova and Yu. A. Dyadin, *Izv. SO AN USSR Ser. Khim. Nauk* **2**, 66 (1986); **12**, 75 (1978).

69. Yu. V. Zephirov and P. M. Zorkij, *Zurn. Strukt. Khim.* **17**, 745, 994 (1976).

70. M. Born and K. Huang, *Dynamical Theory of Crystal Lattices*, Clarendon Press, Oxford, 1954.

71. V. E. Zubkus, A. A. Vlasova, and E. E. Tornau, *Surf. Sci.* **215**, 47 (1989).

72. L. D. Landau and E. M. Lifshitz, *Statistical Physics*, Pergamon Press, Oxford, 1967.

73. L. I. Schiff, *Quantum Mechanics*, McGraw-Hill, New York, 1955.

74. V. R. Belosludov, Yu. A. Dyadin, G. N. Chekhova, S. I. Fadeev, and B. A. Kolesov, *J. Incl. Phenom.* **3**, 243 (1985).

75. V. A. Istomin, *Zh. Phys. Khim.* **61**, 1404 (1987).

76. D. Stroud, *Theory of Alloy Phase Formation*, L. H. Bernett, Ed., Metallurgical Society of AIME, 1980, p. 84.

77. D. de Fontaine, *Solid State Phys.* **34**, 73 (1979).

78. A. A. Vlasova and V. E. Schneider, *Lietuvos Fiz. Rink.* **23**, 61 (1983).

79. C. Domb, *Adv. Phys.* **9**, 149, 245 (1960).

80. V. G. Vaks and V. E. Schneider, *Phys. Stat. Sol. (a)* **35**, 61 (1976).

81. V. E. Schneider and E. E. Tornau, *Chem. Phys.* **98**, 41 (1985).

82. N. G. Parsonage and R. C. Pemberton, *Trans. Faraday Soc.* **63**, 311 (1967).

83. Yu. A. Dyadin, V. R. Belosludov, G. N. Chekhova, and M. Yu. Lavrentiev, *J. Incl. Phenom.* **5**, 195 (1987).

84. G. Leibfried and W. Ludwig, *Solid State Phys.* **12**, 3 (1961).

85. J. C. Ryan and N. M. Lawandy, *Phys. Rev.* **B41**, 2369 (1990).

86. D. E. Williams, *J. Chem. Phys.* **47**, 4680 (1967).

87. W. C. Child, *J. Phys. Chim.* **68**, 1834 (1964).

88. A. A. Gurskas and V. E. Schneider, *Kristallografija* **30**, 1190 (1985) (in Russian).

89. S. R. Gough, S. K. Garg, and D. W. Davidson, *Chem. Phys.* **3**, 239 (1974).

90. L. M. Casey and L. K. Runnels, *J. Chem. Phys.* **51**, 5070 (1969).

91. R. C. Pemberton and N. G. Parsonage, *Trans. Faraday Soc.* **61**, 2112 (1965).

92. D. J. Gannon and N. G. Parsonage, *J. Chem. Thermodyn.* **4**, 745 (1972).

93. F. Jona and G. Shirane, *Ferroelectric Crystals*, Pergamon Press, New York, 1962.

94. K. H. Lau and S. C. Ying, *Phys. Rev. Lett.* **44**, 1222 (1980).

95. V. P. Zhdanov, *Surf. Sci.* **164**, L807 (1985).

96. V. McKoy and O. Sinanoglu, *J. Chem. Phys.* **38**, 2946 (1963).

97. W. I. David, W. T. A. Harrison, A. J. Leadbetter, T. Matsuo, and H. Suga, *Physica* **B156–157**, 93 (1989).

98. H. Ukegawa, T. Matsuo, and K. Suga, *J. Incl. Phenom.* **3**, 261 (1985).

99. J. Q. Broughton and X. P. Li, *Phys. Rev.* **B35**, 9120 (1987).

100. D. Rigby and R. J. Roe, *J. Chem. Phys.* **87**, 7285 (1987).

101. G. Venkataraman and V. C. Sahni, *Rev. Mod. Phys.* **42**, 409 (1970).

102. G. Nielson and S. A. Rice, *J. Chem. Phys.* **80**, 4456 (1984).

103. H. J. C. Berendsen, J. P. M. Postma, W. F. van Gusteren, and J. Herman, in *Intermolecular Forces*, B. Pullman, Ed., Reidel, Dordrecht, 1981.

104. P. L. M. Plummer and T. S. Chen, *J. Chem. Phys.* **87**, 4190 (1983).

105. J. S. Tse, M. L. Klein, and I. R. McDonald, *J. Chem. Phys.* **78**, 2096 (1983).

106. J. S. Tse, M. L. Klein, and I. R. McDonald, *J. Chem. Phys.* **87**, 4198 (1983).

107. J. S. Tse, M. L. Klein, and I. R. McDonald, *J. Chem. Phys.* **81**, 6146 (1984).

108. M. Marchi, J. S. Tse, and M. L. Klein, *J. Chem. Phys.* **85**, 2414 (1986).

109. J. Anderson, J. J. Ullo, and S. Yip, *J. Chem. Phys.* **87**, 1726 (1987).

110. J. S. Tse and M. L. Klein, *Phys. Rev. Lett.* **58**, 1672 (1987).

111. V. R. Belosludov, M. Yu. Lavrentiev, and S. A. Syskin, *Phys. Stat. Sol.* (*b*) **149**, 133 (1988).

112. V. R. Belosludov, M. Yu. Lavrentiev, Yu. A. Dyadin, and S. A. Syskin, *J. Incl. Phenom.* **8**, 59 (1990).

113. J. M. Ziman, *Models of Disorder*, Cambridge University Press, New York, 1979.

114. J. F. Belliveau, *Diss. Abstr.* **B31**, 2591 (1970).

115. D. V. Plant and N. M. Lavandy, *Phys. Lett.* **A138**, 301 (1989).

116. R. L. Deming, T. L. Carlicle, and D. J. Lanerman, *J. Chem. Phys.* **73**, 1762 (1969).

117. D. F. Evans and R. E. Richards, *J. Chem. Soc.* 3932 (1952).

118. D. F. Evans and R. E. Richards, *Proc. Roy. Soc.* (*London*) **A223**, 238 (1954).

119. J. H. Helle, D. Kok, J. C. Platteeuw, and J. H. van der Waals, *Rec. Chim.* 1068 (1962).

120. N. R. Grey and L. A. K. Stavely, *Mol. Phys.* **7**, 83 (1963).

121. G. H. Cady, *J. Phys. Chem.* **87**, 4437 (1983).

122. J. A. Ripmeester and D. W. Davidson, *J. Mol. Struct.* **75**, 67 (1981).

123. R. M. Barrer and D. J. Ruzicka, *Trans. Faraday Soc.* **58**, 2239 (1962).

124. R. M. Barrer and D. J. Ruzicka, *Trans. Faraday Soc.* **58**, 2253 (1962).

125. R. M. Barrer and W. I. Stuart, *Proc. Roy. Soc.* (*London*) **A243**, 172 (1957).

126. J. Priestley, *Versuche und Beobachtungen uber verschiedene Gattungen der Luft*, Wien-Leipzig, 1980.

127. F. Y. Wu, *Rev. Mod. Phys.* **54**, 235 (1982).

128. N. G. Parsonage, *Disc. Faraday Soc.* **48**, 215 (1969).

129. H. M. James and T. A. Keenan, *J. Chem. Phys.* **31**, 12 (1959).

130. K. Maki, Y. Kataoka, and T. Yamamoto, *J. Chem. Phys.* **70**, 655 (1979).

131. A. Koide and T. Kihara, *Chem. Phys.* **5**, 34 (1974).

132. V. A. Slusarev and Yu. A. Freiman, *Phys. Stat. Sol.* (*b*) **62**, K61 (1974).

133. R. Ahlrichs, R. Penco, and G. Scoles, *Chem. Phys.* **19**, 119 (1977).

134. J. Ziman, *Electron and Phonons*, University Press, Oxford, 1960.

135. W. A. Harison, *Pseudopotentials in the Theory of Metals*, Benjamin Press, New York, 1963.

136. R. A. Johnson, *J. Phys.* **F3**, 295 (1973).

137. J. Hafner, *Phys. Rev.* **B21**, 406 (1980).

138. G. A. Neece and J. C. Poirier, *J. Chem. Phys.* **43**, 4282 (1965).

139. M. J. Gillan, *J. Phys.* **C19**, 6169 (1986).

140. I. F. Silvena and V. V. Goldman, *J. Chem. Phys.* **69**, 4209 (1978).

141. V. E. Zubkus and S. Lapinskas, *J. Phys. Cond. Matt.* **2**, 1753 (1990).

SIMULATION AND SYMMETRY IN MOLECULAR DIFFUSION AND SPECTROSCOPY

M. W. EVANS

*Center for Theory and Simulation in Science and Engineering,
Cornell University, Ithaca, New York*

CONTENTS

Advances in Chemical Physics, Volume LXXXI, Edited by I. Prigogine and Stuart A. Rice.
ISBN 0-471-54570-8 © 1992 John Wiley & Sons, Inc.

INTRODUCTION

This chapter deals with the impact of contemporary computer simulation on our understanding of molecular diffusion processes. It argues the case for the classical equations of motion applied to the minute time scales and dimensions of molecular dynamics. It is therefore assumed implicitly, as is the contemporary practice, that classical mechanics is valid in this context, although there is no rigorous proof of why this should be so. The traditional approach to the theory of molecular diffusion, developed at the turn of the century, does not stand up to the data now available. These are obtained contemporaneously from many spectral sources, and from computer simulation using increasingly powerful techniques. These data now show unequivocally that the traditional approach is fundamentally flawed in at least one respect, the assumption that translational and rotational motion are decorrelated. We now have available a set of nonvanishing time-cross correlation functions with which to define and elaborate upon the fundamental physical properties of molecules diffusing in three dimensions.

The chapter is intended to be readily understandable to undergraduates and also to be useful to specialists in molecular diffusion, computer simulation, and several branches of spectroscopy. It opens with a description of the traditional theory, whose mathematical complexity is kept secondary to the essentially simple physical concepts that make up the Langevin equation of diffusion. Einstein's theory of translational Brownian motion, which was used to prove the existence of molecules, and Debye's theory of rotational diffusion, which was used to prove the molecular origin of the dispersion of dielectric permittivity at radio frequencies, are discussed in terms of simple Langevin equations, whose limitations are clearly defined. These include the failure to describe the far infrared (high frequency) part of the dispersion of permittivity and dielectric loss in dipolar molecular liquids, due to a missing inertial term and inadequate description of the intermolecular potential energy. The relation of the Langevin to the Liouville equation shows that the former is a first approximation to the trajectory of a diffusing asymmetric top in a three-dimensional ensemble of similar molecules. The experimental consequences of this approximation are most visible in the far infrared, but also in other types of data, discussed in Section II.

This section makes a survey of the available spectral data on molecular diffusion, and compares the results for consistency from sources such as combined dielectric and far infrared spectroscopy, the relaxation of nuclear magnetic resonance, light scattering, infrared band shape analysis, Raman scattering, ultra and hyper sound relaxation, and inelastic, incoherent, neutron scattering. Correlation times for one test liquid, dichloromethane, are compared from all spectral sources and with their equivalents from computer simulation. The latter brings some self consistency into what remains an imperfect experimental understanding due to inadequate data coverage.

Section III covers the basics of computer simulation with reference to the numerical integration of the classical rotational and translational equations of motion of the three dimensional diffusion of the asymmetric top molecule dichloromethane in the liquid state. This section covers the approximation of the intermolecular potential with the pairwise additive method, the computation of thermodynamic properties, and spectral data, from the individual trajectories of the simulation.

In Section IV some new results from computer simulation are described, and limits set on the validity of traditional linear response theory. These are described in terms of new effects discovered by the simulation technique, such as fall transient acceleration and field decoupling. The importance of cross-correlation functions is illustrated with respect to the effect of external force fields that are shown by simulation to induce such previously unknown correlations direct in the laboratory frame of reference (X, Y, Z).

Section V introduces the use of symmetry laws that govern the existence of cross-correlation functions both in frame (X, Y, Z) and in frame (x, y, z) fixed in the molecule as it diffuses. These include parity and time-reversal symmetry, and group theory in both frames. Tables are provided to guide the reader as to the use of group theory in the context of time correlation functions of all orders, and for representative molecular point groups.

Some theories of the time dependence of members of the set of cross-correlation functions are described briefly in order to illustrate the limits of validity of the traditional approach of Section I.

Section VI describes the challenge to diffusion theory of the existence of many members of this set for all molecular matter. The nature of the set of cross-correlation functions is illustrated with reference to water under a broad range of conditions, with respect to correlations involving simultaneously vibration, rotation, and center of mass translation in a liquid, with respect to rod-like molecules as models for liquid crystal behavior, and to highly anisotropic diffusers in the liquid state. Some indications are given to the meaning of cross-correlation functions in the context of quantum mechanics, where they become expectation values of wave functions involving both rotational and translational energy levels. The conditions are described under which the rototranslational spectral features corresponding to these energy levels may be observed by spectroscopy.

Section VII introduces group theoretical statistical mechanics, and gives the three principles which govern the application of group theory to the process of molecular diffusion. These principles are illustrated with respect to both frames (X, Y, Z) and (x, y, z) in molecular liquids and liquid crystals.

Section VIII extends the scope of group theoretical statistical mechanics to the effect of fields of various kinds on molecular liquids, and introduces the study of non-Newtonian rheology with the principles of group theory. These chapters are filled out with reference tabular material and illustrations which should be useful for a broad cross section of chemical and computational physicists and flow engineers.

Finally, Section IX deals with the symmetry and simulation of new pump–probe spectroscopies, utilizing the principles outlined in Sections VII and VIII.

I. THE TRADITIONAL VIEW

The traditional approach to molecular diffusion in liquids is a first approximation to the problem of dealing with a very large number of moving and interacting molecules.[1-4] Conventional textbooks[5-15] tend to overemphasise the applicability of the turn of the century techniques first used to deal with this problem. Albert Einstein was the first to realize, in a Berne patent office at the dawning of modern physics, that the restless motion of pollen particles known as *Brownian motion* was caused by unending collisions with much smaller particles, *and these he recognised as molecules*. Their dynamical trajectories, he assumed, were governed by classical statistical mechanics. In a paper published in 1905, Einstein[16] estimated the Avogadro Number from a diffusion equation developed to explain the Brownian motion on the basis of molecular collisions with the much heavier and slower moving pollen particles, whose motion could be studied directly through a microscope. This paper provided one of the first definitive proofs of the existence of molecules as the fundamental entities of molecular matter.

Einstein measured his limits, he did not, for example, take into account the rotational motion of each molecule because this was not needed for the estimation of the Avogadro Number from the observation of Brownian motion. He based his theory on well-defined assumptions, and developed it self-consistently. Now, almost a century later we are privileged to have available a vast ocean of data from computer simulation[4] and far infrared spectroscopy,[1] for example, which describe in abundant detail worlds which the pioneer could but dimly perceive. Our understanding of molecular diffusion has been transfigured in the last 20 years.

This chapter attempts to review the significant results of the last decade or so and to reassess critically the early theories of molecular diffusion[16–19] with new data. The chapter argues the case, for the time being, for a classical mechanical approach to the dynamics of molecular liquids, using the Newton and Euler equations, and variations. This is made possible by contemporary supercomputers. There are fewer approximations than in the traditional theory of diffusion, and the new methods are well founded in the laws of classical physics.

A. Einstein's Theory of Translational Brownian Motion

The explanation for Brownian motion given by Einstein[16] makes several assumptions about the nature of the collision between a pollen particle and a molecule of the surrounding liquid. The basic idea is that the effect of collisions produces random jumps in the position of the pollen particle. The velocity change on collision is assumed to be damped out quickly and the role of velocity in Einstein's treatment is ill-defined.[7,20] The random walk of the pollen particle in three dimensions is described by a partial differential equation for the displacement in each dimension. The solution of this equation showed that the mean-square displacement of a Brownian particle should increase linearly with time. This prediction was verified experimentally by Perrin[21] in 1908, and from Einstein's formula a value of Avogadro's Number was deduced which is in satisfactory agreement with the accepted value.

A contemporary description of the methods used by Einstein to derive the formula for the mean-square displacement of the pollen particle is given by W. T. Coffey in Ref. 4. A straightforward description of the reasoning behind the theory was established by Langevin[17] in 1908. This is more transparent to the student of Brownian motion, and also reveals more clearly the limitations of the methods involved in calculating both the mean-square displacement and the Avogadro Number.[1-4]

B. The Langevin Equation for Translational Brownian Motion

Langevin's treatment is limited to the motion of large pollen particles in suspension in a molecular environment. The equation has often been applied uncritically, however, to the motion of a molecule in solution in others. There are fundamental differences in the nature of a collision between two molecules and between a molecule and pollen particle. The transfer of momentum in the former case is roughly equally shared, and the assumption that the change in velocity can be ignored does not hold. This was realized fully as late as 1964, following a computer simulation by Rahman[22] of 864 diffusing argon atoms. In this simulation, the velocity auto correlation function of the ensemble of argon atoms was found to be markedly different in time dependence from the simple exponential of early diffusion theory, based on the Einstein theory of translational Brownian motion.

Langevin wrote the equation of translational Brownian motion as

$$m\frac{d^2x(t)}{dt^2} + m\beta_T\frac{dx(t)}{dt} = F(t) \tag{1}$$

assuming that the forces on the pollen particle could be divided into a systematic part $-m\beta_T\dot{x}(t)$, representing a friction on the pollen particle generated by its molecular collisions, and a random force $F(t)$ generated by a random walk of the position of the pollen particle relative to the surrounding molecules. The frictional force opposes the motion and is therefore given a negative sign, and the random force generates the unending motion of the pollen by collision with molecules which are in constant thermal motion and have constant kinetic energy at constant temperature. Langevin assumed that the frictional term was governed by Stokes's law of macroscopic hydrodynamics, which applies to a spherical particle in a viscous fluid. In contrast, the random force $F(t)$ was assumed to be independent of x and to vary very rapidly compared with any variation in $x(t)$. There is no *statistical correlation* between $F(t)$ and $F(t + \Delta t)$. The picture is therefore a mixture of hydrodynamic and statistical concepts. It has fundamental mathematical limits as described by Doob.[20] Its physical shortcomings were not easily found until the computer simulation by Rahman, using the much older Newton equations of motion. The Langevin equation is really an ad hoc mixture of concepts. In the same way, Einstein's estimate of the Avogadro Number was based on a combination of his result for the mean square displacement

$$\langle \Delta x^2 \rangle = \frac{kTt}{3\pi\eta a} \tag{2}$$

with Stokes's Law for the friction coefficient

$$\beta_T = 6\pi\eta a \tag{3}$$

Here a is the effective radius of the pollen particle, assumed spherical, η is the effective viscosity of the molecular surroundings, and t is the time after an arbitrary initial instant for which the particle has been under the influence of the Brownian random walk. There was no rigorous justification for the mixture of concepts, except that it seemed to work at first, providing a good approximation to the Avogadro Number.

This early work was, however, pivotal, because it was the first to use the concept of molecular dynamics in explaining Brownian motion. It provided evidence for the existence of molecules when such evidence was needed. The role of contemporary computer simulation is very different.

C. Contemporary Criticisms of the Traditional Approach

One of the most powerful arguments against the indiscriminate use of the Langevin equation is that it cannot be derived rigorously from the fundamental equations of motion used in statistical mechanics, for example the Liouville equation. This was demonstrated clearly by Mori[23] in 1965, using projection operators. Mori started from the Liouville equation

$$\dot{\Gamma} = iL\Gamma \tag{4}$$

where L is the Liouville operator and Γ is the phase space of positions and momenta. Mori showed that a dynamical quantity A or $\mathbf{A}$ (i.e., scalar or vector) which obeys the Liouville equation

$$\dot{\mathbf{A}} = iL\mathbf{A} \tag{5}$$

also obeys the equation

$$\dot{\mathbf{A}} = i\Omega_A \mathbf{A}(t) - \int_0^t \phi_A(t-\tau)\mathbf{A}(\tau)d\tau + \mathbf{F}_A(t) \tag{6}$$

which is the same equation as (5). Here Ω_A is the resonance operator, and ϕ_A is the *memory function*. $\mathbf{F}_A$ is a vector of random quantities. Equation (6) reduces to the Langevin equation only when the resonance operator vanishes and when the memory function is a delta function of time. If $\mathbf{A}$ represents linear velocity, $\mathbf{v}$ for example, the original Langevin equation (1) is recovered only with these drastic approximations. *The memory function can never be a delta function for any meaningful molecular dynamical process.* Mori showed that repeated application of projection operators to the Liouville equation produces a string of interrelated equations of the type (6), but never a simple Langevin equation of type (1). Laplace transformation of the string of equations produces the *Mori continued fraction*[1-3] for the statistical correlation between the variable $\mathbf{A}$ at time t and its value at $t = 0$, the *velocity autocorrelation function*. The equivalent result from the simple Langevin equation is an exponential in time, an oversimplification of the rigorous Mori continued fraction. Rahman's computer simulation of 1964[22] showed conclusively that the velocity autocorrelation function was far from being a simple exponential in 864 argon atoms interacting with a realistic (Lennard Jones) model for the interatomic potential.

D. Molecular Translation and Rotation

By restricting consideration to spherical particles the treatment so far has not begun to account for rotation, and for simultaneous rotation and translation in an irregular body, or "asymmetric top," diffusing in three dimensional space (X, Y, Z). The dispersion of the dielectric permittivity at radio frequencies was known in the first decade of this century to be a rotational phenomenon involving the molecular antenna, the permanent molecular dipole moment, $\boldsymbol{\mu}$. In general this rotation occurs on top of the translation in three dimensions. There is a statistical spread of angular velocities which causes the dielectric permittivity of a molecular liquid to be frequency dependent.[1-4] The dipole moment is the result of an asymmetric distribution of charge, due to the distribution of atoms within a molecule. The dispersion of permittivity is always accompanied by a dielectric loss as the frequency of the measuring field is increased. This is governed by a relatively slow motion, the rotation of the whole molecule, and can be observed using conductance/capacitance changes, or by the direct attenuation of radiation, at *far infrared* frequencies.[12,24] The interaction of molecular (and other) matter with electromagnetic radiation is governed by electromagnetic field theory, based classically on Maxwell's equations.

E. Dielectric Relaxation and the Debye Theory of Rotational Diffusion

Essentially speaking, dielectric relaxation involves the partial polarization[9-14] of the molecules of a liquid with an electric field. If this alternates in frequency, the periodic change in direction produces a change in direction of the molecular dipole moments as the molecules attempt to rotate and follow the field direction. The electric field produces only a very slight alignment of the molecules, because the thermal motion is relatively so much more energetic at room temperatures and available electric field strengths. The degree of alignment is governed essentially by the ratio

$$b = \frac{\mu E}{kT} \qquad (7)$$

where E is the electric field, k the Boltzmann constant, and T the temperature in absolute units. The ratio b is the argument of the Langevin function, measuring the degree of alignment produced by the external field, the result of competition between the aligning energy and the thermal energy, kT.

F. Loss of Polarization

The simplest kind of dielectric relaxation occurs when a static electric field is applied initially to a liquid of diffusing molecules, allowing enough time for alignment to occur, and is then switched off instantaneously. The degree of molecular alignment produced by the external electric field is lost, but not immediately. The natural thermal motion that makes alignment a difficult process also prevents its instantaneous loss. The length of time needed to align the sample as fully as possible by an applied static electric field, and conversely, the time needed for loss of alignment, both depend on the nature of molecular diffusion. This fact can be used experimentally[12] to investigate the nature of diffusion. In the alignment process, the degree of orientation of the molecules as a function of time is known as the rise transient, and the loss of alignment as the fall transient. These can only be calculated theoretically by a full consideration of the molecular dynamics. In particular, the rotational dynamics of each molecule in its molecular environment must be known in detail, because the rise and fall transients are both products of the rotational torque generated by the vector product of the electric field and the molecular dipole moment. No progress at all can be made, therefore, with theories of translational diffusion.

G.   The Concept of Rotational Diffusion

Debye was the first to develop a theory[19] of *rotational diffusion*. This was first applied to explain a form of dielectric relaxation which led to a decrease in relative permittivity with the increasing frequency of an applied, alternating, electric field. This had been observed to occur in dipolar molecules whose charge distribution was not symmetrical. The phenomenon could therefore be used to investigate the structure of molecules. Debye reduced the problem[5] to considering the two dimensional Brownian motion of a dipole in an external, time-varying electric field. The original approach[5,19] is summarised in Ref. 4 and involved the use of a Smoluchowski equation[18] with the x coordinate of translational diffusion replaced by the angular coordinate θ. The Langevin equation for Debye's two dimensional rotational diffusion theory is

$$\beta_R \dot{\theta}(t) + \frac{\partial V(\theta, t)}{\partial \theta} = \lambda(t) \tag{8}$$

where λ is the random torque from the Brownian movement of the surroundings, and β_R is the rotational friction coefficient. The potential energy due to the aligning field is

$$V = -\mu E \cos \theta = -\boldsymbol{\mu} \cdot \mathbf{E} \tag{9}$$

We can see immediately from a comparison of the rotational Langevin equation (8) used by Debye[5] and the translational Langevin equation (1) that there is a term missing in the former, that is, $I\dot{\theta}(t)$, the inertial term. The rotational equation ignores from the outset the finite rotational acceleration of each diffusing molecule. Einstein also effectively ignored the inertial term in arriving at his value of the diffusion coefficient, (Ref. 4, pp. 86 to 87).

H.   Rise Transient

The solution of the Langevin equation (8) for the alignment of the molecular dipole moments by an applied static electric field is the mirror image of its solution (the fall transient) after the instantaneous removal of the field. To compute the rise transient we merely solve the Langevin equation for the fall transient, which is Eqn. (8) with

$$V = 0 \tag{10}$$

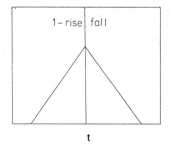

t

Figure 1. Log of rise and fall transients in Debye's theory of rotational diffusion. One is an exponential increase in molecular orientation and the other a mirror image decrease (schematic).

I. The Fall Transient

This is obtained from the Langevin equation as

$$\langle \cos \theta \rangle = \frac{\mu E}{2kT} e^{-t/\tau_D} \tag{11}$$

where τ_D is the Debye relaxation time. The fall and rise transients in the Debye theory of rotational diffusion are therefore both exponentials in time, one the mirror image of the other as sketched in Fig. 1.

J. Polarization and Frequency Spectrum

The polarization decay after the instantaneous removal of the aligning electric field is therefore a simple exponential in this approximation. *If we reinstate the missing inertial term*, however, the polarization decay becomes[25,26]

$$\langle \cos \theta \rangle = \frac{\mu E}{kT} \exp\left\{ - \frac{kTI}{\beta_R^2} \left(\frac{\beta_R t}{I} - 1 + e^{-\beta_R t/I} \right) \right\} \tag{12}$$

where I is the molecular moment of inertia. Equation (12) reduces to Eq. (11) only at very long times, that is, as t goes to infinity. This immediately shows that the Debye rotational diffusion theory, and the model upon which it is based, Eq. (1) with its inertial term missing, work only over relatively long time scales or low frequencies. They are "coarse grained" theories of molecular diffusion and dynamical evolution. Something goes wrong in both theories as the time scale of events becomes shorter and shorter.

To realize these shortcomings is a necessary step towards progress in the theory of molecular diffusion. Short time scales mean high frequencies, long time scales low frequencies conversely. In mathematical terms there is always a rigorous and general relation between a given function of time which is continuous and differentiable in the range of interest and its counterpart in frequency, known as the *frequency spectrum*. This is the Fourier Integral Theorem, which may be written as

$$C(\omega) = \frac{1}{2\pi} \int_{-\infty}^{\infty} C(t)e^{-i\omega t} \, dt \tag{13}$$

$$C(t) = \int_{-\infty}^{\infty} C(\omega)e^{i\omega\tau} \, d\omega \tag{14}$$

Here ω is the angular frequency in radians per second, and if $C(\omega)$ is an angular frequency spectrum, then $C(t)$ is a correlation function.[1] There are many types of correlation function available from modern spectroscopy[1,4] and so there are many types of time correlation. In statistics, "correlation" is the interdependence of quantitative data. The time correlation function is in general the product of two dynamical variables averaged in a special way. It is the internal correlation between two observations in time. The normalized autocorrelation function can be defined in general as

$$C(t) = \lim_{T \to \infty} \left\{ \frac{\int_0^T (1/T)A(t)A(t+\tau) \, dt}{\int_0^T A^2(t) \, dt} \right\} \tag{15}$$

The numerator is the autocovariance, the denominator is the variance, and this type of average is called the running time average. By a basic theorem of statistics[1] it is rigorously equivalent to the Maxwell–Boltzmann ensemble average for a stationary ensemble at reversible thermodynamic equilibrium. The autocorrelation function (acf) of the molecular dipole moment μ is the Fourier transform of the complex dielectric permittivity. The power absorption coefficient of the high frequency far infrared is the Fourier transform[1] of the autocorrelation function of the second time derivative of $\langle \mu(t)\mu(0) \rangle$.

This is known as the rotational velocity acf. Furthermore, the power absorption coefficient of the far infrared, $\alpha(\bar{\nu})$, in neper per cm, is related fundamentally to the dimensionless dielectric loss $[\epsilon(\omega)]$ through Maxwell's equations

$$\alpha(\omega) = \frac{\omega\epsilon''(\omega)}{n(\omega)c} \tag{16}$$

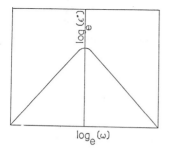

Figure 2. A plot of dielectric loss against the logarithm of frequency. In Debye's theory, this is the well-known bell-shaped curve of dielectric relaxation theory (schematic).

where $n(\omega)$ is the refractive index of the liquid at far infrared frequencies and c is the velocity of light.

There is a chain of relations between the frequency-dependent complex dielectric permittivity

$$\epsilon = \epsilon'(\omega) - i\epsilon''(\omega)$$

the far infrared power absorption coefficient and the acf.'s of the molecular dipole moment and its derivative.

K. Dielectric and Far Infrared Spectrum

Dielectric spectra, the frequency dependence of the dielectric loss,[12,24] and far infrared spectra, that of the power absorption coefficient, are obtained at low frequencies and high, respectively. Although they are observed with entirely different experimental methods, they are always interrelated by Eq. (16). It is obvious that the theory of rotational diffusion of Debye must produce a self-consistent picture of both spectra, which can stretch over a vast frequency range.[1,4,12] This is clear in hindsight, but for over half a century the definitive experimental test of fundamental molecular diffusion theory was delayed by lack of high-frequency data and misunderstandings due to the missing inertial term. The Debye theory seemed to produce results in excellent agreement[14] with the observed angular frequency dependence of complex permittivity at low frequencies, where the well-known bell shaped curve[12,14,24] appears to be the single dominant feature (Fig. 2). This is the Fourier transform of an exponential function of time. According to a theorem of linear response theory,[1–4] known as the fluctuation–dissipation theorem, the fall transient of Eq. (11) has the same time dependence as the acf of the molecular dipole moment

$$\frac{\langle \boldsymbol{\mu}(t) \cdot \boldsymbol{\mu}(0) \rangle}{\langle \mu^2 \rangle} = \frac{\langle \cos \theta \rangle}{\langle \cos \theta(0) \rangle} \qquad (17)$$

when both are normalized to the value of 1 at $t = 0$. The Debye theory of rotational diffusion therefore produces an exponential normalized orientational acf

$$C_\mu(t) = \frac{\langle \boldsymbol{\mu}(t) \cdot \boldsymbol{\mu}(0) \rangle}{\langle \mu^2 \rangle} = e^{-t/\tau_D} \tag{18}$$

where τ_D is the Debye relaxation time. The contemporary theory of statistical mechanics shows that the frequency dependence of the dielectric loss[27] and dispersion is obtained by Fourier transformation of the orientational acf. The well-known bell-shaped curve obtained by Debye is the result of an exponential fall transient, whose time dependence has been taken to be the same as that of the orientational acf at equilibrium. By a comparison of the results of the original Debye theory (Eq. 11) and the same theory corrected for its missing inertial term (Eq. 12) we have seen that the original theory can be valid only at low frequencies. At the time the Debye theory was developed, and for 50 years thereafter, only low-frequency data were available to test the theory. Its apparent success in explaining these data masked its inherent flaws. The theory became accepted uncritically.[9,14]

L. The Debye Plateau

If we neglect the complications of the internal field effect, that is, the difference between the applied electric field and that actually felt by a diffusing molecule at an instant in time, the relation between dielectric loss and the normalized orientational acf $C_\mu(t)$ can be shown[1] to be the following, which are valid in dilute solutions of dipolar molecules in other, symmetrical, (nondipolar) molecules such as carbon tetrachloride or sulphur hexafluoride:

$$\epsilon''(\omega) = \frac{N\mu^2\omega}{3kT\epsilon_0} \int_0^\infty C_\mu(t) \cos \omega t \, dt \tag{19}$$

$$\alpha(\omega) = \frac{N\mu^2\omega^2}{3kT\epsilon_0 c} \int_0^\infty C_\mu(t) \cos \omega t \, dt \tag{20}$$

$$C_\mu(t) = \frac{2}{\pi} \frac{3kT\epsilon_0 c}{N\mu^2} \int_0^\infty \frac{\alpha(\omega)}{\omega^2} \cos \omega t \, d\omega \tag{21}$$

Here ϵ_0 is the dielectric permittivity at static frequencies, N is the molecular number density (the number of molecules per unit volume), and $\alpha(\omega)$ is the far infrared power absorption coefficient in neper cm^{-1}. From Eqs.

(18) and (19) the result is obtaincd that the dielectric loss corresponding to an exponential dipole acf is, in dilute solution

$$\epsilon''(\omega) = \frac{N\mu^2}{3kT\epsilon_0} \frac{\omega\tau_D}{1 + \omega^2\tau_D^2} \tag{22}$$

Combining this equation with (16) it can be seen that the infrared power absoption coefficient from Debye's rotational diffusion theory is

$$\alpha(\omega) = \frac{N\mu^2}{3kT\epsilon_0 n(\omega)c} \frac{\omega^2\tau_D}{1 + \omega^2\tau_D^2} \tag{23}$$

This is well behaved at low frequencies and vanishes with the angular frequency, but at high frequency leads to the Debye plateau[12]

$$\alpha(\omega) \underset{\omega \to \infty}{\to} \frac{N\mu^2}{3kT\epsilon_0 n(\omega)\tau_D c} \tag{24}$$

In Debye's diffusion theory all molecular dipolar liquids must be opaque at all frequencies from the far infrared upwards, including the visible, an absurd result.

M. Far Infrared Interferometry

The development of computers gave great impetus[1,28,29] to the exploration of the far infrared region by Fourier transformation of interferograms obtained by Michelson interferometry. This optical technique produces patterns of light intensity from two interfering beams of far infrared radiation as a function of distance travelled by a mirror in one arm of a Michelson interferometer. The radiation reaching a heat-sensitive Golay detector is recorded as the mirror travels and is known as an interferogram. Fourier transformation of the interferogram (a type of correlation function in distance) produces the far infrared spectrum. Fourier transform spectroscopy now dominates infrared and NMR spectroscopy.

In the late 1960s, about 60 years after Debye proposed his theory of rotational diffusion, it became possible to explore the behavior of molecular liquids at frequencies in the far infrared, which stretches from the upper end of the gigahertz range (microwave) and beyond into the infra-

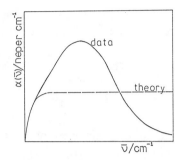

Figure 3. Far infrared power absorption coefficient of liquid dichloromethane at room temperature (solid line) compared with the theoretical prediction of Debye type rotational diffusion (dashed line).

red. Figure 3 illustrates that the far infrared absorption of a dipolar molecular liquid such as dichloromethane[1] in the range from about $2 \, \text{cm}^{-1}$ to about $250 \, \text{cm}^{-1}$ greatly exceeds in intensity of power absorption the Debye plateau computed for that liquid. At very high frequencies the power absorption coefficient once more reaches zero, that is, drops away from the Debye plateau. There is little or no resemblance at these frequencies between theory and experiment.

If however, we compare the same set of data and the same theory in terms of dielectric loss, over the same frequency range, using Eq. (16) to convert the power absorption coefficient to dielectric loss, the result is a deceptively good fit of theory and data, Fig. 4. The theory of rotational diffusion seems to perform well at low frequencies when expressed in terms of dielectric loss, but fails completely at high frequencies when expressed in terms of the far infrared power absorption coefficient in neper per centimeter.

N. The Effect of Including the Inertial Term in Eq. (8)

The procedure of including the inertial term in the rotational Langevin equation used by Debye, Eq. (8), is known as the "inertial correction"

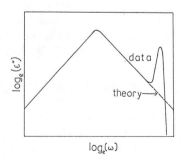

Figure 4. The comparison illustrated in Fig. 3 made in terms of the dielectric loss at lower frequencies than the far infrared (schematic).

to the Debye theory of rotational diffusion. The spectrum in the dielectric and far infrared frequency regions obtained from the inertia-corrected Debye theory is the Fourier transform of the correlation function

$$C_\mu(t) = \frac{\langle \boldsymbol{\mu}(t) \cdot \boldsymbol{\mu}(0) \rangle}{\langle \mu^2 \rangle} = \exp\left\{ -\frac{kTI}{\beta_R} \left(\frac{\beta_R t}{I} - 1 + e^{-\beta_R t/I} \right) \right\} \quad (25)$$

This can be expressed in terms of the continued fraction

$$\frac{\alpha_p(\omega)}{\alpha_p'(0)} = 1 - \cfrac{i\omega/\beta}{i\omega/\beta + \cfrac{\gamma}{1 + i\omega/\beta + 2\cfrac{\gamma}{2 + i\omega/\beta + 3\gamma \ldots}}} \quad (26)$$

where $\alpha_p(\omega)/\alpha_p'(0)$ is the normalized polarizability. This expression was first obtained by Sack[25] from the probability diffusion equation corresponding to the Langevin equation

$$I\ddot{\theta}(t) + \beta_R \dot{\theta}(t) = \lambda(t) \quad (27)$$

Here γ is $kT/I\beta^2$; with

$$\tau_D = \frac{\beta_R}{kT} = \frac{I\beta}{kT}; \qquad \beta = \beta_R/I \quad (28)$$

The original Debye rotational diffusion theory is equivalent to the first convergent[3] of this continued fraction, that is,

$$\frac{\alpha_p(\omega)}{\alpha_p'(0)} = \frac{1}{1 + i\omega\tau_D} \quad (29)$$

with

$$\tau_D = \frac{\beta_R}{kT}$$

The second convergent gives Rocard's equation

$$\frac{\alpha_p(\omega)}{\alpha_p'(0)} = \frac{1}{1 + i\omega\tau_D - \omega^2\tau_D/\beta} \quad (30)$$

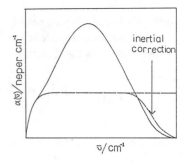

Figure 5. The effect of correcting Debye's theory for the missing inertial term.

which was the earliest attempt (1933) to correct the Debye theory for the missing inertial term. Rocard produced Eq. (30) from a Smoluchowski diffusion equation in which he replaced the term left out by Debye in 1913.

However, the only effect on the theoretical power absorption coefficient in the far infrared of taking any convergent of Sack's continued fraction is to produce a high-frequency return to transparency from the Debye plateau. The inertia-corrected Debye theory fails[25,26] to describe the observed far infrared power absorption coefficient, as illustrated in Fig. 5.

O. Rotational Diffusion in a Potential Well: The Itinerant Oscillator

There is something more fundamental lacking in the simple theory of rotational diffusion than a missing inertial term. The unrealistic nature of the rotational Langevin equation (8) when used to describe data from molecular diffusion is exposed starkly by a result such as that of Fig. 5. Several approaches have been devised[24] to meet this fundamental difficulty, sharing the common feature that the diffusion of an individual molecule was for the first time described theoretically as a process involving torsional oscillation or libration[12,29] with an ever-changing cage of nearest-neighboring molecules. Essentially, the featureless environment of the friction term in Langevin's own treatment was given some shape. It was recognized that a diffusing molecule must simultaneously undergo oscillatory motion of the diffusing center of mass, combined with torsional oscillation about this point in the laboratory frame (X, Y, Z). The influence of Langevin and his contemporaries was so overwhelming, however, that the new approach to diffusion theory which slowly emerged during these years tied itself to the pioneering concept of the Langevin equation, and the equivalent equations for the evolution of probability density developed[1-4] by Smoluchowski, Fokker, Planck, Klein, and Kramers. The new approach still stuck to the concept of rotational diffusion, and left translation out of consideration.

As we shall see, this turns out to be the only tractable way of extending the validity of rotational diffusion, the only way which avoids a morass of mathematical complexity dealing with too many unknown (empirical) parameters. We perceive for the first time very general and profound limitations on the concepts devised at the turn of the century.

One of the most obvious ways of attempting to deal with the failure of the Langevin equation (8) when faced with accurate data from the far infrared was to add a term on the left-hand side to mimic the effect of an extra torque generated by the immediate surroundings of a diffusing molecule. These surroundings are, of course, other diffusing molecules, which form a cage which itself diffuses with time. To put these ideas in simple mathematical form several assumptions were made which can only be justified by working out their consequences mathematically and comparing with experimental data in the complete range of available frequencies, not just in the dielectric range where every model gives closely similar results, difficult to distinguish from the bell-shaped curve, giving an illusion of explanation. The complete range extends from static (zero frequency) to the THz, as much as 14 frequency decades.

P. The Harmonic Approximation

The extra torque on the left-hand side of the Langevin equation is assumed to take the form

$$\frac{\partial V}{\partial \theta} = V_0 \sin \theta \tag{31}$$

If the angle θ is below about five degrees, then the sine can be replaced by the angle itself, so that the rotational Langevin equation becomes

$$I\ddot{\theta} + \beta_R \dot{\theta} + V_0 \theta = \lambda \tag{32}$$

The extra torque comes from assuming that the potential energy generated by torsional oscillation of a diffusing molecule in a cage of neighbors is a simple cosine. Differentiation of the cosinal potential energy with respect to angular displacement gives the torque. For small angles the sine is approximated by its argument, so that torsional oscillation occurs at the bottom of a potential "well". Implicit in the whole exercise are at least two further hidden assumptions, that the rotational diffusion of the molecule is statistically uncorrelated to its own translation, and that the torsional oscillation takes place in a plane, so that the nonlinearities of the Euler equations do not complicate the mathematics. Given all these assump-

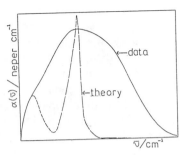

Figure 6. An attempt at reproducing the far infrared power absorption of a dipolar liquid with a harmonic approximation or with the planar itinerant oscillator of Calderwood and Coffey (schematic).

tions, the Langevin equation (32) may be solved by Laplace transformation for the far infrared power absorption coefficient and the dielectric complex permittivity. The mathematical expressions for these optical coefficients as a function of frequency (the spectra) are very complicated[1,4] functions of the friction coefficient β_R and the barrier height or well depth V_0. These two parameters cannot, however, be expressed in terms of the fundamental physical constants and are therefore empirical.

We arrive at the conclusion that the theory of rotational diffusion must be supplemented by an extra torque in order to provide even a qualitative theoretical explanation for experimental data in the electromagnetic frequency range from static to far infrared. This is necessarily an empirical procedure, because there are two parameters which must be varied for best fit of experiment and theory. A typical result of such an approach is illustrated in Fig. 6. The low-frequency part of the data range is matched satisfactorily, as is the case for many models of this type, but in the far infrared the harmonic approximation typically gives much too sharp a result. It provides limited progress from the Debye plateau, and approximants of the Sack continued fraction, the various "inertial corrections" of Debye's theory.

Q. The Itinerant Oscillator

This was developed initially by Hill[30] and Wyllie.[31] In the early 1970s, versions of the theory were first tested against far infrared data.[12,32] A later and simpler planar version was developed[33] which provided a clear explanation of its fundamental principles. All of the various models rely heavily on the theory of rotational diffusion, and decorrelate molecular rotation from molecular translation. The planar version[33] was developed to avoid the mathematical complexities of three-dimensional rotational diffusion.

The itinerant oscillator considers diffusion to be superimposed on oscillation. For rotational motion the latter is torsional oscillation ("libration") in a cage of nearest neighbouring molecules which is assumed to diffuse throughout the medium as a rigid entity. A molecule simultaneously diffuses and librates within the cage. It is assumed that a Langevin equation can be used to describe the diffusion of the outer cage. The theory[33] of Calderwood and Coffey also assumes that the rotational diffusion of both cage and molecule can be approximately planar, and that there is no friction between the cage and inner molecule.

With these assumptions, the Langevin equation of the cage is

$$I_1\ddot{\psi} + I_1\beta\dot{\psi} - I_2\omega_0^2(\theta - \psi) = I_1\dot{W} \tag{33}$$

linked to the oscillatory diffusion of the inner molecule

$$I_2\ddot{\theta} + I_2\omega_0^2(\theta - \psi) = 0 \tag{34}$$

Here I_1 is the effective moment of the cage (the "annulus" in two dimensions); I_2 is the moment of inertia of the inner molecule (the "disk" in two dimensions); θ and ψ are angular variables of the motion defined as follows. θ is the time-dependent angular displacement between the inner molecule's dipole moment and the applied electric field which measures the polarization in the technique of dielectric spectroscopy. ψ is the displacement between a point on the rim of the annulus and this electric field direction. The torque on the outer annulus due to Langevin friction is $I_1\beta\dot{\psi}$ and the random force on the annulus is $I_1\dot{W}$. Note that the equations of this, the simple planar itinerant oscillator, can be written as

$$I_1\ddot{\psi} + I_1\beta\dot{\psi} + I_2\ddot{\theta} = I_1\dot{W} \tag{35}$$

which is a simple Langevin equation with an added inertial term $I_2\ddot{\theta}$ on the left-hand side.

The mathematical solution of Eq. (35) is elaborate and incomplete, despite its physical simplicity. Full descriptions are given in the specialist literature.[1-4,33] Solutions are available, however, for the acf of the angular velocity, $\dot{\theta}$, of the encaged molecule; for the orientational acf, essentially the Fourier transform of the dielectric loss; and for the rotational velocity acf from the model, the Fourier transform of the far infrared spectrum. Using these results, the combined dielectric and far infrared spectra can be modelled with two parameters: β, the friction coefficient on the cage, and the harmonic torsional oscillation frequency, ω_0, of the itinerant oscillator, the encaged molecule.

R. Experimental Testing of the Itinerant Oscillator

The model has been tested thoroughly with far infrared and dielectric data
(see Sections 2–4). These were the first tests carried out on a theory of
rotational diffusion qualitatively able to account for new spectral features
in the far infrared. The theory contains β and ω_0^2 as parameters, and also
the ratio I_1/I_2, which are emipical. This reflects the somewhat superficial
nature of the original Debye approach of 1913, the ad hoc mixture of
concepts of the traditional approach to molecular dynamics. The problem
of comparing the theory with far infrared data was tackled[34–40] in several
ways, using the necessary broad sweep of data over the microwave and
far infrared regions in sample liquids such as dichloromethane, of C_{2v},
point-group symmetry. In one method the three parameters were optimized freely with least-mean-squares simultaneous minimization. This force
fit to the complete frequency sweep of available data exposed a fundamental flaw in the Calderwood Coffey variation of Debye's original theory, in
that the moment of inertia of the annulus had to be about eight times
smaller than that of the disk for most samples of molecular liquid. This
ratio is the wrong way around, because the cage is more massive than the
encaged molecule. For reasonable values of the paramcters β and ω_0^2 the
physically expected ratio $I_1/I_2 \simeq 10$ always produced a very sharp theoretical peak in the far infrared reminiscent of the simple harmonic librator,
Eq. (32). This was much sharper than the experimentally observed absorption. This problem defines the severe limitations of the original theory.[33]
Other methods of comparison provided the same result, for example the
friction coefficient, β, was estimated numerically by making sure that the
theory produced peaks at the right far infrared and microwave frequencies.
This method again produced the unphysical ratio of moments of inertia.

This defect was remedied[41–43] in 1987 by removing the assumption that
the interaction between molecule and cage is harmonic and frictionless.
Unfortunately, the remedy introduces a new parameter, the friction coefficient between the cage and the encaged molecule. However, the use of
four parameters finally manages to provide a fit to observed data over the
complete frequency range with physically acceptable moment of inertia
ratios. The new Langevin equations are

$$I_1\ddot{\psi} + I_1\beta_1\dot{\psi} - \mu F \sin(\theta - \psi) = \lambda_1 \tag{36}$$

$$I_2\ddot{\theta} + I_2\beta_2\dot{\theta} + \mu F \sin(\theta - \psi) = \lambda_2 \tag{37}$$

where β_2 is the new friction coefficient. These are simple linked Langevin
equations, still closely based on rotational diffusion theory, carrying with
them all its original flaws.

1. There are too many empirical parameters for the data available from the microwave and far infrared. It would require the simultaneous use of four independent data sources to define the four parameters, and this exercise has never been attempted.

2. There is a loss of physical realism in an effort to retain mathematical tractability, that is, the equations are written down so that they can be solved.

3. The fundamental dynamics neglect the Euler cross terms, because these are mathematically intractable.

4. These equations retain the drastic assumptions made at the turn of the century that rotation and translation are decorrelated. Contemporary computer simulation has exposed this as an erroneous assumption which produces an illusory understanding of diffusion.

5. Even with these assumptions, Eqs. (36) and (37) can be solved only in restricted special cases, and then only by recourse to the equivalent equations of probability diffusion, the Klein/Kramers equations.[41-43] Solutions are available for equal friction coefficients only.

6. With the free use of four parameters, even if a solution could be found for unequal friction coefficients, the equations fail if exposed to a sufficiently broad range of viscosity. In supercooled molecular liquids, for example, they cannot produce the observed split in the dielectric loss known[1-4] as the α and β processes, whose far infrared adjunct is the γ process. These three processes cover a very broad frequency range and exhibit a complexity of behaviour which no simple diffusion theory could describe.

7. The itinerant oscillator cannot follow phase changes, and neither can any theory based on Debye rotational diffusion.

8. The equations of the itinerant oscillator bear no direct or reasonable relation to the fundamental equations of classical mechanics, or statistical mechanics (the Liouville equation). This is so in any theory of rotational diffusion which does not use the memory function approach developed by Mori in 1965 and described in the specialist literature.[1-4]

The itinerant oscillator theory is an empirical description of N body dynamics which can "force-fit" data over a restricted range of viscosity. It is not tenable as a fundamental theory, and for this we need computer simulation.

S.   Relationship of the Itinerant Oscillator with the Liouville Equation

We consider the Mori equation (6), a form of the Liouville equation. This describes the conditional probability density function for all positions and momenta of the molecules in the ensemble under consideration. It can be written specifically for molecules in a molecular environment, rather than for Brownian motion. In principle, the Langevin equation is less useful to molecular dynamics than the rigorous Liouville equation. However, the latter is more suitable for use in the Mori form than in the form originally devised by Liouville in 1838.

Mori used projection operators to develop equation (6), which is formally the same as the Liouville equation. To illustrate the connection between the Mori equation and the itinerant oscillator of Eq. (35) consider the molecular angular velocity constrained to planar rotational dynamics as implemented by Debye. Assume that the Mori column vector $\mathbf{A}$ can be replaced by the single scalar entry $\dot{\theta}$, the time derivative of the two-dimensional angular displacement. This implies that the Mori resonance operator vanishes, which means physically that the relation between single molecule diffusion and cooperative effects generated by this diffusion is lost. With these assumptions the Mori equation reduces to

$$\ddot{\theta}(t) = - \int_0^t \phi(t - \tau)\dot{\theta}(\tau)\, d\tau + F_{\dot{\theta}}(t) \tag{38}$$

where ϕ is the memory function for two-dimensional Debye-type rotational diffusion. Inspection of this equation reveals that it reduces to the inertia-corrected Debye equation (27) when the memory function is a delta function in time

$$\phi(t - \tau) = \delta(t - \tau) \tag{39}$$

More generally, however, Mori showed tbat the integro-differential equation (38) is the fit in a chain of similar equations, whose first members are

$$\dot{\phi}(t) = - \int_0^t \phi_1(t - \tau)\phi(\tau)\, d\tau + F_1(t) \tag{40a}$$

$$\dot{\phi}_1(t) = - \int_0^t \phi_2(t - \tau)\phi_1(\tau)\, d\tau + F_2(t) \tag{40b}$$

The memory function is itself governed by a like equation, and so on for more indices. The Laplace transformation (L_A) of the chain of equations (40) produces the well known Mori continued fraction expansion of the angular velocity acf

$$L_A\left(\frac{\langle\dot{\theta}(t)\dot{\theta}(0)\rangle}{\langle\dot{\theta}^2\rangle}\right) = C(p) = \cfrac{1}{p + \cfrac{K_0}{p + \cfrac{K_1}{p + K_2 \ldots}}} \tag{41}$$

where K_0, K_1, and K_2, are constants.

There is a relation between this second approximant of this continued fraction and the Calderwood–Coffey itinerant oscillator, which was first derived in 1976.[44] The second approximant of the continued fraction (41) and the equivalent expression from the itinerant oscillator equations[33] for the same acf are formally identical. The itinerant oscillator equations (35) are therefore approximations at an early stage to the Mori equation (38), truncating the continued fraction after only two approximants. In the same way, the 1987 version of the itinerant oscillator is equivalent to a truncation procedure at the same approximant of a matrix continued fraction.

T. The Grigolini Continued Fraction

We have seen that the Debye theory of rotational diffusion, and its close relative, the itinerant oscillator, are special cases of the Liouville equation written in terms of a continued fraction. Grigolini[1-4] has shown that the continued fraction has a deeper significance in classical and wave mechanics.

Grigolini developed the continued fraction from the "Heisenberg equation"

$$\frac{\partial A}{\partial t} = L^\times A \tag{42a}$$

which is equivalent to the Mori equation in statistical mechanics. Equation (42a) is also a Liouville equation for the general dynamical variable A, a stochastic or random variable, not purely deterministic. The equivalent "Schrödinger equation" in this analogy is the Liouville equation applied directly to the probability density $\rho(\mathbf{a}, \mathbf{b}, t)$

$$\frac{\partial \rho(\mathbf{a}, \mathbf{b}, t)}{\partial t} = L\rho(\mathbf{a}, \mathbf{b}, t) \qquad (42b)$$

The operator L is the effective Liouvillian

$$L = L_\mathbf{a} + L_\mathbf{b} + L_1 \qquad (43)$$

made up of three parts, operating on the stochastic variables $\mathbf{a}$ and $\mathbf{b}$ of the equation (42b). The third part, L_1, represents the statistical interaction between the two sets of variables denoted by $\mathbf{a}$ and $\mathbf{b}$.

The starting point for the Grigolini continued fraction is Eq. (42b), which is obtained by working directly with a variable A of the set $\mathbf{a}$, and then building its time correlation function

$$\Phi(t) = \frac{\langle A(0)A(t) \rangle}{\langle A^2 \rangle_{eq}} \qquad (44)$$

where

$$A(t) = e^{\Gamma t}A(0); \qquad \Gamma = L^\times$$

$$\langle \ \rangle_{eq} = \int dA \, d\mathbf{b}(\ldots)\rho_{eq}(A, \mathbf{b})$$

Here ρ_{pq} is the equilibrium probability distribution. Analogously with Heisenberg quantum mechanics, the time-correlation function is the scalar product, a running time, or ensemble, average

$$\Phi(t) = \frac{\langle A \,|\, A(t) \rangle}{\langle A \,|\, A \rangle} \qquad (45)$$

In analogy with quantum mechanics the Mori continued fraction is obtained by choosing the basis set for the expansion of the operator $L^\times$ of Eq. (42a). The basis set is built up of repeated projections on to subspaces of the complete Hilbert space. Grigolini generalises Mori's treatment of the same problem with a biorthogonal basis set. This leads to the integrodifferential equation

$$\frac{d}{dt}\Phi(t) = \lambda_0 - \Delta_1^2 \int_0^t \Phi(\tau)\Phi_1(t - \tau)\,d\tau \qquad (46)$$

for $\Phi(t)$. Here

$$\frac{d}{dt}\Phi_k(t) = \lambda_k\Phi_k(t) - \Delta_k^2 \int_0^t \Phi_k(\tau)\Phi_{k+1}(t - \tau)\,d\tau \qquad (47)$$

and to the continued fraction in Laplace space

$$\Phi(p) = \cfrac{1}{p - \lambda_0 + \cfrac{\Delta_1^2}{p - \lambda_1 + \cfrac{\Delta_2^2}{\cfrac{\cdots}{\cfrac{\Delta_{n-1}^2}{p - \lambda_{n-1} + \Delta_n^2\Phi_n(p)}}}}} \qquad (48)$$

This has the same form as the Mori and Sack continued fractions, but is more general in applicability.[3] In Eqs. (46)–(48) λ is the equivalent of the Mori resonance operator, Φ_n the Grigolini memory function, and Δ_n^2 can be related to spectral moments and determined unequivocally in some cases. There is therefore a well-understood relation between the fundamental equation of motion and the less general diffusion equations, which they approximate. The itinerant oscillator is an approximant of the Grigolini continued fraction.

U. The Statistical Correlation between Rotation and Translation

The development so far has been restricted to separate consideration of translational and rotational diffusion. From first principles it is clear, however, that one form of motion occurs simultaneously with the other in molecular dynamics. Computer simulations of the last 10 years have shown conclusively that there is statistical correlation of many forms between one type of motion and the other in molecular ensembles. There are important hydrodynamic effects caused by the interaction between rotation and translation in fluid materials. On the molecular level none of these effects had been considered in the traditional approach prior to the use of computer simulation. The precise statistical interrelation between the linear and angular velocity of an asymmetric top molecule diffusing in three dimensions is a major unsolved problem of diffusion theory.

There have been several attempts at extending the theory of diffusion to describe the "roto-translation" of molecules rather than separated rotational and translational diffusion of pollen particles. These have to deal with elementary considerations such as the following. Let $\mathbf{u}$ be a unit vector joining the center of mass of a molecule to one of its atoms. Let the centre of mass velocity of the diffusing particle be $\mathbf{v}$ and the linear velocity of the atom $\mathbf{v}_a$. If $\boldsymbol{\omega}$ is the angular velocity of the complete molecule, assumed rigid, then

$$\mathbf{v}_a = \mathbf{v} + \tfrac{1}{2}\boldsymbol{\omega} \times \mathbf{u} \qquad (49)$$

The acf of $\mathbf{v}_a$ therefore contains information on both linear and angular velocities simultaneously, and the acf can be extended as follows

$$\langle \mathbf{v}_a(t) \cdot \mathbf{v}_a(0) \rangle = \langle (\mathbf{v}(t) + \tfrac{1}{2}\boldsymbol{\omega}(t) \times \mathbf{u}(t)) \cdot (\mathbf{v}(0) + \tfrac{1}{2}\boldsymbol{\omega}(0) \times \mathbf{u}(0)) \rangle \quad (50)$$

Using the fundamental kinematic relation[26] between the rotational velocity $\dot{\mathbf{u}}$ and the orientational unit vector $\mathbf{u}$

$$\dot{\mathbf{u}} = \boldsymbol{\omega} \times \mathbf{u} \qquad (51)$$

The expansion of the acf in Eq. (50) contains the *cross*-correlation function (ccf)

$$C_{\dot{u}v} = \langle \dot{\mathbf{u}}(t) \cdot \mathbf{v}(0) \rangle \qquad (52)$$

between the rotational and linear velocities of the diffusing molecule. This ccf exists directly in the laboratory frame (X, Y, Z). An example, from a recent computer simulation of liquid water, is shown in Fig. 7.

Computer simulations have shown[45] that there are many ccf's such as this, involving rotational and translational dynamics simultaneously at the molecular level. These are governed by powerful symmetry laws[1,46,47] of time reversal, parity reversal, and point group theory in frames (X, Y, Z) and (x, y, z). These laws will be developed later in this chapter into group theoretical statistical mechanics (gtsm). They allow the existence of some ccf's but forbid that of others in both frames of reference. One of the ccf's disallowed by parity reversal symmetry in frame (X, Y, Z) is

$$C_{\omega v} = \langle \boldsymbol{\omega}(t) \cdot \mathbf{v}(0) \rangle = 0 \qquad (53)$$

However, if we switch in to the frame (x, y, z) which is fixed in the molecule, and moves with it, this type of ccf becomes visible, and other types of ccf appear[47-52] which we cannot see in frame (X, Y, Z). These

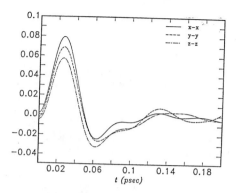

Figure 7. Illustration of the laboratory frame cross-correlation function (ccf) $C_{\dot{u}v}$ obtained from a recent super-computer simulation of liquid water by Evans, Lie and Clementi, IBM, Kingston.

have all been traced in the last few years with the help of computer simulation.[45,47-53] If the frame (x, y, z) is defined as that of the principal molecular moments of inertia, it is simultaneously rotating and translating with respect to frame (X, Y, Z), (Fig. 8). Every molecule has its frame (x, y, z). For each molecule, one frame can be matched with the other by a series of rotations, which define the Euler angles, for example. Any scalar, vector, or tensor quantity can be defined with respect to either frame.

The linear velocity, $\mathbf{v}$, for example, has the components v_X, v_{Y}, and v_Z in frame (X, Y, Z) which can be rotated into frame (x, y, z) with the use of the three unit vectors u_x, u_y, and u_z defined in the axes x, y, and z of the frame (x, y, z) through the rotation equations

$$v_x = v_X u_{xX} + v_Y u_{xY} + v_Z u_{xZ} \tag{54}$$

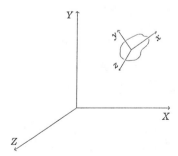

Figure 8. Illustration of frames (X, Y, Z) and (x, y, z) for a molecule diffusing in three dimensions.

$$v_y = v_X u_{yX} + v_Y u_{yY} + v_Z u_{yZ} \tag{55}$$

$$v_z = v_X u_{zX} + v_Y u_{zY} + v_Z u_{zZ} \tag{56}$$

In matrix notation these equations are

$$
\begin{bmatrix} v_x \\ v_y \\ v_z \end{bmatrix} =
\begin{bmatrix} u_{xX} & u_{xY} & u_{xZ} \\ u_{yX} & u_{yY} & u_{yZ} \\ u_{zX} & u_{xY} & u_{zZ} \end{bmatrix}
\begin{bmatrix} v_X \\ v_Y \\ v_Z \end{bmatrix} \tag{57}
$$

so that **v** in one frame is related to **v** in the other by a transformation matrix. Any vector can be transformed from one frame into the other in a similar way. There are major advantages of considering both frames. For example,

1. Group theory can be applied in both frames, revealing a great deal in frame (x, y, z) that is hidden in frame (X, Y, Z). Point-group theory has been highly developed in other branches of chemistry and spectroscopy, and the complete range of results can be used directly to study molecular diffusion, culminating in the three principles of gtsm (Sections VII and VIII).

2. The ccf's allowed by symmetry in either frame can be investigated for their time dependence by computer simulation, and increasingly, by experimental methods. These produce the details of the time dependence of each ccf for any molecular symmetry. The simulation can also be extended to vibrating and flexible molecules. It is generally applicable in all molecular point-group symmetries.

Steps 1 and 2, taken in the past 5 years or so, lead to advances in our understanding of molecular diffusion processes.

V. The Challenge to Traditional Diffusion Theory

The traditional approach has been overtaken by computer simulation and group theory. There has been a sudden increase in understanding based on group theoretical statistical mechanics. The traditional approach is challenged with describing the data now available from computer simulation, and also obtainable experimentally, through processes described in Sections VII and VIII. These are challenges posed by the much older classical equations of motion devised by Newton and Euler:

1. There are many different types of cross correlation that can be used to describe the diffusion of molecules in a molecular environment. These

exist in frames (X, Y, Z) and (x, y, z) and in the latter are governed by the molecular point-group symmetry. They always accompany a range of acf's in both frames which the theory must describe consistently. Some of the acf's and ccf's are experimentally observable. The theory of diffusion must be capable of describing the complete set of correlation functions from simple starting assumptions.

2. The Langevin equation is incapable of more than a general qualitative description, and must be modified to deal with new data from spectra and computer simulation.

3. The traditional rotational and translational diffusion theories are mutually exclusive. The original treatment of Debye left out the inertial term, causing a drastic failure in the far infrared. Reinstating this does not produce the observed far infrared spectrum. There is no rigorous justification for the use of simple rotational diffusion theory in the description of molecular diffusion processes.

4. The various itinerant oscillator theories attempt to extend the validity of rotational diffusion by ad hoc addition of torque terms to the rotational Langevin equation. This is essentially an empirical procedure which runs into difficulties caused by unknown parameters and great mathematical complexity. Some versions are further restricted by deliberate neglect of the nonlinear components of the Euler equations of motion.

5. The Langevin equation, and the diffusion equations in general, are not fundamental equations of motion such as the Liouville equation, and cannot be obtained from the latter without crude approximation.

6. Doob[20] showed as early as 1942 that there are inherent inconsistencies in the simple Langevin equation. The velocity may have no proper time derivatives. The velocity acf from the Langevin equation is an exponential decay, which is not differentiable at $t = 0$. The time expansion of these acf's violates the rules of classical statistical mechanics,[1] which imply that a classical acf must be time-even. Significantly, Doob showed that the Langevin equation is self-consistent only if expressed in the integral form

$$v(b) - v(a) = -\beta_v \int_a^b v(t)\, dt + B(b) \qquad (58)$$

where

$$B(b) - B(a) = \sum_{k=1}^n (B(t_k) - B(t_{k-1}))$$

This is reminiscent of the Mori equation (6). Doob anticipated this equation with its memory function replaced by a constant friction coefficient. The random force B in Eq. (58) must be a Wiener process.[1] The ramifications of this are described in the specialist literature,[14] but the mathematical niceties do not affect the drastic failure of the equation in the far infrared.

7. Most challenging is the failure of contemporary diffusion theory to describe the simultaneous translational and rotational dynamics of a molecule diffusing in three dimensions. This limitation is built in at the most fundamental stage of the theory, and has been exposed to view by computer simulation. Molecular rotation and translation are always correlated, even in spherical tops,[56,57] and the cross-correlation is governed in general by point-group theory,[58] and fundamental symmetry laws of physics. All known generalizations of the Langevin equation run into considerable difficulty when attempts are made to account consistently for the sets of nonvanishing ccf's and acf's generated by the simple fact that a molecule simultaneously rotates and translates. One can question the usefulness of the concept of diffusion at the most basic level. There is no known solution from diffusion theory at the time of writing which is able to describe the data from spectra and computer simulation without approximation, complexity, and overparameterization caused by empiricism. Computer simulation has advanced well beyond the tight boundaries of these theories. Computer simulation is a simpler and more consistent approach, because it relies on fundamental equations of motion. It has advanced to the stage where description of spectra is possible using these equations. Some of these spectra are described in the next section.

II. KEY EXPERIMENTS

A century ago, the Michelson–Morley experiment proved the fallacy of the luminiferous ether (lichtäther), leading to the great advances in the theory of relativity made by Lorentz, Fitzgerald, and Einstein. Einstein's scientific contemporaries based their theories of diffusion on his work of 1905, reviewed in Chapter 1, explaining Brownian motion in molecular terms. The probability diffusion equations from their work foreshadowed the emergence of wave and quantum mechanics in the 1920s.

The opening up of the far infrared by Michelson interferometry had a similar effect on the theory of diffusion. The Michelson interferometer, and the precision it brings to the measurement of the speed of light, destroyed the theory of the luminiferous ether, forcing the great advances in relativity theory. In the late 1960s the same interferometric technique was used to show up severe shortcomings in the received wisdom pertaining to molecular diffusion. In this chapter, the key late twentieth century observations and techniques are described which led to this turning point. Signs of the old theory's limitations showed up rapidly and almost contemporaneously in several different spectroscopic observations. Each of these is described with a view to explaining the need for computer simulation as a guide to further progress in this field.

A. Interferometric Spectroscopy of Molecular Liquids

From its inception in 1913 to the mid-1950s, the data available to test the theory of rotational diffusion were confined to spot frequencies in the range from static to the microwave (GHz). The experimental techniques for obtaining the dielectric loss of a molecular liquid as a function of frequency varied according to the frequency range of interest. A laboratory would be equipped for a total frequency sweep of several decades on the logarithmic scale. This would be accomplished with radio frequency bridges for the Hz to kHz range, Wayne Kerr bridges and so on to the MHz range, microwave apparatus in the GHz range. Microwave measurements were laborious and costly, needing waveguides, klystrons, generators, and other specialized apparatus. Measurements were typically possible at 2, 4, and 8 mm, using frequency doubling. Sweep-frequency apparatus[59] is a recent innovation which allows a spectrum to be measured in the MHz range of frequencies. Before that, spot frequencies only were available. A plot of dielectric loss or permittivity against the logarithm of frequency consisted of isolated points, widely separated on the frequency scale by regions about which nothing was known. Spectra were taken at very low resolution, in other words, in the hope of finding enough information to test a theory of diffusion. The overall objective was the study of molecular dynamics at low frequencies.

These data could not distinguish between different theories of diffusion without information at high frequency and better spectral resolution, obtainable in the GHz to THz (far infrared) frequency range. Spot frequencies in the kHz to MHz range, however accurate, cannot tell the difference between a flawed theory such as rotational diffusion (Section I) and more rigorous descriptions based on the fundamental equations of motion, for example Mori and Grigolini continued fractions and computer simulation. The reasons for this were discussed in Section I and can be traced to the tendency of the orientational acf to become exponential as $t \to \infty$ or at low frequencies. A complete description of the molecular diffusion process in liquids needs the far infrared frequency region as a guideline.

Some of the first indications of the way that molecular liquids absorb in the far infrared (about $2-250 \text{ cm}^{-1}$, where it overlaps with the infrared), were obtained by Poley in 1955.[60] Using spot frequency measurements he found that the effective dielectric loss of dipolar liquids in the very high-frequency end of the GHz range was consistently above the value expected from the Debye rotational diffusion theory, despite the fact that the latter seemed to produce good agreement with dielectric loss and permittivity data at lower frequencies. Poley's measurements were however laborious to repeat in other laboratories and were viewed with uncertainty. The true significance of his data was realized fully after the passage of a decade,[12] which saw the rapid development[1,28,29] of far infrared Michelson interferometry. Computers were harnessed to arrive numerically at the far infrared power absorption spectrum from the interferogram produced by moving a mirror in one arm of the Michelson interferometer. The process of obtaining the spectrum from the interferogram is Fourier transformation (Section I), and for this reason the technique is often known as Fourier transform spectroscopy.

B. The Basic Principles of Fourier Transform Spectroscopy[61] in the Far Infrared

The Michelson interferometer is a simple optical device driven essentially by a light source which produces broad band radiation according to Planck's Law. This is black-body radiation,[29] the intensity of which decreases rapidly with increasing wavelength. The relation between wavelength (λ) and wavenumber ($\bar{\nu}$) is a simple inverse

$$\bar{\nu}\lambda = 1 \tag{59}$$

so that in the far infrared, the intensity of black-body radiation is minute in comparison with, for example, the visible. This simple consequence of Planck's Law means that conventional prism or grating-based spectrometers, of utility in the conventional infrared range just below the visible, become difficult to use with accuracy as the far infrared range is approached. The two important frequency decades from 1 to 10 cm^{-1} and from 10 to 100 cm^{-1} are particularly difficult for grating spectrometers. The great advantage of the Michelson interferometer in the far infrared is that it utilizes the whole of the available radiation from the light source. This radiation is guided after collimation on to a beam splitter which produces two beams at right angles, one by refraction through the beam splitter, and the other by reflection. By positioning the beam splitter at 45° to the incoming radiation from the light source, the refracted and reflected beams travel at right angles to the two mirrors of the Michelson interferometer.[1,28,29] The two beams are reflected back along their paths, which recombine at the beam divider and optically interfere constructively or destructively according to phase difference. The resultant electromagnetic radiation is either refracted through the beam splitter into an optical detector or reflected back into the source.

C. The Interferogram

The interferogram is the intensity of electromagnetic radiation reaching the detector as the function of the distance of one mirror from that point at which both are equidistant from the beam divider. To build up an interferogram, one mirror is therefore displaced in the interferometer, either by stepping it mechanically or electrically, or moving it continuously. If the two mirrors are equidistant from the beam splitter, the two beams from each arm of the interferometer are exactly in phase and interfere constructively. If the mirror to beam divider distance in one arm is displaced by only half a wavelength then destructive interference occurs. With monochromatic radiation entering the interferometer the interferogram is a simple cosine. With polychromatic radiation it is a complicated pattern of maxima and minima, whose Fourier transform gives the spectrum. The power absorption coefficient of the molecular liquid under study is obtained by placing a carefully measured thickness of the liquid just before the detector, measuring its interferogram, and repeating the process with a slightly thicker specimen of liquid. The instrument function of the interferometer is compensated for by taking a ratio of the Fourier transform of the thick to the thinner liquid samples. The far infrared power absorption coefficient is then defined as

$$\alpha(\bar{\nu}) = \frac{1}{d} \log_e \frac{I_0}{I} \tag{60}$$

where d is the increment in liquid thickness, and I_0/I the ratio of radiation intensity at the detector for each frequency.

The technique is now well documented[1,28,29,61] and Fourier transform spectrometers dominate the market.[62] The interested reader is referred to this literature for further details. Some technical steps are necessary to go from the prototype optical set-up to a powerful instrument such as the Bruker IFS 113v[62] or those marketed by Nicolet or Grubb Parsons and several other companies. These steps include the following:

1. The inbuilt computers and software of the contemporary Fourier transform spectrometer are designed to include numerical compensation for discrete sampling of the interferogram, and finite distance travelled by the mirror. These artifacts are treated respectively with the apodisation (sampling) function and the window function. The former is a series of

delta functions and the latter a specially designed mathematical function which compensates for spurious oscillations caused by the unavoidable truncation of the interferogram.

2. The resolution of the spectrometer is determined by the maximum amount (Δd) by which the mirror can be displaced, and is given by

$$\Delta \bar{\nu} = \frac{1}{2\Delta d}$$

Fourier transform spectrometers provide very high resolution across a wide frequency range of four decades.

3. The spectral range of the Fourier transform spectrometer is determined essentially by the stepping distance, and to maximize the range and minimize the problem known as "folding" the mirror stepping distance is minimized. Different spectral ranges require different sources, optical materials, and beam splitters, and spectrometers are automized for different ranges. Most instruments cover the very far infrared to the visible, including the whole of the infrared range.

4. Another major advantage is that the whole range is covered at constant, high-resolution, unlike grating spectrometers.

5. With the use of sensitive liquid helium cooled detectors the upper end of the microwave range can be reached (about 2 cm^{-1}). This is 5 mm in terms of wavelength, or 60 GHz in frequency. Conventional waveguides reach 8 mm typically. The present author has obtained comfortable overlap[63-68] in several different systems using the accurate designs pioneered at the National Physical Laboratory in the U.K. and marketed by Grubb–Parsons and Specac. The overlap was accurate both in terms of frequency and power absorption coefficient (the spectral ordinate in neper cm^{-1}).

D. Key Spectral Data: The Challenge to Diffuion Theory

The challenge is exemplified by the data available for the simple asymmetric top, dichloromethane.[1,3,4] The complete electromagnetic spectrum[1-4] of this asymmetric top can stretch over many frequency decades, the more the greater the viscosity. This signals diffusion processes that evolve over an immense span of time, from picoseconds to years. In the dilute gaseous condition the far infrared spectrum of dichloromethane is a series of rotational lines[69-71] generated by the Schrödinger equation for a freely rotating molecule in an ensemble. The intensity distribution of these lines is governed by the laws of statistical mechanics. The full extent of the challenge to molecular diffusion theory can be gauged when we consider carefully what happens as the gas is condensed into a liquid, and this is cooled and then superooled below its normal freezing point at a given pressure.

E.   Condensation of Gas to Liquid

As a dilute gas of dichloromethane is compressed, the rotational lines broaden and merge.[72-75] The wave functions of the free molecular rotators are affected by the fields of force of other molecules. The energy associated with a particular quantum state of the free rotator is no longer defined sharply at one frequency (energy value) only. The disturbance produced by the fields of neighbouring molecules produces a spread of frequencies around each quantum line of the free rotor. The spectrum begins to merge into a broad band, and there is a transition from quantum mechanical descriptions to those based on statistical mechanics.

One of the problems with the theory of diffusion becomes apparent when the spectrum has merged into a broad band in the far infrared. The original quantum structure has disappeared. The statistical description of the broad band rests on building up an acf (Chapter 1) from the kinematic equation[1]

$$\dot{\mathbf{u}}^{(i)} = \boldsymbol{\omega}^{(i)} \times \mathbf{u}^{(i)} = \mathbf{A}^{(i)}\mathbf{u}^{(i)} \tag{61}$$

written separately for each particle $i = 1, \ldots, N$, where N is the total number of molecules in the liquid. The orientational acf is the running time average

$$\langle \mathbf{u}(t) \cdot \mathbf{u}(0) \rangle = \frac{1}{N} \sum_{i=1}^{N} (\mathbf{u}^{(i)}(0)^T \exp(\mathbf{A}^{(i)}t)) \mathbf{u}^{(i)}(0) \tag{62}$$

When N is of the order of Avogadro's Number (6.023×10^{23}) we can replace the sum by an integral involving two types of averaging, over the initial orientations of the dipoles and the second over the angular velocity probability distribution, which by classical statistical mechanics[76-78] is a Maxwell–Boltzmann distribution

$$P(\omega) = Z \exp\left\{ -\frac{1}{2kT} (I_x \omega_x^2 + I_y \omega_y^2 + I_z \omega_z^2) \right\} \tag{63}$$

Here the I's are the three principal moments of inertia and the ω's are the three components of the molecular angular velocity in the principal moment of inertia frame (x, y, z) fixed in the molecule (Chapter 1). Z is

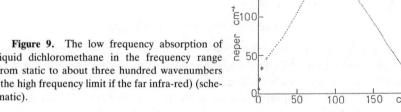

Figure 9. The low frequency absorption of liquid dichloromethane in the frequency range from static to about three hundred wavenumbers (the high frequency limit if the far infra-red) (schematic).

a constant. For the asymmetric top the orientational acf of the classical free rotor follows as[1]

$$\langle \mathbf{u}(t) \cdot \mathbf{u}(0) \rangle = \frac{1}{3} \int d^3\omega P(\omega)(1 + e^{i\omega t} + e^{-i\omega t}) \qquad (64)$$

If the three moments of inertia of the molecule are equal (the spherical top), this expression reduces to

$$\langle \mathbf{u}(t) \cdot \mathbf{u}(0) \rangle = \frac{1}{3} + \frac{2}{3}\left(1 - \frac{kTt^2}{I}\right)\exp\left(-\frac{kTt^2}{2I}\right) \qquad (65)$$

Now this expression, derived from the purely kinematic equation (61), is completely unlike the one derived from the three dimensional diffusion of the spherical top from the rotational Langevin equation (8) after the Debye theory[5] has been corrected[26] for the missing inertial term. Diffusion theory of this kind does not lead to the correct description of the free rotor when the liquid evaporates into a dilute gas.

F.   The Molecular Liquid at Room Temperature and Pressure

A combination of many careful measurements[1,4,79–85] on liquid dichloromethane at room temperature and pressure, using microwave spectroscopy at spot frequencies and Fourier transform spectroscopy for the far infrared region produces the result of Fig. 9. This spectrum stretches over about three frequency decades and is much broader than the envelope of the free rotor absorption, the Fourier transform of Eq. (64). Figure 9 expresses the result both in terms of the far infrared power absorption coefficient and the dielectric loss, using the link provided by Eq. (6). Not only must a complete theory of diffusion describe both the microwave and far infrared data consistently in terms of the fundamental constants but it should also be able to describe all further *spectral moments*.[86–90] The latter are related[1] in classical statistical mechanics to the classical acf's describing the molecular diffusion processes. Among the most prominent of these are the orientational acf, essentially the Fourier transform of the dielectric loss (zero-order spectral moment); and the rotational velocity acf which is the Fourier transform of the second moment, the power absorption coefficient. The next moment is the fourth, which is not observed directly, but which can be obtained[86-90] through the product $\omega^2 \alpha(\omega)$. Similarly the sixth spectral moment is $\omega^4 \alpha(\omega)$ and so on. Odd spectral moments can be constructed, but are of less interest, basically because the time expansion of a classical correlation function contains only even terms.[1]

The self-consistent description of zero and second spectral moments is difficult for contemporary diffusion theory. It is impossible for the original and corrected (Section I) theories of diffusion, and difficult for the itinerant oscillators, even with four parameters. The Mori and Grigolini continued fractions must be truncated at some stage, which introduce empiricism. Contemporary diffusion theory cannot come to grips at all with the many new ccf's from computer simulation[47-57] which indirectly affect the spectra. The fragility of the theoretical approach is exposed by the fourth spectral moment. Approximate analytical expressions for the far infrared power absorption can be obtained[1] from the 1977 itinerant oscillator, an approximant of the Mori continued fraction, and from the improved 1987 version.[41-43] However, if these complicated expressions are simply multiplied by ω^2 to form the fourth spectral moment, they result in plateau absorption which persists indefinitely at high frequencies, an obviously unacceptable result. The fourth moment plateau is the same kind of disaster as the Debye plateau in the second moment.[1,12] The Debye theory fails for the second moment and the itinerant oscillator for the fourth moment. Any force fitting of the itinerant oscillator is bound to unravel at the fourth moment. Similarly, higher approximants of the Mori continued fraction will fail at higher moments, according to the level of truncation.

The only theory discussed so far that maintains integrity for all spectral moments is that of the free rotor ensemble leading to Eq. (64). This is limited to the case where there is no molecular interaction.

G. Molecular Dynamics Simulation

This is a technique[91] that now pervades about 40% of all the literature in physical chemistry and related disciplines. Some of its powerful results have been discussed in Section I. More details of the method will be discussed in Section 3. When faced with spectral moments, however, even this technique runs into well-defined limitations, even though it has left the traditional approach on the blocks. It is as well to describe these limitations here, and to emphasize the fundamental importance of experimental data, accurate and wide ranging, for simple molecular liquids. No

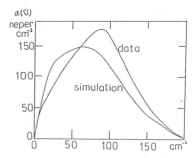

Figure 10. Computer simulation of the experimental absorption of Fig. 9. Simulation; — experimental data.

theory or numerical technique, however powerful, is the last word in natural philosophy.

Contemporary computer simulation methods rely heavily[91-94] on the numerical approximation of the classical equations of motion. These approximations involve expansions of fundamental dynamical variables, for example

$$r(t + \Delta t) = r(t) + \dot{r}(t)\Delta t + \frac{1}{2}\frac{F(t)\Delta t^2}{m} + O(\Delta t^3) \tag{66}$$

$$r(t - \Delta t) = r(t) - \dot{r}\Delta t + \frac{1}{2}F(t)\frac{\Delta t^2}{m} - O(\Delta t^3) \tag{67}$$

illustrating the leapfrog algorithm[91] for translational motion. Here r is the position of the molecular center of mass, $F(t)$ the net force on the molecule, and m its mass. In a digital computer, with a finite storage capacity and speed, these expansions must be truncated at some order. From this it can be shown that only some of the complete set of ccf's (Section I) can be obtained numerically, depending on the order of the truncation of a series expansion such as Eqs. (66) and (67).

Figure 10 is an illustration of the match between simulation[1,4] and far infrared absorption for dichloromethane liquid at 293 K and 1 bar. The simulated power absorption has been obtained by Fourier transforming the rotational velocity acf obtained from the dynamical trajectories generated by the 108 molecules used. The orientational acf, related to the dielectric spectrum at the lower end of the far infrared, can be obtained from the same trajectories. The close match between simulation and spectra, which can now be obtained with model or ab initio potentials, must be tempered with the realization that the simulation involves approximations. These typically include the following:

1. The success of the simulation is limited by its inherent confinement[91]

to the intermediate frequency range covered by the second spectral moment. Accurate definition of the frequency dependence of the fourth moment,[86-90] and of higher moments, would require simulated time-correlation functions of higher derivatives of the molecular dipole moment, and because of the approximations illustrated earlier for the leapfrog algorithm, for example, these are rarely if ever available or are not defined accurately. In exceptional cases, higher-order algorithms have been used[95] to generate the required acf's for higher spectral moments, but little is known in general. Similarly, at long times, conventional computer simulation is limited by several different factors, and does not extend much further than the nanosecond range. Production of the low frequency region of the spectrum of Fig. 10 by simulation is therefore more difficult than the far infrared range.

2. The spectrum of liquid dichloromethane over the three-decade span of Fig. 9 stems from cooperative molecular dynamical processes involving ccf's between different molecules. The dielectric permittivity cannot be built up properly without sufficient consideration[97-100] of these multimolecular effects.

Despite these limitations, computer simulation is able to build up an acceptable representation of the second moment spectrum of liquid dichloromethane without the degree of empiricism of conventional diffusion theory. Increasingly, it is possible to solve the equations of motion of a larger and larger number of moving and interacting molecules interacting with potentials modelled or computed ab initio. An analytical representation of the potential energy between two molecules as a function of intermolecular separation and relative orientation in frame, (X, Y, Z) is known[101-103] as the "pair potential" and will be described in more detail in Chapter 3. Progress is being made rapidly towards constructing the pair potential from fundamental quantum mechanical principles, and towards replacing[104-106] the older model representation of the pair potential based on atom to atom Lennard–Jones parameters.

H. From Liquid to Glass at Constant Pressure

Solutions of dichloromethane in solvents such as decalin can be supercooled[107] well below the normal melting point at room pressure. This has the effect[1] of increasing the viscosity of the solution by several orders of magnitude. In the supercooled liquid solution of dichloromethane in decalin this has the effect[1,107,108] of spreading the dielectric–far infrared spectrum over a vast frequency range. Not only is the peak of the zero moment spectrum (the dielectric loss) split into two and broadened considerably over the simple Debye result, but it is also removed from the peak of the second moment spectrum (the power absorption coefficient) by as much as about 10 frequency decades. The decalin solvent which supports the dichloromethane molecules in their supercooled environment undergoes a transition at a well-defined temperature[1] into a glass, and below this glass transition temperature the splitting of the peak of the very broad zero-order spectral moment disappears. As the temperature of the glassy solution is lowered further the zero moment peak becomes very broad, and becomes separated from the second moment peak in the far infrared by as much as fourteen decades of frequency. The peaks at low frequency are known as the α and β processes, and that in the far infrared γ process, recognized in 1980.[107,108] Clearly, any attempt to explain this development must involve molecular dynamical evolution from picoseconds to years. The zero-moment peak shifts into regions of very low frequency as this cooling process is continued. The peak of the second moment spectrum, the far infrared γ process, shifts upwards in frequency[1,107,108] as the environment in which the dichloromethane molecules librate becomes more and more constricted in the glass at low temperature. This constricting effect is counterbalanced by loss of energy (temperature) but is dominant in the far infrared. This is true generally, the zero- and second-order spectral moments move in opposite directions with temperature in a molecular liquid, above and below the normal melting point in supercooled solutions. This pattern of behaviour can be reproduced qualitatively by the itinerant oscillator, but with the lavish use of empiricism.[1] The diffusion theory available at the time of writing cannot describe, even qualitatively, the α, β, and γ processes and their frequency dependence as a function of temperature and pressure.

I. The Challenge to Computer Simulation

There have been many attempts at investigating the glass transition with computer simulation, but none has been completely successful in representing the multidecade spectral profile. The original, constant volume, algorithms of computer simulation[91] have been successful in many areas, but are not capable of producing phase changes in general. There have been some limited successes with rotator phase to liquid transitions, but the onset of crystallization is not well described by contemporary simulation algorithms. The importance of combined dielectric and far infrared spectroscopy is clear therefore. Even the most powerful contemporary theories and computer simulation algorithms are faced with the challenge of describing self-consistently the first few spectral moments obtained from observation of a simple asymmetric top such as dichloromethane, as it is gradually condensed from gas to liquid and supercooled in solution. Accurate data from these experimental techniques assume key importance when it is realized that even the most advanced contemporary diffusion theory, such as the itinerant oscillator, fails at the fourth moment, and that most contemporary simulation algorithms fail at the fourth or sixth. It follows that it is critically important to study a specimen of molecular liquid such as dichloromethane whose high-frequency power absorption is clearly defined and free of interference from higher frequency infrared vibrational modes. It is possible to construct the fourth, and even the sixth, spectral moments of dichloromethane[86-90] bcause of its clean far infrared spectrum.

A combination of dielectric and far infrared data is a critical and continuing challenge to contemporary theories of molecular diffusion, and has made the most important advances in challenging the traditional approach. Other key experiments have also played a significant role in the advances of the last 20 years, and will be described as follows with reference to the same test liquid, dichloromethane, typical[1,4] of the largest class by far of molecular symmetries, the asymmetric top.

J. Survey of Data for Liquid Dichloromethane

A detailed survey[1] has provided the following picture of investigations into molecular dynamics using many of the available experimental techniques.

1. There is a great deal of data, but not well coordinated. These have been produced using far infrared absorption, acoustic dispersion, and nuclear magnetic resonance relaxation, with a smaller amount of Raman scattering from the molecular liquid.

2. The thermodynamic properties of the molecular liquid at equilibrium are fairly well known, but there is a critical gap when dealing with virial coefficients. There are few data on pressure second and third virial coefficients, and no dielectric virial coefficients.[24]

3. Partly because of this insufficient data base, in particular virial coefficients,[24] the intermolecular potential is not well defined. The existing descriptions are built up empirically, from the Lennard–Jones parameters,[12] for example, and have had limited success in describing experimental data sensitive to the pair potential. There have been no attempts so far to build up the dichloromethane pair potential from first, quantum mechanical, principles. The computer power and basis sets for such a calculation are however available, following the examples set recently for liquid water.[109-112]

4. There is a smaller amount of work available using modern techniques of light scattering, including Brillouin and Rayleigh scat-

tering, and techniques such as coherent anti-Stokes Raman and inverse Raman scattering.

5. There are a few Kerr effect studies,[24] the polarizability of the molecule is known, but its anisotropy less accurately. Some computations are available of the effective charges on the atoms of the dichloromethane molecule from quantum mechanical first principles.

6. There is usually no investigation of the role of pressure in these spectroscopic and thermodynamic observations, although high pressure spectral techniques are well developed.

7. Overall there is a pronounced lack of coordination and interlaboratory cooperation in the study of this simple asymmetric top. Contemporary progress can be made, however, with ingenuity in the face of sparse data.

K. The Relaxation of Nuclear Magnetic Resonance

The archives on nuclear magnetic resonance run into millions of printed pages.[113] However, the conclusions reached with this technique about molecular dynamics are still equivocal.

The physical process of interest is the relaxation of nuclear magnetic resonance, caused by magnetic interactions between nuclei of molecules which are diffusing in a liquid or other environment. Different isotopes or nuclei carried by a diffusing molecule have different resonance frequencies, and this can be used experimentally[113,114] to obtain information that is not available in dielectric–far infrared spectroscopy. This is the anisotropy of the molecular diffusion process, expressed as correlation times[1,4] of different orientation axes in the same molecule. A restriction of NMR relaxation techniques in general is that they can provide only correlation times, not the full time dependence of the correlation function. Frequencies involved in NMR relaxation are relatively low, compared with THz of the far infrared.

In the past 20 years several different types of NMR relaxation have been utilized for the study of anisotropy in molecular diffusion, and the results expressed as different types of correlation time.

1. The longitudinal, or spin-lattice, relaxation time can be measured in a simple ensemble of molecules containing nuclei with a single spin 1/2 placed in an external magnetic field.

2. The transverse, or spin–spin, relaxation time is measured perpendicular to the applied magnetic field. As for the spin–spin relaxation time, it measures the rate of decay of magnetization.

3. A number of NMR relaxation mechanisms can be used to probe

molecular dynamics. Among these are nuclear magnetic dipole–dipole and dipole–quadrupole interactions among nuclei on the same or different molecules, techniques which depend intricately on the details of the molecular diffusion process, both intra- and intermolecular. They can be used to examine diffusion in molecules which carry the interacting nuclei. A third example is spin rotation interaction, which is due to the fact that a moving charge produces a magnetic field, causing the atomic nuclei of molecules in certain rotational states to experience additional magnetic fields. The spin rotation correlation time involves both the molecule's reorientational motion and the change in molecular angular momentum due to torques exerted by other molecules. It is therefore almost uniquely useful to produce information from the angular momentum and orientational correlation function simultaneously.

4. Other relaxation mechanisms such as those involved in the anisotropic chemical shift and technical advances such as nuclear Overhauser enhancement and Fourier transform NMR spectroscopy can also be used to provide information about the overall molecular diffusion process.

In proton relaxation, dipole–dipole interaction usually dominates is molecular liquids, but quadrupole relaxation is overwhelmingly dominant in the relaxation of molecular nuclei with spins greater than 1/2, such as ^{13}C and ^{15}N. In other circumstances, appreciable contributions may result from other mechanisms cited above.

The results from several different studies of this nature have been reviewed for dichloromethane liquid[114] and compared with complementary data from the far infrared and microwave. Brier and Perry[114] found that the relaxation times from different literature sources were at variance with each other but there was enough consistency to show that rotational diffusion could not describe the data. For example, the spin lattice relaxation time was not a third of the dielectric relaxation time, as expected from Debye's inertialess theory of rotational diffusion (Section I).

The opportunity does exist in NMR relaxation to look at different nuclear resonance phenomena. If this is done over a range of temperature and pressure the anisotropy of the molecular diffusion process can be defined. A recent review[4] details the available evidence in this context and summarizes quantitatively with the help of computer simulation of dichloromethane.

L. The Challenge to the Theory of NMR Relaxation

The challenge becomes clear when data from NMR relaxation are used with those from other sources,[24] such as the far infrared, dielectric region, and Kerr effect relaxation. Such a combination,[4,114] used with computer simulation, shows that the traditional approach of (Section I) is unable to match the data quantitatively. The NMR relaxation data are often complicated and difficult to reduce to the form needed for comparison with theory.[114] Despite this and other drawbacks, NMR relaxation is uniquely capable of giving correlation times on the anisotropy of the molecular diffusion process if experimental coordination is achieved among laboratories.

M. Anisotropy of Diffusion in the Asymmetric Top

The nature of this challenge requires a proper description of the theory of anisotropic molecular diffusion in three dimensions in an asymmetric top. Section I has described the rotational diffusion theory for the spherical top, or alternatively in the simplified case of a diffuser in a plane. In both cases the nonlinear Euler cross terms disappear and the mathematics become tractable. If the data indicate however that the diffusion is aniso- tropic, a more complicated theoretical structure is needed, based on the Euler–Langevin equations[1–4]

$$I_x\dot{\omega}_x - (I_y - I_z)\omega_y\omega_z = -I_x\beta_x\omega_x + I_x\dot{W}_x \tag{68}$$

$$I_y\dot{\omega}_y - (I_z - I_x)\omega_z\omega_x = -I_y\beta_y\omega_y + I_y\dot{W}_y \tag{69}$$

$$I_z\dot{\omega}_z - (I_x - I_y)\omega_x\omega_y = -I_z\beta_z\omega_z + I_z\dot{W}_z \tag{70}$$

which are written in frame (x, y, z) of the principal molecular moments of inertia I_i. Here ω_i are the molecular angular velocity components in this frame and β_i are three empirical friction coefficients. Unless there are available three independent data sources, they are not separately definable. The mathematical structure of these equations is complicated, and the solutions, even though approximate, are even more so,[1,4] suffering from the following drawbacks.

1. As for all rotational diffusion theories, the data from the dielectric and far infrared regions are not describable quantitatively.[1,5,12]
2. They always involve the three unknown friction coefficients.
3. They suffer from all the drawbacks of Debye type theory as de- scribed in Chapter I.

Without considerable theoretical modification and additional experimental information, properly coordinated, the evidence for anisotropic diffusion from NMR relaxation must remain qualitative in nature. On the other hand computer simulation can be used directly to provide anisotropic correlation times in dichloromethane.[4]

N. Light Scattering

The elements of the theory of light scattering[46] have been known since the turn of the century through the work of Rayleigh and his contemporaries. In recent years, lasers have brought greater precision to these investigations, especially to those of the depolarised component of light scattered from a liquid.

The scattering of light at right angles is often reported, using VV, VH, HV, and HH configurations, where V stands for vertical and H for horizontal scattered or incident light. These configurations describe changes in the polarization of the scattered light, due to the fact that when electromagnetic radiation is incident on a diffusing molecule it induces the dipole moment

$$\boldsymbol{\mu}(t) = \boldsymbol{\alpha}(t) \cdot \mathbf{n}_0 \cdot \boldsymbol{\epsilon}_0 \tag{71}$$

where $\mathbf{n}_0$ is the polarization vector of the applied field and $\boldsymbol{\epsilon}_0$ is the electric field strength. The tensor $\boldsymbol{\alpha}$ is the molecular polarizability. Equation (71) is written in the linear approximation, as the intensity of the laser is increased higher-order scattering effects become observable through higher-order polarizability tensors.[24] The polarizability can be split into two parts, one is determined by the molecular framework and its point-group symmetry, and the other by the distortions introduced by the bond vibrations within the molecule. The first component gives rise to polarized Rayleigh–Brillouin scattering, which is essentially due to fluctuations in density, and also to a depolarised component dependent on the orientational behavior of the individual molecules as they diffuse and scatter light. The vibration dependent parts are responsible for Raman scattering. Molecular diffusion processes contribute[46] to all four types of scattering in one way or another, and spectra from light scattering contain information on molecular dynamics. Depolarized Rayleigh scattering occurs at THz frequencies, and provides by Fourier transformation the second Legendre polynomial

$$C_{2R}(t) = \tfrac{1}{2}\langle 3(\mathbf{u}(t) \cdot \mathbf{u}(0))^2 - 1 \rangle \tag{72}$$

which involves the orientational acf in the ideal self dynamic limit when all intermolecular cross correlations are ignored. Depolarized Rayleigh scattering contains the same kind of information as that found in far infrared bandshapes, except that the latter is a first-order effect.[1] Theories of molecular diffusion should therefore be capable of describing both types of spectra consistently. Essentially speaking, if the intensity of the depolarized Rayleigh spectrum $[I(\omega)]$ is multiplied by ω^2 the result is very similar in frequency dependence to the far infrared power absorption under the same conditions. The Fourier transform of

$$M_2(\omega) = \omega^2 I(\omega) \tag{73}$$

is the second time derivative of $C_{2R}(t)$.

Both the depolarized Rayleigh spectrum and far infrared absorption contain information about acf's, intermolecular orientational ccf's, and the numerous[47-57] cross correlations between rotational and translational diffusion. The spectra complement each other and should ideally be used in combination in any investigation. Rotational diffusion produces the result[5] that the orientational correlation time from Rayleigh scattering should be three times shorter than that from the first rank equivalent (the first Legendre polynomial).[46] If this is not the case experimentally, then diffusion theory must be modified.

Polarized, Rayleigh–Brillouin, scattering is due to fluctuations in the positions of the diffusing molecules, and is a means of investigating translational processes superimposed on orientational motions. It is an essentially cooperative process, and is described theoretically in terms of the hydrodynamic concept[46] of fluctuation density. The many cross correlations between rotational and translational molecular motion (Section I) contribute to both polarized and depolarized light scattering. Rayleigh–Brillouin scattering involves the creation and destruction of phonons (sound waves), and is the high-frequency adjunct of sound dispersion.[46] This is roughly analogous to the way in which far infrared absorption is the high-frequency adjunct of dielectric relaxation and dispersion. It is observed in VV polarization, but is influenced only indirectly by the motions of individual molecules.[115-120] Depolarized Rayleigh scattering depends in contrast on the reorientation of individual molecules and is observed in VH or HV geometry. In the same way that far infrared spectroscopy and dielectric relaxation should be used to complement each other, so should ultra and hyper sound dispersion and light scattering. A more thorough investigation would also utilize other techniques such as NMR relaxation.

Using the test molecule dichloromethane, it is possible to review the extent to which contemporary investigators have achieved this aim of thoroughly probing the molecular dynamics of one liquid with complementary techniques.[1,4] This review reveals an imbalance of effort. There are many investigations on the hyper and ultra acoustic properties, revealing a dispersion of sound in the GHz range, analogous to dielectric loss in the microwave frequency region. A few reports of Brillouin scattering have deduced hypersonic velocities in the liquid up to 7.2 GHz. There are far fewer measurements involving Rayleigh scattering reported in that review,[1,4] however, and these were carried out under inconsistent experimental conditions, thus blurring the overall picture.

O. Raman Scattering

The frequency shifts of Raman scattering are due essentially to differences in the energy of the molecule due to internal bond vibrations. The molecule absorbs radiation in one state and reemits the radiation in another, via the induced dipole moment. A Raman spectral feature is dependent on the derivative of the polarizability with respect to coordinate. The polarizability tensor of the diffusing molecule fluctuates as the molecule vibrates, and in general, vibrational Raman scattering is influenced by rotational diffusion. Any attempt to separate the two processes is approximate, and in some cases, meaningless. There are always cross correlations between vibration, rotation, and translation in a diffusing molecule which can be investigated with computer simulation but which are intractable with the traditional approach of Section I.

Depolarized Raman scattering is always the result of combined rotation and vibration, superimposed on translation of the molecule's diffusing centre of mass. Individual Raman features in a liquid are broadened by the molecular diffusion process. Depolarization is caused by the rotational motion of the diffusing molecule. If the scattered light is not depolarized, the correlation function is that of the vibrational mode at the frequency at which the light is scattered in VV geometry. The intensities of polarized and depolarized Raman spectra about the vibrational frequencies of interest can therefore be expressed essentially as the Fourier transforms

$$I_{VV}(\omega) \propto \int_{-\infty}^{\infty} \exp(i\omega t) C_{\mathrm{vib}}(t)\, dt \tag{74}$$

$$I_{VH}(\omega) \propto \int_{-\infty}^{\infty} \exp(i\omega t) C_{\mathrm{vib}}(t) C_{2R}(t)\, dt \tag{75}$$

where C_{vib} is the acf of the vibrational dynamics of the mode of scattering under consideration. As the molecule rotates, the polarizability tensor varies with time, giving rise to a rotational band, a broadening centered on the vibrational frequency of each proper mode in the spectrum. Rotational broadening effects predominate for small molecules dissolved in inert solvents and vibrational broadening effects predominate in heavy polyatomic molecules or in the presence of hydrogen bonding. If it is assumed that the rotational and vibrational effects are not cross correlated, that is, that C_{2R} is statistically independent of C_{vib}, then the complete Raman spectrum of a dipolar molecule in the liquid state can be used to obtain information about the nature of molecular diffusion. For each peak in the spectrum the polarized Raman component provides the vibrational correlation function by careful Fourier transformation, and the depolarized spectrum provides according to Eq. (75) a convolution of the rotational and vibrational correlation functions.

A review[1] of seventeen reports on the test liquid dichloromethane revealed information on the Raman and hyper-Raman spectroscopy of the molecular liquid carried out under different experimental conditions. There was no clear indication[121,122] whether the vibration and rotation were accounted for accurately, that is, whether cross correlation was considered or discarded, but there were reported careful corrections for the internal field effect, the amplitude increase in the intensity of scattered radiation caused by differences in the field incident locally on the diffusing molecule, and the original intensity of the incoming laser beam.

There was, however, little attempt at coordinating the results with those from other techniques. Only a few of the available Raman bands had been utilized in these papers.

P. The Challenge to the Traditional Approach to Light Scattering

There are several different ways in which a combination of Rayleigh and Raman scattering can challenge the original theories of rotational diffusion and the itinerant oscillator. Correlation times can be compared from both techniques[4] and with the equivalents from computer simulation. The one from depolarized Rayleigh scattering was 1.85 ps, and equivalents from dielectric relaxation ranged from 0.5 ps in dilute solution to 1.45 ps in the pure liquid. The one from Rayleigh scattering is not three times shorter than that from dielectric relaxation in the pure liquid, as required by the traditional approach. Various other factors contribute which are absent in the theory of rotational diffusion and in related single molecule theories (those which deal with acf's of orientation).

1. Depolarized Rayleigh scattering contains information about statistical cross correlation of the molecular dynamics of one molecule with its diffusing neighbors. In this case the ccf takes the form

$$C_{2Rc}(t) = \frac{1}{2} \left\langle 3 \sum_{i,j} \mathbf{u}_i(t) \cdot \mathbf{u}_j(0) - 1 \right\rangle \tag{76}$$

which reduces at $t = 0$ to the second-order Kirkwood factor

$$K_{2Rc} = \frac{3}{2} \left\langle \sum_{i,j} \mathbf{u}_i \cdot \mathbf{u}_j \right\rangle - \frac{1}{2} \tag{77}$$

In general these cross correlations can be accounted for by implementation of Mori theory.[1]

2. Rayleigh–Brillouin scattering poses the problem of rotation–translation coupling.[46-57] This is imperfectly understood with conventional diffusion theory.

3. Raman scattering involves the cross correlation of intramolecular vibration and molecular rotation, and contemporary computer simulations[123-125] are addressing themselves to this problem through the appropriate ccf in the laboratory frame (X, Y, Z) and the moving frame (x, y, z).

Q. Neutron Scattering[126]

Neutrons are scattered from molecular liquids following the same fundamental equations as light (photon) radiation. The neutron has a finite mass and transfers momentum to the diffusing molecule in an inelastic collision. Wave particle duality applies to the neutron[126] and the equivalent wavelength is in the far infrared range in conventional scattering experiments with neutrons. Relatively low-energy neutrons ("cold" neutrons) are scattered inelastically from molecular liquids. The scattering of laser radiation, on the other hand, is elastic. The hydrogen atom is about as massive as the neutron, and has a large scattering cross section. Neutrons also possess spin,[126] and this adds an extra dimension to the theory of inelastic neutron scattering. The scattered neutron wave contains a coherent and incoherent portion, containing information respectively on individual molecular diffusion and on collective dynamical evolution. This information can be obtained separately for hydrogen-containing molecules because the nucleus of the hydrogen atom has a large spin incoherent cross section. Incoherent, inelastic, neutron scattering provides information on molecular diffusion.

However, there are no selection rules which limit the number of Legendre polynomials which can contribute to the scattering process through the relevant time-correlation functions of orientation. The theory of neutron scattering involves a weighted sum of spherical harmonics of all orders.

This sum is obtainable theoretically[114] only in very simple cases, from the inertialess Debye theory or from the equation (65) of free rotation.

The challenge of neutron-scattering data can be met properly only with computer simulation, which provides a consistent analysis for all the contributing spherical harmonics, and cross refers to other types of data. For the test liquid dichloromethane there has been a thorough analysis by Brier and Perry[114] of the incoherent inelastic neutron scattering spectrum at four scattering angles. The results were compared with NMR data from several sources. The data were sufficient to show up the limitations in the few simple theories considered by Brier and Perry. Neutron scattering would be more incisive, however, if used consistently with far infrared absorption and dielectric relaxation, for example, because neutron scattering data is low in resolution with large instrument corrections. New reactor designs are needed for progress, and this is more expensive than all other techniques put together.

R. Infrared Absorption

Infrared studies of molecular diffusion can be summarized symbolically through the Fourier transform

$$IR(\omega) \propto \int_{-\infty}^{\infty} \exp(i\omega t) C_{\text{vib}}(t) C_{IR}(t)\, dt \tag{78}$$

between the spectrum and the vibrational and rotational acf's. The orientational acf is[1,4] $\langle \mathbf{u}(t) \cdot \mathbf{u}(0) \rangle$ and is assumed to be unaffected by intermolecular correlations in the customary approach. If this assumption is accepted the broadening of the infrared spectrum would be a unique measure of the orientational dynamics of a liquid. However, this is almost never the case in practice, because of statistical crosscorrelation with vibration which cannot be deconvoluted as in Raman scattering, by switching from polarized to unpolarized spectra. Infrared bands are also affected in general by isotopes, Coriolis coupling, and vibrationally excited states (hot bands).

Nevertheless, the information in infrared absorption bands is clean enough to be a significant challenge in its own right to simple diffusion theory. In dichloromethane, for example, the nine fundamental vibrational modes can be used to define the anisotropy of molecular diffusion. The nature of rotational diffusion about each axis of the frame (x, y, z) can be observed in priniple[4] with an appropriate fundamental and the way it has been broadened about its peak frequency. It is assumed that the vibrational acf C_{vib} can be obtained from Raman scattering, but in dichloromethane, this is available only for the totally symmetric vibrational mode. The required information from the Raman is not easily extracted because the Raman band shapes are not related straightforwardly to the reorientation of the symmetry axes of frame (x, y, z). Therefore, despite having available infrared active vibrations parallel to all three axes of frame (x, y, z), the lack of data on vibrational relaxation is a severe constraint. This can however be remedied with contemporary simulation algorithms.

In the 70 or so papers reviewed in Ref. 1 there are many attempts to factorize the vibrational and rotational correlation functions which ignore the cross correlation between these modes. The result is a conflict of conclusions with other sources such as NMR relaxation.[114] There are also discrepancies of up to 50% between results in different papers on the same dynamical process (e.g., diffusion of the dipole axis) using sources from the infrared and NMR relaxation. This limits the usefulness of the data until greater coordination is forged.

S. The Role of Computer Simulation in Data Coordination

Computer simulation is essentially a numerical method[91] of solving the classical equations of motion for a small number of diffusing molecules, typically in the range from 100 to 1000. Depending on the numerical approximations used in integrating the equations of motion, a number of useful time ccf's can be constructed directly from the numerical data by running time averaging. Their Fourier transforms are spectra of various kinds. Both spectra and correlation functions can be used to match consistently experimental data from all sources. This reduces everything to knowledge of the intermolecular pair potential, which can be computed ab initio. Simulation can, in principle, remove empiricism from the interpretation of liquid-state spectra.

T. Dielectric Relaxation and Far Infrared Absorption

The relevant time correlations are respectively the orientational

$$C_{\text{cross}} = \left\langle \sum_i \sum_j \mathbf{u}_i(t) \cdot \mathbf{u}_j(0) \right\rangle \tag{79}$$

of the diffusing dipole moment axis of one molecule and all the others in the ensemble; and its far infrared adjunct,[1] the rotational velocity correlation function, its second time derivative. As with all theories of dielectric relaxation,[9-15] the computation of these time correlation functions involves consideration of intermolecular correlations, as well as cross correlations[46-57] between different dynamical quantities (such as angular and linear velocity) for one particular molecule of the ensemble. The observable spectrum of the molecular liquid presents itself in terms of statistical correlations of orientation vectors, the molecular dipole moments, but a complete understanding of the background dynamics is essential for a proper description of the spectrum. In this respect, computer simulation can contribute in many different ways, despite the fact that it is limited by its inherent numerical approximations to a description of the second spectral moment or power absorption coefficient. In recent years progress has been rapid in the following directions.

1. In the early 1980s, the present author constructed[1-4,127-130] the far infrared power absorption coefficient of liquid dichloromethane from computer simulation of 108 molecules interacting with a model atom–atom representation of the pair potential.[4,12,24] The spectrum was constructed from the auto correlation function, an approximation which neglected intermolecular cross corrrelations because of lack of computer power. This can be improved by using more molecules and by computing the complete cross correlation using an ab initio calculation of the pair potential.[131-133] Not only did the far infrared spectrum from the simulation provide an acceptable representation of the data but the trajectories generated were also used to generate more information about the dynamics. This removed many of the uncertainties caused by empiricism in the customary approach, and showed the existence of many new cross correlations.

2. The anisotropy of diffusion in dichloromethane[1,2,4] was obtained from the relevant correlation functions constructed from fundamental dynamical trajectories. Correlation times for direct comparison with the NMR, Raman, infrared, and neutron-scattering data were obtained from the areas beneath the correlation functions. This is impossible with the Euler–Langevin equations (68) to (70) because the three friction coefficients have never been determined experimentally. This is a clear illustration of the advantage of computer simulation over the theories of Section I. None of these can be used easily without the introduction of ad hoc empiricism, either in the fundamental equations or in closing the continued fractions.

3. Computer simulation was also used to interpret the correlation times from light scattering.[4]

4. The trajectories generated in the same simulation could be used to probe deeply into the nature of statistical ccf's in frames (X, Y, Z) and (x, y, z). The latter were checked against group theory (Sections V, VII,

and VIII), resulting in detailed agreement. A correlation function prohibited by symmetry disappeared also in the simulation, and conversely.

5. The very foundations of the traditional approach were carefully investigated using ccf's in frame (X, Y, Z) and strong externally applied electric fields. The present author showed the existence[134-140] of several ccf's in frame (X, Y, Z) which the customary theories had missed. Several more appeared in response to an applied field such as an electric field. These new ccf's show that molecular diffusion is a much more intricate process than allowed for in the theories of Section I. The ccf's can be built up in computer simulation from the same set of trajectories as those used in the construction of observable spectra. Recent work (Sections VII and VIII) has shown that some of the new ccf's themselves explain fundamental observable phenomena. All this is completely outside the scope of a simple theory of rotational diffusion, however heavily parameterized. The vast majority of the new ccf's signal combined dynamics of rotation and translation, upon which vibration is superimposed. Some of the new ccf's appear only in frame (x, y, z)[141-144] or in frame (X, Y, Z) only in the presence of fields, vanishing at equilibrium. This strikes at the very roots of conventional methods used to describe a spectrum generated in the presence of an external field. This includes the technique of dielectric spectroscopy, which relies on conventional fluctuation–dissipation theory, assuming implicitly that the field generates no new ccf's of its own. This is discussed in greater depth and detail later in this chapter, but it is already apparent from this brief surface scratching exercise that the intensive and detailed work of the 1980s has altered the landscape of diffusion beyond recognition by a combination of key data and computer simulation.

III. COMPUTER SIMULATION

If a diffusing molecule is considered to be a rigid body that simultaneously rotates and translates in three dimensions through an ensemble of similar molecules, then it is assumed in this chapter that the trajectories can be described adequately with classical mechanics. The experimental consequences of this assumption will reveal the limit of its validity.

The numerical technique of computer simulation,[91] introduced in the 1950s and 1960s, solves the classical equations of motion for a small number of molecules, usually in the region from 100 to 1000. The original technique is based on the Newton equation of motion, adapted for rotational dynamics. Newton's equation can be summarised[145] as

$$m\ddot{\mathbf{r}}_i = \mathbf{F}_i = -\sum_j \frac{\partial \phi_{ij}}{\partial \mathbf{r}_i} \tag{80}$$

where m is the molecular mass, $\mathbf{F}_i$ the net force on a molecule i at the instant t, and $\mathbf{r}_i$ the position of the molecular centre of mass in frame (X, Y, Z) of the laboratory observer. Here ϕ_{ij} is the intermolecular pair potential, that between molecule i and j at time t. This equation is dependent on the intermolecular energy ϕ_{ij}. In the simple translational and rotational Langevin equations of Section I the intermolecular potential energy is represented in the first approximation by the friction coefficient in combination with the random force or torque, as the case may be. Langevin's equation is a Newton equation written as

$$m\ddot{\mathbf{r}}_i = -m\beta_T\dot{\mathbf{r}}_i + \mathbf{F}_i(\text{random}) \tag{81}$$

The intermolecular potential has been approximated in a particularly simple way. In one sense, the whole of the theory of molecular diffusion comes down to approximating the intermolecular potential energy generated by the diffusing molecule. The total potential energy $\Phi(\mathbf{r})$ generated by N such diffusing molecules (N is of the order of the Avogadro Number) can be written[145] as the series expansion

$$\Phi(\mathbf{r}) = \frac{1}{2!}\sum \phi_{ij}^2(\mathbf{r}_i, \mathbf{r}_j) + \frac{1}{3!}\sum \phi_{ijk}^3(\mathbf{r}_i, \mathbf{r}_j, \mathbf{r}_k) + \cdots \tag{82}$$

in terms of pair, three-body, four-body and n-body interactions. In an ideal gas of atoms, such as dilute gaseous argon, the higher-order terms can be neglected, and the complete potential energy can be approximated by the first term in the sum, the pair potential. For atomic argon this has customarily been approximated by the Lennard–Jones potential

$$\phi_{ij}(\mathbf{r}_i, \mathbf{r}_j) = 4\epsilon\left[\left(\frac{\sigma}{r_{ij}}\right)^{12} - \left(\frac{\sigma}{r_{ij}}\right)^6\right] \tag{83}$$

where ϵ and σ are adjustable but not purely empirical, because they originate partly in the theory of quantum mechanics.[146] The inverse sixth power comes from the London dispersion energy[146] which has no classical equivalent. This has a negative sign because it is an attractive term, whereas the electronic repulsion between two argon atoms is represented by the empirical inverse twelfth power of the interatomic distance.

In contemporary molecular dynamics computer simulation the pair potential between two molecules is often assumed to be a sum of Lennard–Jones pair potentials for each atom of the two molecules. This is known as the atom–atom Lennard–Jones representation of the pair potential

$$\phi_{ij}^{(12)} \text{ (molecule)} = \sum_1 \sum_2 \phi_{ij}^{(12)} \text{ (atom)} \qquad (84)$$

This is a sum over all the possible atom–atom pair potentials that occur in molecules 1 and 2. If, for example, the molecules are simple diatomics of the type A–B then there are four atom–atom terms: A–A, B–B, A–B, and B–A. The terms A–A and B–B are the same as their equivalents in the isolated atoms A and B, but the cross terms A–B and B–A must be represented by the Lorentz–Berthelot combining rules

$$\sigma_{AB} = \tfrac{1}{2}(\sigma_A + \sigma_B); \qquad \epsilon_{AB} = (\epsilon_A \epsilon_B)^{1/2} \qquad (85)$$

All these assumptions about the nature of the potential energy in an ensemble of diffusing molecules, that it can be represented as a sum of pair potentials between molecules, and that this can in turn be built up from a sum of atom–atom Lennard–Jones potentials, can be justified only by the results, and how they compare with experimental data. The use of the Lennard–Jones approximation, devised in the 1920s in response to the dramatic changes wrought by quantum mechanics, comes from the fact that there is still an inadequate contemporary knowledge of the true intermolecular potential energy. In the test molecule of Section II, dichloromethane, the experimental data[1,4] needed to define the intermolecular pair potential are sparse and inadequate (e.g., pressure and dielectric[24] virial coefficients). The basis sets and computer power for the direct computation of the pair potential ab initio now exist, but there seems to have been no exploration yet along these lines. This typifies the state of the art contemporarily. Furthermore, the whole of classical computer simulation exists only because there are no practical ways, yet, of solving the quantum mechanical equations of motion of an ensemble of diffusing molecules.

Progress can be made against these disadvantages by assuming that the atom–atom pair potential is transferable between molecules to some extent, that is, the quantity A–A in the representation of the intermolecular pair potential between molecule A–B and molecule A–B is about the same as that in any other molecule containing the atom A or B. On this basis, tables of atom–atom Lennard–Jones parameters have been drawn up[101–103] for many of the elements. These literature values are not purely empirical, but are often based on fitting the atom–atom Lennard–Jones form to relevant experimental data, for example, gaseous viscosity,[147] thermal conductivity[148] and, where available, pressure second and third virial coefficients,[1] dielectric second virials,[24] and data such as those on crystal structure.[149] As a rule, the assessment of the dynamical (and spectral) results of a computer simulation should be tempered by the realization that they are obtained with a representation of the true intermolecular potential which may or may not have been adequately tested over a broad enough range of data from all available sources. The survey carried out in 1982 by the present author[1] found that the atom–atom Lennard–Jones potential for dichloromethane could be constructed from data on carbon, chlorine, and hydrogen, the three atomic components of the intermolecular potential, but that there was a serious lack of experimental information with which to assess the potential with any degree of accuracy. In another example, that of the water-water pair potential, a recent survey by Morse and Rice[111,112] using the crystalline structure of the numerous phases of ice found that no potential representation was adequate to reproduce the details available. The most successful water pair potential was not one based on the Lennard–Jones form at all, but from an ab initio computation[150] parameterised for use in a classical simulation algorithm.[151] This method is soundly based, but recquires a large investment of computer time. If inadequate basis sets are used, there may be misleading results for the heavier atoms. However, the ab initio method is destined to replace the models based on Lennard–Jones's approximation.

The technique of molecular dynamics computer simulation, as it stands at present, is not therefore an experimental method. It is a numerical method of solving the classical equations of motion assuming that they can be applied to diffusing molecules over picosecond time scales and angstrom dimensions. The great advantage of the technique over the traditional methods of Chapter 1 include its applicability to three-dimensional diffusion of the asymmetric top, involving simultaneous translation and rotation. Increasingly, simulation methods can also be applied to the diffusion of molecules which are vibrating.[104,105] The quantum mechanical rules governing the various vibrational modes of the molecular framework can be adapted numerically for use in a simulation algorithm.

A. Summary of Numerical Approximations

The method of solving the Newton equation in a collection of, for example, 108 molecules, proceeds[145] on the assumption that the complete potential energy generated by the interaction of any one molecule with all the others can be represented by the sum of $N(N-1)/2$ pair potentials, where N is the number of molecules in the ensemble. The force between any two atoms of two neighbouring and diffusing is approximated by the Lennard–Jones form, as is the contemporary practice in many cases. The complete intermolecular force is the sum of these atom–atom terms. The number of molecules must be restricted in the simulation because of the finite speed and capacity of the computer. This introduces at least two problems.

1. The range of the pair potential must be restricted or truncated at a finite distance, outside of which there is assumed to be no force between the molecules. In other words, the atom–atom potential is truncated at a distance[145] $r_C \simeq 2.5\sigma$.

2. An atom of the diffusing molecule interacts with a finite number of other atoms on other molecules. These are inside the cut-off sphere of radius r_C. There is assumed to be no interaction outside this cut-off sphere.

These approximations are usually basic to any computer simulation algorithm, but the most difficult to accept is the necessity of using periodic boundary conditions.[91,145]

B. Periodic Boundary Conditions

The sample of 108 molecules, for example, dichloromethane molecules, used in the computer simulation has a finite volume, which is usually kept constant (constant volume computer simulation). The kinetic and potential energies of the sample fluctuate considerably in the liquid condition, although the total energy should be constant.[91] In order to preserve this constancy to within a fraction of a percent the 108-molecule sample must be kept from disintegrating by natural diffusion of molecules out of the original sample volume. This is achieved by holding the sample in a cell, usually a cube. If the trajectory of any molecule takes it outside a wall of the cell, periodic boundary conditions ensure that it is not lost from the sample. Usually, a coding device ensures that if the centre of mass moves outside the wall it reenters through the opposing wall with identical velocity. The coordinates of reentry are coded to be the negative of those at which the original centre of mass departed. Periodic boundary conditions can be represented by

$$\dot{\mathbf{r}}(out) = \dot{\mathbf{r}}(in): \qquad (x_i, y_i, z_i)\,(out) = (-x_i, -y_i, -z_i)\,(in) \qquad (86)$$

The effect of equations (86), which apply automatically to any molecule which leaves the cube is to preserve the total energy as a constant. Translational momentum is also conserved, and periodic boundary conditions do not allow net linear flow in the sample.

Their use raises a number of objections, in that the liquid has no periodicity in reality. In practice, however, the molecular dynamics sample is made large enough to eliminate edge and surface effects from the computation of quantities of interest.[145] In particular, time acf's and ccf's do not usually suffer from the effects of imposed periodicity. Numerous simulations[1-4] with different numerical methods[105] for the integration of the classical equations of motion have reinforced this conclusion. Periodic boundary conditions do not conserve the sample's angular momentum, and in consequence, computer simulation has been used to investigate dielectric polarisation due to an applied electric field. Such simulations have already discovered fundamentally new ccf's[141,142] as described in Section II. After 60 years of sporadic development the analytical theory of diffusion failed to anticipate their existence, and is still not capable of describing their time dependence. This is a clear sign of the advantages of computer simulation. The numerous numerical approximations inherent in the technique do not sap its ability to produce original results.

C. Numerical Integration of the Equations of Motion

The essence of the method of computer simulation is to take advantage of the speed of contemporary digital computers to integrate numerically the equations of motion of the sample of molecules under consideration. There are many different ways[91,145] of achieving this objective, with ever-increasing efficiency. The problem is essentially the integration of ordinary differential equations of motion. In this context the nonlinearities of the Euler equations have caused the greatest difficulty due to inherent singu-larities, but these difficulties have bccn circumvented in several different ways, using quaternions,[91] or by methods which directly compute the net molecular torque. The latter is integrated numerically for the molecular angular momentum and angular velocity. Two of the most popular meth-ods for the numerical integration of the translational equations are the Gear predictor–corrector algorithm[145] and the leapfrog algorithm.[91]

D. The Gear Predictor–Corrector

The algorithm uses a Taylor expansion[145] to compute the position of the molecular centre of mass, $\mathbf{r}$, at $t + \Delta t$ given its position at the initial instant t. The Taylor series can be written as

$$\mathbf{r}(t + \Delta t) = \sum_{k=0}^{m} \Delta^k \frac{d^k}{dt^k} \mathbf{r}(t) \tag{87}$$

and is truncated at order m. The differential equation to be solved is a function of the derivatives of the Taylor expansion to order m. This is the "predictor." The "corrector" uses the predicted positions and derivatives to compute the exact mth derivative at the instant $t + \Delta t$. The difference between the exact and predicted mth derivative is then used to correct the derivatives of all orders using a forward differencing scheme.[145] In order to implement this algorithm the forces on the center of mass of the diffusing molecule must be known from a computation of the net forces on each atom. These are computed directly from a knowledge of the interatomic distances between atoms of different molecules by differentiation of the atom–atom potential energy terms. This is implemented for all molecules within the cut-off sphere, and the "forces-loop" is the most time-consuming of the whole algorithm. The cut-off sphere is affected by the cell-code algorithm, which assumes that the pair potential can be neglected beyond the cut-off distance, and by the implementation of periodic boundary conditions.

Usually, the cell-code algorithm is based on the division of the complete cube into subcells, so that the length of one side of the cube is an integer multiple of the sub-cell length, which is not less than the cut-off distance. The molecules are then sorted into sub-cells before the forces loop is implemented, which has the effect of sorting out which molecules are in the nearest-neighboring cells before proceeding with the computation of the net atomic forces for each molecule in the sample. The forces loop is a double loop, because there are $N(N - 1)/2$ interactions. The outer has N steps and the inner $(N - 1)$. For each molecule pair there are several atom–atom terms, so that the use of the nearest-neighbour sub-cells (or "lists") is a great time saver. Parallel implementation[104] on IBM-based architectures shares out the forces loop among different array processors.

Periodic boundary conditions must be applied before the sub-cell code is used.[91,145]

E. The Leapfrog Algorithm

Also known as the Verlet algorithm, this is stable and easy to code and is used to integrate[91] both translational and rotational equations. It uses a Taylor expansion of the center of mass position of each molecule as if it were subjected to a constant net force in the brief time interval Δt. This approximation leads to equations (66) and (67) of Section II. Adding these gives

$$r(t + \Delta t) = r(t) + \delta r(t) \qquad (88)$$

where

$$\delta r(t) = \delta r(t - \Delta t) + F(t) \frac{\Delta t^2}{m} \tag{89}$$

The terms $\delta r(t)$ and $\delta r(t - \Delta t)$ are the position increments. Calculus then leads to the velocity $\dot{r}(t)$ of the center of mass[91]

$$\dot{r}(t) = \frac{r(t + \Delta t) - r(t - \Delta t)}{2\Delta t} \tag{90}$$

$$= \frac{\delta r(t - \Delta t)}{\Delta t} + \frac{1}{2} \frac{F(t) \Delta t^2}{m} \tag{91}$$

It can be shown from a Taylor expansion of the velocity (v) backwards and forwards by the increment $\Delta t/2$ that

$$v^{n + 1/2} = \frac{\delta r(t)}{\Delta t} \tag{92}$$

and

$$v^{n - 1/2} = \frac{\delta t(t - \Delta t)}{\Delta t} \tag{93}$$

where $v^{n + 1/2}$ and $v^{n - 1/2}$ are the half-time step velocities. The force at time $t = n\Delta t$ is calculated from the coordinates at time $n\Delta(t)$. It is then used to update the position increment from time $(n - 1) \Delta t$ to Δt. This new velocity is then used in turn to update the coordinates to time $(n + 1)\Delta t$.

After the coordinates are updated periodic boundary conditions are applied to make sure that all the molecules are inside the cube. Verlet's algorithm is a type of trapezoidal rule numerical integration.

Despite its simplicity, it is more stable than the various orders of Gear algorithm[91] for many applications. Its disadvantages are exposed with spectral moment data, because there are usually no time derivatives of the force, that is, higher-order terms in the Taylor expansions (66) and (67). Higher-order Gear algorithms[145] implement the time derivative of force, and its derivative in turn, but these have a destabilizing effect[91] when using large time steps, probably because of excess fluctuations in the force field. Higher-order Gear algorithms, used with small time steps, can however be tested against spectral moment data. A successful example of the computation of higher-order time correlation functions is given in Ref. 95 for liquid nitrogen. This shows that higher-order time correlation functions and associated spectral moments are much more intricate than allowed for by customary diffusion theory, even when using approximants of the Mori and Grigolini continued fractions.[1-4]

F. Integration of the Rotational Equations

Equations of motion written directly in the Euler angles contain singularities which make them unsuitable for numerical integration. This has led to the quaternion method[91] of representing rotational motion in three dimensions of the diffusing asymmetric top. The rotational equations of motion are expressed as

$$\frac{d\mathbf{J}}{dt} = \mathbf{Tq} \tag{94}$$

and

$$\frac{d\mathbf{q}}{dt} = \mathbf{Q} \tag{95}$$

where $\mathbf{J}$ is the molecular angular momentum and $\mathbf{Tq}$ the torque on the diffusing molecule at time t. Here Eq. (95) is written in four-dimensional quaternion notation, the matrix $\mathbf{Q}$ being defined as

$$\mathbf{Q} = \frac{1}{2}\begin{bmatrix} -\zeta & -\chi & \eta & \xi \\ \chi & -\zeta & -\xi & \eta \\ \xi & \eta & \chi & \zeta \\ -\eta & \xi & -\zeta & \chi \end{bmatrix} \tag{96}$$

The orientation of the asymmetric top is represented in three dimensions through the four-parameter quaternion

$$\mathbf{q} = (\xi, \eta, \zeta, \chi)^T \tag{97}$$

subject to the constraint

$$\xi^2 + \eta^2 + \zeta^2 + \chi^2 = 1 \tag{98}$$

G. Other Molecular Dynamics Simulation Algorithms

There are now available many different types of general and specialist molecular dynamics computer simulation algorithms.[91,145] A library of codes and program descriptions is available from the Daresbury Laboratory of the U.K. Science and Engineering Research Council. The majority of these integrate the classical rotational and translational equations of motion for the asymmetric top molecule in three dimensions, considering this as a rigid body. Increasingly, code is available for flexible molecules, using a method based on the translational equations of motion of each atom of the molecules constrained by a classical representation of the bonds between them.[103,104] An example of time ccf's from this technique can be found in Ref. (104). Exploratory work is in progress in many dimensions, the utilization of higher-order algorithms,[91] such as the Toxvaerd method,[152] the development of algorithms for large molecules such as proteins,[153] the parallel implementation of code for supercomputers,[154] and the rapid development of non-equilibrium simulation.[145] Improvements are also being made in the use of periodic boundary conditions, new link-cell techniques, neighborhood lists, the computation of flow phenomena and hydrodynamic properties, the simulation of the effect of external fields,[155–160] and in many other directions. Volumes of the specialist literature,[161] are increasingly dedicated to new material phases such as the glassy state, and phase transitions. Simulations are also available for the compressed and dilute gas states.[162–166]

This activity has occurred in the last 20 years and has changed most areas of specialization out of recognition. As far as the subject of molecular dynamics is concerned, one of the first signs of the power of the new technique appeared in the computer simulation by Rahman of 864 argon atoms in 1964.

H.   Rahman's Computer Simulation of Liquid Argon

The equations of the translational motion of the 864 atoms were integrated numerically to produce results in conflict with the traditional approach to diffusion theory outlined in Section I. For example, the translational Langevin equation (1) produces an exponential for the velocity acf

$$C_\nu(t) = \langle \mathbf{v}(t) \cdot \mathbf{v}(0) \rangle = \langle \nu^2 \rangle \exp(-t/\tau_\nu) \qquad (99)$$

where τ_ν is the linear velocity correlation time. This result had never been seriously questioned, and never adequately tested experimentally over short enough time scales until the results of this computer simulation appeared in 1964. The prediction of the approximately correct value of the Avogadro Number from consideration of Brownian motion masked the theory's inherent limits. There were no experimental techniques available with which to test the validity of the exponential decay of Eq. (99).

Progress towards such a test had been made slowly, but had resulted in no more than the Green–Kubo relation[167]

$$D_\nu = \frac{1}{3} \int_0^\infty \langle \mathbf{v}(t) \cdot \mathbf{v}(0) \rangle \, dt \tag{100}$$

between the diffusion coefficient and the acf. It had always been assumed by the vast majority that the time dependence itself was exponential. The simulation proved this to be a fallacy.

1. The time acf from the simulation became negative as t increased. It displayed a long negative tail.
2. The acf had a zero slope at $t = 0$, that is, was differentiable at zero time, unlike the simple exponential of the Langevin diffusion equation (1).

These results initiated a vast literature on the corrections necessary[1-4] to account for the numerical results. These attempts are still being made, using many different approaches. This is a clear indication that the simulation leads the theory towards new results.

I. Some Fundamental Properties of Time Correlation Functions

In order to understand more clearly the conflicts between the diffusion theory and the results produced numerically it is necessary to look at the Taylor expansion of the time correlation function, $C(t)$, of two arbitrary and stationary functions of time, $A(t)$ and $B(0)$. This is a real and statistically stationary[1] function of time for a dynamical variable such as linear velocity, implying

$$C_\nu(t) = \langle \mathbf{v}(t) \cdot \mathbf{v}(0) \rangle = \langle \mathbf{v}(-t) \cdot \mathbf{v}(0) \rangle = \langle \mathbf{v}(0) \cdot \mathbf{v}(-t) \rangle \tag{101}$$

Considering the Taylor expansion of this function it can be shown as follows that the odd time derivatives of the classical time correlation function vanish.

Consider the fundamental property

$$C_\nu(t) = \langle \mathbf{v}(t) \cdot \mathbf{v}(0) \rangle = \langle \mathbf{v}(t + \tau) \cdot \mathbf{v}(\tau) \rangle = \langle \mathbf{v}(t - \tau) \cdot \mathbf{v}(\tau) \rangle \tag{102}$$

of the statistically stationary time correlation function at reversible thermodynamic equilibrium. It follows that

$$\langle \dot{\mathbf{v}}(t + \tau) \cdot \mathbf{v}(\tau) \rangle = - \langle \dot{\mathbf{v}}(t - \tau) \cdot \mathbf{v}(\tau) \rangle \tag{103}$$

At $t = 0$ the following results are obtained:

$$\langle \dot{\mathbf{v}}(0) \cdot \mathbf{v}(0) \rangle = - \langle \mathbf{v}(t) \cdot \dot{\mathbf{v}}(0) \rangle \tag{104}$$

and

$$\langle \dot{\mathbf{v}}(0) \cdot \mathbf{v}(0) \rangle = 0 \tag{105}$$

Taking the second derivative,

$$\frac{d^2}{dt^2} \langle \mathbf{v}(t) \cdot \mathbf{v}(0) \rangle = - \langle \dot{\mathbf{v}}(t) \cdot \dot{\mathbf{v}}(0) \rangle \tag{106}$$

and continuing the process

$$\frac{d^{2n}}{dt^{2n}} \langle \mathbf{v}(t) \cdot \mathbf{v}(0) \rangle = (-1)^n \left\langle \frac{d^n}{dt^n} \mathbf{v}(t) \cdot \frac{d^n}{dt^n} \mathbf{v}(0) \right\rangle \tag{107}$$

The classical time acf is therefore an even function of time, with the Taylor expansion

$$C_v(t) = \langle v^2(0) \rangle - \frac{t^2}{2!} \langle \dot{v}^2(0) \rangle + \frac{t^4}{4!} \langle \ddot{v}^2(0) \rangle - \cdots \tag{108}$$

where Taylor coefficients $\langle v^{2n}(0) \rangle$ are spectral moments. The present author showed[44] that the memory functions of the Mori theory of diffusion must also be even functions of time, with well-defined Taylor coefficients related to those of the time correlation function from which the set of memory functions is derived.

At short times, therefore, the time acf of the linear velocity in Rahman's simulation must be

$$C_v(t) \simeq \langle v^2(0) \rangle - \frac{t^2}{2!} \langle \dot{v}^2(0) \rangle \tag{109}$$

that is, the slope at $t = 0$ is zero. The exponential acf from the Langevin equation (1) contravenes this general law because the slope of the exponential is finite and the mean square force $\langle \dot{v}^2 \rangle$ is not definable. The exponential is not differentiable at $t = 0$.

J.   The Long Time Tail

The most telling departure from the exponential is a long negative tail as $t \to \infty$. Intense research into its origin, using simulation and analytical theory, revealed the persistent correlation at long times, the long time tail, to be due to back-flow. The original translation of an atom sets up cooperative motion in neighbouring atoms. An analogy in hydrodynamics is that of spherical particle flowing through a continuous medium. In the same way that a ship marks its path in the water, a translating atom sets up statistically correlated motions of the centers of mass of neighbors which are themselves diffusing. Gradually, a pattern of correlation is built up, which affects the diffusion of the original atom itself. Further research in the wake of Rahman's pioneering work showed[168-172] that both the linear and angular acf's in simulations of molecular diffusion show a time dependence as $t \to \infty$, which is different from the exponential decay expected from the Langevin equations. Explanations of these long time tails were sought in theories[170-172] based on hydrodynamics applied at the molecular level. Some contemporary simulations, using as many as 200,000 atoms, have confirmed[173] that hydrodynamic phenomena, vortices, eddies, and flows, can be generated in a well-defined manner directly from Newton's equation applied to atoms. This closes the gap in our understanding of atomic and molecular phenomena and hydrodynamics, continuum concepts of flow, and provides confidence in the basic applicability and validity of computer simulation in many fields of chemical and theoretical physics and engineering. The traditional approach of Section I cannot explain the long time tails, with the possible exception of some approximants of the Grigolini continued fraction. These appear in several acf's, including the orientational, that of the rotational and angular velocities, as well as that of the linear velocity. Several lines of approximation in recent years have converged to show that when details of the environment (other diffusing molecules) are accounted for properly, using computer simulation, many new correlations appear in the description of the dynamics. A complete understanding of spectral data can be obtained only with computer simulation as a powerful guide to interpretation. This is not to denigrate the approach of Section I, which has provided an important evolutionary base, but one which can be built upon to great advantage.

K.   The Interaction between Rotational and Translational Molecular Diffusion

The presence of long time tails has shown that the rotational and translational diffusion of molecules is statistically correlated in several ways. Once a diffusing molecule is given shape, and once we escape from the confines of separated rotational diffusion, as proposed (with good reason) by Debye, these correlations must be taken into consideration. The theory of hydrodynamics is already richly endowed with concepts which spring from a combination of rotational and translational motions. An example is vortex flow set up downstream of an obstacle. By generalizing the concepts of hydrodynamics, and adapting them for an individual diffusing molecule, it can be shown that its angular velocity acf takes the asymptotic form[168,169,173]

$$C_\omega(t)(t \to \infty) = \frac{d\pi I}{2mn} (4\pi(D + v))^{-(d+2)/2} t^{-(d+2)/2} \tag{110}$$

where I, m, n, v, and D are respectively the molecular moment of inertia, molecular mass, number density, kinematic shear viscosity, self-diffusion coefficient, and dimensionality of the system. For diffusion in two dimensions, $d = 2$, and in three dimensions, $d = 3$. It follows[1] from this relation that the orientational acf

$$\langle \mathbf{u}(t) \cdot \mathbf{u}(0) \rangle \propto \langle u^2(0) \rangle t^{-(d+2)/2} \tag{111}$$

that is, must behave asymptotically as $t^{-(d+2)/2}$. Double differentiation to produce the rotational velocity acf shows that

$$\langle \dot{\mathbf{u}}(t) \cdot \dot{\mathbf{u}}(0) \rangle \propto \langle \dot{u}^2(0) \rangle t^{-(2+(d+2)/2)} \tag{112}$$

By Fourier transformation of the relevant correlation functions (111) and (112) it follows that the low-frequency regions of the dielectric loss and the far infrared power absorption coefficient must contain information on the processes responsible for the long time tails. These spectra contain information on the kinematic shear viscosity, which appears in the relevant asymptotic expressions (110)–(112) for the time correlation functions of the diffusing molecule. This is well known in the theory of light scattering[46] as the "shear doublet" observed at low frequency. No description of the dielectric and far infrared spectral range can be complete without a proper description of the interrelation between rotation and translation. This is discussed in greater detail in Section V, using the new computer simulation results of the last 5 years, obtained for asymmetric tops diffusing in three dimensions. A complete general theory must be capable of generating all these results self consistently, and of interrelating molecular and hydrodynamic concepts.

L. Simulation of Liquid Dichloromethane

The simulation of an asymmetric top diffusing in three dimensions became possible in the early 1980s due to the work of several groups. Within the framework of the European Molecular Liquids Group and the CCP5 Committee of the U.K. Science and Engineering Research Council these algorithms are now freely available to individual scientists who wish to apply them. The emergence of this powerful new technique has meant that the spectral data available from several different sources could be tested numerically for consistency from a representation of the pair potential. This task was undertaken by the present author and co-workers and the results published[1-4] in several review articles for different molecules. For each molecule, the dynamics and available spectral data were probed numerically with an algorithm called TETRA, written by Schofield, Singer, Ferrario, and Evans, full details of which are in the literature.[174] This robust algorithm has now produced a wide variety of new results on several aspects of combined rotational and translational motion in the asymmetric top. This work has pervaded some 150 articles in the specialist literature. A listing and brief explanation of the code is provided in Appendix 1. Video animations of the molecular dynamics of several systems using this algorithm are also available from IBM, Kingston, New York, Dept. 48B/428, and from the author.

In the test liquid dichloromethane, the available experimental data from the sources described in Section II, and others, were tested for consistency with a model potential[175-180] from the atom–atom Lennard–Jones parameters in the literature for chlorine, hydrogen, and carbon. Using these three parameters, which can be refined experimentally, it is possible with TETRA to generate many types of information corresponding to the different spectra. Using the dynamical trajectories generated, during the simulations the variety of statistical ccf's produced by group theory (see Sections V, VII and VIII), in frames (x, y, z) and (X, Y, Z) could be computed entirely self-consistently. The algorithm therefore provides a link between a variety of spectral data and the emerging concepts of hydrodynamics adapted for the motion of individual molecules. The time step used in these computations was 5.0 fs (5.0×10^{-15} sec). This was sufficient to produce stability in the total energy to within $\pm 0.1\%$. A complete description of the results is available in several publications, conveniently collected in Refs. (2) and (4).

M. Comparison of Simulation with Spectral Data

The task of comparing thoroughly the range of available dynamical data on liquid dichloromethane with results from the computer simulation begins with the definition of the intermolecular potential to be used in the simulation. For dichloromethane, with two relatively large chlorine atoms, there is as yet no really adequate ab initio computation of the pair potential as a function of intermolecular distances and relative molecular orientation. Such results are slowly becoming available and it is hoped that the necessary basis sets and methods will soon be used.

This is typical of contemporary knowledge on intermolecular potentials, even for small molecules. The alternative but more empirical Lennard–Jones atom–atom approximation is therefore implemented as a working alternative.

N. The Lennard–Jones Atom–Atom Potential for Dichloromethane

The literature on atom Lennard–Jones parameters usually provides them in the form ϵ/k and σ. For dichloromethane there are several literature sources which do not necessarily provide the same values for these parameters for each atom, in this case H, Cl, and C. This is an unsatisfactory contemporary state of affairs which can only be remedied partially by tuning the Lennard–Jones parameters to given experimental data. In the simulations carried out by Ferrario and Evans,[1,4] which formed the basis for the survey of the properties of dichloromethane, the parameters were taken from the literature but adjusted slightly to provide best agreement with the internal energy of the liquid as measured exprimentally and with the specific heat at constant volume. The Lennard–Jones parameters actually used were

$$\sigma(H - H) = 2.75A \qquad \epsilon/K(H - H) = 13.4\,K$$
$$\sigma(Cl - Cl) = 3.35\,A \qquad \epsilon/k(Cl - Cl) = 175.0\,K$$
$$\sigma(C - C) = 3.20\,A \qquad \epsilon/K(C - C) = 51.0\,K$$

and cross terms were evaluated with Lorentz–Berthelot combining rules.

The representation of the pair potential was completed with electrostatic terms, point charges on each atom. This is a first approximation to the electrostatic attraction between two dichloromethane molecules which does not take into account that the molecule is polarisable.[12,24] The point charges used in the potential were obtained from a quantum mechanical computation which represents the electron cloud of the molecule as fractional charges on each atom, a crude first approximation which continues to be unavoidable in classical approximations to the true quantum potential. The numerical fractional charges were $0.098|e|$ on H; $-0.109|e|$ on Cl; and $0.022|e|$ on C; $e = -1.6 \times 10^{-19}\,C$.

O. Computer Simulation Methods

Having established the approximation to the pair potential, the aim of the computer simulation was to reproduce as accurately as possible as broad a range as possible of dynamical data obtained from observed spectra[1,4] of liquid dichloromethane, and its solutions in non-dipolar molecules such as carbon tetrachloride.

The constant volume algorithm TETRA was initiated by setting up the dichloromethane molecules on a lattice, for convenience a face-centred cubic lattice. The algorithm then generated the necessary Maxwell–Boltzmann distribution of linear and angular velocity with a random number generator. This caused the lattice to melt as the numerical integration routine took over the control and production of the individual molecular dynamical trajectories. The lattice took about 2000 time steps to melt into the liquid. When equilibrium was established the total energy became constant and negative, a sum of negative potential energy terms and positive kinetic energies of rotation and translation. In the liquid the two types of kinetic energy are numerically positive because they are measures of temperature in absolute (Kelvin) units. The potential energy in the liquid state is on the other hand negative because the sum of attractive parts of the pair potentials in the ensemble of 108 molecules is greater than the sum of positive, repulsive parts. The molecular liquid therefore stays together in a cohesive form such as that actually observed in a true liquid.

The next stage after achieving the liquid state and discarding the first 2,000 time steps or so needed for this was to implement further tests as follows.

P. Some Tests for the Liquid State

We have seen that at relatively long times or low frequencies, Eq. (1) should be valid for Brownian motion, the perpetual fluctuations in the position of a massive particle diffusing in an environment of much lighter molecules. One of the predictions[4] of Eq. (1) is that the mean-square displacement of the random walk in the position of the centre of mass of the Brownian particle depends linearly on time. One of the results of the 1964 simulation by Rahman was that this quantity was approximately linear far enough into the dynamical evolution from $t = 0$. Otherwise, shortly after $t = 0$, Eq. (1) is not at all applicable. This pattern was repeated for the simulation of dichloromethane. If the lattice has not melted properly the mean square displacement[181] is oscillatory, and more characteristic of the solid state.

Another test of melting was the computation of the orientational averages $\langle \mathbf{u}_1 \rangle$, $\langle \mathbf{u}_2 \rangle$, $\langle \mathbf{u}_3 \rangle$ over the three unit vectors in the axes of the principal molecular moment of inertia frame. In the isotropic liquid these all vanish, that is, the quantities actually simulated should fluctuate about a zero mean. If there are no special aligning effects caused by a positive Kirkwood factor, for example,[12] alignment caused by the interaction of molecular force fields, these averages should not fluctuate by more than about 0.1 either side of zero.

Finally, the center of mass positions of the molecules in the liquid state should show no solid like ordering, and averages over the center of mass positions in three dimensions should also fluctuate about a zero mean in the simulation.

Q. Thermodynamic Properties

Having made sure of melting the production runs were started, and ran for upwards of 10,000 time steps each. At each time step thermodynamic properties of the liquid were computed and at the end of 10,000 steps, averaged over the complete run for comparison with experimental data if available.

The basic quantities are temperature (T), pressure (P), and volume (V), whose interrelation defines the equation of state.[182] Even in the dilute gas the latter is in general an unknown quantity analytically, although there are numerous empirical representations. In view of this the direct measurement of deviations of real gases from the equation of state of the perfect gas

$$PV/T = \text{constant} \tag{113}$$

is represented in the virial expansion

$$\frac{pV}{T} = A + \frac{B}{V} + \frac{C}{V^2} + \cdots \tag{114}$$

due to Kammerlingh–Onnes. Here, A, B, and C are termed the first, second, and third virial coefficients and if known experimentally, are sensitive measures of the validity of the Lennard–Jones or any other representation of the pair potential and higher-order terms such as the three-body potential. The virial coefficient expansion is also applicable to other thermodynamic and dielectric properties of the dilute gas.[24] Ideally, computer simulation aims at reproducing known virial data from a given representation of the potential. However, there is at present a severe shortage of experimental data on virial coefficients, which is a block to progress.

In the liquid the virial expansion is less useful, because many of its coefficients would be needed for the definition of the equation of the liquid state, and because there are severe fluctuations in the pressure from a molecular dynamics simulation at constant volume. This is a natural consequence of the fact that the mean intermolecular separation in the liquid is much less than in the dilute gas. The fluctuations in the Lennard–Jones potential energy function are inverse sixth and twelfth powers of this distance, and can therefore be very severe. The same is true of temperature, which is computed from the kinetic energies. The translational temperature is

$$T = \frac{1}{3Nk} \sum_{i=1}^{N} \mathbf{v}_i \cdot \mathbf{v}_i \tag{115}$$

The configurational part of the internal pressure, P_c, is computed from the virial coefficient ψ:

$$\langle P_c \rangle = \frac{N}{V} k \langle T \rangle - \frac{\langle \psi \rangle}{V} \tag{116}$$

where

$$\langle \psi \rangle = \left\langle \sum_{i<j} \sum r_{ij} \frac{\partial \phi_{ij}(r)}{\partial r_{ij}} \right\rangle + 2\pi \frac{N^2}{V} \int_{r_c}^{\infty} r^3 \phi'(r) g(r) \, dr \tag{117}$$

Here $\phi_{ij}(r)$ is the functional dependence of the Lennard–Jones pair potential on the interatomic distances (r) of the complete atom–atom potential between two molecules. $g(r)$ is the atom–atom pair distribution function, and is the numerical probability of finding an atom at a distance r from another of the same or different type given the initial position of the second atom. It is a measure of the liquid structure. N is the number of atoms in the sample, either of the same or different type as the case may be. The complete configurational contribution of the internal pressure is built up over all types of individual atom–atom contributions in Eqns. (116) and (117). V is the molar volume. The complete internal pressure of the sample is then given by eqn. (116) and the measurable thermodynamic pressure is

$$P = P_c - T\left(\frac{\partial P_c}{\partial T}\right)_{N/V} \qquad (118)$$

This is the difference between two large and fluctuating numbers, and P as computed is therefore a small number with a large uncertainty. An expected pressure of 1 bar, for example, may be given numerically as something like

$$\langle P \rangle \simeq 1 \pm 300 \text{ bar} \qquad (119)$$

Similarly, the configurational part of the internal energy is computed from

$$\langle U \rangle = \left\langle \sum_{i<j} \sum \phi_{ij} \right\rangle + U_c \qquad (120)$$

where the sum is over all atom–atom pairs for which r_{ij} is less than a critical cut-off distance r_c. This necessitates the correction U_c based on a uniform distribution of molecules beyond:

$$U_c = 2\pi \frac{N^2}{V} \int_{r_c}^{\infty} r^2 \, \phi(r) g(r) \, dr \qquad (121)$$

In the algorithm TETRA[1,2,4] for the simulation of dichloromethane, these principal thermodynamic quantities are computed at each time step, and an average produced at the end of the data production run in SI units. The algorithm also produces estimates of second-order thermodynamic properties such as specific heat, computed from fluctuations in first-order quantities.

R. Thermodynamic Results for Dichloromethane

The simulation algorithm should ideally produce the correct temperature and pressure for a given molar volume

$$V = \frac{M}{\rho} \qquad (122)$$

which is the molecular weight divided by the liquid density. The pressure in particular is extremely sensitive to the individual atom–atom Lennard–Jones parameters σ and ϵ/k. The rotational and translational kinetic energies which produce the temperatures are also fluctuating quantities, and in order to produce the correct temperature for a given input and molar volume (or liquid density), the sample must be thermostatted artificially, especially during the melting phase of the simulation. This is achieved with a routine known as temperature rescaling. The temperature is scaled back to the input value if the natural fluctuations exceed a given maximum, say $\pm 50\ K$. High-quality simulations avoid this procedure in production runs, and the temperature is allowed to fluctuate naturally. In the melting stage, however, very large artificial temperature fluctuations may be unavoidable because of molecules becoming artificially entangled or crushed together during the melt process, as the ensemble finds its optimum equilibrium configuration.

In the simulation of the test dichloromethane molecule[2,4] the optimized Lennard–Jones potential illustrated already produced the following pressure and temperatures for an input molar volume corresponding to 293 K and 1 bar.

1. The mean temperature was 294.5 ± 11 K (50.2% translational and 49.8% rotational), compared with the experimental value expected at 293 K.

2. The mean pressure was 273 ± 300 bar.

3. The computed specific heat at constant volume was $46\ J\ mole^{-1}\ K^{-1}$; compared with the experimental value[2] of $90\ J\ mole^{-1}\ K^{-1}$.

The thermodynamic properties computed for this relatively small sample of 108 dichloromethane molecules are in fair agreement with the true experimental values, but by no means perfect. Given the approximations and uncertainties used throughout the exercise, the agrement is acceptable, and typical of contemporary simulations.

S.   Dynamical and Spectral Properties[1,2,4]

Having produced numerical results for basic thermodynamic properties of liquid dichloromethane in reasonable agreement with the limited amount of experimental data available, the next step is to use the same trajectories to attempt to match the experimental spectral properties of the liquid. These are characterised briefly as in Section II by various types of radiation absorption, dispersion, and scattering.

T.   Microwave and Far Infrared Absorption and Dispersion

These cover a range of about three frequency decades equivalent to the sub-picosecond to nanosecond time range. In comparing the simulations with available experimental data[1,2,4] in pure dichloromethane liquid, there are complications which must be borne in mind. These include the following:

1. The time correlation function from the experimental microwave data is the multiparticle ccf of orientation[183]

$$C_u(t) = \left\langle \sum_i \sum_j \mathbf{u}_i(0) \cdot \mathbf{u}_j(t) \right\rangle \tag{123}$$

which is related through an equation such as (19) of Section I to the Fourier transform of the dielectric loss, and its second time derivative

$$C_{\dot{u}}(t) = \left\langle \sum_i \sum_j \dot{\mathbf{u}}_i(0) \cdot \dot{\mathbf{u}}_j(t) \right\rangle \tag{124}$$

essentially[1] the Fourier transform of the far infrared power absorption coefficient. Limited availability of computer time and storage meant that only the acf's equivalent to Eqs. (123) and (124) were within range of the computation. These are $\langle \mathbf{u}(0) \cdot \mathbf{u}(t) \rangle$ and $\langle \dot{\mathbf{u}}(0) \cdot \dot{\mathbf{u}}(t) \rangle$. Their Fourier transforms will only approximate the observed spectra[1,2] because of the missing intermolecular cross correlations. However, these may be removed by dilution in a nondipolar solvent such as carbon tetrachloride,[1,2,4] which is observed experimentally to shift the complete far infrared power absorption (second spectral moment) to lower frequencies. Proceeding on the assumption that the dilute solution contains less dynamical cross correlation between dichloromethane molecules than in a pure liquid of the latter, a comparison[1,2] with the data from the simulation can be made. The result is sketched in Fig. 11. Here the maxima of the observed and simulated spectra have been adjusted to be the same for ease of comparison of the frequency dependence. The validity of this procedure is discussed in point 2 below. The figure shows reasonably good agreement. In view of the approximations inherent in our computer simulation of the pure liquid, the result is an advance over the theory of Section I. For example, the far infrared absorption peak, clearly defined in the interferometric data obtained experimentally in terms of the far infrared power absorption coefficient, is also clearly present in the computer simulation from a simple Fourier transform of the dipole moment (μ) rotational velocity acf $\langle \dot{\mu}(t) \cdot \dot{\mu}(0) \rangle$. The simulated power absorption coefficient also regains transparency at high frequencies. The simulation is therefore more realistic than the original theory of rotational diffusion, which produces plateau absorption, and also its counterpart corrected for inertial effects,

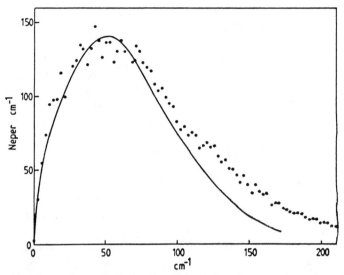

Figure 11. Comparison of simulated and far infrared power absorption of liquid dichloromethane. ● Simulation; — experimental data.

which merely produces a far too gradual fall-off from this plateau. The simulated peak position agrees well with that experimentally observed in a dilute solution of dichloromethane in carbon tetrachloride, chosen to try to remove inter-molecular cross correlations[1,12] between dichloromethane molecules in the neat liquid. Despite the agreement in Fig. 11 it may be objected that the comparison is between a computer simulation of the pure liquid dichloromethane, with cross correlations unaccounted for due to lack of computer availability, and an experimentally derived dilute solution of dichloromethane in carbon tetrachloride. These are valid criticisms which can be met only with increased computer time. The latter is needed both to compute the full intermolecular ccf's (123) and (124) and to simulate directly a solution of dichloromethane in carbon tetrachloride. For a 10% solution with about 108 dichloromethane molecules this would require about 900 carbon tetrachloride molecules. Such large-scale computations would still be faced, of course, with a contemporary lack of ab initio potentials, our limitation to classical equations of motion, and artificial removal of many-body effects in our current approach to computer simulation. In the meantime, of particular interest in Fig. 11 are the high- and low-frequency ends of the spectrum. At high frequencies the observed power absorption coefficient of the dilute solution regains transparency more quickly than the simulated counterpart. This is because the simulation relies on a finite Taylor-series expansion of the type sketched briefly in this chapter. At some point in these Taylor expansions the next higher time derivative is left out as the series is truncated. This implies (Section II) that spectral moments will cease to be definable at some stage. These effects are visible in the simulated second moment (the power absorption coefficient) through the relatively slow decay to transparency of the simulated data in comparison with the experimental observation. The fourth spectral moment is defined in TETRA, because the code defines the time derivative of the mean-square torque, but the sixth moment from the simulation will not regain spectral transparency at high frequencies. This emphasizes that the simulation is an imperfect numerical approximation to the classical equations of motion assumed to govern molecular dynamics. In view of the uncertainty axioms[184–186] of quantum mechanics, operable at subpicosecond time scales, this assumption is itself of limited validity. At low frequencies in Fig. 11 the traditional microwave[12,24] and far infrared ranges overlap. The traditional measure (Section I) of microwave absorption is the dielectric loss and dispersion, and the Debye relaxation time. The dielectric loss is simply another way of expressing the power absorption coefficient through Eq. (16) of Section I. It is also affected by the same cross correlations as those in the far infrared, and these appear implicitly in Eq. (123). A complete computer simulation of the dielectric loss must include these and would proceed over about 50,000 time steps out into the nanosecond range (GHz frequency range). This combination of requirements, together with the need for a 5.0-fs time step to maintain the stability of the total energy in the simulation[91,145] needs an investment of computer time. The original simulation of dichloromethane[2,4] was severely limited by lack of computer availability which restricted the computation of Eq. (123) to the acf, and then only out to about 2.0 ps (400 time steps). At 2.0 ps into its time evolution the acf had fallen[2] to about one-third of its initial value, normalized to 1.0 at $t = 0$, as is the contemporary practice. The time dependence of the acf in this intermediate range (from about 0.2 to the available limit of 2.0 ps) closely approximated an exponential. At longer times (50.0 ps or more) we would expect to see a long time tail (Eq. 111). In the face of these limitations it is reasonable to proceed approximately with the definition of a correlation time from the computer simulation at the instant when the correlation function has decayed to $1/e$ of its value at $t = 0$. This is approximately comparable with the Debye relaxation time, bearing in mind that both times involve autocorrelations only. This procedure gave 1.2 ps from the computer simulation compared with an observed[1] dielectric relaxation time of 1.45 ps. The latter is defined simply as the inverse of the peak frequency of the dielectric loss. The neglect of intermolecular cross correlations in the simulation yields a relaxation time which is too short and a far infrared power absorption peak frequency (55.0 cm^{-1}) which is too low in comparison with the observed peak frequency[1,2] in pure liquid dichloromethane of 80.0 cm^{-1}. In many respects, however, the computer simulation is an improvement over the theory of rotational diffusion. For example, it is carried out in three dimensions, using the proper asymmetric top molecular structure, a fairly realistic model pair potential, and involves both translational and rotational diffusion simultaneously. The trajectories are able to provide a range of thermodynamic and spectral data self-consistently.

2. A further complication not accounted for explicitly in the computer simulation is that of the internal field correction. This occurs[12,24] in all spectral observations of the molecular liquid state for dielectric materials. The correction tries to determine the effective or internal field felt by a diffusing molecule as opposed to the applied field intensity. In dielectric relaxation[12,24] the applied measuring field is often an alternating electric field at a given frequency, or a klystron generator. In the far infrared it is broad-band radiation from a high-pressure mercury lamp, for example. In all cases the diffusing molecule is shielded from the full intensity of the radiation by the fields of other molecules in the sample. The problem of determining the effective internal field is dependent on the molecular dynamics that are being investigated by spectral means. In computer simulation, there is no probe field and time correlation functions are built up directly, without an intervening Fourier transformation. Any comparison of simulation and spectrum must involve, however, a correction of the spectral data for the internal field. There are several literature approaches[1] designed to adjust for the frequency dependence of the internal field correction (the dynamical internal field). The most generally valid approaches are based on the Mori approximation to the Liouville equation developed in the 1970s. The procedures are not entirely unequivocal, however, because use is made of the traditional approach of Section I within the framework of the correction. The argument is circular, the theory of diffusion is used to correct spectra intended to give information on the same theory of diffusion. The circle can be broken, however, as exemplified in Refs. 187 and 188. The micro–macro theorems[1] are easier to apply to practical problems, but at the expense of empiricism introduced by approximants of the Mori continued fraction. An example of one of these theorems[189] is outlined in Ref. 1. As a rule of thumb the various internal field corrections do not have a large effect on the frequency dependence of the far infrared power absorption,[190] but can alter the Debye relaxation time by pushing the peak frequency of the theoretical dielectric loss downwards in frequency by as much as about half a decade.

The correction has a large effect, however, on the intensity of the far infrared and dielectric spectra.[12] This correction is significant whether or not frequency dependent. The typical effect of the Polo–Wilson correction,[1,191] for example, is to reduce the intensity of the spectrum by about 25% over the complete frequency range. This means that the field effective on a diffusing molecule is weaker on the average by about a quarter than that originally applied at the boundary of the sample by the spectroscopist. It is convenient to normalize the far infrared power absorption obtained experimentally to the peak of the same curve obtained in a computer simulation, on the assumption that the internal field correction does not significantly distort the frequency dependence of the spectrum. This allows a comparison such as that of Fig. 11.

3. Another uncertainty in comparing spectrum and simulation is introduced by collision-induced effects coming from the mutual distortion of molecular electron clouds in ensembles. These effects can be dramatic in the far infrared under particular combinations of circumstance. This was shown clearly by Baise[72] in compressed gaseous and liquid nitrous oxide, a weakly dipolar molecule whose electron distribution is easily prone to distortion by other nitrous oxide molecules. The theory of collision induced absorption involves considerations of higher multipole moments of the distribution of electronic charge.[192] These quantities are obtained by a Taylor expansion akin in symmetry to the spherical harmonics.[24] The point charge is a scalar, the dipole moment a vector, the quadrupole moment is a tensor with nine components; the octopole moment has 27 components and the hexadecapole moment 81. The quantum theory of multipole-induced dipole absorption in the far infrared[1] predicts quantum lines spaced by $\Delta J = 2$ for quadrupole-induced dipole absorption, $\Delta J = 3$ for octopole-induced dipole absorption, and so on, where J is the rotational quantum number. In practice the "lines" are almost always broadened into bands which are superimposed on the high-frequency side of the far infrared absorption due to the permanent dipole itself. Baise[72] observed a large effect of multipole induced dipole absorption in nitrous oxide and further effects[65-67] have been isolated due to dipole-, quadrupole-, and octopole-induced dipole absorption in the halogenated hydrocarbons such as fluorotrichloromethane with dipole moments in the intermediate range. The present author isolated the hexadecapole moment of oxygen using its far infrared absorption.[193] Fortunately for the study of diffusion processes characterized solely by the permanent dipole moment of the molecule, the induced effects go through a maximum in the moderately compressed gas, but thereafter decrease due to symmetry,[72] because the arrangement of nearest neighbors on the average leads to a cancella-

tion of multipole-induced absorption. In a molecular crystal, the cancellation is sometimes complete and collision-induced spectra vanish. In preparation for the European Molecular Liquids Group pilot project on dichloromethane, account was taken of the fact that the area beneath the power absorption coefficient in the far infrared of this strongly dipolar species is an experimentally observed linear function of number density (N) in dilute solutions.[1] Multipole effects evidently assume a secondary role, because they are nonlinear in N. Dichloromethane is therefore a good choice for the comparison of data from simulation with those from the far infrared. Nevertheless, collision-induced absorption or scattering is always present, theoretically, and some computer simulations[194] have addressed the problem directly.

U. Infrared Absorption

This technique produces correlation times which are usually assumed to relate directly to the acf of the transition dipole moment of the absorption. For dichloromethane, early work[194] produced a correlation time for the axis of the permanent dipole of only 0.5 ps, much shorter than the 1.2 ps from the simulation by Ferrario and Evans[2,4] and shorter than the dielectric relaxation time of 1.45 ps. This is one illustration of the fact that cross correlation between vibration and rotation must be understood properly in infrared absorption bands broadened by liquid diffusion. Contemporary uncertainty is further illustrated by the fact that an independent investigation[195] of the same infrared information produced a correlation time of about 1.2 ps, in close agreement with the simulation.

V. Relaxation of Nuclear Magnetic Resonance

By collecting data in liquid dichloromethane from a number of different types of nuclear magnetic resonances it is possible to arrive at correlation times[114] for different axes of the molecule fixed frame (x, y, z). For any orientation vector $\mathbf{u}$ the correlation function is a second rank Legendre polynomial, which is most simply expressed as

$$P_2(t) = \tfrac{1}{2}\langle 3(\mathbf{u}(t) \cdot \mathbf{u}(0))^2 - 1 \rangle \tag{125}$$

For the axis defined by the H–H vector in the dichloromethane molecule the correlation time is 0.53 ps from relaxation of resonance of the proton nucleus.[114] From deuterium quadrupole relaxation[114] the correlation time for the C–D axis is 0.80 ps. The correlation time for the diffusion of the C–H axis is 0.70 ps,[114] from C–H dipole nuclear magnetic resonance relaxation. From Cl nuclear magnetic resonance quadrupole relaxation the relaxation time for the diffusion of the C–Cl axis is 1.20 ps. These experimental results were collected by Brier and Perry[114] and show an anisotropy of the molecular diffusion process which was successfully reproduced by Ferrario and Evans[2,4] by computer simulation. The latter produced 0.5 ps for the correlation time of the acf (Eq. 125) of the dipole unit vector, and 0.90 ps for the correlation time of the perpendicular axis of the principal moment of inertia frame. The equivalent of the H–H correlation time from the computer simulation was found to be 0.51 ps. Although there is broad agreement about the anisotropy of the diffusion process, there is internal discrepancy among experimentally derived correlation times.[114] The computer simulation is more effective than the traditional approach to this problem, which relies on three empirical and unknown friction coefficients within an inadequate theoretical framework.

W. Depolarized Rayleigh Scattering

The orientational correlation function from depolarised Rayleigh scattering from liquid dichloromethane complements the data available from a combination of far infrared and microwave spectroscopy. However, the orientational time correlation function from Rayleigh scattering is the second-order Legendre polynomial of Eq. (125). In general, data from depolarised Rayleigh scattering contain contributions from collision-induced effects[195] of the type discussed already in this chapter for the far infrared absorption spectra of dipolar liquids. The process of data reduction to provide a correlation time characteristic of the diffusion of an axis of the dichloromethane molecule must therefore take into account the presence of cross correlations and collision-induced effects.[195] There is a further practical problem in depolarized light scattering as compared with the far infrared in that the intensity of the depolarised scattered light falls very rapidly to the baseline,[117–120] being essentially equivalent to a zero spectral moment. In shape, the frequency dependence of Rayleigh scattered radiation is approximately equivalent[117] to $\alpha(\omega)/\omega^2$, that is, the far infrared power absorption divided by ω^2. This is essentially speaking dielectric loss divided by ω. We have seen in (Section I) that dielectric loss is relatively insensitive to high-frequency molecular dynamics and it follows that raw Rayleigh scattering data are even less so. In actual experimental data the information of most interest to molecular dynamics is contained in the far wing of the Rayleigh scattering profile at frequencies from about 5–300 wavenumbers from the frequency of the exciting line of the laser radiation with which the sample is probed for its light scattering characteristics.[117–120]. Collision-induced effects (dependent on the interaction of two or more molecules) are also more pronounced[195] in Rayleigh scattering than in far infrared absorption because the former depends

TABLE I

Comparison of Observed and Computer Correlation Times for Liquid Dichloromethane

Technique	Axis	Correlation Time (ps)
NMRH	H–H	0.53 ± 0.06
NMRD	C–D	0.80 ± 0.10
NMR C–H	C–H	0.70 ± 0.07
NMR Cl	C–Cl	1.20 ± 1.10
Computer simulation, P_2	$\mathbf{u}_1$	0.5
	$\mathbf{u}_2$	0.9
	$\mathbf{u}_3$	0.5
Neutron scattering	H (cm)	0.56
Dielectric relaxation	$\mathbf{u}_1$	1.45
Infrared	$\mathbf{u}_1$	0.5
	$\mathbf{u}_1$	1.1
Computer simulation, P_1	$\mathbf{u}_1$	1.2
	$\mathbf{u}_1$	3.8
	$\mathbf{u}_3$	1.2
Rayleigh scattering	$\mathbf{u}_1$	1.85

directly on the polarisability tensor of Eq. (71) and the latter on the dipole moment vector.

Despite these complications van Konynenberg and Steele[195] have obtained a correlation time from depolarized Rayleigh scattering at one temperature and pressure. This is the second Legendre correlation time of the dipole moment axis, and was found to be 1.85 ps. This is in marked disagreement, however, with the computer simulation of Ferrario and Evans,[2,4] which produced 0.5 ps for the equivalent correlation time, approximating much more closely to the N.M.R. relaxation time.[114]

X. Incoherent, Inelastic Neutron Scattering

The neutron-scattering data obtained by Brier and Perry[114] were analysed by them in terms of several models of the liquid state, but were insufficiently varied or accurate to produce a reliable picture of the molecular diffusion process. This is a contemporary problem with neutron scattering and the interpretation of such data would benefit from routine computer simulation.

Y. Survey of Correlation Times

A survey of correlation times from the various data sources available is provided in Table I for liquid dichloromethane. The autocorrelation times from the computer simulation and the NMR relaxation produce a picture of anisotropic diffusion, which is corroborated by the results from infrared spectroscopy. A combination of the multiparticle correlation times from dielectric relaxation (1.45 ps) and depolarized Rayleigh scattering (1.85 ps) shows that the theory of rotational diffusion cannot explain the data satisfactorily, because the theory requires the latter to be three times shorter than the former. The equivalent autocorrelation times from the computer simulation are 1.2 ps and 0.5 ps respectively, the opposite trend to that from the dielectric and Rayleigh scattering data.

The anisotropy of the diffusion process is shown by the fact that the correlation time for diffusion of the dipole axis from the simulation is 1.2 ps, while that for the principle moment of inertia axis parallel to Cl–Cl is 3.8 ps. The same trend is observed in the correlation times from infrared analysis, but the absolute values in the latter case are shorter (0.5 and 1.1 ps). From NMR relaxation the same anisotropy is observed from the proton relaxation time, (H–H axis) of 0.53 ps and the chlorine quadrupole relaxation time (1.30 ps, C–Cl axis). The anisotropy in the equivalent second-order relaxation times from the computer simulation is expressed as 0.5 and 1.1 ps, respectively. There is therefore overall agreement but not in detail.

In particular the correlation time from depolarized Rayleigh scattering is anomalous, and the neutron scattering results equivocal. Table 1 shows however that computer simulation is far more effective than the theory of Section I in explaining all the available data coherently. The same simulation trajectories can produce a variety of new information (Sections V, VII, and VIII) unavailable to conventional spectral investigation.

Z. A Survey of Molecular Liquids

The methodology sketched in this section of comparing a range of spectral and thermodynamic data for the same liquid with computer simulation has been applied in several cases. The interested reader is referred to the specialist literature[4] and to source papers for more details:

1. The symmetric tops chloroform,[196] methyl iodide,[197] and acetonitrile[198] in the liquid state.
2. The symmetric tops bromoform[199] and t-butyl chloride[200] in the liquid and rotator phases.
3. The asymmetric tops acetone[201] and ethyl chloride[202] in the liquid and supercooled states.
4. The H-bonded asymmetric tops water[203–206] and methanol.[207]
5. The spherical tops carbon tetrachloride,[56,57] germanium tetrabromide,[208] carbon tetrabromide,[208] phosphorus,[208] and sulphur hexafluoride.[208]
6. The chiral asymmetric tops bromochlorofluoromethane,[209] 1,2-di-

methylcyclopropane,[210,211]   fluorochloroacetonitrile,[212–214]   and others.[215–220]

7. Plate-like and rod-like molecules.[221]

In all cases the computer simulation povided a broad range of data from the same set of trajectories, and also investigated new ccf's of many different types. This has been achieved in the last decade and is an advance over the state of understanding summarized in Section I.

IV. SIMULATION — NEW RESULTS

In the last decade computer simulation has produced many new and fundamental results on molecular diffusion processes. These have made obsolete many of the analytical ideas of Section I. We have seen in Section III that computer simulation can be used to describe a range of experimental information given a representation of the pair potential incorporated in the classical equations of motion. Having tested the technique against experimental data in this way, it is reasonable to apply it to probe the limits of applicability of the basic analytical ideas. Some of the simulation results are fundamentally important, and the analytical approach of Section I is still struggling to come to terms with the new numerical advances. Similarly, the computer simulations have been able to predict the outline results of key spectral data, giving a more profound and specific picture of the molecular liquid state and its dynamical properties. Before embarking on a description of these results, it is necessary to define some basic statistical concepts and techniques of utility.

A. Linear Response Theory and the Fluctuation–Dissipation Theorem

One of the clearest approaches to understanding the linear response approximation in the theory of diffusion is to consider thc diffusion equation used originally in the Debye theory of 1913. This equation was derived from a Smoluchowski equation governing the evolution in time of a conditional probability density function $f_2(x_2, t_2 \mid x_1, t_1)$ of the random variable x. The product $f_2 \, dx_2$ is the probability that $x(t_2)$ has a value in the range $(x_2, x_2 + dx_2)$ given that $x(t_1)$ had the value x_1 at t_1. A random variable[1] is a dynamical quantity that may take any of the values of a specified set with a specified relative frequency or probability. In molecular dynamics it can be a vector quantity such as centre-of-mass linear velocity or, as in the Debye equation, an angular variable. It is defined through a set of permissible values to each of which is attached a probability. The Smoluchowski equation is derived by considering the probability density function (pdf) $f_3(x_3, t_3 \mid x_1, t_1) \, dx_3$ which is the probability that the random variable $x(t_3)$ has the range of values $(x_3, x_3 + dx_3)$ given that $x(t_2)$ has the value x_2 at time t_2 and that $x(t_1)$ has the value x_1 at time t_1.

To find the pdf $f_3(x_3, t_3 | x_1, t_1)$ we have to multiply f_2 and f_3 and average over all possible values taken by x_2. This averaging process can be considered as integration, from the basic laws of integral calculus. This leads directly to the Chapman–Kolmogorov equation[222]

$$f_3(x_3, t_3 | x_1, t_1)\, dx_3 = \int_{-\infty}^{\infty} f_2(x_2, t_2 | x_1, t_1) f_3(x_3, t_3 | x_2, t_2; x_1, t_1)\, dx_2\, dx_3 \tag{126}$$

The Smoluchowski equation is derived from this equation by making the additional assumption that the conditional pdfs

$$f_3(x_3, t_3 | x_2, t_2; x_1, t_1) = f_2(x_3, t_3 | x_2, t_2) \tag{127}$$

The Smoluchowski equation is therefore

$$f_2(x_3, t_3 | x_1, t_1) = \int_{-\infty}^{\infty} f_2(x_2, t_2 | x_1, t_1) f_2(x_3, t_3 | x_2, t_2)\, dx_2 \tag{128}$$

In the context of molecular diffusion, the Markov hypothesis is equivalent to approximating the memory function of Eq. (6) of Section I with a time-independent friction coefficient, which appears in the rotational Langevin equation (Eq. 27). The physical consequence of this "loss of memory" is that Markov theory loses its ability to describe the far infrared power absorption coefficient.

B. Taylor Expansion of the Smoluchowski Equation

An analytical solution of the Smoluchowski equation can be approximated by expanding the pdf's which appear on both sides of Eq. (128) in Taylor series. This is achieved with the variable $z = x_3 - x_2$, fixing x_2. Writing $t_2 = t$ and $t_3 = t + \delta t$, we have

$$f_2(x_3, t_3) = f(x_3, t + \delta t) = \int_{-\infty}^{\infty} f(x_2, t) f(x_3, t + \delta t | x_2, t)\, dx_2 \tag{129}$$

The variable x_3 determines z because x_2 is fixed, that is, given a finite value with probability 1. This implies

$$f(x_3, t + \delta t | x_2, t) = q(z, \delta t | x_2, t) \tag{130}$$

Assuming that q is independent of the value x_2 from which the transition is made and independent of the time t_2,

$$f(x_3, t + \delta t \,|\, x_2, t) = q(z, \delta t) \qquad (131)$$

Equation (129) now becomes

$$f(x, t + \delta t) = \int_{-\infty}^{\infty} q(z, \delta t) f(x - z, t)\, dz \qquad (132)$$

It is now assumed that $f(x, t + \delta t)$ and $f(x - z, t)$ have well-defined Taylor expansions:

$$f(x - z, t) = f(x, t) - z \frac{\partial f(x, t)}{\partial x} + \frac{z^2}{2!} \frac{\partial^2 f(x, t)}{\partial x^2} + \cdots \qquad (133)$$

and

$$f(x, t + \delta t) = f(x, t) + \delta t \frac{\partial f(x, t)}{\delta t} + \frac{(\delta t)^2}{2!} \frac{\partial^2 f(x, t)}{\partial t^2} + \cdots \qquad (134)$$

Finally, the Smoluchowski equation as used by Debye is obtained by assuming that the time interval δt is very small. This leads to

$$\frac{\partial f(x, t)}{\partial t} = \frac{1}{2!} \frac{\langle z^2 (\delta t) \rangle}{\delta t} \frac{\partial^2 f(x, t)}{\partial x^2} \qquad (135)$$

Making use of the averages

$$D = \lim_{\delta t \to 0} \frac{1}{2} \frac{\langle z^2 (\delta t) \rangle}{\delta t} \qquad (136)$$

and

$$\langle z^2 (\delta t) \rangle = \int_{-\infty}^{\infty} z^2 q(z, \delta t)\, dz \qquad (137)$$

finally leads to

$$\frac{\partial f}{\partial t} = D \frac{\partial^2 f}{\partial x^2} \tag{138}$$

which is the form of the Smoluchowski equation used as the starting point of Debye's theory of 1913 and of Einstein's theory of 1905. The diffusion coefficient (Section I) is

$$D = \frac{kT}{6\pi\eta a} = \frac{kT}{m\beta} \tag{139}$$

and Eq. (138) is an approximation to the integral equation (Eq. 129) for the pdf:

$$f(x, t \mid x_1, t_1) = f(x, t) \tag{140}$$

Equation (138) is derived from the very simple framework of the Chapman–Kolmogorov equation (Eq. 126).

Note that Eq. (138) refers to the pdf of position only. A more complete analysis refers to both velocity and position simultaneously, and is known as the Fokker–Planck equation.[1-4] It governs the conditional probability of finding the molecule both with a given position and velocity at time t_1, given both its position and velocity at t.

C. The Concept of Linear Response

If the diffusing molecule is subjected to an external or internal potential energy barrier, the approximation (Eq. 138) must be modified. For the sake of simplicity we consider linear diffusion in one axis. The analytical representation of the molecular diffusion under the influence of a potential $V(x)$ is

$$\frac{\partial f}{\partial t} = \frac{kT}{m\beta} \frac{\partial^2 f}{\partial x^2} + \frac{1}{\beta} \frac{\partial}{\partial x} \left(\frac{\partial V(x)}{\partial x} f \right) \tag{141}$$

D. Rotational Diffusion

Equation (141) was adopted by Debye for rotational diffusion in two dimensions of a molecular dipole moment subjected to an external time-varying electric field $\mathbf{E}$. In this case $f(\theta, t)\, d\theta$ is the number of dipoles whose axes lie in an element $d\theta$ on the circumference of a circle. The pdf is a function of the angle θ between the dipole and the field. The Smoluchowski equation for this process is[1,2,4]

$$\frac{\partial f(\theta, t)}{\partial t} = \frac{\partial}{\partial \theta} \left(\frac{kT}{\zeta} \frac{\partial f(\theta, t)}{\partial \theta} + \frac{1}{\zeta} \frac{\partial V(\theta, t)}{\partial \theta} f(\theta, t) \right) \tag{142}$$

where ζ is the rotational friction coefficient. This equation corresponds to the Langevin equation with the inertial term missing. The statistical process described by the Smoluchowski equation (Eq. 142) involves instantaneous jumps in the angular motion of the dipole axis. The pdf involves position only, and not angular velocity, so that the equation is said to be written in configuration space.

These historical approximations illustrate the meaning of linear response if we write the potential between dipole and field as

$$V = - \mu E \cos \theta = - \boldsymbol{\mu} \cdot \mathbf{E} \tag{143}$$

at Eq. (142) becomes

$$\frac{\partial f(\theta, t)}{\partial t} = \frac{\partial}{\partial \theta} \left(\frac{kT}{\zeta} \frac{\partial f(\theta, t)}{\partial \theta} + \frac{\mu E}{\zeta} \sin \theta f(\theta, t) \right) \tag{144}$$

This is an equation valid for rotation about fixed axes only, but if the dipoles are also free to rotate in space then it is modified for use in spherical polar coordinates, f being now the pdf governing the number of dipoles whose axes point in an element of solid angle Ω. The three dimensional equation is

$$\frac{\partial f(\Theta, t)}{\partial t} = \frac{1}{\sin \Theta} \frac{\partial}{\partial \Theta} \left[\sin \Theta \left(\frac{kT}{\zeta} \frac{\partial f(\Theta, t)}{\partial \Theta} + \frac{\mu E}{\zeta} \sin \Theta f(\Theta, t) \right) \right] \tag{145}$$

where $f(\Theta, t) \sin (\Theta) \, d\Theta$ is the probability that at time t the dipole has an orientation between Θ and $\Theta + d\Theta$ relative to $\mathbf{E}(t)$; and $M(\Theta, t) = \mu E \sin \Theta$ is on the torque on the dipole due to the applied field. Equation (145) is a special case of the Smoluchowski equation:

$$2\tau_D \frac{\partial f}{\partial t} = \text{div} \left[\text{grad} f + \frac{f}{kT} \text{grad} V \right] \tag{146}$$

E. Solution for Arbitrary Field Strength

Analytical solutions for arbitrary field strength are available[1-4] only with simple forms, such as the rectangular pulse

$$\mathbf{E}(t) = \mathbf{E}_0(U(t) - U(t - t_1)) = \mathbf{E}_0 e(t) \qquad (147)$$

where $U(t)$ is the unit step function. The solution of Eq. (145) is assumed to take the general form[223]

$$f(\Theta, t) = \sum_{n=0}^{\infty} a_n(t) P_n(\cos \Theta) \qquad (148)$$

where P_n are the Legendre polynomials and a_n are to be determined. The Legendre polynomials have the orthogonality[224]

$$\int_{-1}^{1} P_n(x) P_m(x) \, dx = \frac{2}{2n + 1} \delta_{m,n} \qquad (149)$$

The ensemble averages defining the system are

$$\langle P_n(\cos \Theta) \rangle = \frac{\displaystyle\int_{-1}^{1} f(\Theta, t) P_n(\cos \Theta) \, d(\cos \Theta)}{\displaystyle\int_{-1}^{1} f(\Theta, t) \, d(\cos \Theta)} = \frac{a_n(t)}{a_0(2n + 1)} \qquad (150)$$

and these can be computed from Eq. (145) to any order in $\cos \Theta$ following Ref. 1, pp. 151 ff. The first two rise transients are then (for $t < t_1$)

$$\langle P_1 \rangle_R = \frac{1}{3} \left[\gamma(1 - e^{-t/\tau_D}) - \frac{1}{5} \gamma^3 \left(\frac{1}{3} - \frac{t e^{-t/\tau_D}}{2\tau_D} - e^{-2t/\tau_D} - \frac{1}{4} e^{-t/\tau_D} \right) + \cdots \right] \qquad (151)$$

$$\langle P_2 \rangle_R = \frac{1}{5} \gamma^2 \left[\left(\frac{1}{3} - \frac{1}{2} e^{-t/\tau_D} + \frac{1}{6} e^{-3t/\tau_D} \right) \right] + \cdots \qquad (152)$$

and the fall transients after switching off the fields instantaneously at $t = t_1$ are

$$\langle P_1\rangle_F = \frac{1}{3}\gamma(e^{-(t-t_1)/\tau_D} - e^{-t/\tau_D})$$

$$-\frac{1}{15}\gamma^3\left(\frac{1}{3}e^{-(t-t_1)/\tau_D} - \frac{t_1}{2\tau_D}e^{-t/\tau_D} - \frac{1}{4}e^{-t/\tau_D} - \frac{1}{12}e^{-(t-t_1)/\tau_D}\right) + O(\gamma^5)$$

$$(153)$$

$$\langle P_2\rangle_F = \frac{1}{5}\gamma^2\left[\frac{1}{3}e^{-3(t-t_1)/\tau_D} + \frac{1}{6}e^{-3t/\tau_D} - \frac{1}{2}e^{-t/\tau_D}e^{-2(t-t_1)/\tau_D} + O(\gamma^4)\right]$$

$$(154)$$

It can be seen from these equations that the rise and fall transients are mirror images (Section I) only when terms linear in $\gamma = \mu E_0/(kT)$ are retained. This is the approximation first given by Debye and is a consequence of the linear response approximation.

Another is that in the linear response limit the fall transient is an exponential, and has the same time dependence as the orientational acf from the equivalent Langevin equation (Eq. 8) of Section I. Finally, the rise and fall transients cease to be functions of the field strength only in the same linear response regime, $\gamma^3 \ll \gamma$.

These linear response approximations are assumed implicitly in the relation of spectral band shapes to time-correlation functions. In the theory of dielectric relaxation, for example,[12] the field strength does not enter into consideration, because implicit in the theory is the fluctuation-dissipation theorem

$$\frac{\langle P_1\rangle_F}{\langle P_1\rangle_0} = \frac{\langle\cos\Theta(t)\cos\Theta(0)\rangle}{\langle\cos\Theta(0)\cos\Theta(0)\rangle}$$

$$(155)$$

that is, the fall transient's time dependence is the same as that of the equilibrium acf in the linear response approximation. The acf is defined at field-free reversible thermodynamic equilibrium. In dielectric relaxation the field is in general frequency-dependent, from subradio frequencies to the far infrared.

F. Computer Simulation of Rise and Fall Transients in an Arbitrarily Strong External Field — Violation of the Linear Response Theory

The theory given above, due originally to Coffey and Paranjape,[223] gave the first indication of what would happen if linear response were inapplicable. Roughly contemporaneously, several experimental indications of the limits of linear response were obtained,[24,225-227] but a full theoretical description remains intractable. This is due essentially to the fact that the Smoluchowski equation (Eq. 145) is for configurational space only, and there is no proper definition of the role of angular velocity. The appropriate equation for the space of configuration and velocity is the Kramers equation.[1-4] The latter corresponds to a Langevin equation with inertial and potential terms on the left-hand side, akin to Eq. (32) of Section I, or the itinerant oscillator equations (Eqs. 36 and 37). In general, the Kramers equation must be solved by numerical differential differencing, and this is possible only in idealized cases.[1-4]

A much more direct and practical approach to the exploration of nonlinear response theory, and of nonlinear problems in general, is computer simulation.

G. Computer Simulation in an Applied External Field

This technique was devised[2,3,160,228] in the early 1980s and prompted fundamental changes in our appreciation of diffusion processes. The basic idea is an adjustment to the forces loop (Chapter (3)) of a standard constant volume algorithm such as TETRA. The extra torque $-\mu \times E$ is coded into the algorithm at the point where the net intermolecular torque is computed from the forces. The extra torque is switched on at the arbitrary initial $t = 0$ and the main simulation algorithm starts to respond. Extra information is provided numerically which can be compared with any available analytical theory, however rudimentary the latter may be, through the careful investigation of (1) rise transients, (2) fall transients, and (3) field-induced acf's and ccf's.

H. Rise Transients

Equilibrium orientation averages such as those represented by Eqs. (151)–(154) are produced in the computer simulation by monitoring the effect of the extra torque on a simple average (over the number of molecules in the sample) of the unit vector u_1 in the axis of the molecular dipole moment μ or any other definable axis in the molecule. In the absence of the aligning field averages such as these vanish, because the sample is isotropic. Optical properties such as the refractive index are the same on average in each axis X, Y, or Z of the laboratory frame. However, when the electric field is applied, and the extra torque between field and dipole takes effect, each molecule becomes partially aligned on average in the field direction: the sample is polarized and becomes birefringent. The complex permittivity (ϵ) and the complex refractive index (n) are linked by

$$\epsilon(\omega) = [n(\omega)]^2 \tag{156}$$

so that any directional phenomenon such as polarization (a field-dependent change in the complex permittivity) is accompanied by an equivalent effect in the complex refractive index. The microscopic quantity equivalent to these observable effects on permittivity and refractive index after the field is switched on is the rise transient, the average $\langle \mathbf{u}_1(t) \rangle_R$ as a function of time t.

I. The Langevin Function

The rise transient eventually reaches a value at which it fluctuates about a mean value greater than zero. This final level is dependent on the torque generated between field and dipole and therefore on the extra energy imparted by the applied electric field. An illustration for several field strengths is given in Fig. 12. An analytical description of the dependence of the level reached by the transient on the applied electric field was first derived by Langevin and is known as the Langevin function. It is expressed in terms of the ratio $b = \mu E/(kT)$ (Section I). For the first-order rise transient $\langle \mathbf{u}_1(t) \rangle_R$ the Langevin function is

$$L_1(b) = \frac{e^b + e^{-b}}{e^b - e^{-b}} - \frac{1}{b} \qquad (157)$$

The second-order rise transient described analytically in Eq. (152) has its equivalent Langevin function also:

$$L_2(b) = 1 - \frac{2L_1(b)}{b} \qquad (158)$$

These analytically derived Langevin functions are plotted against the saturation values of rise transients from computer simulation in Figs. 13 and 14. The computer simulation reproduces the Langevin functions[2,3] for the test molecule dichloromethane treated with strong electric fields, sufficient to saturate the functions. This would be difficult experimentally[24] owing to heating effects, sparking, and difficulty in generating the huge electric field strengths required. The numerical simulation, having been checked out in Section III, produces information on Langevin functions of all orders. This takes us into unknown territory, that of nonlinear dielectric spectroscopy, unexplorable with contemporarily available electric field strengths, and inaccessible to anything but the most rudimentary of analytical theories.

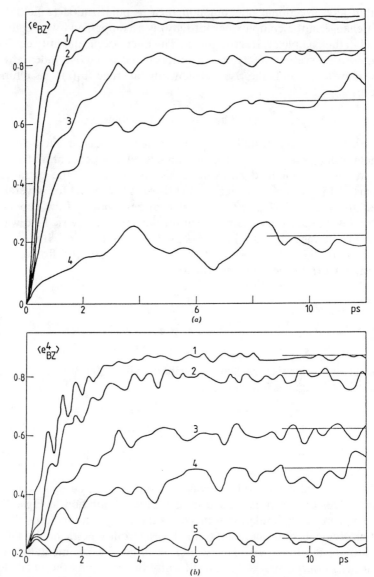

Figure 12. Rise transients as a function of field strength.

J. Time Dependence of the Rise Transient

The Langevin functions of order $n = 1$ to m are available analytically because they are simple thermodynamic averages.[9] The time dependence

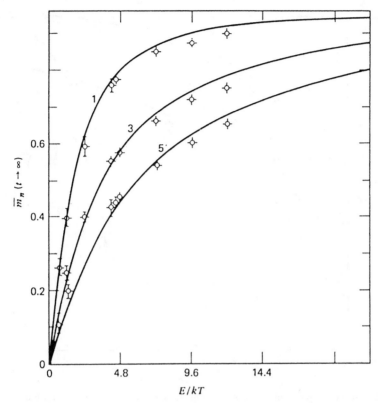

Figure 13. Langevin functions at first, third, and fifth order (full lines) plotted against plateau levels from computer simulation (symbols).

of the transient is much more difficult to obtain. The results of Coffey and Paranjape [223] were among the first to be obtained outside the confines of linear response, but are nevertheless derived in configurational space only, from a simple Smoluchowski equation. They carry with them all the inherent flaws of the customary theory of diffusion. Even in this case Eqs. (151)–(154) are series approximations, sufficient only to show that the rise and fall transients are no longer mirror images. As soon as linear response is left behind, the transients become field-dependent. What little is known analytically shows only that the transients are in general dependent on the approximation employed for the intermolecular potential energy, again a departure from the results in the linear response approximation. The analytical representation of the intermolecular potential in diffusion theory is woefully inadequate. For example, the rise transient may be obtained from a Langevin equation[229]

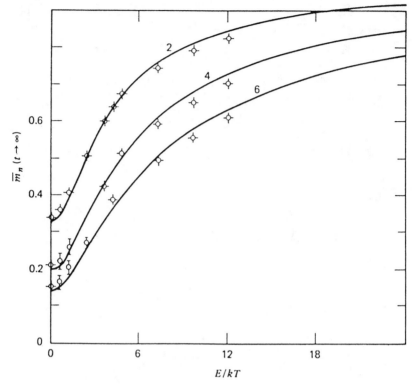

Figure 14. As for Fig. 13, Langevin functions at second, fourth, and sixth order.

$$I\ddot{\theta}(t) + \zeta\dot{\theta}(t) + E_0 \sin\theta(t) = \lambda(t) \qquad (159)$$

The inertial term is present in this equation and makes a startling difference to the time dependence of both the rise and fall transients, as illustrated in Figs. 15 and 16. The Langevin equation (Eq. 159) must be transfomed to the equivalent Kramers equation, however, and is insoluble as it stands. The results from the Kramers equation are obtained numerically, by standard differential differencing, and cannot be obtained analytically. The presence of the inertial term and extra torque term on the left-hand side of Eq. (159) have profound effects on the rise transient, making it markedly oscillatory, and accelerate the decay of the fall transient. These indications have been matched by computer simulation[230] in liquid water, using a more realistic representation of the pair potential. Figures 17 and 18 illustrate the oscillations in the rise transient from the simulation, which considered 1372 water molecules. The simulated oscillations are dependent on the electric field strength and the pattern of oscillations is

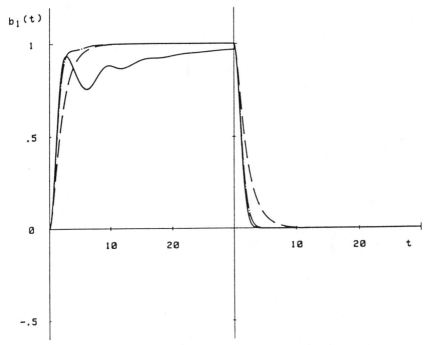

Figure 15. First-order rise and fall transients obtained from the work of Coffey et al., ref. (229).

dependent on the model used of the intermolecular pair potential. This means that experimental information on the potential can be obtained in principle from observations of oscillatory rise transients. This is a critically important simulation result, verifying the analytical indications, and leading to a new method of investigating the pair potential. The field strengths necessary to achieve this are available not in electric fields but in the electric field components of laser fields, applied in trains of very short-lived but extremely powerful pulses. The pair potential in this simulation was modelled with a 5×5 site–site representation devised by the present author.[203–206] It is probable that a different representation of the water pair potential (or triplet and higher-order potentials)[151] would produce a different pattern of oscillations, and thus provide a practical method of differentiating between different model and ab initio potentials.

The restriction to two-dimensional diffusion (Eq. 159) are removed in the simulation. Both the two- and three-dimensional approaches give the same overall indications.

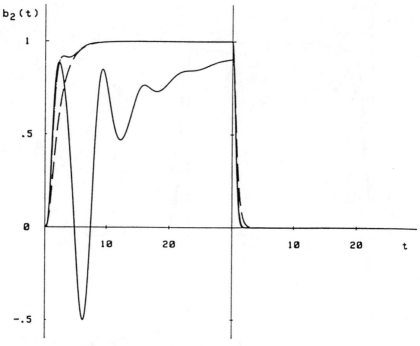

Figure 16. As for Fig. 15, second-order rise transients.

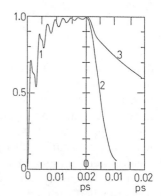

Figure 17. Rise and fall transients from computer simulation of 1372 water molecules, illustrating rise-transient oscillation and fall transient acceleration for a given electric field strength, ref. (230).

K. The Fall Transient

Agreement in principle has also been achieved between computer simulation and analytical theory concerning the time dependence of the fall transient. In linear response theory this is simply the mirror image of the rise transient, both being exponentials which are identical with the normalised equilibrium acf

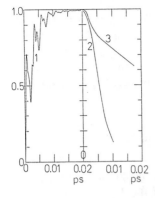

Figure 18. As for Fig. 17, varying the field strength. The oscillation and acceleration effects are dependent on field strength.

$$\frac{\langle \mathbf{u}_1(t) \rangle}{\langle \mathbf{u}_1(0) \rangle_{t=0}} = \frac{\langle \mathbf{u}_1(t) \cdot \mathbf{u}_1(0) \rangle}{\langle \mathbf{u}_1(0) \cdot \mathbf{u}_1(0) \rangle_{E=0}} \qquad (160)$$

Equations (153) and (154), from the relatively simple model of the Debye diffusion equation, show that the fall transient has a complicated time dependence, involving the field strength. The fall transient can decay more quickly to zero than the field-free equilibrium acf. This was first discovered[228] in a computer simulation of liquid dichloromethane in 1982. The simulation was followed by an analytical confirmation.[229] Similar analytical work[231,232] was reported later which again showed the same result. The new effect was identified as fall transient-field acceleration, and later appeared in several corroborative simulations[3,230-232] The most important likely consequence is the use of the effect in the experimental investigation of the intermolecular potential, using intense laser pulse trains. In so doing the individual time dependencies of both the rise and fall transients must be describable consistently by the same representation of the potential energy between molecules. This can only be achieved by computer simulation back-up, because the analytical appreciation is still severely limited. However, analytical theory has managed to show that the effect is produced only when the dependence of the model potential on the angular variable is nonlinear, as for example in Eq. (159). The harmonic approximation $\sin \theta \simeq \theta$ does not produce a fall transient acceleration. This is contrary to the findings of several computer simulations, and the harmonic approximation must be discarded. This conclusion is consistent with the fact that the far infrared power absorption coefficient of a liquid such as dichloromethane needs for its rudimentary description an effectively nonlinear potential representation. Fall-transient acceleration is a fundamental property of the molecular liquid state and any analytical description which fails to describe it must be modified. The same is true of far infrared power absorption. Ideally, both sources of information should be treated by the same computer simulation, as in our work on the test liquid dichloromethane.

Fall-transient acceleration results specifically from the nonlinear nature of the effective potential energy and is also dependent on the applied electric field strength.

L.   Correlation Functions in the Field on Steady State

We havc seen that the Debye–Smoluchowski equation is derived by applying the Markov approximation to the Chapman–Kolmogorov equation. The Markov approximation therefore underlies the Langevin equation, for example, Eq. (8) of Section I. Reinstating the inertial term modifies the nature of the diffusion process, so that the Smoluchowski equation in configuration space must be replaced by the Fokker–Planck equation in the space of both configuration and velocity. The extension of the phase space in this way leaves unaffected, however, thc basic Markov assumption uscd in attempting to solve Eq. (126), and also leaves unaffected the truncation of the Taylor expansions Eqs. (133) and (134). Much of the analytical work in exploring the ramifications of these approximations has occurred in the last decade,[3] and is constantly spurred on by the findings of computer simulation.

One major shortcoming of the Markov approximation is the failure of the resultant Langevin equation, with or without inertial terms, to describe the observed far infrared power absorption coefficient. The Liouville equation does not, furthermore, reduce straightforwardly to the Langevin equation, and this is also due to the Markov approximation. This was made particularly clear by Mori, using projection operators. These produced the Liouville equation in the form Eq. (6) of Section I, a form which resembles the Langevin equation, but reduces to it only if the Mori resonance operator vanishes and the memory function is a delta function, the friction coefficient of Langevin. The latter has no time dependence because it is a consequence of the Markovian exclusion of the statistical influence of historical events upon the time evolution of the molecular dynamical ensemble.

The experimental indications such as far infrared spectra suggest that the liquid state is non-Markovian in nature. Past events affect the statistical nature of future events, and the Markov approximation (Eq. 127) is too severe.

One of the clearest signs of this non-Markovian nature is the field-decoupling effect anticipated by Grigolini[233] and confirmed[3] by the present author with computer simulation.

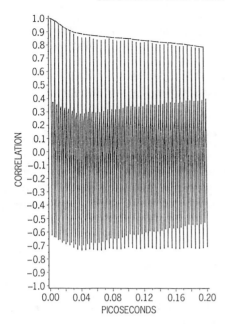

Figure 19. Illustration of the decoupling effect, the field-induced envelope marked by the dashed line decays more slowly than the equivalent field-off acf due to the liquid's inherently non-Markovian statistical nature. (Water, right C.P. laser, first order torque. Rotational velocity ACF'S, F = 400 THZ, field = 6000.)

M. The Decoupling Effect

When a molecular liquid sample is treated with an intense uniaxial electric field, some of the acf's in the ensemble become oscillatory. The frequency of the oscillations is governed by the torque $-\mu \times \mathbf{E}$. They are often most pronounced in the angular velocity acf, and the decoupling effect relates to the envelope of the time decay of the field-induced oscillations. The rate of decay of this envelope, marked by the dashed line in Fig. 19, is considerably slower than the loss of correlation in the field-free acf, and the greater the field strength, the slower the decay of the envelope.

To explain this consistently with fall-transient acceleration and far infrared absorption it is necessary to abandon the Markov approximation completely. The friction coefficient in an equation such as Eq. (159) must be replaccd by a memory function, which has time dependence. The analytical theory must be upgraded once more in response to advances made by computer simulation to the point where it is minimally capable of describing the information available. This new level of description is represented by

$$I\ddot{\theta}(t) + \int_0^t \phi(t-\tau)\dot{\theta}(\tau) + E_0 \sin\theta(t) = \lambda(t) \tag{161}$$

with the memory function ϕ. This equation can describe the decoupling effect qualitatively, and also the fall-transient acceleration, oscillations in the rise transient, and the rudiments of far infrared absorption.

This upgrading of the analytical theory will no doubt continue in the future as more is discovered by computer simulation. The theory of diffusion of Section I is clearly, therefore, not a complete theory which can stand on its own and make predictions in terms of the fundamental physical constants. It is not a quantum theory or a Newton theory of motion in bodies. Numerical simulation is clearly able to provide something more, to anticipate the existence of new physical phenomena. Diffusion theory not only finds itself incapable of prediction, but is also unable to describe available data, especially when judiciously combined. Some examples are given next of the findings of the simulation method.

N. New Type of Birefringence

A material is birefringent if its refractive index in one axis (say X) is different from those in Y and Z. Birefringence can occur naturally, as in a molecular crystal or aligned nematogen[1] or can be induced in a normally isotropic liquid by an external field of force,[9] mechanical, electrical, magnetic, electromagnetic, and others.

O. The Kerr Effect[9]

In the Kerr effect the inducing field is an electric or electromagnetic field, whose effect can now be understood through the imposed torque $-\mathbf{\mu} \times \mathbf{E}$ on each molecule. It can be investigated experimentally with great accuracy.[234] The development of the birefringence as a function of time (the rise transient), and its loss after switching off the field (the fall transient), can be measured. Diffusion theory[223] provides a limited appreciation of these results, which are available usually at relatively low frequencies. Both the traditional experimental[234] and theoretical[9] methods associated with the time-dependent Kerr effect failed completely to detect rise-transient oscillations and fall-transient acceleration. The theoretical approaches failed to realize the importance of inertial and nonlinear terms, took insufficient account of memory effects, and relied too much on low frequency data. There are purely experimental restrictions[24] due to weak aligning fields and limitation to low frequencies. However, if these can be remedied, and the use of computer simulation intensified, Kerr effect relaxation may become a less superficial means of investigation than at present.

P. The Cotton–Mouton Effect

The induction of birefringence in this case takes effect with an external magnetic field, and this is a strong effect in liquid crytals. Much less is reported about the effect of an aligning magnetic field in isotropic, molecular, liquids. The alignment takes effect through the torque generated by the magnetic dipole and magnetic field. The strength of the effect depends on the magnetic dipole, whether this is diamagnetic or paramagnetic. The new superconducting materials[235] offer scope for the generation of very intense magnetic fields with which to develop this technique for use with isotropic molecular liquids.

Q. Birefringence Induced with Electromagnetic and Neutron Radiation

Electromagnetic and neutron radiation induce birefringence, the former is sometimes known as the optical Kerr effect, but neutron-beam induced birefringence is not yet developed experimentally

R. Flow-Induced Bifringence

In a flowing molecular liquid, the birefringence parallel and perpendicular to the axis of flow is measurably different, and this problem can now be tackled by large scale computer simulation.[4,91,145] Similarly, rotational or vortex flow in a molecular liquid leads to birefringence.

S. Birefringence Effects Discovered by Computer Simulation

The development of new computer architectures, at IBM and elsewhere, has allowed the exploration of bigger samples in molecular dynamics computer simulation and improvements in the statistical quality of the runs. Partly as a result of this new found power and computer time, new birefringence phenomena have been discovered numerically. These are described as follows with reference to liquid water, simulated with the author's 5×5 site–site potential.

T. Electric-Field-Induced Translational Anisotropy

Computer simulation takes us outside the limitations of linear response and rotational diffusion, and opens the door to new exploration[235]. The use of a simple torque of the type $-\mu \times E$ has resulted in the numerical discovery of new effects, which the traditional approach of Section I cannot describe. One of the most revealing of these is the induction of birefringence in the linear centre of mass velocity acf in response to an applied electric field. At field strengths outside the range of linear response and conventional dielectric relaxation, this acf develops a different time dependence perpendicular and parallel to the applied electric field. This

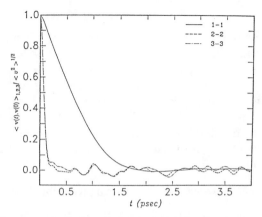

Figure 20. Computer simulation of electric field induced anisotropy in the acf of the molecular centre of mass in liquid water.

was known neither experimentally nor theoretically prior to the computer simulation[235] which produced the result illustrated in Fig. 20. An applied torque of the type $-\mathbf{\mu} \times \mathbf{E}$ results in different rates of translational diffusion parallel and perpendicular to the field, although there is no net flow in the sample. This result is a new measurable phenomenon of the liquid state which marks the boundary beyond which rotational diffusion theory cannot go. A purely rotational Mori theory (Eq. 161) cannot be used to describe this development. Translational variables such as the molecular center of mass linear velocity simply do not appear in our "upgraded" equation (Eq. 161). An imposed *torque* has resulted in anisotropy of *linear* velocity. Nothing in the purely rotational theory of this chapter, dependent on the Debye–Smoluchowski approach, and its more realistic rotational relatives, could have predicted the result obtained numerically and shown in Fig. 20.

This reveals a major flaw in the analytical description of molecular diffusion, one which can be traced directly to the original assumptions made independently by Einstein and Debye. Both assumed that the two fundamental modes of diffusion in a rigid dipolar molecule, rotation and translation, can be dealt with independently. There were sound reasons for this at the turn of the century (Section I), but Fig. 20 shows that rotational and translational effects are interrelated. A purely rotational or purely translational description of the molecular liquid state is insufficient. There are many new things to explore once this is realized, and some of these are described later in this chapter. Computer simulation is in advance of both analytical theory and experimental investigation in being the first technique to produce an unexpected result such as electric field induced anisotropy in the translational diffusion process.

U. Birefringence Induced by a Circularly Polarized Laser

Experimental exploration of the effects of intense electric fields must rely contemporaneously[236] on laser pulses, in which the electric and magnetic field components are described classically by Maxwell's field equations. One of the easiest ways of appreciating how the laser field can interact with a liquid of dipolar molecules is by considering again the simple torque $-\boldsymbol{\mu} \times \mathbf{E}$. This time, the electric field comes from the Maxwell equations of the complete electromagnetic laser field. The electric component of the laser field sets up the torque through interaction with the molecular dipole moment (the antenna which picks up the laser radiation). This introduces new possibilities, for example, the use of a circularly polarized laser field of the type

$$E_X = 0 \tag{162}$$

$$E_Y = E_0 \cos \omega t \tag{163}$$

$$E_Z = E_0 \sin \omega t \tag{164}$$

The components of the torque set up by the laser are then

$$\boldsymbol{\mu} \times \mathbf{E} = \begin{vmatrix} \mathbf{i} & \mathbf{j} & \mathbf{k} \\ \mu_X & \mu_Y & \mu_Z \\ 0 & E_0 \cos \omega t & E_0 \sin \omega t \end{vmatrix} \tag{165}$$

Incorporating these into the forces loop of TETRA produces new information about the way in which an intense laser field can interact with a molecular ensemble to produce new and measurable ccf's in the laboratory frame (X, Y, Z), and from these new information on the molecular dynamics. This information is beyond the boundaries of the traditional approach because a torque of the type in Eq. (165) makes the equations of rotational diffusion completely intractable. Once we depart from linear response theory, there is no closed solution[223] for the Debye–Smoluchowski equation, as described earlier in this chapter. The computer simulation method, on the other hand, deals with the laser field through a simple extra torque coded in to the algorithm. More complicated forms of torque can also be utilized straightforwardly.

A recent simulation of the effect of a strong circularly polarized laser[236] has revealed a range of new results using an intense pulse of radiation. The simulation was carried out on liquid water in three dimensions using 1372 molecules. Good statistics were obtained with 6000 time steps of 0.5 fs each. All 6000 time steps were implemented to build up time correlation functions by running time averaging. These were computed in the steady state obtained after the sample had come to equilibrium in the presence of the laser. In this steady state, new cross-correlation functions appeared directly in the laboratory frame (X, Y, Z) exemplified by

$$\mathbf{C}_1^{ij}(t) = \frac{\langle \boldsymbol{\mu}_i(t)\boldsymbol{\mu}_j(0) \rangle}{\langle \mu_i^2 \rangle^{1/2} \langle \mu_j^2 \rangle^{1/2}} \tag{166}$$

The diagonal elements of this tensor make up the usual acf of the water molecule's dipole moment, that is, the orientational acf related through a Fourier transform such as Eq. (19) to the dielectric loss. The computer simulation showed that this acf becomes anisotropic in the presence of the intense laser field, that is, its components in (X, Y, Z) develop a different time dependence (Fig. 21). This shows the laser inducing a first order birefringence in the molecular liquid. The refractive index in the axis of the applied laser is different from those in the two mutually perpendicular axes. This follows from the relation between dielectric permittivity and refractive index, the former being related by Fourier transformation to the orientational acf.

This birefringence develops without any consideration of the molecular polarizability, which results in the extra torque

$$-(\mathbf{E} \cdot \boldsymbol{\alpha} \cdot \mathbf{E}) \times \mathbf{E} \tag{167}$$

on each molecule, where a is the molecular polarizability tensor. Considerations of the second order torque (Eq. 167) leads to a second order birefringence phenomenon. The effect in Fig. 21 is to first order in the electric field component of the laser, and was unknown prior to the computer simulation. This illustrates the predictive ability of computer simulation in areas where the customary approach has become ineffective due to analytical intractability or where experimental investigation has not been initiated.

The simulation is also capable of providing detailed supplementary information from the same set of molecular trajectories built up over 1372 molecules and 6000 time steps. This information is exemplified in Fig. 22, which shows that the off-diagonal elements of the tensor (Eq. 166) are stimulated to exist directly in frame (X, Y, Z) by the circularly polarized laser. Any future analytical description must take this finding into account,

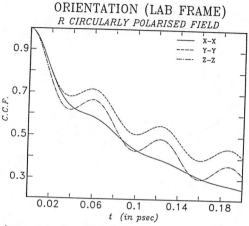

Figure 21. Anisotropy in the dipole orientational acf of liquid water induced by a circularly polarized laser.

and experimentally, these new ccf's are capable of direct observation. Similar results appear in the computer simulation of the tensor of the rotational velocity

$$\mathbf{C}_2^{ij}(t)_{XYZ} = \frac{\langle \dot{\boldsymbol{\mu}}_i(t)\dot{\boldsymbol{\mu}}_j(0)\rangle}{\langle \dot{\mu}_i^2\rangle^{1/2}\langle \dot{\mu}_j^2\rangle^{1/2}} \tag{168}$$

illustrated in Figs. 23 and 24. This means that the far infrared power absorption coefficient of the liquid water sample perpendicular and paral-

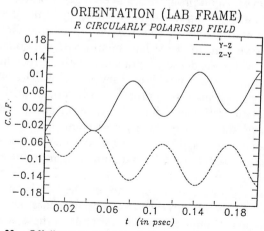

Figure 22. Off-diagonal elements of the tensor illustrated in Fig. 21.

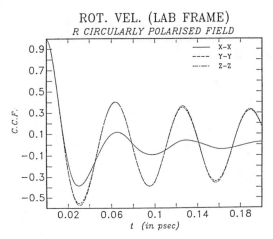

Figure 23. As for Fig. 21, rotational velocity acf.

lel to the applied laser field will become different in frequency dependence, together with the refractive index at these frequencies. There is birefringence both at high (far infrared) and low (dielectric) frequencies, giving plenty of scope for the eventual experimental observation of the phenomenon with laser pulse trains.

V. Simulation of Birefringence due to Circular Flow

Linear flow induced birefringence[237] is well known and described in the literature. However, appreciating its full potential using molecular dynamics as opposed to hydrodynamics is a more difficult task, and this is true in general of the borderline between flow and molecular dynamics. Recently, however, rapid progress has been made towards the unification of both subjects using computer simulation. Many of the flow phenomena of textbooks in hydrodynamics can now be reproduced directly from the fundamental Newtonian and Eulerian equations of motion, and computers are already in widespread use in practical testing procedures involving flow, over aircraft wings, ships' hulls, and so forth. This illustrates how quickly the seemingly abstract approach can be put to work.

Recent simulations[173] using up to 200,000 atoms over many millions of time steps, have shown how the hydrodynamic properties of a liquid flowing past an object (a disk or a cylinder) can be constructed from Newton's equations of motion. Simulation can reproduce the classical phenomena of hydrodynamics.

Having achieved this agreement between the two branches of physics simulation can confidently advance to the point where new flow phenomena can be anticipated numerically. An example of these is vortex induced birefringence.

ROT. VEL. (LAB FRAME)

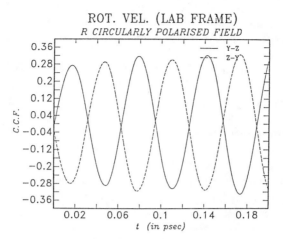

Figure 24. As for Fig. 20, rotational velocity acf.

W.   Field Equations for Circular Flow

Circular flow can be induced in a sample of, for example, 500 water molecules in a simulation cell using the simple equations

$$\mathbf{F} = F_Y\mathbf{j} - F_X\mathbf{i} \tag{169}$$

$$F_X = F_0 Y / (X^2 + Y^2)^{1/2} \tag{170}$$

$$F_Y = F_0 X / (X^2 + Y^2)^{1/2} \tag{171}$$

The X and Y components of the flow field are proportional in each case to the radius vector of each molecule's center of mass and propels all the molecules to which it is applied on a vortex about the Z axis. In order to minimise the effect of periodic boundary conditions on the vortex, Eqs. (170) and (171) are set up in an inner cell insulated from the edges at which the boundary conditions apply. At the edges of this inner cell there are natural boundary conditions, individual flowing water molecules encounter those to which no force field has been applied. The molecules in the outer cell shield the vortex from artificial periodicity which would destroy its development over time. Eventually, the flow set up in the inner cell propagates to the outer, forming the beginnings of macroscopic flow. However this occurs very slowly, over millions of time steps, as the applied force field takes effect like a stirring rod.

On a microscopic scale, however, over picoseconds and angstroms, the cell of 500 water molecules may be used to demonstrate numerically the presence of many new patterns of dynamical correlation at the molecular level itself. An example is anisotropy in the correlation tensor of molecular linear velocity and related dynamical quantities affected by the applied vortex field. The familiar patterns of macroscopic flow, eddies, wakes,a and macroscopic vortices, are always accompanied on the molecular scale by new time cross-correlations at the molecular level. These generate the visible macroscopic phenomena direct from the basic equations of motion, making unnecessary much of the intermediate level of approximation inherent in the subject of hydrodynamics. An example of a measurable effect is vortex induced birefringence.[237] It would be impossible to describe this adequately without the simultaneous consideration of translation and rotation in the diffusing asymmetric top, bringing into consideration both the positional and velocity space, and needing a complicated and insoluble Kramers equation in the approach of Chapter 1. A vortex is made up of many molecules rotating with a definite symmetry in the laboratory frame (X, Y, Z). The theory of Brownian motion in contrast as applied to "purely" rotational or translational diffusion dispenses with adequate consideration of the role of the position vector of center of mass. In rotational diffusion theory there is no consideration of translation and vice versa. To interrelate molecular and flow dynamics it is necessary to identify and define flow fields on a molecular scale. Some preliminary attempts in this direction are described in Refs. 238 and 239.

X. Anisotropy in Linear Diffusion Produced by a Circularly Polarized Laser

This is another new effect[240] and general characteristic of the molecular liquid state discovered by computer simulation. It is also accompanied by a birefringence in the dielectric loss and far infrared power absorption. The effect needs an explanation in terms of the statistical correlation between molecular rotation and translation, exemplified by that between the angular and linear molecular velocities.

Y. Electric or Circularly Polarized Laser Field Applied to a Dilute Gas

We recall from Chapter 1 that the classical-time acf of linear velocity in the infinitely dilute gas is a constant, which can be normalized to one at the arbitrary $t = 0$,

$$\frac{\langle \mathbf{v}(t) \cdot \mathbf{v}(0) \rangle}{\langle v^2(0) \rangle} = 1 \tag{172}$$

On the other hand the equivalent rotational acf of the infinitely dilute gas is[1]

$$\frac{\langle \boldsymbol{\omega}(t) \cdot \boldsymbol{\omega}(0) \rangle}{\langle \omega^2(0) \rangle} = \exp\left(\frac{-kTt^2}{2I}\right) \tag{173}$$

for the spherical top with slightly more complicated expressions for the symmetric and asymmetric top.

If we now consider the imposition of a torque between the dipole moment and the electric field, then the only direct effect is to add a rotational acceleration to each molecule. If there were no molecular interactions, the linear velocities would be unaffected. The linear velocity acf of the dilute gas would remain a constant in time, unaffected by the adjustment to the angular molecular dynamics, and there would be no statistical cross correlation between molecular linear and angular velocities in any frame of reference. We conclude therefore that without the intermediacy of intermolecular interaction there is no way of explaining why a rotational torque generated by an electric field or electromagnetic field through the term $-\boldsymbol{\mu} \times \mathbf{E}$ should result in anisotropy of the center of mass linear velocity acf as illustrated in Fig. 20. The electric or laser field induced linear anisotropy shows, in fact, that in a molecular liquid (as opposed to the infinitely dilute gas) there is always cross correlation between rotational and translational diffusion on the fundamental single molecule level. It is impossible to induce linear anisotropy of the center of mass molecular velocity in the infinitely dilute gas because in this case, and only in this case, the ccf vanishes. In corollary the field induced linear anisotropy of Fig. 20 is a potential experimental method of investigating qualitatively, and perhaps quantitatively, this type of fundamental cross-correlation.

Z.   Correlation between Rotation and Translation Induced by Electric Fields

The above argument is supported conclusively by a molecular dynamics computer simulation of the fundamental ccf:

$$\mathbf{C}_3^{ij}(t)_{XYZ} = \frac{\langle \mathbf{v}_i(t)\boldsymbol{\omega}_j(0)\rangle}{\langle v_i^2\rangle^{1/2}\langle \omega_j^2\rangle^{1/2}} \qquad (174)$$

in the laboratory frame of reference (X, Y, Z) carried out in liquid dichloromethane.[141,142] This result is outside the scope of a purely rotational or translational diffusion theory and therefore of the "up-graded" equation (Eq. 161).

The electric field induces a simple type of ccf directly in frame (X, Y, Z). If the electric field is in the Z axis of the frame (X, Y, Z) it induces the (X, Y) and (Y, X) components of the ccf tensor (Eq. 174), which are equal and opposite. This result can only be described analytically by trying to link together Langevin equations for rotational and translational motion with cross friction coefficients, introducing in the process a plethora of empiricism. The simple static electric field induces anisotropy according to the Kerr effect, anisotropy which is accompanied by the appearance of a new ccf which is inaccessible to the approach of Section I. The conventional theory of the Kerr effect[9] takes no account of the existence, let alone the detailed time dependence, of this ccf. We must try to adapt the theory to meet the unequivocal indications obtained by computer simulation.

AA. Consequences to the Theory of Polarization

The challenge is however very difficult to meet without introducing too much empiricism, too many unknowns. We have met this problem already in the itinerant oscillator approach and in approximants of the continued fractions of Section I. The existence of a set of nonvanishing ccf's strikes at the very foundations of the theory of diffusion and polarization[9]. The challenge is made with the basic equations of classical dynamics. Also at risk is the conventional approach to fluctuation–dissipation which relates field-on to field-off dynamics. The unaccountable existence of new ccf's which appear when an electric field is switched on and disappear again when it is switched off is not compatible with simply equating an orientation transient to an equilibrium acf. Both the rise and fall transients must reflect the fact that new ccf's are transiently appearing or disappearing as the case may be. The fundamentals of the customary approach to dielectric relaxation take no account of cross correlations whatsoever, and continue to rely on the Debye equation (Eq. 145) in many contemporary treatments. The field-on solution of this equation is a fundamentally inadequate description of the complete process of diffusion in molecular liquids irrespective of the strength of the applied electric field and of the linearity or otherwise of the sample response.

The validity of the Green–Kubo relations linking the translational velocity acf to the translational diffusion coefficient and the rotational and angular velocity acf's to distinctly different, rotational type, coefficients of diffusion, is brought into question whenever a molecular liquid is polarized. The induced ccf's mean that it is doubtful whether purely rotational or purely translational diffusion coefficients ever have an independent existence.

Similarly, a laser can induce first- and higher-order types of birefringence and can polarize the liquid in the same way as an electric field. In this case again we must reckon with the induction of fundamentally new types of cross correlation which accompany the appearance of birefringence and polarization. These ccf's will again be present with flow induced birefringence. These conclusions are made on the basis of the (X, Y) and (Y, X) components of the tensor (Eq. 174) induced by a simple electric

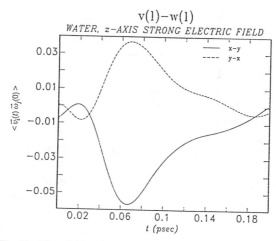

Figure 25. The (X, Y) and (Y, X) elements of the tensor time-correlation function C_3^{ij} in a Z-axis electric field.

field and illustrated in Fig. 25. The strength of cross correlation, and its time dependence, depend on the electric field strength and also on the model used of the effective pair potential. The molecular symmetry determines the cross correlation in the molecule fixed frame (x, y, z). As the electric field strength is increased, the ccf's become more oscillatory and show a decoupling effect. The ccf vanishes at $t = 0$ and $t \to \infty$. With accurate enough computer simulations it is probable that the ccf's would show long time tails which are difficult to observe because of their low amplitude.

BB.   Diffusion Equations for Rotation and Translation

The analytical description of the molecular liquid state needs diffusion equations which are capable of describing the simultaneous rotation and translation of the diffusing molecule in three dimensions. Progress may be attempted either by extending and modifying the old theory or by depending more and more on computer simulation while checking the latter with actual experimental data. Section V deals in greater detail with the attempts made in the last few years to extend the boundary of validity of molecular diffusion, and in this closing part of Section IV we describe and attempt to define the problems associated with this analytical extension.

1. *The Problem of Analytical Tractability*

We have encountered a recurring and increasingly severe general constraint on the traditional theoretical approach to molecular diffusion, that of complexity and mathematical intractability. Within the strictly defined and somewhat artificial limits of rotational diffusion theory the problem is kept under control by ad hoc assumptions about the nature of the intermolecular potential and of the rotational mechanics of a rotating molecule. In this context the rotational diffusion theory has not progressed much beyond the theory of the nonlinear itinerant oscillator, which effectively reduces the rotational problem to two dimensions and uses a very simple approximation to the intermolecular potential. Even then, the Langevin equations (Eqs. 36 and 37) are insoluble directly, and must be put into the form of a Kramers equation which can only be solved numerically,[41,42] frequently consuming as much computer resource as a full computer simulation.

As we have seen in Section I these equations are analytically fully equivalent to an early approximant of the Mori continued fraction, and this fixes their relation to the Liouville equation. The itinerant oscillator equations contain a number of parameters, friction coefficients, moments of inertia, and harmonic force constants, which can only be guessed at, making the description empirical. A historical perspective reveals clearly that the analytical approach has been slow to evolve, and has frequently failed to anticipate new effects discovered numerically or experimentally. The rotational diffusion equation resulted in the physically meaningless Debye plateau and the translational equation missed the existence of long time tails and negative regions in the velocity acf. The theory is now faced with the emergence of a set of new ccf's.

2. *The Problem of Over-Parameterization*

The simplest type of ccf requires for its description a linked Langevin structure such as[141,142]

$$m\dot{\mathbf{v}} + \beta_\nu \mathbf{v} + \beta_{\nu\omega}\boldsymbol{\omega} = \mathbf{F} \tag{175}$$

$$I\dot{\boldsymbol{\omega}} + \beta_\omega \boldsymbol{\omega} + \beta_{\omega\nu}\mathbf{v} + E_0 \sin \Theta t = \mathbf{Tq} \tag{176}$$

Here $\mathbf{v}$ is the molecular center-of-mass linear velocity, $\boldsymbol{\omega}$ the angular velocity of the same molecule, m the molecular mass, I an "effective" moment of inertia, β_ν the translational friction coefficient, β_ω the rotational friction coefficient, and $\beta_{\nu\omega}$ and $\beta_{\omega\nu}$ are cross friction coefficients designed to link the two types of motion purely empirically. $\mathbf{F}$ is a stochastic force and $\mathbf{Tq}$ a stochastic torque. The field term $E_0 \sin \Theta t$ is needed to represent the interaction between the external electric field and molecular dipole, to polarize the liquid for observation by dielectric spectroscopy and to make the sample birefringent for Kerr effect observation. It also induces the existence of new ccf's. To approximate this complicated pair of nonlinear stochastic differential equations would lead back to separated rotational and translational diffusion, thus defeating the purpose of the exercise. In this decoupled form the theory is able to force fit some of the least discriminating of data, such as low frequency dielectric loss, but gets no further below the surface.

A more difficult but necessary approach would be to attempt the direct solution of Eqs. (175) and (176) simultaneously.[141,142] The solution contains four friction coefficients which cannot be expressed in terms of the fundamental physical constants. Given four independent data sources, the four could be found in principle, but in practice this has never been attempted. Such a combination might involve, for example, far infrared spectra, dielectric spectra, Rayleigh scattering data, and various forms of NMR data. These sources would have to be wide ranging and discriminating. Assuming for the sake of argument the availability of such data we would determine four friction coefficients. What would we have achieved?

We would have arrived at four numbers with which to characterize the molecular diffusion process. However, we know independently that the equations upon which these numbers are based are approximate and imperfect descriptions of the molecular dynamical process. The friction coefficients are first approximations to time dependent memory functions, a hierarchy of which is needed to build up the Liouville equation. The equations contain no acceptable description of what is known from quantum mechanics to be the accurate form the pair potential must take. The Langevin equations (Eqs. 175 and 176) would not be able to provide an explanation of effects such as fall transient acceleration and field decoupling. Knowing all this, it would be useless to pretend that our four numbers would be anything more than the parameters of a curve-fiitting exercise.

An additional constraint is that Eqs. (175) and (176) become insoluble without the use of an effective moment of inertia I which approximates the kinematics of an asymmetric top with those of a spherical top. More rigorous consideration leads back to the Euler–Langevin equations (Eqs. 68 to 70) which are analytically insoluble even when no account of translation is made, and which introduce yet more friction coefficients.

The next section attempts to describe how molecular dynamics can be tackled with a combination of group theory and computer simulation. We have arrived at a point where the traditional methodology of natural philosophy, observation followed by description and hypothesis, followed by a repetition of the cyclical gathering of understanding, has been modified to involve consideration of many data sources interpreted with computer simulation. The course of this advance is made smoother with the symmetry principles of the next section.

V. SYMMETRY

The diffusion of a rigid asymmetric top occurs in three dimensions and involves simultaneous rotation and translation. We have seen in Sections I–IV that the available theories of molecular diffusion cannot describe this process adequately because of intractability and empiricism. An experimental understanding needs data from as many sources as possible interpreted with computer simulation. Without theory or data, however, considerable insight may be obtained from considerations of symmetry. These include the fundamental symmetry operations of classical physics, such as parity inversion and time reversal, and the rules of point group theory. Symmetry considerations alone can be used to determine the existence or otherwise of statistical correlation in an ensemble of diffusing asymmetric tops, and they can be applied in the frame (X, Y, Z) of the laboratory or (x, y, z) fixed in the molecule. The symmetry rules to be developed in this chapter and in Sections VII and VIII should be used before starting to simulate the properties of the ensemble. They filter out the ccf's that vanish for all t from those that may exist and which therefore may be detected numerically. They do not provide the detailed time dependence of any correlation function or frequency dependence of any spectrum, these have to be found from equations of motion.

An example of a symmetry rule at work is the use of parity reversal to show that the ccf

$$C_3^{ij}(t)_{XYZ} = \frac{\langle \mathbf{v}_i(t)\boldsymbol{\omega}_j(0)\rangle}{\langle v_i^2\rangle^{1/2}\langle \omega_j^2\rangle^{1/2}} \tag{177}$$

must vanish for all t because the parity symmetry of $\mathbf{v}$ is negative and that of $\boldsymbol{\omega}$ is positive. However the same is not true in frame (x, y, z) and here certain elements of the tensor

$$C_3^{ij}(t)_{xyz} = \frac{\langle \mathbf{v}_i(t)\boldsymbol{\omega}_j(0)\rangle}{\langle v_i^2\rangle^{1/2}\langle \omega_j^2\rangle^{1/2}} \tag{178}$$

exists for $t > 0$, a result first discovered[47] by computer simulation. Again the traditional theory of diffusion failed to predict or anticipate the result.

There are available several fundamental symmetry operations, which should be applied in order of applicability. The most generally valid operation should be applied first, and the less general thereafter. What follows is a brief summary of the most useful symmetry operations.

A. Parity Inversion

The parity inversion operation in frame (X, Y, Z) is defined as

$$\hat{P}:(\mathbf{p}, \mathbf{q}) \rightarrow (-\mathbf{p}, -\mathbf{q}) \tag{179}$$

where $\mathbf{p}$ is the molecular momentum and $\mathbf{q}$ its position. It inverts all positions and momenta, so that, for example $(X, Y, Z) \rightarrow (-X, -Y, -Z)$. The linear velocity is therefore negative to parity inversion but the angular velocity is an axial vector defined by the vector cross product between linear velocity and position, two quantities which are negative to parity reversal. In consequence the angular velocity is positive to parity reversal. Similarly the linear acceleration $\dot{\mathbf{v}}$ is negative to this operation and the angular acceleration positive. None of these considerations apply in the molecule fixed frame (x, y, z)

B. Time Reversal

The time-reversal symmetry rule is

$$\hat{T}:(\mathbf{p}, \mathbf{q}) \rightarrow (-\mathbf{p}, \mathbf{q}) \tag{180}$$

and reverses the sign of time dependent quantities in frame (X, Y, Z). Linear and angular velocity depend on time and are reversed by the operation (Eq. 180). Both are negative to time reversal. The linear and angular accelerations, however, are both positive to time reversal and this may be seen by considering them as velocity divided by time. The latter is given a negative label by time reversal, and accelerations in terms of operation (Eq. 180) are negative quantities divided by negative, being therefore overall positive to time reversal.

C. Application to Time-Correlation Functions

In an isotropic ensemble of molecules at reversible thermodynamic equilibrium, the ensemble average $\langle \mathbf{ABC} \ldots \rangle$ may exist in frame (X, Y, Z) if the product $\hat{P}(\mathbf{A})\hat{P}(\mathbf{B})\hat{P}(\mathbf{C}) \ldots$ is positive. If this product is negative the ensemble average vanishes. If the ensemble average is a time correlation function, the latter vanishes for all t if the product is negative.

This is the most generally applicable rule, and a simple proof is given in Ref. 241.

Note that if the molecular liquid is not isotropic or if the Hamiltonian itself becomes negative to parity inversion, the rule may no longer be applied as it stands, and we must refer to the three principles of group theoretical statistical mechanics developed in Chapter 7.

If the product is positive the ensemble average is subjected to the next test, which is time reversal. This operation must be applied with great care, because it can be misleading when applied to correlation functions. For example, thc acf's $\langle \mathbf{A}(t) \cdot \mathbf{A}(0) \rangle$ or $\langle \mathbf{B}(t) \cdot \mathbf{B}(0) \rangle$ are always positive to time reversal, and the operation does not change the value of the correlation function at $t = 0$ from positive to negative. However, for some ccf's such as $\langle \mathbf{A}(t) \cdot \dot{\mathbf{A}}(0) \rangle$ or $\langle \mathbf{B}(t) \cdot \dot{\mathbf{B}}(0) \rangle$ time reversal also reverses the sign of the correlation function at $t = 0$ from positive to negative. In this case the time reversal rule does not apply, and nothing more can be said about the time existence or otherwise of the cross correlation function. Therefore, the time reversal rule is applicable in general to ccf's which are known by inspection to have a zero slope at the time origin $t = 0$. It is therefore less general in nature than parity reversal symmetry.

D. The Application of Group Theory to Time Correlation Functions

If an ensemble average has passed the tests of parity and time reversal, point group theory may be applied in the frames (X, Y, Z) and (x, y, z). However, in frame (X, Y, Z) the symmetry rules must be applied in the correct order: (1) parity inversion, (2) time reversal, (3) point-group theory. For example, point group theory allows the existence of the cross correlation function $\mathbf{C}_3^{ij}$ frame (X, Y, Z) for some ensembles of chiral molecules, but parity reversal considerations do not. Therefore the ccf must vanish in frame (X, Y, Z) for all t for both enantiomers and the racemic mixture. In frame (x, y, z) point group theory may be used to pinpoint the existence of ensemble averages with reference to *the molecular point group character tables*. Dichloromethane, for example, has C_{2v} molecular point group symmetry, allowing two off-diagonal elements of $\mathbf{C}_3^{ij}$ to exist in frame (x, y, z). These are independent elements with different time evolutions, obtainable[1-4] from computer simulation. The signature of cross correlation between linear and angular velocities is obtained in frame (x, y, z) but vanishes in (X, Y, Z).

E. Point-Group Theory in Frame (X, Y, Z)

Point group theory is a well-developed subject which has recently been applied to ensemble molecular dynamics by Whiffen[242] and the present author, (Sections VII and VIII). The basic assumption in the application of group theory to ensemble molecular dynamics in frame (X, Y, Z) is embodied in the first principle of group theoretical statistical mechanics developed in Section VII. If the point group symmetry representation of the dynamical variable $\mathbf{A}$ is denoted $\Gamma(\mathbf{A})$, then the symmetry representation of the ensemble average $\langle \mathbf{A} \rangle$ is assumed to be the same as that of $\mathbf{A}$. If the representation $\Gamma(\langle \mathbf{A} \rangle)$ contains the totally symmetric representation of the point group of frame (X, Y, Z), then the ensemble average may exist, subject to passing the tests of parity inversion and time reversal.

The theory can be applied to scalars, pseudoscalars, polar and axial vectors, and tensors representing molecular dynamical quantities of interest. Polar vectors such as linear velocity are negative to parity inversion, whereas axial vectors such as angular velocity are positive. Scalars are always positive to both time $\hat{P}$ and $\hat{T}$, and pseudoscalars such as the optical activity coefficient reverse sign between enantiomorphs. A molecular dipole moment and electric field are both polar vectors, while a magnetic dipole moment and a magnetic field are axial.

Molecular polarizability and moment of inertia tensors are second order cartesian quantities. Third order cartesian tensors have $3 \times 3 \times 3 = 27$ elements and so on. Here n is used for the order of the tensor and n_s denotes the number of suffix pairs in which the tensor is symmetrical. If this is nonzero, the number of independent elements, denoted n_i, is less than the number of elements.

If any quantity is positive to $\hat{P}$, it is labelled by the suffix g. If negative it is labelled u. These come from the spectroscopic terms "gerade" or "ungerade".

The friction tensor (whose elements are Langevin friction coefficients), and the molecular moment of inertia tensor are each 3×3 tensors with $n_s = 1$ and $n_i = 6$.

Time correlation functions in this context are valid tensor elements to which symmetry theory can be applied both in frames (X, Y, Z) and (x, y, z). The tensor is usually a 3×3 Cartesian whose elements are individual ccf's. Point group theory can say nothing about the actual time dependence of these elements, but is used to indicate whether or not they vanish by symmetry in frames (X, Y, Z) and (x, y, z). The time dependence of existing elements may be different for the same type of correlation in different frames, and will be nonzero only if they contain the fully symmetric representation of the point group.

The point group of frame (X, Y, Z) is the rotation–reflection group $R_h(3)$ and the symmetries of quantities in this frame are denoted through the irreducible D representations with subscript g or u and superscripts $0, 1, 2, \ldots$ In this notation the scalar quantity is $D_g^{(1)}$, the pseudoscalar is $D_u^{(0)}$, the polar vector is $D_u^{(1)}$, the axial vector is $D_g^{(1)}$; and higher order tensors are subscripted g or u and superscripted $2, 3, 4, \ldots$ This notation is used extensively in Sections VII and VIII.

The product of two or more D representations can be built up with the help of the Clebsch–Gordan Theorem, which can be expressed in the form

$$D^{(n)}D^{(m)} = D^{(n+m)} + \cdots + D^{|(n-m)|} \tag{181}$$

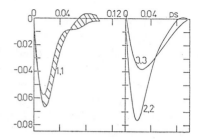

Figure 26. The third-rank tensor $\langle \mathbf{v}(t) \times \omega(t)\mathbf{v}^T(0)\rangle_{xyz}$ for liquid water.

The rule for the product of subscripts is $g \times g = g$; $u \times u = g$; $g \times u = u \times g = u$. If we are looking at the correlation function of a product of linear and angular velocities, the symmetry representation of the correlation function is in general the product of $D_u^{(1)}$ and $D_g^{(1)}$. Equation (181) shows this to be a sum of three parts. All three are however, negative to parity reversal, because each carries the subscript u. This type of ccf does not survive the test of parity inversion, which is more general than group theory, and it vanishes in frame (X, Y, Z). Similarly, the triple ccf $\langle \mathbf{v}(t) \times \omega(t) \cdot \mathbf{v}(0)\rangle$ vanishes in frame $(X, ,Y, Z)$. In frame (x, y, z), however, independent elements exist, and one of these is illustrated in Fig. (26) from a recent computer simulation.[243]

F. Frame (x, y, z), Molecular Point-Group Theory

The point group for the molecule fixed frame is that of the individual molecule.[244] In order to make use of point group theory in this frame the irreducible representations of each quantity have to be mapped from frame ,(X, Y, Z). The existence of an ensemble average in frame (x, y, z) is then determined by the principles of group theoretical statistical mechanics, developed in Section VII. Morc details of the mapping procedures are given in that Section.

G. Ensemble Averages of Scalars

Quantities such as mass and charge are simple scalars which are invariant to frame transformation. In the language of point group theory, this means that the totally symmetric representation in the point group $R_h(3)$ maps on to its equivalent in any molecular point group. The ensemble average of a scalar quantity always exists in both frames (X, Y, Z).

H. Ensemble Averages of Vectors

No vector contains the totally symmetric representation $D_g^{(0)}$ and the ensemble average of any vector quantity at field free thermodynamic equilibrium vanishes in frame (X, Y, Z). This is not necessarily so in frame

(x, y, z), however, and vector ensemble averages may be observable in this frame if also positive to parity inversion and time reversal.

I. Ensemble Average of Tensors

Some tensor quantities such as molecular polarizability contain the totally symmetric representation of $R_h(3)$ at least once. Ensemble averages over these quantities therefore exist in frame (X, Y, Z). Examples are the traces (isotropic diagonal parts) of the molecular polarizability and magnetizability, and moments of inertia. The D representation of a tensor quantity may be mapped from frame (X, Y, Z) to (x, y, z), and the number of independent ensemble averages over tensor elements in each frame may be different. This argument is developed in more detail in Section VII. In general, there may be more independent ensemble averages in frame (x, y, z) than in frame (X, Y, Z), depending on the symmetry of the molecule.

J. Ensemble Average Properties in Frame (x, y, z) — Some Examples

In the C_{2v} point group of dichloromethane, for example, there may be three independent averages of a vector quantity which is positive to parity inversion and time reversal. There are none in frame (X, Y, Z). For a symmetric top of C_{3v} symmetry such as chloroform, two of these averages in frame (x, y, z) are equal and different from the third. In a spherical top such as carbon tetrachloride all three averages in frame (x, y, z) are equal. For all molecular symmetries they vanish in frame (X, Y, Z).

The cross-correlation function between molecular linear and angular velocity is in general a 3×3 tensor, all of whose elements vanish in frame (X, Y, Z) at isotropic reversible equilibrium. In frame (x, y, z) this is not necessarily so. For the molecular point group symmetries C_{2h}, C_{2v}, C_{3v}, and T_d, for example, there are $0, 2, 1, 0$ nonzero elements of this ccf respectively. A C_{2v} asymmetric top such as dichloromethane or water has two independent elements of the ccf in its frame (x, y, z) which can be found using the point group character tables as described in (242). These were originally discovered by computer simulation[3,4] and are illustrated in Fig. 27. The elements found by simulation exactly match those confirmed [242] by group theory. Both simulation and symmetry continually pose new challenges to the original theoretical approach (Chapter 1).

A molecule of C_{3v} symmetry only has one independent element of the ccf tensor, and this is again what is found by simulation.[196] However, for molecules of C_{2h} and spherical top symmetry all elements of the ccf vanish in frame (x, y, z). The point group theory applies to all known molecular

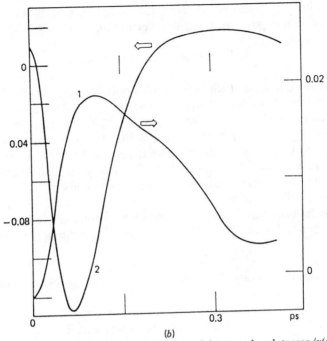

Figure 27. Nonvanishing off-diagonal elements of the second-rank tensor $\langle \mathbf{v}(t)\omega^T(0)\rangle_{xyz}$ for liquid dichloromethane.

symmetries, and is confirmed by available computer simulation results.[47] In chiral molecules all nine elements of the ccf matrix may exist, and all change sign from one enantiomorph to the other and vanish in the racemic mixture.

K. Higher Order Tensors

There are many rank three tensors which characterize simultaneous rotation and translation through the appropriate time ccf's. None of these is yet available in diffusion theory, but both group theory and computer simulation show their existence in frames (X, Y, Z) and (x, y, z). In this case the group theory allows many independent elements, whose identity can be found by detailed reference to the point group character table. In the laboratory frame, however, they must be positive to parity inversion in order to have a time dependence for $\tau > 0$, and if they pass this test, they play a role in the intercorrelation of rotation and translation. An illustration[243] is given in Fig. 28. In physical terms the existence of a time ccf such as this means[242] that the Coriolis acceleration $2\mathbf{v} \times \boldsymbol{\omega}$ plays a direct role in mixing rotational and translational velocities. This is the acceleration which stops a spinning and precessing top from falling over

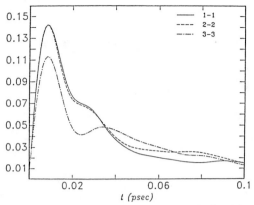

Figure 28. Illustration from computer simulation of the laboratory frame ccf; $\langle \mathbf{v}(t) \times \boldsymbol{\omega}(t) \dot{\mathbf{v}}(0) \times \boldsymbol{\omega}(0))^T \rangle_{XYZ}$.

because it is in opposition to gravity. Point group theory allows, for instance, a statistical correlation to exist in frame (X, Y, Z) between the Coriolis acceleration of a molecule at time t and its own linear acceleration a time t earlier for all molecular symmetries, including the spherical top. It is illustrated for carbon tetrachloride[57] in Fig. 28.

For a molecule of C_{2v} symmetry such as dichloromethane, point group theory allows the possible existence of a g-type ccf such as $\mathbf{C}_4 = \langle \mathbf{v}(t) \times \boldsymbol{\omega}(t) \cdot \mathbf{v}(0) \rangle$ in frame (x, y, z), but this ccf fails the test of time reversal symmetry and vanishes in frame (X, Y, Z) for all t.[203–206] In frame (x, y, z) on the other hand the parity and time reversal rules are not applicable, and point group theory allows the existence of no less than six independent elements. These come from the vector product in the Coriolis acceleration, so that the general symmetry of the ccf is

$$\mathbf{C}_4 = \begin{bmatrix} + & 0 & 0 \\ 0 & + & 0 \\ 0 & 0 & + \end{bmatrix}_{xyz} \tag{182}$$

and the three elements are illustrated in Fig. 29.

For all ranks of ungerade tensors in frame (X, Y, Z) all elements vanish in frame (X, Y, Z). However, in frame (x, y, z), elements may exist depending on the molecular point group symmetry. An example is the general triple product

$$\mathbf{C}_5(t) = \langle \boldsymbol{\omega}(t) \mathbf{v}(t) \boldsymbol{\omega}(0) \rangle \tag{183}$$

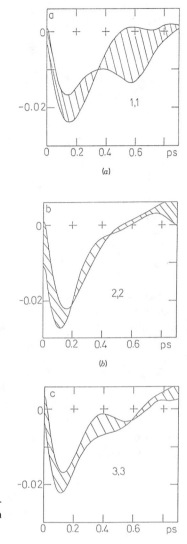

Figure 29. Laboratory frame ccf's for the spherical top carbon tetrachloride in the liquid state, from a recent computer simulation.[57]

in which the several separate elements can be investigated by computer simulation.

In general, in frame (x, y, z), as we increase the point group symmetry from asymmetric to spherical top the number of independent elements decreases, and in this respect computer simulation and point group theory are in detailed agreement. Enantiomorphs containing optically active molecules of low symmetry tend to have the largest number of nonvanishing elements.

The effect of external fields was introduced in Section IV and different types of fields have different effect according to the third principle of group theoretical statistical mechanics, described in Section VII. The electric field, for example, makes visible extra elements of cross correlation between linear and angular velocities in frame (X, Y, Z), elements which are negative to parity inversion in the absence of the field. The electric field itself is negative to parity inversion, and it imparts this symmetry characteristic to certain ensemble averages.

Normally gravitational fields are small in comparison with electric, magnetic and electromagnetic fields, or mechanical force fields such as those encountered in shear or elongational flow (Section VIII). Any of these fields may remove the symmetry of the sample at equilibrium. The interaction of an electric field with the ensemble may occur through different powers of the electric field strength $\mathbf{E}$ with the molecular dipole; E^2 with the molecular polarizability; E^3 with the hyperpolarizability, and so on. In the presence of fields, the general rules on parity and time reversal no longer apply, because the Hamiltonian itself has become affected by the external field. In this case the most general rules available are those provided by point group theory in both frames. These are summarized in the third principle of Section VII.

M. Fundamental Dynamics: The Noninertial Frame of Reference

The intricate relation between rotational and translational variables in molecular diffusion becomes clear if we consider a frame transformation[245] from the laboratory frame (X, Y, Z) to a rotating frame $(1, 2, 3)$ whose origin is the same as that of the laboratory frame. A theorem of elementary dynamics [246] links the differential operator $\hat{D}_f$ in frame (X, Y, Z) to its equivalent $(\hat{D}_m)$ in $(1, 2, 3)$. Considering the position vector of the molecular center of mass the theorem is

$$\hat{D}_f \mathbf{r} \equiv (\hat{D}_m + \boldsymbol{\omega} \times)\mathbf{r} \tag{184}$$

and conversely

$$\hat{D}_m \mathbf{r} \equiv (\hat{D}_f - \boldsymbol{\omega} \times)\mathbf{r} \tag{185}$$

The operator $(\hat{D}_m + \boldsymbol{\omega} \times))$ in frame $(1, 2, 3)$ is equivalent to the operator $\hat{D}_f$ in (X, Y, Z), and conversely in Eq. (185). Rewriting these equations in terms of velocities gives

$$(\mathbf{v})_{XYZ} \equiv (\mathbf{v} + \boldsymbol{\omega} \times \mathbf{r})_{123} \qquad (186)$$

and conversely

$$(\mathbf{v})_{123} \equiv (\mathbf{v} - \boldsymbol{\omega} \times \mathbf{r})_{XYZ} \qquad (187)$$

These identities show that the linear center of mass velocity, $\mathbf{v}$ in frame (X, Y, Z) is rigorously equivalent to the sum $\mathbf{v} + \boldsymbol{\omega} \times \mathbf{r}$ in the rotating frame $(1, 2, 3)$. Carrying out this procedure for each molecule of the ensemble, we note that the axes of the frame $(1, 2, 3)$ are each parallel to those of (x, y, z), but the origin of one frame is displaced with respect to the other. For the purposes of constructing ensemble averages the two frames are interchangeable. We can therefore continue the development with subscripts (x, y, z) substituted for $(1, 2, 3)$.

The noninertial accelerations are generated in both frames (X, Y, Z) and (x, y, z) by the equivalences

$$(\dot{\mathbf{v}})_{XYZ} \equiv (\dot{\mathbf{v}} + 2\boldsymbol{\omega} \times \mathbf{v} + \dot{\boldsymbol{\omega}} \times \mathbf{r} + \boldsymbol{\omega} \times (\boldsymbol{\omega} \times \mathbf{r}))_{xyz} \qquad (188)$$

and conversely

$$(\dot{\mathbf{v}})_{xyz} \equiv (\dot{\mathbf{v}} - 2\boldsymbol{\omega} \times \mathbf{v} - \dot{\boldsymbol{\omega}} \times \mathbf{r} + \boldsymbol{\omega} \times (\boldsymbol{\omega} \times \mathbf{r}))_{XYZ} \qquad (189)$$

which may be written more transparently as[246]

$$\hat{D}_f^2 \mathbf{r} \equiv \hat{D}_f(\hat{D}_f \mathbf{r}) \equiv (\hat{D}_m + \boldsymbol{\omega} \times)(\hat{D}_m + \boldsymbol{\omega} \times \mathbf{r}) \qquad (190)$$

and conversely

$$\hat{D}_M^2 \mathbf{r} \equiv \hat{D}_m(\hat{D}_m \mathbf{r}) \equiv (\hat{D}_f - \boldsymbol{\omega} \times)(\hat{D}_f - \boldsymbol{\omega} \times \mathbf{r}) \qquad (191)$$

These equivalences define the noninertial linear accclerations, which are real in both frames of reference. They are known as the molecular Coriolis acceleration $-2\boldsymbol{\omega} \times \mathbf{v}$, the molecular Eulerian acceleration $-\dot{\boldsymbol{\omega}} \times \mathbf{r}$, and molecular centripetal acceleration $\boldsymbol{\omega} \times (\boldsymbol{\omega} \times \mathbf{r})$.

It follows that acf and ccf may be constructcd involving these accelerations, and also the non inertial linear velocity

$$\mathbf{v}_{\text{noninertial}} \equiv -\boldsymbol{\omega} \times \mathbf{r}$$

which involves the position vector in both frames. The theory of Section I contains no specific mention of correlations such as these, which clearly contain both rotational and translational molecular dynamical quantities. Some of these correlations were described earlier in this section using simulation and symmetry.

The process leading to Eqs. (188) and (189) may be extended to the complete set of time derivatives of noninertial linear accelerations by operating repeatedly in each frame

$$\hat{D}_f^3 \mathbf{r} \equiv (\hat{D}_m + \boldsymbol{\omega} \times)(\hat{D}_m + \boldsymbol{\omega} \times)(\hat{D}_m + \boldsymbol{\omega} \times \mathbf{r}) \qquad (192)$$

and conversely

$$\hat{D}_m^3 \mathbf{r} \equiv (\hat{D}_f - \boldsymbol{\omega} \times)(\hat{D}_f - \boldsymbol{\omega} \times)(\hat{D}_f - \boldsymbol{\omega} \times \mathbf{r}) \qquad (193)$$

The number of terms increases rapidly and there is an infinite number of ccf's describing simultaneous rotation and translation in molecular ensembles. The majority of these vanish by symmetry, but this still leaves an infinite number which does not. All of these, in both frames of reference, contribute to spectra in some way, but clearly, no spectral observation can determine them unequivocally. Computer simulation is able to generate the first few members of the infinite sets in frames (X, Y, Z) and (x, y, z).

N. Consequences for the Theory of Diffusion

The major consequence for the theory of rotational diffusion is to make the appropriate Langevin equations intractable. This is easily demonstrated by considering the diffusion of the asymmetric top in three dimensions. To consider the simultaneous rotational and translational diffusion means supplementing the Euler–Langevin structure Eqs. (68)–(70) with the translational Langevin equation written in frame (x, y, z). It becomes the highly nonlinear stochastic differential equation

$$(\dot{\mathbf{v}} + 2\boldsymbol{\omega} \times \mathbf{v} + \dot{\boldsymbol{\omega}} \times \mathbf{r} + \boldsymbol{\omega} \times (\boldsymbol{\omega} \times \mathbf{r}) + \beta(\mathbf{v} + \boldsymbol{\omega} \times \mathbf{r}))_{xyz} = (\mathbf{F}/m)_{xyz}$$
$$(194)$$

There is no known analytical solution of this equation, even if did not have to be solved simultaneously with the intractable Euler–Langevin equations (68)–(70).

This is a simple illustration of the limits of usefulness of the analytical diffusion theory.[1,247–255] In this situation, limited progress is possible only by relying on computer simulation and symmetry, and by attempting to put the rigorous analytical theory in a form suitable for approximate solution. An example of what is possible is given next.

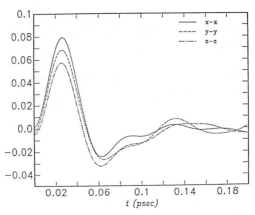

Figure 30. Diagonal elements of the laboratory frame ccf $\langle \dot{\mu}(t)\mathbf{v}^T(0)\rangle_{XYZ}$ from a recent computer simulation of liquid water.

O. A Simple Langevin Theory[256] of ccf's in Frame (X, Y, Z)

The c.c.f. between $\mathbf{v}$ and $\boldsymbol{\omega}$ vanishes for all t in frame (X, Y, Z). It is therefore incorrect to write two Langevin equations in this frame directly linked by cross friction coefficients because this would produce a non-zero value for the ccf. Cross friction coefficients such as those of eqns (175) and (176) must vanish with the applied electric field. (For finite $\mathbf{E}$ however, the ccf exists.)

We have seen from elementary dynamical considerations that the ccf between linear and rotational velocity, $\langle \mathbf{v}(t)\dot{\boldsymbol{\mu}}(0)\rangle$, exists in frame (X, Y, Z) directly, (Eq. (49) to (52)). This is supported by the symmetry rules of this section because the complete D symmetry of this ccf contains the totally symmetric representation of $R_h(3)$. This represents the trace of the ccf, the sum of the diagonal elements. These are illustrated in Fig. 30 from a recent computer simulation.[256] The off-diagonal elements vanish for all t.

Simulation and symmetry lead to a simple Langevin structure

$$\dot{\mathbf{v}} + \beta_T \mathbf{v} + \beta_{Tr}\dot{\boldsymbol{\mu}} = \mathbf{W}_1 \tag{195}$$

$$\ddot{\boldsymbol{\mu}} + \beta_{\dot{\mu}}\dot{\boldsymbol{\mu}} + \beta_{rT}\mathbf{v} = \mathbf{W}_2 \tag{196}$$

where the β's are friction coefficients. The physical meaning of these equations becomes clear by addition, giving

$$\frac{d}{dt}(\mathbf{v} + \dot{\boldsymbol{\mu}}) + \beta_T \mathbf{v} + \beta_\mu \dot{\boldsymbol{\mu}} + \beta_{Tr}\dot{\boldsymbol{\mu}} + \beta_{rT}\mathbf{v} = \mathbf{W}_1 + \mathbf{W}_2 \qquad (197)$$

which is a Langevin equation written for the linear diffusional velocity in three dimensions of an atom of the molecule, displaced from the latter's center of mass by the vector $\boldsymbol{\mu}$. The simultaneous equations (195) and (196) can be solved[1-4] for acf's and ccf's of interest, for example,

$$\langle \mathbf{v}(t) \cdot \dot{\boldsymbol{\mu}}(0)\rangle = \frac{\langle v^2\rangle}{(c - b^2)^{1/2}} \beta_{Tr}e^{-bt}\sin((c - b^2)^{1/2}t); \qquad (c > b^2) \quad (198)$$

or

$$\langle \mathbf{v}(t) \cdot \dot{\boldsymbol{\mu}}(0)\rangle = \frac{\langle v^2\rangle}{(b^2 - c)^{1/2}} \beta_{Tr}e^{-bt}\sinh(b^2 - c)^{1/2}t); \qquad (c < b^2) \qquad (199)$$

where $b = 2(\beta_T + \beta_\mu)$ and $c = \beta_T\beta_\mu - \beta_{Tr}\beta_{rT}$. Furthermore, the friction coefficients β_T and β_μ can be found from the velocity and rotational velocity acf's individually. Therefore, the new ccf's can be found approximately in terms of the coupling parameter β_{rT}. An optimum fit to the computer simulation data at 1 bar and 296 K was found[256] to be with the three friction coefficients $\beta_T = 1.0\,\text{THz}$, $\beta_\mu = 25.0\,\text{THz}$, and $B_{Tr} = 50.0\,\text{THz}$. This very simple theory can therefore provide analytical representations of the velocity, rotational velocity, and cross correlation functions in terms of the three friction coefficients. For example, the translational velocity acf becomes negative at 0.04 ps, as found in the molecular dynamics simulation of water, and the ccf rises to a maximum of about 0.3 at 0.02 ps, as illustrated in Fig. 31. However, the theoretical curves have the usual severe limitation of an essentially Markovian treatment[13] which can only be remedied with memory functions at the cost of introducing empiricism when truncating the continued fractions. The linked Langevin equations do produce a negative overshoot, however, in both acf's, as observed experimentally and numerically. This is not possible with decoupled Langevin equations, which produce simple exponential decays.

P. The Role of the Intermolecular Potential

The linked Langevin equations (195) and (196) can be modified[257] to include an effective potential $V(\mathbf{r} - \boldsymbol{\mu})$, which is the barrier height to translation caused by the simultaneous rotational velocity of the diffusing molecule. The linked Langevin equations then take the form

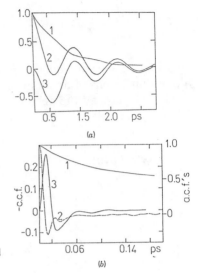

Figure 31. Comparison of theoretical (*a*) and computer-simulated (*b*) ccf's.

$$\ddot{\mathbf{r}} + \beta\mathbf{r} + V'(\mathbf{r} - \boldsymbol{\mu}) = \boldsymbol{\lambda}_1(t) \tag{200}$$

$$\ddot{\boldsymbol{\mu}} + \beta\dot{\boldsymbol{\mu}} - V'(\mathbf{r} - \boldsymbol{\mu}) = \boldsymbol{\lambda}_2(t) \tag{201}$$

$$\ddot{\mathbf{r}} + \ddot{\boldsymbol{\mu}} + \beta(\dot{\mathbf{r}} + \dot{\boldsymbol{\mu}}) = \boldsymbol{\lambda}_1(t) + \boldsymbol{\lambda}_2(t) \tag{202}$$

where **r** is the position vector in frame (X, Y, Z) of the molecular center of mass, and **v** is defined by $\mathbf{v} = \dot{\mathbf{r}}$. In order to solve these Langevin equations simultaneously it is necessary to assume [41–43] that the friction coefficient appearing in Eq. (200) is the same as that in Eq. (201). In this context V is a potential energy generated by the mutual constraints in frame (X, Y, Z) of the molecular rotational velocity on its own center-of-mass velocity. Equations (200) and (201) are then linked pendulum equations with the extra friction and stochastic terms of the Langevin approach. The stochastic terms $\boldsymbol{\lambda}_1$ and $\boldsymbol{\lambda}_2$ are Wiener processes.[1] The use of an equal friction model implies that the rotational velocity acf and the linear center of mass velocity acf must have the same analytical time dependence when both are normalized at $t = 0$. This is implies that Eqs. (200) and (201) are not applicable in the limit of free rotation, a general restriction on all Langevin equations. In general the dependence of V on its argument is intricately non linear, and obtainable from ab initio computation. However, it can be simply assumed that the argument can be expanded in a Taylor series. The harmonic approximation in this context is then

$$V'(\mathbf{r} - \boldsymbol{\mu}) \simeq 2V_0(\mathbf{r} - \boldsymbol{\mu}) \tag{203}$$

The use of a nonlinear potential would introduce into consideration barrier crossing processes superimposed on the overall diffusion process, making the equations insoluble without transformation to the Kramers form.[41-43] If the nonlinear potential is mimicked with a simple cosine, its derivative would be a sine, which can be approximated by the angular displacement itself, the harmonic approximation. This restricts the physical meaning of the theory to torsional oscillation at the bottom of a steep and deep potential well.

In the harmonic approximation the equations do have an analytical solution which is a complicated expression in terms of the friction coefficient and the parameter

$$\omega_1 = (4V_0 - \beta^2/4)^{1/2} \tag{204}$$

The solution gives a variety of results for acf's and ccf's known to exist from simulation and symmetry.[257] Some of these were compared with simulation data in Ref. 257, using simulation data for liquid water over the complete range of its thermodynamic existence.

The overall behavior of the system of Eqs. (200) and (201) is similar to that of the correlation functions from computer simulation, but we must temper this conclusion with what we know to be the shortcomings of the Langevin equations themselves. As a rough guide to understanding, however, these equations are reasonably satisfactory facsimiles of the numerical data, but become intractable if we attempt a frame transformation from (X, Y, Z) to (z, y, z). In the laboratory frame (X, Y, Z) in which solutions are available[257] the analytical rotational and linear velocity acf's become more oscillatory with increasing V, and the ccf that links them in frame (X, Y, Z) becomes stronger (Fig. 32). The barrier height V seems to be directly responsibly in this approximation for the strength of the cross correlation. The latter is therefore directly dependent on intermolecular forces and torques as well as on overall considerations of symmetry. The harmonic approximation (203) is a very simple description of the information available from symmetry and simulation. However, the theory is severely limited in several ways, because otherwise the solutions, if obtainable at all, would be overparameterized and difficult to use.

1. The friction coefficient is assumed to be a scalar, whereas it is more accurately a tensor.

2. It is assumed to be the same in both equations. More generally the friction coefficents in Eqs. (200) and (201) can take different values.

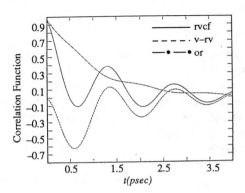

Figure 32.   (*a*).

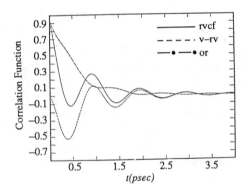

Figure 32.   (*b*).

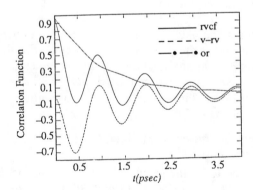

Figure 32.   (*c*).

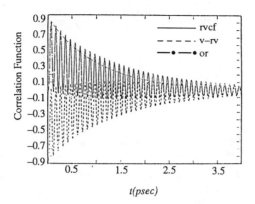

Figure 32. Analytical correlation functions as the barrier height V_0 is varied, (a)–(d).

3. The intrinsic nonlinearities of the Euler equations operate in frame (x, y, z) and in the (X, Y, Z) representation of Eqs. (200) and (201) are assumed to be contained in the potential term. This assumes that Eq. (203) describes the diffusion of a point in the molecule at the end of the axis $\mathbf{r} = 1$ in the laboratory frame.

4. There are general limitations on the use of Langevin equations, described already.

Given these limitations, Eqs. (200) and (201) represent an approach to diffusional dynamics which attempts to retain mathematical tractability and to reduce complexity.

Q. Patterns of Cross Correlations in Frame (X, Y, Z) and (x, y, z)

The set of nonvanishing ccf's obtainable from these fundamental dynamical considerations can be divided for classification into subsets, or patterns, which serve as a guide to the nature of the ccf's themselves. These patterns can be obtained from the operator equations

$$(\mathbf{v})_{XYZ}^{(n+2)} \equiv (\hat{D}_m + \boldsymbol{\omega} \times)_{(1)}(\hat{D}_m + \boldsymbol{\omega} \times)_{(2)} \ldots (\hat{D}_m + \boldsymbol{\omega})_{(n+1)}$$
$$\times (\hat{D}_m + \boldsymbol{\omega} \times \mathbf{r}) \tag{205}$$

and

$$(\mathbf{v})_{xyz}^{(n+2)} \equiv (\hat{D}_f - \boldsymbol{\omega} \times)_{(1)}(\hat{D}_f - \boldsymbol{\omega} \times)_{(2)} \ldots (\hat{D}_f - \boldsymbol{\omega} \times)_{(n+1)}$$
$$\times (\hat{D}_f - \boldsymbol{\omega} \times \mathbf{r}) \tag{206}$$

R. The Omega Pattern in Frame (x, y, z)

By inspection of the terms in the laboratory frame and by application of the standard rotation operation into frame (x, y, z) for any dynamical vector ccf's of the following types exist for $t < 0$ in frame (x, y, z)

$$\langle \boldsymbol{\omega}(t) \times \mathbf{A}(t)\mathbf{A}^T(0)\rangle_{xyz}$$

This is confirmed by point-group theory for all molecular symmetries and for polar vectors $\mathbf{A}$ such as velocity or inertial and noninertial linear accelerations such as the Coriolis, or centripetal terms in frame (x, y, z). This conclusion is confirmed by the available computer simulations of these ccf's.

A subset of the omega pattern is $\langle \boldsymbol{\omega}(t) \times \hat{D}_f\mathbf{A}(t)\boldsymbol{\omega}^T(0)\rangle$, all members of which exist in frame (x, y, z) from inspection of the general operator equation (206). There also exist patterns of ccf's in frame (x,y,z) made up of cross terms of the operator products in Eq. (206). None of these seems to have been explored yet by computer simulation, but are allowed by point group theory in frame (x, y, z) for all molecular symmetries, including that of the spherical top.

S. The $\hat{D}_f$ Pattern in Frame (X, Y, Z)

These results are generated from the subset

$$\langle \dot{\mathbf{A}}(t) \cdot \mathbf{A}(0)\rangle > 0, \qquad t > 0 \tag{207}$$

for all acfs. In Eq. (20) the ccf has been constructed between a dynamical quantity and its time derivative. By applying this result to linear but noninertial velocities and accelerations or to noninertial angular accelerations and higher time derivatives it is possible to see the emergence of many individual ccf's which may exist in frame (X, Y, Z) for all molecular symmetries, including the spherical top. None of these can be described by the theory of Section I, but can be followed approximately by linked Langevin equations of the type used in this chapter. These patterns may be summarized as follows, and emerge from the operator equation (Eq. 205).

The $\hat{D}_f$ pattern is simply

$$\mathbf{C}(t) = \langle \hat{D}_f\mathbf{B}(t)\mathbf{B}^T(0)\rangle_{XYZ} \tag{208}$$

where in general the vector quantity $\mathbf{B}$ is a linear noninertial variable such as velocity or acceleration. One of the simplest examples of this type is

$$\langle \hat{D}_f(\mathbf{v}(t) - \boldsymbol{\omega}(t) \times \mathbf{r}(t))(\mathbf{v}(0) - \boldsymbol{\omega}(0) \times \mathbf{r}(0))^T \rangle$$

which shows the possible existence in frame (X, Y, Z) of six component ccf's for $t > 0$. Each of these must be individually tested with parity and time-reversal symmetry before its exact time dependence is simulated from a model of the pair potential. For ccf's of this nature, time-reversal symmetry must be applied with care because it reverses the sign of each ccf component as it passes through $t = 0$.

Parity inversion, however, allows the existence of each member of the set of six in frame (X, Y, Z), irrespective of molecular symmetry. The first of the group, $\langle \dot{\mathbf{v}}(t) \cdot \mathbf{v}(0) \rangle$, is the inertial, or Newtonian, ccf. The other five all contain rotational variables, such as the molecular angular velocity, and are noninertial in nature. They vanish only when $\boldsymbol{\omega}$ vanishes and therefore distinguish molecular from atomic ensembles.

Repeated application of the above for more complicated vectors $\mathbf{B}$, with more components, produces many more ccf's in frame (X, Y, Z), even for spherical top symmetry. An example is the system

$$\langle \hat{D}_f(\boldsymbol{\omega}(t) \times (\mathbf{v}(t) - \boldsymbol{\omega}(t) \times \mathbf{r}(t)))(\boldsymbol{\omega}(0) \times (\mathbf{v}(0) - \boldsymbol{\omega}(0) \times \mathbf{r}(0)))^T \rangle$$

T. Other Patterns in Frame (X, Y, Z)

The $\hat{D}_f$ pattern just described can be extended to selected cross terms from Eq. (206) provided that the general symmetry rules apply, and the overall symmetry of the ccf's remains that of Eq. (207). These terms can be obtained easily from Eq. (206) by inspection.

U. Symmetry of Some Ccf's

Although the set of diffusion equations represented by the Euler–Langevin components Eqs.(68) to (70) and Eq. (194) is insoluble, its structure may be used by inspection to predict the possible existence in frame (x, y, z) of many ccf's that are among the clearest signatures of the interrelation of the fundamental dynamical variables of molecular diffusion. These ccf's in frame (x, y, z) can be constructed[52,53] by correlating a particular term on the left-hand side of, say, Eq. (194) with counterparts on the left-hand sides of Eqs. (68)–(70), which together comprise the simplest diffusional representation of combined rotation and translation in three dimensions of the asymmetric top molecule.

The rules of parity inversion and time reversal then test the existence of the resulting ccf's in frame (X, Y, Z). These are followed both in frame (X, Y, Z) and (x, y, z) by the rules of point-group theory (Sections VII and VIII). A uniaxial electric field applied to the molecular ensemble may produce new ccf's which may be investigated in turn with symmetry and computer simulation according to the third principle of group theoretical statistical mechanics of Section VII.

The nature of these ccf's, both in the presence and absence of an electric field, is described in more detail in the source papers.[52,53] One of the effects of the electric field can be summarized as follows:

$$\mathbf{C}_{v\omega} = \langle \mathbf{v}(t)\boldsymbol{\omega}(0)\rangle_{XYZ} \equiv \begin{bmatrix} 0 & 0 & 0 \\ 0 & 0 & 0 \\ 0 & 0 & 0 \end{bmatrix} \xrightarrow[E_Z]{} \begin{bmatrix} 0 & + & 0 \\ + & 0 & 0 \\ 0 & 0 & 0 \end{bmatrix} \tag{209}$$

whose off-diagonal elements exist in the presence of an electric field. In frame (x, y, z) the same effect is as follows

$$\langle \mathbf{v}(t)\boldsymbol{\omega}(0)\rangle_{xyz} \equiv \begin{bmatrix} 0 & + & 0 \\ + & 0 & 0 \\ 0 & 0 & 0 \end{bmatrix} \xrightarrow[E_Z]{} \begin{bmatrix} 0 & + & 0 \\ + & 0 & 0 \\ 0 & 0 & 0 \end{bmatrix} \tag{210}$$

The symmetry effect of different types of field, for example a circularly polarized field, is described in the next chapter. Different types of ccf have been simulated recently[52,53] for liquid dichloromethane and water. A complete diffusion theory of the molecular liquid state of matter would provide the time dependence of noninertial ccf's such as this consistently with those of simple acf's. In this context, computer simulation now plays a leading role in linking what we know experimentally to what we can predict on the grounds of symmetry.

In order to illustrate the limits of the analytical theory in the description of field effects on the dynamics of diffusing molecules this section ends with a short description of Mori theory applied to this problem.

V. Analytical Theory[141,142] of Field-Induced ccf's

The following analytical theory is a simple first approximation to the results available from computer simulation but is capable of anticipating an experimental method for the indirect determination of laboratory frame ccf's in the presence of an electric field. The method uses the Kerr effect and its contemporary technology for the measurement of ccf's in liquid, liquid crystals, and other states of matter such as compressed dipolar gases. Another suggested method is the measurement of electric-field induced phenomena in liquid crystals parallel and perpendicular to an applied Z-axis electric field.

The theory is based on the Mori equation

$$\dot{\mathbf{C}}(t) = \boldsymbol{\lambda}(t)\mathbf{C}(t) - \int_0^t \boldsymbol{\phi}(t - \tau)\mathbf{C}(\tau)\,d\tau \tag{211}$$

for the correlation matrix $\mathbf{C}$ in terms of the Mori resonance operator $\boldsymbol{\lambda}$ and memory function matrix $\boldsymbol{\phi}$. It is convenient to write

$$\mathbf{C}(t) = \begin{bmatrix} \langle \mathbf{v}(t)\mathbf{v}^T(0)\rangle\langle \mathbf{v}(t)\boldsymbol{\omega}^T(0)\rangle \\ \langle \boldsymbol{\omega}(t)\mathbf{v}^T(0)\rangle\langle \boldsymbol{\omega}(t)\boldsymbol{\omega}^T(0)\rangle \end{bmatrix} \tag{212}$$

so that the elements of $\mathbf{C}$ are themselves matrices, whose elements in turn are correlation functions. The time evolution of the complete "supermatrix" $\mathbf{C}$ is governed by the Mori equation (Eq. 211), which is the Liouville equation put into a form more suitable for solution.

In order to reduce the problem to the simplest possible form consistent with describing the appearance of the elements of the ccf between linear and angular velocity directly in frame (X, Y, Z), the following assumptions are made:

1. The three molecular moments of inertia of the asymmetric top are approximated by the trace, a scalar average I.

2. It is assumed that the molecule carries a net dipole moment $\boldsymbol{\mu}$ which generates the torque $-\boldsymbol{\mu} \times \mathbf{E}$ described by the Mori resonance operator

$$\boldsymbol{\lambda} = i\omega_1 \begin{bmatrix} 0 & 0 \\ 0 & 1 \end{bmatrix} \tag{213}$$

The scalar frequency in this equation is

$$\omega_1 = \left(\frac{\mu E_Z}{I}\right)^{1/2} \tag{214}$$

With these assumptions we have the definitions

$$C_{vv} = \langle \mathbf{v}(t)\mathbf{v}^T(0)\rangle \tag{215}$$

$$C_{\omega\omega} = \langle \boldsymbol{\omega}(t)\boldsymbol{\omega}^T(0)\rangle \tag{216}$$

$$C_{v\omega} = \langle \mathbf{v}(t)\boldsymbol{\omega}^T(0)\rangle$$

$$\equiv \begin{bmatrix} 0 & \langle v_X(t)\omega_Y(0)\rangle & \langle v_X(t)\omega_Z(0)\rangle \\ \langle v_Y(t)\omega_X(0)\rangle & 0 & \langle v_Y(t)\omega_Z(0)\rangle \\ \langle v_Z(t)\omega_X(0)\rangle & \langle v_Z(t)\omega_Y(0)\rangle & 0 \end{bmatrix} \quad (217)$$

The matrix of memory functions is a supermatrix of the form

$$\boldsymbol{\phi}(t) = \begin{bmatrix} \boldsymbol{\phi}_{vv}(t) & \boldsymbol{\phi}_{v\omega}(t) \\ \boldsymbol{\phi}_{\omega v}(t) & \boldsymbol{\phi}_{\omega\omega}(t) \end{bmatrix} \quad (218)$$

Using the computer simulation result that the electric field produces the elements (X, Y) and (Y, X) of $\mathbf{C}_{v\omega}$ for an electric field, E_Z gives the simplification

$$\mathbf{C}_{v\omega}(t) = \mathbf{C}_{\omega v}(t) = C_{v\omega}^{xy}(t)\begin{bmatrix} 0 & -1 & 0 \\ 1 & 0 & 0 \\ 0 & 0 & 0 \end{bmatrix} \quad (219)$$

The electric field has the dual role of promoting the existence of $\mathbf{C}_{v\omega}$ in frame (X, Y, Z) and of making the sample anisotropic and birefringent, that is,

$$\langle v_X^2\rangle = \langle v_Y^2\rangle \neq \langle v_Z^2\rangle \quad (220)$$

$$\langle \omega_X^2\rangle = \langle \omega_Y^2\rangle \neq \langle \omega_Z^2\rangle \quad (221)$$

Laplace transformation of the Mori equation gives the linear equation

$$(p\mathbf{1} + \boldsymbol{\phi}(p) - i\omega_1\mathbf{1})\mathbf{C}(p) = \mathbf{C}(0) \quad (222)$$

whose supermatrices are

$$\mathbf{C}(0) = \begin{bmatrix} \mathbf{C}_{vv}(0) & 0 \\ 0 & \mathbf{C}_{\omega\omega}(0) \end{bmatrix} \quad (223)$$

where

$$\mathbf{C}_{\nu\nu}(0) = \begin{bmatrix} \langle \nu_X^2 \rangle & 0 & 0 \\ 0 & \langle \nu_Y^2 \rangle & 0 \\ 0 & 0 & \langle \nu_Z^2 \rangle \end{bmatrix} \tag{224}$$

and

$$\mathbf{C}(p) = \begin{bmatrix} \mathbf{C}_{\nu\nu}(p) & \mathbf{C}_{\nu\omega}(p) \\ \mathbf{C}_{\omega\nu}(p) & \mathbf{C}_{\omega\omega}(p) \end{bmatrix} \tag{225}$$

where

$$(p - i\omega_1)\mathbf{1} + \boldsymbol{\phi}(p) = \begin{bmatrix} (p - i\omega_1)\mathbf{1} + \boldsymbol{\phi}_{\nu\nu}(p) & \boldsymbol{\phi}_{\nu\omega}(p) \\ \boldsymbol{\phi}_{\omega\nu}(p) & (p - i\omega_1)\mathbf{1} + \boldsymbol{\phi}_{\omega\omega}(p) \end{bmatrix} \tag{226}$$

Using the symmetry of the computer simulation result, the elements of the memory matrix are themselves matrices defined by

$$\boldsymbol{\phi}_{\nu\omega} = \boldsymbol{\phi}_{\omega\nu} = \phi_{\nu\omega}^{XY} \begin{bmatrix} 0 & -1 & 0 \\ 1 & 0 & 0 \\ 0 & 0 & 0 \end{bmatrix} \tag{227}$$

$$\boldsymbol{\phi}_{\nu\nu} = \begin{bmatrix} \phi_{\nu\nu}^{XX} & 0 & 0 \\ 0 & \phi_{\nu\nu}^{YY} & 0 \\ 0 & o & \phi_{\nu\nu}^{ZZ} \end{bmatrix} \tag{228}$$

and so on.

For an electric field in the Z axis,

$$\phi_{\nu\nu}^{XX}(p) = \phi_{\nu\nu}^{YY}(p) \neq \phi_{\nu\nu}^{ZZ}(p) \tag{229}$$

$$\phi_{\omega\omega}^{XX}(p) = \phi_{\omega\omega}^{YY}(p) \neq \phi_{\omega\omega}^{ZZ}(p) \tag{230}$$

and similarly for $\mathbf{C}_{\nu\nu}$ and $\mathbf{C}_{\omega\omega}$ and the linear and angular velocity matrices.

Comparing scalar elements in Eq. (222) provides the following set of scalar equations in Laplace space, in terms of the Laplace variable p:

$$C_{vv}^{XX}(p) = \frac{\langle v_X^2 \rangle + \phi_{v\omega}^{XY}(p)C_{v\omega}^{XY}(p)}{p + \phi_{vv}^{XX}(p)} \tag{231}$$

$$C_{vv}^{YY}(p) = \frac{\langle v_Y^2 \rangle + \phi_{v\omega}^{XY}(p)C_{v\omega}^{XY}(p)}{p + \phi_{vv}^{YY}(p)} \tag{232}$$

$$C_{vv}^{Z}(p) = \frac{\langle v_Z^2 \rangle}{p + \phi_{vv}^{ZZ}(p)} \tag{233}$$

$$C_{\omega\omega}^{XX} = \frac{\langle \omega_X^2 \rangle + \phi_{v\omega}^{XY}(p)C_{v\omega}^{XY}(p)}{p - i\omega_1 + \phi_{\omega\omega}^{XX}(p)} \tag{234}$$

$$C_{\omega\omega}^{YY}(p) = \frac{\langle \omega_Y^2 \rangle + \phi_{v\omega}^{XY}(p)C_{v\omega}^{YY}(p)}{p - i\omega_1 + \phi_{\omega\omega}^{YY}(p)} \tag{235}$$

$$C_{\omega\omega}^{ZZ}(p) = \frac{\langle \omega_Z^2 \rangle}{p - i\omega_1 + \phi_{\omega\omega}^{ZZ}(p)} \tag{236}$$

The ratios of Eqs. (231)–(233) and of (234)–(236) provide the following approximate but transparent results:

$$\phi_{v\omega}^{XY}(p)C_{v\omega}^{XY}(p) \simeq \langle v_Z^2 \rangle \frac{C_{vv}^{XX}(p)}{C_{vv}^{ZZ}(p)} - \langle v_X^2 \rangle \tag{237}$$

$$\simeq \langle \omega_Z^2 \rangle \frac{C_{\omega\omega}^{XX}(p)}{C_{\omega\omega}^{ZZ}(p)} - \langle \omega_X^2 \rangle \tag{238}$$

and if, in the Markov approximation, we regard the memory function as a constant, that is,

$$\phi_{v\omega}^{XY}(p) \simeq \phi_{v\omega}^{XY} \tag{239}$$

we obtain the following results, which have a simple physical interpretation:

$$C_{v\omega}^{XY}(p) \simeq \frac{\langle v_Z^2 \rangle}{\phi_{v\omega}^{XY}} \frac{C_{vv}^{XX}(p)}{C_{vv}^{ZZ}(p)} - \frac{\langle v_X^2 \rangle}{\phi_{v\omega}^{XY}} \tag{240}$$

$$\simeq \frac{\langle \omega_X^2 \rangle}{\phi_{v\omega}^{XY}} \frac{C_{\omega\omega}^{XX}(p)}{C_{\omega\omega}^{ZZ}(p)} - \frac{\langle \omega_X^2 \rangle}{\phi_{v\omega}^{XY}} \tag{241}$$

Equations (240) and (241) show that it is possible to obtain an indication of the time dependence of the (X, Y) and (Y, X) elements of the ccf by applying an electric field to a molecular liquid sample and observing the anisotropy in the angular velocity acf (and also in the linear velocity acf) with Kerr effect apparatus. Eliminating C_{vv}^{XX} between Eqs. (231)–(236) and the further four relations

$$C_{v\omega}^{XY}(p)(p + \phi_{vv}^{XX}(p)) + \phi_{v\omega}^{XY}(p)C_{\omega\omega}^{XX}(p) = 0 \qquad (242)$$

$$C_{v\omega}^{XY}(p)(p + \phi_{vv}^{XY}(p)) + \phi_{v\omega}^{XY}(p)C_{\omega\omega}^{YY}(0) = 0 \qquad (243)$$

$$C_{v\omega}^{XY}(p)(1 - i\omega_1 + \phi_{\omega\omega}^{XX}(p)) + \phi_{v\omega}^{XY}(p)C_{vv}^{XX}(p) = 0 \qquad (244)$$

$$C_{v\omega}^{XY}(p)(p - i\omega_1 + \phi_{\omega\omega}^{YY}(p)) + \phi_{v\omega}^{XY}C_{\omega\omega}^{YY}(p) = 0 \qquad (245)$$

and eliminating C_{vv}^{XX} between Eqs. (231) and (243) and between Eqs. (234) and (242) provides the results

$$\langle v_X^2 \rangle = \langle \omega_X^2 \rangle \qquad (246)$$

$$\langle v_Y^2 \rangle = \langle \omega_Y^2 \rangle \qquad (247)$$

$$\langle v_Z^2 \rangle \neq \langle \omega_z^2 \rangle \qquad (248)$$

Solving simultaneously Eqs. (231), (234), (242), and (243) finally gives the results

$$C_{v\omega}^{XY} = -\frac{\langle \omega_X^2 \rangle \phi_{v\omega}^{XY} e^{-at/2}}{(b - a^2/4)^{1/2}} \sin((b - a^2/4)^{1/2}t) \qquad (249)$$

$$C_{vv}^{XX}(t) = \langle v_X^2 \rangle e^{-at/2}$$
$$\times \left\{ \cos(b - a^2/4)^{1/2}t) + \frac{(\phi_{vv}^{XX} - a/2)}{(b - a^2/4)^{1/2}} \sin((b - a^2/4)^{1/2}t) \right\} \quad (250)$$

$$C_{\omega\omega}^{XX}(t) = \langle \omega_X^2 \rangle e^{-at/2} \cos((b - a^2/4)^{1/2}t)$$
$$+ \langle \omega_X^2 \rangle \frac{(\phi_{\omega\omega}^{XX} - i\omega_1 - a/2)}{(b - a^2/4)^{1/2}} \sin((b - a^2/4)^{1/2}t) \qquad (251)$$

for the angular, linear, and cross correlations as a function of the electric field strength E_z in terms of the parameters

$$a = \phi_{vv}^{XX} + \phi_{\omega\omega}^{XX} - i\omega_1 \qquad (252)$$

$$b = \phi_{v\omega}^{XY2} + \phi_{vv}^{XX}\phi_{\omega\omega}^{XX} - i\omega_1\phi_{vv}^{XX} \qquad (253)$$

With the test parameters

$$\phi_{vv}^{XX} = \phi_{\omega\omega}^{XX} = 10^{12}\,\text{Hz}$$

the following analytical results emerge:[141,142]

1. The envelope of the oscillations remains constant for constant $\omega_1/\phi_{v\omega}^{XY}$ in the ccf $C_{v\omega}^{XY}$. By parity inversion symmetry $\phi_{v\omega}^{XY}$ must disappear for $\omega_1 = 0$ and, therefore, there should be some link between $\phi_{v\omega}^{XY}$ and ω_1.

2. It is possible to observe the function $C_{v\omega}^{XY}$ even for values of $\phi_{v\omega}^{XY}$ much smaller than ϕ_{vv}^{XX} and $\phi_{\omega\omega}^{XX}$.

3. The ccf vanishes as $t \to 0$ and $t \to \infty$. This is also the result obtained from computer simulation.

The simple theory of Eqs. (249)–(251) produces both the linear and angular velocity acf's observed by computer simulation consistently with a ccf, in terms of a small number of parameters. This is a small analytical step towards explaining the set of nonvanishing ccf's between rotational and translational diffusion produced by numerical simulation and allowed by symmetry. The next section provides examples of such ccf's from recent supercomputer simulations.

VI. TIME-CROSS CORRELATION FUNCTIONS — A MAJOR CHALLENGE TO DIFFUSION THEORY

The first ccf to link statistically the rotational and translational diffusion of a molecule was computed in the early eighties.[47] Prior to that the theory of diffusion had not been able to supply the time dependence of any simple but fundamental ccf. In the few years since then the number of known ccf's has increased dramatically. The symmetry laws controlling their existence (Sections VII and VIII) have been identified for all known molecules. This is clearly a major challenge to the approach of (Section I) posed at a fundamental level within the overall framework of classical statistical mechanics. Theories of rotational or translational diffusion cannot by definition describe the time dependence of the simplest type of ccf involving both motions simultaneously. Contemporary attempts at extending or adapting the fundamental theory (Section V) are first approximations which can describe a small fraction of the data available from computer simulation. A more rigorous approach developed from the principles of traditional diffusion theory[1-4] leads to great complexity and empiricism. The basic dynamical properties of molecular liquids can now be described more successfully and in greater detail by computer simulation. In this context classical dynamics is being applied to the motion of molecules and through the intermediacy of powerful contemporary computers produces a range of results and depth of insight unobtainable with traditional diffusion theory. It has become clear that in comparison with simulation, diffusion theory is restricted severely by its own axioms to a few model approximations, such as the Debye theory of rotational diffusion, which are essentially mathematical approximations of the N-body problem. There are a very few of these that have a closed analytical solution for acf's and these are successful in describing only a very limited amount of spectral data, over a limited frequency range. The use of data from spectral moments soon exposes the limitations of the most elaborate analytical theory. The most accurate computer simulation is capable of describing only the first few moments of, for example, combined dielectric and far infrared spectroscopy.

The weaknesses inherent in the original axioms of diffusion theory are now exposed in several ways, for example, experimentally through the use of several different data sources under consistent conditions[1-4] of investigation, theoretically through the use of projection operators[3] applied to the Liouville equation and memory functions, and most starkly, through the set of fundamental cross-correlation functions now available from many computer simulations. The latter show, for example, that hitherto unknown ccf's exist[55-57] in a liquid of spherical top molecules, both in the laboratory frame (X, Y, Z) and in the molecule fixed frame (x, y, z). These show unequivocally and in many different ways that the rotational diffusion of a spherical top is correlated statistically to its own center-of-mass translation. The fundamental assumption that these motions are statistically uncorrelated is erroneous. Analytical attempts to remedy this error have so far run into difficulties, particularly in modelling realistically the effective pair potential. These appear even under conditions most favorable to the Debye theory, that is, for spherical tops in two or three dimensions.

In considering the diffusion through three dimensions of the most commonplace of molecules, the asymmetric top, the theory of diffusion has never been successful without the use of crude approximation. Even within the restricted confines of rotational diffusion, the Euler–Langevin equations (58)–(60) have no closed solution, and its approximation involves three unknowns, introduced 50 years ago but never determined experimentally. Their independent determination involves the use of three data sources which can be considered as mutually independent. These shortcomings are compounded greatly when account is taken of the new results briefly introduced in Section V. Modern off-shoots of the Debye theory, such as the various itinerant oscillators, have appeared[1-4] throughout the last twenty five years, but carry the same inherent flaws as the original model. There is no itinerant oscillator theory that can describe the set of nonvanishing ccf's in frames (X, Y, Z) and (x, y, z) without excessive complexity and empiricism. There is no itinerant oscillator which is tractable and rigorous in three dimensions. These theories have achieved very little compared with computer simulation, and are empirically descriptive in nature. Even the rotational versions do not address the nonlinearities of Euler's equations, for example, and even were this to be possible the model representation of the effective pair potential is less realistic by far than those used routinely in computer simulation, sometimes taken directly from ab initio quantum mechanics. When the analytical theory of diffusion appears in the contemporary literature as a description, for example, of spectral data, that description must be tempered with the realization that the comparison of data with theory is inevitably limited to that particular context. For example, the traditional range of data from dielectric relaxation does not extend into the far infrared. The fitting of low-frequency data with a simple rotational Debye theory produces a number, the Debye relaxation time, which is no more than the result of a physically flawed curve-fitting technique.

One of the most convincing illustrations of the extent to which computer simulation has out-paced the analytical methods can be derived from a consideration of intramolecular vibration[156] in a rotating and simultaneously translating diffuser, such as a water molecule. It is now possible to construct an effective pair potential for the water molecule that takes into account the internal vibrations of the H_2O framework, and this is adapted for use in a molecular dynamics computer simulation.

A. Ccf's Involving Molecular Vibration

The potential energy between two flexible water molecules can be written as the sum of two contributions arising from inter- and intramolecular motions

$$V(\mathbf{r}_{\alpha\beta}, \mathbf{q}_i) = V_{inter} + V_{intra} \tag{254}$$

where $\mathbf{r}_{\alpha\beta}$ are the intermolecular atomic distances, and $\mathbf{q}_i$ the internal coordinates. The potential is built up[258] from a consideration of the rigid MCY ab initio potential,[151] which is itself built up from first principles. The flexibility of the water molecule is modelled with a further ab initio term for the intramolecular motions taken from a high-quality computation by Bartlett, Shavitt, and Purvis.[259] The original rigid MCY potential[151] contains a negative charge which does not coincide with any atom of the molecule, and its extension to flexible geometry assumes that the charge always resides on the line bisecting the H–O–H angle within the molecular framework. This reduces to the rigid MCY potential when there is no framework distortion. The new flexible potential is known as the MCYL potential, which can be used[156] in a computer simulation.

B.   Computer Simulation of Flexible Water

The system used for the computation[156] of ccf's in flexible water consisted of 343 water molecules subject to periodic boundary conditions (Chapter 3) and a cut-off radius of 10.87 A. Coded into the algorithm was a numerical method for the evaluation of reaction fields, using a method developed by van Gunsteren et al.[260] The forces due to the complete inter- and intramolecular potential energies and the reaction energy were then evaluated using a sixth-order Gear algorithm. To start the run, the atoms of the ensemble were given random velocities with a Maxwell–Boltzmann distribution, the initial spatial configuration being taken from a previous Monte Carlo simulation. The time step was 0.15 fs for all data evaluation and collection. This allowed the simulation of vibrational features such as infrared and Raman fundamentals with frequencies of up to 100 THz. With this time-step the energy was conserved to 0.0033% in 1000 steps. Trajectories were saved every 10 time steps and the total number of configurations collected was 7600, corresponding to a total simulation time of 11.4 ps.

This method is independent of the simulation of rigid water[202–204] carried out by the present author with a 5×5 site-site model potential. This used no reaction fields, and integrated the equations of motion with an entirely different method. The simulation of flexible water, on the other hand, takes each atom as a separate diffuser, the translational motion of which is constrained by potential bond terms linking it to other atoms of the molecule. This method uses no rotational equations of motion. In order to build up correlation functions of molecular dynamical quantities such as angular velocity, and in order to define orientational and rotational velocity acf's, the data must be refined, allowing direct comparison with the simulation of rigid water[202] and with experimental data. Comparison of results from the two algorithms also allows the estimation of the effect of flexibility on the overall molecular dynamics.

C.   The Effect of Internal Vibrations on Correlation Functions

The two methods are capable of producing a set of acf's and ccf's whose individual members contain information on the diffusion in three dimensions of a rigid and flexible model of water. Comparison of the rotational velocity acf from both sources (i.e., the Fourier transform of the far infrared spectrum) exposes the effect of vibration on an essentially rotational and translational motion. In this context the minimum in the rotational velocity acf from the MCYL flexible potential is slightly deeper and more structured up to 0.1 ps, but otherwise there is little difference from the results of the rigid model potential developed by the present author. Vibration does not significantly affect the far infrared spectrum. The results for the rigid water molecule were obtained with as few as 500 saved configurations and 108 molecules. The results from the flexible model were computed for 343 molecules over 3400 configurations. Despite the numerous approximations in both algorithms, and despite the very different models used for the water–water pair potential, the results from both algorithms are very similar.

The same is true, essentially speaking, for other acf's from the simulations, with the exception of the center-of-mass linear velocity acf which is significantly more oscillatory in the model of flexible water due to internal displacements of the water center of mass with respect to the three vibrating atoms. It is significant that the angular velocity acf's from both algorithms are similar in time dependence and also nearly identical in both frames (X, Y, Z) and (x, y, z) with the rotational velocity acf both for the rigid and flexible models of water. The orientational acf's decay far more slowly in both models. In other words, the far infrared librational motion is a far faster process than the low-frequency dielectric adjunct. The peak frequency of the dielectric loss is about 50 times lower than that of the far infrared power absorption coefficient. This is close to what is observed experimentally, and the results from the flexible water model also closely agree with those from the rigid model in producing an anisotropy of angular and linear acf's in frame (x, y, z), anisotropies that are verifiable experimentally in both cases with, for example, NMR relaxation and tracer diffusion analysis. The similarities between rigid and flexible model acf's extends to the noninertial linear velocities and accelerations introduced in Section V. However, a major difference emerges, as expected, in the acf's of atomic linear velocities in frames (X, Y, Z) and (x, y, z). The acf of the velocity of the diffusing hydrogen atom in particular is highly oscillatory over a short time scale in the femtosecond range (infrared frequencies). The rigid model does not produce these oscillations because the H atom does not vibrate with respect to the O atom in the diffusing molecule. The rigid model cannot therefore produce the fundamental vibrational frequencies of the infrared spectrum.

D. Cross-Correlation Functions

Due to the flexible nature of the MCYL pair potential the simulation provides information on the statistical cross correlations between rotation, translation and intramolecular bond vibration in the diffusing water molecule. This is wholly outside the scope of the diffusion theories now available.

E. Vibration–Translation

The ccf between vibration and center-of-mass translation in the water molecule can be defined as follows. The vibration is described through the linear velocity $\mathbf{v}_H$ in frame (x, y, z) of the hydrogen atom, and translation as the linear velocity of the complete molecule, defined as that of the center of mass $(\mathbf{v})$ in the same frame (x, y, z). The latter is fixed by the axes of the molecular principal moments of inertia, which are vibration dependent. The nine elements of the tensor

$$\mathbf{C}_{vt} = \left\{ \frac{\langle \mathbf{v}_H(t)\mathbf{v}(0) \rangle}{\langle v_H^2 \rangle^{1/2} \langle v^2 \rangle^{1/2}} \right\}_{xyz} \tag{255}$$

provide the details of this cross correlation. The overall symmetry of this tensor was found to be

$$\mathbf{C}_{vt} = \begin{bmatrix} + & + & 0 \\ + & + & 0 \\ 0 & 0 & + \end{bmatrix}_{xyz} \tag{256}$$

and the time dependence of the nonvanishing elements is illustrated in Figs. 33 and 34. Both the diagonal and off-diagonal elements are very anisotropic, and it is clear that the translational diffusion of the molecular center of mass is correlated statistically to the individual H-atom translation in several different ways. The coupling in the (yy) component (Fig. 33b) is oscillatory at the frequency of the H-atom vibration, because the y axis is the one that approximately bisects the molecule through the oxygen atom and the centre of mass. The xx component (Fig. 34a) shows the fast oscillations superimposed on slower oscillations of the O-atom vibration. The zz component (Fig. 33c) is free of vibrational oscillations because the z axis is perpendicular to the plane of the molecule. This element is nearly identical with the off-diagonal xz element of the ccf. $\mathbf{C}_{v\omega}$ between the center-of-mass linear velocity and the molecular angular velocity ω in frame (x, y, z). Only the off-diagonal elements (Figs. 34a and b) of $\mathbf{C}_{vt}$ not involving the z axis exist. There is no correlation of the atomic and center-of-mass linear velocity components parallel and perpendicular to the plane of the molecule, but only for those vector components that are both either in or normal to the plane of the molecule.

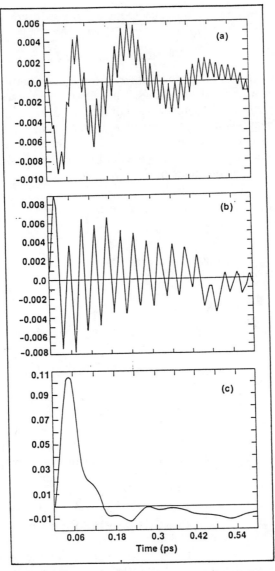

Figure 33. Elements of the ccf $\mathbf{C}_{vt}$: (a) xx, (b) yy, (c) zz.

The effect of vibration on rotation can be looked at with the tensor

$$\mathbf{C}_{H\omega}(t) = \left\{ \frac{\langle \mathbf{v}_H(t)\boldsymbol{\omega}(0) \rangle}{\langle v_H^2 \rangle^{1/2}\langle \omega^2 \rangle^{1/2}} \right\}_{xyz} \tag{257}$$

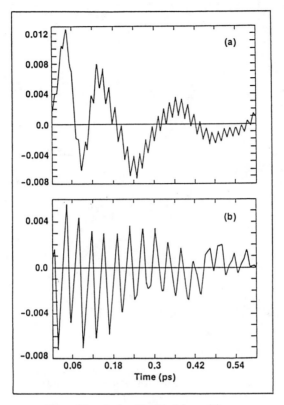

Figure 34.   As for Fig. 34, (a) yx, (b) xy.

Figure 35 shows four off-diagonal elements of this ccf. The diagonal and remaining two off-diagonal elements vanish in the noise of the simulation, so that the overall symmetry of the matrix is

$$\mathbf{C}_{H\omega} = \begin{bmatrix} 0 & 0 & + \\ 0 & 0 & + \\ + & + & 0 \end{bmatrix}_{xyz} \tag{258}$$

This is opposite to the symmetry of the vibration–translation matrix. Figure 35 shows that the zx and zy components of the ccf are essentially perfect mirror images, while there is some residual asymmetry in the yz and xz components. Since the molecule can only vibrate in the molecular

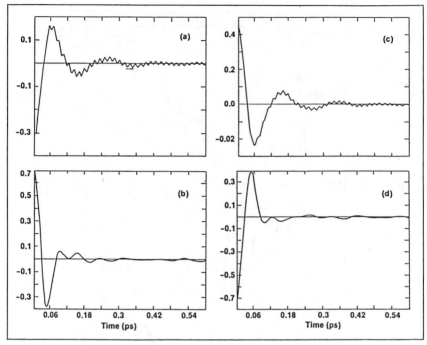

Figure 35.   Elements of $\mathbf{C}_{H\omega}$: (*a*) *xz*; (*b*) *zx*; (*c*) *yz*; (*d*) *zy*.

plane, we see the effect of vibrations in the *xz* and *yz* components as small ripples.

Figures 36*a* and *b* confirm the symmetry of Fig. 35 for the equivalent ccf matrix between the molecular angular velocity in frame (x, y, z) and the linear velocity in this frame, $\mathbf{v}_O$, of the oxygen atom. In this case the whole matrix is dominated by the *zx* element (Fig. 36*a*). Due to the fact that the amplitude of the oxygen vibration is much smaller than that of the hydrogen, no ripples are present here. The *xz* element (Fig. 36*b*) seems to be signal, but with a very small amplitude. All the other seven elements vanish in the background noise of the simulation, so that the overall symmetry is

$$\mathbf{C}_{O\omega} = \begin{bmatrix} 0 & 0 & + \\ 0 & 0 & 0 \\ + & 0 & 0 \end{bmatrix}_{xyz} \tag{259}$$

It is interesting to cross reference Figs. 33*c* and *a*, which illustrate the

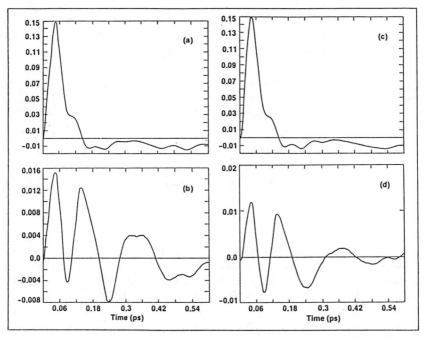

Figure 36. (a) zx element of $\mathbf{C}_{O\omega}$; (b) xz element; (c) xz element of $\mathbf{C}_{v\omega}$; (d) zx element of $\mathbf{C}_{v\omega}$.

time dependence of the zz component of the vibration–translation matrix and the zx component of the oxygen linear translation to molecular angular velocity ccf matrices respectively. The time dependence of these elements is essentially the same, with slightly different maxima. Furthermore, Fig. 36c shows that the xz component of the ccf for molecular angular velocity to center-of-mass linear velocity in frame (x, y, z) is identical with the component in Fig. 36a. This can be explained by the proximity of the oxygen atom to the molecular center of mass. Thus, there is a clear interrelation between the ccf tensors of vibration–translation, vibration–rotation, and in Fig. (36c), rotation–translation. None of these ccf's is clearly and unambiguously available in traditional diffusion theory, which would regard the diffusing water molecule as a "rigid sphere with embedded dipole."

F.   Rotation–Translation

The relevant matrix is the usual tensor product $\mathbf{C}_{v\omega}$ between the molecular angular velocity and the linear centre-of-mass velocity in the flexible and diffusing frame of reference (x, y, z). In this case Figs. 36c and d show that the matrix is dominated, as in the rigid model of water simulated by the present author, by the xz element. However, there is the trace of a signal in the zx element, as allowed by group theory for the point group C_{2v}. The results of Fig. 36c, obtained by simulation from the flexible MCYL potential, is almost identical with the equivalent from the rigid model of water, which is evidence for the reality of the ccf as obtained independently from both simulations. The pattern of the ccf matrix is

$$\mathbf{C}_{\nu\omega} = \begin{bmatrix} 0 & 0 & + \\ 0 & 0 & 0 \\ + & 0 & 0 \end{bmatrix}_{xyz} \tag{260}$$

and this corroborates the simulation of rigid water, together with the predictions of group theory. There is little or no effect on the rotation–translation matrix of vibration, that is, of using a flexible ab initio model potential. This means that the data archive generated[4] from rigid models of several different molecular point groups for the rotation–translation matrix in rigid molecular frameworks is essentially valid as it stands for their flexible counterparts. The theory of Section I is not supported by this finding, however, because of the fundamental decorrelation of rotational and translational diffusion. Computer simulation should be used to interpret data from as wide a variety of sources as possible for the same molecular liquid in terms of an effective pair potential derived ab initio, and analytical theory wherever possible to cement our understanding, as illustrated in Section V. For three rank tensor ccf's, containing the noninertial linear accelerations, the symmetry of the relevant ccf's in frame (x, y, z) is different. For example, nonvanishing diagonal elements exist in frame (x, y, z) of the ccf,

$$\mathbf{C}_{C\nu} = \left\{ \frac{\langle \mathbf{v}(t) \times \boldsymbol{\omega}(t)\mathbf{v}(0) \rangle}{\langle (\mathbf{v} \times \boldsymbol{\omega})^2 \rangle^{1/2} \langle \nu^2 \rangle^{1/2}} \right\}_{xyz} \tag{261}$$

between the molecular Coriolis acceleration in frame (x, y, z) and the linear center-of-mass velocity of the molecule in this frame. In this case also the rigid and flexible models of the water molecule give similar, if not quite identical, results, the time dependence of the flexible potential being slightly more oscillatory.

The overall conclusions from these simulations include the following:

1. Some fundamental characteristics of molecular diffusion are well described with a rigid model of water, for example, the ccf $\mathbf{C}_{v\omega}$.

2. The mutual statistical correlation between vibration and translation or rotation can be described accurately in terms of ccf's simulated from a flexible ab initio model of the water pair potential.

3. Very few of these ccf's are amenable to analytical description with traditional diffusion theory.

4. Algorithms for water dynamics that are based on the rigid model with no reaction field corrections are adequate for a range of properties.

G. Simulation Results for Water from 10 to 1273 K

Water is a well-studied liquid, and the most prevalent. It is of interest to simulate its diffusional dynamics over the complete range of existence, and this has been initiated recently[261] using high-quality runs over 6000 time steps of the rigid model. The thermodynamic properties of liquid and amorphous solid water are known experimentally[262-265] from about 5 K to about 250 kbar at 1043 K, and the critical temperature and pressure are known very accurately. It is particularly interesting to compute time-correlation functions at constant molar volume over a range of state points, data being available experimentally at a constant density of 1 gm/cm^3 from 293 K, 1 bar to 1273 K, 15 kbar in the liquid state. Additionally, experimental data are available from shock wave experiments at a constant inverse density of 0.47 cm^3/gm at two state points: (1) 773 K, 230 kbar and (2) 1043 K, 250 kbar, and these provide further opportunity for simulations at very high pressure and constant molar volume with which to investigate the nature of the molecular dynamics of liquid water through a wide variety of time correlation functions. At the critical point of water (647.02 K, 220.91 bar) the experimentally measured critical molar volume is 56.8 cm^3/mole, and simulations at the critical point are supported in this section from the recent supercomputer runs. To complete the range of conditions available there are experimental data at 10 and 77 K, both at 1 bar and well below the normal melting point of 273.16 K. These low-temperature conditions are obtained experimentally by suddenly quenching, and are mimicked in computer simulation runs by instantaneously dropping the temperature from about 293 K. This is equivalent to removing kinetic energy instantaneously from the system, and is known as slam or splat quenching. This does not result in a phase change, and the configuration in which the system finally equilibrates at low temperature in the computer simulation is amorphous, in the sense that it has no regular ice-like structure. If the rate of cooling is very rapid, as in the computer simulation, the structure of the configuration before the quenching is retained approximately at the much lower temperature, that is, the potential energy is about the same but the kinetic energy greatly diminished. The ensemble is frozen in a vitreous condition that is not a fluid, known as amorphous solid water.

Across this range the simulation algorithm TETRA provides the correct thermodynamic results within the relatively large uncertainty in the computed pressure. The computed pressure at input molar volume tends to increase more rapidly than the experimental pressures, but this is satisfactory in view of the fact that an ab initio potential such as the MCYL produces a pressure which is 6000 times greater than that observed at room temperature and pressure.

H. The ccf's from 10 to 1273 K

The technical difficulties of observing the spectral and dynamical properties of water over this range are such that few reliable data are available. On the other hand, computer simulations are possible over the complete range of existence of liquid and amorphous solid water, providing a range of acf's and ccf's. The same trajectories can be used to provide structural data such as pair distribution functions between the atoms of individual water molecules. Some of the acf's can be linked to spectra by Fourier transformation, but the ccf's in this range are especially significant in marking the limits of applicability of classical diffusion theory in both frames of reference. For example, Fig. 37 illustrates a direct cross correlation between the linear and rotational velocities of the diffusing water molecule in frame (X, Y, Z) of the laboratory. The intensity of the cross correlation increases with liquid density, that is, molar volume decreases, and is approximately constant at constant molar volume. It is therefore impossible for the water molecule to diffuse according to the rules of Section I. The language of cross correlation as illustrated in Fig. 38 seems simpler than the equivalent description in terms of hydrogen bonds,[3] and the large number of cross correlations now known seems capable of providing a more consistent and clearer description of the molecular diffusion process in water.

In the moving frame, for example, the ccf $C_{v\omega}$ between the molecular linear and angular velocities has a time dependence and amplitude which is a sensitive function of density, pressure and temperature over the complete range of existence of liquid and amorphous solid water. At a constant molar volume of 18.0 cm^3/mole the symmetry allowed elements of this ccf remain relatively constant in time dependence and intensity. They are not mirror images because of the lack of symmetry in the water molecule itself. The ccf in frame (x, y, z) is driven primarily by changes in molar

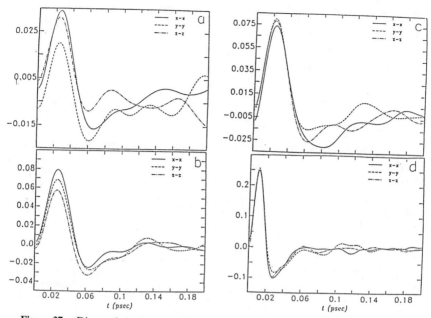

Figure 37.   Diagonal elements of the lab frame ccf $\langle \dot{\mu}(t)\mathbf{v}(0)\rangle$ for liquid water and steam: (*a*) 1800 cm³/mole, 293 K (steam); (*b*) 18.0 cm³/mole, 293 K; (*c*) 18.0 cm³/mole, 15 kbar, 1273 K; (*d*) 8.5 cm³/mole, 250 kbar, 1043 K.

volume, much more so than changes in temperature. At constant density the ccf remains constant, but if density is for example doubled by the application of 250 kbar of pressure at 1043 K, the intensity of each element of the ccf is increased by roughly four times. All the other elements of the matrix remain zero for all t at all state points, in agreement with point-group theory (Section V).

I. Higher-Order ccf's

In the frame (x, y, z) the diagonal elements of such higher order ccf's such as

$$\mathbf{C}_{r\omega} = \frac{\langle \mathbf{r}(t) \times \boldsymbol{\omega}(t)\mathbf{r}(0)\rangle}{\langle (\mathbf{r} \times \boldsymbol{\omega})^2\rangle^{1/2}\langle r^2\rangle^{1/2}} \tag{262}$$

and $\mathbf{C}_{Cv}(t)$ exist at all state points. The intensity of the former is greatest at the critical point, the zz element reaching an intensity of 0.45 (Fig. 39). The time dependence at the other state points is broadly similar, the intensity falling to 0.2 at 1 bar and 293 K and rising to 0.35 at 15 kbar and

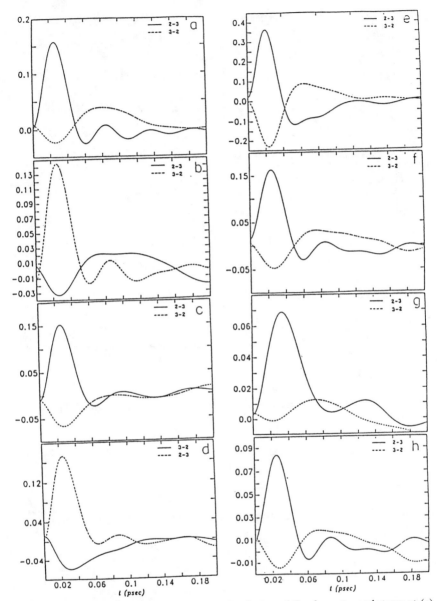

Figure 38. xz and zx elements of the moving frame ccf $\mathbf{C}_{v\omega}$ for water and steam at (a) 1800 cm³/mole, 1 bar, 293 K (steam); (b) 18.0 cm³/mole, 1.0 kbar, 373 K; (c) 18.0 cm³/mole, 9.5 kbar, 773 K; (d) 18.0 cm³/mole, 15.0 kbar, 1273 K; (e) 8.5 cm³/mole, 250.0 kbar, 1043 K; (f) 18.0 cm³/mole, 1.0 kbar, 10 K; (g) 56.8 cm³/mole, 647 K (critical point); (h) 1800 cm³/mole, 20.0 bar, 293 K (steam).

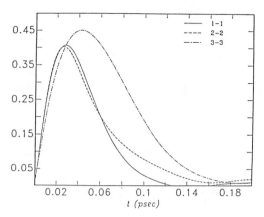

Figure 39. Example of the diagonal elements of $\mathbf{C}_{cv}$ in frame (x, y, z) for liquid water at 221 bar and 647 K (critical point).

1273 K. At 250 kbar and 1043 K it falls to 0.12 with a much shorter time decay. The pattern over all state points is broadly opposite to the simple ccf $\mathbf{C}_{v\infty}(t)$ and the intensity is a minimum at high pressures for the higher order ccf, a maximum for the simple ccf's.

Again, these results are entirely consistent with the symmetry considerations for the C_{2v} point group.

In the frame (X, Y, Z) the higher order ccf's

$$\mathbf{C}_{2v\omega} = \frac{\langle \dot{\mathbf{v}}(t) \times \boldsymbol{\omega}(t)\mathbf{v}(0) \times \boldsymbol{\omega}(0)\rangle}{\langle (\dot{\mathbf{v}} \times \boldsymbol{\omega})^2\rangle^{1/2}\langle \mathbf{v} \times \boldsymbol{\omega})^2\rangle^{1/2}} \tag{263}$$

and

$$\mathbf{C}_{3v\omega} = \frac{\langle \mathbf{v}(t) \times \dot{\boldsymbol{\omega}}(t)\mathbf{v}(0) \times \boldsymbol{\omega}(0)\rangle}{\langle (\mathbf{v} \times \dot{\boldsymbol{\omega}}^2\rangle^{1/2}\langle (\mathbf{v} \times \boldsymbol{\omega})^2\rangle^{1/2}} \tag{264}$$

also exist over the range of state points as diagonal matrices. The off-diagonal elements vanish by symmetry. In frame (X, Y, Z) the intensity behavior is constant up to the kbar range, and there is no sign of a maximum or minimum at the critical point. These laboratory frame ccf's are illustrated in Figs. 40 and 41. The maximum intensity of $\mathbf{C}_{3v\omega}$ remains constant at about 0.5 across the complete range of state points and is illustrated in Fig. 41 at 1.0 bar and 293 K.

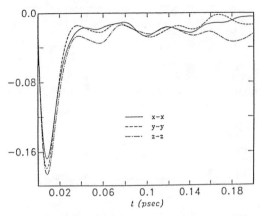

Figure 40. Diagonal elements of $C_{2v\omega}$ in the lab frame at 15 kbar and 1273 K.

J. Solutions of Water in Carbon Tetrachloride

The molecular dynamics of water free from hydrogen bonding is known experimentally to be very different from that of pure water.[266,267] There is a dramatic shortening of the correlation time of orientation in free water diffusing in an organic solvent such as carbon tetrachloride, cyclohexane, or benzene. The far infrared peak frequency shifts from about $700 \, \text{cm}^{-1}$ in pure water to about $200 \, \text{cm}^{-1}$ in dilute solution, clearly showing the effect of H bonding on the far infrared spectrum and on the molecular dynamics of water freed from hydrogen bonds.

The memory function analysis of Ref. 267 can be supplemented by contemporary computer simulation of the various acf's and ccf's which

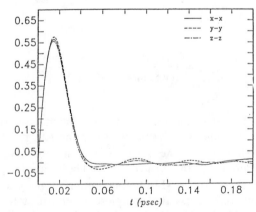

Figure 41. Diagonal elements of $C_{3v\omega}$ in the lab frame at 1 bar, 293 K.

can be used to characterize the molecular dynamics. A solution of water in carbon tetrachloride has been simulated[268] using site–site pair potentials. The most characteristic results were obtained by directly comparing correlation functions for the pure liquids and solutions to isolate the effects of disrupting the H bonds.

The potential energy of the mixture was assumed to be built up of effective molecule to molecule pair potentials obtained from individual atom to atom terms.[268,269] Lorentz–Berthelot combining rules were used to account for cross terms. The algorithm TETRA was adopted to deal with the water–carbon tetrachloride mixtures. and ccf's computed over the water molecules of the mixture. These were used to monitor the dynamical behavior of the system in several ways. For example, in an equimolecular mixture of 54 water and 54 carbon tetrachloride diffusers the H-bonding network appeared already severely disrupted, and the water molecular dynamics evolved differently. In the tail of the angular velocity acf in frame (X, Y, Z), the oscillations are damped out in the equimolecular microemulsion, H bonding thus appears to be the underlying cause of the oscillations in pure water. The effect of microemulsification on the rotational velocity acf is similar in the 0.5–0.15 ps range, implying that the far infrared spectrum shifts to low frequencies, broadening in the process. This is approximately what is observed experimentally.[266] The simulation and observation are not, however, directly compatible because the solutions used experimentally[266] were necessarily very dilute (less than 0.1% mole fraction) and were thermodynamically equilibrated solutions. In the computer simulation the mixture was not equilibratcd because the total time span was restricted to 5.0 ps. It takes many hours for the water and carbon tetrachloride layers to separate in the laboratory.

The effect of removing H bonds appears least in the center-of-mass velocity acf and most in the ccf between linear and angular velocitics in frame (x, y, z) (Fig. 42). Here, curve I shows the ccf in pure liquid water, and curve 2 is the same for the water molecules in the equimolecular microemulsion. The element zx of the ccf has been greatly reduced in amplitude and its sign reversed in the mixture. The T_d symmetry of the carbon tetrachloride molecules does not allow this to exist in a liquid composed only of these molecules, the element in pure carbon tetrachloride must vanish for all t. The much reduced amplitude in the 50/50 mixture indicates the strong effect on this type of ccf of H bonding in pure water, where H bonding and cross correlation appear to be synonymous. The big amplitude decrease occurs even though thc ccf is computed over the water molecules of the mixture. This result comes essentially from the fact that the 54 water molecules in the mixture have a ccf symmetry:

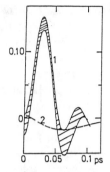

Figure 42. The xz element of $\mathbf{C}_{v\omega}$ in frame (x, y, z) in a microemulsion of water in carbon tetrachloride. (1) Pure water; (2) 50/50
mixture.

$$\mathbf{C}_{v\omega} \equiv \begin{bmatrix} 0 & 0 & + \\ 0 & 0 & 0 \\ + & 0 & 0 \end{bmatrix}_{xyz} \tag{265}$$

and 54 an equivalent in which all elements vanish.

Other types of ccf whose elements are allowed by symmetry both for
water and carbon tetrachloride do not show such a significant amplitude
decrease under the same conditions. These are exemplified by the generic
$\langle \mathbf{A}(t) \times \boldsymbol{\omega}(t)\mathbf{A}(0) \rangle$ where $\mathbf{A}$ is linear velocity or linear noninertial velocity,
for example, in frame (x, y, z). This is shown in Fig. 43. In this type, the
diagonal elements exist both for T_d and C_{2v} symmetries, but with different
strengths of correlation. The 50/50 mixture therefore produces an average
of the amplitude for carbon tetrachloride and water, despite the fact that
the averaging is carried out over the water molecules only. This is an
indication that the dynamics of the water molecules are affected by their
environment in such a way as to reduce the statistical cross correlation
computed in the pure liquid water. Cross correlations on this evidence
appear to be more sensitive to environment than autocorrelations, and

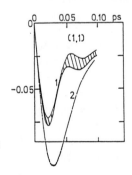

Figure 43. The xx element of the ccf $\mathbf{C}_{Cv}$, for (1) pure
water; (2) for water in carbon tetrachloride.

this is useful in computations of spectra of liquid mixtures, suspensions, and solutions.

K. Liquids of Spherical Top Molecules

Recent computer simulations[55] of T_d and O_h spherical tops have shown that their molecular dynamics are cross correlated in many different ways. These simulations were supported in detail by group theory for both symmetry types. In addition to non-vanishing elements of $\langle \mathbf{v}(t) \times \boldsymbol{\omega}(t)\mathbf{v}(0) \rangle$ in frame (x, y, z), ccf's such as $\langle \mathbf{v}(t)\dot{\boldsymbol{\mu}}(0) \rangle$ appear directly in frame (X, Y, Z). The elements of the simple ccf between linear and angular velocities vanish in both frames, but the diagonal elements of $\langle \mathbf{A}(t) \times \boldsymbol{\omega}(t)\mathbf{A}(0) \rangle$ are equal in time dependence in frame (x, y, z) for both symmetries.

L. The Effect of External Fields in Liquid Water

Various types of applied field (Section V) have appropriate symmetry effects in frames (X, Y, Z) and (x, y, z). In the former the Hamiltonian in the presence of the field may no longer be positive to parity inversion, and will develop a directional property. According to the third principle of group theoretical statistical mechanics (Section VII), new ccf's may appear in the molecular ensemble as a result of applying the field. In frame (x, y, z) the point group theory still applies with the molecular symmetry as reference, but the applied field may change the time dependence of ccf's in this same. Some of the consequences for birefringence theory have been dealt with in Section IV but the following describes changes in symmetry patterns in both frames for several different ccf's.

M. Circularly Polarized Laser Field

A circularly polarized external field has an electric component which rotates in space about an axis of frame (X, Y, Z). The effect of this type of field can be studied by constructing symmetry pattens of correlation functions. The field produces a birefringence effect, as described in Section IV, which is experimentally observable. The ccf's which are available in this context from computer simulation are not describable with the traditional approach of Section I, for example, the effect on the orientational correlation tensor that of the permanent dipole moment) is

$$
\begin{bmatrix} + & 0 & 0 \\ 0 & + & 0 \\ 0 & 0 & + \end{bmatrix} \rightarrow \begin{bmatrix} + & 0 & 0 \\ 0 & + & - \\ 0 & + & + \end{bmatrix}_{(X, Y, Z)} \tag{266}
$$

and on the rotational velocity correlation tensor,

$$
\begin{bmatrix} + & 0 & 0 \\ 0 & + & 0 \\ 0 & 0 & + \end{bmatrix} \rightarrow \begin{bmatrix} + & 0 & - \\ 0 & + & + \\ + & - & + \end{bmatrix}_{(X, Y, Z)} \tag{267}
$$

The indirect effect on the linear center of mass velocity correlation tensor is

$$
\begin{bmatrix} + & 0 & 0 \\ 0 & + & 0 \\ 0 & 0 & + \end{bmatrix} \rightarrow \begin{bmatrix} + & 0 & 0 \\ 0 & + & - \\ 0 & + & + \end{bmatrix}_{(X, Y, Z)} \tag{268}
$$

so that the overall effect in all three cases is antisymmetric, marked by the plus and minus entries which denote mirror images in time dependence. The extent of induced antisymmetry varies from dynamical variable to variable, and most off-diagonals appear in the correlation tensor of molecular angular velocity, where the complete effect is

$$
\begin{bmatrix} + & 0 & 0 \\ 0 & + & 0 \\ 0 & 0 & + \end{bmatrix} \rightarrow \begin{bmatrix} + & + & - \\ - & + & - \\ - & - & + \end{bmatrix}_{(X, Y, Z)} \tag{269}
$$

The correlation tensor of the molecular Coriolis acceleration in frame (X, Y, Z) is affected as follows:

$$
\begin{bmatrix} + & 0 & 0 \\ 0 & + & 0 \\ 0 & 0 & + \end{bmatrix} \rightarrow \begin{bmatrix} + & 0 & 0 \\ 0 & + & + \\ 0 & - & + \end{bmatrix}_{(X, Y, Z)} \tag{270}
$$

and other, higher-order, acf's in frame (X, Y, Z) are similarly affected.

The challenge to diffusion theory is therefore a detailed one, and all these new ccf's underpin the experimentally observable birefringence. There are also pronounced field induced effects on the correlation tensors of the noninertial linear velocity $\omega(t) \times \mathbf{r}(t)$, that is,

$$
\begin{bmatrix} + & 0 & 0 \\ 0 & + & 0 \\ 0 & 0 & + \end{bmatrix} \rightarrow \begin{bmatrix} + & 0 & 0 \\ 0 & + & + \\ 0 & - & + \end{bmatrix}_{(X, Y, Z)} \tag{271}
$$

In the molecule fixed frame (x, y, z) of the principal molecular moments of inertia, the circularly polarized field has a different type of symmetry effect, which can be interpreted consistently with point-group theory. The field in this case has an effect on the time dependence of the ccf elements, but none on the symmetry. The anisotropy produced in the diagonal elements of the acf tensors of the linear and angular velocities is made more pronounced in frame (x, y, z) but the off-diagonal elements vanish. Only the diagonal elements exist of higher order ccf's such as that between the molecular Coriolis acceleration and linear velocity in frame (x, y, z). The overall symmetry effect of the field in these cases is

$$
\begin{bmatrix} + & 0 & 0 \\ 0 & + & 0 \\ 0 & 0 & + \end{bmatrix} \rightarrow \begin{bmatrix} + & 0 & 0 \\ 0 & + & 0 \\ 0 & 0 & + \end{bmatrix}_{(x, y, z)} \tag{272}
$$

The actual time dependence of these elements is different in frame (x, y, z) from the equivalents in frame (X, Y, Z).

The effect of the field on the simple ccf $\langle \mathbf{v}(t)\omega(0) \rangle$ is to distort the time dependence of the nonvanishing elements in frame (x, y, z), but not to change the symmetry

$$
\begin{bmatrix} 0 & 0 & + \\ 0 & 0 & 0 \\ + & 0 & 0 \end{bmatrix} \rightarrow \begin{bmatrix} 0 & 0 & + \\ 0 & 0 & 0 \\ + & 0 & 0 \end{bmatrix}_{(x, y, z)} \tag{273}
$$

This is similar to the effect of a static electric field in this frame, and in general, external fields do not change the overall symmetry in frame (x, y, z), only the time dependence. In contrast, in frame (X, Y, Z) the

(Y, Z) and (Z, Y) elements appear in response to a static electric field in the X axis.

The complete set of results summarized briefly in this section must be used in general to characterize the effect of the external circularly polarized field on the molecular dynamics of water.

N. Chiral Liquids

The equivalent effect of the external laser field on the acf's and ccf's of a chiral liquid such as bromochlorofluoromethane introduces an extra element of symmetry. There are two enantiomorphs (mirror-image pairs) which are physically distinct liquids. These are composed of the R and S enantiomer molecules. If these liquids are mixed in equal proportion, the racemic mixture is formed. The physical properties of each enantiomorph, such as melting point, boiling point, density, refractive index, infrared spectrum, and so on are identical in frame (X, Y, Z). However, the two enantiomers rotate the plane of polarized light in equal but opposite directions, and the well-known phenomenon of circular dichroism results from the interaction of circularly polarized radiation with optically active liquids.

These are all well-known properties of optically active liquids (Section IX) but the use of computer simulation has produced a new and fundamental property of the optically active molecular liquid state discovered by the present author[48] in 1983. This involves the nature of statistical cross correlation in frame (x, y, z) between $\mathbf{v}$ and $\boldsymbol{\omega}$ in the enantiomorphs and racemic mixture of an optically active (chiral) liquid. It was shown that the ccf in frame (x, y, z) was *different* in the enantiomorphs. In general all nine elements of the ccf exist (Sections VII and VIII) and all change sign from one enantiomorph to the mirror image

$$\begin{bmatrix} + & + & + \\ + & + & + \\ + & + & + \end{bmatrix} \rightarrow \begin{bmatrix} - & - & - \\ - & - & - \\ - & - & - \end{bmatrix}_{(x, y, z)} \tag{274}$$

The time dependencies of the elements are equal and opposite in general, and the ccf's cancel in the racemic mixture. The signs of the elements can change from positive to negative or vice versa. In optically active liquids with molecules containing more than one optically active center, it is likely that permutations and combinations of sign changes will become possible. In the simplest type of chiral molecule there is only one optically active center and the switching properties in general depend on the molecular symmetry. In the lowest chiral molecular point group (C_1), the ccf symmetries are in general

$$
\begin{bmatrix} + & + & + \\ + & + & + \\ + & + & + \end{bmatrix} \rightarrow \begin{bmatrix} 0 & 0 & 0 \\ 0 & 0 & 0 \\ 0 & 0 & 0 \end{bmatrix} \leftarrow \begin{bmatrix} - & - & - \\ - & - & - \\ - & - & - \end{bmatrix}_{(x,\,y,\,z)} \tag{275}
$$

This is a fundamental physical property of all chiral liquids and occurs only in frame (x, y, z). It is another substantial challenge to the approach of Section I.

Dichroism (Section IX) can be computer simulated by applying an external circularly polarized laser field to both the R and S enantiomers of, for example, bromochlorofluoromethane. The results[240] in frames (X, Y, Z) and (x, y, z) can be measured through the appropriate set of acf's and ccf's as for water. In this case, however, additional information is provided by the fact that there are three types of liquid available, denoted by (R), (S), and (RS). In the orientational ccf tensor the off-diagonal structure induced by the field in frame (X, Y, Z) is dominated for all three liquids by the (Z, Y), and (Y, Z) elements for a circularly polarized laser applied in the X axis. There is no discernible symmetry change between enantiomers and mixtures and the result can be summarized as

$$
\begin{bmatrix} + & 0 & 0 \\ 0 & + & 0 \\ 0 & 0 & + \end{bmatrix}_R \rightarrow \begin{bmatrix} + & 0 & 0 \\ 0 & + & + \\ 0 & - & + \end{bmatrix}_{R(X,\,Y,\,Z)} \tag{276}
$$

and similarly for the S enantiomorph and the racemic mixture. However, for the rotational velocity correlation tensor a difference develops between the off-diagonal elements of the enantiomers and racemic mixture, a laboratory frame difference between the response of the enantiomers and the racemic mixture. Part of the explanation for the dynamics of circular dichroism lies in this observation, which can be summarized as

$$
\begin{bmatrix} + & 0 & 0 \\ 0 & + & 0 \\ 0 & 0 & + \end{bmatrix}_R \rightarrow \begin{bmatrix} + & - & + \\ - & + & + \\ - & - & + \end{bmatrix}_{R(X, Y, Z)} \tag{277}
$$

and

$$
\begin{bmatrix} + & 0 & 0 \\ 0 & + & 0 \\ 0 & 0 & + \end{bmatrix}_{RS} \rightarrow \begin{bmatrix} + & + & + \\ - & + & + \\ + & - & + \end{bmatrix}_{RS(X, Y, Z)} \tag{278}
$$

There is a different symmetry in the field-on ccf's for the enantiomer and the racemic mixture attributable to the different response of the liquids to the circularly polarized laser field. The symmetry pattern reverses in a left and right circularly polarized field and signals a dichroism in the far infrared and lower frequency adjuncts of the zero to THz range. The birefringence associated with the dichroism manifests itself in the anisotropy of the diagonal element of the rotational velocity and orientational correlation tensors. There is also an interesting indirect effect on the elements of the linear velocity correlation tensor in frame (X, Y, Z). The pattern of field induced cross correlation is further exemplified in the molecular angular velocity correlation tensor in frame (X, Y, Z), but in this case the effect is opposite in the off-diagonals:

$$
\begin{bmatrix} + & 0 & 0 \\ 0 & + & 0 \\ 0 & 0 & + \end{bmatrix}_R \rightarrow \begin{bmatrix} + & + & - \\ + & + & + \\ + & - & + \end{bmatrix}_{R(X, Y, Z)} \tag{279}
$$

and on the racemic mixture

$$
\begin{bmatrix} + & 0 & 0 \\ 0 & + & 0 \\ 0 & 0 & + \end{bmatrix}_{RS} \rightarrow \begin{bmatrix} + & 0 & 0 \\ 0 & + & + \\ 0 & - & + \end{bmatrix}_{RS(X, Y, Z)} \tag{280}
$$

There is a direct laboratory frame difference in the off-diagonal elements for enantiomers and racemic mixture related to the fundamental dynamics of linear and nonlinear circular dichroism.

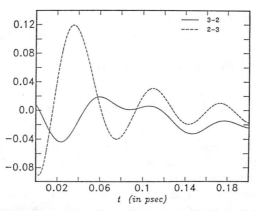

Figure 44. Simple ccf under the influence of a circularly polarized electromagnetic field applied to liquid water, zx and xz elements.

There appears to be no such change, however, in the symmetry of the center of mass linear velocity correlation tensors in frame (X, Y, Z)

$$\begin{bmatrix} + & 0 & 0 \\ 0 & + & 0 \\ 0 & 0 & + \end{bmatrix} \rightarrow \begin{bmatrix} + & 0 & 0 \\ 0 & + & - \\ 0 & + & + \end{bmatrix}_{(X, Y, Z)} \tag{281}$$

both for the enantiomers and the racemic mixture. It is noteworthy that the field induces no new elements of $\langle \mathbf{v}(t)\boldsymbol{\omega}(0) \rangle$ either in the enantiomers or the racemic mixture. The reason for this is that the rotating electric field component of the circularly polarized laser has gerade symmetry, and the ccf has ungerade symmetry to parity inversion.

In the frame (x, y, z) the following symmetry changes were observed when the three liquids are treated in turn with a circularly polarized laser field. The simple ccf just described for frame (X, Y, Z) exists for $t > 0$ and is different in time dependence in frame (x, y, z) for enantiomer and racemic mixture. In the presence of a right circularly polarized field the zy and yz elements in each case are not symmetric in time dependence, but are otherwise similar both for the enantiomer and mixture. Substantial differences begin to appear in the xz and zx elements, the former attaining twice its normalized maximum value for the enantiomer compared with the mixture. The main overall effect of the field on these (x, y, z) frame elements is to remove symmetry difference at field off equilibrium between enantiomer and racemic mixture. These effects are illustrated in Fig. 44.

The circularly polarized field has no effect on the overall symmetry of the ccf's in frame (x, y, z), and this is again a consequence of point group theory described in (Sections VII and VIII). It is an advantage to work directly in the laboratory frame (X, Y, Z) when simulating the effect of external fields, allowing direct comparisons to be made with the available experimental data.

O. Circular Flow in Liquid Water

The conditions for simulating circular flow in liquid water were defined in Section IV, where mention was made of birefringence induced by this effect in frame (X, Y, Z). In this section we examine the effect of the flow on correlation matrices of type $\langle A(t)B(0)\rangle$ where A and B are dynamical quantities in either frame. Circular now effects on the molecular level can be studied through the symmetry changes induced in matrices of this type, which is a key to detecting cross mode couplings in macroscopic flows such as vortices and eddies. There is an infinite number of molecular ccf matrices which may be utilized to explore flow effects, utilizing noninertial and inertial molecular dynamical quantities. Symmetry tests exist to predetermine which members of the set of nonvanishing ccf's may exist for $t > 0$.

In the laboratory frame (X, Y, Z) ccf's of the general type $\langle A(t)\dot{A}(0)\rangle$ are affected by the applied flow field. An example is

$$\frac{d}{dt}\langle v(t) \times \omega(t)v(0) \times \omega(0)\rangle$$

$$= \langle \dot{v}(t) \times \omega(t)v(0) \times \omega(0)\rangle + \langle v(t) \times \dot{\omega}(t)v(0) \times \omega(0)\rangle \quad (282)$$

Both the components on the right-hand side exist in frame (X, Y, Z) and the effect of the flow field is to induce off-diagonal elements with $D_g^{(2)}$ symmetry (Sections VII and VIII)

$$\begin{bmatrix} + & 0 & 0 \\ 0 & + & 0 \\ 0 & 0 & + \end{bmatrix} \rightarrow \begin{bmatrix} + & + & 0 \\ + & + & 0 \\ 0 & 0 & + \end{bmatrix}_{(X,Y,Z)} \quad (283)$$

The rotational flow changes considerably the time dependence of the diagonal elements, inducing a nonzero value at $t = 0$ for each. In the absence of flow each element vanishes at $t = 0$, and the properties of this ccf alone give ample scope for linking rotational flow to molecular dynamics.

In frame (x, y, z) rotational flow also has an effect on the overall symmetry of the simple ccf between molecular linear and angular velocities, the yz element, for example, is amplified to twice its field-free maximum, and the zy element reaches a minimum of -0.12. In contrast, the effect on

$$C_{r\omega} = \left[\frac{\langle \mathbf{r}(t) \times \boldsymbol{\omega}(t)\mathbf{r}(0)\rangle}{\langle(\mathbf{r} \times \boldsymbol{\omega})^2\rangle^{1/2}\langle r^2\rangle^{1/2}} \right]_{(x,y,z)} \tag{284}$$

is to decrease the amplitude of the diagonal elements, and different again is the behavior of the matrix

$$C_{r\omega\omega} = \left[\frac{\langle \boldsymbol{\omega}(t) \times (\boldsymbol{\omega}(t) \times \mathbf{r}(t)\boldsymbol{\omega}(0) \times \mathbf{r}(0)\rangle}{\langle(\boldsymbol{\omega} \times (\boldsymbol{\omega} \times \mathbf{r}))^2\rangle^{1/2}\langle(\boldsymbol{\omega} \times \mathbf{r})^2\rangle^{1/2}} \right]_{(x,y,z)} \tag{285}$$

where there is little change in amplitude or time dependence as a result of applying the field.

Another pattern of symmetry changes is produced in the ccf matrix

$$C_{\omega v} = \left[\frac{\langle \boldsymbol{\omega}(t) \times \mathbf{v}(t)\mathbf{v}(0)\rangle}{\langle(\boldsymbol{\omega} \times \mathbf{v})^2\rangle^{1/2}\langle v^2\rangle^{1/2}} \right]_{(x,y,z)} \tag{286}$$

between the molecular Coriolis acceleration and linear velocity in frame (x, y, z). The overall symmetry effect appears to be

$$\begin{bmatrix} + & 0 & 0 \\ 0 & + & 0 \\ 0 & 0 & + \end{bmatrix} \rightarrow \begin{bmatrix} + & 0 & + \\ 0 & + & + \\ + & + & + \end{bmatrix}_{(x, y, z)} \tag{287}$$

The amplitude of the diagonal elements is increased to about twice the value in the field-free sample, and there appear to be extra off-diagonal elements, which however may be noise, because they are symmetry-disallowed.

Off-diagonal elements are produced by the field of acf's in the laboratory frame (X, Y, Z), whose symmetry is $D_g^{(2)}$. These off-diagonal elements vanish in the field-free isotropic liquid and in consequence there is scope for analysing external flow in terms of field-induced ccf's of this type, using the laboratory frame directly.

P.   Other Types of Applied External Field

The examples so far have shown how the application of external, symmetry-breaking, force fields produces new ccf's in frame (X, Y, Z). The discussion has been restricted to a few examples, and in general there are many types of field which can be considered:

1. A magnetic field has the symmetry breaking properties of an applied axial vector, such as the circularly polarized electromagnetic field, rotating electric field, or rotational flow, and would be expected to induce similar dynamical effects in both frames of reference. In this case there is an interaction via the magnetic field–dipole torque and the Lorentz force on each dipole charge. Linear and nonlinear magnetic field effects can be investigated by molecular dynamics computer simulation of a liquid subjected to a magnetic field.

2. Nonlinear effects may be coded by considering the torque generated by an induced dipole moment and terms to higher order in the electric field strength **E** and magnetic field strength **B** of the electromagnetic field. The same equations may be used within the context of classical physics to investigate the effect of neutron radiation, provided that account is taken of the greater transfer of momentum between neutron and molecule. These considerations would, however, involve tensor traces such as that of the molecular polarizability, hyperpolarizability, and so forth, which arc usually unknown experimentally to any precision, except for some limited number of small molecules. The anisotropy of the molecular polarizability appears in the conventional theory of the Kerr effect, and this can also be investigated by computer simulation provided some way is found of mimicking the anisotropy with pair potentials. This is not, however, a straightforward procedure because polarizability is not pairwise additive.

3. Gravitational fields are already used successfully in computer simulations of flow past a disk in two dimensions or a cylinder in three. These produce all the characteristics of macroscopic flow phenomena as derived from continuum hydrodynamics, effects which are always accompanied by new symmetries in ccf tensors at the fundamental molecular level. Two major branches of physics are related in this way, hydrodynamics and classical molecular dynamics.

4. There is a large and growing area of physics devoted to the molecular dynamics simulation of transport phenomena using new types of algorithm.

5. In chiral media the effects discussed above are supplemented by the

possibility of investigating the three symmetries represented by (R), (S), and (RS).

Q. Removal of H-Bonding, the Computer Simulation[270] of Hydrogen Selenide

The dynamics of hydrogen selenide can be investigated in the liquid state using ccf tensors. The dynamics of hydrogen bonding can be described in liquid water in this way, using an approach which is intrinsically much simpler than conventional theories of hydrogen bonding. A complete picture of an H-bonded entity such as water can be built up using a few members of the complete set of ccf's, each one of which is an individual signature of the complete dynamical process. The available analytical theory can be made to attempt to follow the wealth of data from a single computer simulation, but at the price of over-parameterization and overwhelming mathematical complexity. The greatest single drawback of the analytical theory is that it is not predictive, whereas the computer simulation can produce the time dependence of correlation functions from a model of the pair potential that is much more realistic than its equivalent in diffusion theory. As an illustration of the use of computer simulation to investigate the effective removal of H-bonding from the dynamical environment of a water-like asymmetric top, we describe results from the recent computer simulation[270] of hydrogen selenide, a C_{2v} asymmetric top, H_2Se.

Hydrogen selenide has one moment of inertia much smaller than the other two, and the subsequent anisotropy of the molecular diffusion is greater than that in water. Diffusion theory therefore requires the solution of the Euler–Langevin equations and three rotational friction coefficients, together with the translational Langevin equations in frame (x, y, z). This is mathematically intractable and no further progress can be made without approximation. From basic considerations, however, we know that the selenium atom is almost two orders of magnitude more massive than the two hydrogen atoms, causing pronounced anisotropy of both rotational and translational dynamics. Computer simulation can be used to investigate this in great detail, and also to follow the diffusional dynamics from liquid to compressed gas. This can be achieved by use of the atom–atom Lennard–Jones potential,[270] from which several ccf's are obtainable using good-quality runs over 6000 time steps. The molecular dynamics were simulated in Ref. 270 at several state points and Fig. 45 illustrates a direct laboratory frame ccf, $\langle \mathbf{v}(t)\boldsymbol{\mu}(0)\rangle$, as a function of molar volume. As this is decreased from $180 \, cm^3/mole$ in the compressed gas to $40 \, cm^3/mole$ in the liquid, the amplitude increases by a factor of about 10. The amplitude is plotted against molar volume in Fig. 46, and it is clear that at high

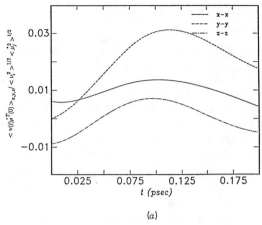

(a)

Figure 45. (a).

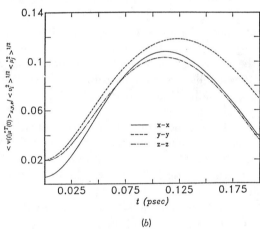

(b)

Figure 45. (b).

number densities there is considerable cross correlation. Figure 46 is a plot of the height of the first peak against molar volume and the transition from compressed gas to liquid is marked with a dashed line. As the line is crossed there is a discontinuity in the dependence of the height of the first peak on the molar volume, that is, there is a pronounced change of slope. The phase transition is measurable through the fact that the total energy of the system changes from positive (compressed gas) to negative (liquid). At constant temperature the change of slope becomes apparent at a molar volume of $150 \text{ cm}^3/\text{mole}$, which is the point at which the total energy of the system becomes negative. In the compressed gas the total

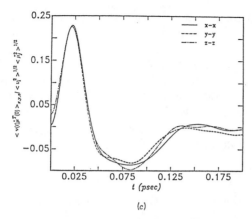

(c)

Figure 45. Ccf between linear and rotational velocity in hydrogen selenide in frame (X, Y, Z) as a function of molar volume at 296 K: (*a*) 180 cm³/mole, compressed gas; (*b*) 40 cm³/mole; (*c*) 20 cm³/mole.

energy is positive, and in the liquid, negative until about 20 cm³/mole is reached, when it once more becomes positive due to the high external pressures (up to 250 kbar) needed to sustain the sample at 20.0 cm³/mole at 243 K. In this state a further phase change has occurred from liquid to amorphous solid. Cross correlation between the linear and rotational velocity is therefore sensitive to molar volume, and to the mean intermolecular separation. At the phase change between the compressed gas and liquid there is a discontinuity in the height of the first peak as a function of molar volume which coincides with the point at which the total energy becomes positive from negative.

The overall pattern is similar for the ccf between orientation (the dipole vector) and linear velocity. The intensity once more increases as a function of pressure.

In the frame (x, y, z) the simple ccf $\langle \mathbf{v}(t)\boldsymbol{\omega}(0) \rangle$ has two nonvanishing elements whose amplitude depends on the molar volume. The fundamen-

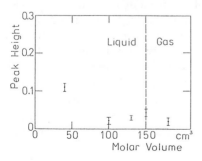

Figure 46. Amplitude of the first peak of the ccf in Fig. 45 as a function of molar volume.

tal molecular dynamics of hydrogen selenide are heavily influenced by cross correlation, which becomes stronger with decreasing molar volume.

The lack of hydrogen bonding in hydrogen selenide is revealed most clearly by higher-order ccf's in frame (x, y, z), whose diagonal elements are anisotropic because the diffusional characteristics are primarily determined by the mass distribution within the molecule itself, and not from intermolecular hydrogen bonding as in water. An extreme example of such anisotropy is methanol,[272] where one diagonal ccf is a thousand times greater in amplitude than the other two.

R. Some Consequences for Quantum Mechanics

For matrix-isolated small molecules such as HD, the rotational spectra generated from the Schrodinger equation are perturbed by mixing with quantized translational lines which appear from the periodicity imposed by the solvent cage. Quantized rotation–translation features occur in the final spectrum. There are shifts in rotational levels, broadening, and disturbance of the rotational selection rules. The existence of the set of classical ccf's in frames (X, Y, Z) and (x, y, z) implies new types of fine and hyperfine structure in the electromagnetic absorption, dispersion, and scattering spectra of liquids and gases, or lattices such as HD in solvents such as argon. These new features could lead to specific and observable information on the nature of rotation to translation to vibration coupling in the liquid state of matter.

The set of nonvanishing ccf's also exists in the compressed and dilute gas states. As dilution proceeds the classical broad-band features typical of a moderately compressed gas give way to quantum absorptions. Under certain conditions, solutions of dipolar molecules in rare gas matrices at low temperatures, the infrared and Raman spectra are quantized. The existence of classical ccf's means that in the dilute gas the translational and rotational wave functions will be mutually perturbed and split the energy levels. Usually, the eigenvalues of linear momentum from Schrodinger's equation form a continuous spectrum from minus infinity to infinity, but in rare gas lattices translational energy levels may become observable at intervals of about $100 \, \text{cm}^{-1}$. The levels have been reproduced using a particle in the box model due to Friedmann et al.[273] and the rotational diffusion is linked to the translational. This means that the corresponding wave functions and inner products must also exist in frames (X, Y, Z) and (x, y, z). Usually this kind of perturbation in a compressed gas is thought to result merely in broadening, but under favorable circumstances translational energy levels may be observed.

S. Analogies between Quantum and Classical Mechanics

Consider a linearly independent set $\{A_j(t)\}$, $j = 1, \ldots, n$ of real-time dependent variables of a given n particle system. The set of all possible dynamical variables is a Hilbert space. The ensemble thermodynamic average in this space, $\langle \ \rangle$, is an inner product, so that

$$(A, B) = \langle A(0)B(t) \rangle \tag{288}$$

where A and B are separate dynamical variables. The equilibrium canonical distribution function is

$$f(\Gamma_0) = \frac{\exp(-H(\Gamma_0)/kT)}{\int \exp(-H/kT)\, d\Gamma_0} \tag{289}$$

where Γ_0 is the phase space variable and H is the Hamiltonian. With this definition the inner product becomes

$$(A^\times, B) = \int d\Gamma f(\Gamma) B(\Gamma) A^\times(\Gamma) \tag{290}$$

The state variable A obeys the Liouville equation

$$\dot{A} = i\hat{L}A \tag{291}$$

which has a formal solution

$$A(\Gamma, t) = \exp(i\hat{L}t)A(\Gamma, 0) \tag{292}$$

providing a definition of the time acf

$$C(t) = (e^{i\hat{L}t}A, A^\times) \tag{293}$$

The scalar product used in quantum mechanics and the time-correlation functions are formally equivalent. More generally, two variables A and B can be used to construct the time ccf

$$C(t) = \int d\Gamma f(\Gamma) A^\times(\Gamma) e^{i\hat{L}t} B(\Gamma) \tag{294}$$

In analogy with quantum mechanics the two variables A and B are orthogonal if $(A^\times, B) = 0$. The propagator e^{Lt} is an orthogonal operator, and by analogy

$$f^{1/2} \equiv \psi_A(\Gamma) \qquad (295)$$

is a wavefunction. The time correlation function $C(t)$ is the expectation value of the propagator in the state defined by the wavefunction. The set of nonvanishing ccf's in classical mechanics implies the existence of wave functions in quantum mechanics.[274]

Friedmann and Kimel[273] have considered the far infrared spectrum of HD in an argon lattice, and have accounted for the quantum mechanical mixing of rotational and translational wave functions in great detail. The effect of the correlation between wave functions is to give zero-point translational energy to the rotational quantum states and to raise the rotational energy levels. The cross correlation doubles the rotational band widths of HD with respect to those of pure hydrogen or deuterium in argon lattices, where the spectrum is Raman active. The rotational selection rules are changed considerably as a result of perturbation from the translational motion. Similar effects are also observed in other small molecules such as water trapped in clathrate cages.

There are quantum mechanical equivalents of the noninertial linear velocities and accelerations, and the resulting spectral effects should be observable directly under the right conditions. In analogy with the well-known Stark and Zeeman effects this fine structure would be split into substructure, signalling the appearance of new wave functions due to loss of isotropy in frame $(X\ Y, Z)$. Extra information would also be available for chiral molecules.

T. Rod-like Molecules and Liquid Crystals

Liquid crystals can be made anisotropic[1] by the application of relatively weak external electric and magnetic fields, and these effects are always accompanied by the induction of a new set of nonvanishing ccf's. Several long-rod diffusion theories are available,[275,276] which have recently been tested against new simulation data for the long rod methyl hexa tri yne[277] using 6000 configurations and a site–site pair potential. The Doi–Edwards theory[275] is relevant for dilute gas long rod ensembles, and was tested against a simulation[277] in the gas phase at 150 cm³/mole. The computed pressure in this condition was 100 ± 200 bar. The usual range of acf's and ccf's was supplemented by the type

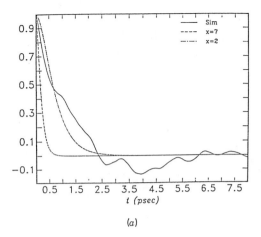

(a)

Figure 47. (a).

$$C_{\text{rod}} = \frac{\langle \mathbf{v}(t) \cdot \mathbf{e}_1(0) \mathbf{v}(0) \cdot \mathbf{e}_1(0) \rangle}{\langle (\mathbf{v}(0) \cdot \mathbf{e}_1(0))^2 \rangle} \tag{296}$$

used by Frenkel and Maguire[276] in their theory of hard-rod diffusion. Here $\mathbf{e}_1$ is the initial orientation vector of each molecule in axis I of the dipole moment. A closed solution for this acf is available for direct comparison with simulation.

A second state point was used in the liquid at 50 K and 130 cm³/mole. Some alignment was observed in the liquid which appears to foreshadow the development of a nematogen. With a small sample of 108 molecules, however, it is not meaningful to see the alignment as characteristic of the nematic phase because swarms in the latter are greater by far than the box size used in the simulation. These swarms define the director axis, which is aligned by an external electric or magnetic field.

Figure 47 is a comparison of C_{rod} from the simulation and the equivalent from the Frenkel–Maguire theory for different x in

$$C_{\text{rod}} = 1/\cosh(xt) \tag{297}$$

The theory fails to reproduce the major features of the simulation, for example, the long negative tail and superimposed oscillations. The root cause of this seems to be that the approach seems to rely on approximation at far too early a stage in the hierarchy of memory functions.

The Doi–Edwards theory is basically a conventional Debye type rotational diffusion approach with the inertial term missing. It is unable to

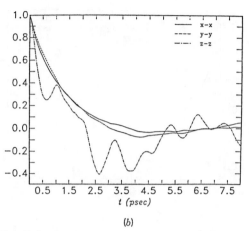

(b)

Figure 47. (a) Comparison of C_{rod} from the computer simulation with two results from the Frenkel Maguire theory of hard rod diffusion for different x (see text). Compressed gas. (b) Individual computer simulations of elements of C_{rod} in frame (X, Y, Z). Compressed gas.

describe the set of nonvanishing ccf's in frames (x, y, z) and (X, Y, Z) and has no means of giving even a qualitative description of the statistical correlation between rotation and translation.

U. Survey of Results and Future Progress

Progress away from the empirical and descriptive approach to the phenomenon of molecular diffusion can be made with an appreciation of the set of nonvanishing ccf's. We have attempted to show in the first six sections of this chapter that the traditional approach is limited in comparison with simulation and data now available. In developing the theory further, symmetry is a fundamental consideration, and the principles of group theoretical statistical mechanics are developed and applied in the next two sections.

VII. GROUP THEORY AND STATISTICAL MECHANICS

At the end of the 19th century, Neumann,[278] in a treatise on the luminiferous ether, recognized that there is a link between the symmetry of properties on the the macroscopic scale and the fundamental. The latter was not accepted universally at that time to be molecular in nature. This link is sometimes known[279] as Neumann's Principle, but more often as Curie's Principle.[280] These principles were recognized well before the emergence of Einstein's molecular explanation for Brownian motion, and preceded the development of correlation functions by more than thirty years. In this section they are extended and applied to molecular dynamics.

We have seen that computer simulation produces a pattern of statistical cross-correlation which is outside the scope of the customary theoretical approach of Section I. Simulation has produced results that the analytical theory had not been able to anticipate. A simple static electric field, for example, induces a statistical cross correlation between the linear and angular velocities of a diffusing molecule. Does the *symmetry* of the electric field vector in the laboratory frame (X, Y, Z) tell us anything about why this particular ccf should appear out of the infinite number possible? We have introduced the frame (x, y, z) fixed in the diffusing molecule, and have seen that the pattern of statistical crosscorrelation is different from that in frame (X, Y, Z). What use can be made of molecular point group theory in this context? What form does the Neumann–Curie Principle take in frame (x, y, z)? Can we devise a form of the principle as a guide to the effect of external force fields in both frames?

Given affirmative answers to these questions we would have constructed guidelines to the symmetry properties[281-290] of molecular diffusion in ensembles, guidelines which can lead to unprecedented experimental insight, new ways of explaining well-established observation, and new experimental techniques.

A.   Point-Group Theory in Frame (X, Y, Z)

In the majority of circumstances use is made of point-group theory[291-295] because the ensemble lacks translational symmetry. (In the molecular crystalline condition we would have to deal with space groups because unit cells repeat themselves periodically.) Molecular liquids at field-free, reversible, thermodynamic equilibrium can be considered as spatially isotropic, properties measured in each axis of frame (X, Y, Z) turn out to be the same. There is no translational symmetry in an isotropic atomic or molecular ensemble, whose symmetry is therefore that of a point group, the three-dimensional rotation–reflection group $R_h(3)$. This realization is the key to the well-developed mathematical theory of point groups and their irreducible symmetry representations.[291-295] In this language, the Neumann–Curie Principle can be written as follows, and becomes the first principle[296-299] of group theoretical statistical mechanics (gtsm).

B.   The First Principle (Neumann–Curie Principle)

The thermodynamic ensemble average $\langle \mathbf{ABC} \ldots \rangle$ exists in the laboratory frame (X, Y, Z) if the product of symmetry representations $\Gamma(\mathbf{A})\Gamma(\mathbf{B})\Gamma(\mathbf{C}) \ldots$ of the molecular quantities $\mathbf{A}, \mathbf{B}, \mathbf{C}, \ldots$ contains at least once the totally symmetric irreducible representation $D_g^{(0)}$ of $R_h(3)$.

Note that this statement links implicitly concepts on the macroscopic scale, characterized by the thermodynamic average $\langle \ \rangle$, with equivalents on the fundamental atomic or molecular level, characterized by the irreducible representations Γ. Both Neumann and Curie achieved this without the help of point-group theory, but for the same reason their ideas could not be built upon very easily in terms of D representations (Section V).

C. D Representation in Frame (X, Y, Z)

The point group $R_h(3)$ is the symmetry group of isotropic three-dimensional space. Its irreducible representations are $D_g^{(0)}, \ldots, D_g^{(n)}$ and $D_u^{(0)}, \ldots, D_u^{(n)}$, respectively. These are even (g) and odd (u) to the parity reversal $(X, Y, Z) \to (-X, -Y, -Z)$. A scalar in this notation has symmetry $D_g^{(0)}$, a pseudo-scalar is $D_u^{(0)}$, a polar vector such as $\mathbf{v}$, the center-of-mass velocity, is $D_u^{(1)}$ and an axial vector such as angular velocity $(\boldsymbol{\omega})$ is $D_g^{(1)}$. The Neumann–Curie Principle implies that in isotropic environments such as those of atomic or molecular ensembles, the symmetry of the ensemble average $\langle \mathbf{A} \rangle$ over the physical property $\mathbf{A}$ will be that of $\mathbf{A}$ itself, either in the laboratory or molecule fixed frames. The symmetry of the ensemble average $\langle \mathbf{AB} \rangle$ is given by the product of representations $\Gamma(\mathbf{A})\Gamma(\mathbf{B})$, where Γ denotes the symmetry representation in the point group of interest. The latter is $R_h(3)$ in frame (X, Y, Z) and the molecular point group in frame (x, y, z). Thus, the average $\langle \mathbf{AB} \rangle$ over the product of molecular physical properties $\mathbf{AB}$ exists in an isotropic molecular ensemble if $\Gamma(\mathbf{A})\Gamma(\mathbf{B})$ contains the totally symmetric representation (tsr) at least once.

In $R_h(3)$ the tsr is $D_g^{(0)}$, and ensemble averages over scalars of this symmetry exist in field-free isotropic equilibrium. Ensemble averages over pseudo-scalars exist in chiral media, and have opposite sign for each enantiomorph. Ensemble averages over both polar and axial vectors vanish at isotropic equilibrium because the symmetry representations of these quantities do not contain the tsr. These considerations can be extended to higher-order tensors and also to products, such as time-correlation functions. Thus, the generic autocorrelation function $\langle \mathbf{A}(0) \cdot \mathbf{A}(t) \rangle$ exists in frame (X, Y, Z) because the product of representations

$$\Gamma(\mathbf{A})\Gamma(\mathbf{A}) = D_g^{(0)} + D_g^{(1)} + D_g^{(2)} \tag{298}$$

contains the tsr once, representing the trace of the matrix product $\langle \mathbf{A}(0)\mathbf{A}(t) \rangle$. In Eq. (298) we have used the Clebsch–Gordan theorem

$$D^{(n)} D^{(m)} = D^{(n+m)} + \cdots + D^{|(n-m)|} \tag{299}$$

to expand the product of symmetry representations on the left-hand side. The expansion on the right-hand side of this equation contains three terms, the scalar, vector, and tensor products, respectively, $D_g^{(0)}$, $D_g^{(1)}$, and $D_g^{(2)}$. The vector product of two vectors is another vector, and can be denoted

Vector product part of $\langle \mathbf{A}(t)\mathbf{A}(0)\rangle = \langle \mathbf{A}(t) \times \mathbf{A}(0)\rangle$

$$= \left\langle \begin{vmatrix} \mathbf{i} & \mathbf{j} & \mathbf{k} \\ A_X(t) & A_Y(t) & A_Z(t) \\ A_X(0) & A_Y(0) & A_Z(0) \end{vmatrix} \right\rangle \quad (300)$$

The overall symmetry of this vector is $D_g^{(1)}$. Here $\mathbf{i}$, $\mathbf{j}$, and $\mathbf{k}$ are unit vectors in the axes X, Y, and Z, respectively. If we examine an individual component of the determinant on the right-hand side of Eq. (300), for example the $\mathbf{j}$-component, we find

$$\{\langle \mathbf{A}(t) \times \mathbf{A}(0)\rangle\}_{\mathbf{j}} = \{\langle A_Z(t)A_X(0)\rangle - \langle A_X(t)A_Z(0)\rangle\}\mathbf{j} \quad (301)$$

with similar results for the other components.

The overall symmetry is however $D_g^{(1)}$, and does not contain the tsr. According to principle 1 these averages all vanish, therefore, at field-free (isotropic) equilibrium, but by principle 3, developed later, may exist in frame (X, Y, Z) in the presence of an external field of the correct symmetry.

Similarly, the tensor product of two vectors is a tensor, a 3×3 matrix, the dyadic product

Tensor part of $\langle \mathbf{A}(t)\mathbf{A}(0)\rangle$

$$= \left\langle \begin{bmatrix} A_X(t)A_X(0)\mathbf{ii}^T & A_X(t)A_Y(0)\mathbf{ij}^T & A_X(t)A_Z(0)\mathbf{ki}^T \\ A_Y(t)A_X(0)\mathbf{ji}^T & A_Y(t)A_Y(0)\mathbf{jj}^T & A_Y(t)A_Z(0)\mathbf{kj}^T \\ A_Z(t)A_X(0)\mathbf{ki}^T & A_Z(t)A_y(0)\mathbf{kj}^T & A_Z(t)A_Z(0)\mathbf{kk}^T \end{bmatrix} \right\rangle \quad (302)$$

where

$$\mathbf{ij}^T = \begin{array}{l} 0 \text{ if } \mathbf{i} = \mathbf{j}; \\ 1 \text{ if } i \neq j \end{array} \quad (303)$$

and so on. The overall symmetry in Eq. (302) is the symmetric second rank $D_g^{(2)}$ and does not contain the tsr, so that these ensemble averages all vanish at field-free equilibrium. The matrix is symmetric (D rank 2) so that there is index symmetry in the off-diagonal elements.

D. Point-Group Theory in Frame (x, y, z)

In this frame the *molecular* point group is the starting point for the symmetry analysis. Principle 1 can be adapted for use in frame (x, y, z) and becomes the second principle of gtsm.

1. *Principle 2*

The thermodynamic ensemble average $\langle \mathbf{ABC} \ldots \rangle$ exists in frame (x, y, z) of the molecular point group if the product of symmetry representations $\Gamma(\mathbf{A})\Gamma(\mathbf{B})\Gamma(\mathbf{C}) \ldots$ in this point group contains at least once the totally symmetric irreducible representation of the molecular point group itself.

The second principle links quantities on a macroscopic scale to those on the fundamental, but unlike principle 1, the averaging is carried out over properties projected onto frame (x, y, z). In order to use molecular point group theory, the D representations of frame (X, Y, Z) are mapped[291–295] from $R_h(3)$ on to the molecular point group. For each D representation there is an equivalent in the molecular point group. For example, the totally symmetric irreducible representation $D_g^{(0)}$ maps on to the representation of a scalar quantity in frame (x, y, z). A collection of such representations is given in Table II for scalars, pseudoscalars, and axial vectors, and for products of vector representations. This table is further explained later in this chapter.

E. The Effect of Fields—Principle 3

The symmetry effect of external fields may be embodied in the third principle, which is as follows.

1. *Principle 3*

If an external field of force is applied to an atomic or molecule ensemble which subsequently reaches a steady state in the presence of that field, new ensemble averages may be created whose symmetry is that of an applied field.

An applied field of force has its own symmetry signature in a given point group, and in frame (X, Y, Z) this is a D representation. By principle 3 ensemble averages with the same D representation may appear in the N-particle ensemble at the field-on steady state. An electric field, $\mathbf{E}$, for example, is a polar vector of $D_u^{(1)}$ symmetry; a magnetic field $\mathbf{B}$ is an axial vector of $D_g^{(1)}$ symmetry, and a shearing rate of the type $\partial v_X/\partial Z$ has the symmetry[300–303] $D_g^{(0)} + D_g^{(1)} + D_g^{(2)}$.

TABLE II

Symmetry Mapping on Selected Molecular Point Groups

Point Group	Scalar		Vector		$D_g^{(1)}D_g^{(1)} = D_g^{(0)} + D_g^{(1)} + D_g^{(2)}$	$D_u^{(1)}D_u^{(1)} = D_u^{(0)} + D_u^{(1)} + D_u^{(2)}$	No.	No.	Local Symmetry
	$D_g^{(0)}$	$D_u^{(0)}$	$D_g^{(1)}$	$D_u^{(1)}$					
C_1 (1)	A	A	$3A$	$3A$	$9A$	$9A$	9	9	Triclinic
$C_i(S_2)$ ($\bar{1}$)	A_g	A_u	$3A_g$	$3A_u$	$9A_g$	$9A_u$	9	0	Triclinic
C_2 (2)	A	A	$A+2B$	$A+2B$	$5A+5B$	$5A+4B$	5	5	Monoclinic
C_{1h} (m)	A'	A''	$2A'+A''$	$2A'+A''$	$5A'+4A''$	$4A'+5A''$	5	4	Monoclinic
C_{2h} (2/m)	A_g	A_u	A_g+2B_g	A_u+2B_u	$5A_g+4B_g$	$5A_u+4B_u$	5	0	Monoclinic
C_{2v} (2mm)	A_1	A_2	$A_1+B_1+B_2$	$A_1+B_1+B_2$	$3A_1+2B_1+2B_2+2A_2$	$2A_1+3A_2+2B_1+2B_2$	3	3	Orthorhombic
D_2 (232)	A_1	A_1	$B_1+B_2+B_3$	$B_1+B_2+B_3$	$3A_1+2B_1+2B_2+2B_3$	$3A_1+2B_1+2B_2+2B_3$	3	0	Orthorhombic
D_{2h} (mmm)	A_{1g}	A_{1u}	$B_{1g}+B_{2g}+B_{3g}$	$B_{1u}+B_{2u}+B_{3u}$	$3A_{1g}+2B_{1g}+2B_{2g}+2B_{3g}$	$3A_{1u}+2B_{1u}+2B_{2u}+2B_{3u}$	3	0	Orthorhombic
C_3 (3)	A	A	$A+E$	$A+E$	$3A+3E$	$3A+3E$	2	2	Trigonal
C_{3v} (3m)	A_1	A_2	A_2+E	A_1+E	$2A_1+A_2+3E$	A_1+2A_2+3E	2	1	Trigonal
D_3 (32)	A_1	A_1	A_2+E	A_2+E	$2A_1+A_2+3E$	$2A_1+A_2+3E$	2	2	Trigonal
D_{3d} ($\bar{3}m$)	A_{1g}	A_{1u}	$A_{2g}+E_g$	$A_{2u}+E_u$	$2A_{1g}+A_{2g}+3E_g$	$2A_{1u}+A_{2u}+3E_u$	2	0	Trigonal
S_6 ($\bar{3}$)	A_g	A_u	A_g+E_g	A_u+E_u	$3A_g+3E_g$	$3A_u+3E_u$	3	0	Trigonal
C_4 (4)	A	A	$A+E$	$A+E$	$3A+2B+2E$	$3A+2B+2E$	3	3	Tetragonal
C_{4v} (4mm)	A_1	A_2	A_1+E	A_1+E	$2A_1+A_2+B_1+B_2+2E$	$A_1+2A_2+B_1+B_2+2E$	2	1	Tetragonal
C_{4h} (4/m)	A_g	A_u	A_g+E_g	A_u+E_u	$3A_g+2B_g+2E_g$	$3A_u+2B_u+2E_u$	2	0	Tetragonal
D_4 (422)	A_1	A_1	A_2+E	A_2+E	$2A_1+A_2+B_1+B_2+2E$	$2A_1+A_2+B_1+B_2+2E$	2	0	Tetragonal
D_{4h} (4/mmm)	A_{1g}	A_{1u}	$A_{2g}+E_g$	$A_{2u}+E_u$	$2A_{1g}+A_{2g}+B_{1g}+B_{2g}+2E_g$	$2A_{1u}+A_{2u}+B_{1u}+B_{2u}+2E_u$	2	0	Tetragonal
D_{2d} ($\bar{4}2m$)	A_1	B_1	A_2+E	B_2+E	$2A_1+A_2+B_1+B_2+2E$	$A_1+A_2+2B_1+B_2+2E$	2	1	Tetragonal
S_4 ($\bar{4}$)	A	B	$A+E$	$B+E$	$3A+2B+2E$	$2A+3B+2E$	3	2	Tetragonal
C_6 (6)	A	A	$A+E_1$	$A+E_1$	$3A+2E_1+E_2$	$3A+2E_1+E_2$	3	3	Hexagonal
C_{6v} (6mm)	A_1	A_2	A_2+E_1	A_1+E_1	$2A_1+A_2+2E_1+E_2$	$A_1+2A_2+2E_1+E_2$	2	1	Hexagonal
C_{3h} ($\bar{6}$)	A'	A''	$A'+E''$	$A''+E'$	$3A'+2E''+E'$	$3A''+3E''$	3	0	Hexagonal
C_{6h} (6/m)	A_g	A_u	A_g+E_{1g}	A_u+E_{1u}	$3A_g+2E_{1g}+E_{2g}$	$3A_u+2E_{1u}+E_{2u}$	3	0	Hexagonal
D_6 (622)	A_1	A_1	A_2+E_1	A_2+E_1	$2A_1+A_2+2E_1+E_2$	$2A_1+A_2+2E_1+E_2$	2	0	Hexagonal
D_{3h} ($\bar{6}2m$)	A_1'	A_1''	$A_2'+E''$	$A_2''+E'$	$2A_1'+A_2'+E'+2E''$	$2A_1''+A_2''+E'+2E''$	2	0	Hexagonal
D_{6h} (6/mmm)	A_{1g}	A_{1u}	$A_{2g}+E_{1g}$	$A_{2u}+E_{1u}$	$2A_{1g}+A_{2g}+2E_{1g}+E_{2g}$	$2A_{1u}+A_{2u}+2E_{1u}+E_{2u}$	2	0	Hexagonal
T_d ($\bar{4}3m$)	A_1	A_2	T_2	T_2	$A_1+T_1+E+T_2$	$A_2+T_2+E+T_1$	1	0	Cubic
O_h (m3m)	A_{1g}	A_{1u}	T_{1g}	T_{1u}	$A_{1g}+T_{1g}+E_g+T_{2g}$	$A_{1u}+T_{1u}+E_u+T_{2u}$	1	1	Cubic
T (23)	A	A	T	T	$A+E+2T$	$A+E+2T$	1	1	Cubic
T_h (m3)	A_g	A_u	T_g	T_u	$A_g+E_g+2T_g$	$A_u+E_u+2T_u$	1	0	Cubic
O (434)	A_1	A_1	T_1	T_1	$A_1+E+T_1+T_2$	$A_1+E+T_1+T_2$	1	1	Cubic
$C_{\infty v}$	Σ^+	Σ^-	$\Sigma^++\Pi$	$\Sigma^++\Pi$	$2\Sigma^++\Sigma^-+2\Pi+\Delta$	$2\Sigma^-+\Sigma^++2\Pi+\Delta$	2	1	Linear dipolar
C_∞	Σ	Σ	$\Sigma+\Pi$	$\Sigma+\Pi$	$3\Sigma+2\Pi+\Delta$	$3\Sigma+2\Pi+\Delta$	3	3	Linear dipolar chiral
$C_{\infty h}$	Σ_g^+	Σ_u^-	$\Sigma_g^++\Pi_g$	$\Sigma_u^++\Pi_u$	$2\Sigma_g^++\Sigma_g^-+2\Pi_g+\Delta_g$	$2\Sigma_u^-+\Sigma_u^++2\Pi_u+\Delta_u$	2	0	Linear nondipolar
D_∞	Σ_g^+	Σ_u^+	$\Sigma_g^-+\Pi_g$	$\Sigma_u^++\Pi$	$2\Sigma_g^++\Sigma_g^-+2\Pi_g+\Delta_g$	$2\Sigma_u^-+\Sigma_u^++2\Pi_u+\Delta$	2	2	Linear nondipolar chiral

In addition to the ensemble averages that may exist at field-free equilibrium, applied fields set up new ensemble averages in frame (X, Y, Z)
whose symmetry, by principle 3, is that of the field. For example, an
electric field sets up averages of the type $\langle \boldsymbol{\omega}(t) \times \mathbf{v}(0) \rangle$ similar to those of
Eq. (300), but whose symmetry is $D_u^{(1)}$. These have been detected by
computer simulation in the form of component ensemble averages which
have the property

$$E_Y \text{ induces } \langle \omega_X(t)v_Z(0) \rangle = -\langle \omega_Z(t)v_X(0) \rangle \tag{304}$$

and similarly for the X and Z components of the applied field. These are
antisymmetric time ccf's which vanish at field free equilibrium, and have
overall symmetry $D_u^{(1)}$. They were found by computer simulation and are
not considered in the conventional theory of diffusion.

Similarly, the response of an ensemble of atoms to shear is described
by a combination of $D_g^{(1)}$ (vector) symmetry and a tensor symmetry $D_g^{(2)}$.
For a strain rate $\partial v_X/\partial Z$, which is traceless (i.e., in which the $D_g^{(0)}$ component is zero) the induced ensemble average is a weighted combination.

F. Symmetries in Frame (x, y, z)

The D symmetry of any quantity in the laboratory frame (X, Y, Z) may
be expressed in the molecule fixed frame (x, y, z) by mapping irreducible
representations from the $R_h(3)$ point group on to the molecular point
group. Table II gives a selection of symmetry mappings on to 36 of the
molecular point groups, ranging from lowest symmetry (C_1) to the high-
symmetry molecular point groups such as O_h. The significance of such
mappings is summarized in principle 2.

Table II describes the symmetry of a scalar, pseudo-scalar, polar, and
axial vector, and the product $\mathbf{AA}$ of two vectors in each molecular point
group. The latter may be a product either of two polar or of two axial
vectors, and includes the scalar, vector, and tensor components as described in the heading to column six of the table. Column 2 of the table
gives the tsr of each point group, and the other columns record the number
of times the tsr occurs in the quantity being described in each column.
Using principle 2 we can see, for example, that the existence or otherwise
of ensemble averages in frame (x, y, z) of each point group is also determined by the number of occurrences of the tsr. Taking the C_{2v} point
group of water as an example the table, used with principle 2, shows that
the ensemble average $\langle \dot{\mathbf{v}} \rangle_{(x,y,z)}$ may exist in frame (x, y, z) because the
molecular linear acceleration is a polar vector whose symmetry representa-
tion in the point group C_{2v} is given in column 5. This includes the tsr
once, and by principle 2, one scalar component of the ensemble average

of linear acceleration may exist in frame (x, y, z). To find which component, we refer to the character table[291-295] for C_{2v} and find that A_1 refers to the z axis. We conclude that the ensemble average $\langle \dot{v}_z \rangle$ may exist by principle 2 in frame (x, y, z). This finding is overlooked by conventional diffusion theory, and it is important to note that the same average vanishes in frame (X, Y, Z) by principle 1. For a complete picture of statistical processes we need both frames of reference. This can be achieved with combined and systematic use of computer simulation and gtsm.

In other molecular symmetries more than one scalar component of this type may exist in frame (x, y, z), depending on the number of occurrences of the tsr. In the chiral group C_1, for example, there are three occurrences of the tsr both for axial and polar vectors (columns 4 and 5 respectively) and this means that the x, y, and z scalar ensemble averages over both types of vector may exist in frame (x, y, z). In each case each average is independent, and each has a different magnitude. By principle 1 they all vanish, however, in frame (X, Y, Z). In a high-symmetry point group such as T_d, for example that of methane, there are no occurrences of the tsr in columns 4 and 5, and in consequence there can be no surviving scalar components of ensemble averages over vector quantities in frame (x, y, z). In C_{4h} the point group's tsr appears once in column 4, but not in column 5, signifying one component ensemble average over an axial vector such as the angular acceleration in frame (x, y, z). Note that in the chiral point groups C_n and D_n there are always occurrences of the tsr in both columns 4 and 5.

The number of occurrences of the tsr of column 2 in column 6 signifies the number of independent scalar elements that may exist in frame (x, y, z) of the ensemble average $\langle \mathbf{AA} \rangle_{(x,y,z)}$. In the molecular point group of lowest symmetry, C_1, nine scalar components of this ensemble average may exist in frame (x, y, z). For the correlation function $\langle \mathbf{A}(t)\mathbf{A}(0) \rangle_{(x,y,z)}$ each may have a different time dependence, even at field-free equilibrium. In frame (X, Y, Z) in contrast, principle 1 implies that only the autocorrelation function may exist, with one independent time dependence. There is only one occurrence of the tsr in frame (X, Y, Z), but no less than nine in frame (x, y, z) for the same diffusion process. These are the three diagonal and six off-diagonal elements, all with different time dependencies. All may change sign from one enantiomorph to the other.

In point groups of higher symmetry, such as C_{2v}, there are fewer occurrences of the tsr in column 6, indicating the existence of fewer independent elements of the ensemble average. In C_{2v} there are three occurrences of the group's tsr in column 6, and a little group theory is needed to interpret their meaning. The result in column 6 for this point group, as for all point groups, is obtained using products of irreducible representations.[291-295] The product of representations for two polar vectors, for example, is

$$(A_1 + B_1 + B_2)(A_1 + B_1 + B_2) = A_1A_1 + A_1B_1 + A_1B_2 + B_1A_1 + B_1B_1$$
$$+ B_1B_2 + B_2A_1 + B_2B_1 + B_2B_2 \quad (305)$$

from which

$$A_1A_1 = A_1; \qquad B_1B_1 = A_1; \qquad B_2B_2 = A_1 \quad (306)$$

using the rules[291-295] for products of irreducible representations. The character table for C_{2v} finally shows that the products in (306) represent the three ensemble averages

$$\langle A_x(t)A_x(0)\rangle; \qquad \langle A_y(t)A_y(0)\rangle; \qquad \langle A_z(t)A_z(0)\rangle \quad (307)$$

each with a different time dependence in frame (x, y, z). This is verified precisely by computer simulation.[1-4] In the laboratory frame there is only one ensemble average, the autocorrelation function with the symmetry of the tsr $D_g^{(0)}$.

Similarly, we may interpret the significance of the tsr occurrences in the other point groups of Table II as they appear in column 6. In some high-symmetry groups such as T_d there is only one occurrence, signifying the one *independent* element

$$\langle A_x(t)A_x(0)\rangle = \langle A_y(t)A_y(0)\rangle = \langle A_z(t)A_z(0)\rangle \quad (308)$$

In other point groups such as C_{3v} there are two occurrences, signifying

$$\langle A_z(t)A_z(0)\rangle \neq \langle A_y(t)A_y(0)\rangle = \langle A_x(t)A_x(0)\rangle \quad (309)$$

that is, two independent elements. In some of the low-symmetry point groups off-diagonal elements may also exist in addition.

G. External Fields: Symmetry

The effect of external fields on averages such as these is to change their individual time dependencies. The overall symmetry, in contrast to frame (X, Y, Z), remains the same, and is given by Table II. It is essential therefore to investigate the overall N-particle ensemble dynamics in both frames for a complete understanding of the diffusional process.

At field-on equilibrium in the presence of shear, for example, we have asymmetric averages induced in frame (X, Y, Z) (Section VIII) which are part of the overall ensemble average $\langle \mathbf{vv} \rangle$. The latter's symmetry in frame (x, y, z), however, remains the same, and is given in column 6 of Table 2. For each molecular symmetry, therefore, the signature of shear in frame (x, y, z) is different. In frame (X, Y, Z) its symmetry signature is always the same, and is the sum of D symmetries at the head of column 6. We conclude that there is a rich variety of behavior in molecular liquids subjected to couette flow. As usual, symmetry patterns should be investigated in both frames of reference.

Similar considerations apply in frame (x, y, z) for molecular liquids under the effect of an electric or magnetic field. The external field changes the time dependencies of the individual ensemble averages in frame (x, y, z), but leaves the symmetry patterns of Table II unchanged. In frame (X, Y, Z), however, the fields induce new ensemble averages according to principle 3.

H.   The Weissenberg Effect

The Weissenberg effect is observed experimentally[304,305] as the pressure (and therefore flow) imparted to a sheared liquid in an axis perpendicular to that of the applied shear plane. It is important in industrial contexts, because the perpendicular flow is caused by a pressure great enough to cause damage, for example, to rollers in the print industry.

The first explanation for the effect has been reported fully elsewhere[300] and is due to new cross correlation functions between elements of the pressure tensor (or stress tensor) in a sheared N-particle ensemble. These can have asymmetry properties described in Section VIII. The existence of these new elements was anticipated by gtsm and confirmed by simulation.[300] They are apparently unknown to conventional rheology and were first characterised by computer simulation. Their description in conventional terms probably requires new constitutive equations and phenomenological parameters. However, it is clear that the underlying cross-correlation functions can be observed experimentally by using computer simulation to match experimental data on Weissenberg flow.

I.   Chirality[279]

In a chiral medium (Section VI), pseudoscalar quantities change sign between enantiomorphs. Barron[279] has recently provided a new definition of a chiral influence, which must be odd to parity inversion and even to time reversal. In magnetochiral dichroism,[306–309] (which we consider in more depth later in this section), the chiral influence is a combination of a static magnetic field $\mathbf{B}$, $D_g^{(1)}$, and the propagation vector $\boldsymbol{\kappa}$ ($D_u^{(1)}$) of unpolarized or linearly polarized electromagnetic radiation. The symmetry signature of this chiral field is therefore

$$\Gamma(\mathbf{B})\Gamma(\mathbf{\kappa}) = D_u^{(0)} + D_u^{(1)} + D_u^{(2)} \tag{310}$$

being again a sum of scalar, vector, and tensor components of the type discussed already. These components are all negative to parity inversion and even to time reversal. If the applied chiral influence is made up of colinear vectors, only the $D_u^{(0)}$ component is expected, otherwise principle 3 implies the appearance of averages of all three types in general in response to a chiral external force field.

J. Nonequilibrium: New Fluctuation–Dissipation Theorems

The three principles imply the existence of new fluctuation–dissipation theorems, an ideal vehicle for the development of which is the recent theory of Morriss and Evans,[310] valid for arbitrary applied field strength. We refer to this as Morriss–Evans theory. It is an important step forward in our understanding of nonequilibrium molecular dynamics and provides many new insights to the way in which molecular and atomic ensembles respond to applied force fields of arbitrary strength.

The three principles are generalized in turn by this theory to nonlinear and transient (nonequilibrium) processes.

The main result of the theory is the relation

$$\langle B(t)\rangle = \langle B(0)\rangle - \frac{F_e}{kT}\int_0^t \langle B(s)J(0)\rangle\, ds \tag{311}$$

where $B(t)$ is an arbitrary time-dependent phase variable of the molecular (or atomic) ensemble, and where F_e and J are, respectively, the applied force field and dissipative flux, defined by

$$\frac{dH_0}{dt} = -JF_e \tag{312}$$

Here H_0 is the Hamiltonian. The integrand on the right-hand side of Eq. (311) is a *transient time-correlation function*, which plays a major role for nonlinear, nonequilibrium statistical mechanics analogous to the partition function in thermodynamics and equilibrium statistical mechanics. Equation (311) generalizes the Green–Kubo relations[311] and links the nonequilibrium value of a phase variable (the left-hand side) to the integral over a time-correlation function between the dissipative flux $J(0)$ in the equilibrium state and B at a time s after the external field F_e has been turned on. One example of the theorem at work (Section VIII) is the relation

$$\eta(t) = -\frac{\langle P_{xy}(t)\rangle}{\dot{\gamma}} = \frac{V}{kT}\int_0^t \langle P_{xy}(s)\,P_{xy}(0)\rangle\,ds \qquad (313)$$

for the viscosity, which reduces to the well-known Green–Kubo relation in the limit $\dot{\gamma}\to 0$. Here $\dot{\gamma}$ is the strain rate and P_{xy} is the off-diagonal component of the pressure tensor. Equation (313) involves a time-correlation function between $P_{xy}(0)$ from the equilibrium system and $P_{xy}(s)$ *from the perturbed system*, and is valid for arbitrary strain rate $\dot{\gamma}$. The Morriss–Evans theorem thus deals indiscriminately and consistently with linear and nonlinear response, thus removing the need for linear response approximation.

K. Application of the Theorem to Dielectric Relaxation and the Dynamic Kerr Effect

We consider a static electric force field E_Z applied to a dipolar molecular ensemble. It is well known that the field interacts with each molecule through the latter's multipole moments, that is, the dipole, quadrupole, octopole, and so on. Without loss of generality we can consider that part of the interaction between field and molecular dipole moment. This is characterized by the torque $-\boldsymbol{\mu}\times\mathbf{E}$, where $\boldsymbol{\mu}$ is the molecular dipole moment. The energy $\boldsymbol{\mu}\cdot\mathbf{E}$ supplements the system Hamiltonian. Thus, Eq. (311) reads

$$\langle B(t)\rangle = \langle B(0)\rangle - \frac{E_Z}{kT}\int_0^t \langle B(s)\,\dot{\mu}_Z(0)\rangle\,ds \qquad (314)$$

where $\dot{\mu}_Z$ is the time derivative of μ_Z, and is known as the "rotational velocity". Equation (311) thus reads, for example,

$$\langle \dot{\mu}_Z(t)\rangle = -\frac{E_Z}{kT}\int_0^t \langle \dot{\mu}_Z(s)\dot{\mu}_Z(0)\rangle\,ds \qquad (315)$$

In Eq. (314), $B(t)$ is in general *any* phase variable of the N-molecule ensemble, and this equation allows us to take account of the set of nonvanishing ccf's induced by E_Z of arbitrary strength, $E_Z \neq 0$. The following are examples of the new fluctuation–dissipation theorems governing dielectric relaxation and the dynamic Kerr effect.

Setting $B(t)$ to μ_Z, we recover a generalization of the fluctuation dissipation of linear response theory as customarily applied to dielectric relaxation,

$$\langle \mu_Z(t) \rangle = -\frac{E_Z}{kT} \int_0^t \langle \mu_Z(s) \dot{\mu}_Z(0) \rangle \, ds \tag{316}$$

relating the orientational transient $\langle \mu_Z(t) \rangle$ to the nonequilibrium time-correlation function which is the integrand on the right-hand side. Traditional experimental methods use $\mu E_Z \ll kT$, but Eq. (316) shows clearly that for any applied electric field strength the orientational fall transient is dependent on the field strength. Linear response theory equates the time dependence of the fall transient and the equilibrium orientational autocorrelation function. The traditional approach also has no method of explaining why time ccf's, such as that between linear and angular molecular velocity exist for $\mathbf{E} \neq \mathbf{0}$ and vanish *if and only if* $\mathbf{E} = \mathbf{0}$. However, this is easily accommodated by the new theorem using, for example, $B(t) = \mathbf{v}(t)|\boldsymbol{\omega}(t)$, giving the relations

$$\langle \mathbf{v}(t) \rangle = -\frac{E_Z}{kT} \int_0^t \langle \mathbf{v}(s) \cdot \dot{\boldsymbol{\mu}}(0) \rangle \, ds \tag{317}$$

and

$$\langle \boldsymbol{\omega}(t) \rangle = -\frac{E_Z}{kT} \int_0^t \langle \boldsymbol{\omega}(s) \cdot \dot{\boldsymbol{\mu}}(0) \rangle \, ds \tag{318}$$

The fluctuation–dissipation theorem of type (317) shows the presence of a nonvanishing velocity transient due to the fact that the time ccf $\langle \mathbf{v}(t) \cdot \dot{\boldsymbol{\mu}}(0) \rangle$ exists in frame (X, Y, Z), both for $\mathbf{E} \neq \mathbf{0}$ *and* $\mathbf{E} = \mathbf{0}$ from the first principle of group theoretical statistical mechanics. The fluctuation dissipation theorem of Eq. (318) shows the existence of a nonvanishing angular velocity transient due to the fact that the time ccf $\langle \boldsymbol{\omega}(t) \cdot \dot{\boldsymbol{\mu}}(0) \rangle$ exists in frame (X, Y, Z) for $\mathbf{E} \neq \mathbf{0}$ (third principle of group theoretical statistical mechanics), but vanishes in this frame when $\mathbf{E} = \mathbf{0}$ (first principle). The transient ccf's exist for $0 < s < t$ in both cases. We can obtain a third expression merely by multiplying Eqs. (317) and (318), that is,

$$\langle \mathbf{v}(t)\rangle\langle \boldsymbol{\omega}(t)\rangle = \left(\frac{E_Z}{kT}\right)^2 \int_0^t \langle \mathbf{v}(s)\cdot\dot{\boldsymbol{\mu}}(0)\rangle\,ds \int_0^t \langle \boldsymbol{\omega}(s)\cdot\dot{\boldsymbol{\mu}}(0)\rangle\,ds \qquad (319)$$

which involves the product of translational and rotational transients.

L. Experimental Observations

The new fluctuation–dissipation theorems can be investigated by computer simulation for all E_Z, but also provide an opportunity for the experimental observation of transient averages caused by cross-correlation functions. An interesting example is $\langle \mathbf{v}(t)\rangle$. This vanishes both at field-off thermodynamic equilibrium and field-on equilibrium (the steady state) because of time-reversal symmetry. However, it may exist as a transient, and should be observable using conventional apparatus to pulse the molecular ensemble with an applied E_Z. A small net translation should be transiently observable, akin to the well-known phenomenon of dielectrophoresis[4] usually attributed to nonuniformities in the applied electric field, that is, to field gradients. Similarly, Eq. (318) shows the existence of a nonvanishing transient angular velocity, which may be observable by techniques sensitive to molecular angular motion, such as far infrared absorption or nuclear magnetic resonance relaxation. The transient angular velocity is intuitively understandable in terms of a removal of external torque.

M. New Dichroic Effects and Absolute Asymmetric Synthesis

The definition of chirality recently provided by Barron[279] and alluded to already in this section implies that the basic requirement for a combination of two vector force fields to be chiral is that one transform as a polar vector and the other as an axial vector and that both be time even or time odd. Thus, at reversible thermodynamic equilibrium,[312] no combination of a static magnetic and static electric field can be chiral. (Note that such a combination can be chiral if one field is made time-dependent, however.) At thermodynamic equilibrium, no combination of static $\mathbf{B}$ and $\mathbf{E}$ can be effective in absolute asymmetric synthesis,[313,314] the preferential production of an enantiomer from a prochiral reaction mixture in what would otherwise be a racemic product.

True chiral influences cause dichroism, an example being magnetochiral dichroism[279] and spin-chiral dichroism.[315] The former is independent of the direction of polarization of the electromagnetic field vector $\boldsymbol{\kappa}$, but reverses if the latter reverses with respect to $\mathbf{B}$. Principle 3 suggests that there is a molecular dynamical mechanism which always accompanies both effects, and which can be observed conversely by the measurement of shifts in the power-absorption coefficient caused by dichroism, or by the associated birefrigerence. This has the same D symmetry at the depicted on the right-hand side of Eq. (310), a symmetry which is also generated by the ccf $\langle \mathbf{v}(t)\boldsymbol{\omega}(0) \rangle$. This sum of D representations includes pseudoscalar D_u^0, which is invariant to proper rotation in the space of hte complete sample,[279] but which has no directional properties. The pseudoscalar is negative to parity inversion, however, and is the scalar product of an axial and polar vector. Natural optical rotation[279] is a time-even pseudoscalar. The symmetry in frame (X, Y, Z) of the ccf $\langle \mathbf{v}(t)\boldsymbol{\omega}(0) \rangle$ is the sum of the pseudoscalar, a polar vector, negative to parity inversion, and an ungerade second-rank tensor. By principle 3 of gtsm these are all generated by magnetochiral dichroism, which is therefore accompanied by molecular rotation–translation coupling, and is a means of its experimental measurement in the laboratory frame.

Note that the Faraday effect (optical rotation caused by a static magnetic field), does not induce the ccf between linear and angular molecular velocity because the magnetic field has $D_g^{(1)}$ symmetry and cannot, by principle 3, induce effects with ungerade parity-reversal symmetry.

N. Spin-Chiral and Other Dichroic Effects from Principle 3

A true chiral influence is capable of absolute asymmetric synthesis, and it is interesting in consquence to look for other types of true chirality by examining appropriate products of D representations in frame (X, Y, Z).

One of these is spin-chiral dichroism, caused by a combination of κ with angular velocity $\boldsymbol{\Omega}$.[315] The angular velocity may be obtained by spinning an electric field around the direction of the unpolarized laser, for example, with the resulting dichroism detected with a Rayleigh refractometer as suggested by Barron and Vrbancich,[316] the laser beam being sent down the arms of this instrument in opposite directions relative to spinning electric fields in each arm. Another experimental alternative is to send circularly polarized laser radiation in opposite directions in each arm.

By Principle 3, a true chiral influence is accompanied in general by all nine elements of the ccf between linear and angular velocity, the "propeller" function. An estimate of the magnitude of magnetochiral dichroism has been given by Barron and Vrbancich[316] who gave a conservative value of about 10^{-8}. A similar effect is expected for spin-chiral dichroism, and can also be investigated by computer simulation in principle.

O. Symmetry Effects in Liquid Crystals

The principles of group theoretical statistical mechanics can be applied[317] to the molecular dynamics of nematogens and cholesterics with point groups

$$C_{\infty v}, D_{\infty h}, C_{\infty}, \text{ and } D_{\infty}$$

and the effect of alignment with an external static electric field can be treated in terms of new ensemble averages that take the symmetry of the applied field and make swarm averages directly visible in the laboratory frame (X, Y, Z). The symmetry arguments lead to experimentally observable effects and to characteristic ensemble averages of swarm dynamics which should be reproducible numerically by computer simulation.

In a shearing field (Section VIII) which causes a shear strain of the type $\partial v_X / \partial Z$, for example, where v_X is the strain velocity, new time asymmetric cross-correlation functions are observed[300] with the (traceless) D symmetry

$$D_g^{(1)} + D_g^{(2)}$$

The shearing field makes visible directly in the laboratory frame (X, Y, Z) time antisymmetric ensemble averages of $D_g^{(1)}$ symmetry and time-symmetric ensemble averages of $D_g^{(2)}$ symmetry. Examples are the velocity cross-correlation functions

$$\langle v_X(0)v_Z(t)\rangle = -\langle v_X(t)v_Z(0)\rangle \quad (D_g^{(1)})$$
$$\langle v_X(0)v_Z(t)\rangle = \langle v_X(t)v_Z(0)\rangle \quad (D_g^{(2)})$$

(320a)

The new cross-correlation function between orthogonal atomic velocity components seen in the computer simulation[300] of an atomic ensemble subjected to this type of shear is a weighted sum of the equations above, giving the new and unexpected result

$$\langle v_X(0)v_Z(t)\rangle \neq \langle v_X(t)v_Z(0)\rangle$$

(320b)

The effect of this type of shear stress on nematogens and cholesterics is of widespread interest and is explored in this section, which is intended to extend the application of gtsm to the molecular dynamics in liquid crystals. In the unaligned nematogenic phase, for example, the director axis[318–321] forms a frame of reference (x_D, y_D, z_D) with two axes of the frame, say x_D, and y_D, mutually perpendicular to the director axis z_D. The swarm axes form a right-handed frame for consistency of definition when dealing with chiral molecules. The extra time cross correlations set up by the presence of molecular alignment along the director vector are calculated with symmetry arguments, both for electrically dipolar and nondipolar liquid crystal molecules. Both in dipolar and nondipolar nematogens, the number of extra cross correlations help to synchronize the molecular dynamics. In the presence of an aligning field, such as $\mathbf{E}$ or E^2, the director frame becomes virtually coincidental with the laboratory frame (X, Y, Z) and the complete aligned liquid crystal specimen ceases to have the isotropic three-dimensional symmetry $R_h(3)$, taking on $C_{\infty v}$ symmetry, for example, for dipolar molecules in a field $\mathbf{E}$. The extra correlation functions and pair-distribution functions that previously existed only in frame (x_D, y_D, z_D), but vanished in the overall isotropic, unaligned, nematogenic sample, now survive ensemble averaging, and accompany the appearance of macroscopic birefringence.

Similar considerations are developed for cholesterics, where the relevant point groups are chiral. They isolate the set of nonvanishing time cross-correlation functions and radial distribution functions which occur exclusively in cholesterics.

A shearing field makes possible the existence in the laboratory frame of new time asymmetric ensemble averages. Shearing would tend to align the director, as with electric fields, but this time would also allow averages of types $D_g^{(1)}$ (antisymmetric in the indices X and Z, and related to shear-induced vorticity) and $D_g^{(2)}$ (symmetric in the indices X and Z, and related to shear-induced deformation) to exist in frame (X, Y, Z). Here we make the simple ansatz that averages equivalent to these D symmetries in the field-free liquid crystal that exist in the absence of the field only in frame (x_D, y_D, z_D) become visible in frame (X, Y, Z) when the liquid crystal is sheared.

P. Basic Symmetry Arguments in Nematogens and Cholesterics

In the unaligned nematogen, or cholesteric phase,[1] group theoretical arguments can be used in the three frames (X, Y, Z), (x_D, y_D, z_D), and (x, y, z), respectively the laboratory, director, and molecule fixed frames of reference. In the frame of reference (X, Y, Z) of the unaligned sample, the relevant point group is $R_h(3)$ of isotropic three-dimensional space. The irreducible representations are $D_g^{(0)}$ (scalar); $D_g^{(0)}$ (pseudoscalar); $D_u^{(1)}$ (polar vector), and $D_g^{(1)}$ (pseudo or axial vector). Higher-order tensors are designated $D_u^{(2)}$, $D_g^{(2)}$, and so on. The point group of the director frame is $C_{\infty v}$ for a dipolar nematogen and $D_{\infty h}$ for a nondipolar nematogen. The irreducible representations are those of these point groups, whichever is appropriate. The director slowly meanders through the laboratory frame but has these point-group symmetries over a well-defined region of three-dimensional space which is large[1] in comparison with molecular dimensions but small in comparison with the volume occupied by the macroscopic sample. In theory, the director point group may have any symmetry, but in nematogens there is alignment in one axis only (z_D). This feature is absent in isotropic molecular liquids such as water, and in nematogens vanishes at the nematic–isotropic transition temperature. Thus, a nematic phase is distinguished by extra ensemble averages (e.g., dynamic time-correlation functions and static, radial pair-distribution functions) in the director frame of reference. Finally, in the molecule fixed frame (x, y, z) of the point group character tables, the relevant point group and irreducible representations are those of the molecular symmetry itself.

The molecular structure and dynamics of the unaligned nematogen are determined by principle 1 in combination with computer simulation and the theory of diffusion.

The director defines a region of three-dimensional space which is described by the point groups $C_{\infty v}$ and $D_{\infty h}$, respectively, depending on whether the individual molecules are dipolar or nondipolar. This region is referred to conveniently as the "swarm." Thermodynamic ensemble averages may be constructed inside the swarm, bringing principle 2 into operation, but in frame (x_D, y_D, z_D).

1. The $C_{\infty v}$ Swarm

This applies to electrically dipolar nematogen molecules, the vast majority. Mapping[291–295] the irreducible representations of a scalar, vector, and so on from $R_h(3)$ to $C_{\infty v}$ gives

$$
\begin{aligned}
&D_g^{(0)} \rightarrow \Sigma^+; \qquad D_u^{(0)} \rightarrow \Sigma^- \\
&D_g^{(1)} \rightarrow \Sigma^- + \Pi; \qquad D_u^{(1)} \rightarrow \Sigma^+ + \Pi \\
&D_g^{(2)} \rightarrow \Sigma^+ + \Pi + \Delta; \qquad D_u^{(2)} \rightarrow \Sigma^- + \Pi + \Delta \\
&D_g^{(3)} \rightarrow \Sigma^- + \Pi + \Delta + \Phi; \qquad D_u^{(3)} \rightarrow \Sigma^+ + \Pi + \Delta + \Phi
\end{aligned}
\tag{321}
$$

The totally symmetric irreducible representation in the $C_{\infty v}$ swarm is Σ^+. Using principle 2 it can be shown that extra cross-correlation functions exist inside the swarm that vanish in (X, Y, Z). For example, one independent element of $\langle \mathbf{v}(t)\boldsymbol{\omega}(0) \rangle$ exists because the product of representations inside the $C_{\infty v}$ point group

$$
\Gamma(\mathbf{v})\Gamma(\boldsymbol{\omega}) = (\Sigma^+ + \Pi)(\Sigma^- + \Pi) = \Sigma^+ + 2\Sigma^- + 2\Pi + \Delta \tag{322}
$$

contains the tsr Σ^+ once. Using this point group's character table[291–295] shows that this is

$$
\langle v_{x_D}(t)\omega_{y_D}(0) \rangle = -\langle v_{y_D}(t)\omega_{x_D}(0) \rangle \tag{323}
$$

This type of ccf vanishes in frame (X, Y, Z). This is one member of the set of nonvanishing ccf's inside the swarm whose members all vanish in (X, Y, Z). Thus any attempt to describe molecular dynamics inside a swarm with computer simulation must produce the result shown in Eq. (323). The swarm volume is bigger than the molecular dynamics "cube" itself, and spontaneous swarm formation under the right conditions can be measured through the spontaneous appearance of the result.[323] This ccf element will vanish for all t if the swarm disappears for some reason, for example, at the nematic (or cholesteric) to isotropic phase change. This gives a simple test for the occurrence of a liquid crystal in computer simulation.

There are many other differences between swarm dynamics and those in (X, Y, Z). For example, time-autocorrelation functions (acf's) become anisotropic

$$\langle \mu_{z_D}(t)\mu_{z_D}(0)\rangle \neq \langle \mu_{x_D}(t)\mu_{x_D}(0)\rangle = \langle \mu_{y_D}(t)\mu_{y_D}(0)\rangle \tag{324}$$

$$\langle \omega_{z_D}(t)\omega_{z_D}(0)\rangle \neq \langle \omega_{x_D}(t)\omega_{x_D}(0)\rangle = \langle \omega_{y_D}(t)\omega_{y_D}(0)\rangle \tag{325}$$

Under some circumstances, thermodynamic averages may exist in the swarm over polar vectors such as $\dot{\mathbf{v}}$ or $\ddot{\boldsymbol{\mu}}$ which have positive time-reversal symmetry. This again can be picked up by computer simulation, and signifies that the swarm can have a net linear acceleration inside the macroscopic unaligned nematogen, measured with respect to the frame (x_D, y_D, z_D). These net accelerations vanish in frame (X, Y, Z).

2. The $D_{\infty h}$ Swarm

This refers to electrically nondipolar molecules which form a nematic phase. An example is hexaphenyl (linear end-to-end arrangement of phenyl groups). In this case the irreducible representations of $R_h(3)$ map on to $D_{\infty h}$ as follows:

$$D_g^{(0)} \rightarrow \Sigma_g^+; \qquad D_u^{(0)} \rightarrow \Sigma_u^-$$

$$D_g^{(1)} \rightarrow \Sigma_g^- + \Pi_g; \qquad D_u^{(1)} \rightarrow \Sigma_u^+ + \Pi_u$$

$$D_g^{(2)} \rightarrow \Sigma_g^+ + \Pi_g + \Delta_g; \qquad D_u^{(2)} \rightarrow \Sigma_u^- + \Pi_u + \Delta_u$$

$$D_g^{(3)} \rightarrow \Sigma_g^- + \Pi_g + \Delta_g + \Phi_g; \qquad D_u^{(3)} \rightarrow \Sigma_u^+ + \Pi_u + \Delta_u + \Phi_u$$

There are fewer new ensemble averages specific to the $D_{\infty h}$ swarm. For example, there can be no ccf's between $\mathbf{v}$ and $\boldsymbol{\omega}$ because the product of symmetry representations

$$\Gamma(v)\Gamma(\omega) = (\Sigma_g^- + \Pi_g)(\Sigma_u^+ + \Pi_u) = \Sigma_u^+ + 2\Sigma_u^- + 2\Pi_u + \Delta_u \quad (326)$$

does not contain the group's tsr Σ_g^+. It is more difficult therefore for the molecular dynamics to be synchronised in a $D_{\infty h}$ swarm, which helps to explain why there are far fewer observed to date. However, the results of Eqs. (324) and (325) are retained in this type of swarm, the relevant products of representations containing in each case two occurrences of the tsr. Computer simulation must be able to pick up this result when attempting to describe the molecular dynamics within a nondipolar swarm, for example, one made up of long rods. In the $D_{\infty h}$ swarm the mean molecular angular acceleration vanishes, but the mean molecular linear center-of-mass acceleration exists. In general, the set of nonvanishing ensemble averages in frame (x_D, y_D, z_D) of this swarm contains fewer members than in the $C_{\infty v}$ swarm.

3.   The $C_{\infty v}$ Swarm

This is found in an unaligned cholesteric phase[1] made up of chiral dipolar molecules. The D representations map onto the following in the point group C_∞

$$D_g^{(0)} \text{ and } D_u^{(0)} \to \Sigma$$

$$D_g^{(1)} \text{ and } D_u^{(1)} \to \Sigma + \Pi$$

$$D_g^{(2)} \text{ and } D_u^{(2)} \to \Sigma + \Pi + \Delta$$

$$D_g^{(3)} \text{ and } D_u^{(3)} \to \Sigma + \Pi + \Delta + \Phi$$

These mappings show that the cholesteric C_∞ swarm has more intrinsic nonvanishing ensemble averages in the frame (x_D, y_D, z_D) than the nematic $C_{\infty v}$ swarm. For example, the product of representations

$$\Gamma(v)\Gamma(\omega) = (\Sigma + \Pi)(\Sigma + \Pi) = 3\Sigma + 2\Pi + \Delta \quad (327)$$

shows that there are three independent elements in the swarm frame of a ccf such as $\langle v(t)\omega(0)\rangle$. Reference to the C_∞ point group character table shows these to be

$$\langle v_{z_D}(0)\, \omega_{z_D}(t)\rangle \neq \langle v_{y_D}(0)\, \omega_{x_D}(t)\rangle = \langle v_{x_D}(0)\omega_{x_D}(t)\rangle \quad (328)$$

$$\langle v_{x_D}(0)\, \omega_{y_D}(t)\rangle = -\langle v_{y_D}(0)\, \omega_{x_D}(t)\rangle$$

This cholesteric swarm also allows the existence of off-diagonal elements of time acf's, the overall symmetry being the same pattern as Eq. (328), two independent diagonal elements and one independent off-diagonal. All ccf elements change sign from one enantiomorph to the other. A computer simulation of a cholesteric swarm would be expected to reproduce these results. At the cholesteric to isotropic phase transition, the set of ccf elements of Eq. (328) all disappear with the director and the swarm frame, and this is another test for computer simulation.

4. The D_∞ Swarm

This is a cholesteric swarm with overall nondipolar symmetry. The relevant mappings are as follows:

$$D_g^{(0)} \text{ and } D_u^{(0)} \to \Sigma^+$$

$$D_g^{(1)} \text{ and } D_u^{(1)} \to \Sigma^- + \Pi$$

$$D_g^{(2)} \text{ and } D_u^{(2)} \to \Sigma^+ + \Pi + \Delta$$

$$D_g^{(3)} \text{ and } D_u^{(3)} \to \Sigma^- + \Pi + \Delta + \Phi$$

again providing the opportunity for several new time ccf patterns to develop in the swarm.

Extra time cross correlations are set up in frame (x, y, z) which are governed by principle 2 and the point-group symmetry of the molecule itself. The molecules making up nematogens and cholesterics are often of quite low symmetry, thus allowing many more thermodynamic ensemble averages to become visible at the molecular level which are different from those at the intermediate swarm level and which vanish at the macroscopic level. A computer simulation must be able to produce numerical results which reproduce self-consistently all the symmetry-predicted thermodynamic averages in all three frames. Thermodynamically and statistically therefore, the existence of unaligned nematogens and cholesterics is accompanied by the appearance of extra time cross correlations at the swarm and molecular level. Properties specific to such liquid crystals can be traced to these cross correlations, which can therefore be related to these observable properties with computer simulation.

5. The Electrically Aligned Nematogen

In the aligned nematogen, the director axis no longer meanders in three dimensions, but is locked in one axis, say Z, of the laboratory frame (X, Y, Z) by the torque generated between the molecular dipole moment μ and the electric field E_z. On the basis of principle 3, the first-order applied electric field of symmetry $D_u^{(1)}$ sets up new ensemble averages of the same symmetry at field-on equilibrium. One example is the time ccf

$$\langle v_X(t)\,\omega_Y(0)\rangle = -\langle v_Y(t)\,\omega_X(0)\rangle \qquad (329)$$

which becomes directly visible in the laboratory. This was first demonstrated, as we have seen in earlier chapters, for molecular liquids using computer simulation, but has yet to be explored in aligned nematogens and other liquid crystals. The ccf (Eq. 329) represents the $D_u^{(1)}$ part of the tensor $\langle \mathbf{v}(t)\,\omega(0)\rangle$ which has

$$D_u^{(0)} + D_u^{(1)} + D_u^{(2)}$$

symmetry. The $D_u^{(1)}$ is the vector part $\langle \mathbf{v}(t) \times \omega(0)\rangle$.

The effect of the static electric field E_z on a nematogen made up of molecules each with the electric dipole moment μ is to make the frames (x_D, y_D, z_D) and (X, Y, Z) indistinguishable as reference frames for molecular dynamics. This is because the director axis z_D is the laboratory axis Z. The extra ensemble averages that have been shown to exist in, for example, a $C_{\infty v}$ swarm now exist in (X, Y, Z) itself. If the nematogen is only partially aligned, the ensemble averages lose some amplitude but are still visible in (X, Y, Z). Similar conclusions are valid for a second-order field E^2, but in this case the field interacts with the polarizability tensor of the molecule, producing orientation, but not alignment. This means that half the aligned molecules have dipoles in one direction, on average, and half in the other. This is described in D language as a sum of $D_g^{(0)}$- and $D_g^{(2)}$-type ensemble averages in frame (X, Y, Z). Mappings of both these D representations on to the $C_{\infty v}$ point group contain the latter's tsr, as is also the case for the other swarm point groups. Thus a field E^2 makes $D_g^{(2)}$ averages of the swarms directly observable in frame (X, Y, Z). (The $D_g^{(0)}$ represents scalars which are frame invariant, i.e., their thermodynamic averages exist in any frame of reference.) In the aligned or oriented state we denote, for convenience, "$C_{\infty v}$ nematogens," constructed from swarms of the same symmetry, and so on.

6. The $C_{\infty v}$ Nematogen

The mappings given already for this swarm can be utilized directly to find the nonvanishing ensemble averages in frame (X, Y, Z) in the $C_{\infty v}$ nematic. For example, whenever a D representation of group $R_h(3)$ maps onto a $C_{\infty v}$ representation of the aligned nematic that contains the latter group's tsr, extra ensemble averages will appear in frame (X, Y, Z) itself.

Thus for example, the averages previously confined to the swarm and described earlier now exist directly in (X, Y, Z):

$$\langle v_X(t)\, \omega_Y(0)\rangle = -\langle v_Y(t)\, \omega_X(0)\rangle \tag{330}$$

$$\langle \mu_Z(t)\, \mu_Z(0)\rangle \neq \langle \mu_X(t)\, \mu_X(0)\rangle = \langle \mu_Y(t)\, \mu_Y(0)\rangle \tag{331}$$

$$\langle \omega_Z(t)\, \omega_Z(0)\rangle \neq \langle \omega_X(t)\, \omega_X(0)\rangle = \langle \omega_Y(t)\, \omega_Y(0)\rangle \tag{332}$$

Equation (330) signals the existence of direct rotation translation coupling in the aligned nematic which has a considerable effect on the molecular dynamics because of the large degree of alignment possible experimentally, even with a weak electric field, of equivalent energy μE_Z much smaller than the thermal energy kT. So far in computer simulations, alignment has been achieved, and ccf's such as Eq. (330) observed, but only through the use of field energies μE_Z about the same as kT. This is because the simulations dealt with isotropic molecular liquids such as water. No complete computer simulation of an aligned nematogen is yet available.

Equations (331) and (332) show that the sample is anisotropic in the laboratory frame under the influence of an aligning first-order electric field E_z. It is therefore birefringent. Our symmetry arguments have therefore consistently reproduced a well-known feature of nemotogenic behavior, especially noticeable in dielectric loss and dispersion. The dielectric complex permittivity is essentially[1] the Fourier transform of Eq. (331) and is by symmetry and gtsm measurably different in Z, being identical in X and Y. The complex permittivity is the square of the complex refractive index, implying birefringence as observed experimentally. We have shown that this is describable in terms of the point group $C_{\infty v}$ and its irreducible representation.

One interesting implication of symmetry is that the ensemble averages over time-even polar vectors, such as $\langle \dot{\mathbf{v}}\rangle$ or $\langle \ddot{\boldsymbol{\mu}}\rangle$ include the tsr of $C_{\infty v}$ and must have the possibility of existence in the aligned nematogen of this symmetry. In this case the $D_u^{(1)}$ representation implies a nonvanishing Σ^+, that is, a nonvanishing linear acceleration or force in the Z axis of the aligned nematogen. The applied E_Z cannot generate the mean acceleration directly, but only through a torque $-\boldsymbol{\mu} \times E_Z$ on each dipolar molecule. The net linear acceleration must come from correlation between molecular rotation and translation, and is an experimental measure of this phenomenon. The net linear acceleration of the aligned sample would result in such effects as a meniscus at the surface of the nematogen as the sample is forced against the electric-field-generating apparatus, normally an electrode. This meniscus should be measurable with a microscope.

Similarly, gtsm predicts the possibility of nonvanishing rotational acceleration, the second time derivative of the dipole moment, implying

$$\langle \ddot{\mu}_Z(t)\,\ddot{\mu}_Z(0)\rangle \rightarrow \langle \ddot{\mu}_Z\rangle^2; \qquad t \rightarrow \infty \tag{333}$$

$$\langle \mu_Z(t)\,\mu_Z(0)\rangle \rightarrow \langle \mu_Z\rangle^2; \qquad t \rightarrow \infty \tag{334}$$

Equation (334) implies the existence of a zero-frequency (infinite time) component of the dielectric spectrum, as observed[1] experimentally in the Z axis of the aligned nematogen. This is the static permittivity component in axis Z, which becomes different from those in X and Y. A similar effect is expected on the fourth spectral moment,[1] which is the Fourier transform, essentially speaking, of Eq. (333).

7. The $D_{\infty h}$ Nematogen

In this case the molecules of the swarm are nondipolar, but if they are polarizable new thermodynamic averages of total symmetry $\Gamma(E^2)$ appear in frame (X, Y, Z) by principle 3. The sample is oriented along the axis of the applied field, for example the Z axis. In D language,

$$\Gamma(E^2) = D_g^{(0)} + D_g^{(2)} \tag{335}$$

More generally

$$\Gamma(\mathbf{EE}) = D_g^{(0)} + D_g^{(1)} + D_g^{(2)} \tag{336}$$

but the notation E^2 implies $\mathbf{E}$ parallel to itself, so that $\mathbf{E} \times \mathbf{E}$ vanishes and by principle 3 no $D_g^{(1)}$ averages appear in (X, Y, Z). This leaves the other two D terms on the right-hand side of Eq. (336), representing averages of the second Legendre polynomial type

$$\tfrac{1}{2}\langle 3(\mathbf{A}(t) \cdot \mathbf{B}(0))^2\rangle - 1 \tag{337}$$

and even-order Langevin functions.[1] In Eq. (336) $\mathbf{A}$ and $\mathbf{B}$ must have the same parity-reversal symmetry.

Averages of this kind that had previously been visible only in the swarm frame become visible under the effect of the field in the laboratory frame. In this case they must all be even to parity reversal symmetry. This time there is no net linear acceleration, therefore, and no net dipole $\langle \mu_Z\rangle$.

Note that the sum (Eq. 335) maps on to

$$2\Sigma_g^+ + \Pi_g + \Delta_g$$

of $D_{\infty h}$. This implies two independent occurrences of this point group's tsr, and two independent types of thermodynamic average in the sample oriented by E^2. This means that the time dependence of the ZZ component of averages such as Eq. (337) is different from those of the YY and XX components, which are the same. In other words, the sample is anisotropic and birefringent, and supports even-order Langevin functions.

8. The C_∞ Cholesteric

In this cholesteric symmetry, alignment with a first-order electric field produces a unidirectional spiral symmetry along the axis of the applied field, say Z. Ensemble averages that map on to symmetry representations in the C_∞ point group that contain the group's tsr at least once now survive in the frame (X, Y, Z). In this point group the tsr appears in the symmetry representations both of polar and axial vectors. If a vector quantity is also even to time-reversal symmetry, its thermodynamic ensemble average might not vanish. The average switches sign from one enantiomer to the other, and must vanish by symmetry only in the racemic mixture. This produces the possibility of observing in the electrically aligned cholesteric net molecular linear acceleration, molecular angular acceleration, and a net molecular rotational acceleration which vanish in the racemic mixture.

9. The D_∞ Cholesteric

In this case similar considerations apply to orientation by an E^2 field, as in the case of $D_{\infty h}$. The second-order field induces averages of type in Eq. (337), but this time the vectors **A** and **B** can have different parity-reversal symmetry.

Q. The Effect of a Shearing Field

The symmetry of a shearing field (Section VIII) consists of a vector component $D_g^{(1)}$ and a tensor component $D_g^{(2)}$. A shear of the type $\partial v_X/\partial Z$ produces by computer simulation[300] velocity ccf's between orthogonal molecular cartesian components X and Z.

These D symmetries of the laboratory frame map differently on to the four swarm point groups considered in this Chapter

$$D_g^{(1)} + D_g^{(2)} \rightarrow \begin{array}{l} \Sigma^+ + \Sigma^- + 2\Pi + \Delta \ (C_{\infty v}) \\ \Sigma_g^+ + \Sigma_g^- + 2\Pi_g + \Delta_g \ (D_{\infty h}) \\ 2\Sigma + 2\Pi + \Delta \ (C_\infty) \\ \Sigma^+ + \Sigma^- + 2\Pi + \Delta \ (D_\infty) \end{array}$$

and there are two independent occurrences in the C_∞ group, one in each of the others. This distinguishes the symmetry effect of shear on a dipolar cholesteric from the other types considered here. Only in the dipolar cholesteric is the shear-induced vorticity transferred to the point group of the swarm. In the other point groups the tsr does not occur in the relevant representation of $D_g^{(1)}$. This will help decide whether shearing an isotropic phase of a liquid crystal induces a phase transition into the nematic or cholesteric phase.

R. Computer Simulation — A Specific Application

One specific application of the symmetry theory is to test for the appearance of predicted averages in a computer simulation of long rods, or ellipsoids, of $D_{\infty h}$ symmetry or $C_{\infty v}$ symmetry. The predicted averages will be immediately useful in deciding whether a transition from an isotropic to a liquid crystal phase has indeed occurred. Long rods are $D_{\infty h}$ symmetry but the addition of a dipolar term in the potential, with charges and asymmetric mass distribution, produces $C_{\infty v}$ symmetry. Liquid crystals and long-rod polymers are known to be highly non-Newtonian in response to shear and work is in progress to simulate linear molecules with applied shearing fields with SLLOD and PUT[300] (Section VIII) and arbitrary applied field strength. This will be a stringent test of the symmetry expectations.

S. Symmetry in Smectic Liquid Crystals

Group theoretical statistical mechanics can be applied to determine the number of nonvanishing ensemble averages in the point groups of the smectic liquid crystals, modelled on the 32 possible crystallographic point groups supplemented by the four linear symmetries (Table II). Assuming that the thermodynamic average exists according to the number of irreducible representations in each point group that are totally symmetric, it is possible to conclude whether or not that average exists at thermodynamic equilibrium. The conclusion is valid within the point-group symmetry of the smectic liquid crystal. The number of ensemble averages supported by the smectic point group, exemplified by the time-correlation functions, decreases from triclinic to monoclinic to orthorhombic to trigonal to tetragonal to hexagonal to cubic. Within each of these major classifications the pattern of nonvanishing correlation functions and other ensemble averages has its own distinctive signature, based on point-group theory. By choosing the correlation functions that are known Fourier transforms of spectra, this type of analysis leads to a convenient method of determining how spectra are affected by the type of smectic point-group symmetry. The analysis leads to an appreciation of the differences in allowed ensemble averages between the various smectic point groups and molecular and other types of liquid. The treatment can be extended straightforwardly to consider the effects of external fields.

1. *Local Smectic Point Groups*

In the smectic liquid crystals, the relevant point-group symmetries are similar to those of the 32 point groups of solid molecular crystals, described by the following seven major classifications: triclinic, monoclinic, orthorhombic, trigonal, tetragonal, hexagonal, and cubic. Each of these major classifications supports a number of point-group symmetries which cover all known crystal symmetries. Molecules crystalize within these point groups, forming an underlying structure classified by 230 space groups. However, the external symmetry and physical appearance of a molecular crystal falls into one of the point groups. Similarly, the physical properties of a smectic liquid crystal may be explored with a point-group description of the local structure which distinguishes it from an isotropic molecular liquid. In principle the point-group symmetry within the smectic phase may be any symmetry, but in order to construct a systematic approach to the problem of smectic local ordering, it is assumed that the smectic symmetry can be described by the 32 crystallographic point groups supplemented by the four point groups of the nematic and cholesteric liquid crystals. We therefore provide symmetry data for the 36 point groups of Table II, any one of which may provide a framework for the application of gtsm and the determination of nonvanishing ensemble averages of the molecular dynamics within the local smectic point-group symmetry. Many such averages exist at the local level, but disappear if the smectic sample is isotropic on the macroscopic scale of a laboratory sample. If the smectic liquid crystal maintains the point-group anisotropy at this level, and is anisotropic overall, then the ensemble averages survive averaging in the laboratory frame (X, Y, Z) as well as in the local frame of the point group, which we denote (x_D, y_D, z_D) as for the nematics and cholesterics.

The differences between solid molecular crystals and liquid crystals with three-dimensional anisotropy is not easy to define. De Vries[322] has pointed out that the most significant difference is that in the solid crystal any alkyl chains at the ends of the molecule have very little disorder, or none at all, whereas in a liquid crystal, which has three-dimensional anisotropy, these chains are slightly disordered. The difference in local point-group symmetry seems to be minimal. For example, the smectic H phase of BBEA (4-*n*-butyloxybenzal-4-ethylaniline) is a liquid and not a solid molecular crystal because there is no transmission of phonon modes through the smectic liquid crystal, which is capable of flow. Translational coupling is much weaker than rotational coupling. When the two compete, as in structures with optically active molecules, the translational correlation is destroyed so as to achieve a more favorable rotational arrangement. A molecular crystal structure of an optically active compound has a single three-dimensional lattice, but that of a three-dimensional smectic phase becomes twisted if the molecules are optically active, similar to a cholesteric phase.[1] Three-dimensional anisotropic liquid crystals may be mixed in all proportions, in contrast to molecular crystals. If translational order disappears, there is no three-dimensional order at all.

Smectic liquid crystals and solid molecular crystals therefore differ essentially in translational order. In the smectic liquid crystals there is still a residual point-group symmetry, however, although translational space-group symmetry may have been partially or completely destroyed.

The difference between a smectic liquid crystal and an isotropic molecular liquid can be expressed in terms of extra ensemble averages supported by the former within definable local point-group symmetries. In the isotropic molecular liquid these are absent, its point group in three dimensions is $R_h(3)$.

2. Mapping from $R_h(3)$ to the Smectic Point Groups

Table II may be used to summarize these mappings. Thus, for cubic (O_h) symmetry, the principles of gtsm state that thermodynamic averages exist in frame (x_D, y_D, z_D), governed by this local cubic symmetry, provided that the symmetry representations of the ensemble averages contain at least once the tsr of this cubic point group. This is the A_{1g} irreducible representation. This result is independent of the molecular symmetry within the point group. [Analogously, gtsm states that the ensemble averages in an isotropic environment exist if their symmetry representations contain the tsr $D_g^{(0)}$ of the point group $R_h(3)$, irrespective of the molecular symmetry.]

In general we must define the irreducible representation of the quantity being averaged in the $R_h(3)$ point group and then in the relevant local smectic point group. If, for example, the latter is the tetragonal D_{2d} (or $\bar{4}\,2m$ in the Hermann–Mauguin international notation), the relevant tsr is the irreducible representation A_1. However, gtsm implies the powerful result that any thermodynamic ensemble average within the local smectic point group D_{2d} of the frame (x_D, y_D, z_D) may exist in this frame if its irreducible representation contains A_1. This is a powerful result because it is wholly independent of any residual space group structure in the smectic liquid crystal. Proceeding on these grounds, it is possible to link the symmetry of ensemble averages in the liquid and smectic liquid crystal by mapping the irreducible representation of the quantities being averaged from the point group $R_h(3)$ to D_{2d}. For ease of development it is convenient as usual to divide quantities into scalars, pseudoscalars, polar and axial vectors, and higher-order tensors. A scalar is characterised by the tsr in any point group. The irreducible representation of the pseudoscalar in $R_h(3)$ is, as we have seen in other contexts, $D_u^{(0)}$, which is odd to parity reversal, but still a zeroth order quantity with no directional property. This representation maps on to B_1 of the point group D_{2d}. In chiral point groups: C_n, D_n, T, O the irreducible representations of both the scalar and pseudoscalar map on to the tsr of the point group, so that ensemble averages over both scalars and pseudoscalars can exist in chiral point groups at the local level in smectic liquid crystals. There are two distinct liquid crystal enantiomorphs.

The polar vector (e.g., velocity) is represented in $R_h(3)$ by $D_u^{(1)}$, meaning that it is odd to parity reversal and has first-order directional properties. The axial vector, on the other hand, is $D_g^{(1)}$, which is even to parity reversal. An example of a polar vector is molecular linear velocity, $\mathbf{v}$. Molecular angular velocity is an axial vector. The irreducible representations $D_u^{(1)}$ and $D_g^{(1)}$, respectively, map on to $B_2 + E$ and $A_2 + E$ of the point group D_{2d}. These do not contain the point group's tsr, and in consequence no ensemble average over a polar or axial vector can exist in the frame (x_D, y_D, z_D) of the local smectic point group D_{2d}.

In local smectic point groups where the tsr appears as part of the irreducible representation in the point group of a polar or axial vector quantity, the thermodynamic average in the local frame of reference (x_D, y_D, z_D) may exist, provided that it is positive to time-reversal symmetry. Examples are listed in Table II.

Second rank tensor quantities are characterized in $R_h(3)$ by $D_g^{(2)}$ or $D_u^{(2)}$ and are second-order directional quantities which are even or odd to parity reversal.

Relations between D products are given as usual by the Clebsch–Gordan theorem, which is also applicable in the point groups of the smectic liquid crystals because the sum of D representations maps on to the same sum in the smectic point group. Thus, if we extend our consideration of ensemble averages to time-correlation functions which are averages over products of vectors, we have, for example,

$$\Gamma(\mathbf{v})\Gamma(\boldsymbol{\omega}) = D_u^{(1)} D_g^{(1)} = D_u^{(0)} + D_u^{(1)} + D_u^{(2)} \quad (R_h(3)) \tag{338}$$

$$= (B_2 + E)(A_2 + E) = A_1 + A_2 + 2B_1 + B_2 + 2E \quad (D_{2d}) \tag{339}$$

This shows that the D representations of the time-correlation tensor generated by the tensor product of molecular linear and angular velocity is a sum of three ungerade D representations in the molecular liquid point group $R_h(3)$. The time-correlation function vanishes for all t in the isotropic liquid because its complete D representation does not include $D_g^{(0)}$. In the D_{2d} smectic point group, however, this is not the case, because the irreducible representation of the ccf $\langle \mathbf{v}(t)\boldsymbol{\omega}(0)\rangle$ is the sum $A_1 + A_2 + 2B_1 + B_2 + 2E$, which includes the tsr A_1 once. Therefore, gtsm implies that one element of $\langle \mathbf{v}(t)\boldsymbol{\omega}(0)\rangle$ may exist in the D_{2d} local point group of the smectic liquid crystal. This result may be checked in principle with molecular dynamics computer simulation.

T. Time-Reversal Symmetry

Time-reversal symmetry is defined as in other contexts as the operation $(\mathbf{q}, \mathbf{p}) \rightarrow (\mathbf{q}, -\mathbf{p})$, which leaves positions unchanged but reverses momenta. Parity reversal is the operation $(\mathbf{q}, \mathbf{p}) \rightarrow (-\mathbf{q}, -\mathbf{p})$. When dealing with scalars and pseudoscalars, the ensemble average over these quantities must be unchanged (i.e., positive) to time reversal. When dealing with vectors, however, some are positive to time reversal, like the electric field $(\mathbf{E})$, the position vector $(\mathbf{r})$, the linear acceleration, $(\dot{\mathbf{v}})$, the angular acceleration, $(\dot{\boldsymbol{\omega}})$, and the acceleration due to gravity, $(\mathbf{g})$. Others are negative to time reversal, such as the magnetic field, $(\mathbf{B})$, the linear velocity $(\mathbf{v})$, the angular velocity, $(\boldsymbol{\omega})$, and the electromagnetic field propagation vector $(\boldsymbol{\kappa})$. A thermodynamic ensemble average over a vector which is negative to time reversal vanishes, but one which is positive may exist if the vector is also positive to parity reversal in the isotropic liquid point group $R_h(3)$. In the local smectic phase point groups, however, the latter requirement is not necessary, and the thermodynamic ensemble average exists, provided it is carried out over a vector which is positive to time reversal and provided the irreducible representation of that vector includes the tsr of the point group. Thus, the average $\langle \mathbf{r}\rangle$, for example, may exist in the point group C_{3v} of the trigonal class, because the vector's irreducible representation in the point group is $A_1 + E$. The latter includes the tsr, A_1 once. From gtsm one independent ensemble average exists over a polar vector with positive time-reversal symmetry. An example is $\langle \mathbf{r}\rangle$. Other examples in this local smectic point group are $\langle \dot{\mathbf{v}}\rangle$ and $\langle \dot{\boldsymbol{\omega}}\rangle$. This may again be investigated with computer simulation. In the tetragonal local smectic point group D_{2d}, however, all these ensemble averages should vanish by gtsm.

When dealing with time-correlation functions, the time-reversal arguments must be applied with care to each individual case. The pitfalls of the procedure may be illustrated, as in other contexts, with reference to the simple time ccf of the type $\langle \mathbf{A}(t)\dot{\mathbf{A}}(0)\rangle$. If $\mathbf{A}$ represents linear velocity, $\mathbf{v}$, for example, the product within the averaging brackets $\langle \; \rangle$, is overall negative to time reversal. However, this type of time-correlation function clearly does not vanish for all t because it is simply the time derivative of

$\langle \mathbf{A}(t)\mathbf{A}(0)\rangle$. It exists according to the elementary theory of correlation functions in reversible thermodynamic equilibrium. It is itself a function which is a time derivative, and is intrinsically negative to time reversal. More generally, the class of time correlation functions which are time derivatives of other correlation functions have an existence for $0 < t < \infty$ despite the fact that the product of the two quantities inside the averaging brackets may in itself appear negative to time reversal.

Bearing in mind these considerations of time reversal, mappings of some D representations from the point group $R_h(3)$ are given in Table II for 36 local smectic point groups. Column 1 of the table contains the name of the point group in Schonflies and Hermann–Mauguin notation (the latter in brackets). Column 2 is the representation of the scalar in the point group, (i.e. the tsr), column 3 contains the symmetry representation of the pseudoscalar, column 4 that of the axial vector, column 5 that of the polar vector. Columns 6 and 7 map products of D representations on to each point group. Columns 8 and 9 give the number of independent ensemble averages expected in the local smectic point group for representative time-correlation functions, column 8 for the rotational velocity correlation tensor, and column 9 for the angular linear velocity cross-correlation tensor. Both tensors are defined in frame (x_D, y_D, z_D) of the local smectic point group. Finally, column 10 records the crystal class of the point group if it were being used to describe solid molecular crystals.

Note that in the isotropic molecular liquid environment only one independent ensemble average (the trace) is expected in column 8, and none in column 9, thus evidencing by gtsm a considerable difference between the local molecular dynamics, of smectic liquid crystals and isotropic molecular liquids.

Some examples of the symmetry of the correlation functions in columns 8 and 9 are given as follows:

EXAMPLE 1: MONOCLINIC $C_{1h}(\mathbf{m})$

1. Rotational Velocity Correlation Tensor

The irreducible representation of this tensor in the point group C_{1h} is

$$\Gamma(\dot{\boldsymbol{\mu}}(t)\Gamma(\dot{\boldsymbol{\mu}}(0)) = (2A^1 + A^{11})(2A^1 + A^{11})$$
$$= 4A^1A^1 + 2A^1A^{11} + 2A^{11}A^1 + A^{11}A^{11} = 5A^1 + 4A^{11} \quad (340)$$

showing five occurrences of the tsr in the product of representations of the correlation tensor. Thus five independent ensemble averages may exist in the local smectic frame of reference (x_D, y_D, z_D). The individual products that give A^1 in Eq. (340) are (1) $4A^1A^1$ and (2) $A^{11}A^{11}$. The others give A^{11}. Referring to the point group character table for C_{1h} we find that A^1 represents the cartesian components x_D and y_D of the local smectic frame of reference. The A^{11} entry represents z_D. The product $4A^1A^1$ therefore represents four independent components of the rotational velocity correlation tensor:

$$\langle \dot{\mu}_{x_D}(t)\, \dot{\mu}_{x_D}(0) \rangle; \quad \langle \dot{\mu}_{y_D}(t)\, \dot{\mu}_{y_D}(0) \rangle; \quad \langle \dot{\mu}_{x_D}(t)\, \dot{\mu}_{y_D}(0) \rangle; \quad \langle \dot{\mu}_{y_D}(t)\, \dot{\mu}_{x_D}(0) \rangle$$

The fifth component $\langle \dot{\mu}_{z_D} \dot{\mu}_{z_D}(0) \rangle$ comes from $A^{11}A^{11}$. The complete symmetry of the correlation tensor in the local smectic frame of reference is therefore

$$\langle \dot{\boldsymbol{\mu}}(t)\, \dot{\boldsymbol{\mu}}(0) \rangle = \begin{bmatrix} a & d & 0 \\ e & b & 0 \\ 0 & 0 & c \end{bmatrix}, \qquad C_{1h}(m)$$

and the five independent elements are recorded in the table.

Similar arguments applied to the generic acf $\langle \mathbf{A}(t)\mathbf{A}(0) \rangle$ show these five independent elements, with the same symmetry pattern. In a molecular crystal this result is related to the number of lattice modes, but in the smectic liquid crystal there are no phonon modes.

2. The Cross Correlation Function $\langle \boldsymbol{v}(t)\boldsymbol{\omega}(0) \rangle$

The irreducible representation is now

$$\Gamma(\mathbf{v})\Gamma(\boldsymbol{\omega}) = (2A^1 + A^{11})(A^1 + 2A^{11})$$
$$= 2A^1A^1 + A^{11}A^1 + 4A^1A^{11} + 2A^{11}A^{11} = 4A^1 + 5A^{11} \qquad (341)$$

showing four occurrences of A^1. Thus, the crystal class supports four independent ensemble averages which are scalar elements of $\langle \mathbf{v}(t)\boldsymbol{\omega}(0) \rangle$. Bearing in mind that the linear velocity $\mathbf{v}$ is referred to by the cartesian components X, Y, and Z of the C_{1h} point-group character table, and the angular velocity component ω by R_X, R_Y, and R_Z, we arrive at the symmetry

$$\langle \mathbf{v}(t)\,\boldsymbol{\omega}(0)\rangle = \begin{bmatrix} 0 & 0 & a_1 \\ 0 & 0 & b_1 \\ C_1 & d_1 & 0 \end{bmatrix} \tag{342}$$

Thus four independent elements exist which all vanish in the equivalent molecular liquid.

EXAMPLE 2: ORTHORHOMBIC D_2 (232)

1. Rotational Velocity acf

The relevant irreducible representation of the time acf is

$$\Gamma(\dot{\boldsymbol{\mu}})\Gamma(\dot{\boldsymbol{\mu}}) = (B_1 + B_2 + B_3)\,(B_1 + B_2 + B_3) = 3A_1 + 2B_1 + 2B_2 + 2B_3 \tag{343}$$

which contains the totally symmetric component three times. From the axioms of gtsm we can expect three independent ensembles in the local smectic point group D_2 (232), a chiral class. From the point group character table for D_2, and using the rules for forming the products of irreducible representations, we have the symmetry

$$\langle \dot{\boldsymbol{\mu}}(t)\,\dot{\boldsymbol{\mu}}(0)\rangle = \begin{bmatrix} a & 0 & 0 \\ 0 & b & 0 \\ 0 & 0 & c \end{bmatrix} \quad (D_2)$$

In this case all the off-diagonal elements vanish, leaving three independent diagonal elements. The far infra-red spectrum in the local smectic frame is different for each element.

2. The ccf $\langle \mathbf{v}(t)\boldsymbol{\omega}(0)\rangle$

The irreducible representation is

$$\Gamma(\mathbf{v})\Gamma(\boldsymbol{\omega}) = (B_1 + B_2 + B_3)(B_1 + B_2 + B_3) = 3A_1 + 2B_1 + 2B_2 + 2B_3 \tag{344}$$

which again contains A_1, three times and is the same as Eq. (343). This is because the D_2 point group is chiral, and $D_g^{(1)}$ and $D_u^{(1)}$ of $R_h(3)$ map on to the same representation in D_2, (i.e., $B_1 + B_2 + B_3$). Using Eq. (344) and the cartesian and R representations in the point-group character table for D_2 leads to the symmetry

$$\langle \mathbf{v}(t)\,\boldsymbol{\omega}(t) \rangle = \pm \begin{bmatrix} a_1 & 0 & 0 \\ 0 & b_1 & 0 \\ 0 & 0 & c_1 \end{bmatrix}$$

Thus, the three diagonal elements of $\langle \mathbf{v}(t)\boldsymbol{\omega}(0) \rangle$ exist in the class D_2 (232). These elements change sign in the opposite enantiomer.

EXAMPLE 3: ORTHORHOMBIC C_{2v} (2 mm)

1. Rotational Velocity a.c.f.
In this achiral orthorhombic crystal class the irreducible representation is

$$\Gamma(\boldsymbol{\mu})\Gamma(\boldsymbol{\mu}) = (A_1 + B_1 + B_2)(A_1 + B_1 + B_2) = 3A_1 + 2A_2 + 2B_1 + 2B_2 \tag{345}$$

leading to the symmetry recorded in Table II. Three independent diagonal elements exist as in the orthorhombic D_2.

2. The ccf $\langle \mathbf{v}(t)\,\boldsymbol{\omega}(0) \rangle$
Here the irreducible representation is

$$\Gamma(\mathbf{v})\Gamma(\boldsymbol{\omega}) = (A_1 + B_1 + B_2)(A_2 + B_1 + B_2) = 2A_1 + 3A_2 + 2B_1 + 2B_2 \tag{346}$$

which contains A_1 twice. Reference to the point-group character tables reveals that the two independent elements are off-diagonals, so that the complete matrix symmetry is

$$\langle \mathbf{v}(t)\,\boldsymbol{\omega}(0) \rangle = \begin{bmatrix} 0 & a_1 & 0 \\ b_1 & 0 & 0 \\ 0 & 0 & 0 \end{bmatrix} C_{2v} \tag{347}$$

This contrasts with the orthorhombic D_2 (232) crystal class where only the diagonals of the ccf are visible.

EXAMPLE 4: TRIGONAL C_{3v} (3m)

1. Rotational Velocity

The irreducible representation in this case is

$$\Gamma(\dot{\boldsymbol{\mu}})\Gamma(\dot{\boldsymbol{\mu}}) = (A_1 + E)(A_1 + E) = 2A_1 + A_2 + 3E \qquad (348)$$

which contains the tsr twice. One comes from the product A_1A_1 and the other from EE, which signifies the product $(x_D, y_D)(x_D, y_D)$ in the Cartesian representation of the point group character table. It is well known that this notation implies the equivalence of x_D and y_D. The product implies four rotational velocity-correlation function elements according to gtsm. These four elements are not independent, and are grouped together, being equivalent to one A_1, generated by the product rule

$$EE = A_1 + A_2 + E \qquad (349)$$

However, from the elementary theory of time correlation functions, we know that

$$\langle \dot{\mu}_{x_D}(t)\, \dot{\mu}_{x_D}(0) \rangle \neq \langle \dot{\mu}_{x_D}(t)\, \dot{\mu}_{y_D}(0) \rangle \qquad (350)$$

because one is an autocorrelation function, with finite value at $t = 0$, and the other a ccf, which vanishes at $t = 0$. This, together with the independent appearance of $\langle \dot{\mu}_{z_D}(t)\, \dot{\mu}_{z_D}(0) \rangle$ from A_1A_1, leads to the final symmetry

$$\langle \dot{\mu}_{x_D}(t)\, \dot{\mu}_{x_D}(0) \rangle = \langle \dot{\mu}_{y_D}(t)\, \dot{\mu}_{y_D}(0) \rangle \neq \langle \dot{\mu}_{z_D}(t)\, \dot{\mu}_{z_D}(0) \rangle \qquad (351)$$

The two independent nonvanishing elements are thus

$$\langle \dot{\mu}_{x_D}(t)\, \dot{\mu}_{x_D}(0) \rangle = \langle \dot{\mu}_{y_D}(t)\, \dot{\mu}_{y_D}(0) \rangle \qquad (352)$$

and $\langle \dot{\mu}_{z_D}(t)\, \dot{\mu}_{z_D}(0) \rangle$. The further result

$$\langle \dot{\mu}_{x_D}(t)\, \dot{\mu}_{y_D}(0) \rangle \neq \langle \dot{\mu}_{y_D}(t)\, \dot{\mu}_{x_D}(0) \rangle \qquad (353)$$

follows from the fact that the irreducible representation (Eq. 348) allows two and only two independent nonvanishing elements. The symmetry of the complete acf matrix is thus

$$\langle \dot{\boldsymbol{\mu}}(t)\,\dot{\boldsymbol{\mu}}(0)\rangle = \begin{bmatrix} a & 0 & 0 \\ 0 & a & 0 \\ 0 & 0 & b \end{bmatrix}; \qquad C_{3\nu}(3m) \tag{354}$$

and is recorded in the table.

2. The ccf $\langle \mathbf{v}(t)\,\boldsymbol{\omega}(0)\rangle$

The relevant irreducible representation is

$$\Gamma(\mathbf{v})\,\Gamma(\boldsymbol{\omega}) = (A_1 + E)(A_2 + E) = A_1 + 2A_2 + 3E \tag{355}$$

allowing one occurrence of A_1, and by gtsm, one independent non-vanishing ccf element. This comes from the product $EE = A_1 + A_2 + E$. The non-vanishing element must therefore be from the four possibilities generated from $(x_D, y_D)(x_D, y_D)$. There is no independent occurrence of the diagonal element $\langle v_{z_D}(t)\,v_{z_D}(0)\rangle$ from (Eq. 355), and therefore the single independent element is

$$\langle v_{x_D}(t)\,\omega_{y_D}(0)\rangle = -\langle v_{y_D}(t)\,\omega_{x_D}(0)\rangle \tag{356}$$

The minus sign comes from the fact that the overall matrix symmetry is odd to parity reversal; the result (Eq. 356) represents the vector cross product symmetry, denoted $D_u^{(1)}$ in the $R_h(3)$ point group. The overall matrix symmetry is thus

$$\langle \mathbf{v}(t)\,\boldsymbol{\omega}(0)\rangle = \begin{bmatrix} 0 & a_1 & 0 \\ -a_1 & 0 & 0 \\ 0 & 0 & 0 \end{bmatrix} \quad C_{3\nu}(3m) \tag{357}$$

as in Table II.

EXAMPLE 5: THE CUBIC CRYSTAL CLASSES

There are five cubic crystal classes of high symmetry. Two of these (T and O) are chiral. Applying the same methods as in Examples 1–4 results in the classification in Table II, which shows that the diagonal elements are supported for the rotational velocity and other time auto-correlation functions. Cross correlation functions are supported only in the chiral classes T and O. No more than one independent element appears in each class, that is, no more than one occurrence of the relevant totally symmetric irreducible representation in each local nematic point group. The far infrared spectrum remains the same along the x_D, y_D, and z_D axes of the cubic point group.

Table II provides a classification scheme for nonvanishing thermodynamic ensemble averages in the given point group classification scheme. Each point group may accommodate molecules of independent symmetry, but the overall thermodynamic average is determined by the point group symmetry alone. We have shown only a few representative thermodynamic ensemble averages in Table II, but in general all such averages may be accommodated. For example, if we wish to consider polarizability of a volume element in the isotropic environment, we take the $R_h(3)$ point group and represent the polarizability with the D symmetry $D_g^{(0)} + D_g^{(2)}$, a symmetric second rank tensor even to parity reversal. The overall macroscopic polarizability of a point group representation of local smectic symmetry may then be investigated according to how many occurrences there are of the tsr in the local point group. The latter's character table may then be used to find out in more detail the nature of the polarizability tensor in the local smectic point group, that is, which ensemble averages over the polarizability vanish and which exist. If we are investigating pyroelectric symmetry, on the other hand, we note that pyroelectricity has $D_u^{(1)}$ symmetry in $R_h(3)$ and then we map this on to the smectic point group representing local symmetry in the liquid crystal as in Table 2. This shows that in some local smectic point groups pyroelectric properties are supported in principle while in others they are not, according to whether the totally symmetric representation occurs in the point group. In some cases the pyroelectricity is different along each smectic axis (three occurrences of the tsr). For a given point group symmetry, care must be taken to examine the time reversal symmetry of the quantity being averaged thermodynamically.

Some D symmetries of physical quantities are listed below.

1. The magnetic dipole is an axial (or pseudo) vector of $D_g^{(1)}$ symmetry.

2. Electric polarizability, thermal and electric conductivity, thermoelectricity, thermal expansion, and magnetic susceptibility each have $D_g^{(0)} + D_g^{(2)}$ symmetry in $R_h(3)$.

3. The quadrupole moment has $D_g^{(2)}$ symmetry.

4. The gyration tensor of optical activity has $D_u^{(0)} + D_u^{(2)}$ symmetry.

5. The first hyperpolarizability has $D_u^{(1)} + D_u^{(3)}$ symmetry.

6. Piezoelectricity and the electrooptic Kerr effect have $2D_u^{(1)} + D_u^{(2)} + D_u^{(3)}$ symmetry.

7. Elasticity is a symmetric fourth rank tensor of $2D_g^{(0)} + 2D_g^{(2)} + D_g^{(4)}$ symmetry.

Thermodynamic ensemble averages over all these quantities vanish in the isotropic liquid environment except for polarizability and elasticity, which contain the $D_g^{(0)}$ representation. In the local smectic point groups, however, new thermodynamic averages may exist which vanish in the laboratory frame if the local (crystal-like) smectic symmetry is not maintained to the macroscopic level. (A key difference between a smectic liquid crystal and a molecular solid crystal is that the same (crystal) point group symmetry is maintained in the latter from the local to the macroscopic level.) Local ensemble averages may exist in the smectic liquid crystal depending on the number of occurrences of the appropriate totally symmetric irreducible representation and on the time reversal symmetry of the quantity being averaged. The molecular electric polarizability and quadrupole moment are both positive to time reversal symmetry, and extra ensemble averages over these quantities might appear in some of the local point groups. For example, in the orthorhombic class of C_{2v} point group symmetry the representation $D_g^{(0)} + D_g^{(2)}$ of the electric polarizability maps on to $3A_1 + A_2 + B_1 + B_2$, showing that there are three independent non-vanishing thermodynamic averages over the molecular polarizability in this local point group. These correspond in the cartesian notation of the point group character table to X^2, Y^2, and Z^2, the three diagonal elements of the polarizability tensor average. All three become equal in the isotropic molecular liquid. Again, this result is independent of the individual molecular symmetry within the C_{2v} point group. In the monoclinic, C_{1h} crystal class $D_g^{(0)} + D_g^{(2)}$ maps on to $4A^1 + 2A^{11}$, meaning that four independent thermodynamic averages over polarizability exist in this class. These are denoted X^2, Y^2, Z^2, and XY in the point-group character table opposite to the A^1 entry, signifying the existence of three independent diagonal thermodynamic averages and one off-diagonal symmetric pair, $XY = YX$. In the molecular liquid, only one average exists, the trace of the diagonal averages, which are the same in the three isotropic laboratory axes in the molecular liquid. Group theoretical statistical mechanics provides a unifying picture of the properties of these ensemble averages together with those of the set of non-vanishing time correlation functions for each local point group.

These results provide a coherent system of predicting the existence of ensemble averages in the point groups of the table. The numerical value of these averages and the time dependence of the correlation functions must be obtained with additional complementary methods, such as band-shape analysis and molecular dynamics computer simulation. Not only would this provide a needed and detailed check on the predictions of gtsm applied to smectic liquid crystals, but it would also be a new era of fruitful investigation of liquid crystal molecular dynamics, extending the range of liquid state computer simulations.

These methods can also be extended to deal with the effect of external fields on smectic liquid crystals, using the third principle. This states that the symmetry of ensemble averages set up in a molecular environment subjected to an externally applied macroscopic force field is the symmetry of the applied field itself. In an ensemble of atoms subjected to a strain rate of overall symmetry $D_g^{(0)} + D_g^{(1)} + D_g^{(2)}$, recent computer simulation[300] has indeed revealed the existence of new types of ensemble average set up by the field and taking its overall symmetry. These ccf's explain the fundamental origin of the well known Weissenberg effect of rheology (Section VIII). Similarly, an electric field of symmetry $D_u^{(1)}$ sets up ensemble averages of this symmetry in the $R_h(3)$ point group, and thus also in the local smectic point groups. The symmetry of the electric-field-induced ensemble average in a given point group is $D_u^{(1)}$, mapped on to its equivalent irreducible representation in the local point group. Similar predictions can be made for other applied macroscopic fields, such as a magnetic field, an electromagnetic field, and strain rate, applied to the smectic liquid crystal in the laboratory axes X, Y, and Z. These methods, used with computer simulation and experimental spectroscopy, for example, will reveal a great deal about fundamental and unknown areas of chemical physics. They can also be extended to deal with n-time correlation functions[323] and higher order angularly resolved pair distribution functions.[324]

VIII. SIMULATION AND SYMMETRY IN NON-NEWTONIAN FLUID DYNAMICS

The past few years have seen the rapid evolution of nonequilibrium molecular dynamics computer simulation[325,326] for the numerical investigation of fluid dynamics, including couette flow, shear, extrusion, turbulence, and the phenomena of non-Newtonian rheology. It is now possible to investigate directly the linear relation between stress and strain, first proposed by Newton, in terms of his fundamental equations of motion. This allows an understanding of the rheology of "simple" liquids with computer simulation, and thereby a better understanding of the flow engineering of colloidal dispersions, flocs, gells, suspensions, polymer shearing and extrusion, new materials such as polymer liquid crystals, the transportation of residue suspensions, and other practical problems.

In general, sheared liquids are "non-Newtonian" in that the linear relation between stress and strain does not hold in general. The subject of non-Newtonian rheology tackles such phenomena as the thinning and thickening of a liquid in response to shear, and the onset of turbulence. Nonequilibrium molecular dynamics (nemd) computer simulation has advanced to the point where these phenomena can be described numerically. It has been shown that shear thinning and thickening (dilatency), viscoelasticity, and related phenomena can all be reproduced in simple ("atomic") liquids, using sufficiently scaled shear rates to correspond with the picosecond time capability of most contemporary computers. Heyes has shown[326] that sedimentation during storage, the levelling of liquids, lubrication, gell stability, and processability are all areas in which non-Newtonian effects are important but largely intractable to conventional rheology. There are several phases[326] in the development of non-Newtonian response which Heyes has classified with the Deborah Number

$$D = \tau_r \dot{\gamma} \tag{358}$$

where τ_r is the structural relaxation time and $\dot{\gamma}$ the shear rate. Newtonian behavior occurs when D is low and the viscosity is a constant. However, if the stress exceeds a critical value, the strain rises more quickly than the linear relation proposed by Newton allows. Viscoelastic effects appear when D is about 1. The material typically exhibits an increasing time dependent stress, which fails to approach a constant value, so that viscosity is not uniquely defined. A further increase in the strain rate results in shear thinning, followed by shear thickening, or dilatency. At this point relaxation effects take place too slowly to be observable, unless high pressures are used. The basics of all these phenomena have now been observed by computer simulation,[325,326] using new concepts and new equations of motion such as those devised by D. J. Evans and co-workers.[325] These have been developed by Heyes and co-workers[326] for many purposes, including couette, laminar, elongational, and sheet flows in atomic ensembles. For laminar flow, for example, the SLLOD equations of motion[326] are

$$\dot{\mathbf{r}} = \mathbf{p}/m + \dot{\gamma} r_Y \hat{X} \tag{359}$$

$$\dot{\mathbf{p}} = \mathbf{F} - \dot{\gamma} P_Y \hat{X} - \alpha \dot{P} \tag{360}$$

where X is a unit vector in the X direction, $\mathbf{F}$ is the force on the particle,

$\mathbf{r}$ is the position, $\mathbf{p}$ the momentum, $\dot{\gamma}$ the strain rate, $\mathbf{P}$ the applied pressure, and α a constant. These are Newton's equations supplemented by extra terms due to laminar shear. They were first devised by D. J. Evans and co-workers[325] and use thermostatting procedures which may be Gaussian isokinetic or profile unbiased.[326] The former has the disadvantage of rigidly fixing the total kinetic energy.

A. D Representation of Simple Couette Flow

In couette flow in an incompressible ensemble[300] a shear is applied with a given stress and strain rate. Newton's relation between the two is

$$\Pi_{XZ} = \eta \, \frac{\partial v_X}{\partial Z} \tag{361}$$

where η is the viscosity, a scalar quantity with $D_g^{(0)}$ symmetry. The Newton relation applies in the limit of vanishing strain rate. Otherwise η becomes a function of the strain rate itself

$$\Pi = 2\eta(\dot{\gamma}) \, \dot{\gamma} \tag{362}$$

In both equations, gtsm (Section VII) applies in the shear-on steady state, through principle 3.

In general, the complete tensor symmetry of the shear may be found by considering that of the strain rate tensor, which is

$$\Gamma(\dot{\gamma}) = \Gamma(\mathbf{v}\mathbf{r}^{-1}) = D_u^{(1)} D_u^{(1)} = D_g^{(0)} + D_g^{(1)} + D_g^{(2)} \tag{363}$$

The shear stress tensor and pressure tensor have the same D symmetry ($\mathbf{P} = -\Pi$), which is made up of scalar, vector, and tensor components positive to parity inversion. In simple couette flow there is no diagonal part to the strain rate, and the scalar part of the D representation vanishes. This leaves an antisymmetric component (Section VII) of $D_g^{(1)}$ symmetry and a symmetric traceless component with D symmetry $D_g^{(2)}$. This symmetry signature applies whatever the complexity of the mathematical treatment, and irrespective of the molecular symmetry in the ensemble, and simplifies the rheological approach to structured fluids, where there are five conservation and eight constitutive equations.[327]

According to principle 3 of Section VII, extra ensemble averages may appear in frame (X, Y, Z) with the symmetry $D_g^{(1)} + D_g^{(2)}$. This D symmetry is a combination of the vector and tensor parts of the generic time correlation function $\langle \mathbf{A}(t)\mathbf{A}(0) \rangle$, which may therefore appear under shear in frame (X, Y, Z). When there is only one component of the strain rate, that is, $\partial v_X/\partial Z$, then only one off-diagonal element of the correlation function is expected to be observable, and is the microscopic characteristic of the macroscopic stress.

In the special case of couette flow in an atomic ensemble[300] a fundamentally new type of ccf is anticipated to exist by gtsm which is a weighted combination of antisymmetric (vector), and symmetric (tensor) components, and which has been described by Eqs. (320a) of Section VII. The weighted combination, Eq. (320b), is in general dissymmetric in time, and this is precisely as observed by nemd computer simulation,[300,301] using both Gaussian isokinetic and profile unbiased thermostatting.

B. Consequences for Langevin Theory

We can attempt to analyze this observation by computer simulation of shear-induced dissymmetric cross correlation functions with linked Langevin equations in the linear, Markovian approximation of Section I. The difference between the analytical results and computer simulation are interpreted in terms of the fact that the simulated cross-correlation functions are nonlinear and non-Markovian, and also seem to be nonstationary, that is, dissymmetric to time displacement or index reversal. In this condition, the Onsager reciprocal relations,[328] which pertain to equilibrium, reversible, linear, and stationary processes, no longer hold, and the simple Langevin equation is no longer able to describe the results of computer simulation with any accuracy.

In order to obtain a qualitative Langevin description, use has to be made of cross friction coefficients which are either symmetric or antisymmetric in the indices X and Z of the laboratory frame (X, Y, Z). However, this produces results which are distinctly different from the simulations[300,301] in the sense that the simulated ccf's are finite at $t = 0$, but the analytical counterparts vanish at $t = 0$.

C. Derivation and Solution of the Langevin Equations

The starting point of the derivation of the Langevin equations is Eq. (3.48) of Ref. (325), the Dolls tensor adaptation of Newton's equation of motion

$$m\dot{\mathbf{v}} = \mathbf{F} - \nabla\mathbf{u} \cdot m\mathbf{v} \tag{364}$$

where $\mathbf{F}$ is the force and $\mathbf{v}$ is the velocity of a particle externally subjected to shear. The latter is represented by the tensor $\nabla\mathbf{u}$ with nine components in general. It is assumed that the shear causes a strain rate response in the N-particle ensemble consisting of $D_g^{(1)}$-type vorticity and $D_g^{(2)}$-type deformation. The former is represented from Eq. (364) as

$$F_X = m\dot{v}_X + m \frac{\partial u_X}{\partial Z} v_Z \qquad (365a)$$

$$F_Z = m\dot{v}_Z - m \frac{\partial u_Z}{\partial X} v_X \qquad (365b)$$

and the latter by the same equations but with a positive sign on the right hand side of Eq. (365b). We assume that these equations can be written with

$$\dot{\gamma}_{XZ} = \frac{\partial u_X}{\partial Z}; \qquad \dot{\gamma}_{ZX} = \frac{\partial u_Z}{\partial X}$$

The deterministic equations (Eq. 365) are developed now into Langevin equations which are solved in the linear Markovian approximation for the dissymmetric cross correlation function, whose components are symmetric and antisymmetric. The Langevin equations corresponding to (Eq. 365) are

$$F_{X \text{ stochastic}} = m\dot{v}_X + m\beta v_X + m\beta_{XZ} v_Z \qquad (366a)$$

$$F_{Z \text{ stochastic}} = m\dot{v}_Z + m\beta v_Z - m\beta_{ZX} v_X \qquad (366b)$$

These equations have been written for $D_g^{(1)}$-type vorticity. For $D_g^{(2)}$-type deformation the minus sign on the right-hand side of Eq. (366b) is replaced by a plus sign. In Eq. (366) the beta's are friction coefficients in the linear, Markovian approximation. It has been assumed that

$$\beta_{XZ} = \frac{\partial u_X}{\partial Z}; \qquad \beta_{ZX} = \frac{\partial u_Z}{\partial X} \qquad (367)$$

that is, that the components of the strain rate response can be identified with cross-friction coefficients in the linear, Markovian approximation. More generally, the friction coefficients are non-Markovian memory functions, as we have seen in earlier sections and the Langevin equation is nonlinear.[3] However, in the linear, Markovian approximation (366) the Langevin equations may be solved for the cross-correlation functions of interest:

$$\langle v_X(t) v_Z(0) \rangle = \frac{\langle v_X(0) v_X(0) \rangle}{(c - b^2)^{1/2}} \beta_{XZ} e^{-bt} \sin\{(c - b^2)^{1/2} t\}; \qquad c > b^2 \qquad (368a)$$

$$\langle v_X(t) v_Z(0) \rangle = \frac{\langle v_X(0) v_X(0) \rangle}{(b^2 - c)^{1/2}} \beta_{XZ} e^{-bt} \sinh\{(b^2 - c)^{1/2} t\}; \qquad b^2 > c$$

$$(368b)$$

where

$$b = 4\beta; \qquad c = \beta^2 - \beta_{XZ} \beta_{ZX}$$

with a similar expression for $\langle v_Z(t) v_X(0) \rangle$ with β_{XZ} replaced by β_{ZX}. For shear-induced vorticity

$$\beta_{XZ} = -\beta_{ZX} \qquad (369a)$$

and for shear-induced deformation

$$\beta_{XZ} = \beta_{ZX} \qquad (369b)$$

The final dissymmetric cross-correlation function assumed to be a *weighted* sum of both types

$$\langle v_X(t) v_Z(0) \rangle = A \langle v_X(t) v_Z(0) \rangle_{\text{vorticity}} + B \langle v_X(t) v_Z(0) \rangle_{\text{deformation}} \qquad (370a)$$

and

$$\langle v_Z(t) v_X(0) \rangle = -A \langle v_Z(t) v_X(0) \rangle_{\text{vorticity}} + B \langle v_Z(t) v_X(0) \rangle_{\text{deformation}} \qquad (370b)$$

where A and B are weighting constants. If $A \ll B$, for example, the cross-correlation functions from Eq. (370) will be slightly dissymmetric, and the same for $A \gg B$. There will be intermediate cases of varying dissymmetry. However, despite being able to explain quantitively the major feature of the simulation, that is, that the cross-correlation functions are dissymmetric, Eq. (370) is not able to show why the simulated ccf's[300,301] remain finite at $t = 0$. Equation (370) produces ccf's which vanish at $t = 0$.

The simple linear, Markovian approach thus fails qualitatively at short times.

The failure of the Langevin equations (Eq. 366) to describe the results from computer simulation is an important indication of the fact that non-Newtonian sheared N-particle ensembles have several features which are fundamentally different from their equilibrium counterparts:

1. The sheared ensemble supports cross correlation functions which are dissymmetric in time displacement and in the indices X, Z of the shear plane. These ccf's have the property[300,301]

$$\langle v_X(0) v_Z(0) \rangle \neq 0 \qquad (371)$$

which is not reproduced by the linear Markovian approximation represented in Eq. (366). This is unlikely to be remedied by developing the friction coefficients into memory functions, thus making the system non-Markovian, and we are led to consider

2. that the system is nonlinear. In one sense it is nonlinear because the stress and the strain rate are not linearly related, as in Newton's law of sheared fluids. In this sense the system is nonlinear because it is non-Newtonian. If we are to attempt an approach to the new ccf's with Langevin equations, we are led to the conclusion that the friction coefficients are no longer simple linear multiples of velocity, as in Eq. (366), because this approach fails qualitatively at $t \to 0$ both for Markovian and non-Markovian approximations to the rigorous Eq. (365). More generally, the Langevin equation can be nonlinear, containing friction coefficients that multiply powers of velocity on the right-hand side. In general the equation would contain a sum of such terms, with interesting analytical implications.

3. The new cross correlation functions are observed by numerical simulation to be dissymmetric in time displacement. They are not therefore stationary[1-4] in the conventional sense, because they are neither symmetric in time displacement nor antisymmetric.

4. This leads directly to the conclusion that in the presence of shear, the N-particle ensemble no longer obeys the Onsager reciprocal relations, which are laws applicable to N-particle ensembles at thermodynamic equilibrium, where the system is reversible.

5. The ccf's are therefore indicative of a dynamical process under shear which is irreversible, in the sense that they are not governed by Onsager's reciprocal relation.

An N particle ensemble in the steady state under shear which is non-Newtonian produces dissymmetric time cross-correlation functions which indicate a statistical process which is non-linear, irreversible, non-Markovian and dissymmetric in time displacement, being in this sense nonstationary. In consequence, a simple linear Markovian description fails qualitatively as $t \to 0$. This leads to an entirely new appreciation of non-Newtonian N-particle dynamics.

D. Shear Induced Structural Effects and GTSM

Angular resolution[326] of the pair-radial distribution function in computer simulations of shear induced thickening in atomic (Lennard–Jones) liquids has revealed the presence of anisotropic local structure, which can be explained on the basis of group theoretical statistical mechanics using principle 3, that the symmetry of allowed ensemble averages in the steady state in the presence of shear is the same as that of the applied strain rate. The computer simulation results[329] can be reproduced from group theory by assuming that the crystal-like lattice arrangement of atoms which appear in the simulations under shear can be described by some of the 32 crystallographic point groups in Table II, hexagonal, trigonal, and triclinic. The hexagonal lattice symmetries C_{3h} and C_{6h}, the trigonal symmetry S_6, and the triclinic symmetry $C_i(S_2)$ are found to support the crystal-like structures necessary to explain the observed[329] angular resolution of the pair-radial distribution function.

The nemd computer simulations of Heyes and co-workers[326] have revealed a number of significant new phenomena of non-Newtonian rheology in atomic liquids using a battery of new numerical techniques. Among the most interesting of these is in the context of shear thinning and thickening. Simulations have shown[329] that as the shear rate is increased the atoms of the liquid ensemble form structurally arrested states with crystal-like symmetries. The point group of the ensemble is therefore changed from $R_h(3)$ to that of the shear induced lattice. Gtsm can be used to to explain the observed symmetry of angularly resolved pair radial distribution functions (rdf's) in non-equilibrium simulations of these atomic ensembles. Only a small number of lattice symmetries support the observed anisotropy under shear, and gtsm is used to explain why shear is able to produce this anisotropy.

We have seen that principle 3 is a statement of how externally applied force fields of given symmetry set up extra ensemble averages at field-applied equilibrium. It is also valid in transient, nonequilibrium regimes. In both cases the overall symmetry of the new ensemble averages is that of the applied field. The complete D symmetry of the strain rate tensor is $D_g^{(0)} + D_g^{(1)} + D_g^{(2)}$, which by principle 3 sets up new ensemble averages, such as pair distribution functions or time correlation functions in frame (X, Y, Z) at the field on steady state. State of the art nonequilibrium computer simulation[300,301] shows the presence of new time ccf's with the overall symmetry of the strain rate. Principle 3 produces similar entirely novel results[300] for the time ccf's of pressure tensor components, revealing the fundamental origin of the Weissenberg effect of macroscopic non-Newtonian rheology, and explaining through ccf's the pressure set up in a sheared liquid in a direction perpendicular to the plane of shear. The new ccf's are also sensitive to the typical macroscopic phenomena of non-Newtonian rheology, including the appearance of shear induced thickening and thinning,[326] the appearance of string phases,[330] and structurally arrested states. These all involve time ccf's in frame (X, Y, Z) for an atomic liquid, and also in frame (x, y, z) for a molecular liquid. Conventional methods of macroscopic rheology have failed to recognize this, in the same way that conventional diffusion theory has failed to recognize the role of ccf's at equilibrium. In both cases they are governed by principle 3.

E. Crystal-Like Arrested States at High Shear Rates—An Excess of Symmetry

Principle 3 may be applied to angularly resolved pair distribution functions, defined[329] by

$$f_{\alpha\beta}(r) = \left(\sum_{i \neq j}^{N} \sum^{N} \langle R_{\alpha ij} R_{\beta ij} / R_{ij}^2 \rangle \right) \bigg/ N \tag{372}$$

$$g_{\alpha\beta}(r) = 15 V f_{\alpha\beta}(r) / (V(r)N) \tag{373}$$

The angular component $f_{\alpha\beta}$ measures the anisotropic dispositions of molecules or atoms, and involves the ensemble average $\langle R_{\alpha ij} R_{\beta ij} \rangle$. Peaks in $g_{\alpha\beta}$ supply information on shear-induced structurally arrested states. In Eq. (373), V is the volume of the shell bounded by $r \pm \delta r/2$

$$V(r) = 4\pi r^2 \delta r \tag{374}$$

for a shear resulting in a strain-rate response of type $\partial v_X / \partial Z$. Principle 3 predicts the existence of the ensemble average $\langle R_{Xij} R_{Zij} \rangle$, but no other off-diagonal elements such as $\langle R_{Xij} R_{Yij} \rangle$ or $\langle R_{Yij} R_{Zij} \rangle$. This is simulated by Heyes[331] and is in satisfactory agreement with numerically derived data for low shear rates, but as the latter increase, off-diagonal elements of the angularly resolved pair rdf appear which are disallowed by principle 3 in $R_h(3)$. Heyes has explained this in terms of slow structural relaxation, outside the time window of the simulation. The applied shear has clearly led to lower symmetry in frame (X, Y, Z), a crystal-like environment has been generated from a shear-induced phase change, taking the ensemble from $R_h(3)$ to some other crystal-like point group of lower symmetry. The problem is how to apply group theoretical methods within this new group to explain the results actually observed by Heyes (Figs. 7 and 8 of Ref. 331).

In order to explain Fig. 8 of Ref. 331 it is necessary to assume that the overall point group of at least some part of the ensemble is no longer $R_h(3)$ of the isotropic liquid, distorted by shear, but is that of a shear-induced crystal-like structure. From the "snapshots" of the simulation provided by Heyes,[331] this appears to be hexagonal, trigonal, or triclinic, an overall triangular lattice which produces

$$f_{XX} \neq f_{YY} \neq f_{ZZ} \neq f_{XZ} \neq f_{XY} \neq f_{YZ} \neq 0 \tag{375}$$

To find the symmetry of the applied shear within each of these crystal-like point groups it is necessary to map $D_g^{(0)} + D_g^{(1)} + D_g^{(2)}$ onto the appropriate irreducible representation within that crystal-like point group. These representations appear in Table II. The following are examples of this procedure for shear-induced crystal-like point groups.

F. Hexagonal C_{3h} (Hermann Mauguin $\bar{6}$)

In this crystal point group, one of the hexagonal crystal symmetries, the symmetry of shear is

$$D_g^{(0)} + D_g^{(1)} + D_g^{(2)} = A^{11}A^{11} + E^1A^{11} + A^{11}E^1 + E^1E^1 \tag{376}$$

which allows ensemble averages of the type

$$\langle R_{Xij}R_{Xij}\rangle \neq \langle R_{Yij}R_{Yij}\rangle \neq \langle R_{Zij}R_{Zij}\rangle; \tag{377a}$$

$$\langle R_{Xij}R_{Yij}\rangle \neq \langle R_{Xij}R_{Zij}\rangle \neq \langle R_{Yij}R_{Zij}\rangle \tag{377b}$$

using principle 3 with the C_{3h} point group rather than the $R_h(3)$ point group. Equation (377) predicts angularly resolved pair-distribution functions with the property Eq. (375). Thus both the diagonal and off-diagonal elements have a different r dependence if the structurally arrested state has the crystal symmetry C_{3h} or $\bar{6}$. Comparison of this result with available computer simulations by Heyes can be made by examining the numerical data in Figs. 7 and 8 of Ref. (331). These show that at strain rates $\dot{\eta} = 110$ and $\dot{\gamma} = 30$ in normalized units the diagonal elements g_{YY} and g_{ZZ} are almost equal but discernably different, the third, g_{XX}, being distinctly different from the other two. In Fig. 8 of Ref. 331 the amplitudes of the off-diagonal elements g_{XY} and g_{YZ} are greater than the third element g_{XZ}, all three being distinctly different.

Our calculations, based on principle 3 applied to the various point groups, show that the hexagonal point group $C_{6h}(6/m)$ also gives this result, along with the trigonal S_6 and the triclinic S_2. Other hexagonal or in general triangular-type lattices are either disqualified on the basis of being chiral, or produce degeneracies, in the sense that one or more angularly resolved pair rdf elements are equal.

We are able to conclude that shear thickening is accompanied by the appearance of structurally arrested states in which the crystal-like symmetry supports six different angularly resolved radial pair distribution functions. These are rationalized with group theoretical statistical mechanics.

G. Shear-Induced Depolarized Light Scattering

The existence by gtsm and nemd computer simulation[300,301] of new dissymmetric ccf's implies that there exists theoretically a hitherto unmeasured depolarized component of light scattered from a sheared N-particle ensemble. In atomic ensembles, and in ensembles of molecules of symmetry higher than T_d, this is the only component, neglecting weak collision-induced effects.

The third principle of gtsm allows the existence of the new current correlation function[332–334]

$$C_{XZ} = \langle v_X(0) v_Z(t) \exp(i\mathbf{q} \cdot (\mathbf{r}(0)) - \mathbf{r}(t))) \rangle \tag{378}$$

in an N particle ensemble subjected to the shear strain $\partial v_X/\partial Z$. This has the same dissymmetry properties as the velocity and pressure tensor ccf's.[300,301] Here $\mathbf{q}$ is the scattering vector and

$$\mathbf{r}(t) - \mathbf{r}(0) = \Delta\mathbf{r}(t) = \int_0^t \mathbf{v}(t)\,dt \tag{379}$$

where $\mathbf{v}$ is the center-of-mass velocity of the diffusing particle (atom or molecule). We have

$$\langle \Delta r^2(\tau) \rangle = \int_0^\tau \int_0^\tau \langle v_X(t_1) v_Z(t_2) \rangle\, dt_2\, dt_1 \tag{380a}$$

$$= 2 \int_0^\tau (t - \tau)\langle v_X(0) v_Z(t) \rangle\, dt \tag{380b}$$

The current ccf (Eq. 378) is related to a self-dynamic structure factor[332–334] by

$$F_{XZ}^{(s)}(\mathbf{q}, t) = \langle \exp(i\mathbf{q} \cdot (\mathbf{r}(0) - \mathbf{r}(t))) \rangle \tag{381}$$

which upon double differentiation provides

$$\frac{d^2}{d\tau^2} F_{XZ}^{(s)}(\mathbf{q}, \tau) = -q^2 C_{XZ}(\tau) \tag{382}$$

Equation (382) gives the result

$$J_{XZ}^{(s)}(\mathbf{q}, \omega) = \frac{\omega^2}{q^2} S_{XZ}^{(s)}(\mathbf{q}, \omega) \tag{383}$$

where J is the temporal Fourier transform of C_{XZ}, and S that of Eq. (381). The latter is the intermediate scattering function in the ideal self-dynamic limit for the sheared N-particle ensemble. Equations (383) shows that this is related to the new current ccf defined by Eq. (378).

H. Light-Scattering Geometry

Integrating Eq. (382) gives the result

$$\lim_{q \to 0} \left(\frac{d}{dt} F_{XZ}^{(s)}(\mathbf{q}, t)_{\tau > \tau_l} \right) = -q^2 \int_0^\infty \langle v_X(0) v_Z(t) \rangle dt \tag{384}$$

where τ_l is the correlation time. This shows that the intermediate scattering function S is related to the new dissymmetric cross correlation functions generated by shear.[300,301] The function S is observable by light scattering where the initial polarization vector is in the X axis of the laboratory frame (X, Y, Z) and where the scattered polarization vector is in the Z axis of this frame. The plane XZ is that of the shear strain $\partial v_X / \partial Z$.

The existence of shear induced ccf's means that there will be depolarized light scattered from a sheared N-particle ensemble with intensity S. This spectrum is related to the temporal Fourier transform of Eq. (378) by Eq. (383).

In atomic ensembles, or in ensembles of molecules of symmetry greater than T_d, this will be the sole contribution to the new type shear-induced light scattering apart from small contributions[332–334] from collision induced polarization anisotropies. At equilibrium in the absence of shear, this spectrum will disappear, because the shear-induced dissymmetric ccf's disappear.

In order to observe the spectrum experimentally, the incident laser

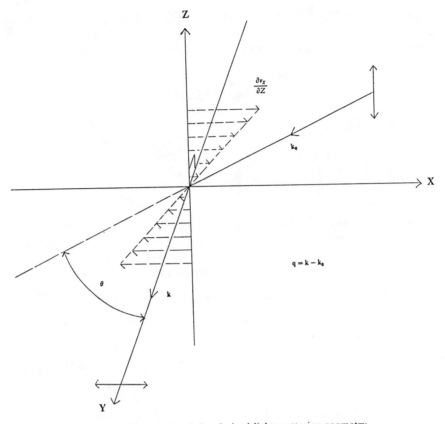

Figure 48. Shear-induced depolarized light-scattering geometry.

beam, polarized in the X axis, is scattered conveniently from an arrangement of coaxial cylinders, and scattered radiation is observed polarized in the axis Z for a given scattering angle, angular frequency and scattering vector. This is the spectrum S which gives J by Eq. (383), and thus the temporal Fourier transform of Eq. (378). The inner cylinder (a stirring rod) rotates rapidly and the outer cylinder is the wall of a round light-scattering cuvette. This arrangement creates a shear on a liquid held between the cylinders. The geometry is illustrated in Fig. 48.

The high frequency wing of the spectrum S is amplified by the multiplication by the square of the angular frequency in Eq. (383) to give the second spectral moment J. The latter is related directly to the new ccf (Eq. 378) and this ccf is in turn a direct measure of non-Newtonian effects in a sheared fluid.

The new depolarized spectrum J is therefore a direct measure of the non-Newtonian nature of sheared N particle ensembles.

Depolarized scattering depends solely on the optical anisotropy of the scattering center, and this may be thought of as a scattering element of polarizability tensor $\boldsymbol{\alpha}$. The scattered electric field vector of the electromagnetic radiation is then

$$E_X = \sum_j \alpha^j_{XZ} \exp(i\mathbf{q} \cdot \mathbf{r}_j(t)) \tag{385}$$

where α^j_{XZ} the XZ component of the polarizability tensor $\boldsymbol{\alpha}$ of the jth scattering element, and $\mathbf{q}$ the scattering vector. Let us situate the center of mass on an atom of our sheared ensemble. On average, we have, for atomic ensembles

$$\alpha^j_{XZ} = \alpha_0 \frac{\langle R_X R_Z \rangle}{R^2} \tag{386}$$

where the vector $\mathbf{R}$ is defined with respect to the center of mass of the polarizability element in frame (X, Y, Z), and where the averaged quantity on the rhs of Eq. (386) is proportional to the angularly resolved radial distribution function introduced by Heyes and Szczepanski.[335] The latter, denoted by ρ_{XZ}, is dependent on the positions of nearest neighbours, next nearest neighbors, and so on, around the center of mass atom, and is assumed here to be approximately the time independent equilibrium average. Therefore, we have

$$\alpha^j_{XZ} = \alpha_0 \rho_{XZ}(\mathbf{R}) \tag{387}$$

where "R" is the argument of the angularly resolved pair-radial distribution function. Without loss of generality, we can assume that

$$\rho_{XZ}(\mathbf{R}) = \rho_{XZ}(\mathbf{R}_0) \tag{388}$$

where $\mathbf{R}_0$ is the position of the first peak of the pair-radial distribution function. Finally, we have

$$\alpha^j_{XZ}(t) = \alpha^j_{XZ}(0) = \alpha_0 \rho_{XZ}(\mathbf{R}_0) \tag{389}$$

so that the depolarized light scattering spectrum is given by

$$S_{VH}(\mathbf{q}, t) \propto \rho^2_{XZ}(\mathbf{R}_0) S^{(s)}_{XZ}(\mathbf{q}, t) \tag{390}$$

In order to provide an idea of the frequency and scattering vector dependence of the spectrum J this can be computed for an ensemble of atoms subjected to shear strain by PUT computer simulation.[300,301] The new current correlation function of type in Eq. (378) is computed directly and its temporal Fourier transform gives J.

I. Polarimetry

The plane of polarization of electromagnetic radiation is expected to be changed a little by shearing on the basis of the argument given above, and this could be detected with a simple polarimeter, providing a direct method of observing the effect of shear on polarized electromagnetic radiation.

J. Shear Induced Dipole Relaxation

In a dipolar molecular ensemble the existence of shear induced dissymmetric ccf's[300,301] implies that of shear-induced polarizability and polarization and direct dissymetric cross correlation between orthogonal components of the permanent molecular dipole moment whose Fourier transform[1–4] is a complex frequency-dependent permittivity. The relaxation of the shear induced permittivity provides a direct method of investigating non-Newtonian phenomena such as shear thinning, thickening, and turbulence in dipolar media. At far infrared frequencies[1] the direct shear-induced cross correlation between orthogonal components of the molecular rotational velocity can be isolated and observed experimentally in principle as a power absorption spectrum with crossed wire grid polarizers.

The shear-induced permittivity spectrum can cover the frequency regions of experimentally obtainable strain rates (up to MHz) and may be used to investigate non-Newtonian effects experimentally. The high-frequency adjunct of this shear induced frequency process, a far infrared power absorption[1–4] likewise has a sheared induced component which is the Fourier transform of the shear induced dissymmetric ccf of the molecular rotational velocity

$$\langle \dot{\mu}_Z(t) \dot{\mu}_X(0) \rangle \neq \langle \dot{\mu}_X(t) \dot{\mu}_Z(0) \rangle \tag{391}$$

It is argued that this can be isolated and observed experimentally and directly with crossed wire grid polarizers, providing another direct probe of non-Newtonian rheology.

K. Shear Symmetry in the Dielectric and Far Infrared

Using the language of irreducible D representations the symmetry of strain rate of type (1) is again assumed to be

$$\Gamma(\dot{\gamma}) = D_g^{(1)} + D_g^{(2)} \tag{392}$$

a traceless, purely off-diagonal, symmetry with a vector part $D_g^{(1)}$ and a tensor part $D_g^{(2)}$. The third principle imparts this symmetry to ensemble averages at the shear-applied steady state, giving rise to dissymmetric ccf's at the atomic (or molecular) level. Conventional rheology does not postulate the existence of atoms and molecules, and in consequence is unable to explain this fundamental result.

In atomic ensembles, the dissymmetric ccf's are exemplified by velocity ccf's (Section VII) and also by related types with the same symmetry, such as the mixed velocity–position ccf's[300,301]

$$\langle v_z(t) r_x(0) \rangle \neq \langle v_x(t) r_z(0) \rangle \tag{393}$$

and the position–position ccf's

$$\langle r_z(t) r_x(0) \rangle \neq \langle r_x(t) r_z(0) \rangle \tag{394}$$

In ensembles of dipolar molecules, the permanent molecular dipole moment μ_0 is always expressible as the vector sum of the position vectors of the atoms of the molecule in frame (X, Y, Z). This immediately implies the existence of the shear induced dissymmetric ccf's of μ_0, and of its time derivative,[1] the rotational velocity $\dot{\mu}_0$. We have the results

$$\langle \mu_{0Z}(t) \mu_{0X}(0) \rangle \neq \langle \mu_{0X}(t) \mu_{0Z}(0) \rangle \tag{395}$$

and

$$\langle \dot{\mu}_{0Z}(t) \dot{\mu}_{0X}(0) \rangle \neq \langle \dot{\mu}_{0X}(t) \dot{\mu}_{0Z}(0) \rangle \tag{396}$$

The Fourier transform of Eq. (395) is a dissymmetric, shear-induced complex permittivity. That of Eq. (396) is a dissymmetric far infrared power absorption accompanied by a dispersion in the referactive index. Thus, the results of Refs. 300 and 301 immediately give new types of observable, shear-induced spectra, which are direct probes of non-New-tonian phenomena.

L. Shear-Induced Molecular Polarizability and Polarization

The terms "polarizability" and "polarization" are usually applied to the response of a dielectric to an applied electric field. However, the exis-tence[300,301] of dissymmetric ccf's of the molecular dipole moment implies that of a *shear-induced molecular polarizability* which is given in the shear applied steady state by

$$\langle \alpha_{XZ} \rangle = \frac{\langle \mu_{0Z} \mu_{0X} \rangle}{\Delta E_{\text{shear}}} \tag{397}$$

where ΔE_{shear} has the units of energy (J), and the polarizability has the units of $C^2 m^2 J^{-1}$. More conventionally, the polarizability is given in units of volume (the "volume definition") by dividing the right-hand side of Eq. (397) by $4\pi\epsilon_0$, where ϵ_0 is the permittivity of free space. The shear induced molecular polarizability then has units of m^3.

The results of Ref. 300 show that at $t = 0$ (the "equilibrium value"),

$$\langle v_Z(0) v_X(0) \rangle = \langle v_X(0) v_Z(0) \rangle \tag{398}$$

so that the shear-on equilibrium value of the polarizability, given by Eq. (397), is not dissymmetric in X and Z. The equilibrium value of the energy in the denominator of Eq. (397) is then the energy of formation of the numerator, the $t = 0$ value of the cross-correlation function of the perma-nent molecular dipole moment in the shear-on steady.

The existence of the shear induced molecular polarizability implies that the sample is *polarized by shear*. This is formally analogous to the polarization caused by an electric field, which is the basis of dielectric spectroscopy, but is due to the field of force caused by shear, the "shearing field." We refer to this as shear-induced polarization. In the same way that dielectric polarization may be expressed as a power series in the applied electric field, shear induced polarization is a power series in the applied shearing field. The coefficients of the series in electric-field-in-duced polarization are: the molecular polarizability (multiplied by the electric field); the molecular hyperpolarizability (multiplied by the electric field squared) and so on. Those in shear-induced polarization are the shear-induced molecular polarizability; shear induced molecular hyper-polarizability, and so on.

The shear-induced polarization may be expressed through a total molecular dipole moment with components X and Z. These are sums of the equivalents for the permanent molecular dipole moment and those induced by shear. (This is again formally analogous to the total dipole moment produced by electric polarization, which is a sum of the permanent dipole, that induced by the product of polarizability and the electric field, and so on.) At shear-on equilibrium the total molecular dipole components are

$$\mu_{X,\text{total}} = \mu_{0X} + \mu_{1X} + \mu_{2X} + \cdots \tag{399a}$$

and

$$\mu_{Z,\text{total}} = \mu_{0Z} + \mu_{1Z} + \mu_{2Z} + \cdots \tag{399b}$$

We now express the shear-induced dipole components μ_{1Z} and so on in terms of integrals over dissymmetric ccf's of type in Eq. (395). This is accomplished using the Morriss–Evans theorem described in Section VII.

M. Adaptation of the Morris–Evans Theorem

The Morriss–Evans theorem is a generalisation of the Green–Kubo relations[1–4] and a fusion of linear and nonlinear response theory, providing a new framework for fluctuation–dissipation theorems in general (Section VII).

Taking the first induced term in Eq. (399), we have, by definition

$$\langle \mu_{1X} \rangle = \frac{F_{\text{shear}}}{\Delta E_{\text{shear}}} \langle \mu_{0X}(0)\mu_{0Z}(0) \rangle = \langle \mu_{1Z} \rangle \tag{400}$$

where F_{shear}, is the shearing field. The structure of the Morriss–Evans theorem allows this to be written as

$$\langle \mu_{1Z} \rangle = -\frac{F_{\text{shear}}}{\Delta E_{\text{shear}}} \int_0^\infty \langle \mu_{0X}(t)\dot{\mu}_{0Z}(0) \rangle \, dt \tag{401}$$

with a similar expression for $\langle \mu_{1X}(t) \rangle$.

In Eq. (401) we have taken the $t \to \infty$ limit of the Morriss–Evans integral, which implies that the nonequilibrium average $\langle\langle \ \rangle\rangle$ becomes the shear-on steady-state average $\langle \ \rangle$, because the external field is applied for an infinite time, allowing the system to reach a steady state.

We note finally that the factor before the integral in Eq. (401) can be expressed in the form

$$\frac{F_{\text{shear}}}{\Delta E_{\text{shear}}} = \frac{\text{Constant}}{\langle \mu_{0X} \rangle} \frac{\partial v_X}{\partial Z} \tag{402}$$

where the "constant" is the shear-induced polarization constant.

N. Shear Induced Dipole Relaxation and Far Infrared Power Absorption

These are three of the many areas of observation affected by the phenomenon implied by Ref. 300, and specifically by the existence of dissymmetric ccf's of the permanent molecular dipole moment and its time derivative in frame (X, Y, Z).

Shear-induced dipole relaxation, a relatively low-frequency process which can cover several frequency decades, is expressed in spectral terms through the Fourier transformation of the shear-induced dipole ccf. In the shear-on steady state it causes polarization, which is the result of statistical correlation between orthogonal X and Z components of the permanent molecular dipole moment. The polarization may be isolated and detected experimentally by a special arrangement of electrodes, one in the XY plane and the other in the ZY plane, one electrode being perpendicular to the other and both being perpendicular to the plane of shear, XZ. In the Hz to kHz frequency region, the relaxation of the shear-induced polarization may be detected with a Wayne–Kerr bridge, and with other types of bridge technique and sweep frequency apparatus up to the MHz range. Direct measurements with orthogonal electrodes of shear induced polarization seem never to have been made, but would isolate the cross-correlation between the X and Z components of the permanent molecular dipole moment.

Analogously, the high-frequency adjunct of the shear induced dipole relaxation process is a far infrared power absorption and accompanying refractive index dispersion. The power absorption spectrum in the far infrared can be isolated in principle by the use of wire grid polarizers. One polarizer is oriented in the Z axis between the exit port of a far infrared interferometer and the sheared sample, the spectrum taken, and the experiment repeated with a polarizer in the X direction between the sample and the detector. The polarizers selectively block out radiation in the presence and absence of shear.

Careful choice of sample and conditions, with accessible laboratory strain rates, leads to this far infrared spectrum, which is a direct measurement of the non-Newtonian molecular rotational velocity ccf. The fundamental reason for this is that the far infrared power absorption is always the high frequency adjunct of a dipole relaxation process which occurs at much lower frequencies, and the high- and low-frequency parts of the overall dynamical process are never independent. This has been demonstrated experimentally by Evans and Reid[1-4] using supercooled liquids and glasses, work which isolated the far infrared gamma process as the high frequency, ever present, adjunct of much lower frequency alpha and beta processes in supercooled liquids and glasses. The gamma process may be separated from the other two by a dozen frequency decades, but is always observable. Similarly, the accessible laboratory strain rate may be only a few kHz at most, but the molecular response extends from Hz to THz frequencies in general. This is simply due to the fact that a molecular diffusion process evolves temporally from the picosecond scale onwards.

In contrast to the absorption processes above, light scattering involves an induced dipole moment, equivalent to our μ_1. Conventionally, this is attributed to the molecular polarizability. However, in the shear applied steady state, the shear induced molecular polarizability can also cause *depolarized* light scattering, as discussed already. Light-scattering apparatus is implemented analogously at high or low frequencies (Rayleigh Brioullin and photon correlation spectroscopy respectively).

O. The D Symmetries of Shear in the Presence of Fields

The important industrial technique of *electrorheology* is based on shear in the presence of an electric field, static or time dependent. The simultaneous application of shear and an electric field, through the use of rotating electrodes for example (Sections IV–VI), leads to the possibility of new ensemble averages whose D symmetries are governed by principle 3. An electric field is a polar vector (Section VII) of $D_u^{(1)}$ symmetry. The combined D symmetry of electrorheology (electric field plus shear) is therefore

$$\Gamma(\mathbf{E})\Gamma(\dot{\gamma}) = D_u^{(1)}D_g^{(1)}D_g^{(1)} = D_u^{(0)} + 3D_u^{(1)} + 2D_u^{(2)} + D_u^{(3)} \qquad (403)$$

using the Clebsch–Gordan theorem. The overall parity inversion symmetry is negative (ungerade), so that by principle (3) the ensemble averages specific to electrorheology must also be negative to parity inversion. The overall time reversal symmetry of the D combination depends on whether the electric field is static or alternating. The former is positive to time reversal and the latter is negative. The time-reversal symmetry of shear is a product of that of position and linear velocity, and is therefore negative to time reversal. The time-reversal symmetry of electroshearing is therefore overall negative in a static and overall positive in an alternating electric field.

The various possible field combinations in electroshearing can be expressed as subsets of the general triple vector product $\mathbf{vrE}$ with 27 elements. In couette flow, there are no scalar $\mathbf{v} \cdot \mathbf{r}$ components, so that the field combinations with $\mathbf{E}$ all involve $\mathbf{v} \times \mathbf{r}$ or $\mathbf{vr}^T$. From the signature (Eq. 403) the possible combinations allowed by symmetry are as follows:

1. The symmetry product $D_u^{(1)} D_g^{(1)}$ gives the three possible combinations $\mathbf{E} \cdot \mathbf{v} \times \mathbf{r}$, $\mathbf{E} \times \mathbf{v} \times \mathbf{r}$, and $\mathbf{E} \times \mathbf{vr}^T$.
2. The product $D_u^{(1)} D_g^{(2)}$ gives the three further possible combinations $\mathbf{E} \times \mathbf{vr}^T$, $\mathbf{E}(\mathbf{vr}^T)^T$, and $\mathbf{Evr}$.

Thus, depending on the relative geometry of the applied fields, different D symmetries are generated. If electroshearing is carried out with an electric field between rotating electrodes, which also serve as shearing plates, then $\mathbf{E}$ is parallel to $\mathbf{r} \times \mathbf{v}$, so that only fewer D signatures are possible than in the general Eq. (403).

The application of principle 3 in this context produces the following, for example.

1. The rotation of plane-polarized light is one possible consequence of applying the field combination $\mathbf{E} \cdot \mathbf{v} \times \mathbf{r}$ to the sheared liquid using an alternating electric field. This particular field combination has positive time-reversal symmetry and negative parity-inversion symmetry and includes the pseudoscalar $D_u^{(0)}$. These satisfy the requirements for the field combination to be chiral.[279] There is the possibility, therefore, from principle 3, of a new kind of effect which which resembles the Faraday effect. The magnetic field which induces optical rotation in the latter is *not* however a true chiral influence in the definition given by Barron[279] because the magnetic field is a $D_g^{(1)}$ axial vector. A chiral influence must be negative to parity inversion and also positive to time reversal[279] (Section VII). In the new effect the magnetic field of Faraday's experiment is replaced by a truly chiral combination of time-dependent electric field and shear. In order to observe the expected optical rotation (signature $D_u^{(0)}$) the electric field vector must be parallel to the vector generated by the cross product $\mathbf{v} \times \mathbf{r}$. In a liquid being sheared by rotating plates, the latter is perpendicular to the plane of shear, and thus parallel to an alternating (or otherwise time-dependent) electric field applied using the rotating plates as electrodes. Principle 3 means that the combined influence of shear and electric field introduces in frame (X, Y, Z) new ccf's as well as the ensemble average of signature $D_u^{(0)}$ of optical rotation angle. If the sample also absorbs, the combination of shear and electric field produces dichroism.

As we have seen in Section VII, the chiral nature of the shear and time-dependent electric field combination also induces the appropriate elements of the ccf $\langle \mathbf{v}(t)\boldsymbol{\omega}(0) \rangle$ between the molecular linear and angular velocities. Ccf's of this type ought to be observable in a computer simulation of electroshearing, together with several other effects anticipated by the D signature on the right-hand side of Eq. (403). Another example follows.

2. Electric field and shear can also result in the generation of thermal conductivity, defined as the Green–Kubo integral[336] over the heat flux tensor **J** of Irving and Kirkwood. The tensor **J** has the same D symmetry as the right-hand side of Eq. (403), and in consequence all possible field combinations in electroshearing may contribute in principle to its appearance in the frame (X, Y, Z). This has recently been confirmed by computer simulation.[337] However, in this case it is necessary to use a static electric field, so that the overall time reversal symmetry is negative. Then, principle 3 shows that the application of a static electric field across the shearing plates will result in measurable changes in the sample's thermal conductivity, a property which may be used as an analytical tool for non-Newtonian electrorheology.

3. In electrorheology the electrodes, which are also the shearing plates, may be built into a circuit to measure the dielectric complex permittivity as a function of frequency over a several decade range. The thermodynamic average over the molecular permanent dipole moment, $\langle \mathbf{\mu} \rangle$, is positive to time reversal and negative to parity inversion with the D signature $D_u^{(1)}$ of a polar vector. This means that the dielectric frequency spectrum is affected by shear, and can be used to measure its non–Newtonian manifestations, reinforcing the conclusions arrived at earlier in this chapter.

4. If a static electric field is used and the overall time reversal symmetry is in consequence negative, principle 3 means that it is possible to induce ensemble averages over quantities such as molecular linear velocity, creating a drift in the sample. This may be used[279] to separate physically the enantiomers of a racemic mixture by applying the electric field to shearing plates immersed in a sample of the racemic mixture. The drift in linear velocity creates a propeller action[279] which is in a different sense for the two enantiomers which drift apart and which can be concentrated in different ends of the vessel.

5. As we have seen, a combination of shear and an alternating electric field is a true chiral influence negative to parity inversion and positive to time reversal. In this sense, the material being electrosheared is a chiral medium, even though the individual molecules in the liquid may be structurally achiral. A beam of chiral radiation incident upon the chiral medium so defined will be attenuated differently according to whether the beam is left- or right-handed, (e.g., left or right circularly polarized electromagnetic radiation). Examples of such chiral radiation are circularly polarized lasers in the ultraviolet, visible, and infrared, including the far infrared

and radio frequency regions (chiral photon beams); spin-polarized electron beams; spin polarized neutrons; and so on. These will result in dichroism or differential scattering which may be attributed to the non-Newtonian effects of electroshearing.

6. Magnetic resonance phenomena may be utilized similarly to investigate non–Newtonian electroshearing. The symmetry argument in this instance constructs a chiral medium using a combination of shear and alternating electric field, and uses a chiral probe consisting of a static magnetic field combined with the same alternating electric field. Both probe and medium are negative to parity inversion and positive to time reversal, and both satisfy Barron's definition[279] of chirality. The apparatus can be imagined to consist of the electroshearing unit with good conducting electrodes such as brass or silver (nonmagnetic) embedded in a solenoid, or put between the pole pieces of a powerful magnet. The chiral probe field is a combination of the static magnetic field and alternating electric field applied between the (counterspinning) silver electrodes. The same ac electric field is used for the electric component of electroshearing, that is, to create the chiral medium. Thus, according to the handedness of the probe and medium, there will be asymmetric attenuation of the magnetic field, a kind of dichroism (and therefore birefringence) which could be detected through magnetic properties such as nuclear magnetic resonance and its relaxation. The NMR features would depend on the non–Newtonian nature of the sample's response to electroshearing. Conventional NMR technology could be implemented, the alternating electric field allowing fine tuning of the chirality of both probe and medium. The expected magnetic effects will be asymmetric in the handedness of the chiral probe, or alternatively of the medium, adding extra observables.

IX. NEW PUMP–PROBE LASER SPECTROSCOPIES: SYMMETRY AND APPLICATION TO ATOMIC AND MOLECULAR SYSTEMS

A. Basic Symmetry Concepts

We have seen in Sections VII and VIII that symmetry is one of the cornerstones of the scientific edifice, and in chemical spectroscopy has a particularly elegant historical facade. By the early 1840s Michael Faraday had convinced himself that there is an ineluctable link between electric and magnetic fields and light. In 1846 he proved this to the world[338] using static magnetic flux density (B) to rotate the plane of polarization of light passing through lead borate glass. In his own words, he had magnetized and electrified a ray of light. Maxwell's equations met the challenge of

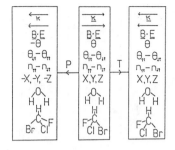

Figure 49. Schematic of motion-reversal symmetry.

Faraday's work, and shed light of their own.[339,340] Much of what we know still relies directly on the work of Faraday and Maxwell, with a little retrospective wisdom from the intervening years.

Some of this allows us to see now that if Faraday had attempted to rotate the plane of his ray of light with static electric field strength (**E**), he would have seen nothing.[341] (A diary entry, reviewed by Bragg[342] suggests that Faraday did indeed try the experiment with a static electric field, and briefly noted "no effect".) This has to do with two of the profound symmetry principles of physics,[343] those of reversality and parity inversion, first proposed in 1927 by Wigner.[344]

B. Complete Experiment Symmetry

1. Wigner's Principle of Reversality (T)

If a complete experiment is realizable in the laboratory frame (X, Y, Z), then it must also be so when all motions are reversed.

2. Wigner's Principle of Parity Inversion (P)

If a complete experiment is realizable in (X, Y, Z), then it must also be so in the frame $(-X, -Y, -Z)$, with all position coordinates reversed.

These deceptively simple statements contain the key to why **B**, but not **E**, rotates the plane of polarization of electromagnetic radiation in an atomic or molecular ensemble. They also underline a central theme of this section, that if we were to find an influence, which we call **Π**, that has the same T and P symmetry as **B**, Wigner's Principles would allow it to rotate plane-polarized light, causing circular birefringence and dichroism.[345] We describe later how **Π** is generated by the conjugate product of a powerful circularly polarized "pump" laser, such as a neodymium-doped yttrium aluminium garnet (Nd:YAG) laser, or dye laser, propagating parallel to the "probe" light beam. The latter may be another, tunable, laser, or broad-band radiation from a contemporary interferometric spectrometer.

Figures 49 and 50 illustrate the application of Wigner's Principles to

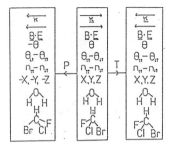

Figure 50.   Schematic of parity-inversion symmetry.

the Faraday Effect. They contain the variables of the complete experiment: the propagation vector ($\boldsymbol{\kappa}$) of the probe laser, parallel or antiparallel to **B**; the laboratory frame of reference (X, Y, Z); the angle, $\Delta\theta$, through which the plane of polarization of the laser is rotated; and the molecular structure, which can be achiral or chiral. The former is represented by water, and the latter by an enantiomer of bromochlorofluoromethane. Figure 49 deals with the effect of T, which is motion reversal. Under T, the ray of light moves in the opposite direction, so that $\boldsymbol{\kappa}$ changes sign. Static magnetic flux density **B** is an axial vector which also reverses sign under T. It is the curl, $\nabla \times \mathbf{A}$, of the vector potential[286] **A**. The angle of rotation, the frame (X, Y, Z), and the molecular structures are unaffected by T, because they are motion-independent. The angle $\Delta\theta$ depends on the circular polarity of the light, a plane-polarized probe laser being made up in equal parts of right- and left-handed components, the mathematical descriptions of which are given later in this section. If a right-handed screw is reversed in motion, its pitch, or screw sense, is not changed by T. However, it is customary to define right circular polarity as clockwise rotation of the electric field vector as the propagation vector travels towards an observer. The effect of motion reversal, T, is to change clockwise motion to anticlockwise, so that T reverses circular polarity from right to left or vice versa. T also reverses the propagation vector, but clearly leaves the product of this with the circular polarity unchanged. It is this product that we have described as the "screw sense" of the electromagnetic wave. The result of applying T is shown in Fig. 49, the motion-reversed variables are relatively in the same configuration. For example, if $\boldsymbol{\kappa}$ had been parallel to **B**, it remains parallel, and has not become antiparallel. The vector product $\boldsymbol{\kappa} \cdot \mathbf{B}$ has not changed sign. Its symmetry representation, denoted $\Gamma(\boldsymbol{\kappa} \cdot \mathbf{B})$, has remained the same under T. Similarly, the sign of $\Delta\theta$ is the same, and also (X, Y, Z) and the achiral and chiral molecular structures. The Faraday effect is an "observable," because the motion-reversed experiment is realizable.

TABLE III
P and I Symmetries of κ, B, and E

	P	T
κ	−	−
B	+	−
E	−	+

In Fig. 50 the operator P is applied to the same variables as those of Fig. 49. P has the following effect on position, $\mathbf{q}$, and momentum, $\mathbf{p}$ (Section V):

$$(\mathbf{q}, \mathbf{p}) \rightarrow (-\mathbf{q}, -\mathbf{p}) \tag{404}$$

and leaves the sign of time, t, unchanged. The propagation vector, κ, being a photon momentum,[346] is reversed by P. The magnetic flux density $\mathbf{B}$ is not, essentially because it is generated[286] by a cylindrical current flow, whose sense of rotation is reversed by T but not by P. (In contrast, $\mathbf{E}$ is generated by two electrodes, one positively and the other negatively charged. Under P the positions of the plates are reversed, but the charges are not, so that $\mathbf{E}$ changes sign. T does not affect the stationary charges, so $\mathbf{E}$ does not change sign under T. It is invariant, i.e., positive, to T). P also makes a right-handed screw into a left-handed screw, and so reversed the circular polarity of the probe laser. It reverses the sign therefore, of κ. It reverses the frame of reference, that is, $(X, Y, Z) \rightarrow (-X, -Y, -Z)$, and inverts the positions of all atoms of a molecule. It generates the opposite enantiomer, therefore, of a chiral molecule. These effects are summarized in Fig. 50. In the P-inverted experiment, everything has reversed in sign, but relative to each other, the variables have remained the same. The Faraday effect is again an "observable," because the P-inverted experiment is realizable.

The P and T symmetries of the variables κ, $\mathbf{B}$, and $\mathbf{E}$ are summarized in Table III. We see that κ is a "time odd, parity odd" variable; $\mathbf{B}$ is "time odd, parity even;" and $\mathbf{E}$ is "time even, parity odd."

It is known[342] that Faraday attempted to substitute $\mathbf{E}$ for $\mathbf{B}$ when examining the lead borate glass and recorded "no effect." How do these symmetry principles explain this? If we examine Figs. 49 and 50, it becomes clear that T has the effect

$$\Gamma(\kappa \cdot \mathbf{E}) \overset{T}{\rightarrow} -\Gamma(-\kappa \cdot \mathbf{E}) \tag{405}$$

on the vector product of κ and $\mathbf{E}$, showing immediately that $\mathbf{E}$ cannot rotate the plane of a light ray without violating the Wigner Reversality Principle. This is because the motion-reversed variables do *not* have the same relative sign, that is, $\kappa \cdot \mathbf{E}$ is reversed, but the others are not. Therefore the *static* $\mathbf{E}$ equivalent of the Faraday effect is said to violate T. A $\Delta\theta$ observed under these conditions would be a "nonobservable", that is, a symmetry-violating observable.[346] Similarly P does not reverse the product $\kappa \cdot \mathbf{E}$ (Table III), but reverses all the other variables in Fig. 50. The $\mathbf{E}$ equivalent of the Faraday effect also violates the Wigner Principle of Parity Inversion. Even if it did not violate T, it could only be observed, in consequence, in chiral ensembles. The group theory introduced later makes this point clearer.

It appears that $\Delta\theta$ has never been observed with $\mathbf{E}$ substituted for $\mathbf{B}$ in this context. If it were, unequivocally, it would signal the presence of a T-violating phenomenon (i.e., nonconservation of reversality), something which has been observed only once, and indirectly, in nuclear physics.[347] In contrast, the Wigner Principles allow the Faraday effect both in achiral and chiral ensembles, and this is what is observed experimentally.[286] P-violating phenomena were first predicted and observed in nuclear physics in the mid-1950s,[348] leading to the unification of the electromagnetic and weak forces,[349] the well-known CERN Experiment,[350] and to the expectation that P-violating phenomena pervade the whole of atomic and molecular spectroscopy,[351] one of the foremost achievements of science in this century. It is part of the purpose of this section to try to prepare theoretically for experiments to observe these P-violating phenomena spectroscopically.

C. The Symmetry of Cause and Effect: Group Theoretical Statistical Mechanics[352]

If an experiment conserves the Wigner P and T Principles, we can proceed with an investigation of the symmetries of cause and effect. These are subtle concepts, needing for clarity the language of group theory,[353] as described in Sections VII and VIII. The Neumann–Curie, or first, principle, as we have seen, holds for thermodynamic equilibrium when entropy is not changing systematically. In the rise transient condition[354] just after a field has been applied, for fall transients just after an applied field has been removed,[355] or in the steady state in the presence of an external field,[356] it is supplemented by a cause–effect principle which we have called the third principle of group theoretical statistical mechanics (gtsm).[352] The second principle is the equivalent of principle one applied[357] to a molecule-fixed frame of reference. The point groups of relevance refer, as we have seen, to the *ensemble* rather than the molecule itself, the group $R_h(3)$ for achiral ensembles, and the group $R(3)$ for ensembles of chiral molecules. The former is the group[352] of all rotations and "reflections" (more accurately parity inversions) about an origin (or "point") in frame (X, Y, Z); and the latter is the group of all rotations only, because reflection in $R(3)$ would result in the opposite enantiomer, a different physical entity. In consequence, "reflection" (i.e., parity inversion) is not a valid group theoretical operation of $R(3)$.

The irreducible representations of $R_h(3)$ are the D symbols,[286] which are subscripted u or g, respectively, negative and positive to parity inversion. They are superscripted by a number ranging from 0 to n, indicating tensor order. The D symbols can be used to denote the symmetry of an ensemble average in $R_h(3)$ (Section VII). For example, the symmetry of $\langle m \rangle$, where m is a scalar quantity (zero rank tensor) such as mass, is $D_g^{(0)}(+)$. This is the totally symmetric irreducible representation (tsr), and principle one indicates that $\langle m \rangle$ is a finite quantity in frame (X, Y, Z). The plus sign in brackets denotes positive to T. The average $\langle v \rangle$, where v is molecular center of mass velocity, vanishes in (X, Y, Z) by principle one, because its symmetry in $R_h(3)$ is $D_u^{(1)}(-)$, denoting an odd parity polar vector (a tensor of rank 1), which is negative to T. The average $\langle \omega \rangle$, where ω is molecular angular velocity, has the symmetry $D_g^{(1)}(-)$, and also vanishes by principle one because it does not contain the tsr of $R_h(3)$.

In the point $R(3)$ of chiral ensembles, the irreducible representations are D symbols without subscripts, because parity inversion in $R(3)$ is not a valid group theoretical operation, as we have seen. The symmetry of $\langle m \rangle$ in $R(3)$ is therefore $D^{(0)}(+)$, which is the tsr. The symmetries of $\langle v \rangle$ and of $\langle \omega \rangle$ are the same in $R(3)$: $D^{(1)}(-)$. This does not contain the tsr of $R(3)$, and in consequence, both $\langle v \rangle$ and $\langle \omega \rangle$ vanish in a chiral ensemble at field-free equilibrium.

D.   The D Symmetries of Natural and Magnetic Optical Activity

The rotation of plane-polarized electromagnetic radiation by $\mathbf{B}$, the Faraday effect, is often referred to as "magnetic optical activity."[286] It occurs both in $R_h(3)$ and $R(3)$, and its symmetry is that of $\mathbf{B}$ itself,[358] that is, $D_g^{(1)}(-)$ in $R_h(3)$ and $D^{(1)}(-)$ in $R(3)$. It is therefore "time odd" and vanishes in the absence of an external influence which is also time odd. This may be $\mathbf{B}$, but, using principle three, may also be the conjugate product $\mathbf{\Pi}$ of a pump laser,[359–362] a simple statement with many consequences, some of which are described later in this review.

Rotation of plane-polarized radiation is observed in a chiral ensemble without $\mathbf{B}$ or $\mathbf{\Pi}$, and is called "natural optical activity." It is therefore a property of $R(3)$ only, and must also pass the test[286] of Wigner's Principles at field-free equilibrium. If we remove $\mathbf{B}$ from Fig. 49 it becomes clear

M. W. EVANS

TABLE IV
Some Combined Field Symmetries

Tenor	$\hat{P}$	$\hat{T}$	$R_h(3)$	$R(3)$
E	$-$	$+$	$D_u^{(1)}$	$D^{(1)}$
Eκ	$+$	$-$	$D_g^{(0)} + D_g^{(1)} + D_g^{(2)}$	$D^{(0)} + D^{(1)} + D^{(2)}$
Bκ	$-$	$+$	$D_u^{(0)} + D_u^{(1)} + D_u^{(2)}$	$D^{(0)} + D^{(1)} + D^{(2)}$
B	$+$	$-$	$D_g^{(1)}$	$D^{(1)}$
EE	$+$	$+$	$D_g^{(0)} + D_g^{(1)} + D_g^{(2)}$	$D^{(0)} + D^{(1)} + D^{(2)}$
BB	$+$	$+$	$D_g^{(0)} + D_g^{(1)} + D_g^{(2)}$	$D^{(0)} + D^{(1)} + D^{(2)}$
EB	$-$	$-$	$D_u^{(0)} + D_u^{(1)} + D_u^{(2)}$	$D^{(0)} + D^{(1)} + D^{(2)}$
EBκ	$+$	$+$	$D_g^{(0)} + 3D_g^{(1)} + 2D_g^{(2)} + D_g^{(3)}$	$D^{(0)} + 3D^{(1)} + 2D^{(2)} + D^{(3)}$

that in the motion-reversed experiment, all variables for natural optical activity are reversed, that is, are relatively unchanged, and Wigner reversality is satisfied. In Fig. 50 with $\mathbf{B}$ missing, P has changed the sign of κ, has produced the opposite enantiomer of bromochlorofluoromethane, and has changed the sign of the angle of rotation. In a chiral ensemble, the P-inverted complete experiment is possible, because the relevant variables are relatively unchanged. In an achiral ensemble, however, P results in a water structure (Fig. 50) which is indistinguishable from the original. The molecular structure is plus to P, while all others are minus. In consequence, Wigner's P principle is not obeyed, and natural optical activity is not observable in an achiral ensemble at field-free equilibrium in the absence of parity violation[363] due to electroweak interactions.

The symmetry of natural optical activity is therefore that of a pseudoscalar in $R(3)$, denoted by $D^{(0)}(+)$, the tsr. By principle one its ensemble average is nonzero, and changes sign between enantiomers. It exists at equilibrium and is positive to T. It has no equivalent in the point group $R_h(3)$ in the absence of parity violation. In consequence, the symmetries of natural and magnetic optical activity are given different signatures.[364]

E. Application of gtsm — Combined Field Symmetries

The basic symmetries of Table III allow the definition of combined field symmetries.[364] Define the combined D symmetry of a tensor product[364] such as $E_i B_j$ as the complete product of their individual symmetry representations in the appropriate point group of the ensemble. This gives the results of Table IV where κ is the laser propagation vector, which is negative to P and T.[365] This may be the propagation vector of a probe or pump laser.

In classical electromagnetic field theory, rank zero natural optical ro-

TABLE V
Summary of Field-Induced Optical Activity

	Occurrence of Signature			
Tensor	$D_u^{(0)}$ (+)	$D_g^{(1)}$ (−)	$D^{(0)}$ (+)	$D^{(1)}$ (−)
E	No	No	No	No
Eκ	No	Yes	No	Yes
Bκ	Yes	No	Yes	No
B	No	Yes	No	Yes
EE	No	No	Yes	No
BB	No	No	Yes	No
EB	No	No	No	Yes
EBκ	No	No	Yes	No

tation is $D^{(0)}(+)$ in $R(3)$ and $D_u^{(0)}(+)$ in $R_h(3)$. The equivalents for magnetic optical activity are $D^{(1)}(-)$ and $D_g^{(1)}(-)$, respectively, the quantities in brackets denoting positive or negative to motion reversal T. Table V is a summary of effects.

Ross et al. have given an equivalent analysis in terms of photon selection rules.[366] Their results can be obtained by this application of gtsm to **E**, **B**, and **κ**.

Table V shows, for example, that the application of a static electric field strength **E** produces neither natural nor magnetic optical activity, leading to no effect, as first noted by Faraday.[342] The magnetic flux density **B** produces magnetic optical activity; the product **Bκ** has been used by Barron[367] and the present author[297] to define the Wagnière–Meier effect (forward–backward birefringence[306–308] due to **B** coaxial with **κ** of an unpolarized probe laser). By reference to Table V this product contains the symmetry of natural optical activity. An interesting example[368] is the combined **EB** symmetry, which produces magnetic optical activity in chiral ensembles only, implying that a type of "Faraday effect" can be obtained in chiral ensembles with a combination of electric field strength and magnetic flux density. This is discussed in detail in Refs. 368 and 369.

F. Application of gtsm to Nonlinear Optical Activity

Nonlinear optics is an important part of chemical physics, and in classical terms, involves quantities nonlinear in the complex conjugates[297,370] of the electric and magnetic field components of the classical electromagnetic field from Maxwell's equations. In general, these have plus (+) and minus (−) complex conjugates, and are right or left circularly polarized:[297, 370]

$$\mathbf{E}_L^- = E_0(\mathbf{i} + i\mathbf{j})e^{-i\theta_L}; \qquad \mathbf{E}_L^+ = E_0\,(\mathbf{i} - i\mathbf{j})e^{i\theta_L}$$

$$\mathbf{E}_R^- = E_0\,(\mathbf{i} - i\mathbf{j})e^{-i\theta_R}; \qquad \mathbf{E}_R^+ = E_0\,(\mathbf{i} + i\mathbf{j})e^{i\theta_R}$$

$$\mathbf{B}_L^- = B_0(\mathbf{j} - i\mathbf{i})e^{-i\theta_L}; \qquad \mathbf{B}_R^- = B_0(\mathbf{j} + i\mathbf{i})e^{-i\theta_R} \qquad (406)$$

$$\mathbf{B}_L^+ = B_0(\mathbf{j} + i\mathbf{i})e^{i\theta_L}; \qquad \mathbf{B}_R^+ = B_0(\mathbf{j} - i\mathbf{i})e^{i\theta_R}$$

$$\theta_L = \omega t - \boldsymbol{\kappa}_L \cdot \mathbf{r}; \qquad \theta_R = \omega t - \boldsymbol{\kappa}_R \cdot \mathbf{r}$$

In a laser propagating in Z of the laboratory frame (X, Y, Z), $\mathbf{i}$ and $\mathbf{j}$ here are unit vectors in X and Y, ω is the frequency of the field, and $\boldsymbol{\kappa}$ its wave vector. The position vector is denoted $\mathbf{r}$.

With these definitions,[297, 371] the electric dipole moment of a molecule in a strong laser field can be expressed as the double Taylor expansion[371]

$$\mu_i = \mu_{0i} + \alpha_{1ij}E_j + \alpha_{2ij}B_j + \frac{1}{2!}$$

$$\times (\beta_{1ijk}E_jE_k + \beta_{2ijk}E_j\beta_k + \beta_{3ijk}B_jE_k + \beta_{4ijk}B_jB_k)$$

$$+ \frac{1}{3!}\,(\gamma_{1ijkl}E_jE_kE_l + \cdots + \gamma_{8ijkl}B_jB_kB_l)$$

$$+ \cdots \qquad (407)$$

in which all quantities are in general complex.[371] Similarly, the molecular magnetic dipole moment can be double Taylor expanded as

$$m_i = m_{0i} + a_{1ij}B_j + a_{2ij}E_j + \frac{1}{2!}$$

$$\times (b_{1ijk}B_jB_k + b_{2ijk}B_jE_k + b_{3ijk}E_jB_k + b_{4ijk}E_jE_k)$$

$$+ \frac{1}{3!}(g_{1ijkl}B_jB_kB_l + \cdots + g_{8ijkl}E_jE_kE_l)$$

$$+ \cdots \qquad (408)$$

For example, the dynamic molecular property tensor α_{1ij} is the complex polarizability, α_{2ij} is the Rosenfeld tensor, and so on. The quantity μ_{0i} is the permanent molecular dipole moment in the absence of the laser field.

The accepted definition of rotational strength in linear optics is the Rosenfeld tensor α_{2ij}, which contains a tensor product of the molecular

TABLE VI
D Symmetries of Molecular Properties and Field Tensors for Optical Rotation: Point Group $R_h(3)^a$

Molecular Property	Field		Part of Dipole Moment	$D_u^{(0)}$	Order
$\left.\begin{array}{l}\alpha_2\\[4pt]\alpha_2\end{array}\right\} D_u^{(0)} + D_u^{(1)} + D_u^{(2)}(-)$	B	$B_g^{(1)}(-)$	μ	1	1
	E	$D_u^{(1)}(+)$	m	1	1
$\begin{array}{l}\beta_1\\ \beta_4\\ b_2\\ b_3\end{array}$ $D_u^{(0)} + 3D_u^{(1)} + 2D_u^{(2)} + D_u^{(3)}(+)$	EE	$D_g^{(0)} + D_g^{(1)} + D_g^{(2)}(+)$	μ	1	2
	BB	$D_g^{(0)} + D_g^{(1)} + D_g^{(2)}(+)$	μ	1	2
	BE	$D_u^{(0)} + D_u^{(1)} + D_u^{(2)}(-)$	m	1	2
	EB	$D_u^{(0)} + D_u^{(1)} + D_u^{(2)}(-)$	m	1	2
$\begin{array}{l}\gamma_2\\ \gamma_3\\ \gamma_5\\ \gamma_8\end{array}$ $\begin{array}{l}3D_u^{(0)} + 6D_u^{(1)} + 6D_u^{(2)}\\ \quad +3D_u^{(3)} + D_u^{(4)}(-)\end{array}$	$\left.\begin{array}{l}\text{EEB}\\ \text{EBE}\\ \text{BEE}\\ \text{BBB}\end{array}\right\}$ $D_g^{(0)} + 3D_g^{(1)} + 2D_g^{(2)} + D_g^{(3)}(-)$		$\begin{array}{l}\mu\\ \mu\\ \mu\\ \mu\end{array}$	$\begin{array}{l}3\\3\\3\\3\end{array}$	$\begin{array}{l}3\\3\\3\\3\end{array}$
$\begin{array}{l}g_2\\ g_3\\ g_5\end{array}$	$\left.\begin{array}{l}\text{BBE}\\ \text{BEB}\\ \text{EBB}\end{array}\right\}$ $D_u^{(0)} + 3D_u^{(1)} + 2D_u^{(2)} + D_u^{(3)}(+)$		$\begin{array}{l}m\\ m\\ m\end{array}$	$\begin{array}{l}3\\3\\3\end{array}$	$\begin{array}{l}3\\3\\3\end{array}$

^{a} For $R(3)$ remove g or u subscripts.

electric and magnetic dipole moments.[372] This can be seen at first order multiplying B_j in the expansion (407) of the electric dipole moment.

Applying gtsm we arrive at the classification scheme of Table VI.

From Table VI it can be deduced[371] that natural optical rotation to a given order in **E** and/or **B** occurs if the mediating property tensor contains the signature $D_u^{(0)}(\pm)$, with either a negative or positive T signature. The equivalent signature in $R(3)$ is $D^{(0)}(\pm)$. The possible nonlinear optical activities are summarized in Table VII.

A similar symmetry analysis can be made for "magnetic" optical activity.

Using Tables VI and VII, different classifications for the natural optical activity induced at different field orders become superfluous. We need only look to see if the relevant molecular property tensor contains $D_u^{(0)}$ with a plus or minus T signature coming from the opposite T signatures[286] of its real and imaginary parts. (The T signature, for example, of the real

TABLE VII
Known and New Optical Rotation Effects to Third Order with Suggested Nomenclature

Effect	Origin	Accompanies	Reference	Status
Rosenfeld optical rotation	α_2 **B** part	Polarization	372	Known
(First-order **B** rotation)	of μ			
First-order **E** rotation	a_2**E**	m Magnetization	—	New
Second-order **EE** rotation	β_1**EE**	μ Polarization	—	New
Magnetochiral birefringence	β_4**BB**	μ Polarization	306	Known
(Second-order **BB** rotation)				
Inverse magnetochiral birefringence	b_2**BE**			
(Second-order **BE** and **EB** rotations)	b_3**EB**	m Magnetization	370	Known
Third-order **EEB** rotation	γ_2**EEB**	μ Polarization		
Third-order **EBE** rotation	γ_3**EBE**	μ Polarization		
Third-order **BEE** rotation	γ_5**BEE**	μ Polarization		
Third-order **BBB** rotation	γ_8**BBB**	μ Polarization		
Third-order **BBE** rotation	g_2**BBE**	m Magnetization	—	New
Third-order **BEB** rotation	g_3**BEB**	m Magnetization		
Third-order **EBB** rotation	g_5**EBB**	*m* Magnetization		
Third-order **EEE** rotation	g_8**EEE**	*m* Magnetization		

part of dynamic polarizability is plus and that of the imaginary part is minus. The P signature of both parts[286] is plus.)

Equations (407) and (408) and Tables VI and VII provide a summary for the unified treatment of the various linear and nonlinear optical effects. Within this framework, the magnetochiral effect[306] for example, is treated through the complex molecular property tensor β_4, which contains $D_u^{(0)}$ (+). The effect is an electric dipole moment induced through β_4 by the tensor product **BB** (or $B_i B_j$ in Einstein notation) of the electromagnetic field. It can be thought of as the "second-order equivalent" of Rosenfeld rotation, characterized by $\alpha_{2ij}B_j$. The third-order equivalent in this sequence is $\gamma_{8ijkl}B_j B_k B_l$, whose γ_8 tensor contains $D_u^{(0)}$ three times, signifying three independent third-order optical rotation effects. This analysis can be repeated for other sequences of optical rotatory effects, involving, for example, the inverse magnetochiral effect, recently proposed by Wagnière.[370]

G. Optical Activity Induced by a Pump Laser

A particularly useful nonlinear property of an intense pump laser is optical rectification, defined through vector cross products of the complex conjugate solutions to Maxwell's equations. These are referred to as conjugate products, and have been defined by Ward[373] in terms of Feynman diagrams of quantum perturbation theory. In this subsection the P and T symmetries of the conjugate product

$$\mathbf{\Pi} = \mathbf{E}_L^+ \times \mathbf{E}_L^- = -\mathbf{E}_R^+ \times \mathbf{E}_R^- = 2E_0^2 i\mathbf{k} \tag{409}$$

are discussed, and shown to be the same as that of magnetic flux density **B**. This leads to the important conclusion that $\mathbf{\Pi}$ of a pump laser can produce, theoretically, *all the effects of* **B**.

The conjugate product defined in Eq. (409) is nonzero *only in a circularly polarized laser*, and changes sign if the polarization is switched from right to left. It is a purely imaginary quantity, which is proportional to the laser electric field strength amplitude squared, E_0^2. Its interaction energy with an atom or molecule is therefore

$$\left.\begin{aligned}
\Delta H &= -\frac{1}{2}\, \alpha_{1ij}\Pi_{ij} \\[6pt]
&= \frac{i}{2}\, \alpha_{1k}''\Pi_k \\[6pt]
(\alpha_{1k} &\equiv \alpha_{1k}' - i\alpha_{1k}'')
\end{aligned}\right] \tag{410}$$

where α_{1k}'' is the axial vector (rank one tensor) representation of the antisymmetric, imaginary part, of the rank two dynamic polarizability tensor. The latter is T-negative from semiclassical theory,[286] and to obtain a T-positive scalar interaction energy, $\mathbf{\Pi}$ must be T-negative. This expectation is reinforced from first principles as follows.

We wish to prove that

$$\mathbf{\Pi} \xrightarrow{T} -\mathbf{\Pi} \tag{411}$$

that is, that the conjugate product is negative to motion reversal. Considering the four electric field strengths in Eq. (406) we expand $\mathbf{E}_L^-$ as

$$\left.\begin{aligned}
\mathrm{Re}(\mathbf{E}_L^-) &= E_0(\mathbf{i}\cos\theta_L + \mathbf{j}\sin\theta_L) \\
\mathrm{Im}(\mathbf{E}_L^-) &= iE_0(\mathbf{j}\cos\theta_L - \mathbf{i}\sin\theta_L)
\end{aligned}\right] \tag{412}$$

and apply T term by term as follows:

$$\left.\begin{aligned}
\mathrm{Re}(\mathbf{E}_L^-) &\xrightarrow{T} E_0(\mathbf{i}\cos\theta_R - \mathbf{j}\sin\theta_R) \\
\mathrm{Im}(\mathbf{E}_L^-) &\xrightarrow{T} iE_0(\mathbf{j}\cos\theta_R + \mathbf{i}\sin\theta_R)
\end{aligned}\right] \tag{413}$$

These follow because

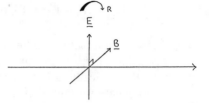

Figure 51. Basic elements of an electromagnetic wave. $\mathbf{E}$ and $\mathbf{B}$ are mutually perpendicular to the propagation vector, which is parallel to the conjugate product.

$$t \overset{T}{\to} -t; \qquad \omega \overset{T}{\to} \omega; \qquad \boldsymbol{\kappa}_L \overset{T}{\to} -\boldsymbol{\kappa}_R; \qquad \mathbf{r} \overset{T}{\to} \mathbf{r} \tag{414}$$

using the basic properties

$$\cos \theta_L \overset{T}{\to} \cos \theta_R; \qquad \sin \theta_L \overset{T}{\to} -\sin \theta_R \tag{415}$$

under T. These in turn follow because ω is a scalar angular frequency (a *number* of radians per unit time unaffected by motion reversal) and $\mathbf{r}$, the position vector, is invariant to T by definition. The time t itself reverses by definition of T as motion reversal, because time is position $\mathbf{r}$ divided by velocity $\mathbf{v}$, and the latter is reversed by the definition of motion reversal. Finally, $\boldsymbol{\kappa}$ the wave vector, is reversed in direction because the laser beam reverses direction with motion reversal. We have seen that T reverses the circular polarity of the laser (right to left or vice versa). Thus $\boldsymbol{\kappa}_L$ becomes $-\boldsymbol{\kappa}_R$ under T. Finally, Eq. (415) follows from the mathematical properties of the cosine and sine. Overall, therefore, T has the effect

$$\left. \begin{array}{cc} \mathbf{E}_L^- \overset{T}{\to} \mathbf{E}_R^+; & \mathbf{E}_L^+ \overset{T}{\to} \mathbf{E}_R^- \\ (\mathbf{E}^+ \times \mathbf{E}^-)_L \overset{T}{\to} -(\mathbf{E}^+ \times \mathbf{E}^-)_R \end{array} \right] \tag{416}$$

and reverses the sign of the conjugate product when the laser is switched from left to right or vice versa. The conjugate product is therefore negative to T, as the Hamiltonian (Eq. 410) indicated.

The conjugate products $\mathbf{\Pi}_L$ and $\mathbf{\Pi}_R$ of a left and right circularly polarized pump laser, such as a Nd:YAG, have opposite signs. The quantity $\mathbf{\Pi}$ derives from the helical motion of conjugate electric field components of the pump laser and is given mathematically by the vector product of orthogonal conjugates. Figure 51 is an illustration of the fact that $\mathbf{\Pi}$ is an axial vector with the same T and P symmetries as $\mathbf{B}$: minus to T and plus to P. Its forward direction is along the axis of the helix, and is therefore the direction of the propagation vector $\boldsymbol{\kappa}$ of the pump laser. The effect of

T is to reverse the direction of the laser beam, thus reversing both $\mathbf{\Pi}_L$ and $\mathbf{\Pi}_R$. However, $\mathbf{\Pi}$ is not reversed by P because it is the vector product of *two* P-negative electric field vectors.

From principle three, we expect $\mathbf{\Pi}$ to produce spectroscopic phenomena akin to $\mathbf{B}$, because it has the same P and T symmetries. In the first instance, we expect an equivalent of the Faraday effect, that is, optical activity with $D_g^{(1)}$ $(-)$ symmetry in $R_h(3)$ and $D^{(1)}$ $(-)$ symmetry in $R(3)$. By substituting $\mathbf{\Pi}$ for $\mathbf{B}$ in Figs. (49) and (50) we can illustrate the conservation of Wigner reversality and parity inversion in achiral and chiral ensembles. Furthermore, wherever $\mathbf{B}$ appears in atomic and molecular spectroscopy, our five symmetry principles allow us to substitute $\mathbf{\Pi}$, *immediately yielding a variety of possible new phenomena*, for example: $\mathbf{\Pi}$-induced Zeeman splitting; $\mathbf{\Pi}$-induced nuclear resonance; and $\mathbf{\Pi}$-induced parity violating spectral features of profound interest. In the same way that $\mathbf{B}$ couples to orbital and spin-angular atomic and molecular properties, we expect $\mathbf{\Pi}$ to couple to fundamental properties of the same symmetry, both electronic and nuclear in nature. Our task is now to evaluate these possibilities theoretically, and, most importantly, to examine the theories experimentally. In this context we have the added advantage that $\mathbf{\Pi}$ of a pump laser which is focused and Q-switched can be enormous, allowing a great deal of latitude in the experimental investigation. Additionally, the frequencies of both pump and probe can be tuned to the frequency of the spectral feature under investigation, allowing resonances due to (1) the probe, (2) the pump, and (3) double resonance with consequent amplification of low-intensity spectral features such as P violating transitions which do not obey Laporte's Rule.[374] The symmetry allows $\mathbf{\Pi}$ to couple in the interaction Hamiltonian to a quantity akin to nuclear spin. This must be a nuclear "spin polarizability" because $\mathbf{\Pi}$ is proportional to pump-laser electric field strength (E_0^2) squared. This would be a fundamentally new nuclear quantity akin to the nuclear magnetic dipole moment which links it to $\mathbf{B}$. It would mediate the phenomenon of "nuclear electromagnetic resonance," a "symmetry clone" of NMR, opening up a multitude of new possibilities, theoretical and experimental.

Part of the purpose of this section is to try to estimate the likely order of magnitude of some of these new induced effects, in order to prepare the way for the all-important experimental investigation of the theory.

H.   Some Expected $\mathbf{\Pi}$-Induced Spectroscopic Effects

In this subsection we mention briefly some expected effects of $\mathbf{\Pi}$, using as a guideline the historical development of one or two of the spectroscopies which rely on $\mathbf{B}$. This sets the theme for the variations that follow in other subsections.

I.   Π-Induced Zeeman Splitting

A good account of conventional Zeeman splitting due to **B** is given by Barron[286] (p. 12 ff.). It was first observed by Zeeman[375] as the broadening of the two lines of the first principal doublet from a sodium flame placed between the poles of an electromagnet. The main features of the conventional (magnetic) Zeeman effect are described in contemporary theory in the *A* term of the electronic Faraday effect owing to **B**,[286] and therefore we will be able in this section to develop an analogous semiclassical theory of the *A* term of optical activity due to **Π**. The latter is expected therefore both in atomic and molecular ensembles, and later in this section an estimate is given of the magnitude of the splitting in terms of **Π** of the pump laser.

The hyperfine (nuclear) part of the optical Zeeman effect gives the exciting prospect of optical nuclear resonance in analogy with NMR. This analogy is developed quantum mechanically and classically later in this section.

J.   Rayleigh–Raman Optical Activity Induced by Π

Radiation scattered from a chiral molecular ensemble is optically active, the Rayleigh and Raman spectral features of which provide specific information about fundamental molecular property tensors.[376] This is analogous with natural optical activity. Similarly, radiation scattered from a chiral or achiral ensemble to which **B** is applied shows magnetic optical activity. This leads to the expectation that radiation scattered from a probe laser will become optically active if **Π** from a pump laser is applied to the molecular ensemble which is the source of the scattered radiation. The semiclassical theory of this effect is developed in this section in terms of new molecular properties.

K.   Forward–Backward Birefringence due to Π

In 1982, Wagnière and Meier[306] proposed another fundamental effect of **B** in the spectroscopy of chiral ensembles: forward–backward (FB) birefringence, the semiclassical theory of which was given later by Barron and Vrbancich[316] in terms of new molecular property tensors. FB birefringence due to **B** is measured with a probe laser whose propagation vector **κ** is parallel or antiparallel to **B**, as in the Faraday effect. However, Wagnière–Meier birefringence is a forward–backward asymmetry,[377] not a circular asymmetry as in the Faraday effect, and is measured with *unpolarized* probe radiation. We can apply the Wigner Principles to **B**-induced FB birefringence with reference, as usual, to Figs. 49 and 50, and replacing the observable of the Faraday effect (the angle of rotation) by that of the Wagnière–Meier effect. The latter[378] is $(n^{\parallel} - n^{\parallel})$, where $n^{\parallel}$ denotes the real part of the refractive index with $\mathbf{B} \parallel \boldsymbol{\kappa}$; and $n^{\parallel}$ denotes $\mathbf{B} \parallel \boldsymbol{\kappa}$. Clearly, since T leaves the dot product $\mathbf{B} \cdot \boldsymbol{\kappa}$ unchanged, it has no effect on $(n^{\parallel} - n^{\parallel})$. The Wagnière–Meier effect therefore conserves Wigner Reversality. The Wigner P operation reverses the sign of the product $\boldsymbol{\kappa} \cdot \mathbf{B}$, and in consequence

$$(n^{\parallel} - n^{\parallel}) \xrightarrow{P} - (n^{\parallel} - n^{\parallel}) \tag{417}$$

that is, P reverses the sign of the variable in $\mathbf{B}$-induced axial birefringence, which is observable in consequence in *chiral* ensembles only. Because of the forward–backward asymmetry of the effect, Barron[279] has suggested that it has the symmetry $D_u^{(1)}(-)$ of a time-odd polar vector in $R(3)$. Note that this has the same D symmetry as that of the Faraday effect (a time-odd axial vector symmetry) in $R(3)$. There is no Wagnière–Meier effect, however, in achiral ensembles, for example of atoms, without parity violation.[377] The optical equivalent with $\mathbf{B}$ replaced by $\mathbf{\Pi}$ therefore gives an excellent opportunity of observing P violation in atomic and achiral molecular ensembles by tuning the pump laser (for example a dye laser) to exact resonance.

Substituting $\mathbf{\Pi}$ for $\mathbf{B}$ we obtain the phenomenon of FB birefringence caused by a pump laser parallel or antiparallel to $\boldsymbol{\kappa}$ of a probe laser, whose semiclassical theory is developed, with order of magnitude estimates, in this section.

L. Parity Violation in Molecular Ensembles due to $\mathbf{\Pi}$

The terra watt power levels achievable with contemporary pump lasers makes $\mathbf{\Pi}$-induced *circular* bifringence (the $\mathbf{\Pi}$-induced analogue of the Faraday effect) a candidate for the attempted observation in molecular ensembles of minute P violating phenomena due to electroweak interactions mediated by the neutral intermediate vector boson[279] whose existence was recently verified by the CERN experiment.[350] It is shown later, for example, that a focused and Q-switched Nd:YAG laser delivering $\mathbf{E}_0^2$ up to 10^{18} volts^2 m^{-2} is capable of rotating the plane of polarization of a probe laser at visible frequencies by about a million radians per meter of sample for $\mathbf{E}_0$ at a modest 10,000 volts cm^{-1}. By tuning the high power Nd:YAG pump to a P-violating transition (e.g., one that violates Laporte's Rule[374]) it may be possible to attain the enormous amplification needed to see clearly the P-violating spectral absorption.

M.   The Optical Zeeman Effect — Quantization of the Imaginary Part of the Atomic or Molecular Polarizability (the Electronic Orbital–Spin Angular Polarizability)

It is well known that quantized angular momentum is described by important commutator relations and coupling coefficients such as those of Clebsch and Gordan.[379] It is proportional through the gyromagnetic ratio to the quantized magnetic dipole moment, which in general has orbital and spin components. Many important spectral phenomena, are induced by applied static magnetic flux density **B**. NMR depends on the availability of magnets of up to 14 Tesla, with a necessarily high degree of homogeneity. Imaging, medical, and Fourier transform NMR and ESR now pervade much of contemporary analytical chemistry. It would be interesting to supplement or replace the NMR magnet with Π of a circularly polarized and inexpensive laser.

The interaction Hamiltonian[410] describes the way in which the quantity Π interacts with an ensemble of atoms or molecules. The quantity Π_i is, as we have seen, a T-negative, P-positive axial vector, proportional to the square of the electric field strength of the laser. As such, it can be expressed mathematically as a rank two antisymmetric polar tensor[380]

$$\Pi_i = \epsilon_{ijk}\Pi_{jk} \tag{418}$$

where ϵ_{ijk} is the Levi Civita symbol (the rank three, totally antisymmetric, unit tensor). The interaction Hamiltonian in this representation is, mathematically, the tensor contraction on to the scalar[380]

$$\Delta H = \frac{i}{2}\alpha''_{1ij}\ \Pi_{ij} \equiv \frac{i}{2}(\alpha''_{1xx}\Pi_{xx} + \alpha''_{1xy}\Pi_{xy} + \cdots + \alpha''_{1yz}\Pi_{yz} + \alpha''_{1zz}\Pi_{zz}) \tag{419}$$

where α''_{1ij} is the rank two tensor representation of the imaginary part of the polarizability.[279,286] On the grounds of symmetry, this must also be an antisymmetric second-rank polar tensor in order to contract on to a scalar energy in Eq. (419) For a pump laser propagating in the Z direction, the tensor representation of its conjugate product is

$$\Pi_z \equiv \begin{bmatrix} 0 & 2iE_0^2 & 0 \\ -2iE_0^2 & 0 & 0 \\ 0 & 0 & 0 \end{bmatrix} \tag{420}$$

with XY and YX components being nonzero. By definition of the mathematical procedure of tensor contraction,[286,380] this implies that the imaginary part of the polarizability is

$$\left.\begin{array}{c} \alpha''_{1i} = \epsilon_{ijk}\alpha''_{1jk} \\[4pt] \alpha''_{1z} = \alpha''_{1xy} - \alpha''_{1yx} \end{array}\right] \tag{421}$$

that is, either an axial vector with one Z component, or an antisymmetric polar tensor with XY and YX components. From semiclassical theory[286] these components are

$$\alpha''_{1xy} = -\alpha''_{1yx}$$

$$= -\frac{2}{\hbar}\sum_{j\neq n}\frac{\omega}{\omega_{jn}^2 - \omega^2}\,\mathrm{Im}(\langle n|\mu_x|j\rangle\langle j|\mu_y|n\rangle) \tag{422}$$

where μ_x and μ_y are orthogonal electric dipole moment components defined by quantum states n and j, with transition frequency

$$\omega_{jn} = \omega_j - \omega_n \tag{423}$$

The angular frequency in radians s^{-1} of the pump laser is ω. It is immediately clear from Eq. (422) that the imaginary part of the dynamic polarizability vanishes as ω goes to zero. If the laser is tuned to the transition frequency,

$$\omega \doteq \omega_{jn} \tag{424}$$

the vector α''_{1i} is amplified enormously.

The axial vector α''_{1i} has the same P and T symmetries as angular momentum, and is therefore quantized in the same way. This allows an important analogy to be drawn between the properties of α''_{1i}, which we name *the electronic orbital-spin angular polarizability* (angular polarizability for short), and the magnetic dipole moment m_i.

The angular polarizability α''_{1i} is a quantized axial vector with positive P and negative T symmetries, the same P and T symmetries as m_i and angular momentum J_i. In relativistic quantum theory, the magnetic dipole moment is proportional through the gyromagnetic ratio (γ_e) to a sum of the orbital (**L**) and spin (2.002**S**) electronic angular momenta, neglecting for the moment the nuclear contribution proportional to the nuclear angular momentum

620

M. W. EVANS

$$m_i = \gamma_e(L_i + 2.002S_i) \tag{425}$$

By a comparison of the Hamiltonians (Eq. 419) and

$$\Delta H_2 = -\gamma_e J_z B_z \tag{426}$$

we can immediately derive

$$\alpha''_{1z} \propto \left(\frac{\gamma_e}{B_0 c^2}\right) \tag{427}$$

showing that the angular polarizability is also proportional to the angular momentum. In analogy with Eq. (425) we have

$$\alpha''_{1i} = \gamma_\pi(L_i + 2.002S_i) \tag{428}$$

where γ_π is a new fundamental atomic or molecular property analogous to the gyromagnetic ratio, a *gyroptic* ratio.

Equation (428) leads to the important conclusion that *the angular polarizability has all the quantum mechanical properties of the angular momentum itself.*

Therefore it has the point-group symmetry in the molecule fixed frame of the well-known R symbols of the point-group character tables.[292] It has the commutator properties

$$[\alpha''_{1x}, \alpha''_{1y}] = i\hbar\alpha''_{1z}$$
$$[\alpha''_{1y}, \alpha''_{1z}] = i\hbar\alpha''_{1x} \tag{429}$$
$$[\alpha''_{1z}, \alpha''_{1x}] = i\hbar\alpha''_{1y}$$

Without having to solve the Schrödinger equation, and using the hermiticity of the commutator, it follows directly that angular polarizability of the commutator, it follows directly that angular polarizability is described by the angular momentum quantum numbers themselves. for orbital momentum these are L and its Z-axis projection, M

$$M = J, J - 1, \ldots, -J \tag{430}$$

and by the spin-angular momentum quantum number S. These angular polarizability quantum numbers couple to others, such as the nuclear angular momentum quantum number N, or the framework angular momentum quantum number O through the Clebsch Gordan, Racah, and Griffith equations.[381]

This leads to the expectation that a circularly polarized pump laser can generate all the spectroscopic properties customarily attributed to static magnetic flux density **B**, with a variety of useful analytical consequences, and with the important advantage of tuning to resonance (Eq. 424) with a natural frequency of the sample. One of these is named here the *optical Zeeman effect*, whose Hamiltonian can be expressed as

$$\Delta H = - \gamma_\pi (L_i + 2S_i) E_{0i}^2 \qquad (431)$$

This puts the Hamiltonian in a form where it can be developed with vector coupling models, as in the conventional (magnetically induced) Zeeman effect.[382] It follows from Eq. (431) that the selection rules for transitions between energy levels in the optical Zeeman effect are

$$\left. \begin{array}{l} \Delta J = 0, \pm 1 \\ \Delta M = 0, \pm 1 \end{array} \right] \qquad (432)$$

where M takes the values $J, \ldots, -J$.

A probe microwave (GHz frequency) field can be used to detect the optical Zeeman effect. When the probe is parallel to Π the selection rule is $\Delta M = 0$, giving the π components; and when the two fields are perpendicular, $\Delta M = \pm 1$, giving the σ components of the optical Zeeman effect. The pump laser is conveniently a circularly polarized narrow-width dye laser, and the sample may be sodium vapor[383–385] as in the original experiment by Zeeman.[375] This combination of GHz frequency probe and circularly polarized visible frequency dye laser can probably be used in the investigation of fine and nuclear (hyperfine) structure in the optical Zeeman effect.[386] The hyperfine structure is conveniently referred to as *optical electronic or nuclear spin resonance*.

For simple diatomic molecules, for example, the Hamiltonian (Eq. 431), initially neglecting hyperfine interactions, can be developed following Townes and Schawlow[382] in terms of the well-known Hund vector-coupling models. In the weak coupling limit, L_i and S_i precess about the molecular axis, which precesses about the total electronic angular momentum J_i. In the presence of the circularly polarized pump dye laser, tuned to resonance[383] with, for example, an atomic beam of diatomic molecules, J_i precesses about Π_i of the dye laser, with projection M_i in the direction of Π_i. This allows the Hamiltonian (Eq. 431) to be rewritten as

$$\Delta H = -\frac{(\Lambda + 2 \cdot 002\Omega)\gamma_\pi M E_{0z}^2}{J(J+1)} \tag{433}$$

with Λ and Ω defined by

$$\Omega = \mathbf{k}_a \cdot \mathbf{J}; \qquad \Lambda = \mathbf{k}_a \cdot \mathbf{L} \tag{434}$$

where $\mathbf{k}_a$ is a unit vector in the molecular axis. We expect $(2J+1)$ equally spaced *optical Zeeman lines* corresponding to the different values of M.

The extent of the splitting is determined by the ratio of the interaction energy $\alpha_{1z}'' E_0^2$ to the reduced Planck constant $h/2\pi$. This ratio is increased dramatically in the resonance condition (Eq. 424).

Hund's case (b) can be written in direct analogy to Eq. (11–5) of Townes and Schawlow[382] as

$$\Delta H = -\frac{1}{2J(J+1)}\left\{\Lambda^2 \frac{(N(N+1) + S(S+1) - J(J+1))}{N(N+1)}\right.$$

$$\left. + 2.002\left[J(J+1) + S(S+1) - N(N+1)\right]\right\} M\gamma_\pi E_0^2 \tag{435}$$

In general, the Hamiltonian of the optical Zeeman effect can be written in terms of the Landé factor g_J:

$$\Delta H = -g_J M \gamma_\pi E_0^2 \tag{436}$$

with g_J of the order of unity for molecules with net angular momentum, as in the conventional, **B**-induced, Zeeman effect. Otherwise g_J is dominated by the nuclear spin quantum number I. In general, g_J depends on the net electronic angular momentum J_i. In the optical Zeeman effect, if the Landé factors in states J_1 and J_2 are g_1 and g_2 respectively, and if the transition frequency (in Hz) between J_1 and J_2 is ν_0, the optical Zeeman spectrum will be a series of lines defined by

$$\nu = \nu_0 + (g_2 - g_1)M\gamma_\pi E_0^2/\hbar \tag{437}$$

for $\Delta M = 0$ (π components), and

$$\nu = \nu_0 + [(g_2 - g_1)M \pm g_1]\gamma_\pi E_0^2/\hbar \qquad (438)$$

for $\Delta M = M_2 - M_1 = \pm 1$ (σ components), with M_1 the lower state quantum number. As the pump laser is swept across resonance with a natural transition frequency of the sample, the angular polarizability α_{1i}'' will increase and decrease dramatically, and the spectral splittings will change accordingly. This pattern of change will be different for each natural transition frequency, giving plenty of scope for the development of useful new analytical methods based on the optical Zeeman effect.

In general, the g factors contain hyperfine (nuclear) and super hyperfine[387] contributions, which cause *nuclear electromagnetic resonance* as the GHz probe is swept across the same frequency as that of a transition frequency between hyperfine states. This is an optical equivalent of NMR.

For observation of the σ lines of the optical Zeeman effect, the electric field of the GHz electromagnetic probe should be perpendicular to Π_i of the circularly polarized pump laser (dye laser, Nd:YAG, CO_2 laser, etc.) so that the direction of propagation of the pump laser is parallel to the length or broadest faces of the waveguide carrying the GHz probe. *No optical Zeeman effect can be observed if the pump laser has not at least some degree of circular polarization.* This is useful in distinguishing it from the well-known Autler–Townes, or optical Stark, effect.[388–391] The latter depends on the *real* part of the polarizability (e.g., Eq. 2.59 of Ref. 388), whose symmetry is *T-positive*[388] from semiclassical theory, and which is a symmetric second-rank tensor with no axial vector equivalent. The optical Stark effect also has a zero frequency (DC) component,[388] whereas the optical Zeeman effect vanishes with vanishing ω from Eq. (422). The optical Stark effect has none of the quantization properties of the optical Zeeman effect, because the real part of the dynamical atomic or molecular polarizability is not proportional to angular momentum, having *opposite* T symmetry and being a *symmetric*-rank two tensor with no rank one axial vector equivalent.

It is important to bear in mind the opposite T and suffix symmetries[286] of the real and imaginary parts of the dynamic electronic polarizabilities. The fundamental difference between the Autler–Townes and optical Zeeman effects is a manifestation of these symmetry differences.

The general appearance of a simple type of optical Zeeman spectrum is expected to be similar to that sketched in Fig. (11-1) of the standard text by Townes and Schawlow[382] but will also depend, as mentioned, on resonance of type in Eq. (424). When nuclear hyperfine structure is considered, a Hamiltonian such as

$$\Delta H_2 = -\frac{1}{2}(\gamma_\pi g_J J_i \Pi_i + \gamma_{\pi n} g_I I_i \Pi_i) \qquad (439)$$

must be used, where $\gamma_{\pi n}$ is the *nuclear gyroptic ratio* and I the nuclear spin quantum number. If this is much smaller than the hyperfine energy, so that Π_i does not disturb the coupling between J_i and I_i, the vector coupling model gives the Landé type Hamiltonian

$$\Delta H_2 = \{-\gamma_{\pi n} g_I [I(I+1) + F(F+1) - J(J+1)]$$

$$-\gamma_\pi g_J [J(J+1) + F(F+1) - I(I+1)]\} \frac{M_F E_0^2}{2F(F+1)} \quad (440)$$

where F is the total angular momentum quantum number and M_F its projection on to Π_i of the circularly polarized pump laser. For a diamagnetic molecule, both terms of the optical Zeeman effect described by Eq. (440) are roughly equal in magnitude, giving considerable extra spectral detail for analytical purposes.

There appears to be another important potential advantage of the optical Zeeman effect over the conventional magnetic Zeeman effect. The pump laser of the former effect puts the molecule into an excited electronic state, in which there is net angular momentum imparted[392] to the molecule. The conjugate product Π_i spins a quantum state of the atom or molecule through the mediacy of α_{1i}''. This extra angular momentum results in a spectrum that is possible in the conventional Zeeman effect only in a molecule such as nitrous oxide, which is in a state with J number 3/2, so that the M_J states are 3/2, 1/2, $-1/2$, and $-3/2$, each of which is split into $M_J = 1$, 0, -1 states. Each M_J state would be expected to show hyperfine structure in optical Zeeman spectroscopy. Another example is that of oxygen, in which there would be an optical Zeeman splitting of the p-type triplets.

Symmetric and asymmetric tops would have more complicated optical Zeeman spectra, the case of HOD, for example, being interesting because its $4\bar{\nu}_{OH}$ state coincides with a circularly polarized Nd:YAG pump frequency.[393,394] The use of such coincidences between circularly polarized pump laser frequencies and natural transition frequencies is reminiscent of the well-developed techniques[395–397] of infrared and infrared–radio frequency double resonance, and of superhigh resolution saturation spectroscopy.[398] In each of these techniques, the circularly polarized pump laser (e.g., a narrow-width circularly polarized dye laser[392] would be used both for resonance, and for generation of Π_i, in analogy with the methods already in existence for multiphoton optical Stark splitting.[397]

One of the most sensitive techniques for optical Zeeman spectroscopy in atomic vapors would be possible with apparatus resembling that of Stroud and co-workers[397] utilizing circularly polarized visible dye lasers and radio-frequency fields focused carefully on to an atomic beam of sodium vapor. Using this apparatus, Molander, Stroud, and Yeazell[398] have characterized what they termed "high angular momentum Stark states" using a process of two-photon absorption by a circularly polarized dye laser, using a circularly polarized radio frequency field to produce quantized angular momentum in the sodium atoms. In this way the sodium atoms were excited to the $n = 25$ manifold, that is, "dressed" by the circularly polarized radio frequency field at 200 MHz, 8 V cm^{-1} equivalent electric field strength. The "dressed" state was then excited by a sensitive resonant two-photon process, reminiscent of the method used by Whitley and Stroud[392] for one of the first unambiguous observations of the optical Autler–Townes effect. For observation of the optical Zeeman effect in sodium vapor, the dye laser of this apparatus would be intense and circularly polarized, possibly Q-switched and focused, and the MHz/GHz probe would not necessarily be circularly polarized. Another possibility would be the use of two radio-frequency fields, one intense and circularly polarized pump to produce Π_i, the other a weaker unpolarized probe. (The same concept of producing Π_i from a radio-frequency field would be potentially very interesting in a conventional NMR spectrometer, using a circularly polarized probe radio-frequency field to produce extra angular momentum, thus *electromagnetically Zeeman splitting the conventional NMR spectrum* and giving a large number of analytical possibilities.)

N. Semiclassical Theory of the Optical Zeeman Effect

It is well known in semiclassical theory[286] that the conventional (**B**-induced) Zeeman effect can be described as the A term of the quantum mechanical description of the Faraday effect,[338] first derived by Serber,[399] and rederived in terms of useful molecular property tensors by Buckingham and Stephens.[400] The semiclassical treatment depends on a Voigt–Born perturbation[286] of the appropriate molecular property by the applied field. In the **B**-induced Faraday effect

$$\alpha''_{1xy}(B_z) = \alpha''_{1xy} + \alpha^{(B)'''}_{1xyz} B_z + \cdots \tag{441}$$

and in the Π-induced Faraday effect

$$\alpha''_{1xy}(\Pi_z) = \alpha''_{1xy} + \alpha^{(\pi)'''}_{1xyz} \Pi_z + \cdots \tag{442}$$

both involving the imaginary part of the dynamic polarizability, the quantity, which, as we have seen, mediates the optical Zeeman effect. This is consistent with the fact that there is also an *optical Faraday effect*, whose A term is the quantum description of the optical Zeeman effect. The first indications of the presence of an optical Faraday effect were obtained[401-403] by measuring the bulk magnetization due to a circularly polarized pulsed giant ruby laser, using[403] a simple inductance coil. The magnetization was easily observable[402,403] in a range of diamagnetic liquids through an electric current generated with the coil during a laser pulse, even though no resonance tuning–amplification was used. The optical Zeeman effect can be thought of as one of the numerous (and unexplored) spectral consequences of this magnetization by the circularly polarized pump laser. With resonance tuning, these effects appear well within the capability of ultrasensitive contemporary apparatus, such as that developed in other contexts by Stroud and co-workers.[397,398]

In semiclassical theory the angle of rotation of the optical Faraday effect can be expressed[286] as

$$\Delta\theta \doteq \frac{1}{2} \omega\mu_0 cl N \langle \alpha''_{1xy}(f) + \alpha^{(\pi)''}_{1xyz}\Pi_z \rangle \tag{443}$$

$$\doteq \frac{1}{2} \omega\mu_0 cl \frac{N}{d_n} E_0^2 \sum_n \left(\langle \alpha^{(\pi)''}_{1xyz}(f) \rangle + \frac{1}{kT} \langle \alpha''_{1zn}\alpha''_{1xy}(f) \rangle \right) \tag{444}$$

where ω is the measuring frequency (radians s^{-1}); μ_0 the permeability in vacuo in SI, c the velocity of light; l the sample length; N the number of molecules per unit volume; d_n the quantum state degeneracy; E_0^2 the square of the pump laser''s electric field strength; and kT the thermal energy per molecule. Here f is the dispersive line-shape function of semiclassical theory.[286] The ellipticity change in the probe is

$$\Delta\eta \doteq \frac{1}{2} \omega\mu_0 cl \frac{n}{d_n} E_0^2 \sum_n \left(\langle \alpha^{(\pi)''}_{1xyz}(g) \rangle + \frac{1}{kT} \langle \alpha''_{1zn}\alpha''_{1xy}(g) \rangle \right) \tag{445}$$

where g is the absorptive lineshape function.[286]

The circular birefringence and dichroism due to Π_i can be written in formal analogy with the magnetic electronic A, B, and C terms, written in the molecule fixed frame, as follows:

$$\Delta\theta \doteq -\frac{\mu_0 cl N E_0^2}{3\hbar} \left[\frac{2\omega_{jn}\omega^2}{\hbar}(f^2 - g^2)A + \omega^2 f\left(B + \frac{c}{kT}\right) \right] \tag{446}$$

$$\Delta\eta \doteq -\frac{\mu_0 cl N E_0^2}{3\hbar} \left[\frac{4\omega_{jn}\omega^2}{\hbar} fgA + \omega^2 g\left(B + \frac{c}{kT}\right) \right] \tag{447}$$

for a quantum transition from n to j, where n is the state of lower energy, usually the ground state. The A, B, and C terms due to the pump laser"s conjugate product are

$$A = \frac{3}{dn} \sum_n (\alpha''_{1jz} - \alpha''_{1nz}) \operatorname{Im}(\langle x|\mu_x|j\rangle\langle j|\mu_Y|n\rangle$$

$$B = \frac{3}{dn} \sum_n \operatorname{Im}\left(\sum_{k \neq n} \frac{\langle k|\alpha''_{1z}|n\rangle}{\hbar\omega_{kn}} (\langle n|\mu_x|j\rangle\langle j|\mu_\gamma|k\rangle - \langle n|\mu_\gamma|j\rangle\langle j|\mu_x|k\rangle) \right.$$

$$\left. + \sum_{k \neq j} \frac{\langle j|\alpha''_{1z}|k\rangle}{\hbar\omega_{kj}} (\langle n|\mu_x|j\rangle\langle k|\mu_y|n\rangle - \langle n|\mu_y|j\rangle\langle k|\mu_x|n\rangle) \right)$$

$$C = \frac{3}{dn} \sum_n \alpha''_{1zn} \operatorname{Im}(\langle n|\mu_x|j\rangle\langle j|\mu_y|n\rangle)$$

which represent a sum over transitions from component states of a degenerate set to an excited state ψ_j which itself could be a member of a degenerate set. Note that the A term, which describes the optical Zeeman effect in semiclassical theory, requires a definition of the angular polarizability α''_{1z} in states n and j. The definition in state n is Eq. (422), and that in state j is

$$\alpha''_{1\alpha\beta,j} = - \alpha''_{1\alpha\beta,j}$$

$$= -\frac{2}{\hbar} \sum_{k \neq j} \frac{\omega}{\omega_{kj}^2 - \omega^2} \operatorname{Im}(\langle j|\mu_\alpha|k\rangle\langle k|\mu_\beta|j\rangle) \tag{448}$$

where k denotes a quantum state higher in energy than j. In writing the A, B, and C terms in this way, weighted Boltzmann averaging[286] is used with the energy ratio $-\alpha''_{1z} E_0^2/kT$. A Q-switched and focused Nd:YAG laser produces E_0 of about 10^9 volt m^{-1}, and for an order of magnitude estimate[404] of about 10^{-41} C^2 m^2 J^{-1} for α''_{1z} the ratio $\alpha''_{1z}E_0^2/kT$ is of the order unity. It appears that this can easily be achieved by tuning to resonance with the transition frequencies ω_{nj} or ω_{jk}.

The A term is responsible for optical Zeeman splitting by the circularly polarized pump laser as measured by a suitable probe. A right circularly polarized pump laser delivers a photon with $-h$ projection in the propagation (κ) axis of the laser, producing a change $\Delta M = -1$ in the atomic or molecular quantum state. Conversely, the left circularly polarized pump laser delivers a photon with projection $+h$, with selection rule $\Delta M = 1$. In a linearly polarized pump there is no optical Zeeman effect, and the selection rule is $\Delta M = 0$. This is accounted for classically in Eq. (409), where the left conjugate product is positive, and the right negative. The sign change produced by switching the pump''s polarity from left to right is equivalent to the change produced in the Faraday effect by switching the direction of **B** relative to the propagation vector of the probe.

As in the conventional, **B**-induced Faraday effect, the A term due to **Π** comes from the splitting of lines by **Π** into right and left circularly polarized components. The B term of the optical Faraday effect originates from the mixing of energy levels due to the pump laser, and the C term is a change of electronic population of the pump-laser-split ground states. In each case the magnetic dipole moment operator of the conventional Zeeman effect is replaced by the angular polarizability vector α_1'' with the same P and T symmetries and M quantum number selection rules. The angular polarizability can be greatly amplified by resonance as we have seen.

In the **Π**-induced A and C terms, the vector polarizabilities in states n and j exist in general in the presence and absence of degeneracy, from the definitions (Eqs. 446 and 447). Therefore the A and C terms should be visible in molecules of lower symmetry than in the conventional **B**-induced equivalents.

O.   Laser-Induced Electronic and Nuclear Spin Resonance

The above discussion leads to the theoretical expectation of electron and nuclear spin resonance due to optical rectification, of great potential value because lasers and circularly polarized radio frequency fields can ultimately[405] supplement interestingly the magnets of NMR and ESR spectrometers. The origin of electron spin resonance with **Π** is found in the quantum nature of the vector polarizability α_1'', which is proportional through Eq. (431) to the sum of the orbital and spin-angular momenta of the electron, the latter taking the values $1/2$ and $-1/2$. If the probe is tuned to the resonance frequency ω_R it is absorbed when

$$\left. \begin{aligned} \hbar\omega_R &= E(1/2) - E(-1/2) \\ &= 2\hbar\gamma_\pi E_0^2 \end{aligned} \right] \tag{449}$$

which is the condition for electron spin resonance due to optical rectification in a circularly polarized electromagnetic field. This depends on the $\Delta M = 1$ transition between the electron spin polarizability states $1/2$ and $-1/2$. If the pump laser inputs the energy $\alpha_{1z}'' E_0^2$, the resonance frequency range is roughly $\alpha_{1z}'' E_0^2 / \hbar$, which is in the MHz for an order of magnitude $10^{-41} \mathrm{C}^2 \, \mathrm{m}^2 \, \mathrm{J}^{-1}$ for the angular polarizability, and a pump laser electric field strength of the order 10^6–10^9 volt m^{-1}. Conventional ESR spectrometers can therefore be adapted for use with the pump laser, using existing microwave probes. High-sensitivity apparatus is available similar to that developed[398] by Stroud and co-workers in another context. Resonance amplification of the angular polarizability can be utilized.

The most useful feature of conventional, **B**-induced, ESR is retained when **B** is substituted by **Π**. This is coupling of electron and nuclear spins, caused by the **Π**-induced transition between electron orientation states by the interaction of the spin-angular momenta of the electron with nuclei which have nonzero spin-quantum numbers I. This hyperfine resonance structure can be induced, under the right conditions, by **Π** of a pump laser, such as a dye laser. In the triphenyl methyl radical, for example, the **B**-induced hyperfine structure of one resonance peak contains no less than 196 lines, and similar detail is expected from **Π** used in place of **B**.

If a circular polarized pump laser is used to *supplement* the magnetic flux density of a conventional NMR spectrometer, the result is a Hamiltonian of the form

$$\Delta H_3 = -\gamma_n I_i B_i - \tfrac{1}{2}\gamma_\pi J_i \Pi_i \qquad (450)$$

where γ_N is the nuclear gyromagnetic ratio, and the angular momentum J_i is the sum

$$J_i = L_i + 2.002 S_i \qquad (451)$$

This Hamiltonian can be rewritten in the Landé form:

$$\Delta H_3 = -\gamma_N \left(1 + \frac{I(I+1) - J(J+1)}{2J_T(J_T+1)}\right) J_{Ti} B_i$$

$$-\frac{\gamma_\pi}{2}\left(1 - \frac{I(I+1) - J(J+1)}{2J_T(J_T+1)}\right) J_{Ti} \Pi_i \qquad (452)$$

where

$$J_T = J + I \qquad (453)$$

is the total angular momentum quantum number. For $I \neq 0$ the customary NMR line, defined through the selection rule

$$\Delta M_I = \pm 1 \tag{454}$$

is split into a new pattern of lines dependent on the selection rule

$$\Delta M_{J_T} = 0, \pm 1 \tag{455}$$

and on the individual values of I and J. A convenient way of doing this is to increase the intensity and to circularly polarize the MHz probe radio-frequency field of the NMR spectrometer. *This has great potential application in analytical laboratories, because modifications to include Π_i can be made in standard NMR instruments, including 2-D (imaging), and Fourier transform NMR.*

P. Rayleigh–Raman Light-Scattering Optical Activity due to Optical Rectification

Magnetic Rayleigh–Raman optical activity was developed theoretically by Barron and Buckingham[406] and experimentally by Barron and co-workers. The radiation scattered from a probe laser becomes optically active by applying a magnetic field to the sample parallel to the incident probe laser. For 90° scattering

$$\Delta_X(90°) = \frac{2\,\text{Im}(\alpha_{1XY}\,\alpha_{1XX}^*)}{\text{Re}(\alpha_{1XX}\,\alpha_{1XX}^* + \alpha_{1XY}\,\alpha_{1XY}^*)} \tag{456}$$

and

$$\Delta_Z(90°) = \frac{2\,\text{Im}(\alpha_{1ZY}\,\alpha_{1ZX}^*)}{\text{Re}(\alpha_{1ZX}\,\alpha_{1ZX}^* + \alpha_{1ZY}\,\alpha_{1ZY}^*)} \tag{457}$$

where

$$\Delta = \frac{I^R - I^L}{I^R + I^L} \tag{458}$$

is the dimensionless circular intensity difference.[286] The scattering is described in Eqs. (456) and (457) by laboratory frame components of complex molecular polarizability tensors and complex conjugates described by a superscripted asterisk. The magnetic field **B** activates optical activity in several different ways, and in consequence so does **Π**. The latter activates the polarizability through a Voigt–Born expansion to first-order in Π_Z. Consequently, in analogy with the effect of **B**, there are several new optically active scattering phenomena due to Π of a pump laser parallel to the incident probe. These can be subclassified into Π-induced Rayleigh and Raman effects associated with diagonal scattering transitions; and with off-diagonal transitions which probe the analogue of ground-state optical Zeeman splitting due to Π. There is also optically active resonance, as well as transparent, Raman scattering due to Π, together with the interesting prospect of double resonance, when both the pump and probe are tuned simultaneously. There is the additional advantage that Π is expected to have a much more direct influence on vibrational spectra than $\mathbf{B}$, because Π is electromagnetic in origin, and Raman scattering in general is a phenomenon which depends on electronic states excited by electromagnetic fields.

Optical activity in scattered probe radiation due to Π of the pump laser (or circularly polarized radio-frequency field) conserves parity and reversality in all molecular ensembles (chiral and achiral) and the main contribution in Rayleigh scattering is due to interference between waves generated by polarizability tensor components respectively perturbed and unperturbed by Π of the pump laser. It is measured by scattered probe radiation at any scattering angle, but the theory simplifies considerably[407] for scattering at 90°.

The same considerations apply for Raman optical activity due to Π, but the interference is now between unperturbed symmetric transition polarizability components, $(\alpha_{ij})^s_{mn}$, and antisymmetric components $(\alpha_{ij})^a_{mn}$ perturbed by Π, and vice versa.[286]

In both Rayleigh and Raman contexts the Voigt–Born expansion in Π of the complex dynamic polarizability is

$$\alpha_{1ij}(\Pi_k) = \alpha'_{1ij} + \alpha^{(\pi)'}_{1ijk}\,\Pi_k - i(\alpha''_{1ij} + \alpha^{(\pi)''}_{1ijk}\Pi_k) \tag{459}$$

and products such as $\alpha_{1xx}\alpha^*_{1ixy}$ are Boltzmann-averaged[286] with the potential energy $\alpha''_{1Z}E_0^2/kT$. We obtain expressions for optically active scattering due to Π analogous with those for $\mathbf{B}$ given in Eqs. (3.5.45) and (3.5.53) of Ref. 286. The most general expression for the Stokes parameters for optically active scattered radiation due to Π are the analogues of those due to $\mathbf{B}$[286] in Barron's Eq. (3.5.51), and in Barron, Meehan, and Vrbancich.[407] Here the superscripts R and L refer to the scattered probe radiation, whose electric field intensity is denoted $E_0^{(p)}$, and R is the distance from the scattering center:

$$\Delta\alpha = (I_\alpha^R - I_\alpha^L)/(I_\alpha^R + I_\alpha^L) \tag{460}$$

where

$$A = (\omega^4 \mu_0 E_0^{(p)2})/(16\pi^2 c R^2) \tag{461}$$

$$I_X^R - I_X^L = A \operatorname{Im}(\alpha_{1XX} \alpha_{XX}^*) \tag{462}$$

$$I_Z^R - I_Z^L = A \operatorname{Im}(\alpha_{1ZY} \alpha_{1ZX}^*) \tag{463}$$

$$I_X^R + I_X^L = \frac{A}{2} \operatorname{Re}(\alpha_{1XX} \alpha_{1XX}^* + \alpha_{1XY} \alpha_{1XY}^*) \tag{464}$$

$$I_Z^R + I_Z^L = \frac{A}{2} \operatorname{Re}(\alpha_{1ZX} \alpha_{1ZX}^* + \alpha_{1ZY} \alpha_{1ZY}^*) \tag{465}$$

The products of polarizabilities in these expressions are perturbed by $\boldsymbol{\Pi}$ of the pump laser. After Boltzmann averaging[286] for $\boldsymbol{\Pi}$ in the Z direction of the incoming probe beam, we have:[408-411]

$$
\begin{aligned}
I_X^R - I_X^L = 2AE_{0Z}^2 \langle & \alpha_{1XX}' \alpha_{1XYZ}^{(\pi)''*} \\
& - \alpha_{1XX}'' \alpha_{1XYZ}^{(\pi)'*} + \alpha_{1XXZ}^{(\pi)'} \alpha_{1XZ}''^* - \alpha_{1XXZ}^{(\pi)''} \alpha_{1XY}'^* \\
& + \frac{1}{kT}(\alpha_{1XX}' \alpha_{1XY}''^* \alpha_{1XY}'' - \alpha_{1XX}'' \alpha_{1XY}'^* \alpha_{1XY}'') \rangle
\end{aligned}
\tag{466}
$$

for the numerator of ΔX. It is seen that this is proportional to the square of the electric field intensities both of the pump and of the probe. In consequence, it appears that the effect can be easily large enough for observation with a suitable pump dye laser.

These Stokes parameters contain cross terms between and $(\alpha_{1ij}')_{mn}^a$ which are responsible for resonance Raman scattering from the probe due to $\boldsymbol{\Pi}$ of the circularly polarized pump. "Resonance" in this context refers to the probe frequencies. The Stokes parameters switch sign if the pump is switched from right to left for a given probe circular polarity. The effect can also be generated by a pump laser at any suitable angle to the probe, automatically removing the need to filter off scattered pump radiation.

In developing averages of the type shown in Eq. (466) use is made[408] of Boltzmann weighted averaging techniques to produce results in the laboratory frame such as

$$I_X^R - I_X^L = 2AE_0^2 \langle \alpha_{1XX} \alpha_{1XY}^* \rangle_\pi \tag{467}$$

which transform into the molecule fixed frame[408] as follows:

$$\langle \alpha_{1XX}' \alpha_{1XYZ}^{(\pi)''*} \rangle = \alpha_{1\gamma\delta}' \alpha_{1\epsilon\alpha\beta}'' \langle j_\alpha k_\beta i_\gamma i_\delta i_\epsilon \rangle$$

$$= \frac{1}{30}(2\alpha'_{1\alpha\beta}\epsilon_{\alpha\gamma\delta}\alpha''^{*}_{1\beta\gamma\delta} + \alpha'_{1\alpha\alpha}\epsilon_{\beta\gamma\delta}\alpha''_{1\gamma\delta\beta}) \qquad (468)$$

using Greek subscripts to refer to molecule fixed-frame quantities. Stokes parameters such as

$$S_0^\alpha(0^0) = -2KE_{0z}^2[2\alpha'_{1\alpha\beta}\,\epsilon_{\alpha\gamma\delta}\alpha''^{*}_{1\beta\gamma\delta}$$

$$+ \alpha'_{1\alpha\alpha}\epsilon_{\beta\gamma\delta}\alpha''^{*}_{1\gamma\delta\beta} + 2\alpha''_{1\alpha\beta}\epsilon_{\alpha\gamma\delta}\alpha'^{*}_{1\gamma\beta\delta}$$

$$+ \alpha''_{1\alpha\beta}\epsilon_{\alpha\beta\gamma}\alpha'^{*}_{1\delta\delta\gamma} + \frac{1}{kT}(2\alpha'_{1\alpha\beta}\epsilon_{\alpha\gamma\delta}\alpha''^{*}_{1\beta\gamma\delta}\alpha''_{1\delta\mu}$$

$$+ \alpha'_{1\alpha\alpha}\epsilon_{\beta\gamma\delta}\alpha''^{*}_{1\gamma\delta}\alpha''_{1\beta\mu})]\rho\sin 2\eta \qquad (469)$$

can then be expressed conveniently in this frame of reference. These equations are formally identical with Barron''s Eqs. (3.5.45–3.5.47) of Ref. 286, but implement Π in place of $\mathbf{B}$ and the angular polarizability α''_1 in place of the magnetic dipole moment $\mathbf{m}$. As in scattered optical activity due to $\mathbf{B}$, that due to Π does not lead to a circularly polarized component in the light scattered at 90° if the pump and probe lasers are parallel. It can be generated when the circularly polarized pump laser is parallel with the scattered beam, and the intensity of the scattered probe radiation depends on the degree of circularity of the incident probe only when the pump is parallel with the probe.

The circular intensity differences for scattering of probe radiation at 0°, 180°, and 90° due to Π in a pump parallel to the probe are found in analogy with the theory of Barron and Buckingham[406] as

$$\Delta(0^\circ) = \Delta(180^\circ)$$

$$= -2E_0^2[2\alpha'_{1\alpha\beta}\epsilon_{\alpha\gamma\delta}\alpha''^{*}_{1\beta\gamma\delta} + \alpha'_{1\alpha\alpha}\epsilon_{\beta\gamma\delta}\alpha''^{*}_{1\gamma\delta\beta}$$

$$+ 2\alpha''_{1\alpha\beta}\epsilon_{\alpha\gamma\delta}\alpha'^{*}_{1\gamma\beta\delta} + \alpha''_{1\alpha\beta}\epsilon_{\alpha\beta\gamma}\,\alpha'^{*}_{1\delta\delta\gamma} + \frac{1}{kT}(2\alpha'_{1\alpha\beta}\epsilon_{\alpha\gamma\delta}\alpha''^{*}_{1\beta\gamma}\alpha''_{1\delta\eta}$$

$$+ \alpha'_{1\alpha\alpha}\,\epsilon_{\beta\gamma\delta}\alpha''^{*}_{1\gamma\delta}\alpha''_{1\beta\eta})]/(7\alpha'_{1\lambda\mu}\alpha'^{*}_{1\lambda\mu} + \alpha'_{1\lambda\lambda}\alpha'^{*}_{1\mu\mu} + 5\alpha''_{1\lambda\mu}\alpha''^{*}_{1\lambda\mu})$$

$$(470)$$

where the molecular property tensors are all expressed in the molecule fixed frame and where the tensor summation convention has been applied to repeated indices.

Q. Forward–Backward Birefringence due to Optical Rectification

Forward-backward birefringence due to a static magnetic field was introduced in 1982 by Wagnière and Meier,[306–308] and is another fundamentally important effect of **B** in atomic and molecular spectroscopy. From the third principle of gtsm we immediately have the possibility of an analogous effect due to **Π**, an effect which can be developed theoretically[409] in terms of the zeta tensor of semiclassical theory,[286] or alternatively,[408–411] directly from the Maxwell equation.

Forward–backward birefringence can be measured with *unpolarized* probe radiation. It is generated in the Wagnière–Meier effect, for example,[306,316] by reversing the direction of **B** with respect to **κ**, the propagation vector of the unpolarized probe. It is sustained *only in chiral ensembles*, and therefore, if observed in atoms, would be an indication of parity nonconservation.[290,412] It has been shown recently that forward–backward birefringence can be generated by a pump laser in at least two ways, called class one and two *spin-chiral birefringence*.[315] Class 1[412,413] is observed by switching the polarity of the pump from left to right, keeping the direction of its propagation constant. Class 2[412,413] keeps the circular polarity constant and reverses the direction of propagation. In both cases, **Π** plays the role of **B** of the Wagnière–Meier effect, and spin-chiral birefringence is sustained only in chiral ensembles, giving another good opportunity of investigating parity nonconservation in achiral ensembles. Both in class 1 and 2 spin-chiral effects, amplification by resonance is feasible by sweeping the frequency of the pump through natural transition frequencies of the chiral ensemble.

The semiclassical theory of spin-chiral birefringence for pump and probe directed in the Z axis relies on the following scalar elements of the zeta tensor:[316]

$$\zeta'_{XXZ} = \frac{2}{c}\left(\frac{\omega}{3}A''_{XXZ} + \alpha'_{2XY}\right) \tag{471}$$

and

$$\zeta'_{YYZ} = \frac{2}{c}\left(\frac{\omega}{3}A''_{YYZ} - \alpha'_{2YX}\right) \tag{472}$$

and on Voigt–Born perturbations linear in Π_z of the Rosenfeld tensor[316] and electric dipole – electric quadruple tensor[413]

$$\alpha'_{2XY}(\Pi_Z) = \alpha'_{2XY} + \alpha^{(\pi)'}_{2XYZ}\Pi_Z + \cdots \qquad (473)$$

$$\alpha'_{2YX}(\Pi_Z) = \alpha'_{2YX} + \alpha^{(\pi)'}_{2YXZ}\Pi_Z + \cdots \qquad (474)$$

$$A''_{XXZ}(\Pi_Z) = A''_{XXZ} + A^{(\pi)''}_{XXZZ}\Pi_Z + \cdots \qquad (475)$$

$$A''_{YYZ}(\Pi_Z) = A''_{YYZ} + A^{(\pi)''}_{YYZZ}\Pi_Z + \cdots \qquad (476)$$

subjected to Boltzmann averaging with the interaction energy

$$V(\Omega) = -E_0^2 \alpha''_{1Z} \equiv -E_0^2(\alpha''_{1XY} - \alpha''_{1YX}) \qquad (477)$$

The forward–backward birefringence of the class 1 effect is the difference

$$n'(\Pi_Z \uparrow\uparrow \kappa_Z) - n'(\pi_Z \uparrow\downarrow \kappa_Z) \qquad (478)$$

where κ is the propagation vector of the probe laser. In a dilute solution for $E_0^2 \alpha''_{1XY} \ll kT$ we have[413]

$$
\begin{aligned}
(n^{\parallel} - n^{\parallel}) &\doteq 2\mu_0 c N E_0^2 \left\{ \frac{1}{3} \epsilon_{\alpha\beta\gamma}\alpha^{(\pi)'}_{2\alpha\beta\gamma}(f) \right. \\
&\quad + \frac{1}{30} kT \left[(4\delta_{\alpha\beta}\delta_{\gamma\delta} - \delta_{\alpha\gamma}\delta_{\beta\delta} - \delta_{\alpha\delta}\delta_{\beta\gamma})(\alpha'_{2\alpha\gamma}\alpha''_{1\beta\delta} \right. \\
&\quad \left. - \alpha'_{2\beta\gamma}\alpha''_{1\alpha\delta}) \right] + \frac{\omega}{45} (3A^{(\pi)''}_{\alpha\alpha\beta\beta}(f) - A^{(\pi)''}_{\alpha\beta\beta\alpha}(f) + \cdots \left. \right\}
\end{aligned}
\qquad (479)
$$

with tensor components defined in the molecule fixed frame. An order-of-magnitude estimate of this effect[413] produces

$$n^{\parallel} - n^{\parallel} \doteq 10^{-23}E_0^2 \qquad (480)$$

which for a Q-switched and focused Nd:YAG laser delivering 10^{18} (volt/m)^2 is of the order 10^{-5}, even without the added advantages of resonance tuning, in which condition the effect is amplified greatly. If a highly polarizable chiral material is chosen, such as a helical biomacromolecule or a cobalt complex, it is probable that the forward–backward birefringence can be increased to the point where it is easily observable even in a transparent part of the spectrum.

R. The Optical Faraday Effect — Order-of-Magnitude Estimate of the Angle of Rotation of a Plane-Polarized Probe

An expressison for the optical Faraday effect in the laboratory frame of reference can be obtained from the XY element of the perturbed polarizability, Eq. (441), which parallels the standard Voigt–Born perturbation in the semiclassical theory of the Faraday effect. The rotation of the plane of polarization of the probe laser can be derived[413] from Eq. (441) as the laboratory frame expression

$$\Delta\theta \doteq \frac{1}{2}\omega\mu_0 c |NE_0^2 \left(\langle \alpha_{xyz}^{(\pi)''}(f) \rangle + \frac{\langle \alpha_{1zn}'' \alpha_{xy}''(f) \rangle}{kT} \right) \tag{481}$$

and where $\alpha_{xy}''(f)$ is the absorptive[286] part of the tensor component XY of the perturbed angular polarizability. In the molecule fixed frame, Eq. (481) becomes

$$\Delta\theta \doteq \frac{1}{6}\omega\mu_0\, clN\, E_0^2\, \alpha_{1\alpha\beta}'' \frac{\alpha_{1\alpha\beta}''}{kT} + \cdots \tag{482}$$

for $\alpha_{1z}'' E_0^2 \ll kT$ in dilute solution.

For a conservative order of magnitude $10^{-41}\,\mathrm{J}^{-1}\,\mathrm{C}^2\,\mathrm{m}^2$ for $\alpha_{1\alpha\beta}''(f)$, we obtain[413]

$$\Delta\theta \doteq 10^{-15} E_0^2 \text{ radian} \tag{483}$$

at 300 K for the angle of rotation due to the component of Eq. (482). This is easily within range of a contemporary laser spectropolarimeter,[414] even for an unfocused, CW dye laser operating out of resonance. As in the conventional Faraday effect, there will be an accompanying dichroism and optical rotatory dispersion in the visible and infrared frequency ranges. This type of spectrum provides unique and potentially useful analytical information for atomic and molecular ensembles, both chiral and achiral.

S. Electric Circular Birefringence and Dichroism

Faraday noted "no effect" when he attempted to see circular birefringence due to a static electric field. It is now known that such an effect would imply nonconservation of reversality. However, a nonzero time derivative of an electric field is negative to T, and can produce electric circular birefringence and dichroism in chiral ensembles. Recently, this effect has

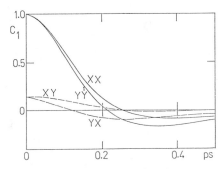

Figure 52. Diagonal (auto) and two off-diagonal (cross) correlation elements of the rotational velocity correlation function of the electric Faraday effect. The Fourier transform of the autocorrelation function is far infrared dichroism.

been supercomputer-simulated,[415] following a semiclassical treatment.[412] Electric circular dichroism was observed in the simulation through the Fourier transform of frequency spectra in the far infrared and dielectric range (Section I of the S enantiomer of bromochlorofluoromethane, using a Voigt–Born expansion of electric dipole moments:

$$\mu_{iR} = \mu_{0i} + \left(\alpha_{1ij} \pm \alpha_{1ijZ} \frac{\partial E_Z(t)}{\partial t} \right) E_j^R + \cdots \tag{484}$$

$$\mu_{iL} = \mu_{0i} + \left(\alpha_{1ij} \pm \alpha_{1ijZ} \frac{\partial E_Z(t)}{\partial t} \right) E_j^L + \cdots \tag{485}$$

induced respectively by a right and left circularly polarized probe field parallel to the electric field derivative. The latter was assumed to be cosinusoidal.[415] The electric circular dichroism observed in the far infrared range during the course of this simulation is illustrated in Fig. 52 for an input electric field derivative equivalent in energy to 7.0 kJ/mole. Figure 52 illustrates the difference in the rotational velocity correlation tensor (Sections I–VII) for right and left circularly polarized probe electromagnetic radiation. Electric-field-induced birefringence and dichroism is therefore accompanied by the appearance of asymmetric cross-correlation functions of the type seen in another context in Section VIII. It is accompanied, also, by anisotropy in the diagonal elements of the cross-correlation function. This produces spectral differences which are observable in principle with a modified Fourier transform spectrometer (Section II) with electrodes with central apertures to allow alternating left and right circularly polarized probe radiation from a piezoelectric modulator[416] or wire grid beam dividers[417] to pass through the chiral sample in the Z axis of the laboratory frame. The electrodes are used to apply the AC field derivative.

The orientational acf (Sections I to VII) exhibits the same type of

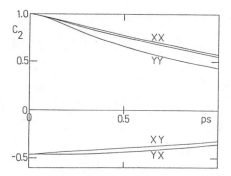

Figure 53. As for Fig. 52, orientational correlation function.

behavior (Fig. 53). However, in this case, the cross-correlation functions (*XY* and *YX* elements) are symmetric in time dependence. These are accompanied by the development of interesting new nonvanishing diagonal elements of the cross-correlation function between the linear and angular molecular momentum (Sections I–VII). These are much smaller in magnitude than the *XY* and *YX* ccf's of orientation and rotational velocity shown in Figs. 52 and 53, but nevertheless exist above the noise of the simulation. The *ZZ* element has a different time dependence from the *XX* and *YY* elements, which are approximately equal.

T. Frequency-Dependent Electric Polarization due to Optical Rectification in Chiral Ensembles

It has been shown recently by Evans and Wagnière[418] that Π of the optical rectification effect can produce frequency-dependent electric polarization in chiral ensembles through the mediation of the angular polarizability. (This is the optical equivalent of **B**-induced electric polarization in a chiral ensemble capable of supporting the Rosenfeld tensor.) It should be carefully noted that this effect is again different from the well-known[388] A.C. Stark effect, or Autler–Townes effect, because the latter is mediated by the *T*-positive symmetric real part of the atomic or molecular polarizability and has a zero-frequency (DC) component.

It is convenient to discuss the new effect by Evans and Wagnière[418] in terms of the quantum mechanical expressions for optical rectification introduced by Ward[373] and used recently by Wagnière[370] to derive the inverse magnetochiral effect.

Optical rectification as discussed by Ward[373] leads to an expression for the DC electric polarization induced to second order by the electromagnetic field, and consists of double sums over all eigenstates of the unperturbed molecular system. The individual terms in these sums contain in the numerators products of matrix elements of the system field interaction. In the denominators appear the transition energies of the system, the frequency of the radiation field, and appropriate damping factors. It is sufficient for our purposes to consider the numerators, which are of the general form

$$\mu(\mu' \cdot E^-)(\mu'' \cdot E^+) \tag{486}$$

where μ, μ', and μ'' designate matrix elements of the electric dipole moment operator and E^-, E^+ are the complex conjugate electric field strength vectors of the laser. Using isotropic averaging,[286,419] Eq. (486) splits up into a real part:

$$(\mu \cdot \mu' \times \mu'')(E^- \times E^+) \tag{487}$$

The product $E^- \times E^+$ vanishes if the radiation is linearly polarized, and is purely imaginary (Eq. 409) for circularly polarized lasers. The induced DC electric polarization proportional to the product (Eq. 487) is therefore not directly observable, because it is purely imaginary.

However, at finite laser frequency ω in Eq. (422) the imaginary part of the polarizability is nonzero, and multiplies the imaginary product $E^+ \times E^-$ to give a *real* electric polarization which is mediated by a *P*-negative molecular property tensor.[418] *This means that a circularly polarized electromagnetic field produces electric polarization in ensembles of chiral molecules.* A particularly interesting aspect of this is the potential utilization of a circularly polarized radio-frequency field (from a waveguide) to produce linear and nonlinear[419] dielectric relaxation (Sections I–VII) in chiral liquids.

The first supercomputer simulation[418] and video animation[420] of this effect has recently been pursued with a torque (Sections IV–VI) of the type[421]

$$\langle T_q \rangle = \langle \beta_{ijk} E_j E_k \times E_j \rangle \tag{488}$$

and approximating the electromagnetic phase factor by

$$\theta_L \doteq \theta_R \doteq \omega t \tag{489}$$

The real part of this torque was incorporated into the code of the program TETRA (see Appendix) for 108 molecules of *S*-bromochlorofluoromethane, a chiral structure. The orientation and rotational velocity time-correlation functions of the ensemble were evaluated in the field-applied

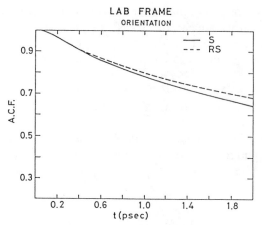

Figure 54. Orientational acf for (*S*)-bromochlorofluoromethane, (S), and the racemic mixture (RS), at field-free equilibrium.

steady state using two far infrared field frequencies of 10.0 and 1.0 THz respectively.

Figure 54 illustrates the orientational acf under field-free conditions for the *S* enantioner and racemic mixture. There are no orientational cross correlations. Figure 55 is the same correlation function for an applied field

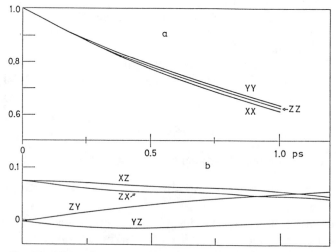

Figure 55. The effect of the conjugate product at a field-frequency of 10.0 THz. (*a*) Autocorrelation functions; (*b*) cross-correlation functions.

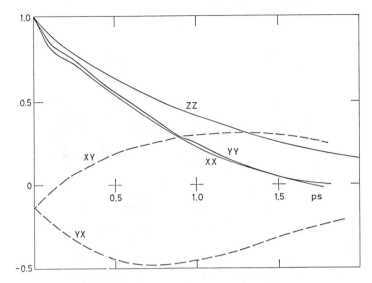

Figure 56. As for Fig. 55, frequency 0.01 THz.

of 1.0 THz, and Fig. 56 for 10.0 THz, each in the S enantiomer. At the lower frequency, anisotropy in the orientational acf components is clearly visible for a laser beam propagating in the Z axis. This is accompanied by the development of off-diagonal cross-correlation functions of orientation. At the higher frequency, Fig. 56, the anisotropy is lessened considerably, but the time dependencies of the orientational autocorrelation functions are clearly different from those at field-free equilibrium in Fig. 54. The Fourier transform of these acf elements is dielectric loss,[1,2] and this shows the presence of *laser-induced dielectric relaxation*[418] a potentially very useful phenomenon. In practice the THz fields of the simulation are replaced by GHz circularly polarized fields from a klystron or waveguide. Figure 55 also shows the presence of four off-diagonal orientational time-cross correlation functions in the presence of the circularly polarized electromagnetic field.

Figure 57–59 show the effect of the laser on the Fourier transform of the far infrared power absorption of the chiral liquid. There are interestingly asymmetric rotational velocity cross-correlation functions reminiscent of those simulated for shear (see Section VIII) by Evans and Heyes.[422] This shows that a circularly polarized radio-frequency field is capable, in a chiral liquid, of producing bandshape changes in the far infrared, providing new information on the molecular dynamics (Sections I–VII) of the liquid. The far infrared is of course the high-frequency adjunct (Section I) of the frequency range in which relaxational behavior has been investigated

(S) ENANT LAB FRAME
ROTATIONAL VELOCITY

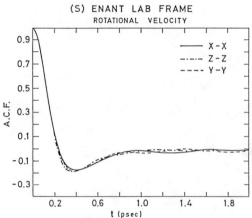

Figure 57. As for Fig. 55, rotational velocity acf's.

historically, and these results show the presence of *electromagnetically induced far infrared relaxational effects*.

The precise molecular dynamical nature of these have been animated on video at the Cornell National Supercomputer Facility and video copies are available on request from the Cornell Theory Center.

U. Symmetry of Laser-Induced Electric Polarization in Chiral Single Crystals

It is possible to express the induced electric polarization just described in terms of a *P*-negative molecular property tensor X_{ij} defined through

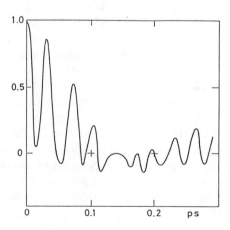

Figure 58. As for Fig. 57, 10.0 THz.

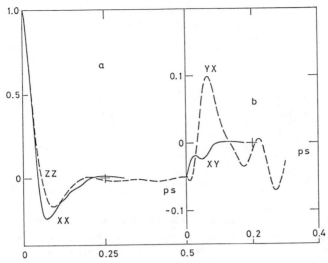

Figure 59.   As for Fig. 57, 0.01 THz

$$\mu_i = X_{ij}\Pi_j \qquad (490)$$

where μ_i is the induced electric dipole moment. It has been shown that X_{ij} is defined through the time-dependent Schrödinger equation by[432]

$$X_{\alpha\beta} = -\frac{2}{\hbar}\sum_{j\neq n}\frac{\langle n|\mu'_\alpha|j\rangle\langle j|\alpha''_{1\beta}|n\rangle}{\omega_{jn}} \qquad (491)$$

for a transition between quantum states n and j. Here μ'_α is the transition electric dipole moment. The angular polarizability is conveniently expressed by

$$\alpha''_{1\gamma} = \alpha''_{1\alpha\beta} - \alpha''_{1\beta\alpha} \qquad (492)$$

through the tensor definition[422] of the T-negative, antisymmetric, part of the imaginary polarizability.

In this subsection, we consider the symmetry characteristics of X_{ij} in the 11 major chiral crystal point groups. This is important for ab initio computations and experimental investigations of X_{ij} in individual chiral molecules, and in chiral single crystals of material with useful nonlinear optical characteristics. In general, the symmetry of X_{ij} is

$$\Gamma(X_{ij}) = \Gamma(\mu'_i)\,\Gamma(\alpha''_{1j}) \qquad (493)$$

TABLE VIII
Symmetry of X_{ij} in the Chiral Crystal Point Groups

Point Group		Finite Components of $X_{\alpha\beta}$	Orientation	Symmetry
Triclinic	$C_1 (1)$	All	Any	9A
Monoclinic	$C_2 (2)$	XX, YY, ZZ, XY, YX	$C^2 \parallel Z$	5A + 4B
Orthorhombic	D_2	XX, YY, ZZ	$C_2 \parallel X \parallel Y$	$3A_1 + 2B_1 + 2B_2 + 2B_3$
Trigonal	$C_3 (3)$	XX = YY, ZZ, XY − YX	$C_3 \parallel Z$	3A + 3E
	$D_3(32)$	XX = YY, ZZ	$C_3 \parallel Z; C_2 \parallel Y$	$2A_1 + A_2 + 3E$
Tetragonal	$C_4 (4)$	XX = YY, ZZ, XY − YX	$C_4 \parallel Z$	3A + 2B + 2E
	D_4	XX = YY, ZZ	$C_4 \parallel Z; C_2 \parallel Y$	$2A_1 + A_2 + B_1 + B_2 + 2E$
Hexagonal	$C_6 (6)$	XX = YY, ZZ, XY − YX	$C_6 \parallel Z$	$3A + 2E_1 + E_2$
	$D_6 (622)$	XX = YY, ZZ	$C_6 \parallel Z; C_2 \parallel Y$	$2A_1 + A_2 + E_2 + 2E_1$
Cubic	$T (23)$	XX = YY = ZZ	$C_4 \parallel Z; C_4 \parallel Y$	A + E + 2T
Chiral liquid		XX = YY = ZZ		$D^{(0)} + D^{(1)} + D^{(2)}$

which is a product of those of the electric dipole moment μ'_i and the angular polarizability α''_{1j}. In the point group R(3) of all rotations, the point group of a liquid ensemble of chiral molecules (Section VII), Eq. (493) can be expressed in terms of the irreducible D representations and D symmetries of Section VII as follows:

$$\Gamma(X_{ij}) = (D^{(0)} + D^{(1)} + D^{(2)})(-) \qquad (494)$$

It is T- and P-negative, and must be activated by the T-negative influence Π to be observed, following the third principle of Section VII. It disappears at field-free equilibrium because the angular polarizability is essentially a projection of the imaginary part of the dynamic polarizability (Eq. 422) on to the Z axis. This is a finite ensemble average in a liquid ensemble only in the presence of Π. It mediates a P-negative electric dipole moment μ_i produced by the P-positive influence Π_j, and in consequence can exist only in chiral media, such as chiral single crystals (Table VIII).

Note that this table can be used to characterize the equivalent molecular point groups, with the laboratory frame (X, Y, Z) replaced by the molecule fixed frame of the point group character tables.

In the triclinic crystal C_1 for example, there are nine independent components of X_{ij}, that is, the crystal supports all components in the laboratory frame (X, Y, Z). This is summarized by nine occurrences of the totally symmetric representation of the C_1 point group in the last column. Similarly, in a molecule of C_1 symmetry, all elements of X_{ij} exist in the molecule fixed frame. In the monoclinic there are five independent elements, in the trigonal C_3 there are three, and so on, thus summarizing the symmetry of the tensor X_{ij}, mediating laser-induced electric polarization in chiral crystals.

V. Electrodynamics of a Rotating Body — Some Spectral Consequences of the Lorentz Transformation

In the relativistic theory of classical electromagnetic fields[424] the Lorentz transformations relate **E** and **B** in the rest frame to their equivalents in a moving frame. The latter may translate or rotate with respect to the former. The Lorentz transformations show that the electromagnetic radiation reaching the rest frame of an earth-bound observer from a source in a distant galaxy, for example, receding along the X axis of the rest frame contains information about the velocity v at which the source is receding from the earth. The further away the galactic source in an expanding universe, the closer the fraction v/c approaches unity, where c is the constant velocity of light. This manifests itself in such well-known spectral phenomena as the stellar–galactic red shift.

The red shift is only one out of many optical phenomena which can be observed in the rest frame of the earth-bound observer with a contemporary laser spectropolarimeter, such as the one constructed by R. V. Jones,[414] attached to a telescope. It is shown in this subsection that circular birefringence in a chiral earth-bound sample is amplified relativistically when observed with the telescope–polarimeter. It is shown in Ref. 425 that the optical activity observed in this way with a source of electromagnetic radiation receding with a velocity v_Z (e.g., a star in a far galaxy) is amplified by the factor

$$R = (c + v_Z)/(c - v_Z) \qquad (495)$$

the relativistic amplification.[425] This appears to be of interest in the amplification of the tiny parity nonconserving optical activity recently observed in atomic vapors.[426–430] The measurement proceeds in principle by gathering the stellar radiation with a powerful telescope, possibly in orbit, plane polarizing it, and analyzing the angle of rotation with the probe laser spectropolarimeter to microradian accuracy. This simultaneously provides information on the velocity of recession v_Z of the source, and amplifies relativistically the natural optical activity of the earth-bound sample.

Elegant use was made of a laser spectropolarimeter by R. V. Jones[414] in the first experimental demonstration of *rotational ether drag*, a circular birefringence of pure relatively origin generated in an achiral rotating glass rod by the Lorentz transformation.

It appears that J. J. Thompson[431] was the first to analyze ether drag, when, a few years after the first Michelson–Morley experiment, he considered light passing through a medium that is rotating about an axis parallel with the propagation axis of the electromagnetic beam. The angular drag per unit path length (the angle of rotation) was later obtained by Fermi[432] and was proportional to the ratio of the angular frequency of the rotating body (Ω_Z) to the velocity of light. Player[433] later extended Fermi's analysis to dispersive ether drag in a transparent rotating rod, which was measured meticulously by R. V. Jones, giving a result in good agreement with theory.[414]

The present author has extended the consideration by Player in several directions[434] to include the magnetization term of the Lorentz transformation, second-order effects, and polarization effects in a rotating rod of absorbing chiral material. In each case relativistic terms were found to give theoretical contributions to the angular rotation measurable in principle by a laser spectropolarimeter.[414]

These new effects were based on the Maxwell equation

$$\frac{1}{\mu_0} \nabla \times \mathbf{B} = \frac{\partial}{\partial t} (\epsilon_0 \mathbf{E} + \mathbf{P}) \tag{496}$$

Equation (496) is written in the approximation $\mathbf{M} \ll \mathbf{P}$, and $\mathbf{J} = 0$, where $\mathbf{M}$ is the bulk magnetization, $\mathbf{P}$ the polarization, and $\mathbf{J}$ the current density in the rest frame of the observer. Here ϵ_0 and μ_0 are respectively the frame invariant permittivity and permeability in vacuo. The time t is measured in the observer frame.

Consider the rod rotating at an angular velocity Ω_Z about the Z axis of the observer frame while radiation from a source in this frame propagates through it in the same Z axis. The quantities $\mathbf{P}$, $\mathbf{E}$, and $\mathbf{B}$ in Eq. (496) are now defined in the frame (x, y, z) which rotates with the rod. The inverse of the Lorentz transformation must be used to relate $\mathbf{P}$, $\mathbf{E}$, and $\mathbf{B}$ in frame (x, y, z) to their equivalents in the frame (X, Y, Z) of the static observer, the frame in which t is defined. In SI units,

$$[\mathbf{B}]_{(x,y,z)} = \left[\beta \left(\mathbf{B} - \frac{1}{c^2} \mathbf{v} \times \mathbf{E} \right) \right]_{(X,Y,Z)} \tag{497}$$

$$[\mathbf{E}]_{(x,y,z)} = [\beta(\mathbf{E} + \mathbf{v} \times \mathbf{B})]_{(X,Y,Z)} \tag{498}$$

$$[\mathbf{P}]_{(x,y,z)} = [\beta(\mathbf{P} - \epsilon_0\mu_0\mathbf{v} \times \mathbf{M})]_{(X,Y,Z)} \tag{499}$$

where

$$\beta = \left(1 - \frac{v^2}{c^2}\right)^{-1/2} \tag{500}$$

and where the velocity v is defined[433] by

$$v_x = -\Omega_z y; \qquad v_y = \Omega_z X; \qquad v_z = 0 \tag{501}$$

where **R** is the radius of the rod. Note that **v** has the components

$$\mathbf{v} = \mathbf{\Omega} \times \mathbf{R}; \qquad \mathbf{\nabla} \cdot \mathbf{v} = 0; \qquad \mathbf{\nabla} \times \mathbf{v} = 2\mathbf{\Omega} \tag{502}$$

Therefore, the Maxwell equation (Eq. 496) becomes

$$\frac{1}{\mu_0}\left(\mathbf{\nabla} \times \mathbf{B} - \frac{1}{c^2}\mathbf{\nabla} \times (\mathbf{v} \times \mathbf{E})\right) = \epsilon_0 \frac{\partial}{\partial t}(\mathbf{E} + \mathbf{v} \times \mathbf{B})$$

$$+ \frac{\partial}{\partial t}(\mathbf{P} - \epsilon_0\mu_0\mathbf{v} \times \mathbf{M}) \tag{503}$$

with all quantities defined in the frame of the observer, the static laboratory frame (X, Y, Z).

Equation (503) has three extra terms of purely relativistic origin.[434] One of these is the Lorentz magnetization, $-\epsilon_0\mu_0\mathbf{v} \times \mathbf{M}$, which introduces a contribution to the Thompson–Fermi–Player angle drag as described in Ref. 434, and also produces an interesting *relativistic forward–backward birefringence* which is measurable[435] in the Z axis with unpolarized light.

Player[433] considered a quasi-monochromatic optical disturbance in an isotropic medium of low absorption, where the displacement **D** was put directly proportional to the electric field strength **E**, with no explicit consideration of the molecular nature of the polarization **P** and Lorentz magnetization. Molecular property tensors were employed in Refs. 434 and 435 to consider the relativistic effects of rotating a rod composed of molecular material. The displacement was accordingly

$$\mathbf{D} = \epsilon_0\mathbf{E} + \mathbf{P} \tag{504}$$

with **E** and **P** defined respectively by Eqs. (498) and (499).

As in Player's analysis, the angular frequencies of the right and left circularly polarized components of the electromagnetic plane wave are affected by equal and opposite Doppler shifts, so that their frequencies in frame (X, Y, Z) appear to the observer to be[435]

$$\omega_R = \omega + \omega_1; \qquad \omega_L = \omega - \omega_1; \qquad \omega_1 = \left(\frac{\Omega_z R}{2c}\right)\omega \qquad (505)$$

respectively, for right and left circular polarization of the probe laser. Accordingly, the frequency-dependent molecular property tensors are functions not of ω, but of these Doppler-shifted frequencies.

Without the Lorentz magnetization, the circular birefringence due to ether drag is found from Eq. (503) to be

$$n'_{LZ} - n'_{RZ'} = 3\Omega_z\left(\frac{1}{\omega - \Omega_D} + \frac{1}{\omega + \Omega_D}\right) \qquad (506)$$

so that the angle of rotation is proportional to the ratio of Ω_z to the velocity of light, as found by both Player[433] and Fermi.[432] When Lorentz magnetization is accounted for in an absorbing chiral rod this result is supplemented[435] by

$$\langle n'_L - n'_R\rangle_{LM} = 2\Omega_Z Y \langle\alpha_{2ZY}\rangle\mu_0 N \qquad (507)$$

which is proportional to the number of molecules per unit volume of the rotating rod N, and to the ensemble average over the Rosenfeld molecular property tensor component α_{2ZY}. There are also interesting second-order effects[434] which can be treated with molecular property tensors.

Relativistic forward–backward birefringence is present, furthermore, both in chiral and achiral rotating rods.[435] This effect is interestingly much larger in magnitude that ether drag, and is proportional to nonvanishing diagonal scalar components of ensemble-averaged molecular property tensors, giving information on them in a rotating chiral rod. In general, it can be expressed in terms of diagonal and off-diagonal elements of molecular property tensors through the magnetization term of the Lorentz transformation. It also depends on the X and Y velocity components of the rotating rod. If this is a chiral crystal, the Lorentz magnetization also makes a contribution, just described, to the rotational ether drag.

Defining $\mathbf{k}$ as a unit vector in the Z axis of the static observer frame, relativistic forward–backward birefringence is generated from the equation

$$\nabla(\mathbf{E} \cdot \mathbf{v}) = -\frac{\partial}{\partial t}(\mathbf{v} \times \mathbf{B}) + \mu_0 \frac{\partial}{\partial t}(\mathbf{v} \times \mathbf{M}) \tag{508}$$

by comparing $\mathbf{k}$ coefficients in Eq. (503). Using the approximation

$$\nabla \times (\mathbf{v} \times \mathbf{E}) \sim \nabla(\mathbf{E} \cdot \mathbf{v}) - 2\mathbf{E} \times \mathbf{\Omega} \tag{509}$$

Ref. 435 derives the results

$$n'_{av} = \tfrac{1}{2}(\langle n'_{LZ}\rangle + \langle n'_{RZ}\rangle) \doteq 1 + \tfrac{1}{2}\mu_0 Nc\langle \alpha'_{2YY}\rangle \tag{510}$$

and

$$A''_{av} = \frac{\omega}{c}(\langle n''_{LZ}\rangle + \langle n''_{RZ}\rangle) \doteq \mu_0 Nc\langle \alpha''_{2YY}\rangle \tag{511}$$

for the average real refractive index n'_{av} and power absorption coefficient A''_{av} along the axis of the rotating rod in terms of the frequency-dependent diagonal elements $\langle \alpha'_{2yy}\rangle$ and $\langle \alpha''_{2yy}\rangle$ of the Rosenfeld tensor. Both $\langle \alpha'_{2yy}\rangle$ and $\langle \alpha''_{2yy}\rangle$ are visible therefore to *unpolarized* radiation, an interesting new effect of relativity. The latter also enters into consideration through the Doppler shifts of the frequency in the molecular property tensors. For example, the real and imaginary parts of the Rosenfeld tensor

$$\alpha'_{2YY} = \frac{\partial}{\hbar}\sum_{j \neq n}\frac{(\omega \pm \omega_1)_{jn}}{(\omega \pm \omega_1)_{jn}^2 - (\omega \pm \omega_1)^2}\,\mathrm{Re}(\langle n|\mu_Y|j\rangle\langle j|m_Y|n\rangle) \tag{512}$$

and

$$\alpha''_{2YY} = -\frac{\partial}{\hbar}\sum_{j \neq n}\frac{(\omega \pm \omega_1)}{(\omega \pm \omega_1)_{jn}^2 - (\omega \pm \omega_1)^2}\,\mathrm{Im}(\langle n|\mu_Y|j\rangle\langle j|m_Y|n\rangle) \tag{513}$$

Note that the former is negative to T and the latter is positive to T from semiclassical theory in a nonrotating medium. The rotation itself provides a vehicle for the "activation" of the real part of the Rosenfeld tensor, which becomes observable (Eq. 510) through the average refractive index measured by unpolarized radiation directed along the Z axis of the rotating rod. Further discussion of the role of bulk angular momentum in a nonrelativistic context is given in Ref. 436.

The above rotating rod method appears to be useful therefore in the study of vibrational dichroism[437] without recourse to circularly polarized radiation.

W. Parity Nonconservation in New Laser Spectroscopies

Parity nonconservation is well known[279,438] to cause tiny optical rotations is atomic ensembles which are otherwise achiral. Several of the new laser spectroscopies developed in this section are mediated only in a chiral ensemble, examples being forward–backward birefringence due to (1) static magnetic flux density (the Wagnière–Meier effect), (2) the conjugate product Π,[412,413] and, as we have just seen, (3) relativity in a rotating chiral rod. If any of these effects are observed in an achiral medium, such as water, it signals the presence of parity nonconservation in molecular matter.[438] Parity nonconservation applies whenever the Wigner Principle of Parity conservation is broken; another example would be the observation of circular birefringence and dichroism caused in an achiral medium by the time derivative of an electric field.

This is probably one of the most profound applications of the new spectroscopies introduced in this section, and further details of the ideas behind P and T nonconservation may be found in the interesting articles by Barron[279,438] and Mason.[439]

Acknowledgments

Part of this research was supported by the Center for Theory and Simulations in Science and Engineering (the Cornell Theory Center), which receives major funding from NSF (US); IBM (US), New York State, and Members of the Corporate Research Institute. Funding and support is also acknowledged from: IBM Kingston, New York State; the Universities of Zurich, London, Lancaster, and Wales; SERC (UK); the Leverhulme Trust; the Nuffield Foundation; and the Cornell National Supercomputer Facility.

Special thanks are due to Dr. Laura J. Evans.

References

1. M. W. Evans, G. J. Evans, W. T. Coffey, and P. Grigolini, *Molecular Dynamics and the Theory of Broad Band Spectroscopy*, Wiley Interscience, New York, 1982.

2. M. W. Evans, W. T. Coffey, and P. Grigolini, *Molecular Diffusion*, Wiley Interscience, New York, 1984; M.I.R., Moscow, 1988.

3. M. W. Evans, P. Grigolini, and G. Pastori-Parravicini, Eds., *Memory Function Approaches to Dynamical Problems in Condensed Matter*, vol. 62 of *Advances in Chemical Physics*, I. Prigogine and S. A. Rice, Eds., Wiley Interscience, New York, 1985.

4. Ibid., vol. 63, M. W. Evans, Ed., *Dynamical Processes in Condensed Matter*.

5. P. Debye, *Polar Molecules*, Chem. Catalog Co., New York, 1929.

6. A. Einstein, *Investigations on the Theory of the Brownian Movement*. R. Furth, Ed., Dover, New York, 1956.

7. N. Wax, Ed., *Selected Papers on Noise and Stochastic Processes*, Dover, New York, 1954.

8. J. Goulon, *Theories Stochastiques des Phenomènes de Transport*, Cas des Mouvements de Reorientations Moleculaires, Nancy, 1972.

9. P. Bordewijk and C. J. F. Böttcher, *Theory of Electric Polarization*, vols. 1 and 2, Elsevier, Amsterdam, 1973, 1979.

10. H. Fröhlich, *Theory of Dielectrics*, Oxford University Press, 1958.

11. J. B. Hasted, *Aqueous Dielectrics*, Chapman and Hall, London, 1973.

12. N. E. Hill, W. E. Vaughan, A. H. Price, and M. Davies, *Dielectric Properties and Molecular Behaviour*, van Nostrand, London, 1969.

13. B. K. P. Scaife, Ed., *Complex Permittivity*, English University Press, London, 1971.

14. C. P. Smyth, *Dielectric Behavior and Structure*, McGraw-Hill, New York, 1965.

15. A. D. Buckingham, Ed., *Organic Liquids, Structure, Dynamics, and Properties*, Wiley, New York, 1979.

16. A. Einstein, *Ann. Phys.* **17**, 549 (1905); **19**, 371 (1906).

17. P. Langevin, see Ref. 7.

18. M. von Smoluchowski, see Ref. 7.

19. P. Debye, Ref. 5.

20. J. L. Doob, *Stochastic Processes*, Wiley, New York, 1963.

21. J. Perrin, see Ref. 7.

22. A. Rahman, *Rev. Nuovo Cim.* **1**, 315 (1969).

23. H. Mori, *Prog. Theor. Phys.* **33**, 423 (1965); **34**, 399 (1965).

24. M. Davies Ed., *Dielectric and Related Molecular Processes*, vols. 1 to 3, Chemical. Society., London, 1972–1977.

25. R. A. Sack, *Physica* **22**, 917 (1956); *Proc. Phys. Soc.* **70**, 402, 414 (1957).

26. J. T. Lewis, J. R. McConnell, and B. K. P. Scaife, *Proc. R. Irish Acad.* **76A**, 43 (1976).

27. See C. Brot and G. Wyllie in Ref. 24.

28. K. D. Muller and W. D. Rothschild, *Far Infra red Spectroscopy*, Wiley Interscience, New York, 1971.

29. G. Chantry, *Submillimetre Spectroscopy*, Academic, New York, 1971.

30. N. E. Hill, *Proc. Phys. Soc.* **82**, 723 (1963).

31. G. Wyllie, reviewed in Ref. 24.

32. I. Larkin and M. W. Evans, *J. Chem. Soc., Faraday Trans.* **2**, 70, 477 (1974).

33. J. H. Calderwood and W. T. Coffey, *Proc. R. Soc., A* **356**, 269 (1977).

34. M. W. Evans, Ref. 24, **3**, 1 (1977).

35. M. W. Evans, W. T. Coffey, G. J. Evans, and G. Wegdam, *J. Chem. Soc., Faraday Trans.* **2**, 74, 1310 (1977).

36. W. T. Coffey and M. W. Evans, *Mol. Phys.* **35**, 975 (1978).

37. M. W. Evans, W. T. Coffey, and G. J. Evans, ibid. **38**, 477 (1978).

38. E. Kestermont, F. Hermans, R. van Loon, G. J. Evans, and M. W. Evans, *Chem. Phys. Lett.* **52**, 521 (1978).

39. M. W. Evans, A. R. Davies, and G. J. Evans, in *Advances in Chemial Physics*, vol. 44, I. Prigogine and S. A. Rice, Eds., Wiley Interscience, New York, 1980, pp. 255–481.

40. M. W. Evans, W. T. Coffey, and G. J. Evans, *Adv. Mol. Rel. Int. Proc.* **20**, 11 (1981).

41. W. T. Coffey, P. Corcoran, and M. W. Evans, *Proc. R. Soc.*, *A* **410**, 61 (1987).

42. W. T. Coffey, P. Corcoran, and M. W. Evans, *Mol. Phys.* **61**, 1 (1987).

43. Ibid., p. 15.

44. M. W. Evans, *Chem. Phys. Lett.* **39**, 601 (1976).

45. M. W. Evans and G. J. Evans, *J. Mol. Liq.* **36**, 293 (1987).

46. B. J. Berne and R. Pecora, *Dynamical Light Scattering with Reference to Physics, Chemistry and Biology*, Wiley Interscience, New York, 1976.

47. J. P. Ryckaert, A. Bellemans, and G. Ciccotti, *Mol. Phys.* **44**, 979 (1981).

48. M. W. Evans, *Phys. Rev. Lett.* **50**, 351 (1983).

49. For a review, see Ref. 4, M. W. Evans and G. J. Evans.

50. M. W. Evans, *Phys. Rev. Lett.* **55**, 1551 (1985).

51. M. W. Evans and G. J. Evans, ibid., p. 818.

52. M. W. Evans, *Phys. Rev. A* **34**, 468 (1986).

53. M. W. Evans and G. J. Evans, ibid. **33**, 1903 (1985).

54. D. H. Whiffen, *Mol. Phys.* **63**, 1053 (1988).

55. M. W. Evans, K. N. Swamy, G. C. Lie, and E. Clementi, *Mol. Sim.* **1**, 187 (1988).

56. M. W. Evans, *Phys. Rev. A* **34**, 2302 (1986).

57. M. W. Evans, ibid. **35**, 2989 (1987).

58. M. W. Evans, *Physica* **168**, 9 (1991).

59. A. H. Price and P. Maurel, *J. Chem. Soc.*, *Faraday Trans.* 2, **69**, 1486 (1973).

60. For a review, see Ref. 12.

61. J. Chamberlain, *The Principles of Interferometric Spectroscopy*, Wiley, Chichester, 1978.

62. For example, Bruker Spectrospin, *Quart. Rev.*

63. M. W. Evans, M. Davies, and I. W. Larkin, *J. Chem. Soc.*, *Faraday Trans.* 2, **69**, 1011 (1973).

64. J. Goulon, D. Canet, G. J. Davies, and M. W. Evans, *Mol. Phys.* **30**, 973 (1975).

65. G. J. Davies and M. W. Evans, *J. Chem. Soc.*, *Faraday Trans.* 2, **71**, 1275 (1975).

66. Ibid. **72**, 727 (1976).

67. Ibid., p. 1206.

68. G. J. Davies, G. J. Evans, and M. W. Evans, ibid., p. 1901.

69. M. W. Evans, *Spectrochim. Acta* **30**, 79 (1974).

70. M. W. Evans, *J. Chem. Soc.*, *Faraday Trans.* 2, **69**, 763 (1973).

71. Ibid. **71**, 843 (1975).

72. A. I. Baise, ibid. **69**, 1907 (1972).

73. J. H. Colpa and J. A. A. Ketelaar, *Mol. Phys.* **1**, 873 (1958).

74. M. W. Evans, *Mol. Phys.* **29**, 1345 (1975).

75. M. W. Evans, *Spectrochim. Acta* **32**, 1259 (1976).

76. R. Balescu, *Equilibrium and non-Equilibrium Statistical Mechanics*, Wiley, New York, 1975.

77. D. A. McQuarrie, *Statistical Mechanics*, Harper and Row, New York, 1975.

78. R. C. Tolman, *The Principles of Statistical Mechanics.*, Oxford University Press, 1938.

79. G. J. Evans, C. J. Reid, and M. W. Evans, *J. Chem. Soc., Faraday Trans.* 2, 74, 343 (1978).
80. C. J. Reid, R. A. Yadav, G. J. Evans, and M. W. Evans, ibid. 74, 2143 (1978).
81. C. J. Reid, G. J. Evans, and M. W. Evans, *Chem. Phys. Lett.* 56, 529 (1975).
82. C. J. Reid and M. W. Evans, *J. Chem. Soc., Faraday Trans.* 2, 75, 1218 (1975).
83. Ibid. 76, 286 (1980).
84. Ibid., *Mol. Phys.* 40, 1357 (1980).
85. Ibid., in *Molecular Interactions*, W. J. Orville-Thomas, Ed., Wiley, Chichester, 1982.
86. M. W. Evans, *Spectrochim. Acta* 36A, 929 (1980).
87. M. W. Evans, *Acc. Chem. Res.* 14, 253 (1981), The Meldola Lecture.
88. M. W. Evans, *Chem. Phys.* 62, 481 (1981).
89. M. W. Evans, G. J. Davies, and M. Veerappa, ibid. 61, 73 (1981).
90. G. J. Evans and M. W. Evans, *Infra red Phys.* 18, 455 (1978).
91. D. M. Heyes and D. Fincham, in Ref. 4.
92. J. P. Hansen and I. R. McDonald, *Theory of Simple Liquids*, 2nd. ed., Academic, New York, 1986.
93. L. Verlet, *Phys. Rev.* 159, 98 (1967).
94. K. Singer, J. V. L. Singer, and A. J. Taylor, *Mol. Phys.* 37, 1239 (1979).
95. M. W. Evans, G. J. Evans, and G. Wegdam. *Mol. Phys.* 33, 180 (1977).
96. M. W. Evans, *Chem. Phys. Lett.* 58, 518 (1978).
97. C. Brot, G. Bossis, and C. Hesse-Bezot, *Mol. Phys.* 40,1053 (1980).
98. D. Kivelson and P. Madden, *Mol. Phys.* 30, 1749 (1975).
99. B. K. P. Scaife, in Ref. 24, vol. 1.
100. J. G. Kirkwood, *J. Chem. Phys.* 7, 911 (1939).
101. A. M. Stoneham, *Handbook of Interatomic Potentials*, A.E.R.E., Harwell, 1981.
102. *The Information Quarterly for MD and MC Simulations*, S.E.R.C., Daresbury, U.K.
103. E. K. Eliel, N. L. Allinger, S. J. Angyal, and G. A. Morrison, *Conformational Analysis*, Wiley, New York, 1965.
104. G. C. Lie and E. Clementi, *Phys. Rev. A* 33, 2679 (1986).
105. M. Wojcik and E. Clementi, *J. Chem. Phys.* 85, 3544 (1986).
106. M. W. Evans, G. C. Lie, and E. Clementi, *J. Chem. Phys.* 89, 6399 (1988).
107. M. W. Evans and C. J. Reid, ibid. 76, 2576 (1982).
108. Ibid., *Spectrochim. Acta* 38A, 417 (1982).
109. M. W. Evans, G. C. Lie, and E. Clementi, *J. Chem. Phys.* 87, 6040 (1987).
110. G. C. Lie and E. Clementi, ibid. 64, 2314 (1976).
111. M. D. Morse and S. A. Rice, ibid. 76, 650 (1982).
112. M. G. Sceats and S. A. Rice, in *Water, A Comprehensive Treatise*, vol. 7, F. Francks, Ed., Plenum, New York, 1982.
113. For example the Specialist Periodical Reports of the London Chemical Society.
114. P. N. Brier and A. Perry, *Adv. Mol. Rel. Int. Proc.* 13, 46 (1978).
115. P. van Konynenberg and W. A. Steele, *J. Chem. Phys.* 56, 4775 (1972).
116. A. M. Amorim da Costa, M. A. Norman, and J. H. P. Clarke, *Mol. Phys.* 29, 191 (1975).

117. P. A. Lund, O. Faurskov-Nielsen, and E. Praestgaard, *Chem. Phys.* **28**, 167 (1978).

118. J. F. Dill, T. A. Litovitz, and J. A. Bucaro, *J. Chem. Phys.* **62**, 3839 (1975).

119. C. K. Cheung, D. R. Jones, and C. H. Wang, *J. Chem. Phys.* **64**, 3567 (1976).

120. S. Claesson and D. R. Jones, *Chim. Scripta* **9**, 103 (1976).

121. A. E. Boldeskal, S. S. Esman, and V. E. Pogornelov, *Opt. Spectrosc.* **37**, 521 (1974).

122. V. M. Baranov, *Vopt. Mol. Spektrosk.* **89**, 348 (1974).

123. M. W. Evans, W. Luken, G. C. Lie, and E. Clementi, *J. Chem. Phys.* **88**, 2685 (1988).

124. J. Anderson, I. J. Ullo, and S. Yip, *J. Chem. Phys.* **86**, 4078 (1987).

125. M. W. Evans, G. C. Lie, and E. Clementi, *J. Mol. Liq.* **40**, 89 (1989).

126. P. A. Egelstaff, *An Introduction to the Liquid State*, Academic, New York, 1967.

127. M. W. Evans and M. Ferrario, *Adv. Mol. Rel. Int. Proc.* **22**, 245 (1982).

128. Ibid., p. 75.

129. Ibid., *Chem. Phys.* **72**, 141 and 147 (1982).

130. M. W. Evans, M. Ferrario, P. Marin, and P. Grigolini, *J. Mol. Liq.* **26**, 249 (1983).

131. M. W. Evans, G. C. Lie, and E. Clementi, *Phys. Lett A* **130**, 289 (1988).

132. M. W. Evans and A. A. Hasanein, *J. Mol. Liq.* **29**, 45 (1984).

133. C. Brot and I. Darmon, *Mol. Phys.* **21**, 785 (1971).

134. M. W. Evans and M. Ferrario, *Adv. Mol. Rel. Int. Proc.* **24**, 139 (1982).

135. M. W. Evans, G. J. Evans, and V. K. Agarwal, *J. Chem. Soc., Faraday Trans.* 2, **79**, 153 (1983).

136. M. W. Evans and G. J. Evans, ibid. **79**, 767 (1983).

137. M. W. Evans, *J. Mol. Liq.* **25**,149 (1983).

138. M. W. Evans and G. J. Evans, ibid. **26**, 63 (1983).

139. M. W. Evans, *Phys. Rev. A* **30**, 2062 (1984).

140. M. W. Evans, *J. Chem. Soc., Faraday Trans.* 2, **81**, 1463 (1985).

141. M. W. Evans, *Phys. Scripta* **31**, 419 (1985).

142. M. W. Evans, *Physica* **131B&C**, 273 (1985).

143. M. W. Evans, G. C. Lie, and E. Clementi, *Chem. Phys. Lett.* **138**, 149 (1987).

144. M. W. Evans, G. C. Lie, and E. Clementi, *Phys. Rev. A* **36**, 226 (1987).

145. D. J. Evans, *Computer Phys. Rep.* **1**, 299 (1984).

146. A van der Avoird, F. Mulder, P. E. Wormer, and P. M. Berns, in *Topics in Current Chemistry*, M. J. S. Dewar, Ed., Springer Verlag, Berlin, 1980.

147. Ref. 1. p. 818.

148. H. S. Sandhu, *J. Am. Chem. Soc.* **97**, 6284 (1975).

149. Ibid., p. 830.

150. M. Woijik and E. Clementi, *J. Chem. Phys.* **85**, 3544 (1986).

151. O. Matsuoka, E. Clementi, and M. Yoshimine, *J. Chem. Phys.* **64**, 1351 (1976).

152. S. Toxvaerd, *Phys. Rev. Lett.* **51**, 1971 (1983).

153. K. N. Swamy and E. Clementi, IBM Technical Report, Dept. 48B/428, Kingston, New York, 1987.

154. E. Clementi and K. N. Swamy, in *Structure and Dynamics of Proteins, Nucleic Acids and Membranes*, E. Clementi and S. Chin, Ed., Plenum, New York, 1986.

155. Summarized in Chapter 5 of Ref. 3.

156. M. W. Evans, K. N. Swamy, K. Refson, G. C. Lie, and E. Clementi, *Phys. Rev. A*, **36**, 3935 (1987).

157. M. W. Evans, *Phys. Rev. A* **31**, 3947 (1985).

158. M. W. Evans, *Phys. Scripta* **30**, 94 (1984).

159. M. W. Evans, *J. Mol. Liq.* **26**, 49 (1983).

160. M. W. Evans, *J. Chem. Phys.* **79**, 5403 (1983); **78**, 925 (1983).

161. For example, volumes in *Advances in Chemical Physics*, I. Prigogine and S. A. Rice, Eds., Wiley Interscience, New York.

162. M. W. Evans, G. Wegdam, and G. J. Evans, *Adv. Mol. Rel. Int. Proc.* **11**, 295 (1977).

163. T. Occelli, B. Quentrec, and C. Brot, *Mol. Phys.* **36**, 257 (1978).

164. B. J. Alder, H. Strauss, and G. Weiss, *J. Chem. Phys.* **59**, 1002 (1973).

165. M. W. Evans and G. J. Evans, *J. Mol. Liq.* **39**, 25 (1988).

166. M. W. Evans, G. C. Lie, and E. Clementi, ibid. **40**, 89 (1989).

167. R. Zwanzig, *Ann. Rev. Phys. Chem.* **16**, 67 (1965).

168. B. J. Alder and T. E. Wainright, *Phys. Rev.* **1A**, 18 (1970).

169. J. J. Erpenbeck and W. W. Wood, *J. Stat. Phys.* **24**, 455 (1981).

170. W. W. Wood, *Fundamental Problems in Statistical Mechanics, Part 3*, E. G. D. Cohen, Ed., Plenum, New York, 1975, p. 331.

171. D. J. Evans and G. P. Morriss, *Phys. Rev. Lett.* **51**, 1776 (1983).

172. Summarized in Chapter 5 of Ref. 1.

173. E. L. Hannon, G. C. Lie, and E. Clementi, *Phys. Lett. A* **119**, 174 (1986).

174. M. W. Evans, *J. Chem. Soc. Faraday Trans. 2*, **82**, 653 (1986).

175. M. W. Evans, *J. Mol. Struct.* **80**, 389 (1982).

176. M. W. Evans, *Adv. Mol. Rel. Int. Proc.* **22**, 1 (1982).

177. M. W. Evans, J. K. Vij, C. J. Reid, G. J. Evans, and M. Ferrario, *Adv. Mol. Rel. Int. Proc.* **22**, 245 (1982).

178. M. W. Evans and M. Ferrario, ibid. **24**, 139 (1982).

179. Ibid. **24**, 139 (1982).

180. M. W. Evans, ibid. **23**, 113 (1982).

181. J. S. Rowlinson and M. W. Evans, *Chem. Soc. Ann. Rep.* **72**, 5 (1975).

182. A. D. Buckingham, *Chem. Soc. Quart. Rev.* **63** (1959).

183. C. Brot, in Ref. 24, vol. 2, 1975.

184. L. D. Landau and E. M. Lifshitz, *Quantum Mechanics, Non-Relativistic Theory*, Pergamon, Oxford, 1958.

185. C. A. Chatzidimitriou–Dreismann in M. W. Evans, Ed., *J. Mol. Liq.* Orville–Thomas Issue, 1988.

186. M. F. Herman and E. Kluck in Ref. 4.

187. See Ref. 1, p. 235.

188. E. Fatuzzo and P. R. Mason, *Proc. Phys. Soc.* **90**, 729 (1967).

189. D. Kivelson and P. A. Madden, *Mol. Phys.* **30**, 1749 (1975).

190. A. Gerschel, I. Darmon, and C. Brot, ibid. **33**, 527 (1977).

191. M. W. Evans, *Spectrochim. Acta* **31A**, 609 (1975).

192. S. Kielich in Ref. 24, vol. 1, 1972.

193. M. W. Evans, *J. Chem. Soc. Faraday Trans.* 2, **71**, 843 (1975).

194. P. Madden, *Mol. Phys.* **36**, 365 (1978).

195. P. van Konynenberg and W. A. Steele, *J. Chem. Phys.* **56**, 4776 (1972).

196. M. W. Evans, *J. Mol. Liq.* **25**, 211 (1983).

197. M. W. Evans and G, J. Evans, ibid. **25**, 177 (1983).

198. Ibid. **25**, 149 (1983).

199. M. W. Evans, G. J. Evans, and V. K. Agarwal, *J. Chem. Soc. Faraday Trans.* 2, **79**, 137 (1983).

200. M. W. Evans and G. J. Evans, ibid. **79**, 153 (1983).

201. M. W. Evans, ibid. **79**, 719 (1983).

202. M. W. Evans, G. J. Evans, C. J. Reid, P. Minguzzi, G. Salvetti, and J. K. Vij, *J. Mol. Liq.* **34**, 269 (1987).

203. M. W. Evans, ibid. **34**, 269 (1987).

204. M. W. Evans, G. C. Lie, and E. Clementi, *Phys. Rev. A* **36**, 226 (1987).

205. M. W. Evans, K. N. Swamy, G. C. Lie, K. Refson, and E. Clementi, ibid. **36**, 3935 (1986).

206. M. W. Evans, G. C. Lie, and E. Clementi, *Phys. Lett. A* **130**, 289 (1988).

207. M. W. Evans, J. K. Vij, and C. J. Reid, *Mol. Phys.* **50**, 1247 (1983).

208. M. W. Evans, K. N. Swamy, G. C. Lie, and E. Clementi, *Mol. Sim.* **1**, 187 (1988).

209. M. W. Evans, *J. Chem. Soc. Faraday Trans.* 2, **79**, 1811 (1983).

210. M. W. Evans, *J. Mol. Liq.* **27**, 19 (1983).

211. M. W. Evans, P. L. Roselli, and C. J. Reid, ibid. **29**, 1 (1984).

212. M. W. Evans, *Phys. Scripta* **39**, 94 (1984).

213. M. W. Evans, C. J. Reid, P. L. Roselli, and J. K. Vij, *J. Mol. Liq.* **29**, 11 (1984).

214. M. W. Evans, *Phys. Rev. A* **30**, 2062 (1984).

215. M. W. Evans, G. J. Evans, and J. Baran, *J. Mol. Liq.* **25**, 261 (1983).

216. M. W. Evans and G, J. Evans, *Chem. Phys. Lett.* **96**, 416 (1983).

217. M. W. Evans, G. J. Evans, and J. Baran, *J. Chem. Soc. Faraday Trans.* 2, **79**, 1473 (1983).

218. M. W. Evans, *J. Mol. Liq.* **26**, 229 (1983).

219. M. W. Evans, *J. Chem. Soc. Faraday Trans.* 2, **81**, 1463 (1985).

220. D. Hennequin, P. Glorieux, E. Arimondo, and M. W. Evans, ibid. **83**, 463 (1987).

221. M. W. Evans, G. C. Lie, and E. Clementi, IBM Technical Report, KGN 142, Kingston, New York, 1987.

222. R. C. Tolman, *The Principles of Statistical Mechanics*, Oxford University Press, 1938.

223. W. T. Coffey and B. V. Paranjape, *Proc. R. Irish Acad.* **78A**, 17 (1978).

224. M. Morse and H. Feschbach, *Methods of Theoretical Physics*, McGraw-Hill, New York, 1953.

225. M. S. Beevers, J. Crossley, D. C. Garrington, and G. Williams, *J. Chem. Soc. Faraday Trans.* 2, **72**, 1482 (1976).

226. M. Gregson and J. Parry–Jones in Ref. 24.

227. M. S. Beevers and J. Khanarian, *Aust. J. Chem.* **32**, 263 (1979); **33**, 2585 (1980).

228. M. W. Evans, *J. Chem. Phys.* **76**, 5473, 5480 (1982); ibid. **77**, 4632 (1983).

229. W. T. Coffey, C. Rybarsch, and W. Schroer, *Phys. Lett. A* **88**, 331 (1982); ibid., *Chem. Phys. Lett.* **92**, 245 (1982).

230. M. W. Evans, G. C. Lie, and E. Clementi, *Phys. Lett. A* **130**, 289 (1988).

231. M. W. Evans, P. Grigolini, and F. Marchesoni, *Chem. Phys. Lett.* **95**, 544 (1983).

232. Ibid., p. 548.

233. P. Grigolini, *Mol. Phys.* **31**, 1717 (1976).

234. M. S. Beevers and D. A. Elliott, *Mol. Cryst. Liq. Cryst.* **26**, 411 (1979).

235. M. W. Evans, G. C. Lie, and E. Clementi, *Phys. Rev. A* **36**, 226 (1987).

236. M. W. Evans, G. C. Lie, and E. Clementi, IBM Technical Report, KGN 153, (1988).

237. M. W. Evans, G. C. Lie, and E. Clementi, *Phys. Rev. A* **37**, 2551 (1988).

238. M. W. Evans, *J. Mol. Liq.* **38**, 175 (1988).

239. M. W. Evans and D. M. Heyes, *Mol. Phys.* **69**, 241 (1990).

240. M. W. Evans, G. C. Lie, and E. Clementi, *Z. Phys. D* **7**, 397 (1988).

241. B. J. Berne and R. Pecora, *Dynamical Light Scattering with Reference to Physics, Chemistry and Biology*, Wiley Interscience, New York, 1976.

242. D. H. Whiffen, *Mol. Phys.* **63**, 1053 (1988).

243. M. W. Evans, *J. Chem. Phys.* **86**, 4096 (1987).

244. J. A Salthouse and M. J. Ware, *Point Group Character Tables*, Cambridge University Press, 1972.

245. L. D. Landau and E. M. Lifshitz, *Mechanics*, Pergamon Press, Oxford, 1976.

246. M. R. Spiegel, *Vector Analysis*, Schaum, New York, 1959.

247. D. W. Condiff and J. S. Dahler, *J. Chem. Phys.* **44**, 3988 (1966).

248. L. P. Hwang and J. H. Freed, *J. Chem. Phys.* **63**, 118, 4017 (1975).

249. P. G. Wolynes and J. M. Deutch, *J. Chem. Phys.* **67**, 733 (1977).

250. G. T. Evans, *Mol. Phys.* **36**, 1199 (1978).

251. U. Steiger and R. F. Fox, *J. Math. Phys.* **23**, 296 (1982).

252. G. van der Zwan and J. T. Hynes, *Physica* **121A** 224 (1983).

253. E. Dickinson, *Ann. Rep. Chem. Soc.* **140**, 421 (1985).

254. N. K. Ailawadi and B. J. Berne, *Faraday Symp.* **11** (1976).

255. B. J. Berne and J. A. Montgomery, *Mol. Phys.* **32**, 363 (1976).

256. M. W. Evans, G. C. Lie, and E. Clementi, *IBM Tech. Rep.*, KGN 131 (1988).

257. M. W. Evans, W. Luken, G. C. Lie, and E. Clementi, *J. Chem. Phys.* **88**, 2685 (1988).

258. G. C. Lie and E. Clementi, *Phys. Rev. A* **33**, 2679 (1986).

259. R. J. Bartlett, 1. Shavitt, and G. D. Purvis, *J. Chem. Phys.* **71**, 281 (1979).

260. W. F. van Gunsteren, H. J. C. Berendsen, and J. A. C. Rullman, *Faraday Disc. Chem. Soc.* **66**, 58 (1978).

261. M. W. Evans, G. C. Lie, and E. Clementi, *J. Chem. Phys.* **88**, 5157 (1988).

262. M. G. Sceats and S. A. Rice, in *Water, a Comprehensive Treatise*, vol. 7, F. Franks, Ed., Plenum, New York, 1982.

263. C. A. Angell, ibid., Chapter 1.

264. O. C. Bridgman and E. W. Aldrich, *J. Heat Transfer* **87**, 26 (1965).

265. M. H. Price and J. M. Walsh, *J. Chem. Phys.* **26**, 824 (1957).

266. G. W. F. Pardoe, Ph.D. Thesis, University of Wales, 1969.

267. M. W. Evans, *J. Chem. Soc., Faraday Trans.* 2, **72**, 2138 (1976).
268. M. W. Evans, *J. Chem. Phys.* **86**, 4096 (1987).
269. F. H. Stillinger and A. Rahman, *J. Chem. Phys.* **60**, 1545 (1974).
270. M. W. Evans, G. C. Lie, and E. Clementi, *J. Chem. Phys.* **89**, 6399 (1988).
271. M. W. Evans, G. C. Lie, and E. Clementi, *J. Chem. Phys.* **87**, 6040 (1987).
272. M. W. Evans, *J. Chem. Soc. Faraday Trans.* 2, **82**, 1967 (1986).
273. H. Friedmann and S. Kimel, *J. Chem. Phys.* **47**, 3589 (1967).
274. G. Ewing, *Acc. Chem. Res.* 2, 168 (1969).
275. M. Doi and S. F. Edwards, *J. Chem. Soc., Faraday Trans.* 2, **74**, 918 (1978).
276. D. Frenkel and J. F. Maguire, *Mol. Phys.* **49**, 503 (1983).
277. M. W. Evans, G. C. Lie, and E. Clementi, *J. Mol. Liq.* **39**, 1 (1988).
278. F. E. Neumann, *Vorlesungen uber die Theorie Elastizitat der Festen Korper und des Lichtathers*, Teubner, Leipzig, 1885.
279. L. D. Barron, *Chem. Soc. Rev.* **15**, 189 (1986).
280. P. Curie, *J. Phys. (Paris)* **3**, 393 (1894).
281. S. F. Mason, *Molecular Optical Activity and the Chiral Discriminations*, Cambridge University Press, 1982.
282. L. Pasteur, *Rev. Sci.* **7**, 2 (1884).
283. H. Primas, *Chemistry, Quantum Mechanics, and Reductionism*, Springer-Verlag, Berlin, 1981.
284. E. P. Wigner, *Z. Phys.* **43**, 624 (1927).
285. R. P. Feynman, R. B. Leighton, and M. Sands, *The Feynman Lectures in Physics*, Addison-Wesley, Reading, MA, 1964.
286. L. D. Barron, *Molecular Light Scattering and Optical Activity*, Cambridge University Press, 1982.
287. R. R. Birss, *Symmetry and Magnetism*, North-Holland, Amsterdam, 1966.
288. A. V. Shubnikov and V. A. Koptsik, *Symmetry in Science and Art*, Plenum, New York, 1974.
289. W. Heisenberg, *Introduction to the Unified Theory of Elementary Particles*, Wiley, New York, 1966.
290. T. D. Lee, *Particle Physics and Introduction to Field Theory*, Harwood, Chur, 1981.
291. E. Bright-Wilson, Jr., J. C. Decius, and P. G. Cross, *Molecular Vibration*, McGraw-Hill, New York, 1955.
292. R. L. Flurry, Jr., *Symmetry Groups, Theory and Applications*, Prentice-Hall, Englewood Cliffs, NJ, 1980.
293. J. A. Salthouse and M. J. Ware, *Point Group Character Tables*, Cambridge University Press, 1972.
294. D. S. Urch, *Orbitals and Symmetry*, Penguin, Harmondsworth, 1970.
295. F. A. Cotton, *Chemical Applications of Group Theory*, Wiley Interscience, New York, 1963.
296. M. W. Evans, *Phys. Lett. A* **134**, 409 (1989).
297. M. W. Evans, *Phys. Rev. A* **39**, 6041 (1989).
298. M. W. Evans, G. C. Lie, and E. Clementi, *J. Mol. Liq.* **37**, 231 (1988).
299. M. W. Evans, *Mol. Phys.* **67**, 1195 (1989).
300. M. W. Evans and D. M. Heyes, *Mol. Phys.* **65**, 1441 (1988).

301. M. W. Evans and D. M. Heyes, *Phys. Rev. B.* **42**, 4363 (1990).

302. J. Harris, *Rheology of non-Newtonian Flow*, Longmans, London, 1977.

303. W. R. Schowalter, *Mechanics of non-Newtonian Fluids*, Pergamon, Oxford, 1978.

304. D. M. Heyes, *J. non-Newtonian Fluid Mech.* **27**, 47 (1988).

305. Ibid., *J. Chem. Soc. Faraday Trans.* 2, **82**, 1365 (1986).

306. G. Wagnière and A. Meier, *Chem. Phys. Lett.* **93**, 78 (1982).

307. G. Wagnière, *Z. Naturforsch* **39A**, 254 (1984).

308. G. Wagnière and A. Meier, *Experientia* **39**, 1090 (1983).

309. D. Radulesu and J. Moga, *Bull. Soc. Chim. Romania* **1**, 2 (1939).

310. G. P. Morriss and D. J. Evans, *Phys. Rev. A* **35**, 792 (1987).

311. D. J. Evans and G. P. Morriss, *Computer Phys. Rep.* **1**, 297 (1984).

312. L. D. Barron, *Chem. Phys. Lett.* **135**, 1 (1987).

313. H. Zocher and C. Torok, *Proc. Natl. Acad. Sci., U.S.A.* **39**, 681 (1953).

314. P. G. de Gennes, *C. R. Hebd. Seances Acad. Sci. Ser. B* **270**, 891 (1970).

315. M. W. Evans, *Chem. Phys. Lett.* **152**, 33 (1988).

316. L. D. Barron and J. Vrbancich, *Mol. Phys.* **51**, 715 (1984).

317. M. W. Evans, *Chem. Phys.* **135**, 187 (1989).

318. W. Maier and A. Saupe, *Z. Naturforsch.* **13A**, 564 (1958).

319. P. G. de Gennes, *The Physics of Liquid Crystals*, Oxford University Press, 1974.

320. G. R. Luchurst and G. Zannoni, *Proc. Roy. Soc.* **343A**, 389 (1975).

321. J. H. Freed, *J. Chem. Phys.* **66**, 3428 (1977).

322. A. de Vries, *Mol. Cryst. Liq. Cryst.* **49**, 1 (1978).

323. M. W. Evans, *Chem. Phys.* **127,** 413 (1988).

324. M. W. Evans and D. M. Heyes, *J. Chem. Soc. Faraday Trans.* 2, **86**, 1041 (1990).

325. D. J. Evans and G. P. Morriss, *Computer Phys. Rep.* **1**, 297 (1984).

326. D. M. Heyes, ibid. **8**, 71 (1988).

327. D. J. Evans, *Mol. Phys.* **42**,1355 (1981).

328. L. D. Landau and E. M. Lifshitz, *Statistical Physics*, Pergamon, Oxford, 1978.

329. D. M. Heyes, *J. Chem. Soc. Faraday Trans.* 2, **84**, 705 (1988).

330. D. J. Evans and G. P. Morriss, *Phys. Rev. Lett.* **56**, 2172 (1985).

331. D. M. Heyes, *J. Chem. Soc. Faraday Trans.* 2, **82** 1365 (1986).

332. P. N. Pusey and R. J. A. Tough, in R. Pecora, *Dynamical Light Scattering, Applications of Photon Correlation Spectroscopy*, Ed., Plenum. New York, 1985.

333. P. A. Madden, *Mol. Phys.* **36**, 365 (1978).

334. B. J. Berne and R. Pecora, *Dynamical Light Scattering with Reference to Physics, Chemistry and Biology*, Wiley Interscience, New York, 1976.

335. D. M. Heyes and R. Szczepanski, *J. Chem. Soc. Faraday Trans.* 2, **83**, 319 (1987).

336. D. J. Evans and J. F. Ely, *Mol. Phys.* **59**, 1043 (1986).

337. M. W. Evans and D. M. Heyes, *Mol. Sim.* **4**, 339 (1990).

338. M. Faraday, *Phil. Mag,* **28**, 294 (1846).

339. P. W. Atkins, *Molecular Quantum Mechanics*, Oxford University Press, 1982.

340. L. D. Landau and E. M. Lifshitz, *The Classical Theory of Fields*, Pergamon, Oxford, 1975.

341. G. E. Stedman, *Diagrammatic Group Theory*, Cambridge University Press, 1990.

342. W. H. Bragg, *Rev. Mod. Phys.* **3**, 449 (1931).

343. E. P. Wigner, *Group Theory*, Academic, New York, 1959.

344. E. P. Wigner, *Z. Phys.* **43**, 624 (1927).

345. M. W. Evans, *Phys. Lett. A* **146**, 475 (1990).

346. M. W. Evans, *Phys. Lett. A* **147**, 364 (1990).

347. J. H. Christenson, J. W. Cronin, V. L. Fitch, and R. Turlay, *Phys. Rev. Lett.* **13**, 138 (1964).

348. C. S. Wu, E. Ambler, R. W. Hayward, D. D. Hoppes, and R. P. Hudson, *Phys. Rev.* **105**, 1413 (1957).

349. E. Fermi, *Z. Phys.* **88**, 161 (1934).

350. K. Gottfried and V. F. Weisskopf, *Concepts of Particle Physics*, Clarendon, Oxford, 1984.

351. M. A. Bouchiat and L. Pottier, *Sci. Am.* **250(6)**, 76 (1984).

352. M. W. Evans and D. M. Heyes, *Phys. Scripta* **42**, 196 (1990).

353. M. W. Evans, *Mol. Phys.* **71**, 193 (1990).

354. M. W. Evans and D. M. Heyes, *Phys. Rev. B*, **42**, 4363 (1990).

355. M. Ferrario, P. Grigolini, A. Tani, R. Vallauri, and B. Zambon, in Ref. 3, Chapter 6, p. 225.

356. M. W. Evans and D. M. Heyes, *Comp. Phys. Comm.*, Thematic Issue, **62**, 249 (1991).

357. M. W. Evans and D. M. Heyes, *J. Mol. Liq.* **44**, 27 (1989).

358. E. Verdet, *Comp. Rendues* **39**, 548 (1854).

359. G. H. Wagnière and J. B. Hutter, *J. Opt. Soc. Am. B* **6**, 693 (1989).

360. N. Bloembergen, *Non-Linear Optics*, Benjamin, New York, 1965.

361. Y. R. Shen, *The Principles of Non-Linear Optics*, Wiley, New York, 1984.

362. J. A. Giordmaine, *Phys. Rev.* **138**, 1599 (1965).

363. M. W. Evans, *J. Mod. Opt.*, **37**, 1655 (1990).

364. M. W. Evans, *Phys. Lett. A* **146**, 185 (1990).

365. G. E. Stedman, *Adv. Phys.* **34**, 513 (1985).

366. H. J. Ross, B. S.. Sherbourne, and G. E. Stedman, *J. Phys. B, At. Mol. Opt. Phys.* **22**, 459 (1989).

367. L. D. Barron, *Bio-Systems* **20**, 7 (1987).

368. M. W. Evans, *J. Chem. Phys.* **93**, 2328 (1990).

369. M. W. Evans, *Int. J. Quant. Chem.*, Clementi Issue, in press, (1991).

370. G. Wagnière, *Phys. Rev. A* **40**, 2437 (1989).

371. M. W. Evans, *Phys. Rev. A* **41**, 4601 (1990).

372. G. Wagnière, *Chem. Phys. Lett.* **110**, 546 (1984).

373. J. F. Ward, *Rev. Mod. Phys.* **37**, 1 (1965).

374. O. Laporte, *Z. Phys.* **51**, 512 (1924).

375. P. Zeeman, *Phil. Mag*, **43**, 226 (1986).

376. L. D. Barron, *Mol. Phys.* **31**, 129 (1976).

377. G. Wagnière, *Z. Phys. D* **8**, 229 (1988).

378. M. W. Evans, *J. Mod. Opt.*, in press (1991).

379. J. S. Griffith, *The Irreducible Tensor Method for Molecular Symmetry Groups*, Prentice Hall, Englewood Cliffs, NJ, 1962.

380. U. Fano and G. Racah, *Irreducible Tensorial Sets*, Academic, New York, 1959.

381. S. B. Piepho and P. N. Schatz, *Group Theory in Spectroscopy with Applications to Magnetic Circular Dichroism*, Wiley, New York, 1983.

382. C. H. Townes and A. L. Schawlow, *Microwave Spectroscopy*, McGraw-Hill, New York, 1955.

383. J. E. Borkholm and P. F. Liao, *Opt. Commun.* **21**, 132 (1977).

384. A. Schabert, R. Keil, and P. E. Toschek, *Opt. Commun.* **13**, 265 (1975).

385. C. Delsart and J. C. Keller, *J. Phys. B.* **9**, 2769 (1978).

386. M. W. Evans, *J. Mol. Spect.*, **146**, 143 (1991).

387. Ch. Salomon, CH. Breant, A. van Lerberghe, G. Camy, and C. U. Bordé, *Appl. Phys. B* **29**, 153 (1982).

388. D. C. Hanna, M. A. Yuratich, and D. Cotter, *Non-Linear Optics of Free Atoms and Molecules*, Springer, New York, 1979.

389. S. H. Autler and C. H. Townes, *Phys. Rev.* **100**, 703 (1955).

390. S. Feneuille, *Rep. Prog. Phys.* **40**, 1257 (1977).

391. S. E. Moody and M. Lambropoulos, *Phys. Rev. A* **15**, 1497 (1977).

392. H. R. Gray and C. R. Stroud, Jr., *Opt. Commun.* **25**, 359 (1978).

393. F. Fleming Krim, A. Sinha, and M. C. Hsiao, *J. Chem. Phys.* **92**, 6333 (1990).

394. W. Worthy, *Chem. Eng. News* **68**(26), 24 (1990).

395. C. Reiser, J. I. Steinfeld, and H. W. Galbraith, *J. Chem. Phys.* **74**, 2189 (1981).

396. E. Arimondo, P. Glorieux, and T. Oka, *Phys. Rev. A* **17**, 1 375 (1978).

397. R. M. Whitley and C. R. Stroud, Jr., *Phys. Rev. A* **14**, 1488 (1976).

398. W. A. Molander, C. R. Stroud, Jr., and J. A. Yeazell, *J. Phys. B.* **19**, L461 (1986).

399. R. Serber, *Phys. Rev.* **41**, 489 (1932).

400. A. D. Buckingham and P. J. Stephens, *Ann. Rev. Phys. Chem.* **17**, 399 (1966).

401. P. S. Pershan, *Phys. Rev.* **130**, 19 (1963).

402. J. P. van der Ziel, P. S. Pershan, and L. D. Malmstrom, *Phys. Rev. Lett.* **15**, 190 (1965).

403. P. S. Pershan, J. P. van der Ziel, and L. D. Malmstrom, *Phys. Rev.* **143**, 574 (1966).

404. S. Kielich, in M. Davies (Sen. Rep.), *Dielectric and Related Molecular Processes*, vol. 1, Chem. Soc., London, 1972.

405. M. W. Evans, *J. Mol. Liq.* in press (1991).

406. L. D. Barron and A. D. Buckingham, *Mol. Phys.* **20**, 1111 (1971).

407. L. D. Barron, C. Meehan, and J. Vrbancich, *J. Raman Spect.* **12**, 251 (1982).

408. M. W. Evans, *J. Mol. Spect.*, **143**, 327 (1990).

409. M. W. Evans, *Spectrochim. Acta*, **46A**, 1475 (1990).

410. M. W. Evans, *Chem. Phys.*, **150**, 197 (1991).

411. M. W. Evans, *Phys. Lett. A*, **149**, 328 (1990).

412. M. W. Evans, *Phys. Rev. Lett.* **64**, 2909 (1990).

413. M. W. Evans, *Opt. Lett.* **15**, 836 (1990).

414. R. V. Jones, *Proc. Roy. Soc.* **349A**, 423 (1976).

415. M. W. Evans, *J. Mol. Liq.*, **47**, 109 (1990).

416. T. B. Freedman, M. Germana Paterlini, Nam Soo Lee, and L. A. Nafie, *J. Am. Chem. Soc.* **109**, 4727 (1987).

417. P. L. Polavarapu, P. G. Quincey, and J. R. Birch, *Infra red Phys.* **30**, 175 (1990).

418. M. W. Evans and G. Wagnière, *Phys. Rev. A*, **42**, 6732 (1990).

419. Ref. 404, vols. 2 and 3.

420. Video animation (with narration) by C. Pelkie, B. Land, and M. W. Evans, Cornell National Supercomputer Facility, based on the code TETRA (see Appendix).*

421. See appendix of Ref. 371 for more details.

422. M. W. Evans and D. M. Heyes, Proceedings of the NATO Conference on Complex Flows, Brussels, 1989, published 1991.

423. M. W. Evans, *J. Phys. Chem.*, **95**, 2256 (1991).

424. E. B. Cullwick, *Electromagnetism and Relativity*, Longmans, Green, and Co., London, 1957.

425. M. W. Evans, *J. Mod. Opt.*, in press (1991).

426. P. G. H. Sandars, in K. Crowe, J. Ducios, G. Fiorentini, and G. Torelli, Eds., *Fundamental Interactions and Structure of Matter*, Plenum, New York, 1980.

427. E. N. Fortson and L. Wilets, *Adv. At. Mol. Phys.* **16**, 319 (1980).

428. L. D. Barron, in *New Developments in Molecular Chirality*, P. Mezey, Ed., Reidel, Netherlands, 1990.

429. M. A. Bouchiat and C. Bouchiat, *J. Phys. (Paris)* **35**, 899 (1974).

430. M. Quack, *Angew. Chem. Int. Ed. Engl.* **28**, 571 (1989).

431. J. J. Thompson, *Proc. Camb. Phil. Soc.* **5**, 250 (1886).

432. E. Fermi, *Rend. Lincei* **32**, 115 (1923).

433. M. A. Player, *Proc. Roy. Soc.* **349A**, 441 (1976).

434. M. W. Evans and A. Lakhtakia, *Phys. Rev. A*, in press (1991).

435. M. W. Evans, *Phys. Lett. A*, in press (1991).

436. M. W. Evans, *Spec. Sci. Tech.*, in press (1991).

437. S. J. Cianciosi, K. M. Spencer, T. B. Freedman, L. A. Nafie, and J. E. Baldwin, *J. Am. Chem. Soc.* **111**, 1913 (1989).

438. L. D. Barron, in *Theoretical Models of Chemical Bonding*, Z. B. Maksic, Ed., Springer, Berlin, 1990.

439. S. F. Mason, *Bio Systems* **20**, 27 (1987).

* Judged best animation in the Natural Sciences and Mathematics Category of the IBM 1990 Supercomputer Competition and Conference; to be distributed by "Media Magic," California in video cassette format; see also M. W. Evans and C. R. Pelkie, Conference Proceedings and *J. Opt. Soc. Am., B*, in press.

APPENDIX: MOLECULAR DYNAMICS SIMULATION
ALGORITHM "TETRA"

This appendix provides FORTRAN code for the molecular dynamics simulation algorithm "TETRA," evolved gradually from work by Schofield, Singer, Ferrario, and Evans. It integrates the classical equations of motion for an ensemble of asymmetric tops diffusing in three dimensions. The version shown is for liquid water, with a facility for applying a right-handed circularly polarized laser.

This code formed the basis for much of the simulation work reported in this article, and copies on magnetic tape are available from the author upon request.

A library of molecular dynamics simulation algorithms has been set up at the United Kingdom's Science and Engineering Research Council Daresbury Laboratory, CCP5 Group, Daresbury, Warrington WA4 4AD, UK. These algorithms are available for individual research scientists and are described regularly in the CCP5 Quarterly Newsletter. They are supplied with detailed descriptive documentation, and are available for many different areas of computer simulation. Algorithms for the test molecule dichloromethane are available from the same source, and were set up during the pilot project of the European Molecular Liquids Group (EMLG).

The following TETRA code is not meant as a substitute for a comprehensive description, but illustrates the stages involved in the computer simulation of an asymmetric top molecule diffusing in three dimensions.

```
      PROGRAM TETRA
C---- MOL DYNAMICS PROGRAM FOR WATER ------
C
      IMPLICIT REAL*8 (A-H,O-Z)
C
C ATOMIC COORDINATES: XAT,YAT,ZAT, 1.LE.IA.LE 432
C
      REAL*8 M(6),JX(108),JY(108),JZ(108),IN(6),TM
      REAL*8 KB,NAV,JCON,NINE,NTEN,NITF,NTTF
      INTEGER TITLE(80),INDEX(25),BIND(6)
      CHARACTER*4 TIND(6),TII(8)
      DIMENSION
     1    TXA(108),TYA(108),TZA(108),EXO(3,108),EYO(3,108),EZO(3,108)
     2,RI(6),ODDT(6),EDDDX(6),EDDDY(6),EDDDZ(6)
      DIMENSION O(6),OSQ(6),OM(6),OMSQ(6),ODOT(6)
     1,SQE(3),CO(6),EEX(6),EEY(6),EEZ(6),EXN(6),EYN(6)
     2,EZN(6),ELX(6),ELY(6),ELZ(6),EDDX(6),EDDY(6),EDDZ(6)
     3,TE(6),OP(101),EDOX(6),EDOY(6),EDOZ(6)
      DIMENSION  SIG(6,6),EPS(6,6),BSIG(36),EP(36),BSIGSQ(36),
     *ACR(36),CHA(6),BCHA(36)
      DIMENSION G(6,100),GR(6,100)
      COMMON /ATT/ XAT(648),YAT(648),ZAT(648),XA(648),YA(648),ZA(648)
```

```
C
C CENTRE OF MASS COORD.: XC,XCN(NEW),XCO(OLD);C.O.M. VEL. VXC...
C
    COMMON /CMT/ XC(108),YC(108),ZC(108),XCO(108),YCO(108),ZCO(108),
   &XCN(108),YCN(108),ZCN(108),VXC(108),VYC(108),VZC(108)
C
C UNIT VECTORS ALONG THE PRINCIPAL AXES: EX,EY,EZ; EXN; EXO,ETC.
C EX(L,IC) L=1,2,3 SPECIFIES THE AXIS; IC(1-108)
C SPECIFIES THE MOLECULE
C
    COMMON /CMO/ EX(3,108),EY(3,108),EZ(3,108)
C
C ANGULAR MOMENTUM JX,...
C
    COMMON /CMJ/ JX,JY,JZ,IN,TM,KB,NAV
C
C FORCES: FXC,FYC,FZC ION C.O.M.,FAX,FAY,FAZ: ON ATOMS.
C TORQUES: TX.TY,TZ; TXN(NEW)........TXO(OLD)....
C
    COMMON /FOR/ FXC(108),FYC(108),FZC(108),TX(108),TY(108),TZ(108),
   &FAX(648),FAY(648),FAZ(648),TXO(108),TYO(108),TZO(108)
C
C   PHYSICAL CONSTANT ...
C
    COMMON /N/ NOM,NOMM1,NORM,NT,NOFST,NTINC,INOF
C
                                              COMMON          /PHY/
TEMP,VOL,DT,BOXL,FACTOR,CONFAC,CUT,RTKTM,RKTF,FF,DTF
C
    COMMON /NTR/ TRIG
C
C   LENNARD JONES POTENTIAL PARAMATERS ...
C
C   SIG=MATRIX OF DISTANCE PARAMETERS
C   EP=MATRIX OF ENERGY PARAMETERS
C
C PRINCIPAL MOMENTS OF INERTIA IN(1-3);THEIR RECIPROCALS RI(1-3)
C
C
C O=ANGULAR VELOCITIES ABOUT PRINCIPAL AXES OSQ=O*O
C.......LOCAL VARIABLES=EX; EDDX, ETC. SECOND DERIVATIVES OF
ELX......
C TE=SCALAR PRODUCTS OF TORQUES WITH UNIT VECTORS E,....
C SUMI(1)=I(1)+I(2)-2I(3) ETC.
C
C   PAIR DISTRIBUTION FUNCTION ... GR(SITE)
```

```
      C
            DATA G,GR/1200*0.0D0/
      C
      C     ----- IBM ERRSET -----
      C
            CALL ERRSET(201,256,-1,1,1,1)
            CALL ERRSET(208,  0,-1,1,1,1)
      C
            KB = 1.3807D-00
            NAV = 6.0223D +23
            ELSQ = 2.3071138D +05
            NATM = 5

            DATA INDEX/1,2,1,3,3,2,4,2,5,5,1,2,1,3,3,3,5,3,6,6,
           *3,5,3,6,6/
            DATA TI1/'CH2 ','CL2 ','OUT ','LJ +C','HARG','T =  ','293K','VOL ='/
            DATA TIND/'H-H ','H- O','H-Q ','O- O','O- Q','Q-Q '/
            DATA BIND/4,4,8,1,4,4/
            DATA ONE,TWO,THREE,SIX,PTFI/1.0D0,2.0D0,3.0D0,6.0D0,0.5D0/
            DATA FOUR,ELEV,TWLE,EITE/4.0D0,11.0D0,12.0D0,18.0D0/
            DATA FIVE,SEVE,NINE,NTEN,TWFO/5.0D0,7.0D0,9.0D0,19.0D0,24.0D0/
                                                                            DATA
      ZOFST,RSTKE,RSRKE,RTQTE,RRQTE,RTPE,RTPR,RTQP,RQTEN,RTEN,
           &RTMO,RTJM,RQEK,RVT,RVIR,RTRTE/16*0.0D0/
            DATA  ZER0,SSTKE,SSRKE, TQTE, RQTE, TPE, TPR, TQP,QTOTE,TTEN,
           &TTMO,TTJM,TQEK,TVT,TVIR,TTRTE/16*0.0D0/
            CALL DATE(DAT)
            CALL TIME(TRIG)


            X = RAND(1)
            READ(5,3) (TITLE(I),I = 1,80)
          3 FORMAT(80A1)
            READ(5,4) NOM,NMAX,NT,NTINC,NDUMP,MM,MODE,MTIME,IPRINT
          4 FORMAT(10I8)
            READ(5,10) TEMP,VOL,DT,FCC,CUT,DFFT,TMAX
         10 FORMAT(8D10.3)
          6 FORMAT(10X,5X,80A1////)
          8 FORMAT(1X,' NUMERICAL SIMULATION RUN CONDITION'/10X,
           &'NUMBER OF MOLECULES = ',I5,' INTEGRATION TIME STEP = ',G12.5
           &,' POTENTIAL CUT-OFF DISTANCE = ',G12.5/)
          7 FORMAT(1X,' THERMODYNAMICAL CONDITION '/3X,' TEMPERATURE
        = '
           &,F10.5,'  MOLAR VOLUME = ',G12.5//)
            NZ = 0
            PYE = (DACOS(-1.0D0))
```

```
AK1 = TWO*PYE/FCC
ANINT = 100.0D0
ROOT2 = DSQRT(TWO)
RR2 = ONE/ROOT2
ROOT3 = DSQRT(THREE)
CONFAC = NAV/(DFLOAT(NOM))
BOXL = (VOL/CONFAC)**(ONE/THREE)
FACTOR = TWO/BOXL
DELGR = CUT/(FACTOR*ANINT)
TKIN = ZERO
TROT = ZERO
RDT = ONE/DT
DTSQ = DT**2
DTCU = DT**3
DTF = DT*FACTOR
NAT = NOM*NATM
NOMM1 = NOM-1
GRFAC = BOXL**3/(TWO*PYE*NOM*NOMM1*DELGR)
INOF = 0
FF = FACTOR**2
FF24 = TWFO*FF
FITW = FIVE/TWLE
TWTH = TWO/THREE
ONTW = ONE/TWLE
ONSI = ONE/SIX
ONTF = ONE/TWFO
FITF = FIVE/TWFO
NITF = NINE/TWFO
NTTF = NTEN/TWFO
STWT = SEVE/TWLE
VCON = 1.0D +02
PCON = 1.0D-25
FCON = 1.0D-13
ECON = 1.0D-23
JCON = 1.0D-35
CUTF = CUT/FACTOR
PRINT 6,(TITLE(I),I = 1,80)
PRINT 7,TEMP,VOL
PRINT 8,NOM,DT,CUTF
CUTSQ = CUT**2
CALL LENJO(SIG,EPS,BSIG,BSIGSQ,EP,KB,FACTOR)
CALL CHARGE(CHA,BCHA,ELSQ,FACTOR,TWFO)
CALL RANGE(PSI,PELR,CUT,NOM,ACR,EP,BSIGSQ)
CALL KINET(NAV,FACTOR,M,TM,IN,RI)
RKTF = KB*TEMP*FF
RMTT = DTSQ/TM
```

```
      RTKTM = DSQRT(KB*TEMP/TM)
      IF(MM .EQ. 1) GO TO 89
      XX = DFLOAT(MM)
      PRINT 2345,DFLOAT(MM)
 2345 FORMAT(G14.6)
      CALL LATFCC(FCC)
      GO TO 90
   89 CONTINUE
C
C    ***    READ INITIAL CONDITION FROM DISK FILE    ***
C
      IF(MODE.NE.3)  GO TO 731
      ZOFST = ZER0
      READ(7)
      READ(7)
      GO TO 734
  731 CONTINUE
      READ(7) ZOFST,SRKE,STKE,SSTKE,SSRKE,TQTE,RQTE
     &      ,TPE,TPR,TQP,QTOTE,TTMO,TTJM,TQEK,TVIR,TVT
      READ(7) G
  734 CONTINUE
      READ(7) XCN,YCN,ZCN,XC,YC,ZC,VXC,VYC,VZC
      READ(7) JX,JY,JZ,EX,EY,EZ
      READ(7) TX,TY,TZ,TXO,TYO,TZO,TXA,TYA,TZA
C
C    ***    ------------------------------------    ***
C
      PRINT 432,ZOFST
  432 FORMAT(1X,' SIMULATION RUN START FROM TIME STEP  = ',F10.3)
      MM = 1
      DO 9 I = 1,2
      PRINT 431,XCN(I),YCN(I),ZCN(I),EX(1,I),EY(1,I),EZ(1,I),
     &JX(I),JY(I),JZ(I),XC(I),YC(I),ZC(I)
    9  CONTINUE
      CALL ZERO(FCC)
      IF(MODE.NE.2) GO TO 90
      PRINT 433,MTIME,TEMP
  433 FORMAT(1X//,3X,' C.O.M. VELOCITIES AND ANGULAR MOMENTUM
RESCA',
     &'LED INITIALLY AND EVERY ',I3,' STEPS. TEMPERATURE = ',F10.3)
      CALL TSCAL(1,TRTE,ROTE)
   90  CONTINUE
      PRINT 431,DTSQ,DTCU,FACTOR,CONFAC,RTKTM
      PRINT 431,FF,EP(1),BSIG(2),TM
      NOFST = INT(ZOFST)
      IF(NOFST.LT.NT) WRITE(1) TI1,VOL
```

```
     IF(NOFST.LT.NT) GO TO 300
     NSTO = (NOFST-NT)/NDUMP + 10
     DO 737 J = 1,NSTO
     KJ = J
     READ(1,END = 847)
737 CONTINUE
847 PRINT 849,KJ
849 FORMAT(10X,'   NUMBER OF RECORDS SKIPPED = ',I10,' READY ',
    &' TO WRITE THE NEW ONES',1X///)
300  CONTINUE
     PE = ZER0
     VIR = ZER0
     VIRT = ZER0
C  START OF LOOP
     SUMEX = 0.0
     SUMEY = 0.0
     SUMEZ = 0.0
     SUMVX = 0.0
     SUMVY = 0.0
     SUMVZ = 0.0


     INOF = INOF + 1.
     NOFST = NOFST + 1
     ZOFST = ZOFST + ONE
C
C
     DO 369 IC = 1,NOM
     UX = TWO*DINT(XCN(IC))
     UY = TWO*DINT(YCN(IC))
     UZ = TWO*DINT(ZCN(IC))
     XCO(IC) = XC(IC)-UX
     YCO(IC) = YC(IC)-UY
     ZCO(IC) = ZC(IC)-UZ
C
C CENTRE OF MASS POSITIONS COMPUTED NEXT
C
     XC(IC) = XCN(IC)-UX
     YC(IC) = YCN(IC)-UY
     ZC(IC) = ZCN(IC)-UZ
C
C
C
369 CONTINUE
     CALL ATPOS(NATM)
C
```

```
C      ATPOS DOES NOT AFFECT CENTRE OF MASS POSITIONS XC,YC,ZC
C
C   FORCES  LOOP
C
      DO 150 I = 1,NOM
      FXC(I) = ZER0
      FYC(I) = ZER0
      FZC(I) = ZER0
      DO 151 IA = 1,NATM
      I4 = NATM*(I-1)
      L = I4 + IA
      FAX(L) = ZER0
      FAY(L) = ZER0
151   FAZ(L) = ZER0
150   CONTINUE
      SK1 = ZER0
      DO 152 IC = 1,NOM
      AK = TWO*AK1*(XC(IC) + YC(IC) + ZC(IC))
152   SK1 = SK1 + DCOS(AK + PYE)
      DO 350 IC = 1,NOMM1
      XCI = XC(IC)
      YCI = YC(IC)
      ZCI = ZC(IC)
      I4 = NATM*(IC-1)
      J = IC + 1
      DO 340 JC = J,NOM
      J4 = NATM*(JC-1)
C    COORD DIFFERENCES BETWEEN C.O.M.S
C
C      POTENTIAL CUT-OFF ON C.O.M DISTANCES TO HELP WITH CHARGES
C
      DCX = XCI-XC(JC)
      DCY = YCI-YC(JC)
      DCZ = ZCI-ZC(JC)
      DCX = DCX-DINT(DCX)*TWO
      DCY = DCY-DINT(DCY)*TWO
      DCZ = DCZ-DINT(DCZ)*TWO
      RCSQ = DCX*DCX + DCY*DCY + DCZ*DCZ
      IF(RCSQ.GT.CUTSQ) GO TO 340
      DO 351 IA = 1,NATM
      K = I4 + IA
      IA5 = 5*(IA-1)
      XATK = XAT(K)
      YATK = YAT(K)
      ZATK = ZAT(K)
      FAXK = ZER0
```

```
      FAYK = ZERO
      FAZK = ZERO
      DO 341 JA = 1,NATM
      L = J4 + JA
      DATX = XATK-XAT(L)
      DATY = YATK-YAT(L)
      DATZ = ZATK-ZAT(L)
      DAAX = DCX + DATX
      DAAY = DCY + DATY
      DAAZ = DCZ + DATZ
      DAAX = DAAX-TWO*(DFLOAT(IDINT(DAAX)))
      DAAY = DAAY-TWO*(DFLOAT(IDINT(DAAY)))
      DAAZ = DAAZ-TWO*(DFLOAT(IDINT(DAAZ)))
      RSQ = DAAX**2 + DAAY**2 + DAAZ**2
      RDIJ = DSQRT(RSQ)
      RRSQ = ONE/RSQ
      IJ = IA5 + JA
      NDF = IDINT(RDIJ*ANINT) + 1
      IF(NDF .GT. 100) GO TO 342
      NIND = INDEX(IJ)
      GR(NIND,NDF) = GR(NIND,NDF) + ONE
  342 CONTINUE
      ALJ = (RRSQ*BSIGSQ(IJ))**3
      BLJ = ALJ*ALJ
      AEL = BCHA(IJ)/RDIJ
      FLJ = EP(IJ)*(BLJ + BLJ-ALJ) + AEL
      PE = PE + EP(IJ)*(BLJ-ALJ) + SIX*AEL
      VIR = VIR-FLJ
      A = FLJ*RRSQ
C     A = A-ACR(IJ)
      VIRT = VIRT + (DAAX*DATX + DAAY*DATY + DAAZ*DATZ)*A
      FAAX = A*DAAX
      FAAY = A*DAAY
      FAAZ = A*DAAZ
      FAX(L) = FAX(L)-FAAX
      FAXK = FAXK + FAAX
      FAY(L) = FAY(L)-FAAY
      FAYK = FAYK + FAAY
      FAZ(L) = FAZ(L)-FAAZ
      FAZK = FAZK + FAAZ
  341 CONTINUE
      FAX(K) = FAXK + FAX(K)
      FAY(K) = FAYK + FAY(K)
      FAZ(K) = FAZK + FAZ(K)
  351 CONTINUE
  340 CONTINUE
```

```
350  CONTINUE
     DO 355 IC=1,NOM
     I4=NATM*(IC-1)
     DO 356 IA=1,NATM
     K=I4+IA
     FXC(IC)=FXC(IC)+FAX(K)
     FYC(IC)=FYC(IC)+FAY(K)
     FZC(IC)=FZC(IC)+FAZ(K)+CHA(IA)*0.0
356  CONTINUE
355  CONTINUE
     DO 354 IA=1,NAT
     FAX(IA)=FAX(IA)*FF24
     FAY(IA)=FAY(IA)*FF24
     FAZ(IA)=FAZ(IA)*FF24
354  CONTINUE
     PE=TWO*TWO*PE
     VIR=TWFO*VIR
     VIRT=VIRT*TWFO
     SPE=PE+PELR
     SVIR=VIR
     SVIRT=VIRT
     RTPE = RTPE + SPE
C
C  PE AND VIR SHOULD BE S.I. UNITS;FORCES CONTAIN FACTOR
C
     DO 353 IC=1,NOM
     FXC(IC)=FXC(IC)*FF24
     FYC(IC)=FYC(IC)*FF24
     FZC(IC)=FZC(IC)*FF24
353  CONTINUE
C  THE ABOVE LOOP COMPUTES NET FORCES
     TKE=ZER0
     XMO = ZER0
     YMO = ZER0
     ZMO = ZER0
C  MD LOOP
C    XCI ETC. UPDATED FROM XC(I) ETC TO COMPUTE VELOCITIES
C    XC(I), YC(I), ZC(I) DUMPED TO WRITE (1)
C
     DO 360 I=1,NOM
     XCI=TWO*XC(I)-XCO(I)+FXC(I)*RMTT
     YCI=TWO*YC(I)-YCO(I)+FYC(I)*RMTT
     ZCI=TWO*ZC(I)-ZCO(I)+FZC(I)*RMTT
     VXC(I)=PTFI*(XCI-XCO(I))
     VYC(I)=PTFI*(YCI-YCO(I))
     VZC(I)=PTFI*(ZCI-ZCO(I))
```

```
C
C  CENTRE OF MASS LINEAR VELOCITIES, LAST POINT BEFORE DUMP
C
      XMO = XMO + VXC(I)
      YMO = YMO + VYC(I)
      ZMO = ZMO + VZC(I)
      TKE=TKE+VXC(I)**2+VYC(I)**2+VZC(I)**2
      XCN(I)=XCI
      YCN(I)=YCI
      ZCN(I)=ZCI
360   CONTINUE
      TMO = DSQRT( XMO*XMO + YMO*YMO + ZMO*ZMO )*TM/DTF
      STKE=(TKE/(DTF**2))*PTFI*TM
      TRTE=TWTH*STKE/(KB*NOM)
      RTMO = RTMO + TMO
      RSTKE = STKE + RSTKE
      RTQTE = RTQTE + TRTE*TRTE
C        ROTATION
C        PRINCIPAL MOMENTS OF INERTIA ARE IN(1)-IN(3)
C        RECIPROCALS  RI(1)-RI(3)
C        CALCULATE ANGULAR MOMENTUM
C        IF NOFST=2 NO INITIAL MODIFICATION NECESSARY
C
      XJI=ZERO
      YJI=ZERO
      ZJI=ZERO
      ROTKE=ZERO
      DO 361 IC=1,NOM
      I4=NATM*(IC-1)
      TXI=ZERO
      TYI=ZERO
      TZI=ZERO
      DO 362 IA=1,NATM
      K=I4+IA
      TXI=TXI+YAT(K)*FAZ(K)-ZAT(K)*FAY(K)
      TYI=TYI+ZAT(K)*FAX(K)-XAT(K)*FAZ(K)
362   TZI=TZI+XAT(K)*FAY(K)-YAT(K)*FAX(K)
      E0=100.00
      OMX= 0.05
      WT=OMX*NOFST
      TXI=E0*(DSIN(WT)*EY(1,IC)-DCOS(WT)*EZ(1,IC))+TXI
      TYI=-E0*DSIN(WT)*EX(1,IC)+TYI
      TZI=E0*DCOS(WT)*EX(1,IC)+TZI
C      INITIALISATION REQUIRED.
      IF(NOFST.GT.3) GO TO 374
      IF(NOFST.EQ.3) GO TO 373
```

```
       IF(NOFST.EQ.2) GO TO 372
  C    FIRST STEP  ZERO ORDER ALGORITHM FOR J
       JX(IC) = JX(IC) + DT*TXI
       JY(IC) = JY(IC) + DT*TYI
       JZ(IC) = JZ(IC) + DT*TZI
       GO TO 371
  C    SECOND STEP FIRST ORDER ALGORITHM FOR J
  372  JX(IC) = JX(IC) + PTFI*(TXI + TX(IC))*DT
       JY(IC) = JY(IC) + PTFI*(TYI + TY(IC))*DT
       JZ(IC) = JZ(IC) + PTFI*(TZI + TZ(IC))*DT
       DTORX = (TXI-TX(IC))*RDT
       DTORY = (TYI-TY(IC))*RDT
       DTORZ = (TZI-TZ(IC))*RDT
       GOTO 375
  C    THIRD STEP SECOND ORDER ALGORITHM FOR J
  373  JX(IC) = JX(IC) + DT*(FITW*TXI + TWTH*TX(IC)-ONTW*TXO(IC))
       JY(IC) = JY(IC) + DT*(FITW*TYI + TWTH*TY(IC)-ONTW*TYO(IC))
       JZ(IC) = JZ(IC) + DT*(FITW*TZI + TWTH*TZ(IC)-ONTW*TZO(IC))
       DTORX = (THREE*TXI + TXO(IC)-FOUR*TX(IC))*PTFI*RDT
       DTORY = (THREE*TYI + TYO(IC)-FOUR*TY(IC))*PTFI*RDT
       DTORZ = (THREE*TZI + TZO(IC)-FOUR*TZ(IC))*PTFI*RDT
       GO TO 376
  C    NORMAL ALGORITHM THIRD ORDER INTEGRATION
                                                                374
JX(IC) = JX(IC) + DT*(NITF*TXI + NTTF*TX(IC)-FITF*TXO(IC) + ONTF*TXA(IC))

JY(IC) = JY(IC) + DT*(NITF*TYI + NTTF*TY(IC)-FITF*TYO(IC) + ONTF*TYA(IC))

JZ(IC) = JZ(IC) + DT*(NITF*TZI + NTTF*TZ(IC)-FITF*TZO(IC) + ONTF*TZA(IC))

DTORX = (ELEV*TXI-EITE*TX(IC) + NINE*TXO(IC)-TWO*TXA(IC))*ONSI*RDT

DTORY = (ELEV*TYI-EITE*TY(IC) + NINE*TYO(IC)-TWO*TYA(IC))*ONSI*RDT

DTORZ = (ELEV*TZI-EITE*TZ(IC) + NINE*TZO(IC)-TWO*TZA(IC))*ONSI*RDT
  376 TXA(IC) = TXO(IC)
      TYA(IC) = TYO(IC)
      TZA(IC) = TZO(IC)
  375 TXO(IC) = TX(IC)
      TYO(IC) = TY(IC)
      TZO(IC) = TZ(IC)
  371 TX(IC) = TXI
      TY(IC) = TYI
      TZ(IC) = TZI
  361 CONTINUE
  370 CONTINUE
```

```
C       THE ANGULAR MOMENTUM HAS BEEN ADVANCED
        SUMC=ZER0
        DO 377 IC=1,NOM
        DO 345 L=1,3
        ELX(L)=EX(L,IC)
        ELX(L+3)=ELX(L)
        ELY(L)=EY(L,IC)
        ELY(L+3)=ELY(L)
        ELZ(L)=EZ(L,IC)
        ELZ(L+3)=ELZ(L)
        OM(L)=(JX(IC)*ELX(L)+JY(IC)*ELY(L)+JZ(IC)*ELZ(L))*RI(L)
        OM(L+3)=OM(L)
        OMSQ(L)=OM(L)**2
        OMSQ(L+3)=OMSQ(L)
        TE(L)=TX(IC)*ELX(L)+TY(IC)*ELY(L)+TZ(IC)*ELZ(L)
345     CONTINUE
        XJI=XJI+JX(IC)
        YJI=YJI+JY(IC)
        ZJI=ZJI+JZ(IC)
        ROTKE=ROTKE+PTFI*(IN(1)*OM(1)**2+IN(2)*OM(2)**2+IN(3)*OM(3)**2)
        DO 346 L=1,3
        O(L)=OM(L+1)*OM(L+2)
        OSQ(L)=OMSQ(L+1)+OMSQ(L+2)
        O(L+3)=O(L)
        OSQ(L+3)=OSQ(L)
        ODOT(L)=(TE(L)+(IN(L+1)-IN(L+2))*O(L))*RI(L)
        ODOT(L+3)=ODOT(L)
        EDOX(L)=OM(L+1)*ELX(L+2)-OM(L+2)*ELX(L+1)
        EDOX(L)=-EDOX(L)
        EDOX(L+3)=EDOX(L)
        EDOY(L)=OM(L+1)*ELY(L+2)-OM(L+2)*ELY(L+1)
        EDOY(L)=-EDOY(L)
        EDOY(L+3)=EDOY(L)
        EDOZ(L)=OM(L+1)*ELZ(L+2)-OM(L+2)*ELZ(L+1)
        EDOZ(L)=-EDOZ(L)
        EDOZ(L+3)=EDOZ(L)
346     CONTINUE
C       TE(1)-(3)=T.EA
C       O(1)-(3)=OMA,OMB/IC ETC.
        DO 385 L=1,3
C
C MIXED ALGORITHM,ONLY OMEGA DOT TERMS IN ACCELERATION
C OF E VECTORS
        EDDX(L)=ODOT(L+2)*ELX(L+1)-ODOT(L+1)*ELX(L+2)
       &+EDOX(L+1)*OM(L+2)-EDOX(L+2)*OM(L+1)
        EDDY(L)=ODOT(L+2)*ELY(L+1)-ODOT(L+1)*ELY(L+2)
```

```
    & + EDOY(L+1)*OM(L+2)-EDOY(L+2)*OM(L+1)
      EDDZ(L) = ODOT(L+2)*ELZ(L+1)-ODOT(L+1)*ELZ(L+2)
    & + EDOZ(L+1)*OM(L+2)-EDOZ(L+2)*OM(L+1)
      EDDX(L+3) = EDDX(L)
      EDDY(L+3) = EDDY(L)
      EDDZ(L+3) = EDDZ(L)
      EDDDX(L) = ZER0
      EDDDY(L) = ZER0
      EDDDZ(L) = ZER0
385 CONTINUE
      IF(NOFST.EQ.1) GOTO 386
      DO 388 L=1,3
      DTE = DTORX*ELX(L) + DTORY*ELY(L) + DTORZ*ELZ(L)
      TEDOT = TX(IC)*EDOX(L) + TY(IC)*EDOY(L) + TZ(IC)*EDOZ(L)
      TEDD = JX(IC)*EDDX(L) + JY(IC)*EDDY(L) + JZ(IC)*EDDZ(L)
      ODDT(L) = DTE + TWO*TEDOT + TEDD
      ODDT(L+3) = ODDT(L)
388 CONTINUE
      DO 382 L=1,3

EDDDX(L) = EDDX(L+1)*OM(L+2)-EDDX(L+2)*OM(L+1) + TWO*(EDOX(L+1)*ODOT

&(L+2)-EDOX(L+2)*ODOT(L+1)) + ELX(L+1)*ODDT(L+2)-ELX(L+2)*ODDT(L+1)

EDDDY(L) = EDDY(L+1)*OM(L+2)-EDDY(L+2)*OM(L+1) + TWO*(EDOY(L+1)*ODOT

&(L+2)-EDOY(L+2)*ODOT(L+1)) + ELY(L+1)*ODDT(L+2)-ELY(L+2)*ODDT(L+1)

EDDDZ(L) = EDDZ(L+1)*OM(L+2)-EDDZ(L+2)*OM(L+1) + TWO*(EDOZ(L+1)*ODOT

&(L+2)-EDOZ(L+2)*ODOT(L+1)) + ELZ(L+1)*ODDT(L+2)-ELZ(L+2)*ODDT(L+1)
382 CONTINUE
386 DO 384 L=1,3

EXN(L) = ELX(L) + DT*EDOX(L) + PTFI*DTSQ*EDDX(L) + ONSI*DTCU*EDDDX(L)

EYN(L) = ELY(L) + DT*EDOY(L) + PTFI*DTSQ*EDDY(L) + ONSI*DTCU*EDDDY(L)

EZN(L) = ELZ(L) + DT*EDOZ(L) + PTFI*DTSQ*EDDZ(L) + ONSI*DTCU*EDDDZ(L)
      EX(L,IC) = EXN(L)
      EY(L,IC) = EYN(L)
      EZ(L,IC) = EZN(L)
      EXN(L+3) = EXN(L)
      EYN(L+3) = EYN(L)
      EZN(L+3) = EZN(L)
      EEX(L) = EXN(L)
```

```
      EEX(L+3) = EEX(L)
      EEY(L) = EYN(L)
      EEY(L+3) = EEY(L)
      EEZ(L) = EZN(L)
      EEZ(L+3) = EEZ(L)
384  CONTINUE
C  ORTHONORMALISATION
      IF (NOFST.LE.4) GOTO 387
      IF (NOFST.NE.(5*NOFST/5)) GOTO  377
387  CONTINUE
C  ORTHOGONALISATION
      DO 391 L=1,3
      CO(L) = EEX(L+1)*EEX(L+2) + EEY(L+1)*EEY(L+2) + EEZ(L+1)*EEZ(L+2)
391  CO(L+3) = CO(L)
      DO 393 L=1,3
      EXN(L) = EEX(L)-PTFI*(CO(L+2)*EEX(L+1) + CO(L+1)*EEX(L+2))
      EYN(L) = EEY(L)-PTFI*(CO(L+2)*EEY(L+1) + CO(L+1)*EEY(L+2))
      EZN(L) = EEZ(L)-PTFI*(CO(L+2)*EEZ(L+1) + CO(L+1)*EEZ(L+2))
      EEX(L) = EXN(L)
      EEY(L) = EYN(L)
      EEZ(L) = EZN(L)
393  CONTINUE
C     NORMALISATION
397    DO 390 L=1,3
      SQE(L) = EXN(L)**2 + EYN(L)**2 + EZN(L)**2-ONE
      CORR = SQE(L)*(THREE*PTFI/TWO*SQE(L)-PTFI) + ONE
      EEX(L) = EXN(L)*CORR
      EEY(L) = EYN(L)*CORR
      EEZ(L) = EZN(L)*CORR
      EEX(L+3) = EEX(L)
      EEY(L+3) = EEY(L)
      EEZ(L+3) = EEZ(L)
390  CONTINUE
392  DO 394 L=1,3
      EX(L,IC) = EEX(L)
      EY(L,IC) = EEY(L)
      EZ(L,IC) = EEZ(L)
394  CONTINUE
      SUMEX = SUMEX + EX(1,IC)
      SUMEY = SUMEY + EY(1,IC)
      SUMEZ = SUMEZ + EZ(1,IC)
      SUMVX = SUMVX + VXC(IC)
      SUMVY = SUMVY + VYC(IC)
      SUMVZ = SUMVZ + VZC(IC)
377  CONTINUE
      TJMI = DSQRT( XJI*XJI + YJI*YJI + ZJI*ZJI )/FACTOR
```

```
ROTKE = ROTKE/FF
RTJM  = RTJM + TJMI
SRKE = ROTKE
ROTE = TWTH*SRKE/(NOM*KB)
TOTEN = STKE + SPE + SRKE
RTEN = RTEN + TOTEN
RSRKE = RSRKE + SRKE
RQEK = RQEK + (SRKE + STKE)*(SRKE + STKE)
RRQTE = RRQTE + ROTE*ROTE
RTRTE = RTRTE + ROTE*TRTE
RVIR = RVIR + SVIR + SVIRT + PSI
RVT = RVT + (SVIR + SVIRT + PSI)*(TRTE + ROTE)
PRESS = (STKE + SRKE-SVIR-SVIRT-PSI)/THREE
PRESS = PRESS/(.101325D-01*(VOL/CONFAC))
RTPR  = RTPR + PRESS
RTQP  = RTQP + PRESS*PRESS
RQTEN = RQTEN + TOTEN*TOTEN
IF (NOFST .LT. NT) GO TO 379
NTT = NOFST-NT
IF(NTT .NE. NDUMP*(NTT/NDUMP)) GO TO 379
ZDUMP = DFLOAT(NTT/NDUMP) + ONE
WRITE(1) ZDUMP,XC,YC,ZC,VXC,VYC,VZC,JX,JY,JZ,EX,EY,EZ,
AFXC,FYC,FZC,TX,TY,TZ,XA,YA,ZA


379 CONTINUE



SUMEX = SUMEX/108.0
SUMEY = SUMEY/108.0
SUMEZ = SUMEZ/108.0
SUMVX = SUMVX/108.0
SUMVY = SUMVY/108.0
SUMVZ = SUMVZ/108.0
PRINT 513,SUMEX,SUMEY,SUMEZ,SUMVX,SUMVY,SUMVZ
513 FORMAT(6E14.7)
IF(NOFST.NE.(NOFST/IPRINT)*IPRINT) GO TO 378
PRINT 501,NOFST,STKE,SRKE,SPE,TOTEN,PRESS,TMO,TJMI,SK1
501 FORMAT(1X,/,' TIME STEP = ',I5,' KIN.EN = ',G12.5,' ROT.EN = ',G12.5
&' POT.EN = ',G12.5,' TOT.EN = ',G12.5,' PRESSURE = ',G12.5/
&' TOT.P = ',G12.5,' ANG.MOM. = ',G12.5,'     S(KX) = ',G12.5)
500 FORMAT(' TR.TEMP = ',F12.2,5X,'ROT.TEMP = ',F12.2/)
PRINT 500 ,TRTE,ROTE

IF(NOFST.NE.NT) GO TO 233
```

```
      NZ = INOF
      DO 222 L2 = 1,6
      DO 222 L1 = 1,100
 222 GR(L2,L1) = ZERO
      RSTKE = ZERO
      RSRKE = ZERO
      RQEK  = ZERO
      RTQTE = ZERO
      RRQTE = ZERO
      RTRTE = ZERO
      RTPE  = ZERO
      RTPR  = ZERO
      RTQP  = ZERO
      RQTEN = ZERO
      RTMO  = ZERO
      RTJM  = ZERO
      RVIR  = ZERO
      RVT   = ZERO
      RTEN  = ZERO
 233 CONTINUE
C
C  TEMP.SCALING
C
 378 CONTINUE
C
      IF((INOF-NZ).NE.((INOF-NZ)/NTINC)*NTINC) GO TO 415
      IF (NOFST .LE. NT) GO TO 417
      SUBAVK = (RSTKE-TKIN)*TWTH/(NTINC*KB*NOM)
      SUBAVR = (RSRKE-TROT)*TWTH/(NTINC*KB*NOM)
      IF(DABS(SUBAVR+SUBAVK-TWO*TEMP) .LT. DFFT) GO TO 417
      CALL TSCAL(3,SUBAVK,SUBAVR)
 417 TKIN = RSTKE
      TROT = RSRKE
      GO TO 416
 415 IF(MODE.NE.2) GO TO 416
      IF(NOFST.NE.MTIME*(NOFST/MTIME)) GO TO 416
      CALL TSCAL(1,TRTE,ROTE)
      CALL ZERO(FCC)
 416 CONTINUE


 431  FORMAT(2X,3E13.5,1X,3E13.5/2X,3E13.6,1X,3E13.5)
      IF (NOFST.GT.NMAX) GO TO 400
      IF (NOFST.LT.NMAX) GOTO 300

 400 CONTINUE
```

```
      INOF  = INOF - NZ
      RNOF  = ONE / ( DFLOAT ( INOF ) )
      IF(NOFST .LT. NT) GO TO 334
      SSTKE = SSTKE + RSTKE
      SSRKE = SSRKE + RSRKE
      TQEK  = TQEK  + RQEK
      TQTE  = TQTE  + RTQTE
      RQTE  = RQTE  + RRQTE
      TTRTE = TTRTE + RTRTE
      TPE   = TPE   + RTPE
      TPR   = TPR   + RTPR
      TQP   = TQP   + RTQP
      QTOTE = QTOTE + RQTEN
      TTMO  = TTMO  + RTMO
      TTJM  = TTJM  + RTJM
      TVIR  = TVIR  + RVIR
      TVT   = TVT   + RVT
      DO 333 L2 = 1,6
      DO 333 L1 = 1,100
333   G(L2,L1) = G(L2,L1) + GR(L2,L1)
334   CONTINUE
      DO 223 L2 = 1,6
      R = -DELGR/TWO
      DO 223 L1 = 1,100
      R = R + DELGR
223   GR(L2,L1) = GR(L2,L1)*RNOF*GRFAC/(BIND(L2)*R**2)
C
C  ***    DUMP FINAL CONFIGURATION TO DISK FOR RESTART  ***
C
      WRITE(8) ZOFST,SRKE,STKE,SSTKE,SSRKE,TQTE,RQTE
     &        ,TPE,TPR,TQP,QTOTE,TTMO,TTJM,TQEK,TVIR,TVT
      WRITE(8) G
      WRITE(8) XCN,YCN,ZCN,XC,YC,ZC,VXC,VYC,VZC
      WRITE(8) JX,JY,JZ,EX,EY,EZ
      WRITE(8) TX,TY,TZ,TXO,TYO,TZO,TXA,TYA,TZA

      ***  ------------------------------------- ***

      RSTKE = RSTKE * RNOF * (ECON*CONFAC)
      RSRKE = RSRKE * RNOF * (ECON*CONFAC)
      RQEK  = RQEK  * RNOF * (ECON*CONFAC)**2
      RTQTE = RTQTE * RNOF
      RRQTE = RRQTE * RNOF
      RTRTE = RTRTE * RNOF
      RTPE  = RTPE  * RNOF * (ECON*CONFAC)
      RTPR  = RTPR  * RNOF
```

```
      RTQP  = RTQP  * RNOF
      RTEN  = RTEN  * RNOF * (ECON*CONFAC)
      RQTEN = RQTEN * RNOF * (ECON*CONFAC)**2
      RTMO  = RTMO  * RNOF * (PCON*CONFAC)
      RTJM  = RTJM  * RNOF * (JCON*CONFAC)
      RVIR  = RVIR  * RNOF * (ECON*CONFAC)
      RVT   = RVT   * RNOF  * (ECON*CONFAC)
      TRTE  = RSTKE * TWTH / (KB*NAV*ECON)
      ROTE  = RSRKE * TWTH / (KB*NAV*ECON)
      PRINT 988,INOF
  988 FORMAT(1H1,' RUN AVERAGED QUANTITIES. ',I8,'-TIME STEPS FOR
     THIS
     &,' JOB.'//)
                                                        PRINT
  989,RSTKE,TRTE,RSRKE,ROTE,RQEK,RTQTE,RRQTE,RTPE,RTPR,RTQP,
     &RTEN,RQTEN,RTMO,RTJM,RVIR,RVT,RTRTE
  989 FORMAT(1X,' < KIN.TRA.EN. > = ',G12.5,20X,' < TRA.TEMP. > = ',
     &G12.5,//,1X,' < KIN.ROT.EN. > = ',G12.5,20X,' < ROT.TEMP. > = ',
     &G12.5,//,1X,' < (KIN.TOT.EN.)**2 > = ',G12.5,/,1X,' < (TRA.TEMP.)
     &'**2 > = ',G12.5,/1X,' < (ROT.TEMP.)**2 > = ',G12.5,/,1X,
     &' < POT.EN. > = ',G12.5,/,1X,' < PRESSURE > = ',G12.5,20X,
     &' < (PRESSURE)**2 > = ',G12.5,//,1X,' < TOTAL ENERGY > = ',G12.5,
     &20X,' < (TOT.EN.)**2 > = ',G12.5,//,1X,' < C.O.M. IMPULSE > = ',
     &G12.5,10X,' < ANG. MOM. > = ',G12.5//1X,
     &' < VIRIAL > = ',G12.5,' < VIR.*TEMP. > = ',G12.5,
     &' < TRA.TEMP*ROT.TEMP > = ',G12.5/)
      PRINT  999,(TIND(IJ),IJ=1,6)
      R = -DELGR/TWO
      DO 244 L1=1,100
      R=R+DELGR
  244 PRINT 1999,L1,R,(GR(J,L1),J=1,6)
 1999 FORMAT(4X,I4,4X,G10.3,6(2X,G12.5))
  999  FORMAT(1H ,' ATOM-ATOM PAIR DISTRIBUTION G(R) '/15X
     &,'R',4X,7(9X,A5)/)
      IF( NOFST .LT. NT ) GO TO 7000
      RZOF = ONE / ( DFLOAT ( NOFST - NT ) )
      IF(NOFST.LT.NT) RZOF=ONE/(DFLOAT(NOFST))
      SSTKE = SSTKE * RZOF * (ECON*CONFAC)
      SSRKE = SSRKE * RZOF * (ECON*CONFAC)
      TQEK  = TQEK  * RZOF * (ECON*CONFAC)**2
      TQTE  = TQTE  * RZOF
      RQTE  = RQTE  * RZOF
      TTRTE = TTRTE * RZOF
      TPE   = TPE   * RZOF * (ECON*CONFAC)
      TPR   = TPR   * RZOF
      TQP   = TQP   * RZOF
```

```
    TTEN = (TPE+SSTKE+SSRKE)
    QTOTE = QTOTE * RZOF * (ECON*CONFAC)**2
    TTMO = TTMO * RZOF * (PCON*CONFAC)
    TTJM = TTJM * RZOF * (PCON*CONFAC)
    TVIR = TVIR * RZOF * (ECON*CONFAC)
    TVT  = TVT  * RZOF * (ECON*CONFAC)
    TRTE = SSTKE * TWTH / (KB*NAV*ECON)
    ROTE = SSRKE * TWTH / (KB*NAV*ECON)
    PRINT 688,NOFST,NT
688 FORMAT(1H1,' RUN AVERAGED QUANTITIES. ',I8,'-TIME STEPS FOR
THIS
   &,' SEGMENTS. OF WHICH ',I5,'-STEPS USED FOR EQUILIBRATION.'//)
    PRINT 689,SSTKE,TRTE,SSRKE,ROTE,TQEK, TQTE, RQTE, TPE, TPR, TQP,
   &TTEN,QTOTE,TTMO,TTJM,TVIR,TVT,TTRTE
689 FORMAT(1X,' < KIN.TRA.EN. > = ',G12.5,20X,' < TRA.TEMP. > = ',
   &G12.5,//,1X,' < KIN.ROT.EN. > = ',G12.5,20X,' < ROT.TEMP. > = ',
   &G12.5,//,1X,' < (KIN.TOT.EN.)**2 > = ',G12.5,/,1X,' < (TRA.TEMP.)
   &,'**2 > = ',G12.5,/1X,' < (ROT.TEMP.)**2 > = ',G12.5,/,1X,
   &' < POT.EN. > = ',G12.5,/,1X,' < PRESSURE > = ',G12.5,20X,
   &' < (PRESSURE)**2 > = ',G12.5,//,1X,' < TOTAL ENERGY > = ',G12.5,
   &20X,' < (TOT.EN.)**2 > = ',G12.5,//,1X,' < C.O.M. IMPULSE > = ',
   &G12.5,20X,' < ANG.MOM. > = ',G12.5,//,1X,
   &' < VIRIAL > = ',G12.5,' < VIR.*TEMP. > = ',G12.5,
   &' < TRA.TEMP*ROT.TEMP > = ',G12.5/)
    PRINT 999,(TIND(IJ),IJ=1,6)
    R = -DELGR/TWO
    DO 247 L1=1,100
    R = R + DELGR
    DO 245 L2=1,6
245 G(L2,L1)=G(L2,L1)*RZOF*GRFAC/(BIND(L2)*R**2)
247 PRINT 1999,L1,R,(G(J,L1),J=1,6)
7000 CONTINUE
    STOP
    END
    SUBROUTINE LATBCC(A)
    IMPLICIT REAL*8 (A-H,O-Z)

    DIMENSION
   &XC(108),YC(108),ZC(108),XCO(108),YCO(108),ZCO(108),
   &XCN(108),YCN(108),ZCN(108),VXC(108),VYC(108),VZC(108)
   &,EX(3,108),EY(3,108),EZ(3,108),IN(6)
   &,JX(108),JY(108),JZ(108)
    COMMON /CMT/ XC,YC,ZC,XCO,YCO,ZCO,
   &XCN,YCN,ZCN,VXC,VYC,VZC
    COMMON /CMO/ EX,EY,EZ
    COMMON /CMJ/ JX,JY,JZ,IN,TM,KB,NAV
```

```
                                        COMMON          /PHY/
TEMP,VOL,DT,BOXL,FACTOR,CONFAC,CUT,RTKTM,RKTF,FF,DTF
      COMMON /N/ NOM,NOMM1,NORM,NT,NOFST,NTINC,INOF
      COMMON /NTR/ TRIG
      LI=0
      DO 603 L1=1,25
      DO 603 L2=1,25
      DO 603 L3=1,25
      LM1=L1-8
      LM2=L2-8
      LM3=L3-8
      XAA=(LM1-LM2+LM3)*A/2.D0-1.D0+A/4.D0+(RAND(0)-0.5D0)*A/8.D0
      YAA=(LM1+LM2-LM3)*A/2.D0-1.D0+A/4.D0+(RAND(0)-0.5D0)*A/8.D0
      ZAA=(-LM1+LM2+LM3)*A/2.D0-1.D0+A/4.D0+(RAND(0)-0.5D0)*A/8.D0
      IF(DABS(XAA).GT.1.0D0.OR.DABS(YAA).GT.1.0D0.OR.DABS(ZAA).GT.1.0D0
     & GO TO 603
      LI=LI+1
      XCN(LI)=XAA
      YCN(LI)=YAA
      ZCN(LI)=ZAA
  603 CONTINUE
      PRINT 33,RTKTM,RKTF,TM
   33 FORMAT(1X,G12.5,1X,G12.5,1X,G12.5)
      DO 3 IC=1,NOM
      DO 2 L=1,3
      EX(L,IC)=0.0D0
      EY(L,IC)=0.0D0
    2 EZ(L,IC)=0.0D0
      EX(1,IC)=1.0D0
      EY(2,IC)=1.0D0
      EZ(3,IC)=1.0D0
      VXC(IC)=GRAND(0)*RTKTM*DTF
      VYC(IC)=GRAND(0)*RTKTM*DTF
      VZC(IC)=GRAND(0)*RTKTM*DTF
      AJ1=GRAND(0)*DSQRT(RKTF*DFLOAT(IN(1)))
      AJ2=GRAND(0)*DSQRT(RKTF*DFLOAT(IN(2)))
      AJ3=GRAND(0)*DSQRT(RKTF*DFLOAT(IN(3)))
      JX(IC)=AJ1*EX(1,IC)+AJ2*EX(2,IC)+AJ3*EX(3,IC)
      JY(IC)=AJ1*EY(1,IC)+AJ2*EY(2,IC)+AJ3*EY(3,IC)
    3 JZ(IC)=AJ1*EZ(1,IC)+AJ2*EZ(2,IC)+AJ3*EZ(3,IC)
C NOTE:VX,VY,VZ, ARE MULTIPLIED BY DT,JX,JY,JZ, ARE NOT
      PRINT 66, LI,A
   66 FORMAT(1H1,' STARTING FROM A RANDOM QUASI-BCC LATTICE
CONFIGURATI
     &N',//,10X,I5,'-RETICOLAR SITES',F10.4,'-LATTICE CONST.'//)
      DO 5 IC=1,NOM
```

```
  5  PRINT 610,IC,XCN(IC),YCN(IC),ZCN(IC)
610   FORMAT(10X,I4,3(4X,D10.4))
      CALL ZERO(A)
      CALL TSCAL(1,ET,ER)
      RETURN
      END
      SUBROUTINE LATFCC(A)
      IMPLICIT REAL*8 (A-H,O-Z)
      REAL*8 IN(6),JX(108),JY(108),JZ(108)
      DIMENSION
     &XC(108),YC(108),ZC(108),XCO(108),YCO(108),ZCO(108),
     &XCN(108),YCN(108),ZCN(108),VXC(108),VYC(108),VZC(108)
     &,EX(3,108),EY(3,108),EZ(3,108)

      COMMON /CMT/ XC,YC,ZC,XCO,YCO,ZCO,
     &XCN,YCN,ZCN,VXC,VYC,VZC
      COMMON /CMO/ EX,EY,EZ
      COMMON /CMJ/ JX,JY,JZ,IN,TM,KB,NAV
      COMMON                                          /PHY/
TEMP,VOL,DT,BOXL,FACTOR,CONFAC,CUT,RTKTM,RKTF,FF,DTF
      COMMON /N/ NOM,NOMM1,NORM,NT,NOFST,NTINC,INOF
      COMMON /NTR/ TRIG
      LI=0
      DO 603 L1=1,25
      DO 603 L2=1,25
      DO 603 L3=1,25
      LM1=L1-8
      LM2=L2-8
      LM3=L3-8
      XAA=(LM1+LM3)*A/2.D0 -1.D0 + A/4.D0 + (RAND(0)-0.5D0)*A/8.D0
      YAA=(LM1+LM2)*A/2.D0 -1.D0 + A/4.D0 + (RAND(0)-0.5D0)*A/8.D0
      ZAA=(LM2+LM3)*A/2.D0 -1.D0 + A/4.D0 + (RAND(0)-0.5D0)*A/8.D0
      IF(DABS(XAA).GT.1.0D0.OR.DABS(YAA).GT.1.0D0.OR.DABS(ZAA).GT.1.0D0
     &  GO TO 603
      LI=LI+1
      XCN(LI)=XAA
      YCN(LI)=YAA
      ZCN(LI)=ZAA
603   CONTINUE
      PRINT 33,RTKTM,RKTF,TM
 33   FORMAT(1X,G12.5,1X,G12.5,1X,G12.5)
      PRINT *, IN(1),IN(2),IN(3)
      DO 3 IC=1,NOM
      DO 2 L=1,3
      EX(L,IC)=0.0D0
      EY(L,IC)=0.0D0
```

```
 2   EZ(L,IC) = 0.0D0
     EX(1,IC) = 1.0D0
     EY(2,IC) = 1.0D0
     EZ(3,IC) = 1.0D0
     VXC(IC) = GRAND(0)*RTKTM*DTF
     VYC(IC) = GRAND(0)*RTKTM*DTF
     VZC(IC) = GRAND(0)*RTKTM*DTF
     AJ1 = GRAND(0)*DSQRT(RKTF*IN(1))
     AJ2 = GRAND(0)*DSQRT(RKTF*IN(2))
     AJ3 = GRAND(0)*DSQRT(RKTF*IN(3))
     JX(IC) = AJ1*EX(1,IC) + AJ2*EX(2,IC) + AJ3*EX(3,IC)
     JY(IC) = AJ1*EY(1,IC) + AJ2*EY(2,IC) + AJ3*EY(3,IC)
 3   JZ(IC) = AJ1*EZ(1,IC) + AJ2*EZ(2,IC) + AJ3*EZ(3,IC)
C NOTE:VX,VY,VZ, ARE MULTIPLIED BY DT,JX,JY,JZ, ARE NOT
     PRINT 66, LI,A
 66 FORMAT(1H1,' STARTING FROM FCC LATTICE CONFIGURATION ',/,
    &10X,I5,'-RETICOLAR SITES',F10.4,'-LATTICE CONST.'//)
     DO 5 IC = 1,NOM
 5   PRINT 610,IC,XCN(IC),YCN(IC),ZCN(IC)
610  FORMAT(10X,I4,3(4X,E10.4))
     CALL ZERO(A)
     CALL TSCAL(1,ET,ER)
     RETURN
     END
     SUBROUTINE LATSCC(A).
     IMPLICIT REAL*8 (A-H,O-Z)
     DIMENSION
    &XC(108),YC(108),ZC(108),XCO(108),YCO(108),ZCO(108),
    &XCN(108),YCN(108),ZCN(108),VXC(108),VYC(108),VZC(108)
    &,EX(3,108),EY(3,108),EZ(3,108)
    &,JX(108),JY(108),JZ(108),IN(6)

     COMMON /CMT/ XC,YC,ZC,XCO,YCO,ZCO,
    &XCN,YCN,ZCN,VXC,VYC,VZC
     COMMON /CMO/ EX,EY,EZ
     COMMON /CMJ/ JX,JY,JZ,IN,TM,KB,NAV
                                          COMMON        /PHY/
TEMP,VOL,DT,BOXL,FACTOR,CONFAC,CUT,RTKTM,RKTF,FF,DTF
     COMMON /N/ NOM,NOMM1,NORM,NT,NOFST,NTINC,INOF
     COMMON /NTR/ TRIG
     LI = 0
     DO 603 L1 = 1,25
     DO 603 L2 = 1,25
     DO 603 L3 = 1,25
     LM1 = L1-8
     LM2 = L2-8
```

```
      LM3 = L3-8
      XAA = LM1*A-1.D0 + A/2.D0 + (RAND(0)-0.5D0)*A/8.D0
      YAA = LM2*A-1.D0 + A/2.D0 + (RAND(0)-0.5D0)*A/8.D0
      ZAA = LM3*A-1.D0 + A/2.D0 + (RAND(0)-0.5D0)*A/8.D0
      IF(DABS(XAA).GT.1.0D0.OR.DABS(YAA).GT.1.0D0.OR.DABS(ZAA).GT.1.0D0
     &  GO TO 603
      LI = LI + 1
      XCN(LI) = XAA
      YCN(LI) = YAA
      ZCN(LI) = ZAA
  603 CONTINUE
      PRINT 33,RTKTM,RKTF,TM
   33 FORMAT(1X,G12.5,1X,G12.5,1X,G12.5)
      DO 3 IC = 1,NOM
      DO 2 L = 1,3
      EX(L,IC) = 0.0D0
      EY(L,IC) = 0.0D0
    2 EZ(L,IC) = 0.0D0
      EX(1,IC) = 1.0D0
      EY(2,IC) = 1.0D0
      EZ(3,IC) = 1.0D0
      VXC(IC) = GRAND(0)*RTKTM*DTF
      VYC(IC) = GRAND(0)*RTKTM*DTF
      VZC(IC) = GRAND(0)*RTKTM*DTF
      AJ1 = GRAND(0)*DSQRT(RKTF*IN(1))
      AJ2 = GRAND(0)*DSQRT(RKTF*IN(2))
      AJ3 = GRAND(0)*DSQRT(RKTF*IN(3))
      JX(IC) = AJ1*EX(1,IC) + AJ2*EX(2,IC) + AJ3*EX(3,IC)
      JY(IC) = AJ1*EY(1,IC) + AJ2*EY(2,IC) + AJ3*EY(3,IC)
    3 JZ(IC) = AJ1*EZ(1,IC) + AJ2*EZ(2,IC) + AJ3*EZ(3,IC)
    C NOTE:VX,VY,VZ, ARE MULTIPLIED BY DT,JX,JY,JZ, ARE NOT
      PRINT 66, LI,A
   66 FORMAT(1H1,' STARTING FROM A RANDOM QUASI-SCC LATTICE
   CONFIGURATI
     &N',//,10X,I5,'-RETICOLAR SITES',F10.4,'-LATTICE CONST.'//)
      DO 5 IC = 1,NOM
    5 PRINT 610,IC,XCN(IC),YCN(IC),ZCN(IC)
  610 FORMAT(10X,I4,3(4X,E10.4))
      CALL ZERO(A)
      CALL TSCAL(1,ET,ER)
      RETURN
      END
      FUNCTION GRAND(N)
      IMPLICIT REAL*8 (A-H,O-Z)
      COMMON/RANDNO/R3(127),R1,I2
      COMMON /NTR/ TRIG
```

```
      PI = 4.D0*DATAN(1.D0)
      GRAND = DSQRT(-2.*DLOG(R1))*DCOS(2.D0*PI*R2)
      RETURN
      END
      SUBROUTINE ZERO(Z)
      IMPLICIT REAL*8 (A-H,O-Z)
      REAL*8 JX(108),JY(108),JZ(108),IN(6)
      DIMENSION
     &XC(108),YC(108),ZC(108),XCO(108),YCO(108),ZCO(108),
     &XCN(108),YCN(108),ZCN(108),VXC(108),VYC(108),VZC(108)
     &,EX(3,108),EY(3,108),EZ(3,108)

      COMMON /CMT/ XC,YC,ZC,XCO,YCO,ZCO,
     &XCN,YCN,ZCN,VXC,VYC,VZC
      COMMON /CMO/ EX,EY,EZ
      COMMON /CMJ/ JX,JY,JZ,IN,TM,KB,NAV
                                              COMMON          /PHY/
     TEMP,VOL,DT,BOXL,FACTOR,CONFAC,CUT,RTKTM,RKTF,FF,DTF
      COMMON /N/ NOM,NOMM1,NORM,NT,NOFST,NTINC,INOF
      SVX = 0.0D0
      SVY = 0.0D0
      SVZ = 0.0D0
      SXX = 0.0D0
      SXY = 0.0D0
      SXZ = 0.0D0
      DO 3 IC = 1,NOM
C NOTE:VX,VY,VZ, ARE MULTIPLIED BY DT,XX,XY,XZ, ARE NOT
      SVX = SVX + VXC(IC)
      SVY = SVY + VYC(IC)
      SVZ = SVZ + VZC(IC)
  3   CONTINUE
      RNOM = 1.0D0/(DFLOAT(NOM))
C SET THE MEAN VALUES TO ZERO
      SVX = SVX*RNOM
      SVY = SVY*RNOM
      SVZ = SVZ*RNOM
      SJX = 0.0D0
      SJY = 0.0D0
      SJZ = 0.0D0
      PRINT 66
  66  FORMAT(1H ,' TOTAL TRASLATIONAL MOMENTA AND C.O.M. POS.',
     &' SET TO ZERO ')
      DO 4 IC = 1,NOM
      VXC(IC) = VXC(IC)-SVX
      VYC(IC) = VYC(IC)-SVY
      VZC(IC) = VZC(IC)-SVZ
```

```
      XC(IC) = XCN(IC)-VXC(IC)
      YC(IC) = YCN(IC)-VYC(IC)
      ZC(IC) = ZCN(IC)-VZC(IC)
      SXX = SXX + XC(IC)
      SXY = SXY + YC(IC)
      SXZ = SXZ + ZC(IC)
      SJX = SJX + JX(IC)
      SJY = SJY + JY(IC)
      SJZ = SJZ + JZ(IC)
    4 CONTINUE
      SXX = SXX*RNOM
      SXY = SXY*RNOM
      SXZ = SXZ*RNOM
      SJX = SJX*RNOM
      SJY = SJY*RNOM
      SJZ = SJZ*RNOM
      DO 44 IC = 1,NOM
      XC(IC) = XC(IC)-SXX
      YC(IC) = YC(IC)-SXY
      ZC(IC) = ZC(IC)-SXZ
      XCN(IC) = XCN(IC)-SXX
      YCN(IC) = YCN(IC)-SXY
      ZCN(IC) = ZCN(IC)-SXZ
      JX(IC) = JX(IC)-SJX
      JY(IC) = JY(IC)-SJY
      JZ(IC) = JZ(IC)-SJZ
   44 CONTINUE
      RETURN
      END
      SUBROUTINE TSCAL(MODE,TTEMP,RTEMP)
      IMPLICIT REAL*8 (A-H,O-Z)
      REAL*8 JX(108),JY(108),JZ(108),IN(6),KB,NAV
      DIMENSION
     &XC(108),YC(108),ZC(108),XCO(108),YCO(108),ZCO(108),
     &XCN(108),YCN(108),ZCN(108),VXC(108),VYC(108),VZC(108)
     &,EX(3,108),EY(3,108),EZ(3,108),OM(6)

      COMMON /CMT/ XC,YC,ZC,XCO,YCO,ZCO,
     &XCN,YCN,ZCN,VXC,VYC,VZC
      COMMON /CMO/ EX,EY,EZ
      COMMON /CMJ/ JX,JY,JZ,IN,TM,KB,NAV
                                            COMMON      /PHY/
 TEMP,VOL,DT,BOXL,FACTOR,CONFAC,CUT,RTKTM,RKTF,FF,DTF
      COMMON /N/ NOM,NOMM1,NORM,NT,NOFST,NTINC,INOF
      KB = 1.3807D-00
      NAV = 6.0223D + 23
```

```
      ELSQ = 2.3071138D + 05
      IF(MODE.NE.1) GO TO 1
      TTEMP = 0.0D0
      RTEMP = 0.0D0
      DO 2 I = 1,NOM
      TTEMP = TTEMP + VXC(I)**2 + VYC(I)**2 + VZC(I)**2
      DO 3 L = 1,3
    3 OM(L) = (EX(L,I)*JX(I) + EY(L,I)*JY(I) + EZ(L,I)*JZ(I))/IN(L)
      RTEMP = RTEMP + 0.5D0*(IN(1)*OM(1)**2 + IN(2)*OM(2)**2 + IN(3)*OM(3)**2)
    2 CONTINUE
      PRINT 3456,KB,NAV,TM,IN(1),IN(2),IN(3)
 3456 FORMAT (7G14.6)


      TTEMP = TTEMP*0.5D0*TM/DTF**2
      TTEMP = 2.0D0*TTEMP/(3.0D0*NOM*KB)
      RTEMP = RTEMP*2.0D0/(3.0D0*NOM*KB*FF)
    1 UT = DSQRT(TEMP/TTEMP)
      UR = DSQRT(TEMP/RTEMP)
      DO 413 I = 1,NOM
      VXC(I) = UT*VXC(I)
      VYC(I) = UT*VYC(I)
      VZC(I) = UT*VZC(I)



      XC(I) = XCN(I)-VXC(I)
      YC(I) = YCN(I)-VYC(I)
      ZC(I) = ZCN(I)-VZC(I)
      JX(I) = UR*JX(I)
      JY(I) = UR*JY(I)
  413 JZ(I) = UR*JZ(I)
      PRINT 412,TTEMP,RTEMP
  412 FORMAT(1X,' TEMPERATURE SCALING. OLD TEMPERATURES : ',
     &1X,' TR.TEMP = ',G12.5,' RO.TEMP = ',G12.5)
      RETURN
      END
      SUBROUTINE RANGE(PSI,PELR,CUT,NOM,ACR,EP,BSIGSQ)
      IMPLICIT REAL*8 (A-H,O-Z)
      REAL*8 PSI,PELR,EP(36),ACR(36),BSIGSQ(36),
     &CUT,CUTSQ,A,PGR
      CUTSQ = CUT**2
      PGR = (DACOS(-1.0D0))
      PSI = 0.0D0
      PELR = 0.0D0
      DO 19 J = 1,25
```

```
      A = (BSIGSQ(J)/CUTSQ)**3
      PSI = PSI + EP(J)*(A-2.0D0/3.0D0*A*A)
      PELR = PELR + EP(J)*(A*A/3.0D0-A)/3.0D0
   19 ACR(J) = EP(J)*(2.0D0*A*A-A)/CUTSQ
      PSI = PSI*2.0D0*PGR*(DFLOAT(NOM*NOM))*CUT**3
      PELR = PELR*PGR*(DFLOAT(NOM*NOM))*CUT**3
      RETURN
      END
      SUBROUTINE KINET(NAV,FACTOR,M,TM,IN,RI)
      IMPLICIT REAL*8 (A-H,O-Z)
      REAL*8 FACTOR,NAV,TM,DCH,DCCL,IN(6),RI(6),M(6)
      \,ROOT3
      D1 = 1.0
      D2 = 1.0
      DOC = 1.0
      M(1) = 1.0D0
      M(2) = 16.0D0
      M(3) = 1.0D0
      M(4) = 0.0D0
      M(5) = 0.0D0
      TM = 0.0D0
      DO 15 I = 1,5
      M(I) = M(I)/(NAV*1.0D-24)
   15 TM = TM + M(I)
      PI = 3.1415927
      ZET = PI*54.5/180.0
      X1 = DSIN(ZET)
      Y1 = -8.0/9.0*DCOS(ZET)
      Z1 = 0.00
      X2 = 0.00
      Y2 = DCOS(ZET)/9.0
      Z2 = 0.00
      X3 = -DSIN(ZET)
      Y3 = Y1
      Z3 = 0.00
      IN(1) = M(1)*(X1*X1 + Z1*Z1) + M(2)*(X2*X2 + Z2*Z2)    AM(3)*(X3*X3 + Z3*Z3)
                        IN(2) = M(1)*(X1*X1 + Y1*Y1) + M(2)*(X2*X2 + Y2*Y2)
AM(3)*(X3*X3 + Y3*Y3)
      IN(3) = M(1)*(Y1*Y1 + Z1*Z1) + M(2)*(Y2*Y2 + Z2*Z2)    AM(3)*(Y3*Y3 + Z3*Z3)
C  MOMENTS OF INERTIA IN BOX UNITS
C  IN(3) IS PARALLEL TO Z-AXIS ,IN(2) PARALLEL TO 3-4 VECTOR,
C  IN(1) PARALLEL TO 1-2 VECTOR DOC=SIST. FROM CENTRAL ATOM TO
COM
      IN(1) = IN(1)*FACTOR**2/(NAV*1.0D-24)
      IN(2) = IN(2)*FACTOR**2/(NAV*1.0D-24)
      IN(3) = IN(3)*FACTOR**2/(NAV*1.0D-24)
```

```
      DO 18 L=1,3
      IN(L+3)=IN(L)
  18  RI(L)=1.0D0/IN(L)
      PRINT 20,IN(1),IN(2),IN(3),RI(1),RI(2),RI(3),FACTOR


  20  FORMAT(1X//2X,' INERTIA(1,2,3)  =',3G12.5,' INV.MOM.INE.(1,2,3)  ='
     &,3G12.5,1X//2X,' FACTOR  =',G12.5,1X//)


      RETURN
      END
      SUBROUTINE LENJO(SIG,EPS,BSIG,BSIGSQ,EP,KB,FACTOR)
      IMPLICIT REAL*8 (A-H,O-Z)
      REAL*8 SIG(6,6),EPS(6,6),BSIG(36),BSIGSQ(36),EP(36)
     &,KB,FACTOR
      SIG(1,1)=2.25
      SIG(2,2)=2.8
      SIG(3,3)=2.25
      SIG(4,4)=0.0
      SIG(5,5)=0.0
      EPS(1,1)=21.1
      EPS(2,2)=58.4
      EPS(3,3)=21.1
      EPS(4,4)=0.0
      EPS(5,5)=0.0
      SIG(1,2)=0.5*(SIG(1,1)+SIG(2,2))
      SIG(1,3)=0.5*(SIG(1,1)+SIG(3,3))
      SIG(1,4)=0.0
      SIG(1,5)=0.0
      EPS(1,2)=DSQRT(EPS(1,1)*EPS(2,2))
      EPS(1,3)=DSQRT(EPS(1,1)*EPS(3,3))
      EPS(1,4)=DSQRT(EPS(1,1)*EPS(4,4))
      EPS(1,5)=DSQRT(EPS(1,1)*EPS(5,5))
      SIG(2,1)=SIG(1,2)
      SIG(2,3)=0.5*(SIG(2,2)+SIG(3,3))
      SIG(2,4)=0.0
      SIG(2,5)=0.0
      EPS(2,1)=EPS(1,2)
      EPS(2,3)=DSQRT(EPS(2,2)*EPS(3,3))
      EPS(2,4)=DSQRT(EPS(2,2)*EPS(4,4))
      EPS(2,5)=DSQRT(EPS(2,2)*EPS(5,5))
      SIG(3,1)=SIG(1,3)
      SIG(3,2)=SIG(2,3)
      SIG(3,4)=0.0
      SIG(3,5)=0.0
```

```
      EPS(3,1) = EPS(1,3)
      EPS(3,2) = EPS(2,3)
      EPS(3,4) = DSQRT(EPS(3,3)*EPS(4,4))
      EPS(3,5) = DSQRT(EPS(3,3)*EPS(5,5))
      SIG(4,1) = SIG(1,4)
      SIG(4,2) = SIG(2,4)
      SIG(4,3) = SIG(3,4)
      SIG(4,5) = 0.0
      EPS(4,1) = EPS(1,4)
      EPS(4,2) = EPS(2,4)
      EPS(4,3) = EPS(3,4)
      EPS(4,5) = DSQRT(EPS(4,4)*EPS(5,5))
      SIG(5,1) = SIG(1,5)
      SIG(5,2) = SIG(2,5)
      SIG(5,3) = SIG(3,5)
      SIG(5,4) = SIG(4,5)
      EPS(5,1) = EPS(1,5)
      EPS(5,2) = EPS(2,5)
      EPS(5,3) = EPS(3,5)
      EPS(5,4) = EPS(4,5)
      DO 14 I = 1,5
      I5 = 5*(I-1)
      DO 14 J = 1,5
      IJ = I5 + J
      BSIG(IJ) = FACTOR*SIG(I,J)
      BSIGSQ(IJ) = BSIG(IJ)**2
      EP(IJ) = KB*EPS(I,J)
   14 CONTINUE
      RETURN
      END
      SUBROUTINE ATPOS(NATM,FACTOR)
      IMPLICIT REAL*8 (A-H,O-Z)
      DIMENSION
     &    XAT(648),YAT(648),ZAT(648),XA(648),YA(648)
     &,ZA(648),XC(108),YC(108),ZC(108),XCO(108),YCO(108),ZCO(108)
     &,XCN(108),YCN(108),ZCN(108),VXC(108),VYC(108),VZC(108)
     &,EX(3,108),EY(3,108),EZ(3,108)

      COMMON /ATT/ XAT,YAT,ZAT,XA,YA,ZA
      COMMON /CMT/ XC,YC,ZC,XCO,YCO,ZCO,
     &XCN,YCN,ZCN,VXC,VYC,VZC
      COMMON /CMO/ EX,EY,EZ
      COMMON /N/ NOM,NOMM1,NORM,NT,NOFST,NTINC,INOF
      TWO = 2.0D0
      ROOT2 = 70.5*3.1415927/180.0
      FACTOR = 0.13508
```

```
    PI = 3.1415927
    ZET = PI*54.5/180.0
    X1 = DSIN(ZET)
    Y1 =-8.0/9.0*DCOS(ZET)
    Z1 = 0.00
    X2 = 0.00
    Y2 = DCOS(ZET)/9.0
    Z2 = 0.00
    X3 =-DSIN(ZET)
    Y3 = Y1
    Z3 = 0.00
    X4 = 0.00
    Y4 = (1.0/9.0 + 0.8)*DCOS(ZET)
    Z4 = 0.8*DSIN(ZET)
    X5 = 0.0
    Y5 = Y4
    Z5 =-Z4
    DO 369 IC = 1,NOM
    IA = NATM*(IC-1) + 1
    XAT(IA) = FACTOR*(X1*EX(3,IC) + Y1*EX(1,IC) + Z1*EX(2,IC))
    YAT(IA) = FACTOR*(X1*EY(3,IC) + Y1*EY(1,IC) + Z1*EY(2,IC))
    ZAT(IA) = FACTOR*(X1*EZ(3,IC) + Y1*EZ(1,IC) + Z1*EZ(2,IC))
    XAT(IA + 1) = FACTOR*(X2*EX(3,IC) + Y2*EX(1,IC) + Z2*EX(2,IC))
    YAT(IA + 1) = FACTOR*(X2*EY(3,IC) + Y2*EY(1,IC) + Z2*EY(2,IC))
    ZAT(IA + 1) = FACTOR*(X2*EZ(3,IC) + Y2*EZ(1,IC) + Z2*EZ(2,IC))
    XAT(IA + 2) = FACTOR*(X3*EX(3,IC) + Y3*EX(1,IC) + Z3*EX(2,IC))
    YAT(IA + 2) = FACTOR*(X3*EY(3,IC) + Y3*EY(1,IC) + Z3*EY(2,IC))
    ZAT(IA + 2) = FACTOR*(X3*EZ(3,IC) + Y3*EZ(1,IC) + Z3*EZ(2,IC))
    XAT(IA + 3) = FACTOR*(X4*EX(3,IC) + Y4*EX(1,IC) + Z4*EX(2,IC))
    YAT(IA + 3) = FACTOR*(X4*EY(3,IC) + Y4*EY(1,IC) + Z4*EY(2,IC))
    ZAT(IA + 3) = FACTOR*(X4*EZ(3,IC) + Y4*EZ(1,IC) + Z4*EZ(2,IC))
    XAT(IA + 4) = FACTOR*(X5*EX(3,IC) + Y5*EX(1,IC) + Z5*EX(2,IC))
    YAT(IA + 4) = FACTOR*(X5*EY(3,IC) + Y5*EY(1,IC) + Z5*EY(2,IC))
    ZAT(IA + 4) = FACTOR*(X5*EZ(3,IC) + Y5*EZ(1,IC) + Z5*EZ(2,IC))
369 CONTINUE
    DO 30 IC = 1,NOM
    I5 = NATM*(IC-1)
    DO 31 IA = 1,NATM
    K = I5 + IA
    XA(K) = XC(IC) + XAT(K)
    XA(K) = XA(K)-TWO*(DFLOAT(IDINT(XA(K))))
    YA(K) = YC(IC) + YAT(K)
    YA(K) = YA(K)-TWO*(DFLOAT(IDINT(YA(K))))
    ZA(K) = ZC(IC) + ZAT(K)
    ZA(K) = ZA(K)-TWO*(DFLOAT(IDINT(ZA(K))))
 31 CONTINUE
```

```
 30  CONTINUE
     RETURN
     END
     SUBROUTINE GRAF(F,F1,F2,KMAX,IK,IL)
C
C
C  PLOTTING 3 CURVES F,F1,F2
C  KMAX = NUMBER OF POINTS
C  IK  = 0 (NORMALIZATION AT FIRST POINT), = 1 ( NO NORMALIZATION

C  IL  = 0 ( 'F' CURVE ONLY ), = 1 ('F'&'F1' CURVES), = 2 ( ALL CURVES
C
C
     IMPLICIT REAL*8 (A-H,O-Z)
     DIMENSION F(200),F1(200),F2(200),O(101)
     DATA B/1H /,TA/1HI/,T1/1H*/,T2/1H + /,T3/1H#/
     FF1 = F(1)
     FF11 = F1(1)
     FF21 = F2(1)
     X = 0.D0
     DO 1 I = 1,KMAX
     IF(IK.EQ.0) F(I) = F(I)/FF1
     X1 = DABS(F(I))
     IF(X1.GT.X)X = X1
     IF(IL.EQ.0)GOTO 1
     IF(IK.EQ.0) F1(I) = F1(I)/FF11
     X1 = DABS(F1(I))
     IF (IL.EQ.1) GOTO 1
     IF (IK.EQ.0) F2(I) = F2(I)/FF21
     X1 = DABS(F2(J))
     IF(X1.GT.X)X = X1
   1 CONTINUE
     DELT = X /50.D0
     DO 2 I = 1,KMAX
     DO 3 J = 1,101
   3 O(J) = B
     O(1) = TA
     O(51) = TA
     O(101) = TA
     M = (F(I) + X )/DELT + 1.5D0
     O(M) = T1
     IF(IL.EQ.0)GOTO 4
     M = (F1(I) + X )/DELT + 1.5D0
     O(M) = T2
     IF(IL.EQ.1) GO TO 4
     M = (F2(I) + X)/DELT + 1.5D0
```

694 M. W. EVANS

```
      O(M)=T3
  4 CONTINUE
    IF (IL.EQ.2) PRINT 997,I,F(I),F1(I),F2(I),O
    IF(IL.EQ.1) PRINT 998,I,F(I),F1(I),O
    IF(IL.EQ.0) PRINT 999,I,F(I),O
  2 CONTINUE
997 FORMAT(1H ,I3,3(1X,G9.3),1X,101A1)
998 FORMAT(1H ,I5,2(1X,G10.3),1X,101A1)
999 FORMAT(1H ,I5,D15.7,101A1)
    RETURN
    END
    SUBROUTINE CHARGE(CHA,BCHA,ELSQ,FACTOR,TWFO)
    IMPLICIT REAL*8 (A-H,O-Z)
    DIMENSION CHA(6),BCHA(36)
    CHA(1)=0.23
    CHA(2)=0.00
    CHA(3)=0.23
    CHA(4)=-0.23
    CHA(5)=-0.23
    DO 14 I=1,5
    I5=5*(I-1)
    DO 14 J=1,5
    IJ=I5+J
    BCHA(IJ)=CHA(I)*CHA(J)*ELSQ*FACTOR/TWFO
 14 CONTINUE
    RETURN
    END
c          RANDOM NUMBER GENERATOR
C
    FUNCTION RAND(N)
    IMPLICIT REAL*8 (A-H,O-Z)
    COMMON /RANDNO/ R3(127),R1,I2
    DATA S,T,RMC/0.D0, 1.D0, 1.D0/
    DATA IW/-1/
    IF((R1.LT.1.D0).AND.(N.EQ.0)) GO TO 60
    IF (IW.GT.0) GO TO 30
 10 IW = IW + 1
    T = 0.5D0*T
    R1 = S
    S = S + T
    IF ((S.GT.R1).AND.(S.LT.1D0)) GO TO 10
    IKT = (IW - 1)/12
    IC = IW - 12*IKT
    ID = 2**(13 - IC)
    DO 20 I = 1, IC
 20 RMC = 0.5D0*RMC
```

```
      RM = 0.015625D0*0.015625D0
30  I2 = 127
      IR = MOD(IABS(N), 8190) + 1
40  R1 = 0.D0
      DO 50 I = 1, IKT
      IR = MOD(17*IR, 8191)
50  R1 = (R1 + DFLOAT(IR/2))*RM
      IR = MOD (17*IR, 8191)
      R1 = (R1 + DFLOAT(IR/ID))*RMC
      R3(I2) = R1
      I2 = I2 - 1
      IF (I2.GT.0) GO TO 40
60  IF (I2.EQ.0) I2 = 127
      T = R1 + R3(I2)
      IF (T.GE.1D0) T = (R1 - 0.5D0) + (R3(I2) - 0.5D0)
      R1 = T
      R3(I2) = R1
      I2 = I2 - 1
      RAND = R1
      RETURN
      END
      SUBROUTINE DATE(DAT)
      IMPLICIT REAL*8 (A-H,O-Z)
      RETURN
      END
      SUBROUTINE TIME(TRIG)
      IMPLICIT REAL*8 (A-H,O-Z)
      RETURN
      END
      SUBROUTINE SECOND(PTIME)
      IMPLICIT REAL*8 (A-H,O-Z)
      RETURN
      END
```

Code for Computation of Cross Correlation Functions

The following is a program for the efficient computation of cross correlation functions of all types written by Keith Refson and the present author. It uses running time averaging over N time steps and M molecules, and is capable of computing a c.c.f. over 6,000 time steps and 108 molecules in about one minute of IBM 3090 time.

```
      PROGRAM  CRLATE
      PARAMETER (NSL=1000,MAXMOL=108,MAXCCF=1000)
      PARAMETER (IR=0,IV=1, IJ=2, IE=3 ,IF=4 ,IT=5, IJ1=6,
1          IED=7, IW=8 , IP=9 , ITOTR=10)
      REAL X(0:3*MAXMOL*NSL-1,2)
```

```
                                                              REAL*8
RCCF(0:MAXCCF),ACCF(0:MAXCCF),DENOM,DENOM1,DENOM2,DOT
   LOGICAL LINTER,LPLOT,FIRST
   NAMELIST /ACFCCF/ NSLICE,NCF,ID1,ID2,IX1,IX2,NMOLS,NBYTE,
   1           FIRST, NSLT, LINTER, NORM, LPLOT, TSLICE, ITOUT
   COMMON /BYTE/ NBYTE
   DATA NSLICE, NCF, ID1, ID2, IX1, IX2, NMOLS,
   1    NSLT, LINTER, NORM, LPLOT, TSLICE, ITOUT
   2    / 900,200, 1, 1, 0, 0, 108, 1000, F, 0, F, 0., 6/
   INDX(NMOL,IX,ISL) = NMOL - 1 + (IX-1)*NMOLS + ISL*3*NMOLS
   NBYTE=8
   FIRST=.FALSE.  C C C
   READ(4,ACFCCF)

OPEN(UNIT=80,ACCESS='DIRECT',RECL=3*NMOLS*NBYTE,ACTION='READ')
   IF(IX1 .EQ. 0) IX2 = 0
   IF( IX1.LT.0 .OR. IX1.GT.3 .OR. IX2.LT.0 .OR. IX2.GT.3) THEN
     WRITE(6,*) ' Invalid values of ccf components IX1, IX2'
     STOP
   ELSE IF(IX1 .EQ. 0) THEN
     NORM = 0
   ELSE
     IF(NORM .NE. 1 .AND. NORM .NE. 2) THEN
       WRITE(6,*) ' Invalid normalisation type - must be 1 or 2'
       STOP
     END IF
   END IF
   IF(ID1 .EQ. ID2 .AND. NORM .EQ. 2) NORM = 0 C C C
   DENOM = 0.D0
   DENOM1 = 0.D0
   DENOM2 = 0.D0
   DO 600 I=0,NCF
600    ACCF(I) = 0:D0

   IF(ID1 .EQ. ID2) THEN

     CALL READAT(X(0,1),ID1,NCF,0,NSLT,NMOLS)
     DO 2000 IBLOCK = NCF, NSLICE-1, NSL-NCF
       NN = MIN(NSL-NCF,NSLICE-IBLOCK)
       CALL READAT(X(3*NMOLS*NCF,1),ID1,NN,IBLOCK,NSLT,NMOLS)
       WRITE(6,*) ' ****** DATA READ IN'

       CALL CCF(X,X,NMOLS,NN+NCF,RCCF,NCF,IX1,IX2)
       DO 1000 ICF = 0,NCF
1000        ACCF(ICF) = ACCF(ICF) + RCCF(ICF)
```

```
            IF(NORM .EQ. 0) THEN C              WRITE(6,*) ' CORRELATION
    FROM',IBLOCK-NCF,' FOR',NN
            DENOM = DENOM + DOT(X,X,3*NMOLS*NN) C              WRITE(6,*)
    ' DENOM =',DENOM
          ELSE IF(NORM .EQ. 1) THEN
            DO 2108 ISL=0,NN
            DENOM1 = DENOM1
   1          +DOT(X(INDX(1,IX1,ISL),1),X(INDX(1,IX1,ISL),1),NMOLS)
            DENOM2 = DENOM2
   1          +DOT(X(INDX(1,IX2,ISL),1),X(INDX(1,IX2,ISL),1),NMOLS)
2108        CONTINUE
          END IF

C         WRITE(6,*) ' MOVING',NCF,' FROM',NSL-NCF,' TO',0
          CALL MOVE(X(3*NMOLS*(NSL-NCF),1),X,3*NMOLS*NCF)
2000    CONTINUE

      ELSE IF(IX1.GE.1 .AND. IX1.LE.3 .AND. IX2.GE.1 .AND. IX2.LE.3)THE

      CALL READAT(X(0,1),ID1,NCF,0,NSLT,NMOLS)
      CALL READAT(X(0,2),ID2,NCF,0,NSLT,NMOLS)
      DO 4000 IBLOCK = NCF, NSLICE-1, NSL-NCF
       NN = MIN(NSL-NCF,NSLICE-IBLOCK)
       CALL READAT(X(3*NMOLS*NCF,1),ID1,NN,IBLOCK,NSLT,NMOLS)
       CALL READAT(X(3*NMOLS*NCF,2),ID2,NN,IBLOCK,NSLT,NMOLS)
       WRITE(6,*) ' ****** DATA READ IN'

      CALL CCF(X(0,1),X(0,2),NMOLS,NN+NCF,RCCF,NCF,IX1,IX2)
      DO 3000 ICF = 0,NCF
3000    ACCF(ICF) = ACCF(ICF) + RCCF(ICF)

      IF(NORM .EQ. 1) THEN
        DO 4108 ISL=0,NN
        DENOM1 = DENOM1
   1      +DOT(X(INDX(1,IX1,ISL),1),X(INDX(1,IX1,ISL),1),NMOLS)
        DENOM2 = DENOM2
   1      +DOT(X(INDX(1,IX2,ISL),2),X(INDX(1,IX2,ISL),2),NMOLS)
4108    CONTINUE
      ELSE IF(NORM .EQ. 2)THEN
        DENOM1 = DENOM1 + DOT(X(0,1),X(0,1),3*NMOLS*NN)
        DENOM2 = DENOM2 + DOT(X(0,2),X(0,2),3*NMOLS*NN)
      END IF

      CALL MOVE(X(3*NMOLS*(NSL-NCF),1),X(0,1),3*NMOLS*NCF)
      CALL MOVE(X(3*NMOLS*(NSL-NCF),2),X(0,2),3*NMOLS*NCF)
4000  CONTINUE
```

```
      ELSE
        WRITE(6,*) ' IX1(2) > 3 OR IX1(2) < 0 ', IX1,IX2
        STOP
      END IF

C   CALL READAT(R,IR,NSLICE,NT,NMOLS)

C     NMOLN=NMOLS  C      CALL COMPCT(V,R,NMOLS,NMOLN,NSLICE) C
WRITE(6,*) ' ****** COMPACTED'

C   CALL FOR 'DOT PRODUCT' ACF
      IF(NORM .NE. 0) DENOM = SQRT(DENOM1*DENOM2)
      WRITE(ITOUT,100) DENOM,ACCF(0)
  100 FORMAT(' DENOMINATOR = ',1P,D16.8,'   <V(0).V(0)> = ',D16.8)
      DO 1080 I=0,200
 1080    ACCF(I) = ACCF(I)/DENOM

      CALL PRINT(ACCF,ID1,ID2,IX1,IX2,NCF,ITOUT)
      IF(LPLOT) THEN
       IF(FIRST) THEN
         REWIND 9
         ENDFILE 9
       ENDIF
      REWIND 9
      DO 98 I=1,10000
      READ (9,END=99)
   98 CONTINUE
   99 BACKSPACE 9
      WRITE (9) ACCF,NCF,TSLICE,NMOLS,ID1,ID2,IX1,IX2
      REWIND 9
      ENDIF
      STOP
      END

      FUNCTION CCFNOR(X1,X2,IC1,IC2,NMOL,NSL)
        REAL*4 X1(1:NMOL,1:3,0:NSL-1),X2(1:NMOL,1:3,0:NSL-1)
        REAL*8 X1X1,X2X2, CCFNOR, DOT
        X1X1 = 0.D0
        X2X2 = 0.D0
        DO 6 ISL = 0,NSL-1
          X1X1 = X1X1+DOT(X1(1,IC1,ISL),X1(1,IC1,ISL),NMOL)
          X2X2 = X2X2+DOT(X2(1,IC2,ISL),X2(1,IC2,ISL),NMOL)
    6   CONTINUE
        CCFNOR = SQRT(X1X1*X2X2)
        RETURN
```

```
      END

      SUBROUTINE READAT(X,IX,NSL,NSL0,NSLMAX,NMOL)
      REAL*4 X(0:3*NMOL*NSL-1)
      PARAMETER (MAXMOL=108)
      REAL*8 BUFF(0:3*MAXMOL-1)
      COMMON /BYTE/ NBYTE C          WRITE(6,*) ' READING SLICES ',NSL0,'
TO ',NSL+NSL0-1

      IF(NMOL .GT. MAXMOL) THEN
        WRITE(6,*) 'READ BUFFER TOO SMALL -INCREASE NMOL'
        STOP
      END IF

      J0 = 0 C        WRITE(6,*) ' READING RECORDS',IX*NSLMAX+NSL0+1,'
TO', C   1            IX*NSLMAX+NSL+NSL0
      IF(NBYTE .EQ. 8)THEN
        DO 1000 IREC=IX*NSLMAX+NSL0+1,IX*NSLMAX+NSL+NSL0
          CALL RBLOCK(BUFF,IREC,2*NMOL)
          DO 1010 J=3*NMOL-1,0,-1
          X(J+J0) = BUFF(J)
1010      CONTINUE
          J0 = J0 + 3*NMOL
1000    CONTINUE
      ELSE
        DO 2000 IREC=IX*NSLMAX+NSL0+1,IX*NSLMAX+NSL+NSL0
          CALL RBLOCK(X(J0),IREC,NMOL)
          J0=J0+3*NMOL
2000    CONTINUE
      END IF
      RETURN
      END

      SUBROUTINE RBLOCK(BUFF,IREC,NMOLS)
      REAL*4 BUFF(0:3*NMOLS-1)
      READ(UNIT=80,REC=IREC) BUFF
      RETURN
      END

C       FUNCTION DOT(A,B,N) C      REAL*4 A(N),B(N) C          REAL*8
SUM,DOT C       SUM = 0.0D0 C     DO 1000 I = 1,N C        SUM = SUM
+ A(I)*B(I) C1000    CONTINUE C     DOT=SUM C      RETURN C    END

      SUBROUTINE MOVE(FROM,TO,LEN)
      REAL*4 FROM(LEN),TO(LEN)
      DO 1000 I=1,LEN
```

```
1000      TO(I) = FROM(I)
         RETURN
      END

      SUBROUTINE CCF(X,Y,NMOL,NSL,RCCF,NCCF,IX1,IX2)
         PARAMETER (NMOLS = 108,NSLICE = 900,NT = 1000)
         REAL*4 X(1:NMOL,1:3,0:NSL-1),Y(1:NMOL,1:3,0:NSL-1)
         REAL*8 DOT,RCCF(0:NCCF)

         DO 1020 I = 0,NCCF
           RCCF(I) = 0.0
1020     CONTINUE
         IF(IX1 .LE. 0 .OR. IX2 .LE. 0) THEN
           IX = 1
           JX = 1
           N = 3
         ELSE IF(IX1 .LE. 3 .AND. IX2 .LE. 3) THEN
           IX = IX1
           JX = IX2
           N = 1
         ELSE
           WRITE(6,*) IX1, ' AND ',IX2 ,' ARE INVALID COMPONENTS'
           STOP
         END IF

         DO 1000 IT0 = 0,NSL-NCCF-1
           IF(MOD(IT0,10).EQ.0) WRITE(6,*) ' WORKING ON SLICE', IT0
           DO 1010 IDT = 0,NCCF
             RCCF(IDT) = RCCF(IDT) + DOT(X(1,IX,IT0),
     1                        Y(1,JX,IT0 + IDT),N*NMOL)
1010       CONTINUE
1000     CONTINUE
         RETURN
      END

      SUBROUTINE COMPCT(X,R,NMOL,NMOLN,NSL)
         PARAMETER(NMOLS = 108,NSLICE = 900)
         PARAMETER (CUBE = 0.5)
         REAL R(1:NMOL,1:3,0:NSL-1),X(NMOLS*3*NSLICE)
         INTEGER MOL(NMOLS)
         LOGICAL IN
         INDX(IM,IK,IS) = (IS*3 + IK-1)*NMOL + IM

         NLIST = 0
         DO 1000 ISL = 0,NSL-1
           DO 1010 IMOL = 1,NMOL
```

```
        IF(  ABS(R(IMOL,1,ISL)).LT.CUBE
  A      .AND.ABS(R(IMOL,2,ISL)).LT.CUBE
  B      .AND.ABS(R(IMOL,3,ISL)).LT.CUBE) THEN
        DO 2000 ILIST=1,NLIST
         IF(IMOL .LE. MOL(ILIST)) GOTO 2600
2000     CONTINUE
        NLIST = NLIST + 1
        MOL (NLIST) = IMOL
        WRITE(6,200) ISL,(MOL(IL),IL=1,NLIST)
2600     IF(IMOL.NE.MOL(ILIST)) THEN
         NLIST=NLIST+1
         DO 3000 IL1=NLIST,ILIST,-1
3000        MOL(IL1)=MOL(IL1-1)
         MOL(ILIST)=IMOL
         WRITE(6,200) ISL,(MOL(IL),IL=1,NLIST)
        END IF
       END IF
1010    CONTINUE
1000 CONTINUE

    WRITE(6,100) NLIST,(MOL(ILIST),ILIST=1,NLIST)
100   FORMAT(' NUMBER OF MOLECULES WHICH ENTER INNER CUBE =
',I6/(12I6)
200   FORMAT(' LIST UPDATED AT TIMESLICE',I5,/' NEW LIST....',(10I6))
    IX=0
    DO 4000 ISL = 0,NSL-1
      DO 4010 I = 1,3
        DO 4020 IMOL = 1,NMOL
         IN = .FALSE.
         DO 4030 ILIST = 1,NLIST
          IF(IMOL .EQ. MOL(ILIST)) IN = .TRUE.
4030      CONTINUE
         IF(.NOT. IN) THEN
          IX = IX + 1
             X(IX) = X(INDX(IMOL,I,ISL)) C          WRITE(6,*)
IX,INDX(IMOL,I,ISL)
         END IF
4020     CONTINUE
4010     CONTINUE
4000 CONTINUE
    NMOLN=NMOL-NLIST
    RETURN
    END

    SUBROUTINE PRINT(RCF,IX,IY,IC1,IC2,NRCF,ITOUT)
      REAL*8 RCF(0:NRCF)
```

```
      CHARACTER*1 XYZ(1:3)
      CHARACTER*127 DESCR
      CHARACTER*15 WHICH(0:9)
      DATA WHICH /'R(123)  ','V(123)  ','VXW(123)     ','RXW(123)
     1        ','(RXW)XW','VXW(L)','FXW(L)       ','TQXV(L)    ',
     1           'W(123)    ','RXW(L)      '/
      DATA XYZ /'X','Y','Z'/

      IF (IX .EQ. IY .AND. IC1 .EQ. 0 .AND. IC2 .EQ. 0) THEN
        DESCR = WHICH(IX)//' AUTOCORRELATION FUNCTION'
      ELSE
        DESCR = WHICH(IX)//'('//XYZ(IC1)//')-'//WHICH(IY)//'('//XYZ(IC2
     1  //') CROSS-CORRELATION FUNCTION'
      END IF
      WRITE(ITOUT,*) DESCR
      WRITE(ITOUT,100) RCF
100   FORMAT(1P,(1X,5D16.8))
      RETURN
      END
```

VIBRONIC INTERACTIONS IN POLYNUCLEAR MIXED-VALENCE CLUSTERS

I. B. BERSUKER and S. A. BORSHCH

Laboratory of Quantum Chemistry, Institute of Chemistry, Academy of Sciences of SSRM, Kishinev, USSR

CONTENTS

Advances in Chemical Physics, *Volume LXXXI*, Edited by I. Prigogine and Stuart A. Rice.
ISBN 0-471-54570-8 © 1992 John Wiley & Sons, Inc.

I. INTRODUCTION

Mixed-valence (MV) compounds are at present very important to many areas of chemistry, physics, and biology. Their most significant feature is the process of electron transfer between equivalent centers that are parts of the same molecule. Although the electron transfer takes place in many other systems, MV compounds are exceptional. Distinguished from, say, the case of electron transfer in the liquid state, the MV centers are fixed, the process of electron transition is not complicated by superposition of other phenomena (as in the case of chemical reactions).[1]

However, the importance of MV compounds is not limited by their ability to serve as models for reactions of electron transfer. Many of these compounds are active centers in metalloenzymes (ferredoxins, manganese complexes in the photosynthetic chain[2]), and are active catalysts. In the last 2–3 years the interest in MV compounds increased in connection with the widespread discussion of the role of the heterovalent ions in the metalloxide ceramics possessing high-temperature superconductivity,[3] as well as works on their use in construction of devices for molecular electronics.[4]

MV compounds, as it follows from their denotion, contain two or more ions of the same chemical element in different valent states. Provided the environments of the ions with different valency are equivalent, it is convenient to consider the MV system as containing equivalent centers with one (or more) excess electrons. In this chapter we consider mostly systems with transition metal ions in different valent states. However, there are such systems also among pure organic compounds, and some of the results of this review are valid also for them. If the near-neighbor environments of the ions differ from each other by the coordination number and/or chemical composition, then the energy of the excess electron at the different valence centers is essentially different. In these systems the electron transfer is possible, but under external perturbation (e.g., photoexcitation).

The situation is different when the centers (without the excess elecrons) are completely equivalent. In these cases there may be different electron charge distributions from localized at one of the centers to uniform delocalization at several or all the centers of the system dependent on the molecular structure, the nature of the bridge ligands, and temperature. Here it is worthwhile to emphasize that the notions "localized" and "delocalized" in the case under consideration are to a certain extent conventional and depend on the means of observation.[5] The same MV compounds in the same conditions may exhibit localized valences in experiments with small characteristic "time of measurement" (e.g., in Mossbauer spectroscopy) and delocalized ones on more "slow" experiments (e.g., in X-ray diffraction).

In most cases when there is a strong intercenter interaction in the same molecule, resulting in the delocalization of the excess electron, the traditional theoretical approach to the description of the electron transfer based on calculations of the probability of electron transitions[6] becomes invalid. Indeed, this approach is based on the assumptions that there are well-defined localized initial and final states and the perturbation theory can be used that can hardly be founded. In these cases the description based on the states determined for the system as a whole (vibronic states) seems to be more adequate for the problem. Here the different possibilities of electron distribution are included in the dependence of the states of the system on its parameters.

The localization of the excess electron at one center of the MV compound results in the distortion of the center of localization, and this in turn, making the centers nonequivalent, instantly lowers the symmetry of the system. This local distortion of the system arising because of the interaction of the electron with the low-symmetry nuclear displacements is one of the most significant manifestations of the vibronic interactions. During the last 30 years the theory of vibronic interactions grew to a whole trend in physics and chemistry of molecules and crystals including many effects and regularities and new areas of development,[5,7] one of its particular demonstrations being the well-known Jahn–Teller effect.

One of the significant last achievements in this area is the proof of the statement (theorem) that all the instabilities of high-symmetry configurations of molecules and crystals are of vibronic origin, that is, the vibronic mixing of the ground electronic state of any polyatomic system in a given high-symmetry configuration with its excited ones (or the mixing of different electronic states of the degenerate ground term) by nuclear displacements is the only possible source of dynamic instability of this configuration.[5,7–9] This statement has many important consequences. In particular, it means that so far as a certain nuclear configuration of the molecular system is assumed, the correct general treatment of its possible stability (instability) with respect to nuclear displacements should be based on the vibronic theory. Another conclusion is that if the nuclear configuration of a polyatomic system in its ground electronic state is unstable, there must be close-in-energy excited states of a certain symmetry (determined by direction of instability) that admix to the ground under the appropriate nuclear displacements. As shown below, the methods of the theory of vibronic interactions allow an adequate description of the properties of

MV compounds from a unique point of view. On the other hand, the development of the theory of MV compounds requires further extension of the general theory of vibronic interactions.

The necessity to take into account the vibronic interactions in the consideration of MV dimers was firstly demonstrated in the works of Mayoh and Day[10] and Hush.[11] In particular, it was shown that the alternative problem "localization–delocalization" for MV dimers is equivalent to the one of the pseudo Jahn–Teller effect, that is, to the problem of two nondegenerate electronic energy levels that mix strongly under a nontotally symmetric vibration; the condition of electron localization in the MV dimer coincides with the condition of instability of the high-symmetry configuration in the pseudo Jahn–Teller effect.

The analogy with the pseudo Jahn–Teller effect was most completely employed in the model of Piepho, Krausz, and Schatz (PKS) (e.g., see Ref. 12). The authors used the methods of calculation of the vibronic states in the pseudo Jahn–Teller effect for evaluation of the vibronic states in the MV dimer. The knowledge of the vibronic states allowed them to calculate the electronic spectra in the region of the intervalent transfer band, the energies and intensities of the low-frequency tunneling transitions, resonance Raman spectra, and so on.

However, as shown by further investigations, the PKS model in its initial form is valid for a rather limited number of cases. The model is based on the assumption that only one nondegenerate state of each center participates in the process of electron transfer. However, in transition metal complexes having a high density of electronic states and high enough symmetry of the near-neighbor environment, two and more electronic degenerate or pseudodegenerate (close-in-energy; see discussion below) electronic states mixing under the nuclear displacements and thus participating strongly in the electron transfer process are most probable. Besides, the transition metal ions in these complexes sometimes possess a nonzero spin in both valent states, and hence the ions with different valency are linked not only by electron transfer but also by exchange coupling. In general, in the situation mentioned above, the resulting energy spectrum cannot be reduced to a couple of nondegenrate states as in the PKS model. Accordingly, the vibronic problem for such systems doesn't reduce to the pseudo Jahn–Teller one.

An essential complication of the energy spectrum, as compared with that assumed in the PKS model, results also from a simple increase of the number of centers among which the electron transfer takes place. This circumstance is important not only when one passes from dimers to trimers, tetramers, and other clusters of higher nuclearity, but even for the dimers themselves when the bridge ligand takes part in the electron transfer by its states directly.

This review article is devoted to the mainly vibronic interactions in MV compounds with an energy spectrum of the centers differing in valency containing more than two electronic states of the PKS approach. As mentioned above, the complication of the basic (active) electronic spectrum beyond the simplest PKS model allows us to consider most real MV systems containing transition metals as different valence centers. The complication of the electronic spectrum certainly complicates also the vibronic problem, requiring new methods for its solution. The results reveal and predict qualitatively new effects and regularities, some of which have already been observed experimentally.

Note that the complication of the electronic spectrum results also in the complication of the vibrational aspects of the vibronic problem. Indeed, in the case of more than one nondegenerate electronic state on each center, not only the local totally symmetric vibration is active, but also other low-symmetry modes (Jahn–Teller or pseudo Jahn–Teller active[5,7]) become essential. For some cases the full vibronic problem is solved numerically. In other cases the semiclassical approach considering the adiabatic potential surfaces allows us to obtain quite reasonable qualitative results.

In this chapter we shall consider only effects existing in isolated MV clusters noninteracting with other molecules. If required we shall refer the reader to original papers dealing with cooperative effects in molecular crystals containing MV clusters as structural units.

The plan of the review is as follows. After a brief presentation of the PKS model and introduction of the appropriate notations, the one-center interactions in the case of more than one electronic state are considered in more detail and the vibronic theory of dimers, taking into account exchange coupling, is given. Then follows the problem of electron delocalization in tricenter MV systems and its vibronic theory including magnetic interactions. A discussion of clusters of higher nuclearity and concluding remarks are also given.

II. THE VIBRONIC MODEL FOR MV DIMERS
(THE PKS MODEL)

The PKS model is reviewed in several articles.[12,13] We review here briefly some of the main statements about this model important to the understanding of the more sophisticated vibronic models for MV systems.

It is assumed that the system consists of two equivalent molecular subunits, containing closed shells of electrons, 1 and 2, and one excess electron that can be localized at each of the monomers. At each center (each subunit) the electron occupies only one nondegenerate one-electron molecular orbital (MO), φ_1 on φ_2, above the fulfilled shell of the core and only two MO are active in the electron transfer, the other possible MO not being taken into account explicitly. The full Hamiltonian of the system can be presented as follows:

$$\hat{H} = \hat{H}_1 + \hat{H}_2 + \hat{W} \tag{2.1}$$

where $\hat{H}_{1,2}$ are the Hamiltonians of the two subunits and $\hat{W}$ is the interaction between them. The Hamiltonian $\hat{H}$ is convenient to write on the basis of the multielectron determinant states ϕ_1 and ϕ_2 that describe the system (including the core states) with the excess electron localized at the appropriate center. The energy is read off from the energy level of the electronic subsystem of the two equivalent monomers with the localized excess electron (without taking account of the local distortion).

Since occupying a nondegenerate MO, the electron interacts only with the totally symmetric (in the local symmetry group of the monomer) displacements Q_i ($i = 1, 2$) of the near-neighbor environment. Therefore the vibronic matrix Hamiltonian contains linear terms in Q_i only,

$$\langle \phi_i | \hat{H} | \phi_i \rangle = H_n(Q) + AQ_i \tag{2.2}$$

where $\hat{H}_n(Q)$ is the vibrational Hamiltonian of the system as a whole, A is the constant of vibronic interactions with the displacements, and Q in $\hat{H}_n(Q)$ stands for all the other nuclear coordinates of the system. In these notations

$$\langle \phi_1 | \hat{H} | \phi_2 \rangle \equiv \langle \phi_1 | \hat{W} | \phi_2 \rangle = w \tag{2.3}$$

This parameter is most important in the MV theory characterizing the strength of the intercenter interaction.

If we neglect all the elastic interactions between the monomers, the terms in Q_1 and Q_2 only remain in the Hamiltonian (Eq. 2.1) and in the presentation of the states ϕ_1 and ϕ_2 it can be written in the matrix form:

$$\hat{H} = \hat{T}_n(Q_1) + \hat{T}_n(Q_2) + \begin{pmatrix} \dfrac{\omega_e^2 Q_1^2}{2} + \dfrac{\omega_n^2 Q_2^2}{2} + AQ_1 & w \\[2mm] w & \dfrac{\omega_n^2 Q_1^2}{2} + \dfrac{\omega_e^2 Q_2^2}{2} + AQ_2 \end{pmatrix}$$

$$(2.4)$$

where $\hat{T}_n(Q)$ is the operator of kinetic energy of the nuclei, ω_e and ω_n being the frequencies of the appropriate nuclear vibrations with and without the excess electron. Usually it is assumed also that $\omega_e = \omega_n = \omega$. This assumption seems to be grounded when the orbital of the excess electron is not very strong bonding or antibonding. Indeed, in many cases the force constants $(\sim \omega^2)$ for different valence states differ by no more than 10%. The frequency effect essentially influences the resonance Raman spectra.[14]

Let us pass to the electronic basis $\phi_{\pm}$ and coordinate $Q_{\pm}$ that transform after the even and odd representations of the system as a whole:

$$\phi_{\pm} = (1/\sqrt{2})(\phi_1 \pm \phi_2) \tag{2.5}$$

$$Q_{\pm} = (1/\sqrt{2})(Q_1 \pm Q_2) \tag{2.6}$$

and introduce dimensionless Q and A as follows:

$$(\omega/\hbar)^{1/2} Q \to Q, \qquad (2\hbar\omega^3)^{-1/2} A \to A \tag{2.7}$$

Under the transformation (Eq. 2.7) the energy becomes presented in $\hbar\omega$ units.

Finally, the vibronic Hamiltonian in the PKS model is

$$\hat{H} = \hat{T}_n(Q_+) + \hat{T}_n(Q_-) + \frac{1}{2}(Q_+^2 + Q_-^2)\hat{I} + \begin{pmatrix} w + AQ_+ & AQ_- \\ AQ_- & -w + AQ_+ \end{pmatrix}$$

$$(2.8)$$

where $\hat{I}$ is the unit matrix. By means of a change of the energy read off, the linear terms in Q_+ can be eliminated and hence the Q_+ vibrations become simple harmonic. The remaining vibronic Hamiltonian contains only one Q_- coordinate:

$$\hat{H} = \hat{T}_n(Q_-) + \frac{1}{2}Q_-^2 \hat{I} + \begin{pmatrix} w & AQ_- \\ AQ_- & -w \end{pmatrix} \tag{2.9}$$

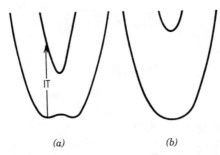

Figure 1. Adiabatic potentials of MV dimers in PKS model: (*a*) strong vibronic coupling (localization), (*b*) weak coupling (delocalization) (IT — intervalence transfer).

and coincides with the pseudo Jahn–Teller one.[5,7] It describes the mixing of two nondegnerate electronic states ϕ_+ and ϕ_- by one nontotally symmetric odd vibration. As shown below, the even mode Q_+ participates also in the electron transfer process when other possible electronic states are also involved. The adiabatic potential of the system is the solution of the Schrodinger equation with the Hamiltonian (Eq. 2.9) for fixed nuclei $[T_n(Q_-) = 0]$

$$\mathcal{E}_\pm(Q_-) = \tfrac{1}{2}Q_-^2 \pm (w^2 + A^2Q_-^2)^{1/2} \tag{2.10}$$

It contains two curves. The lower one describing the ground state under the condition

$$|w| < A^2 \tag{2.11}$$

has two equivalent minima (Fig. 1). At $Q_- = 0$ this curve has a maximum, and therefore Eq. (2.11) is considered as a criterion of instability of the high-symmetry configuration of the system. If $|w| > A^2$, then the ground state curve, as the excited one, has a minimum at this point.

The analysis of the electronic wave function at the minima of the adiabatic potential shows that in the case under consideration, if Eq. (2.11) is fulfilled, the minima states correspond to the localization of the excess electron at one of the two MV centers. In the opposite case with only one minimum at $Q_- = 0$, this electron is uniformly delocalized over the two centers. Thus for MV dimers Eq. (2.11) serves as the condition of localization of the excess electron, and it is used as a base in the classification of MV compounds after Robin and Day.[15]

The semiclassical consideration of MV compounds given above based on the analysis of their adiabatic potentials, gives qualitative results concerning with the possible types of electronic distribution. For quantitative evaluation of observable magnitudes the knowledge of the vibronic states — solutions of the equation of nuclear motions with Eq. (2.10) — are needed. The full equation can be written as follows:

$$\hat{H}\begin{pmatrix} X_n^+(Q_-) \\ X_n^-(Q_-) \end{pmatrix} = E_n \begin{pmatrix} X_n^+(Q_-) \\ X_n^-(Q_-) \end{pmatrix} \tag{2.12}$$

the wave functions taking the form

$$\Psi_n = \phi_+ X_n^+(Q_-) + \phi_- X_n^-(Q_-) \tag{2.13}$$

There are several methods of solutions of vibronic equations.[7] In the PKS model the method suggested by Fulton and Gouterman[16] based on the expansion of the wave function $X_n(Q_-)$ over the ones of the nondisplaced harmonic oscillators $\chi_\nu(Q_-)$ is used:

$$X_n^\pm(Q_-) = \sum_\nu \alpha_{n\nu}^\pm \chi_\nu(Q_-) \tag{2.14}$$

The functions Ψ_n should transform after the even (Ψ_n^+) or the odd (Ψ_n^-) representations of the symmetry group of the system. This means that

$$\Psi_n^+ = \phi_+ \sum_{\nu=0,2} \alpha_{n\nu}\chi_\nu(Q_-) + \phi_- \sum_{\nu=1,3} \alpha_{n\nu}\chi_\nu(Q_-) \tag{2.15}$$

$$\Psi_n^- = \phi_+ \sum_{\nu=1,3} \beta_{n\nu}\chi_\nu(Q_-) + \phi_- \sum_{\nu=0,2} \beta_{n\nu}\chi_\nu(Q_-) \tag{2.16}$$

The coefficients $\alpha_{n\nu}$ and energies E_n^+ are given by the secular equation

$$\sum_\nu \alpha_{n\nu}(H_{\mu\nu} - \delta_{\mu\nu}E_n^+) = 0, \qquad n, \mu, \nu = 0, 1, 2, \ldots \tag{2.17}$$

where

$$H_{\mu\nu} = [\mu + \tfrac{1}{2} + (-1)^\mu w]\delta_{\mu\nu} + A[\sqrt{\mu/2}\,\delta_{\mu,\nu+1} + \sqrt{(\mu + 1)/2}\,\delta_{\mu,\nu-1}] \tag{2.18}$$

Similar equations for $\beta_{n\nu}$ and E_n^- can be obtained by the substitution $w \rightarrow -w$. In concrete calculations of observable characteristics of MV dimers it is enough to keep only several tenths of vibrational states in the expansion (Eq. 2.14). For systems with localized electrons, a faster convergence of the calculations can be obtained if several oscillator functions with displaced equilibrium positions are introduced in this expansion, but this essentially complicates the algorithm and computation procedure.

The main physical magnitude calculated by the PKS model is the intervalent transfer band. In the semiclassical approach it can be regarded as resulting from the transition between the two adiabatic potential curves (Fig. 1) under the influence of the external electromagnetic wave. In a more rigorous quantum–mechanical treatment the intervalent transfer band can be considered as describing the envelope of the summarized individual transitions between the vibronic states (Eq. 2.13).

The problem in the PKS model is reduced to the one-mode one own to the simplifications introduced above, namely, to the employment of only one totally symmetric mode on each center (based on the assumption of only one nondegenerate electronic state on each of them for the excess electron) and to the neglect of the elastic interaction between the monomers. A question arises: To what extent is this approach justified for real systems? To answer this question the experience gained in studies of vibronic systems can be used. For instance, it was shown that the broad optical band of the singlet–doublet $A \rightarrow E$ transition can be described with a high accuracy even when the several active e modes interacting with the E state are substituted by one effective mode of the same symmetry.[17,18] However, such a substitution cannot describe statisfactorily the results on the vibronic fine structure of the band.

Apparently, the vibrations Q_- and Q_+ introduced above have to be considered as some effective modes of appropriate symmetry, and should not necessarily be identified with a certain vibration in the real system. Therefore this approach cannot be used for the interpretation of, say, IR spectra, whereas broad, "nonstructured" intervalence transition bands can be described well.

III. ONE-CENTER INTERACTIONS IN MV DIMERS

As mentioned above, a significant extension of the PKS model for MV dimers is possible by means of employment of more than one state for the excess electron on each center. The additional excited states can take part in the electron transfer between the centers if either they are temperature populated, or they admix to the ground state by some special interaction. This interaction can be effective either in the electronic subsystem only, or it can be an additional vibronic interaction. In this section we consider both these possibilities when the admixture of additional one-

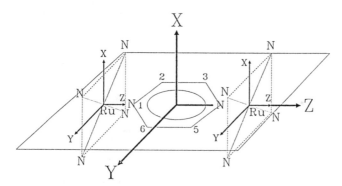

Figure 2.   Structure of the CT ion.

electronic states is due to one-center interaction only. Exchange coupling between centers will be considered in subsequent sections.

A.   Low-Symmetry Crystal Fields and Spin-Orbital Interactions

For complexes of 4d and 5d transition elements the spin-orbital interaction plays a significant role in the formation of the energy spectrum. Even in the case of full delocalization of the excess electron (large w values) the spin-orbital interaction can be of the same order of magnitude as w. The different spin-orbital states may be manifested also in the spectra of intervalent transfer.[19]

We consider here the role of the spin-orbital interaction using as an example the Creutz–Taube (C–T) ion, $[(NH_3)_5Ru(pyz)Ru(NH_3)_5]^{5+}$ (pyz = pyrazine). This ion is a classical object of polynuclear MV systems. Its relatively simple composition and structure allows us to use this system as a convenient example for elucidation of the role of different factors influencing the intramolecular electron transfer.

At present it can be considered definitely that in the ground state of the C–T ion the excess electron is uniformly delocalized over the two ruthenium ions. However, the theoretical models suggested for the description of the various experimental data on this ion from a unique point of view encounter a series of difficulties.

The structure of the C–T ion is shown schematically in Fig. 2. In the octahedral environment the ground state is $^1A_1$ for the bivalent ruthenium ion and $^2T_{2g}$ for the trivalent one. In the real structure with fields of lower symmetry the orbital triplet is split into three orbital singlets that are mixed by the spin-orbital interaction. The electron transfer can take place from the $^1A_1$ state to any of these three states.

The first work considering the low-symmetry crystal field and spin-orbital interactions in the C–T ion was performed by Neuenschwander,

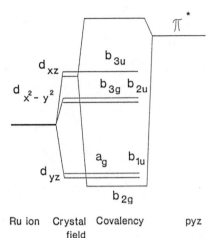

Figure 3.   MO scheme for the CT ion.

Piepho, and Schatz.[20] These authors assumed that the vibronic constants for all the three electronic states are the same and equal to that of the $Ru(NH_3)_6^{2+/3+}$ complex and the parameters of the electronic structure were used as adjustable ones in order to fit the experimental data on the ESR spectra and the intervalence transfer band. This approach results in close values of the resonance interaction (electron transfer probability) w for all three orbital states, but this conclusion does not agree with the data on the electronic structure of this ion. The analysis shows that only one of the three ruthenium orbitals overlaps effectively with the appropriate MO of pyrazine that transfers the resonance interaction.

As mentioned above, the vibrations involved in the interactions with the electronic states have to be considered as forming an effective "interaction mode" (see also Ref. 7), and hence it is difficult to identify real vibronic constants of this interaction with experimental values. Therefore, the approach in which the vibronic constants are taken as adjustable ones and the parameters of electronic structure are evaluated by quantum–chemical calculations seems to be more appropriate to the problem. In the one-electron energy level scheme of the ruthenium ion, the splitting of t_{2g} orbitals by the tetragonal ligand field and the rhombic component of the pyrazine bridge field has to be taken into account. The d_{yz} and $d_{x^2-y^2}$ orbitals (the latter belongs to the t_{2g} representation due to the chosen coordinate system, see Fig. 2) do not overlap with the valent MO of pyrazine and hence are not subject to strong resonance alterations. The orbital d_{xz} overlaps with the empty π^* MO of pyrazine, resulting in the energy level scheme of Fig. 3. In the ground state the unpaired excess electron occupies the nonbonding orbital b_{3u}, resulting in the multielectron state $^2B_{3u}$ (symmetry C_{2v}) of the system as a whole.

The first excited states can also be constructed. Using symmetrized combinations we have:

$$\varphi(a_g) = (1/\sqrt{2})(d^1_{x^2-y^2} + d^2_{x^2-y^2})$$

$$\varphi(b_{1u}) = (1/\sqrt{2})(d^1_{x^2-y^2} - d^2_{x^2-y^2})$$

$$\varphi(b_{3g}) = (-i/\sqrt{2})(d^1_{yz} + d^2_{yz})$$

$$\varphi(b_{2u}) = (-i/\sqrt{2})(d^1_{yz} - d^2_{yz})$$

(3.1)

The occupation of these one-electron orbitals as well as the antibonding orbital $\varphi(b_{2g})$ (Fig. 3) by the unpaired electron results in the excited states ^2A_g, $^2B_{1u}$, $^2B_{3g}$, $^2B_{2u}$, and $^2B_{2g}$, respectively. These states are similar to that of Eq. (2.5) in the PKS model with the significant distinction that in the case at hand we have three pairs of resonance multielectron states (instead of one pair in the PKS approach), and they are mixed by the spin-orbital interaction. Neglecting the covalency contribution to the latter we can present the matrix of the electronic Hamiltonian for the even states as follows:

$$\hat{H}^g_{el} = \begin{pmatrix} E_{A_g} & \pm\xi/2 & -\xi/2 \\ \pm\xi/2 & E_{B_{2g}} & \pm\xi/2 \\ -\xi/2 & \pm\xi/2 & E_{B_{3g}} \end{pmatrix}$$

(3.2)

where E_Γ is the energy of the multielectron states obtained by quantum–chemical calculations without taking account of the spin-orbital interaction, and ξ is the constant of the latter. In Eq. (3.2) the upper and lower signs describe, respectively, two states corresponding to two projections of the spin. The matrix for the odd states can be obtained from Eq. (3.2) by substitution of $E_{B_{2g}}$ by $E_{B_{3u}}$; the energies of the remaining two odd states coincide with that of the similar even states. The relative energies of the multielectron states can be estimated using the data of electronic structure calculations in the DVM-X_α approximation.[22]

Now we pass to the vibronic part of the problem. Apparently the vibronic constants of interactions of the electronic states with the totally symmetric vibrations on each center cannot be estimated by the data for the one-center hexaamine complexes of ruthenium, as it is done in, for example Refs. 20 and 23. When passing from $Ru(NH_3)_6^{3+}$ to $Ru(NH_3)_6^{2+}$

the distance Ru–N increases with 0.04 Å, whereas in $Ru(NH_3)_5(pyz)^{2+}$ as compared with $Ru(NH_3)_5(pyz)^{3+}$ the distance Ru–N(pyz) decreases by 0.07 Å.[2+] Therefore one can suppose that the vibronic constants given above are different for the three lcoalized electronic states. Then the potential energy can formally be written as a six-order matrix (three states per each center)

$$\hat{H}_p = \frac{1}{2}(Q_+^2 + Q_-^2)\hat{I}^+ \begin{pmatrix} \hat{H}^g(Q_+) & \hat{V}(Q_-) \\ \hat{V}(Q_-) & \hat{H}^u(Q_+) \end{pmatrix} \tag{3.3}$$

where

$$\hat{H}^{g,u}(Q_+) = \hat{H}_{el}^{g,u} + \hat{V}(Q_+) \tag{3.4}$$

$$\hat{V}(Q) = \begin{pmatrix} A_1 Q & 0 & 0 \\ 0 & A_2 Q & 0 \\ 0 & 0 & A_3 Q \end{pmatrix} \tag{3.5}$$

and A_i $(i = 1, 2, 3)$ are the vibronic constants for the three localized states A_g, B_{2g}, and B_{3g} respectively. While A_1 and A_3 characterize the interaction of "pure" d states with the nuclear displacement, A_2 is the constant of interaction for an effective localized MO. Here, distinguished from the PKS model and the one that takes into account the spin-orbital interaction but neglects the difference in the A_i values, the Q_+ mode cannot be separated, and the adiabatic potential has to be considered in the space of two coordinates, Q_- and Q_+. Its evaluation can be carried by solving the equation with the 6×6 matrix Hamiltonian (Eq. 3.3).

If we take the value of the spin-orbital constant the same as for the free-ion Ru, $\xi \approx 10^3$ cm^{-1}, and the stretching vibration frequency Ru–N as equal to 500 cm^{-1}, then the vibronic constants can be estimated from the EPR data. In the electron delocalized state the minimum of the lower curve of the adiabatic potential is at $Q_- = 0$, and Eq. (3.3) reduces to a 3×3 matrix. The electronic wave function at the minimum reads

$$\phi_0 = a\phi(^2A_g) + b\phi(^2B_{2g}) + c\phi(^2B_{3g}) \tag{3.6}$$

with this function taken at the minimum $Q_- = 0$ the expresions for the g-factors are

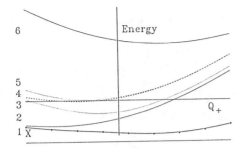

Figure 4. Extremal cross section $(Q_- = 0)$ of the adiabatic potential of the CT ion. X is a saddlepoint between the two minima at $Q_- \neq 0$.

$$g_x = 2(2\kappa ac + a^2 - b^2 + c^2)$$
$$g_y = 2(a^2 + b^2 - c^2 - 2\kappa ab) \qquad (3.7)$$
$$g_z = 2(2\kappa bc + a^2 - b^2 - c^2)$$

where κ is the orbital reduction factor. The constants A_i are here adjustable parameters, and they are chosen to reproduce the experimental values $g_x = 1.346$, $g_y = 2.799$, $g_z = 2.487$.[25] In this way the following values were obtained: $A_1 = 1.44$, $A_2 = -1.19$, and $A_3 = 0.76$. These results show first of all that, as expected, the three values of A_i are different (although they are relatively not very large and not very much different from the value $A = 1.1$ for the hexaamine complex). The A_2 value for the orbital that takes part in the covalent binding to the pyrazine molecule becomes even negative. This is not surprising since, as mentioned above, the excess electron shortens the Ru–N(pyz) distance. The difference between A_1 and A_3 is also understandable since the d_{yz} orbital is more influenced by the pyrazine molecule.

For not very large vibronic constants (as compared with w), the lower curve in the PKS model is not essentially different from the harmonic one, and the intervalence transfer band at low enough temperatures is symmetrical. Figure 4 illustrates the extreme cross sections (at $Q_- = 0$) of the adiabatic potential obtained in our approach. The form of the electronic wave functions allows us to conclude that the most intensive transition from the ground state is that to the sheet 6. In the semiclassical approximation the maximum of the band coincides with the distance between the surfaces 1 and 6 which in our calculation is about $6600\ \mathrm{cm}^{-1}$. This value is close to the experimental one for the frequency of the maximum of the intervalence transfer band in the C–T ion, $6200\ \mathrm{cm}^{-1}$.[25] One of the features of this band is that it is not symmetrical, it slumps slower at higher frequencies, and this takes place also at low temperatures down to 4.2 K. This means that there are intensive enough transitions from the ground state of sheet 1 to not only the ground state of sheet 6, but also to its excited states.

Such transitions can take place when the ground state has a strong vibronic anharmonism. In the case at hand this anharmonism arises, in spite of the relatively small values of the vibronic constants, due to the spin-orbital admixing of the excited states to the ground one. In the region where sheets 1 and 2 are drawn together, the lower one has two minima at $Q_- \neq 0$ divided by a saddle-point X. The surface as a whole is a triminima one, the barriers between the minima being small. The ground vibronic state is not harmonic, resulting in the asymmetry of the optical spectrum mentioned above.

The existence of three minima of the ground-state adiabatic potential surface does not contradict the statement of full delocalization of the excess electron. Indeed, the deepest minimum corresponds to a delocalized electronic distribution, and the probability of localization of the system in other minima (where the electronic cloud can be localized) is hardly possible because of the small barriers between them. Note that a triminima potential energy surface cannot, in principle, be obtained in the two-level model of the PKS type, even if one introduces artificially different vibronic constants for the two states (Eq. 2.5).

B. Electron Delocalization in MV Dimers in the Presence of a Local Pseudo Jahn–Teller Effect

The problem of electron transfer when the local vibronic interactions are essential is under continuous discussion in solid-state physics in connection with the problem of the Jahn–Teller polaron (e.g., see Refs 26 and 27).

In simple cases of crystals with Jahn–Teller MV centers it is usually assumed[26] that the excess electron on each center occupies a twofold degenerate electronic orbital transforming after the E representation of the site symmetry group D_{4h} and interacts with $b_{1u}(D_{4h})$ nuclear displacements resulting in the $E * b_1$ Jahn–Teller problem.[7]

In MV dimers the local symmetry is usually lower, and the case of two close-in-energy electronic energy levels mixed by low-symmetry nuclear displacements (the pseudo Jahn–Teller problem)[6,7] is most probable.[28] Denote these two states, ground and excited, by ϕ_i^g and ϕ_i^e, respectively, $i = 1, 2$ being the numbers of the two centers, the energy distance between them 2Δ, the local vibrations mixing these states by q_i, the corresponding vibronic constant being λ, the vibronic constants of their interaction with the totally symmetric local vibrations by A_g and A_e, respectively, and the appropriate parameters of the resonance interaction between the centers by w_g and w_e. The case is considered when the overlap of ϕ_i^g and ϕ_j^e of different centers can be neglected. Then the vibronic Hamiltonian can be presented similarly to Eq. (2.4):

$$\hat{H} = H_n(Q, q) + \begin{pmatrix} A_g Q_1 & w_g & \lambda q_1 & 0 \\ w_g & A_g Q_2 & 0 & \lambda q_2 \\ \lambda q_1 & 0 & A_e Q_1 & w_e \\ 9 & \lambda q_2 & w_e & A_e Q_2 \end{pmatrix} \tag{3.8}$$

The problem can be simplified by means of symmetry considerations. All the vibronic states of the dimer can be divided into two nonoverlapping sets — odd and even — with respect to the transmutation of the two centers. The equation for these states can be obtained similarly to the work in Ref. 29. Let us introduce the permutation operator of the monomers $\hat{g}$. The operator

$$\hat{Q} = \begin{pmatrix} 0 & \hat{g} & 0 & 0 \\ \hat{g} & 0 & 0 & 0 \\ 0 & 0 & 0 & \hat{g} \\ 0 & 0 & \hat{g} & 0 \end{pmatrix} \tag{3.9}$$

commutes with the Hamiltonian (Eq. 3.8).

The eigenfunction of Eq. (3.9) can be projected from an arbitrary bispinor by means of the following operators of projection:

$$\hat{P}_{\pm} = \frac{1}{\sqrt{2}} \begin{pmatrix} 1 & \pm\hat{g} & 0 & 0 \\ \pm\hat{g} & 1 & 0 & 0 \\ 0 & 0 & 1 & \pm\hat{g} \\ 0 & 0 & \pm\hat{g} & 1 \end{pmatrix} \tag{3.10}$$

The vibronic functions can be written in the form

$$\Psi_\nu^\pm = \frac{1}{\sqrt{2}} [(\phi_1^g \pm \phi_2^g \hat{g}) \chi_\nu^\pm \pm (\phi_1^e \pm \phi_2^e \hat{g}) \xi_\nu^\pm] \tag{3.11}$$

$\chi_\nu^\pm$ and $\xi_\nu^\pm$ being the vibrational wave functions that obey the equation

$$\begin{pmatrix} \hat{H}_n + A_g(Q_+ + Q_-) \pm w_g\hat{g} - \epsilon_\nu & \lambda(q_+ + q_-) \\ \lambda(q_+ + q_-) & \hat{H}_n + A_e(Q_+ + Q_-) + 2\Delta \pm w_e\hat{g} - \epsilon_\nu \end{pmatrix}$$

$$\times \begin{pmatrix} \chi_\nu^\pm \\ \xi_\nu^\pm \end{pmatrix} = 0 \tag{3.12}$$

and

$$q_\pm = \frac{1}{\sqrt{2}}(q_1 \pm q_2) \tag{3.13}$$

Using the expansion of $\chi_\nu^\pm$ and $\xi_\nu^\pm$ into a series of oscillator wave functions, one can reduce Eq. (3.12) to a system of algebraic equations. Its solution allows us to obtain the vectors of the states that can be used to calculate the spectra of the dimer.

To obtain the adiabatic potential of the system we have to go back to Eq. (3.8) and neglect the kinetic energy of the nuclei in the expression of $\hat{H}_n$. The problem of arbitrary parameter relations is complicated; we shall consider it, making several additional simplifications.

Consider the case when the pseudo Jahn–Teller effect on each center is strong enough. Then at the points near minima of the adiabatic potentials of the two noninteracting centers the energy distance to the excited states is large enough and the resonance interaction of only the lowest sheets, which are solutions of Eq. (3.8) with $w_{g,e} = 0$, can be taken into account, all the others being neglected. The two adiabatic potentials are

$$\mathscr{E}_{1,2} = \tfrac{1}{2}\omega^2(q_1^2 + q_2^2) + \tfrac{1}{2}\Omega^2(Q_1^2 + Q_2^2)$$
$$- (\Delta^2 + \lambda^2 q_{1,2}^2)^{1/2} + A(q_{1,2})Q_{1,2} \tag{3.14}$$

where

$$A(Q) = \frac{1}{2}(A_g + A_e) + \frac{\Delta}{2}(A_g - A_e)(\Delta^2 + \lambda^2 q^2)^{-1/2} \tag{3.15}$$

and ω and Ω are the frequencies of q and Q vibrations, respectively. Here mass-weighted (and not dimensionless) coordinates are used.

Consider first the case when the vibronic constants for the ground and excited state $A_g = A_e = A$. Then the ground-state adiabatic potential becomes as follows:

$$\mathscr{E} = \frac{1}{2}\omega^2(q_1^2 + q_2^2) + \frac{1}{2}\Omega^2(Q_+^2 + Q_-^2) + \frac{1}{\sqrt{2}}AQ_+$$

$$- \tfrac{1}{2}(\Delta^2 + \lambda^2 q_1^2)^{1/2} - \tfrac{1}{2}(\Delta^2 + \lambda^2 q_2^2)^{1/2}$$

$$- \{[\sqrt{2}AQ_- - (\Delta^2 + \lambda^2 q_1^2)^{1/2} - (\Delta^2 + \lambda^2 q_2^2)^{1/2}]^2 + 4W^2(q_1, q_2)\}^{1/2}$$

$$(3.16)$$

where $W(q_1, q_2)$ is a function of w_g, w_e, q_1, q_2 that can be obtained from Eq. (3.8). The interaction with the Q_+ vibration can be easily separated into a displaced oscillator, and thus the behavior of the adiabatic potential with respect to Q_-, q_1, and q_2 (or Q_-, q_+, and q_-) has to be considered. A nonuniform (over the two centers) electron distribution occurs when there are minima at Q_-, $q_- \neq 0$. Therefore we shall investigate the instability of the adiabatic potential with respect to this odd vibrations at the q_+ value corresponding to the minimum of the adiabatic potential of the noninteracting pseudo Jahn–Teller centers:

$$q_\pm^0 = \left(\frac{2\lambda^2}{\omega^4} + \frac{2\Delta^2}{\lambda^2}\right)^{1/2} \tag{3.17}$$

Here it is assumed that the pseudo Jahn–Teller distortions of the two centers have the same sign (in-phase distortions). In the opposite case q_+ and $q-$ have to be interchanged. Since the pseudo Jahn–Teller effect was assumed to be strong. Since the psuedo Jahn–Teller effect was assumed to be strong, the value q_+^{min} for the potential (Eq. 3.16) will not differ much from that of Eq. (3.17). Expanding Eq. (3.16) into a power series of q_- and Q_- and keeping the terms up to the second power, we have

$$\mathscr{E} = \mathscr{E}_0 + \frac{1}{2}\left(\omega^2 - \frac{R}{2P}\right)q_-^2 + \frac{A\lambda}{P}q_-Q_- + \frac{1}{2}\left(\Omega^2 - \frac{A^2}{P}\right)Q_-^2 \tag{3.18}$$

where

$$R = 2\lambda^2 + (w_g^2 - w_e^2)\frac{\Delta\Omega^6}{\lambda^4} \tag{3.19}$$

$$P^2 = (w_g^2 - w_e^2) + 2(w_g^2 - w_e^2)\frac{\Delta\Omega^2}{\lambda^2} \tag{3.20}$$

The condition of the minimum of the adiabatic potential at $Q_- = q_- =$

0 (and hence of a delocalized state) is that the coefficient at the squares of displacements in the appropriate quadratic form are positive. Thus the condition of delocalization of the electron distribution is here different from that in the case of a two-level situation (Eq. 2.12) (in the units of this section it reads $|w| > A^2/\Omega^2$). In the case under consideration in this section the excess electron has two "channels" of localization: (1) the interaction with the totally symmetric nuclear displacements (similar to the two-level case) and (2) the interaction with nontotally symmetric displacements active in the local pseudo Jahn–Teller effect. Therefore when the former is small (rigid totally symmetric vibrations, small A values), the probability of localization is small in the two-level model, while it can be large enough due to the psuedo Jahn–Teller effect (soft nontotally symmetric vibrations, large λ values). On the other hand the inclusion of the excited state in the process of electron transfer increases the probability of delocalization, provided the w_e value is nonzero. Hence in the case of large enough w_e values, the pseudo Jahn–Teller effect can also increase the delocalization, distinguished from the Jahn–Teller effect which leads always to the increase of the "effective mass" of the polaron.[26,27]

The direction of the lowest curvature of the adiabatic potential in the (Q_-, q_-) plane forms a nonzero angle determined by the relation

$$\cos^2 \varphi = \tfrac{1}{2} + \tfrac{1}{2}[1 + 4A^2\lambda^2/(\omega^2 P - \Omega^2 P - R + A^2)]^{-1/2} \qquad (3.21)$$

Thus when Q_- is nonzero (localized distribution), q_- is also nonzero. This means that the electron localization results also in out-of-phase pseudo Jahn–Teller low-symmetry distortions of the two centers, and the latter differ not only by the metal–ligand distances but also by the symmetries of the environment.

Apparently, the observed different symmetries of the two centers $Mn^{3+} - Mn^{4+}$ in the MV complex $[Mn_2O_2(bipy)_4]^{3+}$ with identical ligands[30] can be explained as due to the pseudo Jahn–Teller influence on the electron transfer of the kind discussed above. Indeed, the ion Mn^{3+} in the high-spin state and in the near-neighbor environment of C_{2v} symmetry has an empty orbital a_2 which is close-in-energy to the highest occupied by one-electron a_1 MO. These two orbitals originate from the twofold degenerate e_g one in the O_h symmetry, and they mix strongly under the a_2 nuclear displacements, giving the appropriate pseudo Jahn–Teller distortions.

A procedure similar to that leading to Eq. (3.18) can be carried out when the two electronic states on each center have different constants of vibronic coupling with the totally symmetric local vibrations $A_g \neq A_e$, while the parameters of resonance interactions are the same, $w_g = w_e = w$. In this case the Q_+ coordinates are no more separable. However, since we assume that the intracenter interactions are much stronger than the intercenter ones, the Q_+ coordinate, as well as q_+, can be substituted by their equilibrium values for the noninteracting centers. All the qualitative conclusions of the previous case remain valid here, and the condition of an electron-delocalized distribution is

$$\Omega^2 > \frac{1}{2w}\left(A_+^2 + 2\frac{\Delta\omega^2}{\lambda^2}A_+A_-\right) \tag{3.22}$$

$$\omega^2 > \frac{1}{2w}\left(\lambda^2 - \frac{\Delta\omega^4}{\lambda^2\Omega^2}A_+A_-\right)$$

where

$$A_\pm = \tfrac{1}{2}(A_g \pm A_e) \tag{3.23}$$

Consider now the case when the overlap of the electronic states of different centers is large enough and $|w_{g,e}| \gg \Delta$. In this case, for the two lowest electronic states one can take into account first the interaction with the totally symmetric vibrations and obtained their adiabatic potentials,

$$\mathscr{E}_1 = \frac{1}{2}\omega^2(q_+^2 + q_-^2) + \frac{1}{2}\Omega^2(Q_+^2 + Q_-^2) + \frac{1}{\sqrt{2}}A_gQ_+$$

$$-(w_g^2 + \frac{1}{2}A_g^2Q_-^2)^{1/2} \tag{3.24}$$

$$\mathscr{E}_2 = \frac{1}{2}\omega^2(q_+^2 + q_-^2) + \tfrac{1}{2}\Omega^2(Q_+^2 + Q_-^2) + \frac{1}{\sqrt{2}}A_eQ_+$$

$$+ 2\Delta - (w_e^2 + \tfrac{1}{2}A_e^2Q_-^2)^{1/2}$$

and then take into account their mixing by the vibrations q. The resulting lowest sheet of the adiabatic potential with an accuracy up to the second order in Q_- and q_- reads

$$\mathscr{E}(q, Q) = \mathscr{E}_0(q_+, Q_+) + F_1Q_-^2 + F_2q_-^2 + F_3Q_-q_- \tag{3.25}$$

where

$$\mathscr{E}_0(q_+, Q_+) = \tfrac{1}{2}\omega^2 q_+^2 + \tfrac{1}{2}\Omega^2 Q_+^2 + \frac{1}{\sqrt{2}} A_+ Q_+$$

$$- \tfrac{1}{2}[(w_e + 2\Delta - w_g - \sqrt{2} A_- Q_+)^2 + 2\lambda^2 q_+^2]^{1/2}$$

(3.26)

and F_1, F_2, F_3 are complicated functions of Q_+ and q_+, F_3 being proportional to q_+.

The stability of the adiabatic potential with respect to Q_- and q_- has to be investigated at the extreme (minima) points of Q_+ and q_+. The latter can be estimated from the surface $\mathscr{E}_0(q_+, Q_+)$. Its extreme points are

$$q'_+ = 0$$
$$q''_+ = \pm \left[\frac{\lambda^2}{2\omega^4} - \frac{1}{2\lambda^2}(w_g + 2\Delta - w_e - \sqrt{2} A_- Q'_+)^2 \right]^{1/2}$$

(3.27)

where

$$Q'_+ = -\left[A_- \frac{\omega^2}{\lambda^2}(w_g + 2\Delta - w_e) + A_+ \right] \Big/ \sqrt{2}\left(\Omega^2 - \frac{\omega^2}{\lambda^2} A_-^2 \right)$$

(3.28)

If the minimum is at $q_+ = 0$, the cross term in Eq. (3.25) vanishes, and the potential surface is most "soft" along Q_- or q_-. This means that in the electron-localized state the two centers differ by either the totally symmetric normal coordinate or by the nontotally symmetric one. If $q_+^{\min} \neq 0$, both coordinates are different and it is implied that F_1 and F_2 are negative. The condition for the minimum to be at $q_+ = 0$ can be obtained from Eq. (3.26):

$$\omega^2 > \lambda^2 / [w_g + 2\Delta - w_e + A_-(A_+ + A_-)/\Omega^2]$$

(3.29)

To summarize, the vibronic mixing results in a series of effects that cannot be described in the PKS model.

IV. THE VIBRONIC THEORY OF EXCHANGE-COUPLED
MV DIMERS

In the previous section we considered MV systems with one excess electron that occupies a MO above the two closed shells of electrons of the cores. Beside these systems, there are many MV clusters in which the centers of electron transfer themselves contain unpaired electrons. It is obvious that these centers, in addition to the electron transfer, are coupled also by the interaction of their spin moments (exchange interaction). In these cases the resulting electronic energy spectrum is different from that obtained in the one-electron two-level PKS model. Special features of the vibronic effects that are due to the exchange coupling in MV dimers are discussed in this section.

A. Electronic Energy Spectrum

The energy spectrum of spin multiplets of a system with two paramagnetic centers in orbitally nondegenerate states can be described in the model of Heisenberg–Dirac–Van Vleck (HDVV) by the exchange Hamiltonian

$$\hat{H}_{\mathrm{ex}} = -2J\mathbf{S}_1\mathbf{S}_2 \tag{4.1}$$

where $2J$ is the exchange parameter ($J > 0$ for a ferromagnetic interaction, and $J < 0$ for an antiferromagnetic one), $\mathbf{S}_1$ and $\mathbf{S}_2$ are the operators of the spin of the ions. Sometimes additional, more complicated forms of exchange coupling between magnetic ions have to be taken into account [and corresponding terms added to the Hamiltonian (Eq. 4.1)].[31]

The HDVV model is based on the Heitler–London approach to the description of the electronic structure of the system assuming that the unpaired electrons are localized at a certain center. In MV systems one of the electrons takes part in the formation of the magnetic moments of the centers and simultaneously migrates over different centers. This situation cannot be described directly by the Heitler–London scheme and hence by the HDVV model. An adequate approach to the problem of magnetic interactions in a system with migrating electrons is provided by the concept of double exchange suggested by Zener[32] for explanation of the ferromagnetism of magnetic semiconductors. Subsequently, the theory of double exchange in application to crystals was worked out by Anderson and Hasegawa,[33] De Gennes,[34] and in latter works.[35–38]

The idea of double exchange is as follows. The spin of the excess electron localized at the magnetic center is lined up parallel to the spin of the ionic core by the strong intraatomic exchange interaction (the electronic shell of the core is assumed to be less than half filled). Since the resonance interaction responsible for the electron transfer between the centers is independent of the spin, the energy gain by the delocalization of the excess electron is greater when the spins of the two ionic cores are parallel. Hence the electron transfer favors ferromagnetic ordering of the spin centers.

The application of this idea to limited molecular systems, (i.e., MV clusters) was carried out recently.[39-42] The theoretical results obtained earlier for crystals cannot be used directly for molecules for two reasons: (1) in application to crystals the usual HDVV interaction between the magnetic ions (direct or indirect) is not taken into account since, for example, in magnetic semiconductors the parameter of intercenter exchange interaction is always much smaller than the parameter of resonance interaction (electron transfer), whereas for molecular MV systems their ratio can be arbitrary (see below); (2) in the earlier theories of double exchange the multielectron effects due to the identity of the localized and migrating electrons are neglected.[38]

Consider first the energy spectrum of exchange-coupled MV dimers using as an example the simplest problem of two centers plus three electrons. Besides the general theoretical interest, this problem can serve as a model for real MV dimers containing Ni(I)–Ni(II) pairs.[43,44] Consider two one-electron states φ_i and φ_i' for each of the two equivalent centers, $i = 1, 2$. We assume that these states are nondegenerate, their orbital moment is frozen, and the orbitals that belong to different centers are orthogonal. The nonorthogonality of these orbitals can be taken into account at the expense of some complication of the resulting formulas. The two orbitals φ_1 and φ_2 contain one by one two electrons, while the third one can occupy the other two orbitals φ_1' and φ_2' with equal probability. The Coloumb repulsion of two electrons with opposite spins on the same orbital (the u value in the Hubbard model) is assumed to be large enough, so that the appropriate atomic excited states can be neglected.

The electronic Hamiltonian has the usual form:

$$\hat{H}_e = \sum_{i=1}^{3} \hat{h}_i + \sum_{i<j} \hat{g}_{ij} \tag{4.2}$$

where

$$\hat{h}_i = -\frac{1}{2}\nabla_i^2 + \sum_{k=1,2} v(|\mathbf{r}_i - \mathbf{R}_k|) \tag{4.3}$$

$$\hat{g}_{ij} = |\mathbf{r}_i - \mathbf{r}_j|^{-1} \tag{4.4}$$

$v(|\mathbf{R}|)$ is the ion potential. Let us choose as a basis the eigenfunctions of the operators S^2, S_z, and $\tilde{S}^2_{1(2)}$, where S is the operator of the total spin of the system, S_z is its projection, and $\tilde{S}_{1(2)}$ is the operator of the (intermediate) spin of the ion on which the excess electron is localized. These functions can be presented as linear combinations of Slater determinants:

$$|S, M_S, \tilde{S}_1\rangle = \sum_{\text{all } m} C^{SM_s}_{S_1\bar{M}_1,1/2m_3} C^{S_1\bar{M}_1}_{1/2m_1,1/2m_2} |\varphi_1 m_1, \varphi_1' m_2, \varphi_2 m_3|$$

$$|S, M_S, \tilde{S}_2\rangle = \sum_{\text{all } m} C^{SM_s}_{S_2\bar{M}_2,1/2m_1} C^{S_2\bar{M}_2}_{1/2m_2,1/2m_3} |\varphi_1 m_1, \varphi_2 m_2, \varphi_2' m_3|$$

(4.5)

where $C^{SM_s}_{S_1M_1,S_2M_2}$ are the Clebsch–Gordah coefficients, and the determinants are composed by the corresponding one-electron spin-orbitals, for example,

$$|\varphi_1 m_1, \varphi_1' m_2, \varphi_2 m_3| \equiv |\varphi_1 \rho_{m_1}, \varphi_1' \rho_{m_2}, \varphi_2 \rho_{m_3}| \qquad (4.6)$$

ρ_m being the spin function with the projection m.

Since the Hamiltonian (Eq. 4.2) does not contain operators dependent on the total spin and its projection, it does not mix the states with different S and M_S and the appropriate nondiagonal matrix elements vanish. Therefore, the matrix of the secular equation in the basis of the functions (Eq. 4.5) decomposes into four identical matrices of the second rank for $S = \frac{3}{2}$, and two identical matrices of the fourth rank for $S = \frac{1}{2}$. Denote the matrix elements by

$$J_0 = [\varphi_1 \varphi_1' \mid \varphi_1' \varphi_1] = [\varphi_2 \varphi_2' \mid \varphi_2' \varphi_2] \qquad (4.7)$$

$$w = [\varphi_1' \mid \varphi_2'] + [\varphi_1' \varphi_2' \mid \varphi_1 \varphi_1] + [\varphi_1' \varphi_2' \mid \varphi_2 \varphi_2] \qquad (4.8)$$

$$\Delta = 2[\varphi_1 \varphi_2' \mid \varphi_1' \cdot \varphi_1] = 2[\varphi_2 \varphi_2' \mid \varphi_1' \varphi_2] \qquad (4.9)$$

where

$$[\varphi_1' \mid \varphi_2'] = \int d\mathbf{r}_i \, \varphi_1'^*(\mathbf{r}_i) \hat{h}_i \, \varphi_2'(\mathbf{r}_i) \qquad (4.10)$$

$$[\varphi_m \varphi_n \mid \varphi_k \varphi_l] = \int d\mathbf{r}_i \, d\mathbf{r}_j \varphi_m^*(\mathbf{r}_i) \varphi_k^*(\mathbf{r}_j) \hat{g}_{ij} \varphi_n(\mathbf{r}_i) \varphi_l(\mathbf{r}_j) \qquad (4.11)$$

The parameters J_0 and w have the physical meaning of the intraatomic exchange integral and the matrix element of intercenter transfer, respectively. The integral of interatomic exchange interaction J can also be introduced, and as usual in the HDVV model (Eq. 4.1), it is assumed that J is the same for all occupied states:

$$J = [\varphi_1\varphi_2 \mid \varphi_2\varphi_1] \approx [\varphi_1'\varphi_2 \mid \varphi_2\varphi_1'] = [\varphi_2'\varphi_1 \mid \varphi_1\varphi_2'] \qquad (4.12)$$

In the basis of the functions (Eq. 4.5) for $S = \frac{1}{2}$ and arbitrary M_S ($\tilde{S}_i$ acquires the values 0, 1), the matrix Hamiltonian reads

$$\hat{H}_e = \begin{vmatrix} -J_0 + J & 0 & \frac{1}{2}(w - \Delta) & \frac{\sqrt{3}}{2}w \\ 0 & J_0 - J & \frac{\sqrt{3}}{2}w & -\frac{1}{2}(w + \Delta) \\ \frac{1}{2}(w - \Delta) & \frac{\sqrt{3}}{2}w & -J_0 + J & 0 \\ \frac{\sqrt{3}}{2}w & -\frac{1}{2}(w + \Delta) & 0 & J_0 - J \end{vmatrix} \qquad (4.13)$$

The states with a spin $S = \frac{3}{2}$ can be realized only when the intermediate spin $\tilde{S}_i = 1$. Therefore, for them the matrix Hamiltonian has the dimensions 2×2,

$$\hat{H}_e = \begin{pmatrix} -J_0 - 2J & w - \Delta \\ w - \Delta & -J_0 - 2J \end{pmatrix} \qquad (4.14)$$

The analysis[45] shows that the terms proportional to Δ can be neglected. Then for the energies we have

$$E(S = \tfrac{3}{2}) = -(J_0 + 2J) \pm w \qquad (4.15)$$

$$E(S = \tfrac{1}{2}) = \pm[(J_0 - J)^2 + w^2 \pm (J - J_0)w]^{1/2} \qquad (4.16)$$

and all the possible combinations of the signs have to be used.

In our approach presented above we take into account the same interactions (electron transfer and intraatomic exchange) that have been considered in the earlier work of Anderson and Hasegawa[33] plus the intercenter exchange interaction. Therefore the energy levels in Eqs. (4.15) and (4.16), with an accuracy up to a nonessential constant, coincide with the energy levels of the model Hamiltonian, suggested in Ref. 33, provided the additional terms of intercenter exchange interaction are introduced:

$$\hat{H}_{AH} = \begin{pmatrix} -2J_0\mathbf{S}_1\mathbf{s} - 2J(\mathbf{S}_1 + \mathbf{s})\mathbf{S}_2 & w\hat{I} \\ w\hat{I} & -2J_0\mathbf{S}_2\mathbf{s} - 2J\mathbf{S}_1(\mathbf{S}_2 + \mathbf{s}) \end{pmatrix} \tag{4.17}$$

Here $S_1 = S_2 = \frac{1}{2}$ is the spin of the ionic cores, and $s = \frac{1}{2}$ is the spin of the migrating electron. The Hamiltonian (Eq. 4.17) is given in the basis of the states with a definite localization of the excess electron and a definite local spin. If one passes to the basis of states with definite total spin of the system as a whole, $\hat{H}_{AH}$ decomposes into blocks corresponding to different S values. The parts of each of these blocks (subblocks) containing exchange terms transform to diagonal matrices, while the matrix elements of the nondiagonal subblocks are proportional to w, with coefficients proportioanl to $6j-$ symbols.

It is reasonable to assume that the Hamiltonian (Eq. 4.17) is valid also for the description of the energy spectrum of MV dimers with an arbitrary core spin S_0 and indirect mechanism of exchange interaction. Similar to the case of $S_0 = \frac{1}{2}$, the Hamiltonian (Eq. 4.17) for any value of S results in four energy levels for all S values except $S_{\max} = 2S_0 + \frac{1}{2}$, for which two energy levels occur. These energies are

$$E(S < S_{\max}) = J_0/2 + J(S_0 + \frac{1}{2}) - JS(S + 1)$$

$$\pm [(S_0 + \tfrac{1}{2})^2(J_0 - J)^2 + w^2 \pm (J_0 - J)w(S + \tfrac{1}{2})]^{1/2} \tag{4.18}$$

$$E(S_{\max}) = -J_0S_0 + J(2S_0 + 1) - JS(S + 1) \pm w \tag{4.19}$$

For transition metal compounds the orders of magnitude of the parameters entering Eqs. (4.18) and (4.19) are as follows: $J_0 \sim 1\text{--}10\,\text{eV}$, $w \lesssim 1\,\text{eV}$, $J \sim 10^{-1}\text{--}10^{-3}\,\text{eV}$. As shown below, the w values due to the vibronic effects can become comparable in magnitude with J.

It can easily be seen that Eq. (4.18) contains two pairs of close-in-energy states with an energy separation between them of about $2J_0$. The states of the upper pair are often called non-Hund ones because they arise from the states with a local spin of $S_0 - \frac{1}{2}$ (again, it is assumed that the electronic shell is less than half filled). In most cases (exceptions are considered below) the non-Hund states can be neglected. Then, expanding Eq. (4.18) in powers of w/J_0 and keeping linear terms in w only we obtain for the energies of the Hund states as functions of S

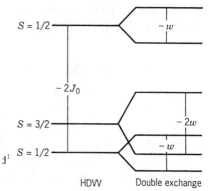

Figure 5. Electronic energy spectrum of a d^1–d^2 exchange-coupled MV dimer (the real ratios between J_0, J, and w are not obeyed).

$$E(S) = -JS(S + 1) \pm \frac{S + \frac{1}{2}}{2S_0 + 1} w \qquad (4.20)$$

Note that here the terms not containing S are omitted and hence the energy read-off is changed as compared with Eqs. (4.18) and (4.19).

Equation (4.20) is also valid in the case of a more than half-filled electronic shell when there are no non-Hund states and terms containing J_0 are not present. This equation emerges also from other theories of energy spectra of MV dimers. In Ref. 39 Eq. (4.20) is obtained within the Hubbard model.[46] In Refs. 41 and 45 the energy spectrum of the dimer is calculated by means of built-up correct multielectron wave functions; when the electrons occupy the degenerate t_{2g} orbital, an essential larger number of spin multiplets were obtained. However, for real systems the model of an energy spectrum based on the Anderson–Hasegawa Hamiltonian (Eq. 4.17) seems to be acceptable, the energy spectrum of model calculations agrees with that obtained in quantum–chemical investigations for concrete MV dimers [e.g., Fe(II)–Fe(III) pairs in ferredoxin[47] and in oxides[48]].

Using the relation between the w and J parameters, Eq. (4.20) allows us to obtain a qualitative picture of the energy spectrum of the systems under consideration. For each value of the total spin there are two resonance states. The spacing of the centers of gravity of these doublets is determined by the intercenter exchange, while their splitting is linearly dependent on the spin value (Fig. 5). Thus the double exchange favors ferromagnetic spin ordering. However distinguished from the case of MV dimers with HDVV ferromagnetic exchange interaction, the double exchange does not necessarily result in a ground state with a maximal spin when $J < 0$. The condition of such a ground state is rather hard, $w > (2n + 1)(n + 1)J$, where n is the number of electrons in the ionic core.

In general, it is clear that the migration of the electron results in an energy spectrum that is essentially different from that expected in the HDVV model.

In molecular crystals MV dimers are coupled by dipole–dipole interaction, which can lead to the transition in the charge-ordered state. If the dimers also feature double exchange, this transition is accompanied by the reduction of magnetic moment. Arising effects and phase diagrams (in the absence of vibronic interaction) are considered in Ref. 49.

B. Vibronic States and Magnetic Characteristics

The interaction with nuclear displacements in MV dimers having several unpaired electrons can be investigated following the same scheme as in the PKS model for MV systems with one electron. Consider the trielectron system discussed in the previous subsection. If we assume that the localization of the excess electrons at the ith center due to the vibronic coupling changes the φ'_i state only,

$$\langle \varphi'_i | \hat{H}_{en} | \varphi_i \rangle = AQ_i \tag{4.21}$$

then it can be shown that the constant of vibronic interactions with the totally symmetric vibrations A is the same for all the multielectron states (Eq. 4.5). Passing to the coordinates $Q_\pm$ and observing that the Q_+ vibrations, as above, can be separated, we can write the vibronic Hamiltonians for the state with $S = \frac{1}{2}, \frac{3}{2}$ in the basis of localized states as follows:

$$\hat{H}(S = \tfrac{1}{2}) = \begin{pmatrix} -J_0 + J + AQ_- & 0 & w/2 & \sqrt{3}w/2 \\ 0 & J_0 - J + AQ_- & \sqrt{3}w/2 & -w/2 \\ w/2 & \sqrt{3}w/2 & -J_0 + J - AQ_- & 0 \\ \sqrt{3}w/2 & -w/2 & 0 & J_0 - J - AQ_- \end{pmatrix}$$

$$+ [\tfrac{1}{2}Q_-^2 + T_n(Q_-)]\hat{I} \tag{4.22}$$

$$\hat{H}(S = \tfrac{3}{2}) = \begin{pmatrix} -J_0 - 2J + AQ_- & w \\ w & -J_0 - 2J - AQ_- \end{pmatrix}$$

$$+ [\tfrac{1}{2}Q_-^2 + T_n(Q_-)]\hat{I} \tag{4.23}$$

Neglecting the non-Hund states and passing to even and odd combinations of the wave functions,

$$\phi_\pm^{SM_S} = \frac{1}{\sqrt{2}} (|S, M_S, \tilde{S}_1 = S_0 + \tfrac{1}{2}\rangle \pm |S, M_S, \tilde{S}_2 = S_0 + \tfrac{1}{2}\rangle) \tag{4.24}$$

one can obtain the vibronic Hamiltonian for all the spin states, analogous to Eq. (2.9). For an arbitrary dimer it reads

$$\hat{H}_S = [\tfrac{1}{2}Q_-^2 + T_n(Q_-) - JS(S+1)]\hat{I} + \begin{pmatrix} \Delta_S & AQ_- \\ AQ_- & -\Delta_S \end{pmatrix} \qquad (4.25)$$

where

$$\Delta_S = w \frac{S + \tfrac{1}{2}}{2S_0 + 1} \qquad (4.26)$$

Equation (4.25) means that we again come to the pseudo Jahn–Teller problem with the distinction that here the parameters of the problem are different for different spin states. While the vibronic constants were assumed to be the same for all the states, the energy gaps Δ_S are different. Besides, the energy read-off for the vibronic states of different spin states is different.

Consider first the adiabatic potential of the system. It follows from Eq. (4.25) that

$$\mathscr{E}_S(Q_-) = \tfrac{1}{2}Q_-^2 - JS(S+1) \pm (\Delta_S^2 + A^2 Q_-^2)^{1/2} \qquad (4.27)$$

From Eq. (4.26) it is seen that a situation can arise when the condition of instability at $Q_- = 0$ of the type shown in Eq. (2.12),

$$|\Delta_S| < A^2 \qquad (4.28)$$

is valid for some of the spin states and invalid for others. This means that the excess electron can be localized in some of the spin states with a lower spin and delocalized in other ones having a higher spin. Since for a two-minima potential the minima depths are larger for the states with a lower spin, we can state that the vibronic interaction weakens the ferromagnetic properties.

If the condition (Eq. 4.28) is valid for all the states, then from Eq. (4.27) one can deduce the energies of the minima of the lower curve of the adiabatic potential:

$$\mathscr{E}_S(Q_0^S) = -\frac{1}{2}A^2 - \frac{1}{2}\frac{\Delta_S^2}{A^2} - JS(S+1) \qquad (4.29)$$

Note that for the differences of the minima energies,

$$\mathscr{E}_S(Q_0^S) - \mathscr{E}_{S-1}(Q_0^{S-1}) = -2S\left(J + \frac{w^2}{2A^2(2S_0 + 1)^2}\right) \tag{4.30}$$

the Lande rule of energy intervals is valid. Based on this formula a conclusion was made[39] that the dimers with a double-minimum potential behave as HDVV dimers with a modified exchange integral

$$\bar{J} = J + \frac{w^2}{2A^2(2S_0 + 1)^2} \tag{4.31}$$

This expression can be used for analysis of magnetic properties in the case of strong vibronic coupling, provided the A values are not very small.[51]

For a more rigorous description of the magnetic properties the spectrum of the spin-vibronic states has to be calculated. For these states, as in Eq. (2.16), we have

$$\Psi_n^{+S} = \phi_+^S \sum_{\nu = 0,2,4,\ldots} \alpha_{n\nu}^S \chi_\nu(Q_-) + \phi_-^S \sum_{\nu = 1,3,5,\ldots} \alpha_{n\nu}^S \chi_\nu(Q_-)$$

$$\Psi_n^{-S} = \phi_+^S \sum_{\nu = 1,3,5,\ldots} \beta_{n\nu}^S \chi_\nu(Q_-) + \phi_-^S \sum_{\nu = 0,2,4,\ldots} \beta_{n\nu}^S \chi_\nu(Q_-) \tag{4.32}$$

The constants $\alpha_{n\nu}^S$ and $\beta_{n\nu}^S$ were obtained for each S value from the secular equation of the type (Eq. 2.17) in which w is substituted by Δ_S.

In Fig. 6 several calculated low-lying vibronic levels of a MV dimer with three unpaired electrons as a function of the vibronic constant A (for $J = 0$) are given.[52] As could be expected, the increase of vibronic interaction decreases the energy gap between the appropriate vibronic levels originating from the doublet and quadruplet electronic states, that is, it quenches the double exchange. Note that the vibronic function of the ground state at the right end of Fig. 6 is a quasi-localized one[53] since the energy separation characterizing the electron transfer probability is very small and any perturbation, always present in the crystal, localizes the electron [in this case the wave function becomes an adiabatic one, its electronic component being one of the functions (Eq. 4.5)]. Hence by spectral methods with a short enough "time of measurements" one can observe states of localized valencies. On the other hand the doublet–quadruplet splitting at this limit is not completely quenched. Thus the HDVV model of the energy spectrum has to be used with great care when describing magnetic characteristics of MV systems.

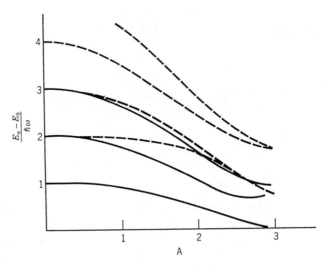

Figure 6. Spin-vibronic energy levels for a d^1-d^2 dimer with $w = 3$ and $J = 0$ as a function of the vibronic constant. The full lines are for $S = \frac{3}{2}$, while the dashed ones stand for $S = \frac{1}{2}$. The zero energy is taken at the lowest level with $S = \frac{3}{2}$.

In Fig. 7 the temperature dependence of the effective magnetic moment determined by the expression

$$\mu^2 = g^2 \beta^2 \frac{\Sigma_{S,n}\, S(S+1)(2S+1) \exp[-E_n(S)/\kappa T]}{\Sigma_{S,n}\, (2S+1) \exp[-E_n(S)/\kappa T]} \qquad (4.33)$$

is shown. It is seen that in the case of antiferromagnetic exchange the vibronic interaction may change the spin of the ground state. This phenomenon is similar to the well-known "crossover" one, the intersection and change of the spin state of the ground term as a function of the crystal field parameter in transition metal complexes.[54] Figure 8 shows that the spin transition is not catched by the semiclassical model of the energy spectrum based on the substitution of the exchange integral (Eq. 4.30).

In conclusion, let us discuss briefly the non-Hund states. By the Hamiltonian (Eq. 4.22) the adiabatic potentials of the dimer, including the interactions of the Hund and non-Hund states with $S = \frac{1}{2}$ can be obtained:

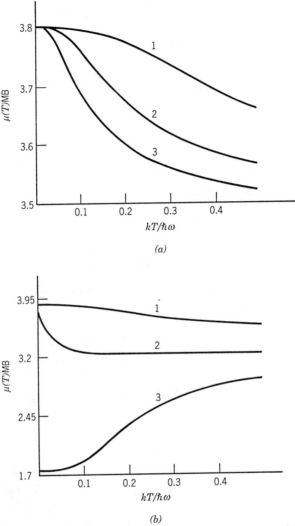

Figure 7. (a) Temperature dependence of the magnetic moment of a d^1–d^2 dimer with $J = 0.15$, $w = 1$; $1 - A = 0$, $2 - A = 1.8$, $3 - A = 3$. (b) The same as in (a) with $J = -0.15$, $w = 2$.

$$\mathscr{E}_{1-4}(Q_-) = \tfrac{1}{2}Q^2_- \pm [(J_0 - J)^2 + w^2 + A^2Q^2_-$$
$$\pm 2(J_0 - J)^2(\tfrac{1}{4}w^2 + A^2Q^2_-)^{1/2}]^{1/2} \qquad (4.34)$$

By means of expansion into a power series near the point $Q_- = 0$ one can be certain that for two curves for excited states (Fig. 9) there is a minimum

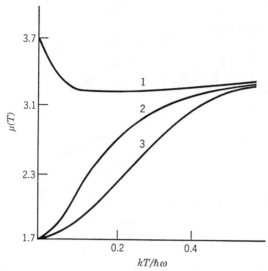

Figure 8. Temperature dependence of the magnetic moment of a d^1–d^2 dimer with $J = -0.2$, $w = 1$, and $A = 1.8$ obtained by: (1) calculations with spin-vibronic levels; (2) the HDVV model with J from Eq. (4.31); (3) without taking account of electron transfer.

at this point if $J_0 > w/2$ (delocalized distribution), while for the other two (including the ground state) alternative electron distribution, either localized or delocalized, is possible. The criterion of localization for the ground state reads as follows (see Eq. 4.35):

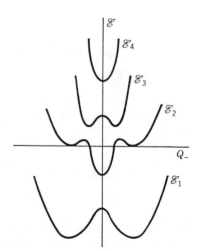

Figure 9. Adiabatic potentials for states with $S = \frac{1}{2}$ taking into account the interaction with the non-Hund states.

$$A^2 > [(J_0 - J)^2 + w^2 + (J - J_0)w]^{1/2} \left(2\frac{J_0 - J}{w} + 1\right)^{-1} \quad (4.35)$$

while for the excited one ($\mathscr{E}_3$ in Fig. 9)

$$A^2 > [(J_0 - J)^2 + w^2 - (J - J_0)w]^{1/2} \left(2\frac{J_0 - J}{w} + 1\right)^{-1} \quad (4.36)$$

Owing to the interaction with non-Hund states a situation is possible when one of the adiabatic potential curves for the excited state becomes a triminima one ($\mathscr{E}_2$ in Fig. 9).

The adiabatic potentials with more than two minima also appear in dimers with an even number of electrons, provided the interelectronic and vibronic interactions as well as the electron transfer are taken into account simultaneously.[55] In such systems a mixed-valence state is possible with valencies differing in two. The detailed vibronic theory of dimers with valency disproportionation analogous to a PKS model is worked out in Ref. 56.

In the next subsection the complications of the energy spectrum in exchange-coupled MV dimers are considered as influencing the spectra of intervalence transfer.

C. The Features of Optical Spectra

The energy spectrum of MV dimers is certainly manifested in their optical spectra. We consider first the influence of exchange effects on the intervalence transfer band because the latter is a distinguished feature of MV compounds.

To begin with we neglect the resonance splitting of all the spin states. In this case the nonadiabaticity effects vanish and the intervalence transfer band is produced by the transitions between the adiabatic states of each pair of parabolic adiabatic potentials, and these transitions are the same for different values S. The electronic matrix element of such a transition, $\langle S, M_S, \tilde{S}_1 | \mathbf{d} | S, M_s, \tilde{S}_2 \rangle$, where $\mathbf{d}$ is the operator of the dipole moment, is proportional to the coefficients of coupling between two schemes of momenta coupling that are different for different spin states. Because of different orderings of these in the cases of ferromagnetic and antiferromagnetic HDVV exchanges the temperature dependence of the transition intensities at the maximum of the band will be also different.[57] If one takes into account the double exchange and the inequality (Eq. 4.28) remains valid but the analysis of the band shape is carried out within the semiclassical approximation, all the features of the spectrum are again reduced to the temperature dependence of the maximum of the band.[39] Indeed, it can be deduced from Eq. (4.27) that the frequency of the Franck–Condon transition from the minima of the lowest sheet of the adiabatic potential to the upper one is the same for all the spin states, and it is equal to $2A^2$.

A more rigorous analysis of the optical spectrum requires its calculation based on the vibronic states (Eq. 4.32).[58] Consider the intervalence transfer band as an envelope of the individual vibronic transitions within the set of states with spins $\frac{1}{2}$ and $\frac{3}{2}$. The restriction of the spin values does not necessarily mean that we restrict the treatment by trielectron MV dimers (for more electrons states with higher spins are possible). The results obtained below are valid for any dimer with antiferromagnetic HDVV exchange, provided the lowest spin-vibronic energy levels with indicated spins $(\frac{1}{2}, \frac{3}{2})$ only are populated. As it follows from Eq. (4.20), for any core spin the splitting of the electronic levels with $S = \frac{3}{2}$ is twice that with $S = \frac{1}{2}$.

For electric dipole transitions considered here transitions between states having the same spin only $\Psi_n^{\pm S} \to \Psi_m^{\mp S}$ are allowed, their probability being

$$f_{n \to m}^s = (Z)^{-1}[\exp(-E_n(S)/\kappa T) - \exp(-E_m(S)/\kappa T)]$$
$$\times \sum_\nu (\alpha_{n\nu}^S \beta_{m\nu}^S)^2 |\langle \phi_+^S | \hat{d}_z | \phi_-^S \rangle|^2 \qquad (4.37)$$

where Z is the statistical sum. As shown in the PKS model,[11] the main contribution to the electronic matrix element in Eq. (4.37) comes from the on-center diagonal matrix elements, that is,

$$\langle \phi_+^S | \hat{d}_z | \phi_-^S \rangle \cong \frac{1}{2}(\langle S, \tilde{S}_1 = S_0 + \frac{1}{2} | \hat{d}_z | S, \tilde{S}_1 = S_0 + \frac{1}{2} \rangle$$
$$- \langle S, \tilde{S}_2 = S_0 + \frac{1}{2} | \hat{d}_z | S, \tilde{S}_2 = S_0 + \frac{1}{2} \rangle \qquad (4.38)$$

(we omitted here the unimportant dependence of the wave functions on M_S). The contribution of the nondiagonal (with respect to the two centers) matrix element can be essential but in the limit of the nonmigrating electron, that is, in the absence of double exchange.

Provided Eq. (4.38) is valid, the same result for the electronic matrix element, as in the PKS model, can be obtained,

$$\langle \phi_+^S | \hat{d}_z | \phi_-^S \rangle = \frac{eR}{2} \qquad (4.39)$$

where R is the distance between the two centers.

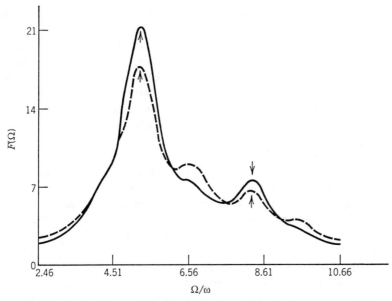

Figure 10. Intervalence transfer band for the d^1–d^2 dimer with $J = -0.15$, $w = 2$, $A = 2$, $kT/\hbar\omega = 0.1$ (full line) and 0.3 (dashed line). The arrows indicate the transitions for $S = \frac{3}{2}$.

To construct the band shape the lines of individual transitions obtained from Eq. (4.37) are substituted by Gauss-shaped bands having a half-width of the order of 1 (i.e., $\hbar\omega$). Approximately the same half-widths of elementary bands were used for adjusting the results otained in the PKS model to the experimentally observed optical spectra of Ruthenium dimers.[59]

In Fig. 10, the calculated spectra of intervalence transfer for different values of the exchange integral and temperature are given. The parameters of electron transfer and vibronic interactions are taken to satisfy the criterion (Eq. 4.28) for both spin states. It is seen that the spectra have a multiplet structure, that is, the components of different spin states are present. In the PKS model it is shown[12] that the intervalence transfer band is the more intensive, the stronger the intercenter interaction w. In our case the increase of w leads to an increase of the intensity of the component with greater Δ_S values, that is, with greater spin values (provided the population of different spin states is the same). To observe this effect experimentally the smaller value of the vibronic matrix element of the transition for states with $S = \frac{1}{2}$ has to be compensated for by their slightly larger population, and this is possible only in the case of antiferro-magnetic exchange.

In the studies cited[39,57] an increase of the maximum intensity of the band with temperature is predicted. As seen from Fig. 10, this prediction is not confirmed. In fact the temperature behavior of the intervalence transfer band is determined by the ordering of spin-vibronic energy levels. In particular, if the experiment is performed within a limited interval of temperatures, so that only one pair of vibronic energy levels of a given S value is populated, then the maximal intensity of the band increases with temperature only for the ground state $S = \frac{1}{2}$.

So far we have neglected the interaction with the non-Hund states. On the other hand, these distort the adiabatic potential (Fig. 9) and hence the band shape. Consider the latter in the semiclassical approximation often used in the calculation of vibronic spectra.[60] The conditions of applicability of this approximation are (1) high enough temperatures and (2) strong vibronic coupling, in our case $A^2 \gg \Delta_S$. The latter condition is close to that resulting in the potentials of Fig. 9.

In the semiclassical approximation the form–function for the part of the spectrum generated from the states with $S = \frac{1}{2}$ is given by the following expression:

$$F(\Omega) = \sum_{i=2,3} \left[\int dQ_- \exp(-\mathscr{E}_1(Q_-)/\kappa T) |\langle \phi_1(Q_-) | \hat{d}_z | \phi_i(Q_-) \rangle|^2 \right.$$
$$\left. \times \delta(\mathscr{E}_i - \mathscr{E}_1 - \hbar\Omega) \middle/ \left[\int dQ_- \exp\left(-\mathscr{E}_1(Q_-)/\kappa T\right) \right] \right. \qquad (4.40)$$

where the $\mathscr{E}_i(Q_-)$ are given by Eq. (4.34), the transition to the curve 4 being forbidden, and the wave functions $\phi_i(Q_-)$ are eigenfunctions of the operator $\hat{H}(S = \frac{1}{2}) - \hat{T}_n(Q_-)$ with $\hat{H}$ given by Eq. (4.22). The integration in Eq. (4.40) yields

$$F(\Omega) = w^2 \exp\left\{ \left[A^4 + \frac{w^2}{4} - (\kappa - A^2)^2 \right] \middle/ 2A^2\kappa T \right\} \middle/$$
$$4\mathscr{H}A(4\kappa^2 - w^2)^{1/2} \int dQ_- \exp\left\{ \left[Q_-^2 - 2\left(\frac{w^2}{4} + A^2Q_2^2\right)^{1/2} \right] \middle/ 2\kappa T \right\}$$
$$\qquad (4.41)$$

where

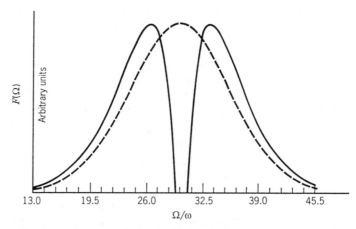

Figure 11. Intervalence transfer band with (——) and without (– – –) taking account of the interaction with the non-Hund states.

$$\kappa = [\Omega^2 - 2\Omega(J_0 - J) - 3w^2/4]/[(2\Omega - 4(J_0 - J)] \qquad (4.42)$$

Here the interaction between the most distant curves 1 and 4 is neglected. If we assume that $\kappa = \Omega/2$, then Eq. (4.41) results in the same Gauss-shaped optical band as in the PKS model,[61] provided it is taken into account also that ϵ^{61} is equivalent to the parameter of resonance splitting of the $S = \frac{1}{2}$ state, $w/2$. In the case considered in this section, κ has a more complicated frequency dependence (Eq. 4.42). Besides, this parameter should be positive since it substitutes Q_-^2 in the integration with the δ function in (Eq. 4.41), and this condition creates a forbidden gap in the frequency spectrum (Fig. 11). In principle, this effect can occur in any part of the spectrum, but it is most observable when falling near the band maximum. Since the interaction with non-Hund states results in a repulsion of the vibronic energy levels near the quasi-crossing of the adiabatic potentials, the region of lowered intensity has to be present also in the full quantum–mechanical calculations.

In general, it has to be noted that both the magnetic and spectral characteristics of exchange-coupled MV dimers are determined by the parameters of electronic and vibronic structure. Therefore, it seems incorrect to extract the exchange parameters separately from the magnetochemical data, while the electron transfer parameters and vibronic constants are determined from the optical spectra; both types of experimental data have to be analyzed simultaneously.

From other possible manifestations of the exchange coupling in MV dimers we mention here the doubling of the hyperfine structure in the

Mössbauer spectra.[62,63] The Mössbauer spectroscopy is widely used in the investigation of MV compounds, mostly ones containing iron. However, the Mössbauer spectra of MV dimers in the case of electron delocalized states are not very informative with respect to the electron transfer parameters. Indeed, at low enough temperatures, when the lowest two tunneling vibronic states $\Psi_0^{\pm S}$ are populated, the magnetic fields on the nucleus generated by these two states are the same (for the same S and fixed M_s values $\langle S_2 \rangle$ is the same) and hence the tunneling splitting characterizing the electron transfer will not be manifest in the hyperfine splitting. The situation can be changed, if one applies an external magnetic field that admixes the next excited states with a spin $S + 1$ (created by the HDVV interaction) to the ground ones. Since the tunneling splitting is different for different spin states, the amount of the above admixture is different for Ψ_0^{+S} and Ψ_0^{-S}. This effect becomes even more essential, if one takes into account that the ordering of even and odd states alternates for consequent spin values.[46]

It follows that in an external magnetic field the two tunneling states, + and −, produce different hyperfine structure, and hence the latter will be doubled. In this case the tunneling spliting can be deduced from the relative intensities of the components of the two hyperfine spectra and their temperature dependence. This effect is possible if the HDVV exchange coupling is small enough. The most suitable systems for its observation seem to be the dimers of Fe(III)–Fe(IV),[64] demonstating electron delocalized states in the Mössbauer spectra and having small HDVV exchange interactions.

D. Kinetics of Electron Transfer

One of the most interesting applications of the PKS model is the calculation of the probability of the intercenter electron transfer. In this subsection we consider the innovations to the kinetics of this process introduced by the effect of exchange coupling.

There are many different methods of calculation of the electron transfer, many of them using as a starting point the Fermi "golden rule," that is, the perturbation theory for transition probabilities. We employ here the Weiner method[65] used earlier in the PKS model,[12] which allows us to overcome the limits of weak intercenter interaction; as in the PKS model, we can separate the effects of exchange coupling neatly.[66]

Assume that at the initial moment the system is localized at one of the minima of the double-minimum potential that is realized for all the spin states. Weiner obtained an approximate relation between temperature-averaged probability of electron transfer between the two minima in an arbitrary double-minimum potential and the energy spectrum of the system with this potential. According to the study[65] the transition frequency is given by the following temperature-averaged magnitude:

$$P = \sum_{n=0}^{\infty} p_n \exp(-E_n/\kappa T) \Big/ \sum_{n=0}^{\infty} \exp(-E_n/\kappa T) \qquad (4.43)$$

where p_n is a function of the tunneling splitting of the nth vibronic energy level ΔE_n

$$p_n = \frac{\pi \omega}{2} \left(\frac{\Delta E_n}{\hbar \omega} \right)^2 \qquad (4.44)$$

In the case under consideration the summation in Eq. (4.43) has to be carried out over all the spin multiplets and over all the vibronic states of a given spin state,

$$P = \frac{\pi \omega}{2} \sum_{S} \sum_{n=0}^{\infty} \left(\frac{\Delta E_{Sn}}{\hbar \omega} \right)^2 \exp(-E_n(S)/\kappa T) \Big/ \sum_{S} \sum_{n=0}^{\infty} \exp(-E_n(S)/\kappa T)$$

$$(4.45)$$

where

$$\Delta E_{Sn} = E_n^+(S) - E_n^-(S) \qquad (4.46)$$

and $E_n(S)$ is the energy level to which the two levels $E_n^+(S)$ and $E_n^-(S)$ tend at $w \rightarrow 0$. Consider first the case of strong vibronic coupling for which the analytical expressions for tunneling vibronic energy levels can be obtained within the perturbation theory, using as a basis the electron-vibrational wave functions of displaced oscillators.[16] The appropriate expressions can be generalized in order to take into account the difference in the positions of the center of gravity and of the splittings of different spin multiplets. Then we have

$$E_n^{\pm}(S)/\hbar \omega = -JS(S+1) + n + \frac{1}{2} - \frac{A^2}{2} \pm (-1)^n \Delta_s e^{-A^2} L_n^0(2A^2) \qquad (4.47)$$

where $L_n^0(x)$ is a Laguerr polynomial. Substituting Eq. (4.47) into Eq. (4.45) and using the formulas for power series of Laguerr polynomials[67] we obtain

$$P = 2\pi^2 \omega F_M I_0 \left(\frac{2A^2}{\sinh x}\right) \exp\left(-2A^2 - e^{-x}\frac{2A^2}{\sinh x}\right) \quad (4.48)$$

where $x = \hbar\omega/2\kappa T$, $I_0(y)$ is the modified Bessel function, and

$$F_M = \sum_S \Delta_s^2 \exp(-JS(S+1)/\kappa T) \bigg/ \sum_S \exp(-JS(S+1)/\kappa T) \quad (4.49)$$

Equation (4.48) for the temperature-averaged tunneling rate has been obtained earlier by different authors,[68,69] but without the factor F_M. The latter thus characterizes the dependence of the frequency of electron transfer on the magnetic exchange interaction in the system. In the magnetochemistry of homovalent exchange-coupled clusters there are examples when the structure of the bridge between the metal atoms of two systems is the same, but the value of the exchange integral varies and even changes its sign dependent on the out-of-sphere ligands.[70] Assume that there are two MV systems of the same metal with approximately the same absolute values of their exchange integrals having opposite signs. Then for the ratio of the electron transfer frequencies for these two systems at low temperatures we have

$$P_F/P_{AF} = (2S_0 + 1)^2 \quad (4.50)$$

It follows that in these cases ceteris paribus for a Ni(II)–Ni(I) pair the appropriate rates of electron transfer in the ferromagnetic and antiferromagnetic cases are different four times, while for the Fe(III)–Fe(II) pair the difference is 36 times.

To calculate the rate of electron transfer in the more general cases the wave functions (Eq. 4.32) have to be employed. In Fig. 12 the temperature dependence of the probability of electron transfer obtained in this way for the cases when up to high temperatures the states with $S = \frac{1}{2}$ and $S = \frac{3}{2}$ only are populated and for different values of the exchange parameters are shown. In most cases this dependence has a characteristic form with a transition from the regime of tunneling to the one of temperature activation. For small J values the curves for P_F and P_{AF} practically coincide due to the fact that in this case intensive transfer makes the ferromagnetic state lowest independent of the sign of J. In the case when the ground vibronic doublet for $S = \frac{1}{2}$ is not very much higher from the same doublet for $S = \frac{3}{2}$, an unusual temperature dependence of the electron transfer occurs when it decreases with temperature.

For large $|J|$ values P_F and P_{AF} at the low-temperature limit differ

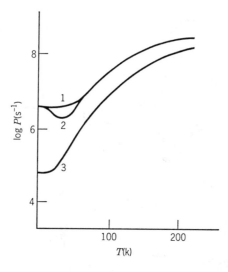

Figure 12. Temperature dependence of the probability of electron transfer at $w = 2$, $A = 3$, $\hbar\omega = 500\,\mathrm{cm}^{-1}$. (1) $J = 0.3$; (2) $J = -0.2$; (3) $J = -0.3$.

much more than follows from Eq. (4.50). Apparently, for $S_0 > \frac{1}{2}$ this effect has to be even more significant.

For experimental verification of the effects considered in this subsection one has to study the regularities that take place in series of similar compounds which differ by the value of the exchange interaction. As a source of information about electron transfer rates various spectral methods, such as Mössbauer spectra, ESR, and so on can be employed.

V. ELECTRON DELOCALIZATION IN TRICENTER MV COMPOUNDS

Tricenter MV clusters exist for many transition metals. There are analogous systems also among ion-radical molecules.[71-73] The presence of three centers results in the fact that the energy spectrum contains at least three energy levels and this, as will be shown below, completely changes the main ideas about the electron delocalization worked out for two-center MV systems.

A. Tricenter Clusters as Equilateral Triangles

Tricenter clusters in which the metal ions form an equilateral triangle are most widespread [e.g., see the well-known series of compounds of the type $[M_3O(RCOO)_6L_3$ (Fig. 13), as well as other molecular systems[74-76]]. Apparently, in its simplest version the theory of such a system has to be based on assumptions similar to that used in the PKS theory for dimers.

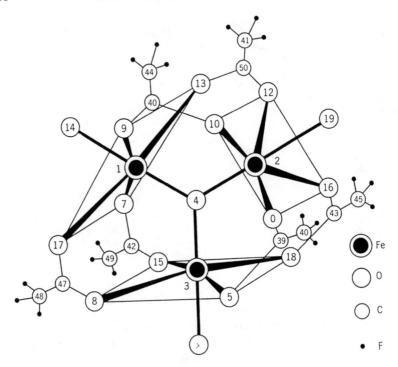

Figure 13. Structure of an MV trimer.

Already the semiclassical version of the theory applied to these system yields some new effects[77-79] considered below.

Consider the behavior of one excess electron that can localize at the ith center, resulting in the multielectron state of the system as a whole ϕ_i. Taking into account the interaction of this electron with the totally symmetric nuclear displacements of the near-neighbor environment of each center Q_i we can, analogously to Eq. (2.4) and neglecting the frequency effect, present the expression for the operator of the potential energy of a MV system with symmetry axis of the third order, as follows:

$$\hat{H}_p = w \begin{pmatrix} 0 & 1 & 1 \\ 1 & 0 & 1 \\ 1 & 1 & 0 \end{pmatrix} + A \begin{pmatrix} Q_1 & 0 & 0 \\ 0 & Q_2 & 0 \\ 0 & 0 & Q_3 \end{pmatrix}$$
$$+ \tfrac{1}{2}(Q_1^2 + Q_2^2 + Q_3^2)\hat{I} \tag{5.1}$$

Let us pass from ϕ_i to symmetrized wave functions ϕ_Γ transforming after the totally symmetric representation $\Gamma = A$ (not to be confused with the vibronic constant) and the twofold degenerate one $\Gamma = E$ of the point-symmetry group of the system

$$\phi_A = (1/\sqrt{3})(\phi_1 + \phi_2 + \phi_3)$$

$$\phi_{E\theta} = (1/\sqrt{6})(2\phi_1 - \phi_2 - \phi_3) \qquad (5.2)$$

$$\phi_{E\epsilon} = (1/\sqrt{2})(\phi_2 - \phi_3)$$

and to the same symmetrized nuclear displacements:

$$Q_A = (1/\sqrt{3})(Q_1 + Q_2 + Q_3)$$

$$Q_{E\theta} = (1/\sqrt{6})(2Q_1 - Q_2 - Q_3) \qquad (5.3)$$

$$Q_{E\epsilon} = (1/\sqrt{2})(Q_2 - Q_3)$$

In the new basis and new coordinates the operator $\hat{H}_p$ reads

$$
\hat{H}_p = w \begin{pmatrix} 2 & 0 & 0 \\ 0 & -1 & 0 \\ 0 & 0 & 0 \end{pmatrix}
$$

$$
+ \frac{A}{\sqrt{3}} \begin{pmatrix} Q_A & Q_{E\theta} & Q_{E\epsilon} \\ Q_{E\theta} & Q_A + \dfrac{1}{\sqrt{2}} Q_{E\theta} & -\dfrac{1}{\sqrt{2}} Q_{E\epsilon} \\ Q_{E\epsilon} & -\dfrac{1}{\sqrt{2}} Q_{E\epsilon} & Q_A - \dfrac{1}{\sqrt{2}} Q_{E\theta} \end{pmatrix}
$$

$$
+ \frac{1}{2}(Q_A^2 + Q_{E\theta}^2 + Q_{E\epsilon}^2) \qquad (5.4)
$$

This Hamiltonian describes two electronic levels, a singlet and a doublet, interacting with the totally symmetric A and twofold degenerate E vibrations. The latter are both active, mixing the states within the electronic E doublet (the Jahn–Teller $E * e$ problem) and mixing the electronic singlet A with the doublet E [the pseudo Jahn–Teller $(A + E) * e$ problem], note that the vibronic constant of the A–E mixing is $\sqrt{2}$ larger than for the E–E one.[77] Thus, distinguished from the case of the dimer where the vibronic problem is reduced to the pseudo Jahn–Teller one, $(A + B) * b$, in the case of the trimers under consideration the problem is a superposition of the Jahn–Teller and pseudo Jahn–Teller ones. Note that an analogous situation arises in the widely discussed problem of the origin of the excited state of the complex $Ru(bpy)_3^{2+}$.[79]

As in the case of the dimer, the interaction with the totally symmetric vibration can be separated by an appropriate choice of their read-off. In the space of the remaining displacements Q_θ and Q_ϵ, the adiabatic potentials can be presented by the expression

$$\mathscr{E}_i(Q_\theta, Q_\epsilon) = \tfrac{1}{2}(Q_\theta^2 + Q_\epsilon^2) + \epsilon_i(Q_\theta, Q_\epsilon) \tag{5.5}$$

where ϵ_i are the eigenvalues of the electronic part of the Hamiltonian (Eq. 5.4). To analyze the shape of these surfaces it is convenient, as usual in the vibronic problems with E vibrations (Eq. 5.7), to pass to polar coordinates

$$Q_\theta = \rho \cos \varphi, \qquad Q_\epsilon = \rho \sin \varphi \tag{5.6}$$

Then $\epsilon_i(\rho, \varphi)$ are the roots of the secular equation[78]

$$\epsilon^3 - \left(\tfrac{1}{2}A^2\rho^2 + 3w^2\right)\epsilon - 2w^3 - \frac{\sqrt{2A^3\rho^3}}{\sqrt{3}}\cos 3\varphi = 0 \tag{5.7}$$

Since $Q_\theta^2 + Q_\epsilon^2 = \rho^2$ is independent of φ, the adiabatic potentials $\mathscr{E}_i(\rho, \varphi) = \rho^2/2 + \epsilon_i(\rho, \varphi)$, which follow from Eq. (5.7), are periodic functions of φ with a period of $2\pi/3$. It can also be shown that the extrema points of the adiabatic potential are at $\varphi = \pi n/3$, $n = 0, 1, \ldots, 5$. If the vibronic constant $A > 0$, then for the lowest sheet of the adiabatic potential the even values of n correspond to maxima along φ (and saddlepoints, in general; see below), while the odd ones give minima, and vice versa for $A < 0$.

Consider the radial dependence of the adiabatic potential in the extremal cross section $\varphi = 0$ (or, equivalently, $Q_\epsilon = 0$). The solutions (Eq. 5.7) in this case are

$$\epsilon_1 = -(1/\sqrt{6})A\rho - w \tag{5.8}$$

$$\epsilon_{2,3} = \tfrac{1}{2}[w - (1/\sqrt{6})A\rho \pm 3(\tfrac{1}{6}A^2\rho^2 - 2w\rho/3\sqrt{6} + w^2)^{1/2}]$$

Further investigation of this expression requires the knowledge of the sign of the parameter of intercenter electron transfer w. While in the case of dimers the sign of w is unimportant, for the trimers it is very essential. Different possibilities of realization of both positive and negative w values are considered in Ref. 77.

If $w > 0$, the electronic doublet is lowest. Its adiabatic potential shape is then similar to that in the Jahn–Teller $E * e$ problem, taking account of the quadratic terms of the vibronic interaction.[5,7] However, in the case at hand the appropriate warped adiabatic potential is obtained as a result of simultaneous Jahn–Teller and pseudo Jahn–Teller distortion already in the linear approximation with respect to vibronic interaction terms. The three minima of the adiabatic potential describe the three possibilities of localization of the excess electron at each of the three centers. Since there is no minimum at $Q_\theta = 0$ (where the three centers are equivalent), it follows that for $w > 0$ delocalized electron distributions are not possible in the trimer MV compounds under consideration.

The case of $w < 0$ (negative value of the parameter of intercenter electron transfer) seems to be more interesting. Indeed, in this case the ground electronic term is the singlet one, and the adiabatic potential shape, as in the case of the dimer, is completely determined by the parameter $|w|/A^2$. Performing the substitution

$$\rho = \sqrt{\frac{2}{3} \frac{|w|}{A}} (2\sqrt{2} \sinh t - 1) \tag{5.9}$$

we obtain the following transcendent equation for the extrema points of the adiabatic potential:

$$4\sqrt{2} \sinh 2t + \left(1 - \frac{|w|}{A^2}\right) \cosh t - 3 \sinh t = 0 \tag{5.10}$$

Figure 14 illustrates the calculated shapes of the extremal cross sections of the adiabatic potential for different $|w|/A^2$ values. Its behavior at $\rho = 0$ can be investigated by means of the expansion of the potential function into a power series of S keeping the terms up to ρ^2. If $|w|/A^2 < \frac{2}{9}$, the point $\rho = 0$ is a local maximum, and the potential has three absolute minima at $\rho \neq 0$. The other minimum at $\rho \neq 0$ on Fig. 14a (and two other equivalents at $\varphi = 2\pi/3$, and $4\pi/3$) are in fact saddlepoints. If $|w|/A^2 > \frac{2}{9}$, a minimum occurs at $\rho = 0$, but this is not necessarily accompanied by the disappearance of the minima at $\rho \neq 0$. In the interval $\frac{2}{9} < |w|/A^2 < 0.255$,

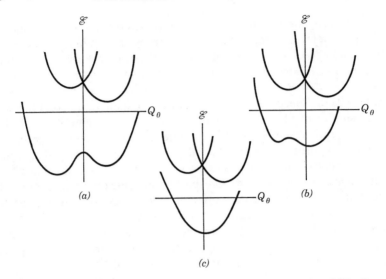

Figure 14. Extremal cross sections of the adiabatic potentials of an MV trimer. (*a*) $|w|/A^2 < 2/9$, (*b*) $2/9 < |w|/A^2 < 0.255$, (*c*) $|w|/A^2 > 0.255$.

both types of minima coexist. Finally, when $|w|/A^2 > 0.255$, the only minimum exists at $\rho = 0$.

Remembering that the minimum at $\rho = 0$ describes the state of the excess electron uniformly delocalized over the three centers, whereas the minima at $\rho \neq 0$ correspond to electron-localized (at one of the three centers) states, we come to the conclusion that for tricenter MV compounds in a certain interval of parameter values a coexistence of localized and delocalized electron distributions is possible. The region of parameter values required for this coexistence of two alternative electron distributions is very small. As stated in Ref. 81 this region increases when the frequency effect (the difference in the frequencies of the local one-center totally synmmetric vibration in two valence states) is taken into account. The coexistence localized and delocalized electron distributions were observed first in a series of compounds of the type $[Fe(II)Fe_2(III)(CH_3COO)_6L_3]$.[82–85] In these works it is shown that the Mössbauer spectra of these MV compounds, along with the lines corresponding to the Fe(II) and Fe(III) ions having an intensity ratio $1:2$, also contain quadrupole doublets that are characteristic for iron ions in the intermediate oxidation state.

The MV systems considered in this subsection form, as a rule, molecular crystals in which the interaction between the molecules depends on the intramolecular electron distribution and hence nuclear configuration distortions. It follows that the cooperative properties of MV compounds in the crystal state depend on their electron localization or delocalization state. The potential curves of the type, considered above, were used to analyze the possible types of phase transitions in crystals of MV trimers.[86–87] It was shown that there may be three types of such phase transitions: (1) order–disorder transition owing to the ordering of the distortions produced by the excess electron localization. Note that here the excess electron remains localized in both phases; (2) static localized–delocalized transition for which the system at low temperatures coexists in the localized and delocalized states of the two types of minima, and at high temperatures it is delocalized coherently; (3) dynamical localization–delocalization transition in which the delocalization results from fast electron hopping and the distortion of each moelcule changes from static to dynamical.

The results obtained in this subsection are based on the assumption that there is only one excess electron above the three zero-spin cores of the three centers, and a question arises whether these results are valid for clusters containing Fe(II), Fe(III), and so on in the high-spin state. This topic is discussed in more detail in Section VI. For each value of the total spin there are several A and E energy levels. But if one takes into account the exchange anisotropy [i.e., the difference in the exchange parameters in the pairs Fe(II)–Fe(III) and Fe(III)–Fe(III)], then these levels are relatively displaced to each other, and it may be possible to consider separately one set containing the pair of A and E levels, as was done above. This possibility has to be considered additionally in each concrete case.

B.   Mössbauer Spectra

In the case when at low temperatures the MV system is in the electron-localized state and the tunneling is very small, the electron transfer can take place with a certain rate at higher temperatures due to temperature-activated transitions over the barrier between the minima. The possibility to check these transitions and hence to observe the averaged valence of the centers with delocalized electrons depends on the means of observation, that is, on the "time of measurement;" it has to be smaller than the lifetime of the excess electron at each center. In this subsection we consider the process of valency averaging in MV trimers of iron centers as observed by Mössbauer spectra.

One of the most convenient systems to study the dynamic delocalization of the electron in Mössbauer spectra are the basic iron carboxylates (Fig. 13). At low temperatures their spectra represent a superposition of partial spectra of iron ions in different valence states, while at high enough temperatures an averaged picture is seen.[88,89] Since the quadrupole splitting for Fe(II) is approximately twice as that for Fe(III) (this ratio is determined from the spectra at low temperatures), their gradients of the electric field having opposite signs, it is expected that at high temperatures the quadrupole splitting taking account of the natural line width will not be observed at all. However, this expectation is not always confirmed experimentally; the averaged spectrum at room temperatures is a well-resolved doublet.

The minima of the adiabatic potential obtained in the previous subsection are defined in the space of some normal coordinates of the E type. Their evaluation requires the knowledge of the vibrational frequencies and vibronic constants. Appropriate full calculations for large systems under consideration encounter great difficulties. In general we can assume that the transition between the minima is accompanied by both electron charge redistribution and nuclear configuration distortion resulting in the change of both absolute parameter values and main axes of the tensor of electric field gradient. This circumstance has to be taken into account in the theory of Mössbauer spectra for the MV systems under consideration.[90]

To calculate the line shape of absorption of γ-quanta by the Mössbauer nucleus of the trimer MV cluster, it is convenient to use the nonsecular stochastic theory of line shapes.[91] This approach takes into account the changes of the axes of quantization that accompany the fluctuation of the hyperfine fields, and describes the whole region of rates of electron transitions.

The fluctuating Hamiltonian of the Mössbauer nucleus reads

$$\hat{H}(t) = \sum_{i=1}^{3} f_i(t)\hat{H}_i \qquad (5.11)$$

where the Hamiltonians $\hat{H}_i$ correspond to three possible positions of the electron in the trimer, and the coefficients $f_i(t)$ are expressed by a random function $f(t)$ that acquires the values 0, 1 and describes a stationary Markoff process:

$$\begin{aligned} f_1(t) &= 1 - f(t) \\ f_{2,3}(t) &= \tfrac{1}{2}f(t)[1 \pm f(t)] \end{aligned} \qquad (5.12)$$

Taking the energy read off at the undisplaced Mössbauer line for the Fe^{57} nucleus, one can write the Hamiltonians $\hat{H}_i$ as follows:

$$\hat{H}_i = \delta_i \hat{I} + \hat{H}_{qi} \tag{5.13}$$

where δ is the isomer shift, $\hat{I}$ is a unit matrix, and $\hat{H}_{qi}$ is the Hamiltonian of the quadrupole splitting when the electron is localized at the ith center. It can be presented as a scalar product of two tensors,

$$\hat{H}_{qi} = \sum_{m=-2}^{2} q^m B_i^{-m} \tag{5.14}$$

where q is the tensor of the quadrupole moment of the nucleus and B_i is the tensor of the gradient of the electric field at the ith nucleus. The components of these tensors in an arbitrary system of coordinates are

$$q^{\pm 2} = \frac{eq}{I(2I-1)} \frac{\sqrt{6}}{4} \hat{I}_{\pm}^2, \qquad q^0 = \frac{eq}{2I(2I-1)} (3\hat{I}_z^2 - I^2) \tag{5.15}$$

$$q^{\pm 1} = \frac{eq}{I(2I-1)} \frac{\sqrt{6}}{4} (\hat{I}_z \hat{I}_{\pm} + \hat{I}_{\pm} \hat{I}_z)$$

$$B^{\pm 2} = \frac{1}{2\sqrt{6}} (V_{xx} - V_{yy} \pm 2iV_{xy}), \qquad B^0 = \frac{1}{2} V_{zz} \tag{5.16}$$

$$B^{\pm 1} = \frac{1}{\sqrt{6}} (V_{xz} \pm iV_{yz})$$

where q is the value of the quadrupole moment of the nucleus and V_{ij} are the Cartesian components of the tensor of the electric field gradient.

With the Hamiltonian (Eq. 5.11) we obtain (further details of calculation are given in Ref. 90) the following expression for the probability of absorption of γ quanta with the frequency Ω by a polycrystalline sample containing MV iron trimers in an equilateral triangle configuration:

$$F(\Omega) \sim \text{Re} \sum_{m_0 m_1} |D_{m_0 m_1}|^2 \text{Re} \sum_{k,l=1}^{3}$$

$$\times \left\langle m_1, k \left| \left[\left(-i\Omega + \frac{\Gamma}{2} \right) \hat{I} - \hat{W} - i \sum_{j=1}^{3} \hat{H}_{qj} \times \hat{F}_j \right]^{-1} \right| m_1, l \right\rangle \tag{5.17}$$

Here $\hat{D}$ is the operator of the Mössbauer transition, m_0 and m_1 are the magnetic quantum numbers of the ground and excited states of the nucleus, respectively (for Fe^{57} $m_0 = \pm\frac{1}{2}$, $m_1 = \pm\frac{1}{2}$, $\pm\frac{3}{2}$), and Γ is the natural line width. The matrix elements of the $\hat{W}$ operator are given by the temperature-dependent probabilities of electron transfer between the centers p_{ij}

$$\langle m_1, k | \hat{W} | m_1', l \rangle = p_{kl} \delta_{m_1 m_1'} \qquad (k \neq l)$$

$$\langle m_1, k | \hat{W} | m_1', k \rangle = -\delta_{m_1 m_1'} \sum_{l \neq k} p_{kl} \qquad (5.18)$$

$\hat{F}_j$ is a 3×3 matrix with the only nonzero elements $\langle j | \hat{F}_j | j \rangle = 1$, the sign $\times$ in $\hat{H}_{qj} \times \hat{F}_j$ denoting the direct product.

Consider the case of D_{3h} symmetry, as in the basic iron acetates.[88,89] Then $p_{kl} = p$. Since the direction of the main axes of the electric field gradient are determined by the local symmetry, one of them is perpendicular to the plane of the triangle containing the metal ions, while at the electron-localized Mössbauer nucleus another main axis is along the one of twofold symmetry of the triangle. When the excess electron passes to one of the other centers the latter main axis rotates by an angle α; the electric field gradient on the remaining center can be easily obtained by symmetry considerations. Thus the reorientations of the main axis of the electric field gradient by the electron transfer are determined by one angle.

Equation (5.17) results in characteristic changes of the spectrum dependent on the electron transfer probability p. Similar changes are always present when there are fluctuations of the hyperfine interactions. At low temperatures (small p values) the Mössbauer spectrum is a superposition of the quadrupole spectra of Fe(II) and Fe(III), while at high enough temperatures (large p values) it changes to a single quadrupole doublet with averaged parameters. However, the quadrupole splitting of the latter depends not only on the ones for Fe(II) and Fe(III), but also on the angle α, as well as on the parameter of asymmetry of the electric field gradient $\eta = (V_{xx} - V_{yy})/V_{zz}$. The appropriate functions are illustrated in Fig. 15.

Thus the effective vibrational mode describing the electron localization in MV trimers results in complex structural distortions producing nontrivial fluctuations of the hyperfine field on the nuclei that are manifest in special temperature-dependent Mössbauer spectra.

C. Electron Delocalization in Linear MV Trimers

The cases of linear or quasilinear (isosceles triangle) tricenter MV systems are of interest from at least two points of view: (1) they really exist[92] and (2) they describe many cases of dimers in which it is necessary to consider

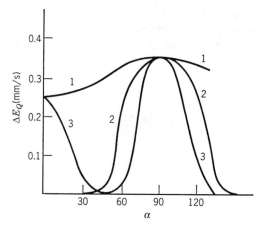

Figure 15. Averaged quadrupole moment as a function of α, the angle of the rotation of the system of principal axes of electric field gradient for different values of the asymmetry parameter of the divalent (η_1) and trivalent (η_2) iron ion. (1) $\eta_1 = 0.9$; $\eta_2 = 0.1$. (2) $\eta_1 = \eta_2 = 0.5$. (3) $\eta_1 = 0.1$; $\eta_2 = 0.9$.

the bridge ligand as a third center of electron localization.[23,93,94] Recently such tricenter MV compounds were considered as possible prototypes of molecular switches for molecular electronic devices.[4,95-97] The idea of using MV compounds as switches is based on the possibility of influencing selectively one of the cluster fragments that controls the electron transfer. One of such possibilities is provided by the tunneling microscope.[98] Other possibilities are based on changes of charge distributions on the near-neighbor molecular groups produced by induction influence.[99] It is assumed that these changes can produce a sharp enough change in electron transfer properties.

The switch of the charge distribution in MV systems can be of two types: (1) transition from localization at one center to that of another one, and (2) transition from the localized state to the delocalized one over all the centers. Provided the two terminal centers are equivalent, the first type of switching can be obtained by influencing one of them (in MV compounds with localized electron distribution), whereas the second one depends on the influence on the bridging center (fragment).

Consider the situation when the central (bridging) fragment is slightly different from the terminal ones. In particular, assume that the energy of the excess electron, when localized at the central fragment, differs by an amount of δ from that of the terminal one. The appropriate two vibronic constants are denoted by A_1 and A_2. The difference in the frequencies of totally symmetric vibration on these slightly different centers is neglected for the sake of simplicity; its inclusion does not essentially change the results.[99] If we neglect also the direct interaction between the terminal centers, the operator of potential energy, analogously to Eq. (5.1), is

$$\hat{H}_p = w \begin{pmatrix} 0 & 1 & 0 \\ 1 & 0 & 1 \\ 0 & 1 & 0 \end{pmatrix} + \begin{pmatrix} A_1 Q_1 & 0 & 0 \\ 0 & A_2 Q_2 + \delta & 0 \\ 0 & 0 & A_3 Q_3 \end{pmatrix}$$
$$+ \tfrac{1}{2}(Q_1^2 + Q_2^2 + Q_3^2)\hat{I} \tag{5.19}$$

In the normal coordinates

$$Q_{1+} = \tfrac{1}{2}(Q_1 + \sqrt{2}Q_2 + Q_3)$$
$$Q_{2+} = \tfrac{1}{2}(Q_1 - \sqrt{2}Q_2 + Q_3) \tag{5.20}$$
$$Q_- = \frac{1}{\sqrt{2}}(Q_1 - Q_3)$$

(the indices + and −, as above, determine the even and odd vibrations, respectively)

$$\hat{H}_p = \tfrac{1}{2}(Q_{1+}^2 + Q_{2+}^2 + Q_-^2)\hat{I}$$
$$+ \begin{pmatrix} A_1\left(\tfrac{1}{2}(Q_{1+} + Q_{2+}) + \tfrac{1}{\sqrt{2}}Q_-\right) & w & 0 \\ w & \tfrac{1}{2}A_2(Q_{1+} - Q_{2+}) + \delta & w \\ 0 & w & A_1\left(\tfrac{1}{2}(Q_{1+} + Q_{2+}) - \tfrac{1}{\sqrt{2}}AQ_-\right) \end{pmatrix}$$
$$\tag{5.21}$$

It can be seen that here, distinguished from many cases above, neither of the coordinates can be separated, and hence the problem is a trimode one. However, the two modes of the same symmetry can still be converted into one, using the method described in Ref. 99. After a shift transformation

$$q_{1+} = Q_{1+} + A_1/2$$
$$q_{2+} = Q_{2+} + A_1/2 \tag{5.22}$$
$$q_- = Q_-$$

we have

$$\hat{H}_p = [\tfrac{1}{2}(q_{1+}^2 + q_{2+}^2 + q_-^2) - \tfrac{1}{4}A_1^2]\hat{I}$$

$$+ \begin{pmatrix} \dfrac{A_1}{\sqrt{2}}q_- & w & 0 \\[2ex] w & \Delta - \dfrac{A_1}{2}(q_{1+} + q_{2+}) + \dfrac{A_2(q_{1+} + q_{2+})}{\sqrt{2}} & w \\[2ex] 0 & w & -\dfrac{A_1}{\sqrt{2}}q_- \end{pmatrix} \tag{5.23}$$

where

$$\Delta = \delta + \tfrac{1}{2}A_1^2 \tag{5.24}$$

Now we perform a transformation of rotation in the space of the coordinate (Eq. 5.22),

$$q_+ = \frac{1}{B}\left[\left(-\frac{A_1}{2} + \frac{A_2}{\sqrt{2}}\right)q_{1+} - \left(\frac{A_1}{2} + \frac{A_2}{\sqrt{2}}\right)q_{2+}\right]$$

$$q'_+ = \frac{1}{B}\left[\left(\frac{A_1}{2} + \frac{A_2}{\sqrt{2}}\right)q_{1+} + \left(-\frac{A_1}{2} + \frac{A_2}{\sqrt{2}}\right)q_{2+}\right] \tag{5.25}$$

where B has the meaning of the vibronic constant for the modified vibration

$$B = (\tfrac{1}{2}A_1^2 + A_2^2)^{1/2} \tag{5.26}$$

Finally we have

$$\hat{H}_p = [\tfrac{1}{2}(q_-^2 + q_+^2 + q_+'^2) - \tfrac{1}{4}A_1^2]\hat{I}$$

$$+ \begin{pmatrix} \dfrac{A_1}{\sqrt{2}}q_- & w & 0 \\ w & \Delta + Bq_+ & w \\ 0 & w & -\dfrac{A_1}{\sqrt{2}}q_- \end{pmatrix} \qquad (5.27)$$

Here the interaction with q_+' is separable, and for the adiabatic potential we obtain a two-mode problem

$$\hat{H}_p = \tfrac{1}{2}(q_-^2 + q_+^2)\hat{I} + \begin{pmatrix} Aq_- & w & 0 \\ w & \Delta + Bq_+ & w \\ 0 & w & -Aq_- \end{pmatrix} \qquad (5.28)$$

where a substitution $A = A_1/\sqrt{2}$ and an energy read-off shift are performed.

Consider the shape of the adiabatic potential surface for different parameter values. If $w = 0$, the potential has the form of three intersecting paraboloids with displaced minima positions. Two equivalent minima lay on the q_- axis at $q_- = \pm A$, their energy being $-A^2/2$. They describe the electron localization at the terminal centers. The third minimum is on the q_+ axis at $-B^2/2$. If $\Delta = A^2$ and $B = \sqrt{3}A$, the potential coincides with that of the Jahn–Teller $T * e$ problem[7] or, equivalently, with the potential obtained from Eq. (5.4) at $w = 0$.

When the intercenter interaction is taken into account, the sheets of the adiabatic potential split and the ground state one distorts, passing gradually (by increasing interaction) from the triminima surface to the double-minimum one and then to the one-minimum potential.

Consider now the variation in the excess electron distribution as a function of the parameters of the system. If the height of the barriers between the minima are high enough and the tunneling splitting is smaller than the possible random strain (in crystals) or other random deformations, the system is mainly localized at the bottom of the deepest minimum, and the distribution of the excess electron is well defined by the coefficients of its expansion over the electronic states of all the centers. Assume that this condition of high barriers is fulfilled. Then the region where all the important parameters have the same order of magnitude, $A \sim B$, $w \sim \Delta \sim A^2$, seems to be most interesting. Indeed, in this case all the minima have almost the same depths, the states with the localized and

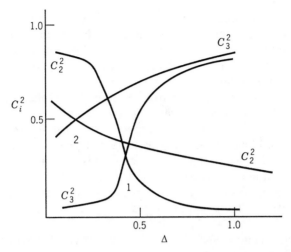

Figure 16. The probability of electron localization C_i^2 (C_i is the LCAO coefficient of the electronic wave function at the minimum of the adiabatic potential) as a function of the parameter Δ: (1) $B/A = \sqrt{3}$, $w/A^2 = 0.9$; (2) $B/A = \sqrt{3}/2$, $w/A^2 = 1.2$.

delocalized excess electron occur at close values of the parameters, and the localization–delocalization transition is sensitive to the variation of the Δ parameters.

Figure 16 illustrates such a transition as a function of the latter. Two qualitatively different situations are shown: (1) the transition from the localization at the central fragment (C_2^2) to the localization at the terminal centers (C_3^2), and (2) the transition from the delocalized (over the whole system) state to the localized one on the central fragment. In both cases at $\Delta \sim 0$, the lowest sheet of the adiabatic potential has one minimum on the q_- axis (the left-hand part of Fig. 16). When Δ increases, the energy of the minimum where the electron is localized at the central fragment increases and the influence of the electron transfer parameter w on the shape of the adiabatic potential decreases. As a result, its lowest sheet in the cross section along q_- changes from a parabolic curve to a double-minimum one with a hump at $q_- = 0$ (the right-hand part of Fig. 16). This transformation can be interpreted as due to the reduction of the effective value of w and hence the quenching of the barrier between the equivalent minima produced by the electron transfer. Thus the increase of the value of Δ increases the barrier between the minima and hence the localization of the excess electron at the terminal centers. Note that in the region of transition between the minima the semiclassical approximation used above is apparently not a good one. In this region there is a delocalization of the excess electron due to its strong tunneling between the equivalent minima. However, in the two limit cases of the left- and right-hand sides in Fig. 16 the criterion of the semiclassical approximation is fulfilled, and the electron distribution is well described by the squares of the appropriate coefficients of the adiabatic wave function.

Apparently, the region of transition from the localized to the delocalized electron distributions obtained above is most suitable for the realization of molecular switches. It is expected that a more exact quantum-mechanical calculation (not semiclassical) will produce a more smooth transition, but this will not change the conclusion that in the transition region, revealed above, the strongest changes of electron conductivity[100] and spectroscopic (especially in the region of the intervalence transfer band) properties of the system take place.

VI. THE VIBRONIC THEORY OF MV TRIMERS

So far we have considered mainly the adiabatic potentials of MV trimers that revealed a series of important results, especially those concerning the electron charge distribution. However, a more detailed determination of the observable characteristics of these compounds requires quantum-mechanical calculation of the vibronic states of the system with such adiabatic potentials. The first step to the solution of this problem is to build up a model for MV trimers analogous to the PKS one for dimers.

A. Vibronic States

For quantum-mechanical calculations of the vibronic states, the full Hamiltonian, including the kinetic energy of the nuclei, has to be employed. For a trimer in the form of an equilateral triangle it reads (see (Eq. 5.4):

$$\hat{H} = \frac{1}{2}\left[\left(-\frac{\partial^2}{\partial Q_\epsilon^2} - \frac{\partial^2}{\partial Q_\epsilon^2}\right) + (Q_\theta^2 + Q_\epsilon^2)\right]\hat{I}$$

$$+ w\begin{pmatrix} 2 & 0 & 0 \\ 0 & -1 & 0 \\ 0 & 0 & -1 \end{pmatrix} + \frac{A}{\sqrt{3}}\begin{pmatrix} 0 & Q_\theta & Q_\epsilon \\ Q_\theta & \frac{1}{\sqrt{2}}Q_\theta & \frac{-1}{\sqrt{2}}Q_\epsilon \\ Q_\epsilon & \frac{-1}{\sqrt{2}}Q_\epsilon & -\frac{1}{\sqrt{2}}Q_\theta \end{pmatrix} \quad (6.1)$$

This Hamiltonian is known in the theory of vibronic interactions as the one for the $(A + E) * e$ problem. Many studies are devoted to the solution of this problem,[101-104] but in most of these cases the mixing A and E

states are excited electronic ones of the system, and the goal of the calculation, as a rule, is to study the transition from the nondegenerate ground state to the A and E combinations. For not very large vibronic constants, this can be done by calculating several lowest-in-energy vibronic states originating from the electronic A and E levels.

However, for MV clusters under consideration the intervalence transfer band includes the transtitions within an essentially larger group of vibronic states of the A and E levels, and hence the larger region of the energy spectrum has to be determined.

Let us pass from the basis functions (Eq. 5.2) to a more convenient complex basis $\psi_\mu(\mu = 0, \pm 1)$:

$$\psi_0 = \phi_A, \qquad \psi_\pm = \frac{1}{\sqrt{2}}(\phi_\theta \pm i\phi_\epsilon) \qquad (6.2)$$

Then the part of the Hamiltonian (Eq. 6.1) describing the vibronic interaction is

$$H_{en} = \frac{A}{\sqrt{3}}\begin{pmatrix} 0 & Q_- & Q_+ \\ Q_+ & 0 & Q_- \\ Q_- & Q_+ & 0 \end{pmatrix} \qquad (6.3)$$

where

$$Q_\pm = \frac{1}{\sqrt{2}}(Q_\theta \pm iQ_\epsilon) \qquad (6.4)$$

The eigenfunctions of the Hamiltonian (Eq. 6.1) can be sought in the form of an expansion:

$$\Psi_n = \sum_{\mu,v,l} C^{(n)}_{\mu,v,l}|\mu, v, l\rangle \qquad (6.5)$$

over the wave functions of the Hamiltonian $\hat{H} - \hat{H}_{en}$

$$|\mu, v, l\rangle = \psi_\mu \chi_{v,l} \qquad (6.6)$$

where $\chi_{v,l}$ is the wave function of a two-dimensional isotropic harmonic oscillator. Substituting Eq. (6.5) into Eq. (6.1) we come to the following systems of linear equations for the C coefficients:

$$\sum_{\substack{\lambda \neq \mu \\ v',l'}} [\langle v, l | \hat{H}'_{\mu\lambda} | v', l' \rangle C^{(n)}_{\lambda,v',l'}] + [H'_{\mu\mu} + v + 1 - E^{(n)}] C^{(n)}_{\mu,v,l} = 0 \quad (6.7)$$

$$\mu = 0, \pm 1; \qquad v = 0, 1, 2, \ldots, \qquad l = -v, -v + 2, \ldots, v.$$

where $E^{(n)}$ is energy of the vibronic state and

$$\hat{H}' = w \begin{pmatrix} 2 & 0 & 0 \\ 0 & -1 & 0 \\ 0 & 0 & -1 \end{pmatrix} + \hat{H}_{en} \qquad (6.8)$$

The dimensionality of the secular equation for $E^{(n)}$ [from Eq. (6.7)] can be reduced by means of the group-theoretical classification of the vibronic states. Denote the quantum number of the electron-vibrational momentum by

$$j = l - \mu \qquad (6.9)$$

Presenting j in the form

$$\xi = 3m - \Lambda \qquad (6.10)$$

where m is an integer, one can see that the states with $\Lambda = 0$ belong to the representation A (A_1 or A_2), while those with $\Lambda = +1$ and $\Lambda = -1$ are, respectively, the E_+ and E_- components of the E representation of the C_{3v} group of the system.[102,103]

For practical reasons, we have to restrict the vibrational basis $\chi_{v,l}$ by some limit value $v_{\max}$. The adaptation of these functions to the irreducible representations allows us to decompose the secular equation matrix into smaller ones, and to diagonalize two matrices of the dimension $(v_{\max} + 1)(v_{\max} + 2)/2$ instead of one matrix with a three times larger dimension (the matrices for E_+ and E_- coincide). Symbolically these matrices can be written as follows:[102]

$$\hat{H}_\Lambda = \begin{pmatrix} 0 & & 0 & & \\ & \hat{h}(j-3,j-3) & \hat{h}(j-3,j) & 0 & \\ 0 & \hat{h}(j,j-3) & \hat{h}(j,j) & \hat{h}(j,j+3) & 0 \\ & 0 & \hat{h}(j+3,j) & \hat{h}(j+3,j+3) & \\ & & 0 & & \end{pmatrix}$$

$$(6.11)$$

The eigenfunctions of Eq. (6.11) are

$$\Psi_\Lambda^{(n)} = \sum_\mu \sum_{k=0,1} \sum_{m=0,\pm1,\dots} C_{\mu,m,k}^{(n)\Lambda} |\mu, |3m - \Lambda + \mu| + 2k, 3m - \Lambda + \mu\rangle$$

$$(6.12)$$

where the summation is limited by the condition

$$|3m - \Lambda + \mu| + 2k \le v_{\max} \qquad (6.13)$$

B. Intervalence Transfer Band

As in the case of dimers, the shape of the intervalence transfer band is presented as the envelope of the individual transitions between the vibronic states. Consider absorption of light polarized in the plane of the triangle of the MV trimer (σ_1 polarization). Then for the transitions between the vibronic states the following selection rules hold:

$$\Lambda = 0 \leftrightarrow \Lambda' = 1$$
$$\Lambda = -1 \leftrightarrow \Lambda' = -1 \qquad (6.14)$$

As in the case of dimers, we assume that the optical transitions take place as a result of the interaction with the electronic dipole moment $\hat{d}_{\mathrm{el}}$ only. In Ref. 13 the neglect of the vibrational dipole moment is based on the fact that the totally symmetric vibrations of the two subunits of the dimer are nonpolar ones. However, as indicated in Section IV.C, for a dimer (and a similar statement is true also for trimers) the electronic dipole moment of the transition is expressed by the dipole moment of the system as a whole in the state with the excess electron localized at one of the equivalent centers. Taking the origin of the coordinate system at the center of the equivalent triangle of the three equivalent centers, we have

$$|\langle\phi_i|\hat{\mathbf{d}}_{\mathrm{el}}|\phi_i\rangle| = e\left|\sum_{n=1}^{3}\sum_k \mathbf{r}_k^{(n)}\right| = \frac{eR}{\sqrt{3}} \qquad (6.15)$$

where $\mathbf{r}_k^{(n)}$ is the coordinate of the kth valence electron localized at the nth center, and R is the distance between the centers. In the region of the intervalence transfer band this matrix element is much larger than the matrix element of the nuclear dipole, even when the appropriate totally symmetric vibrations change the dipole moment. This is not true for the tunneling IR transition,[13] for which both the electronic and nuclear contributions have to be included.

The form–function of the shape of the intervalence transfer band can be presented as follows:

$$F(\Omega) = \frac{1}{Z} \sum_{\substack{\Lambda,\Lambda' \\ nn'}} \exp(-E_\Lambda^{(n)}/kT) |\langle \Psi_\Lambda^{(n)} | \hat{d}_+ | \Psi_{\Lambda'}^{(n')} \rangle|^2$$

$$\times f(\Omega, E_\Lambda^{(n)}, E_{\Lambda'}^{(n')}) \qquad (6.16)$$

where $E_\Lambda^{(n)}$ is the energy of the vibronic state $d_+ = d_x + id_y$ (the metal ions are in the xy plane), and the form–function for individual vibronic transitions $f(\Omega, E_\Lambda^{(n)}, E_{\Lambda'}^{(n')})$ is assumed to be of a Gaussian type. Further on

$$\langle \Psi_\Lambda^{(n)} | \hat{d}_+ | \Psi_{\Lambda'}^{(n')} \rangle = \sum_{\mu,\mu'} \sum_{m,m'} \sum_{k,k'} C_{\mu,m,k}^{(n)\Lambda} C_{\mu',m',k'}^{(n')\Lambda'} \langle \psi_\mu | \hat{d}_+ | \psi_{\mu'} \rangle \quad (6.17)$$

and for the electronic matrix element $\langle \psi_\mu | \hat{d}_+ | \psi_{\mu'} \rangle$ the same selection rules for μ as in Eq. (6.14) hold. Using Eqs. (5.2), (6.2), and (6.15), we obtain

$$|\langle \psi_0 | \hat{d}_+ | \psi_+ \rangle| = |\langle \psi_- | \hat{d}_+ | \psi_- \rangle| = \frac{eR}{\sqrt{3}}$$

Several possible types of intervalence transfer spectra for a variety of parameter values were calculated.[105] The number of vibrational quanta in the basis of the wavefunction expansion (Eq. 6.5) was taken equal to $v_{max} = 20$ (note that the number of basis states is much larger, it is equal to 231 for each of the A, E_+ and E_- representations). The validity of this basis dimension for concrete values of the vibronic constants A was determined by a comparison of the results for the value v_{max} with that of $v_{max} - 1$. This allows to show that the results are reliable for $v_{max} = 20$ when $A \lesssim 3$.

Let us discuss first the qualitative behavior of the spectrum for limit values of the parameters. If the electronic coupling between the centers is negligible ($w = 0$) the spectrum coincides with that expected from intersheet transitions in the Jahn–Teller $T * e$ problem.[7] In this case the adiabatic potential surfaces are presented by three intersecting (but not interacting) parabolloids, while the wavefunctions for each of them has a usual adiabatic form. The theory of the band shapes due to transitions between such surfaces is well known,[106] the band has a shape sometimes called "Pekarian" (after the name of its first author S. Pekar[107]). In the case when the three parabolloid minima are well displaced, which means strong localization of the excess electron, the band has a Gaussian shape, its parameters obeying the Hush relationships for dimers.[108]

In the opposite limit case of strong delocalization ($w \gg A^2$) the pseudo Jahn–Teller mixing of the A and E states can be neglected. Then the shape of the intervalence transfer band coincides with the one of the envelope of the $A \rightarrow E$ or $E \rightarrow A$ (dependent of the sign of w) optical transition in Jahn–Teller systems.[7,109] Depending on the magnitude of the vibronic constant A and the temperature, this band has either one or two maxima. This result is quite different from that for dimers in the delocalized state where the band is rather narrow and has no more than one maximum.[12,13]

In Fig. 17 the shape of the intervalence transfer spectrum is shown for different w values. The dependence of the type of spectrum on the sign of w is seen explicitly. For negative values the band narrows and its intensity increases by increasing the intercenter interaction. For a given A value the double-humped structure of the band takes place for w values close to that for which the minima of localized and delocalized electron distribution coexist. This does not mean that the two maxima of the band correspond to transitions from two minima of the adiabatic surface; in this region of transitions the semiclassical approximation is invalid, in principle. At $A = \sqrt{3}$ and large negative w values, the spectrum has one maximum; this result coincides with that of an appropriate Jahn–Teller system with weak vibronic coupling.[109] The double-humped structure of the band arises again when the vibronic constant A increases.

The intervalence transfer spectra with two maxima were observed in some iron MV trimers.[110]

C. Magnetic Properties

Another interesting use of the MV trimer vibronic states, obtained above, is in the calculation of their magnetic characteristics. Experimentally the latter were obtained in quite a number of works (e.g., see Refs. 89 and 111–113), but their interpretation is always carried out within the HDVV model without taking account of the vibronic interactions. The results of Section IV show that this approach may be incorrect.

As in the case of dimers, before taking into consideration the vibronic interaction, the energy spectrum itself has to be evaluated. There are several approaches to this problem. One is based on the construction of a basis of multielectron wave functions for a definite total spin and localized excess electron with the subsequent diagonalization of the matrix of

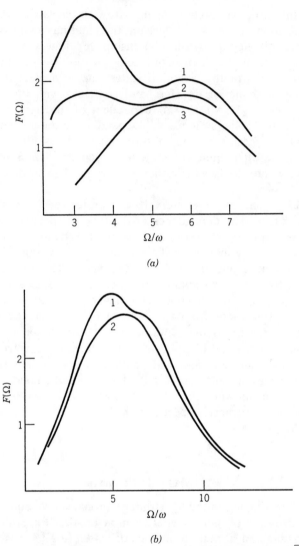

Figure 17. (a) The shape of the intervalence transfer band with $A = \sqrt{3}$ and positive w values: (1) $w = 0.7$; (2) $w = 0.5$; (3) $w = 0.3$. (b) The same as in (a) with negative w values: (1) $w = -0.7$; (2) $w = 0.3$.

the intercenter Coulomb interactions in this basis.[114] However, a simpler approach is based on the Anderson–Hasegawa Hamiltonian (Eq. 4.17) adopted for the case of trimers. For the latter the generalized Hamiltonian can be taken as follows:[115]

$$\hat{H}_{AH} = \begin{pmatrix} -2J_0\mathbf{S}_1\mathbf{s} - 2J_2\mathbf{S}_2\mathbf{S}_3 \\ 2J_1(\mathbf{S}_1 + \mathbf{s})(\mathbf{S}_2 + \mathbf{S}_3) & w\hat{I} & w\hat{I} \\ \\ w\hat{I} & \begin{array}{c} -2J_0\mathbf{S}_2\mathbf{s} - 2J_2\mathbf{S}_1\mathbf{S}_3 \\ 2J_1(\mathbf{S}_2 + \mathbf{s})(\mathbf{S}_1 + \mathbf{S}_3) \end{array} & w\hat{I} \\ \\ w\hat{I} & w\hat{I} & \begin{array}{c} -2J_0\mathbf{S}_3\mathbf{s} - 2J_2\mathbf{S}_1\mathbf{S}_2 \\ 2J_1(\mathbf{S}_3 + \mathbf{s})(\mathbf{S}_1 + \mathbf{S}_2) \end{array} \end{pmatrix}$$

$$(6.18)$$

where J_1 is the exchange integral for the interaction of the heterovalent ions, while J_2 is the one for homovalent ions without the excess electron, the remaining parameters having the same meaning as in Eq. (4.17).

The method of diagonalization of Eq. (6.18) is the same as that used above for Eq. (4.17). If we assume that Eq. (6.18) is written on the basis of states with certain total spin and two intermediate spins (the spin of the core with the excess electron and the sum of the spins of the remaining two i and j centers S_{ij}, respectively), then its matrix decomposes into blocks with different S values. The blocks on the diagonal line of Eq. (6.18) become diagonal matrices, while the ones proportional to w are expressed by $9j$ symbols. Note that for the trimers the interaction with non-Hund states is more essential for magnetic characteristics than in the case of dimers.[115] Nevertheless, they are neglected in this section. We mention also the works[116,117] in which the energy spectrum of the MV trimer is obtained by means of perturbation theory applied to a Hubbard-type Hamiltonian.

Let us consider the vibronic effects in the simple example of a MV trimer with D_{3h} symmetry and the electron configuration d^1-d^1-d^2 on the three centers. In this four-electron system with a fixed localization of the excess electron four spin multiplets are possible: by one with spins $S = 0$, $S = 2$ and two multiplets with $S = 1$. The latter differ by the value of the intermediate spin equal to 0 and 1.

The electron transfer splits each of these multiplets into A and E states (cf. Section V.A). It also mixes states with the same total spin but a different intermediate one, but we neglect this mixing in the first approximation of our consideration.

Assuming that the system is exchange-isotropic, $J_1 = J_2 = J$ and using the known values of $9j$ symbols, for the energies of the spin-resonance states, we obtain

$$E(^1A) = 2J - w, \qquad E(^1E) = 2J + w/2$$
$$E(^5A) = -4J + 2w, \qquad E(^5E) = -4J - w$$
$$E(^3A') = -w, \qquad E(^3E') = w/2 \qquad\qquad (6.19)$$
$$E(^3A'') = E(^3E'') = 0$$

Using the method described in Section VI.A, one can solve the problem of calculation of the vibronic states separately for each pair of A and E levels, and then take into account the mixing of states with $S = 1$ having different intermediate spin by electron transfer. Certainly the vibronic basis used in the calculation of this mixing has to be limited. The basis dimensionality depends on the accuracy to which the magnetic moment has to be determined. The group-theoretical analysis shows that the electron transfer mixes only the components of spin triplets that belong to the conjugated components of the complex representation E. The appropriate matrix elements are

$$\langle ^3\Psi_{+1}^{(n)} | \hat{H}_{AH} | ^3\Psi_{-1}^{(n')} \rangle$$

$$= i\sqrt{\frac{3}{2}} w \sum_{m,k} (C_{+1,m,k}^{+1(n)} C_{-1,m,k}^{-1(n')} - C_{-1,m,k}^{+1(n)} C_{+1,m,k}^{-1(n')}) \qquad (6.20)$$

The finally obtained spin-vibronic states of the system allow us to calculate its magnetic moment after Eq. (4.33). Calculation details are given in Ref. 118. Distinguished from the case of dimers, the magnetic characteristics of trimers depend not only on the sign of the exchange integral, but also on the sign of the parameter of electron transfer w. In Fig. 18 four types of possible manifestation of the vibronic effects in the temperature dependence of the effective magnetic moment are illustrated. Let us consider these situations in more detail.

1. $w > 0$, $J > 0$ (Fig. 18a). In this case the vibronic interaction does not change the spin of the ground state, but weakens the ferromagnetic properties of the system, bringing them closer to that of the HDVV model.

2. $w > 0$, $J < 0$ (Fig. 18b). In this case the Hamiltonian (Eq. 6.18) always leads to a spin-singlet ground state, as in the HDVV model.[114] However for the first excited state with $S = 1$ the energy gap between the A and E states is smaller than that for the ground one, and hence the vibronic interaction is stronger. This results in a situation when for a certain value of the vibronic constant A the crossover phenomenon mentioned above (Section IV.B), that is, the change of the ground-state spin

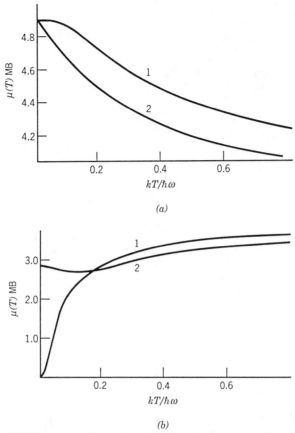

Figure 18. (*a*) Temperature dependence of the magnetic moment of a trimer $d^1-d^1-d^2$ without (1) and with (2) taking account of the vibronic interaction ($A = 2.8$, $J = 0.1$, $w = 1$). (*b*) The same as in (*a*) but with $J = -0.1$. (*c*) The same as in (*a*) but with $w = -0.3$. (*d*) The same as in (*b*) but with $w = -0.3$.

multiplicity, takes place, and the function $\mu(T)$ is nonmonotonous. Note that while for dimers the vibronic interaction always promotes a transition from the ground state with a higher spin to that of a lower spin, that is, it enhances the antiferromagnetic properties, in the case of trimers it amplifies the ferromagnetic ones.

At further increases of the vibronic constant A, an opposite transition from the triplet ground state to the singlet one is expected. At $A \gg 1$ the energies of the spin-vibronic levels become independent of w (the system is localized) and the ground state is again a singlet one, as follows from

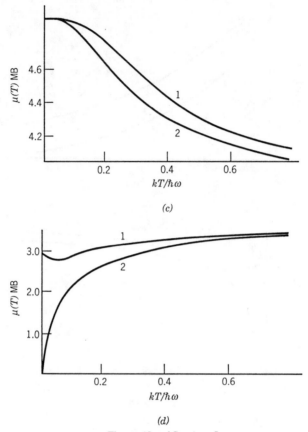

(c)

(d)

Figure 18. (*Continued*).

the HDVV model. However, to reach this region by numerical calculations an essential increase of the vibrational basis is required.

3. $w < 0, J > 0$ (Fig. 18c). Here the vibronic interaction, similar to case 1, does not change the multiplicity of the ground state that has the maximum spin. As in case 1, the magnetic moment by the increase of the vibronic interaction decreases more sharply with temperature.

4. $w < 0, J < 0$ (Fig. 18d). In this case of an antiferromagnetic cluster, the resonance interaction, having a negative matrix element w, may produce ground states of different multiplicity. It can be a singlet (when $|w|/J < 2$) or a triplet (when $2 < |w|/J < 8$), and at $|w|/J > 8$ the ground state is a quintet one. The vibronic interaction lowers the spin of the

ground state in the consequence $2 \rightarrow 1 \rightarrow 0$ or $1 \rightarrow 0$, that is, it enhances the antiferromagnetic behavior.

The analysis of the experimental data on the magnetic properties of trimer MV compounds shows that their interpretation may be rather complicated. In any case one has to be careful when the models based on the consideration of only the electronic energy spectrum, without taking account vibronic interactions, are used. This statement is especially true concerning MV systems that contain ions with large core spins. The number of states in these systems is great, and the vibronic interaction can radically change their ordering. It is obvious that the neglect of the vibronic interactions in the description of the experimental data for such MV systems may yield exchange parameters that have nothing to do with the real systems.

D. Electron Delocalization with Double Exchange — Application to Ferredoxins

The delocalization of the excess electron in trimer MV systems was considered in Section V.A for the case of zero-spin cores, and a question arises: In what measure are these results valid for systems with double exchange (nonzero spin cores) having a more complicated electronic energy spectrum? The trinuclear iron–sulfur cluster in the ferredoxin of the sulfate-reducing bacteria *Desulfovibrio gigas*,[119] for which the presence of double exchange is revealed definitively, seems to be a good subject for solving the questions posed above.

In the reduced form this ferredoxin cluster has the valence composition Fe(III)–Fe(III)–Fe(II). The ESR and Mossbauer spectra show that the excess electron is delocalized over two from three centers.[120] Based on these data a model was proposed considering the dobule exchange between the pair Fe(II)–Fe(III) and the HDVV exchange between all three ions.[120] This approach implies that one of the ions is strongly distinguished from the other two, so its resonance interaction with the two other ions is negligible. However, recent X-ray structural data[121] show that structural differences between the three ions are very small, if any.

In Ref. 122 a model based on the simultaneous influence of magnetic and vibronic effects is suggested, explaining the origin of the peculiar electron distribution in the ferredoxin cluster under consideration. The interaction with the vibrations can be included in the generalized Anderson–Hasegawa Hamiltonian (Eq. 6.18). Since the electron configuration of the three centers is d^5–d^5–d^6 and all the iron ions are in the high-spin states, the terms containing the intraatomic exchange can be omitted. Then

$$\hat{H}_{AH}(Q) =$$

$$\begin{pmatrix} AQ_1 - 2J_2S_2S_3 & w\hat{I} & w\hat{I} \\ -2J_1(S_1 + s)(S_2 + S_3) & AQ_2 - 2J_2S1S_3 & w\hat{I} \\ w\hat{I} & -2J_1(S_2 + s)(S_1 + S_3) & AQ_3 - 2J_2S_1S_2 \\ w\hat{I} & w\hat{I} & -2J_1(S_3 + s)(S_1 + S_2) \end{pmatrix}$$

$$(6.21)$$

As in the previous sections we can pass to symmetrized coordinates and separate out the totally symmetric Q_A vibrations. It is known from the experimental data that the ground state is a quintet $S = 2$, and the excited states with other $S \neq 2$ spins lay higher than the ground one by not less than 80 cm^{-1}.[123] Therefore, for approximately low enough temperatures we can consider in Eq. (6.21) only the quintet states. With the localized electron there are five such states that differ by the intermediate spin $S_{ij} = 0-4$. With the three poossible localizations of the excess electron taken into account, the matrix (Eq. 6.21) has the dimensionality 15×15.

Let us neglect the differences in the exchange integrals, that is, we assume, as in the previous subsection, that the system is exchange-isotropic. Then the analytical expressions for the electron energies at $Q_\theta = Q_\epsilon = 0$ can be obtained:

$$(^5A^{(1)}) = E(^5E^{(1)}) = E(^5E^{(2)}) = -w$$

$$E(^5A^{(2)}) = E(^5A^{(3)}) = E(^5E^{(3)}) = 0 \qquad (6.22)$$

$$E(^5A^{(4)}) = E(^5A^{(5)}) = E(^5E^{(4)}) = E(^5E^{(5)}) = \tfrac{5}{6}w$$

In Eq. (6.21) the energies are read off from the position of nonsplitted HDVV multiplet with $S = 2$. For an arbitrary relation between the exchange parameters J_1 and J_2 and arbitrary values of vibrational coordinates, the electronic energies can be obtained by numerical calculations.

With the known electronic energies at fixed nuclei one can determine the adiabatic potential of the system in the space of the active Q_θ and Q_ϵ or ρ and φ (see Eq. 5.6) coordinates. In Section V.A. it was shown that the adiabatic potential shape depends on the sign of w. In the case at hand, although as shown from Eq. (6.22) the energy-level scheme depends on the sign of w, the shape of the lowest sheet of the adiabatic potential undergoes qualitatively the same transformations with the $|w|/A^2$ parameter for both positive and negative w values.

For small w values the minima of the lowest sheet are along $\varphi = \pi/3$, π, $5\pi/3$, whereas at $\varphi = 0$, $2\pi/3$, $4\pi/3$ there are saddlepoints. As a result its extremal cross sections have the same shape as in Fig. 14a. Similar to the case of one-electron trimer, considered in Section V.A, these minima correspond to the localization of the excess electron at one center. For some values of $|w|/A^2$ a pair of minima occurs instead of each of them, the two minima in the pair occupying symmetrical positions with respect to the directions $\varphi = \pi/3$, π, $5\pi/3$, the minima on the latter thus becoming saddlepoints. The value $|w|/A^2$ for which the number of minima in doubling depends on the relation between the exchange integrals; for $|J_1 - J_2| = 0$–0.1 it lays within the inequality $0.7 < w/A^2 < 0.8$.

The possible existence of six equivalent minima of the adiabatic potential in the space of E vibration (active in the vibronic interaction) was predicted by group-theoretical consideration.[124] To our knowledge so far such a potential surface has been obtained in the only case of hexafluorine d^0 transition metal complexes.[125] The essential feature making similar these two otherwise physically different systems is the presence of more than two electronic states mixing the E-type vibronic interactions.

At further increase of the $|w|/A^2$ value the minima pairwise convert to the values $\varphi = 0$, $2\pi/3$, $4\pi/3$, the latter hence becoming minima (instead of saddlepoints), and the surface as a whole becomes again a triminima one. But distinguished from the previous cases, the excess electron in these minima is delocalized over two centers of the trimer.

If $|w|/A^2 \to \infty$ the energies of the minima and saddlepoints coincide, and we obtain a continuous set of minima, a trough.[5,7] The motion along the trough corresponds to a dynamical delocalization of the excess electron that is not possible in the previously studied MV systems, in dimers, or in one-electron trimers.

The observed charge distribution in the iron–sulfur ferredoxin can thus be explained as due to the joint influence of vibronic and interelectronic (exchange) interactions in a multilevel system with a special relation of parameter values resulting in minima of the adiabatic potential describing partial two-center delocalization of the excess electron.

Obviously, one can expect that the number of novel effects arising due to the combined vibronic and double-exchange interactions will increase with the number of new MV trimers studied. Even more essential effects may be expected in four-nuclear MV ferredoxins that are known to have a delocalized electronic distribution and double exchange.

VII. MV SYSTEMS OF HIGHER NUCLEARITY

The variety of MV clusters is not limited by the bi- and trinuclear ones discussed above. There is quite a number of polynuclear systems with higher than three nuclearity filling the gap between the relatively small clusters considered above and infinite chain structures with alternating valencies[126] or three-dimensional crystals, for example, magnetit Fe_3O_4.[127] To our knowledge there are no published theoretical works devoted to the vibronic interactions in systems with more than three MV centers. However, one may hope that the ideas and methods developed for MV compounds with slower nuclearity discussed in this chapter can be expanded toward clusters with a larger number of centers.

A special group of such systems is formed by the tetranuclear clusters in which the metal ions occupy the apices of a regular tetrahedron. In particular, the cubane iron–sulfur clusters serving as active centers of a series of proteins (including nitrogenaza) and their models[128] belong to such tetranuclear systems. These systems are among the most important of those that provide increasing interest to the whole class of MV compounds.

In this section we intend to show that for tetramers a theoretical treatment of the vibronic interactions is possible similar to that carried out above for clusters with lower nuclearity. Analogous to Eq. (5.1) the operator of the potential energy is

$$\hat{H}_p = \tfrac{1}{2}(Q_1^2 + Q_2^2 + Q_3^2 + Q_4^2)\hat{I}$$

$$+ w \begin{pmatrix} 0 & 1 & 1 & 1 \\ 1 & 0 & 1 & 1 \\ 1 & 1 & 0 & 1 \\ 1 & 1 & 1 & 0 \end{pmatrix} + A \begin{pmatrix} Q_1 & 0 & 0 & 0 \\ 0 & Q_2 & 0 & 0 \\ 0 & 0 & Q_3 & 0 \\ 0 & 0 & 0 & Q_4 \end{pmatrix} \tag{7.1}$$

For a tetrahedron the symmetrized basis of the electronic wave functions contains the ones belonging to the A and T representations expressed by those of the localized basis ϕ_i ($i = 1\text{–}4$):

$$\phi_A = \tfrac{1}{2}(\phi_1 + \phi_2 + \phi_3 + \phi_4)$$
$$\phi_{T\xi} = \tfrac{1}{2}(\phi_1 + \phi_2 - \phi_3 - \phi_4)$$
$$\phi_{T\eta} = \tfrac{1}{2}(\phi_1 - \phi_2 + \phi_3 - \phi_4) \tag{7.2}$$
$$\phi_{T\zeta} = \tfrac{1}{2}(\phi_1 - \phi_2 - \phi_3 + \phi_4)$$

Similarly, the symmetrized displacements of the cluster of A and T representations are expressed by the local totally symmetric ones Q_i,

$$Q_A = \tfrac{1}{2}(Q_1 + Q_2 + Q_3 + Q_4)$$

$$Q_{T\xi} = \tfrac{1}{2}(Q_1 + Q_2 - Q_3 - Q_4)$$

$$Q_{T\eta} = \tfrac{1}{2}(Q_1 - Q_2 + Q_3 - Q_4)$$ (7.3)

$$Q_{T\zeta} = \tfrac{1}{2}(Q_1 - Q_2 - Q_3 + Q_4)$$

In the symmetrized basis the interaction with the totally symmetric (for the system as a whole) displacements Q_A can be separated and the Hamiltonian reads:

$$\hat{H}_p = \tfrac{1}{2}(Q_\xi^2 + Q_\eta^2 + Q_\zeta^2)\hat{I}$$

$$+ w \begin{pmatrix} 3 & 0 & 0 & 0 \\ 0 & -1 & 0 & 0 \\ 0 & 0 & -1 & 0 \\ 0 & 0 & 0 & -1 \end{pmatrix} + \frac{A}{2} \begin{pmatrix} 0 & Q_\xi & Q_\eta & Q_\zeta \\ Q_\xi & 0 & Q_\zeta & Q_\eta \\ Q_\eta & Q_\zeta & 0 & Q_\xi \\ Q_\zeta & Q_\eta & Q_\xi & 0 \end{pmatrix}$$ (7.4)

It can be shown that, analogous to the case of a trimer, we have here a combination of the Jahn–Teller effect (on the T states) and the pseudo Jahn–Teller one (the A–T mixing). In general this is a $(A + T) * t$ problem; for the trimer it was an $(A + E) * e$ problem.[78] It is important that the vibronic constant A is the same for both the Jahn–Teller mixings within the T states and the pseudo Jahn–Teller A–T one. A similar problem, but with nonequal vibronic constants, was considered for the nitrogen atom as an impurity in diamond.[129]

The general investigation of the adiabatic potentials for the Hamiltonian (Eq. 7.4) resulting in fourth-order equations encounters certain difficulties. However, as it is known from the theory of vibronic interactions,[7] the extrema points and curvatures of the adiabatic potentials can be revealed without solving the equations, for example, by using the method of Opik and Pryce. Employing the latter we come to the conclusion that in the space of the T coordinates Q_ξ, Q_η, and Q_ζ there are four minima along the directions $(1, 1, 1)$, $(\bar{1}, \bar{1}, 1)$, $(\bar{1}, 1, \bar{1})$, and $(1, \bar{1}, \bar{1})$; 1 and $\bar{1}$ denote the directions of positive and negative value of the corresponding coordinate, respectively). The displacements in these minima describe the four equivalent trigonal distortions of the tetrahedron. To study them it is convenient to pass to the trigonal basis of the wave functions, in which they belong to the A and E representations (by trigonal distortion $T \rightarrow A + E$):

$$\phi_A^{(1)} = \phi_A$$

$$\phi_A^{(2)} = \frac{1}{\sqrt{3}}(\phi_{T\xi} + \phi_{T\eta} + \phi_{T\zeta})$$

$$\phi_{E\epsilon} = \frac{1}{\sqrt{2}}(\phi_{T\xi} - \phi_{T\eta}) \tag{7.5}$$

$$\phi_{E\theta} = \frac{1}{\sqrt{6}}(\phi_{T\xi} + \phi_{T\eta} - 2\phi_{T\zeta})$$

and to similar A and E symmetrized displacements $Q_A^{(1)}$, $Q_A^{(2)}$, $Q_{E\epsilon}$, and $Q_{E\theta}$. Since the interaction with $Q_A^{(1)}$ is separated, we consider now the cross section of the adiabatic potential along the totally symmetric coordinate in the trigonal group

$$Q = Q_A^{(2)} = \frac{1}{\sqrt{3}}(Q_{T\xi} + Q_{T\eta} + Q_{T\zeta}) \tag{7.6}$$

As a function of this coordinate the four curves can be obtained from the matrix Hamiltonian written in Eq. (7.5), as follows:

$$\mathscr{E}_{1,2}(Q) = \frac{1}{2}Q^2 - w - \frac{1}{2\sqrt{3}}AQ \tag{7.7}$$

$$\mathscr{E}_{3,4}(Q) = \frac{1}{2}Q^2 + w + \frac{1}{2\sqrt{3}}AQ_{\pm} + 2\left(w^2 - \frac{1}{2\sqrt{3}}AQ + \frac{1}{12}A^2Q^2\right)^{1/2}$$

By direct calculation one can easily make certain that if $w > 0$ there is always a minimum of the lowest curve at $Q \neq 0$ corresponding to the excess electron localized at one center. For negative w values three possibilities, similar to that for the trimers, occur: (1) if $|w|/A^2 < 0.125$, the lowest curve has two minima, in the extended space one of them being a saddlepoint; (2) if $0.125 < |w|/A^2 < 0.152$, there are two minima at $Q = 0$ and $Q \neq 0$, respectively; (3) if $|w|/A^2 > 0.152$ the only minima point occurs at $Q = 0$. Concerning the electron charge distribution corresponding to these three cases, they are: (1) electron-localized states, (2) coexistence of localized and delocalized states, and (3) electron delocalization.

Thus for the tetrahedral MV tetramers we again come to the possibility of coexistence of states with localized and delocalized excess electron distributions. Apparently this effect has a more general meaning and can be expected also in clusters with larger numbers of MV centers. So far, to our knowledge there is no experimental evidence of the coexistence effect in tetramers.

The lowest way between the adiabatic potential minima corresponding to the localized electron goes through the saddlepoints along the $Q_{T\xi}$, $Q_{T\eta}$, and $Q_{T\zeta}$ directions. The numerical calculation shows that the barrier height for the electron transfer through these saddlepoints in tetramers is always lower than the appropriate barrier in dimers for the same values of w and A. Perhaps this result explains the fact that in the same conditions, the same structure of the near-neighbor environment of the iron ions and the same bridges between them as well as approximately the same iron–iron distances, the iron–sulfur dimers manifest localized electronic states, whereas in similar tetramers the excess electron is delocalized.

It is obvious that the other problems solved above for dimers and trimers by means of the vibronic theory can be analogously considered also for tetramers and other MV systems with $n > 4$. However, for the latter some new problems may occur. In particular, for these systems the case of two and more excess electrons migrating among the MV centers becomes essential (in the trimers the case of two excess electrons is equivalent to one excess hole and yields similar results). For two and more excess electrons the electron distribution and dynamics is determined by the competition of vibronic, intraatomic, and intercenter interactions. Examples of MV systems with two excess electrons are known among cubane,[128] and quadratic[130] tetramers, six-nuclear clusters,[131] and so on.

It can also be noted that the coexistence of localized and delocalized electronic distributions is polynuclear clusters is analogous to the coexistence of localized and delocalized excitonic states in crystals which is well known in solid-state physics.[132,133]

VIII.   CONCLUDING REMARKS

The vibronic approach based on a more detailed consideration of the coupling of the electronic motion with the nuclear displacements than in the simple adiabatic approximation results in special vibronic states of the system that give a much more adequate description of its properties.[5,7] Employed to reveal the characteristics of MV compounds, the vibronic approach, as shown in this chapter, explains the origin of experimental data and predicts a series of new effects. Attention was paid mostly to MV dimers and trimers that are so far most studied. But even for them approach, are just beginning, and we discuss here just some general results illustrating the possibilities of this approach.

Even for dimers and trimers the vibronic theory needs some further development. Indeed, the main parameters of the theory, such as the electron energy spectrum and vibronic constants, are not calculated from first principles. Such calculations for real MV compounds, which are usually relatively large, may be rather difficult. Nevertheless, for logical accomplishment of the theory such calculations are needed, at least for most simple systems. Full calculations for a MV compound (ab initio or semiempirical) may reveal some new questions, such as the ability to choose local states, vibrations, and vibronic constants.

Among the other assumptions of the vibronic theory of MV compounds that need further theoretical foundation we note here the neglect of the nonadiabatic dependence of the electronic wave functions on the active nuclear displacements and the consideration of only one local active vibration included in the vibronic interaction (i.e., the neglect of the multimode nature of the vibronic problem). While the former effect, the nonadiabacity of the wave functions, seems less important, at least when the properties near the minima of the adiabatic potentials are considered, the multimode nature of the vibronic interaction can be essential in many effects, especially in those that depend on the details of the vibronic energy spectrum (e.g., the fine structure of optical IR and Raman spectra). The problem of vibronic coupling with several vibrational modes of the same symmetry was considered recently for systems with disproportionating valency.[134] The analogous researches are needed for other MV systems with complicated electronic energy spectrum.

ACKNOWLEDGMENTS

We would like to thank our colleagues Drs. I. N. Kotov and L. F. Chibotaru for their valuable help in our studies of MV clusters. It is also a pleasure to thank Prof. B. S. Tsukerblat, Dr. M. I. Belinskii, Prof. K. I. Turta, Dr. R. D. Cannon, Dr. J. J. Girerd, Dr. J. P. Launay, Dr. K. Prassides, Prof. D. N. Hendrickson, and Prof. P. N. Schatz for stimulating discussions.

REFERENCES

1. *Mixed-Valence Compounds*, D. B. Brown, Ed., Reidel, Dordrecht, 1980.
2. S. J. Lippard, in Ref. 1, p. 427.

3. A. R. Bishop, R. L. Martin, K. A. Muller, and Z. Tesanovic, *Zeit. fur Phys. B.* **76**, 17 (1989).

4. J. P. Launay, in *Molecular Electronic Devices II*, F. L. Carter, Ed., Marcel Dekker, New York, 1987, p. 39.

5. I. B. Bersuker, *The Jahn–Teller Effect and Vibronic Interactions in Modern Chemistry*, Plenum, New York, 1984.

6. K. V. Mikkelsen and M. A. Ratner, *Chem. Rev.* **87**, 113 (1987).

7. I. B. Bersuker and V. Z. Polinger, *Vibronic Interactions in Molecules and Crystals*, Springer, Berlin, 1989.

8. I. B. Bersuker, *Pure and Appl. Chem.* **60**, 1167 (1988).

9. I. B. Bersuker, *Fiz. Tverd. Tela (Russ.)* **30**, 1738 (1988).

10. B. Mayoh and P. Day, *J. Am. Chem. Soc.* **94**, 2885 (1972).

11. N. S. Hush, *Chem. Phys.* **10**, 361 (1975).

12. K. Y. Wong and P. N. Schatz, *Progr. Inorg. Chem.* **28**, 369 (1981).

13. P. N. Schatz, in Ref. 1, p. 115.

14. Y. Mikami, in *Proc. 9th Int. Conf. Raman Spectrosc.*, Tokyo, 1984, p. 696.

15. M. B. Robin and P. Day, *Adv. Inorg. Chem. Radiochem.* **10**, 247 (1967).

16. R. L. Fulton and M. Gouterman, *J. Chem. Phys.* **35**, 1059 (1961).

17. L. S. Cederbaum, E. Haller, and W. Domcke, *Sol. State Commun.* **35**, 879 (1980).

18. H. Koppel, W. Domcke, and L. S. Cederbaum, *Adv. Chem. Phys.* **57**, 59 (1984).

19. E. M. Kober, K. A. Goldsby, D. N. S. Narayana, and T. J. Meyer, *J. Am. Chem. Soc.* **105**, 4303 (1983).

20. K. Neuenschwander, S. B. Piepho, and P. N. Schatz, *J. Am. Chem. Soc.* **107**, 7862 (1985).

21. M. J. Ondrechen, D. P. Ellis, and M. A. Ratner, *Chem. Phys. Lett.* **109**, 50 (1984).

22. L. T. Zhang, J. Ko, and M. J. Ondrechen, *J. Am. Chem. Soc.* **109**, 1666 (1987).

23. K. Y. Wong, *Chem. Phys. Lett.* **125**, 485 (1986).

24. M. E. Gress, C. Creutz, and C. O. Quicksoll, *Inorg. Chem.* **20**, 1522 (1981).

25. U. Furholz, H. B. Burgi, P. E. Wagner, A. Stebler, J. H. Ammeter, E. Krausz, R. J. H. Clark, M. Stead, and A. J. Ludi, *J. Am. Chem. Soc.* **106**, 121 (1964).

26. K. I. Kugel and D. I. Khomskii, *J. Exp. Theor. Phys. (Russ.)* **79**, 987 (1980).

27. K. H. Hock, H. Nickisch, and H. Thomas, *Helv. Phys. Acta* **56**, 237 (1983).

28. S. A. Borshch, *J. Strukt. Chem. (Russ.)* **28**, N4, 36 (1987).

29. M. Z. Zgierski and M. Pawlikowski, *J. Chem. Phys.* **79**, 1616 (1983).

30. P. M. Plaksin, R. C. Stoufer, M. Methew, and G. J. Palenik, *J. Am. Chem. Soc.* **94**, 2121 (1972).

31. *Theory and Applications of Molecular Paramagnetism*, E. A. Boudreaux and L. N. Mulay, Eds., Wiley, New York, 1976.

32. C. Zener, *Phys. Rev.* **82**, 403 (1951).

33. P. V. Anderson and H. Hasegawa, *Phys. Rev.* **100**, 675 (1956).

34. P. G. de Gennes, *Phys. Rev.* **118**, 141 (1960).

35. B. V. Karpenko, *J. Magnet. & Magnet. Mater.* **3**, 267 (1976).

36. K. Kubo and N. Ohata, *J. Phys. Soc. Jpn.* **33**, 21 (1972).

37. M. Cieplak, *Phys. Rev.* **B18**, 3470 (1978).

38. N. L. H. Liu and D. Emin, *Phys. Rev. Lett.* **41**, 71 (1979).

39. J. J. Girerd, *J. Chem. Phys.* **79**, 1766 (1983).

40. S. A. Borshch, I. N. Kotov, and I. B. Bersuker, *Khim. Fis. (Russ.)* **3**, 667 (1984) [see English translation, *Sov. J. Chem. Phys.* **3**, 1009 (1985)].

41. M. I. Belinskii, B. S. Tsukerblat, and N. V. Gerbeleu, *Sov. Phys. Solid State* **25**, 497 (1983).

42. M. Drillon, G. Pourroy, and J. Darriet, *Chem. Phys.* **88**, 27 (1984).

43. L. Sacconi, C. Mealli, and D. Gatteschi, *Inorg. Chem.* **13**, 1985 (1974).

44. A. Bencini, D. Gatteschi, and L. Sacconi, *Inorg. Chem.* **17**, 2670 (1973).

45. M. I. Belinskii and B. S. Tsukerblat, *Fis. Tverd. Tela (Russ.)* **26**, 758 (1984).

46. J. Hubbard, *Proc. Roy. Soc.* **A276**, 238 (1963).

47. L. Noodleman and E. J. Baerends, *J. Am. Chem. Soc.* **106**, 2316 (1984).

48. D. M. Sherman, *Solid State Commun.* **58**, 719 (1986).

49. S. I. Klokishner and B. S. Tsukerblat, *Chem. Phys.* **125**, 11 (1988).

50. S. A. Borshch, I. N. Kotov, and I. B. Bersuker, *Chem. Phys. Lett.* **111**, 264 (1984).

51. S. A. Borshch and I. N. Kotov, *Teor. Eksperim. Khim. (Russ.)* (1990) (in press).

52. S. A. Borshch, I. N. Kotov, and I. B. Bersuker, *Teor. Eksperim. Khim. (Russ.)* **20**, 675 (1984).

53. R. Friesner and R. Silbey, *J. Chem. Phys.* **74**, 1166 (1981).

54. N. Sasaki and T. Kambara, *J. Chem. Phys.* **74**, 3472 (1981).

55. Y. Toyozawa, *J. Phys. Soc. Jpn.* **50**, 1861 (1981).

56. K. Prassides, P. N. Schatz, K. Y. Wong, and P. N. Day, *J. Phys. Chem.* **90**, 5588 (1986).

57. P. A. Cox, *Chem. Phys. Lett.* **69**, 340 (1980).

58. S. A. Borshch and I. N. Kotov, *Teor. Eksperim. Khim. (Russ.)* **23**, 211 (1987).

59. M. Tanner and A. Ludi, *Inorg. Chem.* **20**, 2348 (1981).

60. Y. Toyozawa and M. Inoue, *J. Phys. Soc. Jpn.* **21**, 1663 (1966).

61. K. Y. Wong, P. N. Schatz, and S. B. Piepho, *J. Am. Chem. Soc.* **101**, 2793 (1979).

62. S. A. Borshch, *Sov. Phys. Solid State* **26**, 1142 (1984).

63. S. A. Borshch, I. N. Kotov, and I. B. Bersuker, in *Applications of Mössbauer Effect*, Vol. 2, Yu. M. Kagan and I. S. Lyubutin, Eds., Gordon and Breach, New York, 1985, p. 511.

64. R. G. Wollman and D. N. Hendrickson, *Inorg. Chem.* **16**, 723 (1977).

65. J. H. Weiner, *J. Chem. Phys.* **69**, 4743 (1978).

66. S. A. Borshch, *Dokl. AN SSSR (Russ.)* **280**, 652 (1985).

67. I. S. Gradshtein and I. M. Ryznik, *Tables of Integrals, Series and Products*, Academic Press, New York, 1965.

68. J. Jortner, *J. Chem. Phys.* **64**, 4860 (1976).

69. J. P. Laplante and W. Siebrand, *Chem. Phys. Lett.* **59**, 433 (1970).

70. R. J. Butcher, C. J. O'Konnor, and E. Sinn, *Inorg. Chem.* **20**, 3486 (1981).

71. T. Sunduresan and S. C. Wallwork, *Acta Cryst.* **B28**, 491 (1972).

72. V. M. Yartsev, *Phys. Stat. Sol. (b)* **112**, 279 (1982).

73. A. Muller, R. Jostes, and F. A. Cotton, *Angew. Chem. Int. Ed. Engl.* **19**, 875 (1980).
74. C. E. Strause and L. F. Dahl, *J. Am. Chem. Soc.* **93**, 6032 (1971).
75. G. Longoni, M. Manassero, and M. Sansoni, *J. Am. Chem. Soc.* **102**, 7973 (1980).
76. J. L. Walsh, J. A. Bauman, and T. J. Meyer, *Inorg. Chem.* **19**, 2145 (1980).
77. J. P. Launay and F. Babonneau, *Chem. Phys.* **67**, 2591 (1984).
78. S. A. Borshch, I. N. Kotov, and I. B. Bersuker, *Chem. Phys. Lett.* **89**, 381 (1982).
79 R. D. Cannon, L. Montri, D. B. Brown, K. M. Marshall, and C. M. Elliott, *J. Am. Chem. Soc.* **106**, 2591 (1984).
80. M. A. Collins and E. Krausz, in *Photochemistry and Photophysics of Coordination Compounds*, H. Yersin and A. Vogler, Eds., Springer, Berlin, 1987, p. 85.
81. M. A. Collins, E. Krausz, and M. Riley, in *Abstracts of Xth International Symposium on the Jahn–Teller Effect*, Kishinev, 1989, p. 66.
82. S. M. Oh, D. N. Hendrickson, K. L. Hassett, and R. E. Davis, *J. Am. Chem. Soc.* **107**, 8009 (1985).
83. S. E. Woehler, R. J. Witterbort, S. M. Oh, D. N. Hendrickson, D. Inniss, and C. E. Strouse, *J. Am. Chem. Soc.* **108**, 2938 (1986).
84. S. M. Oh, S. R. Wilson, D. N. Hendrickson, S. E. Woehler, R. J. Witterbort, D. Innis, and C. E. Strouse, *J. Am. Chem. Soc.* **109**, 1073 (1987).
85. H. G. Jang, S. J. Geib, Y. Kaneko, M. Nakano, M. Sorai, A. L. Rheingold, B. Monter, and D. N. Hendrickson, *J. Am. Chem. Soc.* **111**, 173 (1989).
86. T. Kambara, D. N. Hendrickson, M. Sorai, and S. Oh, *J. Chem. Phys.* **85**, 2895 (1986).
87. R. M. Stratt and S. H. Adachi, *J. Chem. Phys.* **86**, 7156 (1987).
88. R. A. Stukan, K. I. Turte, A. V. Ablov, and S. A. Bobkova, *Koordin. Khim. (Russ.)* **5**, 95 (1979).
89. J. T. Wrobleski, C. T. Dziobkowski, and D. B. Brown, *Inorg. Chem.* **20**, 684 (1981).
90. S. A. Borshch and I. N. Kotov, *Sov. Phys. Solid State* **24**, 1187 (1982).
91. N. Blume, *Phys. Rev.* **174**, 351 (1968).
92. M. J. Powers, R. W. Callahan, D. J. Salmon, and T. Meyer, *Inorg. Chem.* **15**, 894 (1976).
93. J. Ko and M. J. Ondrechen, *Chem. Phys. Lett.* **112**, 507 (1984).
94. M. J. Ondrechen, J. Ko, and L. T. Zhang, *J. Am. Chem. Soc.* **109**, 1672 (1987).
95. C. Joachim and J. P. Launay, *Nouv. J. Chim.* **8**, 723 (1984).
96. C. Joachim and J. P. Launay, *Chem. Phys.* **109**, 93 (1986).
97. O. Kahn and J. P. Launay, *Chemtronics* **3**, 140 (1988).
98. A. Aviram, C. Joachim, and M. Pomerantz, *Chem. Phys. Lett.* **146**, 490 (1988).
99. I. B. Bersuker, S. A. Borshch, and L. F. Chibotaru, *Chem. Phys.* **136**, 379 (1989).
100. I. B. Bersuker, S. A. Borshch, and L. F. Chibotaru, in *Proceedings of All-Union Conference on Molecular Electronics*, Odessa, 1988 (Russ.) (in press).
101. M. H. Perrin and M. Gouterman, *J. Chem. Phys.* **46**, 1019 (1967).
102. J. H. van der Waals. A. M. D. Berghius, and M. S. de Groot, *Mol. Phys.* **13**, 301 (1967); **21**, 497 (1971).
103. M. Z. Zgierski and M. Pawlikowski, *J. Chem. Phys.* **70**, 3444 (1979).
104. P. Lacroix, J. Weber, and E. Duval, *J. Phys. C.: Sol. State Phys.* **12**, 2065 (1979).

105. I. N. Kotov, S. A. Borshch, and I. B. Bersuker, Manuscript deposited in VINITI (N2064–B87), Moscow, 1987 (in Russian).

106. Yu. E. Perlin and B. S. Tsukerblat, *Effects of Electron-Vibrational Interactions in the Optical Spectra of Impurity Paramagnetic Ions*, Shtiintsa, Kishinev, 1974 (In Russian).

107. S. I. Pekar, *Zh. Exp. Teor. Fys. (Russ.)* **20**, 267 (1950).

108. N. S. Hush, in *Progress in Inorganic Chemistry*, vol. 8, F. A. Cotton, Ed., Wiley, New York, 1967.

109. S. Muramatsu and N. Sakamoto, *J. Phys. Soc. Jpn.* **36**, 839 (1974).

110. R. D. Cannon, unpublished results.

111. D. Lupu, *Rev. Roumaine Chim.* **15**, 417 (1970).

112. D. Lupu, D. Barb, G. Filoti, M. Morariu, and D. Tarina, *J. Inorg. Nucl. Chem.* **34**, 2803 (1972).

113. K. I. Turta, S. A. Bobkova, R. A. Stukan, A. V. Dorogan, and M. E. Veksel'man, *Koordin. Khim.* **8**, 794 (1982).

114. M. I. Belinski, *Molec. Phys.* **60**, 793 (1987).

115. S. A. Borshch, *Fiz. Tverd. Tela* **29**, 1561 (1987) (in

116. G. Pourroy, E. Coronado, M. Drillon, and R. Georges, *Chem. Phys.* **104**, 73 (1986).

117. J. J. Girerd, V. Papaefthymiou, K. K. Surerus, and E. Munck, *Pure and Appl. Chem.* **61**, 805 (1989).

118. I. B. Bersuker, S. A. Borshch, and I. N. Kotov, *Rev. Roumaine Chim.* **32**, 1075 (1987).

119. B. H. Huynh, J. J. C. Moura, I. Moura, T. A. Kent, J. Le Gall, A. X. Xavier, and E. Munck, *J. Biol. Chem.* **255**, 3242 (1980).

120. V. Papaefthymiou, J. J. Girerd, I. Moura, J. J. G. Moura, and E. Munck, *J. Am. Chem. Soc.* **109**, 4703 (1987).

121. C. P. Kissinger, E. T. Adman, L. C. Sieker, and L. H. Jensen, *J. Am. Chem. Soc.* **110**, 8721 (1988).

122. S. A. Borshch and L. F. Chibotaru, *Chem. Phys.* **135**, 375 (1989).

123. A. J. Thomson, A. E. Robinson, M. K. Johnson, J. J. G. Moura, I. Moura, A. V. Xavier, and J. Le Gall, *Biochem. Biophys. Acta* **670**, 93 (1981).

124. R. S. Dagys and I. B. Levinson, *Opt. i Spektroskop. (Russ.)* **3**, 3 (1967).

125. S. A. Borshch, I. Ya. Ogurtsov, and I. B. Bersuker, *Zh. Strukt. Khim. (Russ.)* **23**, 7 (1983).

126. P. Day in *Organic and Inorganic Low-Dimensional Crystalline Materials*, M. Delhaes and M. Drillon, Eds., Plenumd, New York, 1987.

127. J. B. Goodenough in Ref. 1, p. 413.

128. *Iron–Sulfur Proteins*, T. G. Spiro, Ed., Wiley-Interscience, New York, 1982.

129. J. Koppitz, O. F. Schirmer, and M. Seal, *J. Phys. C: Solid State Phys.* **19**, 1123 (1986).

130. J. J. Girerd and J. P. Launay, *Chem. Phys.* **74**, 217 (1983).

131. D. Coucouvanis, M. G. Kanatzidis, A. Salifoglou, W. R. Dunham, A. Simopoulos, J. R. Sams, V. Papaefthymiou, A. Kostikas, and C. E. Strouse, *J. Am. Chem. Soc.* **109**, 6863 (1987).

132. H. Sumi and Y. Toyozawa, *J. Phys. Soc. Jpn.* **31**, 342 (1971).

133. V. V. Hizhnyakov and A. V. Sherman, *Phys. Stat. Sol. (b)* **92**, 77 (1979).

134. K. Prassides and P. N. Schatz, *J. Phys. Chem.* **93**, 83 (1989).

AUTHOR INDEX

Numbers in parentheses are reference numbers and indicate that the author's work is referred to although his name is not mentioned in the text. Numbers in *italics* show the pages on which the complete references are listed.

SUBJECT INDEX